HERBIG–HARO FLOWS AND THE BIRTH OF LOW MASS STARS

INTERNATIONAL ASTRONOMICAL UNION

UNION ASTRONOMIQUE INTERNATIONALE

HERBIG–HARO FLOWS AND THE BIRTH OF LOW MASS STARS

PROCEEDINGS OF THE 182ND SYMPOSIUM OF THE INTERNATIONAL ASTRONOMICAL UNION, HELD IN CHAMONIX, FRANCE, 20–26 JANUARY 1997

EDITED BY

BO REIPURTH

Observatoire de Grenoble, France

and

CLAUDE BERTOUT

Institut d'Astrophysique de Paris, France

KLUWER ACADEMIC PUBLISHERS

DORDRECHT / BOSTON / LONDON

Library of Congress Cataloging-in-Publication Data

ISBN 0-7923-4660-2 (HB)

Published on behalf of
the International Astronomical Union
by
Kluwer Academic Publishers, P.O. Box 17, 3300 AA Dordrecht, The Netherlands.

Sold and distributed in the U.S.A. and Canada
by Kluwer Academic Publishers,
101 Philip Drive, Norwell, MA 02061, U.S.A.

In all other countries, sold and distributed
by Kluwer Academic Publishers Group,
P.O. Box 322, 3300 AH Dordrecht, The Netherlands.

Printed on acid-free paper

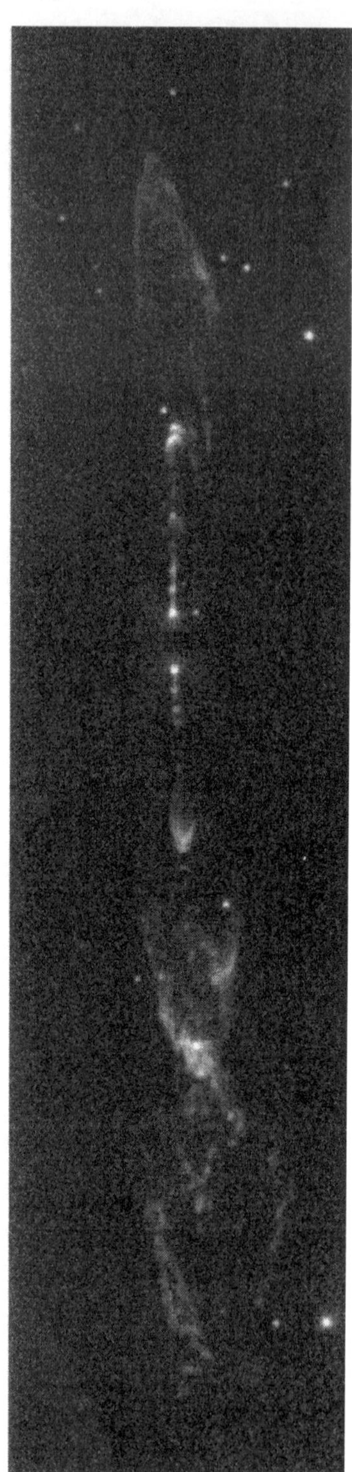

HH 212

This molecular hydrogen image (v=1–0 S(1) line at 2.122 μm) shows the highly symmetric jet HH 212 in Orion, which emanates from a deeply embedded young source, IRAS 05413−0104, and drives an associated molecular outflow. The symmetric and periodic nature of the inner knots and outer bow shocks point to fluctuations at the source as their origin. The image covers 0.15×0.61 parsec at a distance of 450 pc, and has been rotated to place the jet vertically; its true position angle is 24 degrees E of N. Further details can be found in the paper by Zinnecker *et al.* in the poster proceedings of this meeting.

Data taken by Mark McCaughrean using MAGIC on the Calar Alto 3.5-m; 0.7 arcsec seeing; 14 minutes integration time; 1.1×4.6 arcmin field; continuum not subtracted.

TABLE OF CONTENTS

I. HERBIG-HARO OBJECTS, H_2 FLOWS AND RADIO JETS

50 Years of Herbig-Haro Research

B. Reipurth & S. Heathcote

HST Observations of the L1551 IRS5 Jet

C.V.M. Fridlund, M. Huldtgren & R. Liseau

Giant Herbig-Haro Flows

J. Bally & D. Devine

Herbig-Haro Objects in the Orion Nebula Region

C.R. O'Dell

Spectroscopic Properties of Herbig-Haro Flows

K.-H. Böhm & A.P. Goodson

Spectroscopic Signatures of Microjets

J. Solf

The Ionization State along the Beam of Herbig-Haro Jets

F. Bacciotti

Thermal Radio Jets

L.F. Rodríguez

Molecular Hydrogen Emission in Embedded Flows

J. Eislöffel

The Molecular Outflow and CO Bullets in HH 111

J. Cernicharo, R. Neri & B. Reipurth

Shock Chemistry in Bipolar Molecular Outflows

R. Bachiller & M. Pérez Gutiérrez

Models of Bipolar Molecular Outflows

S. Cabrit, A.C. Raga & F. Gueth

The Physics of Molecular Shocks in YSO Outflows

D.J. Hollenbach

The Physical and Chemical Effects of C-Shocks in Molecular Outflows

G. Pineau des Forêts, D.R. Flower & J.-P. Chièze

Herbig-Haro Objects as Searchlights for Dense Cloud Chemistry

S.D. Taylor

III. THEORETICAL MODELS

Protostellar X-Rays, Jets, and Bipolar Outflows

F.H. Shu & H. Shang

Energetics, Collimation and Propagation of Galactic Protostellar Outflows

M. Camenzind

Numerical Simulations of Jets from Accretion Disks

R.E. Pudritz & R. Ouyed

Asymptotic Structure of Rotating MHD Winds and Its Relation to Wind Boundary Conditions

J. Heyvaerts & C.A. Norman

Hydrodynamic Collimation of YSO Jets

A. Frank & G. Mellema

IV. DISKS, WINDS, AND MAGNETIC FIELDS

Hubble Space Telescope Imaging of the Disks and Jets of Taurus Young Stellar Objects

K. Stapelfeldt, C.J. Burrows, J.E. Krist & the WFPC2 Science Team

Disks and Outflows as seen from the IRAM Interferometer

S. Guilloteau, A. Dutrey & F. Gueth

NMA Imaging of Envelopes and Disks around Low Mass Protostars and T Tauri Stars

Y. Kitamura, M. Saito, R. Kawabe & K. Sunada

The Observational Evidence for Accretion

L. Hartmann

The Radiative Impact of FU Orionis Outbursts on Protostellar Envelopes

K.R. Bell & K.M. Chick

Properties of the Winds of T Tauri Stars

N. Calvet

Magnetospherically Mediated Accretion in Classical T Tauri Stars

S. Edwards

Jets, Disk Winds, and Warm Disk Coronae in Classical T Tauri Stars

J. Kwan

Thermal Structure of Magnetic Funnel Flows

S.C. Martin

Magnetic Fields of T Tauri Stars

E.W. Guenther

Evidence for Magnetic Fields in the Outflow from T Tau S

T.P. Ray, T.W.B. Muxlow, D.J.Axon, A. Brown, D.Corcoran, J.E. Dyson & R. Mundt

V. LOW- AND HIGH-MASS PROTOSTARS AND THEIR ENVIRONMENT

The Evolution of Flows and Protostars

P. André

PREFACE

Herbig-Haro objects were discovered 50 years ago, and during this half century they have developed from being mysterious small nebulae to becoming an important phenomenon in star formation. Indeed, HH flows are now recognized not only as fascinating astrophysical laboratories involving shock physics and chemistry, hydrodynamics and radiation processes, but it has gradually been realized that HH flows hold essential clues to the birth and early evolution of low mass stars.

IAU Symposium No. 182 on *Herbig-Haro Flows and the Birth of Low Mass Stars* were held from January 20 to 24, 1997 in Chamonix in the french alps. A total of 178 researchers from 26 countries met to discuss our present level of understanding of Herbig-Haro flows and their relation to disk accretion events and T Tauri winds and other outflow phenomena like molecular outflows, embedded molecular hydrogen flows and radio jets. The present book contains the manuscripts from the oral contributions of the symposium. The poster papers were printed in a separate volume *Low Mass Star Formation – from Infall to Outflow*, edited by Fabien Malbet and Alain Castets, which was distributed at the beginning of the meeting. Together these two books document the vigorous state and the scientific appeal which research into Herbig-Haro flows and related issues in star formation enjoys today, observationally as well as theoretically.

To organize a major symposium like the present one requires the generous support of many people and organisations. We gratefully acknowledge the financial support of the International Astronomical Union, the Grenoble Observatory and its Astrophysics Laboratory (LAOG), the Institute for Millimetric Radioastronomy (IRAM), IBM, the Joseph Fourier University at Grenoble (UJF), the National Center for Scientific Research (CNRS), the French Ministry for Foreign Affairs (MAE), the regional (Région Rhône-Alpes) and local authorities (Conseil Général de Haute-Savoie and Mairie de Chamonix-Mont Blanc). The practical support provided by the Chamonix Tourist Office and the Grenoble Observatory was indispensable. Finally, the hard work of the Local and Scientific Organizing Committees contributed greatly to the success of the Symposium. To all, our warmest thanks.

Bo Reipurth and Claude Bertout

Scientific Organizing Committee

Claude Bertout (Co-chair), Institut d'Astrophysique de Paris, France
Karl-Heinz Böhm, University of Washington, Seattle, USA
Nuria Calvet, Centro de Investigaciones de Astronomia, Mérida, Venezuela
Max Camenzind, Landessternwarte Königstuhl, Heidelberg, Germany
John Dyson, The University of Leeds, England
Suzan Edwards, Smith College, Northampton, USA
George Herbig, Institute for Astronomy, Honolulu, USA
Alex Raga, Instituto de Astronomia, UNAM, México D.F., México
Bo Reipurth (Co-chair), European Southern Observatory, Santiago, Chile
Luis Felipe Rodríguez, Instituto de Astronomia, UNAM, México D.F., México

Local Organizing Committee

Alain Castets (Chair), Observatoire de Grenoble, France
Fabien Malbet, Observatoire de Grenoble, France
Françoise Bouillet, Observatoire de Grenoble, France

I. Herbig-Haro Objects, H_2 Flows and Radio Jets

50 YEARS OF HERBIG-HARO RESEARCH

From discovery to HST

BO REIPURTH
European Southern Observatory
Casilla 19001, Santiago 19, Chile

AND

STEVE HEATHCOTE
Cerro Tololo InterAmerican Observatory
Casilla 603, La Serena, Chile

Abstract. We review the events leading to the discovery of Herbig-Haro objects half a century ago, and the early efforts to understand the nature of these enigmatic objects. The recognition in the mid-seventies of the shocked nature of HH objects heralded a burst of observational and theoretical efforts, and further impetus was soon after provided by the discovery of high proper motions, and by detailed optical, infrared and ultraviolet spectroscopic studies. The recognition in the early eighties of HH jets was the starting point for the increasingly intense studies during the last 15 years which we discuss in this Symposium. In the second half of our review, we summarize the insights into the nature of HH jets provided by analyzing high resolution images obtained with the *Hubble Space Telescope* of two of the finest known HH jets, HH 47 and HH 111.

1. The Discovery of HH Objects

The first example of what we now call Herbig-Haro objects was seen more than a century ago, when Burnham (1890, 1894) looked with the naked eye through the Lick Observatory 36-inch refractor towards T Tauri and saw the faint glimmer of what has come to be known as Burnham's Nebula, or HH 255. The matter rested there for another half century until in the late forties George Herbig and Guillermo Haro independently discovered some curious semi-stellar objects in Orion. A series of four brief papers resulted, which uncovered the basic properties of these objects. In 1950, Herbig drew attention to the peculiar spectrum of Burnham's Nebula with

B. Reipurth and C. Bertout (eds.), Herbig–Haro Flows and the Birth of Low Mass Stars, 3–18.

its strong emission lines of [S II], [O II] and H (Herbig, 1950). This was followed in 1951 by another paper in which Herbig presented a photograph of HH 1, 2 and 3, and noted their spectral similarity to Burnham's Nebula (Herbig, 1951). In 1952 and 1953, Haro presented the discovery of several more objects, and noted that they were invisible on "infrared" plates. Herbig had postulated the existence of either faint blue high-temperature stars or late-type dwarfs inside the HH objects, and Haro concluded from his observations that if the stars were cool he should have detected them (Haro, 1952, 1953). Ambartsumian (1954,1957) suggested that these peculiar nebulous objects, which he called Herbig-Haro objects, might represent the early stages of newly formed T Tauri stars, based on their co-existence with nearby nebulous or emission-line stars (e.g. Herbig, 1946, Haro, 1950).

We have asked George Herbig to describe the circumstances of discovery of HH objects, and he has kindly sent us the following:

To the best of my recollection, and without going through all my early records and correspondence, it went about like this. While looking around for new T Tauri stars as part of my thesis, I ran across BD -6° 1253 (now V380 Ori), which illuminates NGC 1999. A note on this was published in PASP in 1946. In 1946-47, I took some direct photographs of the region of NGC 1999 with the Crossley reflector at Lick, and noticed some odd little fuzzy blobs nearby; these later became HH-1, -2 and -3. According to my notes, the first such plate was taken on 1946 Jan 24, followed by 2 others in Jan. and Feb. 1946. I have among my papers only an enlargement of a plate taken the next year, Jan. 20, 1947, with the same telescope and exposure time. This shows the 3 HHs.

I paid no serious attention to these Objects at the time, but in December 1949 I met Haro at the AAS meeting in Tucson. He gave a paper on his objective-prism discoveries of emission-H-alpha stars around the Orion Nebula, an abstract of which appeared in AJ in 1950, and called attention to the emission-line spectra of these Objects near NGC 1999. He published details later in ApJ in 1952 and 1953. This re-ignited my interest in these spectra, because during the winter of 1948-49 at McDonald I had obtained spectra of Burnham's Nebula at T Tauri, which had the same odd combination of emission lines including [S II] and [O II]; a paper on this appeared in ApJ in 1950.

So at Lick in 1950, I obtained slit spectra of HH-1 and -2, from which came the note in ApJ in 1951, in which attention was drawn to the similarity to Burnham's Nebula. It was probably this connection with T Tauri that gave rise to the conjecture that Herbig-Haro Objects, as they were named by Ambartsumian, had something to do with early stages of star formation.

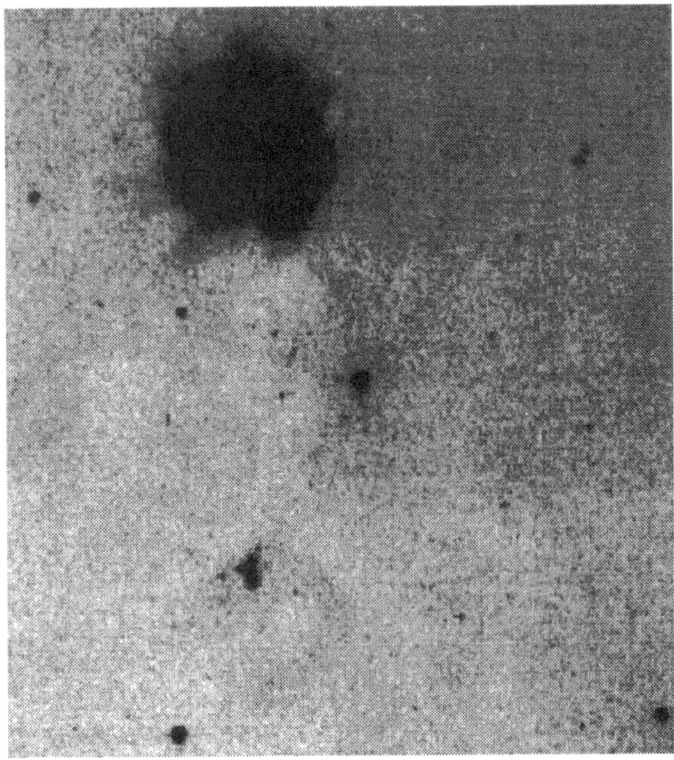

Figure 1. HH 1, 2 and 3 as seen in an enlargement of the Jan 20, 1947 plate which was published in Herbig (1951). The plate was taken in the blue spectral region with the Crossley reflector at Lick Observatory. Courtesy George Herbig.

In Figure 1 we show the Jan. 20, 1947 plate of Herbig, which appeared in his 1950 paper. As our present IAU Symposium started on Jan. 20, 1997, we were thus able to celebrate the 50th anniversary of this first published photograph of HH objects.

2. The Early Years

In 1956, Karl-Heinz Böhm published his spectrophotometric analysis of HH 1, and concluded that the nebula was non-uniform with a mean electron temperature of $T_e = 7500$ K and electron densities between a few times $10^3 cm^{-3}$ to a few times $10^4 cm^{-3}$. In the absence of any known energy source and faced with the relatively high ionization levels observed, Böhm speculated that a low luminosity hot star could be hidden in the nebular condensation. Hoyle (1956), on the other hand, argued that one should not exclude accretion energy as the source of excitation.

In 1958, a short paper appeared by Don Osterbrock in which he argues against photoionization from a hot source inside the HH objects. In the case of Burnham's Nebula he suggested that, as already surmised by Herbig in his two discovery papers, high velocity mass loss ejectae from T Tauri might deposit energy in the nearby nebula. Magnan & Schatzman (1965) proposed that even more energetic protons (100 MeV) hitting a neutral medium could produce the observed ionization.

Haro & Minkowski (1960) demonstrated, based on new and deeper plate material, that no star down to very faint limits could hide in the HH nebulae and argued that the condensations might contain protostars.

An important observational fact about HH objects, namely their variability, was documented by Herbig (1957, 1958, 1968, 1973), who showed that individual HH knots could gradually fade or brighten on timescales of a few years, and new ones could even appear.

Herbig (1974) compiled a list of all HH objects known up to then, a total of 43 objects. For comparison, the latest listing contains about four hundred HH objects (Reipurth, 1997), and new objects are discovered all the time.

3. Shock Physics, Kinematics and Multi-Wavelength Data

After the relative hiatus of the sixties, Herbig-Haro research experienced a true avalanche of papers in the mid-seventies, and only a small number of key studies, especially observational ones, can be mentioned here.

An important discovery was made by Strom *et al.* (1974a), when they detected an embedded infrared source displaced from HH 100 in the dark clouds of Corona Australis. This led to the proposal that HH objects are small patches of reflected light projected onto cloud material through cavities around deeply embedded newborn stars (Strom *et al.*, 1974b).

An alternative model, which had a dramatic effect on Herbig-Haro research, appeared in 1975, when Dick Schwartz proposed that a supersonic stellar wind from T Tauri would create radiating shocks where the flow encounters the ambient medium. The basis for this idea was the observation of lines with supersonic radial velocities around T Tauri, and the similarity of HH spectra to those of certain supernova remnants.

In 1978, three papers appeared which came to have a profound influence on our way of thinking of HH objects.

Dick Schwartz elaborated on his shock interpretation, and proposed a more specific model in which small cloudlets are run over by strong stellar winds from T Tauri stars. In this way stationary shocks are formed on the side towards the star (Schwartz, 1978).

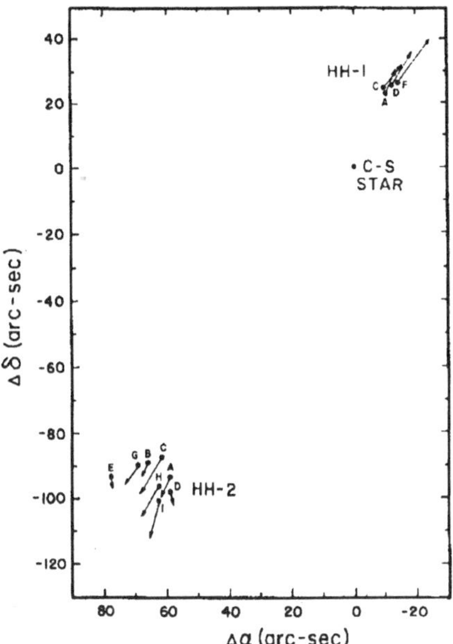

Figure 2. Proper motions of HH 1 and 2, after Herbig & Jones (1981). The Cohen-Schwartz star was earlier thought to be the driving source of the HH complex (Cohen & Schwartz 1979) until the discovery of the VLA source midway between the two HH objects by Pravdo *et al.*(1985).

Karl-Heinz Böhm noted that there are two possibilities for creating such shocks, the quasi-stationary shocks of Schwartz, and alternatively running shocks, similar to blast waves. He demonstrated that the time scales of variability, and the sizes, radial velocities and small filling factors of individual knots could be understood in terms of such shocks (Böhm, 1978).

Finally, Mike Dopita presented a large compilation of spectrophotometry of many HH objects, and a comparison with shock models. In the case of the HH 46/47 objects, he noted a discrepancy between the large radial velocities observed and the lack of [OIII] emission, which implied much lower shock velocities. He concluded that the HH objects were moving into a co-moving medium, and suggested that their sources would be eruptive, similar to FU Orionis eruptions (Dopita, 1978a).

In 1979, Gary Schmidt & Joe Miller published their spectropolarimetry of HH 24, an object known for its large polarization. They concluded that the emission lines are unpolarized, while it is the continuum that is responsible for the observed polarization. This is consistent with the lines forming in shock waves, and the continuum being due to reflected light. The dispute over models was thus closed in favor of the shock model.

Two papers appeared in 1979 and 1981 by Cudworth & Herbig and by Herbig & Jones which had a tremendous impact on the field. In these papers it was established that HH 28/29 and HH 1/2 have very large proper motions of the order of several hundred km/sec. Figure 2 shows the proper motion vectors of HH 1 and 2 from Herbig & Jones (1981),which established the bipolar nature of these objects. In those days the Cohen-Schwartz star was assumed to be responsible for these HH objects (Cohen & Schwartz, 1979), and only later was the central VLA source discovered (Pravdo *et al.*, 1985). For those who have entered the field of HH research within the last 15 years it is perhaps difficult now to imagine the profound effect this figure had on us then. As a curiosity it can be mentioned that the large proper motions of HH 28/29 was noted already back in 1963 by Luyten, who assumed that they were faint red stars.

1980 was the year when unexpected spectroscopic characteristics of HH objects were discovered.

First, Böhm and collaborators drew attention to some rare low excitation HH objects with peculiar line ratios, including very strong forbidden S II lines. These objects could not be explained by the shock models employed in those days, a first sign that plane-parallel shock models were inadequate for dealing with HH objects, thus anticipating the need for bow shock models (Böhm *et al.*, 1980).

1980 was also the year when Jay Elias published his important paper on the detection of molecular hydrogen emission lines by infrared spectroscopy of a number of HH objects. Already in this early paper he pointed out that the shocks producing these H_2 lines could have shock velocities of not more than 20 km/sec, and that this was discrepant with the shock velocities determined from optical lines and plane-parallel models (Elias, 1980).

At the other side of the visible spectrum, Ortolani & D'Odorico (1980) published their ultraviolet spectrum of HH 1 between 1000 and 2000 Å, showing a surprisingly strong UV continuum and emission lines. In a more detailed study shortly afterwards, Böhm *et al.* (1981) noted that HH 1 emits about 20 times more energy in the ultraviolet than in the whole visible range, and pointed out the inadequacy of the then existing shock models to account for the lines detected of very high ionization, like for example C IV.

During the early eighties, the locations and properties of the energy sources of HH objects were gradually established. As an example, L1551 IRS 5, which had been discovered at near-infrared wavelengths by Strom *et al.* (1976), was detected at two far-infrared wavelengths in a balloon-borne experiment by Fridlund *et al.* (1980), who thus derived a total luminosity of this HH source of 25 $L_\odot$ and documented the importance of cold circumstellar dust emission. In a large study, Cohen & Schwartz (1983) surveyed the surroundings of numerous HH objects at near-infrared wavelengths and detected many of the sources with which we are today familiar.

In 1980, Snell, Loren and Plambeck published what is certainly among the most cited papers related to low mass star formation, namely their discovery of the bipolar molecular outflow emanating from L1551 IRS 5, and the presence of HH 28 and 29 in the blue lobe. The relation of HH objects to molecular outflows has become a vital field of research, and is discussed in detail by Cabrit *et al.* and Padman *et al.* elsewhere in this volume.

And now we finally get to the jets. As early as 1978, Bart Bok published an excellent photograph of the HH 46/47 objects in a globule in the Gum Nebula. But it was not until the spectroscopic work of Dopita *et al.* (1982) and Graham & Elias (1983) that it was established as a fine bipolar HH complex, and indeed Dopita *et al.* (1982) realized that the HH 46/47 objects form a bipolar jet emanating from a newborn star. The following year, Mundt & Fried published their highly influential paper about jets from young stars, in which they presented the discovery of four HH jets. This quickly lead to the idea that many, perhaps most, HH objects might be a manifestation of highly collimated outflows.

This is approximately where we stood at the time of the first Herbig-Haro symposium in Mexico City in 1983. This second symposium devoted to Herbig-Haro flows and their relation to star formation documents the enormous advances which the field has experienced in the intervening 14 years. We cannot even begin, in the limited space available, to summarize the enormous body of observational and theoretical work which has been done during this period. Instead, in the remainder of this review, we will focus on what has been learned about two of the best studied HH jets, HH 111 and HH 46/47. One of the most exciting recent developments has been the availability of extremely high resolution images of these and other jets obtained with the *Hubble Space Telescope* and we will place especial emphasis on what such images have to tell us about the nature of HH flows.

4. The HH 111 Jet

The HH 111 jet, together with the very similar HH 34 jet, is the object which comes closest to the text book idealization of how a jet should look. Consequently, it has been the subject of intensive study using the full gamut of observational techniques and has become one of the favorite bench marks against which theoreticians test their models.

The HH 111 outflow is driven by a young star deeply embedded within a compact molecular core in the L 1617 cloud complex in Orion (Reipurth, 1989). In ground based images (see Figure 3) the jet appears as a bony

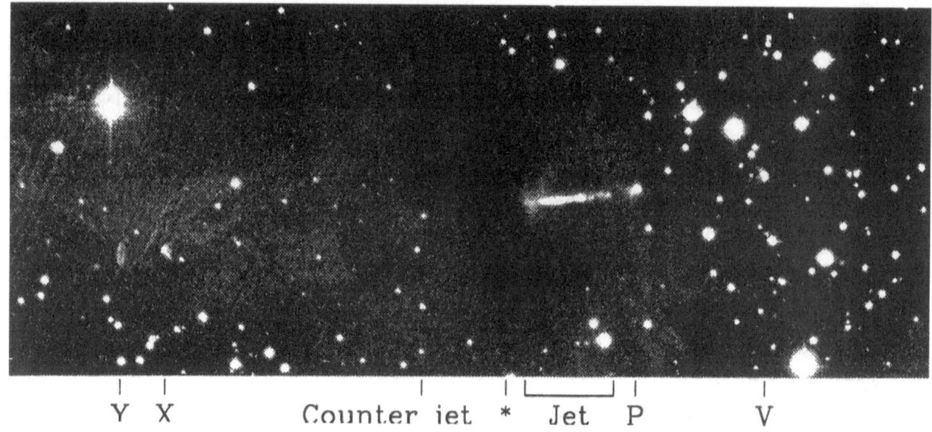

Figure 3. The HH 111 complex as seen on a Gunn-r CCD image obtained at the ESO 3.6m telescope.

finger pointing away from the location of the (optically invisible) source towards an outlying bow shock HH 111V. The jet and V are both blue shifted. Proper motion and radial velocity measurements show that they recede from the source with a space velocity of about 400 $\mathrm{km\,s^{-1}}$ and that the flow is inclined at an angle of only 10° to the plane of the sky (Reipurth *et al.*, 1992). On the opposite side of the source from V, a counter jet and a pair of bow shocks, HH 111 X and Y, trace the red shifted lobe of the outflow. At optical wavelengths the counter jet is highly obscured and consequently faint. However, infrared observations show that the two sides of the flow are in reality highly symmetric (Gredel & Reipurth, 1994).

The most violent action in a jet should occur at the "working surface" where the supersonic flow slams into the surrounding gas, so it is natural to identify high excitation, bow shaped objects like HH 111V with such crash sites. In Figure 4 we give a close up view of this feature obtained using the WFPC2 imager on *HST* (Reipurth *et al.*, 1997a). This object should consist of two shocks, a "reverse shock" (or Mach disk) which decelerates the fast moving jet gas and a "forward shock" (or bow shock) which accelerates the ambient material. At the spatial resolution of *HST* (a WFPC2 pixel subtends 46 AU or 1.5×10^{13} cm at the distance of HH 111) it is possible, for the first time, to resolve the cooling zone behind these radiative shocks and examine their structure.

In the Hα image (Figure 4a), the brightest emission comes from an unresolved filament which wraps around the leading edge of knot V. A broader arc of [S II] bright material runs parallel to this Hα rim, but is displaced $0\farcs5$ (230 AU) closer to the source (Fig. 4b). This stratification, already detected in ground-based images (Reipurth *et al.*, 1992), is especially clear in

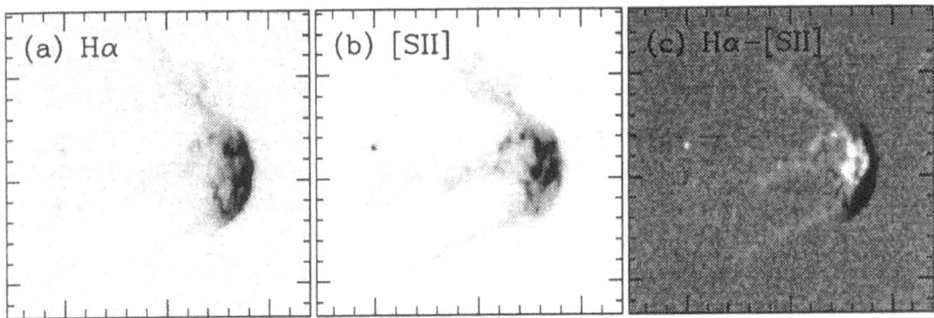

Figure 4. A detailed view of the HH 111V working surface as seen with *HST*: (a) Hα; (b) [S II]; and (c) an Hα − [S II] difference image in which Hα-bright regions are shown in black and [S II]-bright regions in white.

Fig. 4c which shows the result of subtracting the [S II] image from the Hα image. The region enclosed by these leading arcs has a complex structure of clumps and filaments with characteristic sizes of 100-250 AU.

The leading Hα bright filament traces the layer immediately following the "forward shock" where neutral hydrogen atoms are collisionally excited giving rise to strong Balmer emission (Chevalier & Raymond, 1978; Heathcote *et al.*, 1996). This thin skin of collisionally excited Hα wraps around the apex of the bow shock and appears as a bright rim where the line of sight runs tangent to its surface. Hα filaments seen projected against the core of knot V probably arise where corrugations in the surface of the shock front bring it close to the line of sight. The arc of [S II] emission behind the shock front arises in the extended cooling zone where gas heated by passage through the "forward shock" cools and recombines emitting strong forbidden lines of various metals as well as recombination lines of hydrogen. The clumped structure of this region may result from instabilities in the cooling flow. There is no obvious Balmer filament which might delineate the "reverse shock" so its location is less clear. Some of the condensations in the interior of HH 111V emit both [S II] and Hα and are probably material cooling behind the bow shock. Others have strong [S II] emission but little Hα and are probably shocked jet material.

The spectrum of HH 111V presents us with the same puzzle faced by Dopita (1978b) in the case of HH 46/47. Although the space velocity of V is $\sim 400\ \mathrm{km\,s^{-1}}$ no [O III] emission is detected implying that the shock velocity must be $< 100\ \mathrm{km\,s^{-1}}$. These facts can only be reconciled if HH 111V advances into gas which is flowing away from the driving source at high velocity. Comparison between the observed two dimensional velocity field of V and that predicted by bow shock models also suggests that the pre-shock medium moves outward from the source at $300\ \mathrm{km\,s^{-1}}$ (Morse *et al.*, 1993a). This implies that, rather than being the terminal shock of the

HH 111 flow, V is an internal working surface, following in the wake of material ejected earlier. Indeed Reipurth *et al.* (1997b) have demonstrated that the driving source is bracketed by two more distant working surfaces, HH 113 and HH 311, so that in Figure 3 we are only seeing the core of a giant outflow with a total extent of more than 7 pc! The HH 111 complex is only one of several cases where multiple working surfaces are found within a single outflow, confirming the idea first suggested by Dopita (1978b) that the mass loss from new born stars is episodic with major eruptions occurring on time scales of a few centuries.

From the moment of their recognition, HH jets have posed a difficult conundrum; how to account for their characteristic low excitation spectra, indicative of a shock velocity of at most several tens of $\mathrm{km\,s^{-1}}$, in face of their supersonic space velocities of several hundred $\mathrm{km\,s^{-1}}$. Two competing explanations have been advanced. On the one hand, Ray *et al.* (1988) proposed that Kelvin-Helmholtz instabilities at the boundary of the jet beam could excite oblique (hence weak) shocks within it, which would appear as low excitation emission knots. Various observations of jets have been interpreted in the framework of this model (Bührke *et al.*, 1988; Eislöffel & Mundt, 1992). On the other hand, Dopita (1978b) and Reipurth (1989) have argued that jets are highly transient, intermittent events driven by eruptions of their driving sources. Variations in the velocity of the flow would generate internal working surfaces propagating along the jet and having shock strengths comparable to the amplitude of the velocity perturbation (Raga *et al.*, 1990; Stone & Norman, 1993). Observations of knots in jets have been interpreted in light of this model by several authors (e.g. Heathcote & Reipurth, 1992; Reipurth *et al.*, 1992; Morse *et al.*, 1993b).

These two models predict very different internal structures for the knots, which can easily be distinguished at the spatial resolution of *HST*. Fig. 5 provides a detailed view of the bright section of the HH 111 jet (Reipurth *et al.*, 1997a) which is seen to consist of a chain of well resolved knots embedded in a diffuse, gently wiggling sheath of faint [S II] emission. The bright knot L at the tip of the jet is very obviously an internal working surface having a morphology similar to that of V. This result was foreshadowed by the ground based images of Reipurth *et al.* (1992). Further support for this interpretation comes from the fact that this knot has a large velocity dispersion comparable to that of V (Reipurth *et al.*, 1997a) and additionally shows weak [O III] emission (Morse *et al.*, 1993b) implying that it has a shock velocity higher than that of V. Most of the other bright knots in the jet also have clear bow shock shapes. Some are centered on the axis of the jet and form complete arcs with wings swept back symmetrically on both sides, while others are displaced slightly from the axis of the jet and are one-sided with a trailing wing on the "outside" of the jet. Typically these

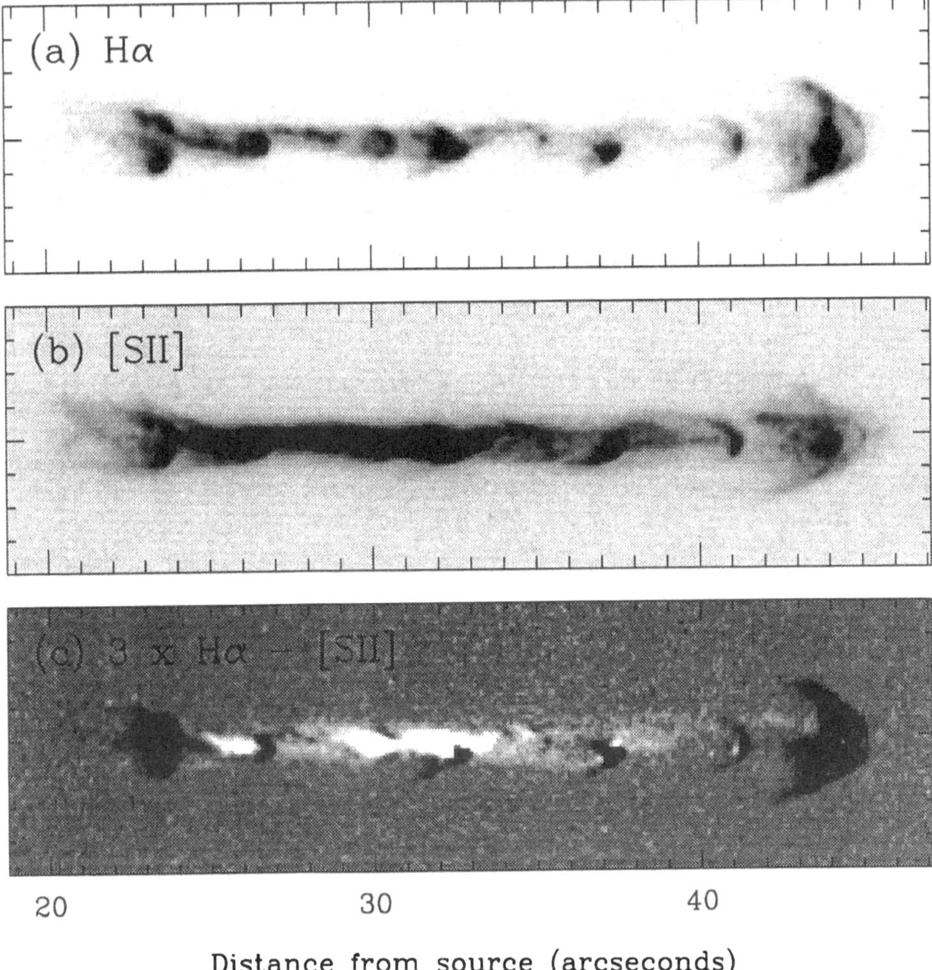

Figure 5. *HST* images of the HH 111 jet: (a) Hα; (b) [S II]; and (c) 3×Hα − [S II].

knots show the same kind of excitation stratification as HH 111V with an Hα filament wrapped around a [S II] bright core. This morphology strongly suggests that, at least in the HH 111 jet, the knots are internal working surfaces caused by variations in the ejection velocity of the jet. If a new impulse emerges precisely along the axis of the jet, then we see a symmetrical bow, whereas if the ejection occurs at a slight angle to the axis of the jet a one-sided feature results. The leading Hα arcs mark the shock front while the [S II] emission arises in the post-shock cooling zone. The spacing of the knots and the proper motion of the jet imply that the internal working surfaces are the result of eruptions which occur at intervals of a few decades.

The HH 111 complex is associated with a molecular outflow (Reipurth & Olberg 1991, see also Cernicharo *et al.*, this volume). High spatial resolution millimeter-wave interferometric data show that the low velocity molecular gas is unusually well collimated and has the form of a cylinder surrounding the HH jet (Cernicharo *et al.*, 1997). This system thus provides the perhaps most persuasive example of a molecular flow driven by an HH jet. The extended wings of a bow shock are effective in sweeping up and accelerating the ambient gas. Thus the passage of a succession of internal working surfaces along the jet could well, over the course of time, have dragged material from the dense molecular core to form the observed CO flow. The wings of the faint bow shocks beyond knot L have a lateral extent comparable to the radius of the molecular cylinder, and may interact with, and be confined by the molecular gas. In addition to the slow molecular flow, Cernicharo & Reipurth (1996) also discovered several high velocity CO bullets in the gap between the tip of the jet and HH 111V and in the region beyond it. These bullets have space velocities of 400-500 $\mathrm{km\,s^{-1}}$ comparable to that of the jet. They are thus unlikely to be entrained material, but rather must be jet material ejected from the driving source. Infrared observations also reveal H_2 knots embedded in the working surfaces L and V (Gredel & Reipurth, 1994). Thus the bullets are probably dense material trapped and squeezed between the radiative shocks in the working surfaces which has cooled sufficiently to form molecules.

5. The HH 47 Jet

Next we turn to HH 46/47, the object which got the entire HH jet bandwagon rolling (Dopita *et al.*, 1982). It emanates from an isolated Bok globule which harbors a highly obscured infrared source. The flow is bipolar with the northeastern lobe approaching us and the southwestern lobe receding. The HH 47 jet itself, located in the approaching lobe, emerges from a reflection nebula illuminated by the hidden source and terminates in the bright bow shock HH 47A. The jet is surrounded by an elongated bubble of faint shock excited emission, HH 47D. In Hα and [S II] only a cap of this bubble is visible. However, in [O II] the wings of HH 47D extend all the way back to the reflection nebula completely enclosing the jet (Hartigan *et al.*, 1990). In the optical, the receding lobe is traced by a faint stubby counter jet which points towards a short arc of emission HH 47C. However, as in the case of HH 111 this asymmetry is in large part the result of overlying obscuration. In the infrared, HH 47C is seen to mark the apex of a complete bubble of H_2 emission very similar to HH 47D (Eislöffel *et al.*, 1994). The total extent of the complex from HH 47D to HH 47C is about 0.57pc. Combining the radial velocity measurements of Morse *et al.* (1994) with

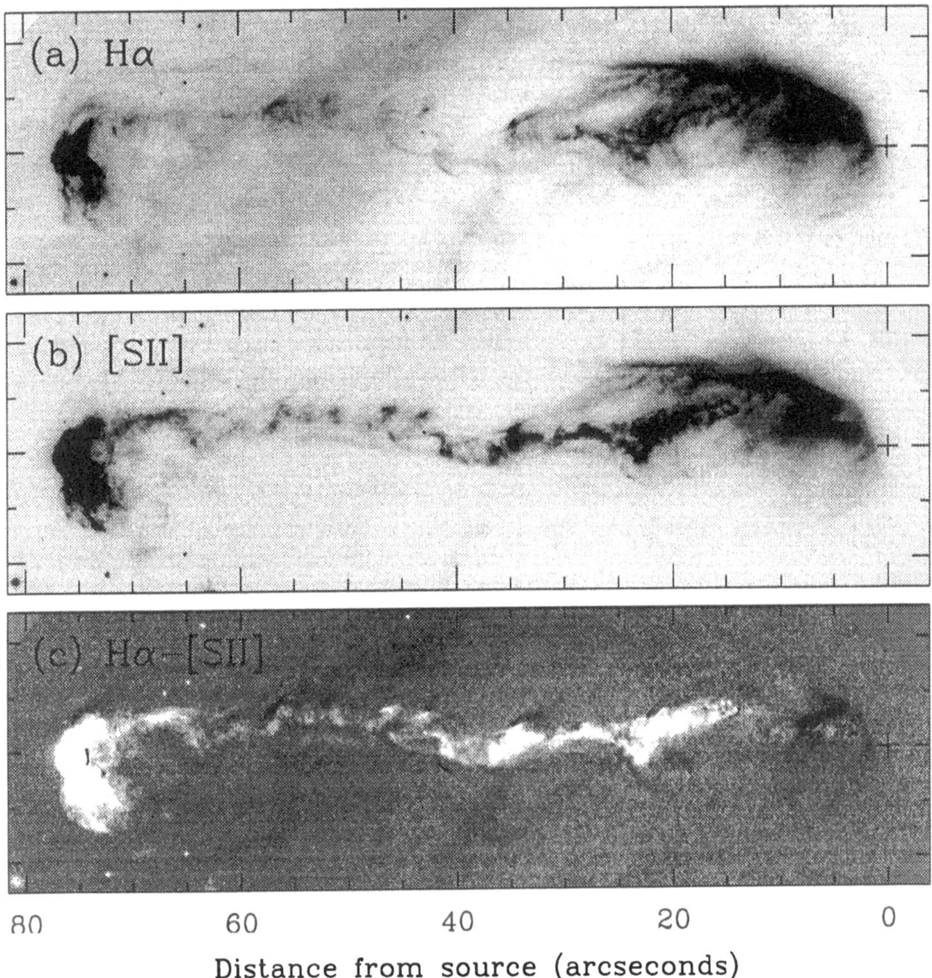

Figure 6. *HST* images of the HH 47 jet: (a) Hα; (b) [S II]; and (c) Hα - [S II].

the proper motions derived by Eislöffel & Mundt (1994), and assuming a distance of 450 pc, one finds that the axis of the flow makes an angle of 28° to the plane of the sky. The jet recedes from the source with a space velocity of 300 km s^{-1}, while space velocities of similar magnitude are seen in the counter flow.

In Figure 6 we show the HH 47 jet as seen with *HST*. At first glance, the strongest impression given by these images is the very different appearance of the jet in Hα (Fig. 6a) and [S II] (Fig. 6b). While in [S II] we see a winding chain of bright knots, the Hα frame shows a delicate tracery of narrow filaments. The relationship between these two structures becomes clear in the Hα – [S II] difference image shown in Fig. 6c. Here we see that the [S II]-bright knots delineate the "core" of the jet, while the Hα emission comes predominantly from the zone along the edges of the jet. This segregation of [S II]-bright and Hα-bright material was already apparent in the ground based images of Reipurth & Heathcote (1991). However, the *HST* images reveal the remarkable thinness of the Hα filaments, several of which are unresolved even with *HST*. The most prominent Hα filaments trail behind knots in the core of the jet like the wings of miniature bow shocks. Many of the other knots appear to be cloaked in a thin skin of Hα emission. Especially in [S II], the HH 47 jet shows considerable wiggling with several abrupt changes in direction. Many of the best Hα filaments occur at these apparent bends in the jet, and invariably lie on the "outside" of the corner.

It is very unlikely that the flow truly follows the tortuous path traced out by the [S II] emission. These images only show the gas where it is radiating. Rather, each fluid element in such a highly supersonic flow will move ballistically away from the source. Thus the apparent sinuous structure of the HH 47 jet is most likely the consequence of changes in the *direction* of ejection of the jet where it leaves the source. Again the Hα filaments are the result of collisional excitation at shock fronts and the Balmer arcs serve to mark the location of internal working surfaces in the flow. The knotty structure of the HH 47 jet is thus also the result of variability of the driving source. The very different appearance of the HH 111 and HH 47 jets results from the differing importance of directional variations in the two systems. In the case of HH 111, the amplitude of the directional variations is small so that an almost straight, narrow jet with primarily axially symmetric internal working surfaces result. Conversely, the HH 47 jet has undergone larger changes in the direction of ejection leading to a broader, wiggling and more chaotic jet with predominantly one-sided internal working surfaces. Although on short time scales changes in the ejection direction dominate in the HH 47 jet, the multiple working surfaces found in this flow attest to the occurrence of massive eruptions of the source on time scales of several hundred years.

The ground based Fabry-Perot observations of Hartigan *et al.* (1993) showed that the material at the core of the HH 47 jet moves more rapidly than that at the edges, suggesting that the jet is transferring momentum to the surrounding medium and entraining it. This together with the fact that the Hα/[S II] ratio is greater at the edges of the jet than at its core was initially thought to be evidence for turbulent entrainment (Raymond *et al.*, 1994). Larger Hα/[S II] ratios translate into higher shock velocities, consistent with a turbulent entrainment model in which the strongest shocks should occur in the boundary layer between the jet and the ambient medium. However, at the higher resolution of the *HST* images it is apparent that the edges of the jet emit Hα along several well defined shock fronts rather than in a chaotic boundary layer. This suggests that, instead, entrainment is occurring as the result of the collective effect of a succession of internal working surfaces which sweep up and accelerate the gas along the edges of the jet. The HH 47 flow is also accompanied by a molecular flow (Chernin & Masson, 1991; Olberg *et al.*, 1992). However, this flow is effectively unipolar, with a well developed receding lobe, but only a weak approaching lobe. This is most likely because the approaching lobe of the HH 47 flow advances into the medium outside the dense globule so that there is relatively little material for it to sweep up. The reflection nebula at the base of the jet has a complex filamentary structure with the filaments extending parallel to the jet. This perhaps suggests that, at least near its base, the HH flow is dragging dense dusty material out of the cloud.

Acknowledgements. We thank George Herbig for providing his recollections of the years of discovery of HH objects and for a print of Fig 1. We also thank our collaborators John Bally, Pat Hartigan and Jon Morse for many stimulating conversations.

References

Ambartsumian, V.A. 1954, Comm. Byurakan Obs. No. 13
Ambartsumian, V.A. 1957, in IAU Symp. No. 3, ed. G.H. Herbig, Cambridge Univ. Press, p.177
Bok, B.J., 1978 PASP, 90, 489
Böhm, K.-H. 1956, ApJ, 123, 379
Böhm, K.-H. 1978, A&A, 64, 115
Böhm, K.-H., Brugel, E.W., Mannery, E. 1980, ApJ, 235, L137
Böhm, K.-H., Böhm-Vitense, E., Brugel, E.W. 1981, ApJ, 245, L113
Bührke, T., Mundt, R., & Ray, T.P. 1988, A&A, 200, 99
Burnham, S.W. 1890, MNRAS, 51, 94
Burnham, S.W. 1894, Pub. Lick Obs. 2, 175
Cernicharo, J., Reipurth, B., 1996, ApJ, 460, L57
Cernicharo, J., Neri, R, Reipurth, B. 1997, *in preparation*
Chernin, L.M., & Masson, C.R. 1991, ApJ, 382, L93
Chevalier, R. A., & Raymond, J. C. 1978, ApJ, 225, L27
Cohen, M., & Schwartz, R.D. 1979, ApJ, 233, L77
Cohen, M., & Schwartz, R.D. 1983, ApJ, 265, 877
Cudworth, K.M., & Herbig, G.H. 1979, AJ, 84, 548
Dopita, M.A. 1978a, ApJ Suppl., 37, 117
Dopita, M.A. 1978b, A&A, 63, 237
Dopita, M.A., Schwartz, R.D., & Evans, I. 1982, ApJ, 263, L73
Eislöffel, J., & Mundt, R. 1992, A&A, 263, 292
Eislöffel, J., & Mundt, R. 1994, A&A, 284, 530
Eislöffel, J., Davis, C.J., Ray, T.P., & Mundt, R. 1994, ApJ, 422, L91
Elias, J.H. 1980. ApJ, 241, 728
Fridlund, C.V.M. et al. 1980, A&A, 91, L1
Graham, J.A., & Elias, J.H. 1983, ApJ, 272, 615
Gredel, R., & Reipurth, B. 1994, A&A, 289, L19

Haro, G. 1950, AJ, 55, 72
Haro, G. 1952, ApJ, 115, 572
Haro, G. 1953, ApJ, 117, 73
Haro, G., & Minkowski, R. 1960, AJ, 65, 490
Hartigan, P., Raymond, J., & Meaburn, J. 1990, ApJ, 362, 624
Hartigan, P., Morse, J.A., Heathcote, S., & Cecil, G. 1993, ApJ, 414, L121
Heathcote, S., & Reipurth, B. 1992, AJ, 104, 2193
Heathcote, S., Morse, J.A., Hartigan, P., Reipurth, B., Schwartz, R.D., Bally, J., & Stone, J.M. 1996, AJ, 112, 1141
Herbig, G.H. 1946, PASP, 58, 163
Herbig, G.H. 1950, ApJ, 111, 11
Herbig, G.H. 1951, ApJ, 113, 697
Herbig, G.H. 1957, in IAU Symp. No. 3, ed. G.H. Herbig, Cambridge Univ. Press, p.3
Herbig, G.H. 1958, in *Stellar Populations*, ed. D.J.K. O'Connell, Vatican Observatory, p.127
Herbig, G.H. 1968, in Non-Periodic Phenomena in Variable Stars, IAU Coll., ed. L. Detre, Reidel, p.75
Herbig, G.H. 1973, Inf. Bull. Var. Stars No. 832
Herbig, G.H. 1974, Lick Observatory Bulletin No. 658
Herbig, G.H., Jones, B.F. 1981, AJ, 86, 1232
Hoyle, F. 1956, ApJ, 124, 484
Luyten, W.J. 1963, Harvard Annu. Card No. 1589
Magnan, C., & Schatzman, E. 1965, C.R. Acad. Sci. Paris 260, 6289
Morse, J.A., Heathcote, S., Cecil, G., Hartigan, P., & Raymond, J.C. 1993a, ApJ, 410, 764
Morse, J.A., Heathcote, S., Hartigan, P. & Cecil., G. 1993b, AJ, 106, 1139
Morse, J.A., Hartigan, P., Heathcote, S., Raymond, J.C., & Cecil, G. 1994, ApJ, 425, 738
Mundt, R., & Fried, J.W. 1983, ApJ, 274, L83
Olberg, M., Reipurth, B., & Booth, R.S. 1992, A&A, 259, 252
Ortolani, S., & D'Odorico, S. 1980, A&A, 83, L8
Osterbrock, D.E. 1958, PASP, 70, 399
Pravdo, S. et al. 1985, ApJ, 293, L35
Raga, A.C., Cantó, J., Binette, L., & Calvet, N. 1990, ApJ, 364, 601
Ray, T.P., Bührke, T., & Mundt, R. 1988, in NATO ASI on *Formation and Evolution of Low Mass Stars*, ed. A.K. Dupree & M. Lago, (Kluwer), p.281
Raymond, J. C., Morse, J.A., Hartigan, P., Curiel, S. & Heathcote, S. 1994, ApJ, 434, 232
Reipurth, B. 1989, Nature, 340, 42
Reipurth, B. 1997, *A general catalogue of Herbig-Haro objects*, electronically published via anon. ftp to ftp.hq.eso.org, directory /pub/Catalogs/Herbig-Haro, 2.edition, available Aug 1997
Reipurth, B., & Heathcote, S. 1991, A&A, 246, 511
Reipurth, B. & Olberg, M. 1991, A&A, 246, 535
Reipurth, B., Raga, A.C., & Heathcote, S. 1992, ApJ, 392, 145
Reipurth, B., Hartigan, P., Heathcote, S., Morse, J.A., & Bally J. 1997a AJ, *submitted*
Reipurth, B., Bally J., & Devine, D. 1997b AJ, *submitted*
Schmidt, G.D., & Miller, J.S. 1979, ApJ, 234, L191
Schwartz, R.D. 1975, ApJ, 195, 631
Schwartz, R.D. 1978, ApJ, 223, 884
Snell, R.L., Loren, R.B., Plambeck, R.L. 1980, ApJ, 239, L17
Stone, J.M., & Norman, M.L. 1993, ApJ, 413, 21
Strom, K.M., Strom, S.E., Grasdalen, G.L. 1974a, ApJ, 187, 83
Strom, S.E., Grasdalen, G.L., Strom, K.M. 1974b, ApJ, 191, 111
Strom, K.M., Strom, S.E., Vrba, F.J. 1976, AJ, 81, 320

HST OBSERVATIONS OF THE L1551 IRS5 JET

MALCOLM FRIDLUND
Astrophysics Division, Space Science Department
European Space Agency
ESTEC, P.O. Box 299, NL 2200AG, Noordwijk
The Netherlands

MONICA HULDTGREN
Stockholm Observatory
S-133 36, Saltsjöbaden, Sweden

AND

RENE LISEAU
Stockholm Observatory
S-133 36, Saltsjöbaden, Sweden

Abstract. The jet emanating from the young stellar object, known as IRS5, located in the molecular cloud L1551 in Taurus has been imaged with the Hubble Space Telescope. The observations and a preliminary interpretation of them is presented. The relevant background history for this important object is briefly reviewed.

1. Introduction & background

The L1551-IRS5 Young Stellar Object (YSO) has been in the focus of front line research for the last 20 years or more, at least as what concerns low mass star formation. It is often the first object where a new phenomenon is observed, or where a new technique has been tried out. When we now report the first Hubble Space Telescope (HST) observations, it is also appropriate to briefly review this source. A more general review also exists in the form of the paper of Staude & Elsässer (1993).

The compact molecular cloud L1551 is located in the southern part of the Taurus molecular cloud region. It shows abundant signs of ongoing star formation, and it has received great attention by those who study this pro-

B. Reipurth and C. Bertout (eds.), Herbig–Haro Flows and the Birth of Low Mass Stars, 19–28.

cess. It is roughly spherical with a diameter of 1°, located at a distance of 140 pc (Elias 1978, Kenyon *et al.* 1994) and has an estimated mass of 100 $M_{\odot}$ (Sandqvist & Bernes, 1980). Optical nebulosity, superposed on the surface of the cloud, is visible on photographic plates of the region. This was classified as a HII region (S239) by Sharpless (1959). The number 1551 is in the listing of Lynds (1962). Felli & Perinotti (1974) did, however, not detect any radio continuum emanating from this region and suggested that the visible emission emanated from a planetary nebula. Two detached small nebulous knots appear in the compilation by Herbig (1974) as Herbig-Haro objects number 28 and 29. The same objects were noted by Luyten (1963, 1971) to have very high proper motions. In studies of regions containing Herbig Haro objects (Strom *et al.* 1974, Strom *et al.* 1976) it was shown, that the S239 nebula possesses the spectroscopic characteristics of HH objects, and as a consequence this object was 're-designated' HH102. In their 2μm data they found one embedded object associated with the nebulosity - number 5 in their listing (IRS5). A first study of the molecular gas in the region, utilising the mm radio lines of CO and ^{13}CO was carried out by Knapp *et al.* (1976). They found large mass motions as evidenced by very broad line wings, but their spatial resolution was too poor to discern the structure of the flow. They suggested that ordered free-fall collapse of gas onto a newborn stellar cluster was responsible for the broad wings detected in their spectra. Cudworth & Herbig (1979) did a proper motion study of objects in the region confirming the high velocities of HH28 & HH29 detected by Luyten (1963, 1971). They found that transversal velocities of the order of 150 km s^{-1} were indicated, and even more importantly, that when extending the proper motion vectors backwards in time, they intersected near IRS5 – but at different epochs (timescales of 600 and 2000 years for HH29 and HH28, respectively). Sandqvist & Bernes (1980) mapped the cloud in the 6cm, 2cm and 2mm transitions of formaldehyde (H_2CO). They estimated the total cloud mass to be $\approx$ 100 $M_{\odot}$, while they found a very dense core, centered also on IRS5, with a calculated mass of $\geq$ 0.7 $M_{\odot}$. The same year IRS5 was detected at the far infrared wavelengths of 83μm and 155μm (Fridlund *et al.*, 1980). The total luminosity inferred from these observations was surprisingly low - only 25 $L_{\odot}$, which suggested that the object was a low mass stellar object in an early stage of evolution. This conclusion – which was confirmed by the near and mid infrared measurements by Beichman & Harris (1981) – was even more surprising when Snell, Loren and Plambeck (1980) at about the same time detected the first bipolar molecular outflow. This was found centered on IRS5, and these authors found that the outflow was ordered into spatially well separated diametrically opposite lobes – one receeding and one approaching. A lower limit to the mass involved in these motions could also be derived, and the low luminosity found in the IR observations then immediately indicated that the outflow could not be radiatively driven. Later, IRAS and KAO results increased the bolometric luminosity somewhat to $30L_{\odot}$ to $40\ L_{\odot}$, but changed neither this conclusion, nor the one that IRS5 is a low mass star (Fridlund *et al.*, 1980).

In a series of studies (Cohen *et al.* 1982, Bieging *et al.* 1984, Bieging & Cohen 1985), radio emission was detected from the immediate vicinity of IRS5 and aligned along the major axis of the CO outflow in a 'jet-like' morphology. It was quite clear given the low luminosity of IRS5 that it could not provide the UV flux necessary to photoionize the gas to a degree sufficient to provide the observed radio continuum. Taken together with the morphology, this indicated instead an origin in shock heated gas.

Cohen *et al.* (1982) noted the coincidence between the 'radio jet' in the VLA data, and a jet-like object visible on a plate taken by Strom *et al.* (1974). This object was imaged in detail by Mundt & Fried (1983), who also detected a number of other optically visible jets, associated with YSO's and with other outflow phenomena. A number of studies now concentrated on detecting IRS5 itself. At one point it was actually believed that the innermost knot visible in the jet was IRS5 itself. It was eventually realised that IRS5 was in fact well hidden behind a very large amount ($\geq$ 100 magnitudes) of visual extinction, albeit relatively near the optically visible inner edge of the jet. VLA & MERLIN observations (Curiel, 1995) show how the knotty structure continues inwards, closer to the source where no optical emission is visible, and the appearence of new radio knots are now being traced into the optical (Curiel, 1995, 1997 private communication; Fridlund & Liseau 1994). At this time, it was realised, that the reflection nebulosity character of S239 (alias HH102) allowed a possibility of determining the spectroscopic type of IRS5. Mundt *et al.* (1985) detected P Cygni-profiles of a type observed in FU Orionis type stars, and assigned a spectral type of G to K, also consistent with IRS5 belonging to this class of objects. Direct 2μm spectroscopy (Carr *et al.*, 1987) provided further evidence of this classification, as well as suggesting the presence of shocked molecular gas very close to IRS5.

Imaging (in the Gunn r band) of the jet was carried out at three epochs between 1983 and 1987 (Neckel & Staude, 1987). Their results showed the lifetime of the knots along the jet to be in excess of 4 years, the V_{tan} for the knot system to be 190 km s^{-1}, and a new knot that had appeared at the inner end of the jet during the epoch of the observations. This study was continued by Fridlund & Liseau (1994), who performed imaging at 3 intervals between 1989 and 1993, thus expanding the total time base to 10 years. These data were collected through R, I, Hα and [SII] filters. The images were acquired during seeing conditions between 0.6 arcsec and

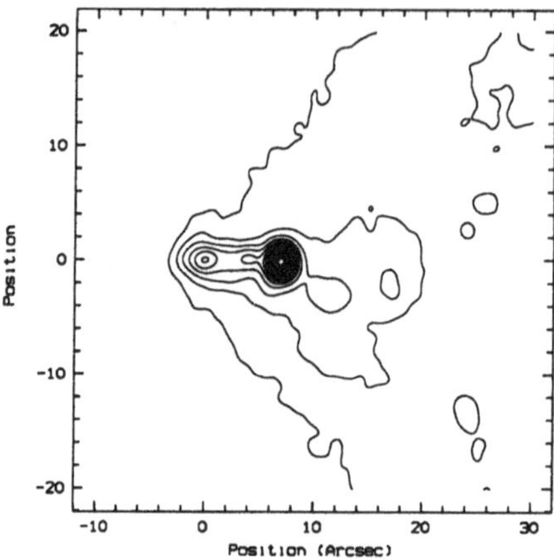

Figure 1. Knots B – F shown on a R band image. The data was obtained with the Nordic Optical Telescope. This Figure is from Fridlund & Liseau (1994). Note that the orientation in this Figure is with the jet horizontally and not like it really is – with 17° inclination to the E – W direction.

0.8 arcsec. Although the knots of Neckel & Staude (1987) were clearly identifiable, their positions had changed significantly (see Figure 1). One knot appeared to be merging with the end point ('working surface') of the jet, and a new knot had appeared, which was also identified in radio data obtained during the same epoch (Curiel, 1995). No less than three transversal velocity systems were detected, and an estimate of the force required to accelerate the individual knots were found to be too small – by two orders of magnitude – than that required to accelerate the bulk of the molecular gas near IRS5 (see Fridlund & Knee, 1993). This is of course for an ionisation fraction of one.

The present ground based capabilities had thus been brought to its limits, and in order to further study the kinematics - particularly the acceleration/deceleration of individual knots – a multi cycle series of observations with HST appeared well motivated. This is the astrophysical jet closest to the Earth, and thus the shock structure along the jet and at the 'working surface' can be studied in unprecedented detail.

2. HST Observations

The area surrounding the L1551 IRS5 jet was imaged using the Wide Field and Planetary Camera 2 (WFPC2) aboard the Hubble Space Telescope (HST) (Trauger *et al.*, 1994). This camera consists of 4 adjacent 800 × 800 Loral CCD's. Three of these arrays have an image scale of 0.1″ per pixel, while the fourth – the planetary camera (PC) – has an image scale of 0.046″ per pixel. The FOV in this last camera is 25 × 25 arcsec^2, and since the jet visible in the ground imagery is less than 15 arcsec long, we centered the PC on the position of IRS5. This will allow us to resolve the smallest elements in any known astrophysical jet. Then, however, we unfortunately loose the multiplexing advantage since with this orientation we can not simultaneously image HH29 or any of the other HH objects within the SW outflow lobe (there are none known within the NE lobe). We obtained two exposures each through the F656N (Hα) and F673N ([SII] λ6717,6731Å) filters on 1996.093. Total exposure time for each filter was 2500s. On 1996.210 we obtained 3 exposures each through the F675W (R) and F814W (I) filters with total observing times of 1800s and 2400s respectively. The data were reduced using methods described e.g. in Holtzman *et al.*, (1995). The calibration of the data into units of erg cm^{-2} s^{-1} were done according to Holtzman *et al.*, (1995), Heathcote *et al.*, (1996) and the WFPC2 handbook. We compare the results with those of Fridlund & Liseau (1994) in Table 1.

TABLE 1. Absolute fluxes for knot D and the total jet in this work and in Fridlund & Liseau (1994 – FL94)

Feature	Hα	[SII]	Source
D	2.29E-14	2.33E-14	HST
Σ_{Alljet}	5.17E-14	4.93E-14	HST
D	1.8E-14	1.9E-14	FL94
Σ_{Alljet}	3.3E-14	2.9E-14	FL94

3. Results

Figure 1 is a contour plot of part of a R-band image obtained from the ground (Nordic Optical Telescope) in 1989 under 0.6 arcsec seeing conditions (Fridlund & Liseau, 1994). In this Figure we have identified the features F to B from that paper. One can interpret feature D as constituting a 'working surface' shock at the end of the visible jet. This notion is strengthened by the apparent ellipticity of this feature. At low intensity levels a 'bow-shock' appearance is also clearly visible in such images. In Figure 2 we show greyscale images of the I-band and R-band HST images. The resolution of these images are then 12 times higher than in the best ground based data available. This is also immediately apparent. The

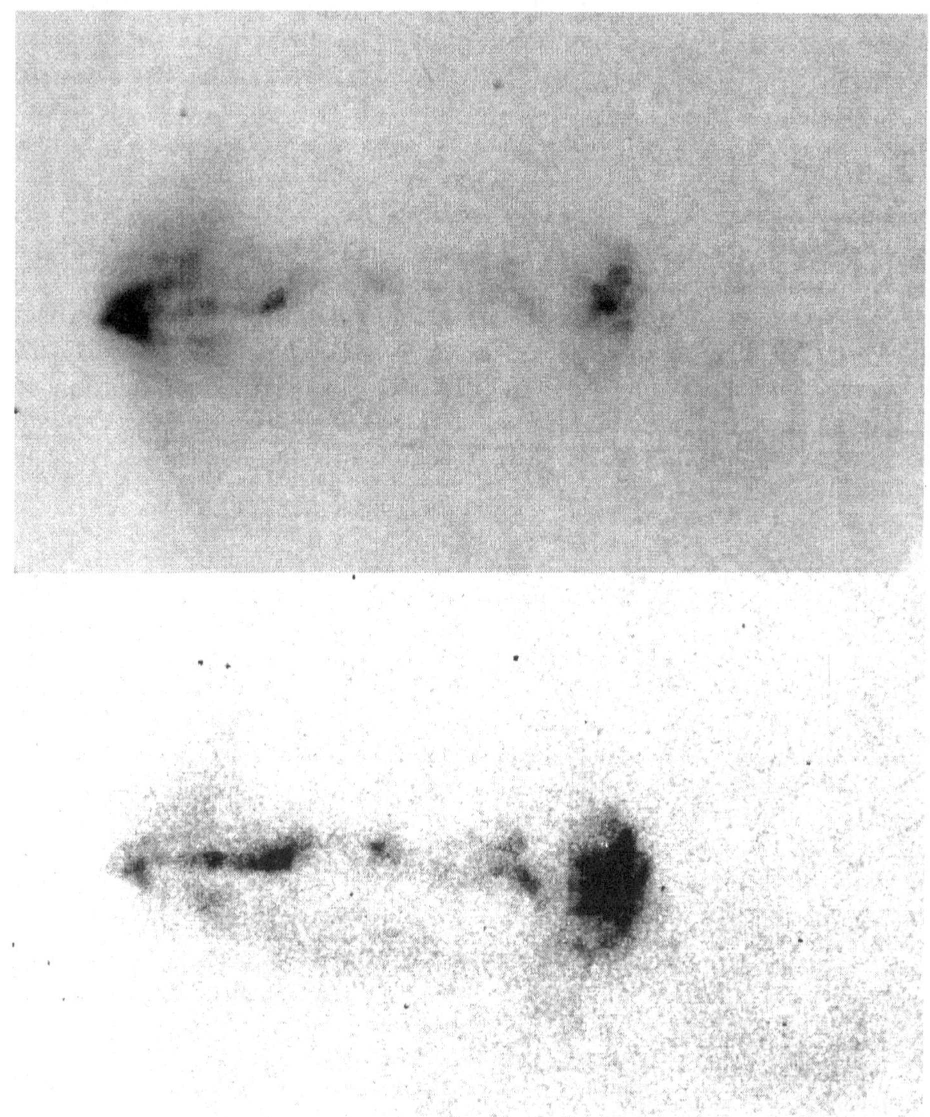

Figure 2. The HST/PC I-band (top) and R-band images. The images have been rotated to match the orientation in Figure 1.

body of the jet itself is much narrower and shows a clear 'wiggling' as it progresses outwards. Features F & D can be readily identified in the HST data – but with much more substructure. Knot D is very different from

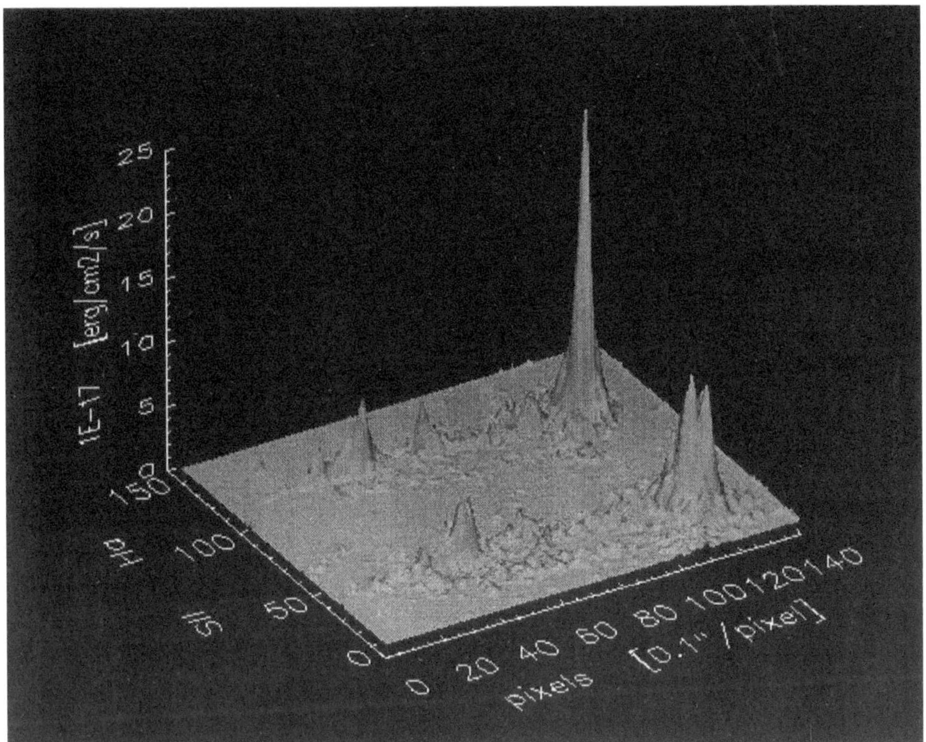

Figure 3. Intensity along the jet in Hα and [SII] as 3D plots. Images have been smoothed to 0.1 arcsec resolution, and have been rotated to match the orientation in Figure 1.

the ground based picture, in that it has a much more irregular appearence with a small bright core instead of the smooth elliptic, 'bow-shock' like shape visible from the ground. In Figure 3 we display the Hα and the [SII] data as 3-D plots. These images have been smoothed to a resolution of 0.1 arcsec. On this image, as well as in the raw data one can clearly see two jets. One brighter towards the north, and one fainter towards the south. It is also obvious that feature D is much brighter in Hα than in [SII]. There is also more structure in the latter. To display this we show contour plots of the line emission images in Figure 4, expanded so that only the 2 arcsec by 2 arcsec surrounding knot D are shown. The inherent resolution in these plots is 6.5 AU.

We have measured the cross section of the jet along its extent and we find that there is no apparent expansion of the visible material. The FWHM is found to be almost constant at 0.42±0.13 arcsec from where the signal/noise becomes high enough to get a reasonable measurement, until it 'disappears' into knot D. This corresponds to a 'jet-diameter' of

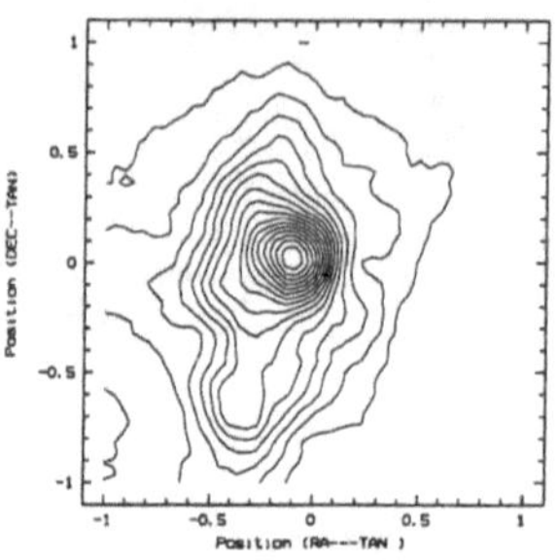

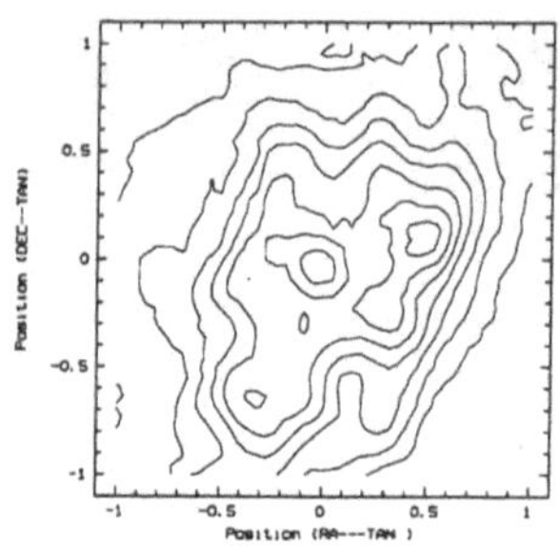

Figure 4. Hα and [SII] contour representations of knot D. Offsets are in arcseconds, and the direction towards IRS5 is in the upper left.

60AU±20AU. As mentioned above, and as can be seen in Figure 2, the brighter portion of the jet 'wiggles' as it progresses outwards. If one plots the peak brightness of the jet, one can see that the area on the sky onto which this 'wiggling' takes place, is cone-shaped with the apex just where the jet first emerges (with a FWHM of 0.4 arcsec). The cone defined by the 'jet-wiggling' expands to $\approx$ 2 arcsec just as the jet reaches knot D. This could conceivably be representative of the projected shape of a cavity within which the jet moves. The nodes defined by the undulation of the jets are separated by $\approx$ 500AU or 7.5×10^{15} cm. The opening angle defined by the expanding envelope is 15°. Comparing the Hα data with the I-band image in Figure 2 we see that the well collimated jet emerges out of a cone shaped reflection nebula which possesses an opening angle of 70°. This cone immediately starts to bend 'inwards' towards the jet axis on both edges, and after only about 1.1 arcsec it appears to have become 'collimated', and there are two bright, more or less parallell streams with a separation of 1.3 arcsec. These features are all seen superposed on top of a fainter nebulosity with a larger opening angle ($\approx$ 150°), which is identical to the cone shaped nebula seen in ground based data (e.g. Fridlund & Liseau, 1994, Figures 1c & 1d), and which is assumed to trace the molecular outflow cavity. The emission from the cone shaped nebula is almost certainly pure reflection, since no trace of it is visible on either the Hα or the [SII] images – each of which traces high excitation (high velocity shocks) and low excitation (low velocity shocks) respectively. The same is true for a bright spot (diameter $\approx$ 0.3 arcsec) which is visible in the I-band image (and very faintly also in the R-band data) very near the apex of the cone shaped nebula. Although very tentative, a possible interpretation of this signature is that it is due to light reflected from a rift in the extinction towards IRS5.

No crossing shocks are immediately evident from the Hα and the [SII]

Figure 5. Difference between Hα (white) and [SII] (black) images for the bright feature in knot D. The area displayed is $1'' \times 1''$. The direction towards IRS5 is in the upper left

images. A careful examination of the emission maxima show that the ionic species are slightly displaced from each other, and positioned close to the 'nodes' mentioned above. This suggest an interpretation where they actually represent shocks at the edge of the cavity walls, which arise due to interaction with the undulating jet.

The feature called knot D is resolved into one very bright Hα peak and several fainter [SII] intensity maxima, each of which have a FWHM of 0.3 arcsec (see Figure 4). It is seen that what was interpreted as a bow shock in the ground based data now is only the fainter envelope surrounding these peaks. The single bright core seen in the Hα image, clearly has the shock on the side *away* from IRS5, as defined by the steepness of the intensity gradient. Taking the difference between the Hα and [SII] we can clearly see (Figure 5) that they are well separated with [SII] on the outside (relative to IRS5). This would indicate that the Mach disk is the shock with the higher excitation, and thus that the jet is less dense than the ambient material that it encounters. On the other hand, the steep gradient on the side *away* from IRS5 is indicative of a jet denser than the ambient medium. The elongation of the bright Hα feature has its long axis well pointed towards IRS5.

4. Conclusions

We have observed the jet emanating from L1551 IRS5 with the Hubble Space Telescope. We find that the brightest knot has a pronounced small scale structure, and that those small features ($\leq$ 40AU) have a high excitation character. The shock is located on the side *furthest away* from the originating source on the Mach disk, but the highest excitation is to be found on the side of the Mach disk *towards* IRS5.

At the origin of the visible jet, it emerges already well collimated to a diameter of $\approx$ 40AU. It appears to 'wiggle' within a cavity of diameter $\approx$ 100AU, until it reaches the assumed 'working surface', i.e. knot D. The nodes of the undulation of the jet are separated along the jet axis by $\approx$ 500AU.

Acknowledgements: The authors gratefully acknowledge the support of STSCI staff. We are pleased to acknowledge the excellent arrangements for the symposium in Chamonix, and finally MF thanks the organisers for the opportunity to present these results.

References

Beichman, C., Harris, S., 1981 ApJ 245, 589
Bieging, J.H., Cohen, M., Schwartz, R.D., 1984, ApJ 282, 699
Bieging, J.H., Cohen, M., 1985 ApJ 289, L5
Carr, J.S., Harvey, P.M., Lester, D.F. 1987 ApJ 321, L71
Cohen, M., Bieging, J.H., Schwartz, R.D., 1982, ApJ 253, 707
Cudworth, K.M., Herbig, G., 1979 AJ 84, 548
Curiel, S., 1995 RevMexAASC Series de conferencias, volumen 1 abril 1995, p59
Elias, J., 1978 ApJ 224, 857
Felli, M., Perinotto, M., 1974 Ap&SS 26, 11
Fridlund, C.V.M., Nordh, H.L., van Duinen, R.J., Aalders, J.W.G., Sargent, A.I., 1980 A&A 91, L1
Fridlund, C.V.M., Sandqvist, Aa., Nordh, H.L., Olofsson, G., 1989 A&A 213, 310
Fridlund, C.V.M., Knee, L.B.G., 1993 A&A 268, 245
Fridlund, C.V.M., Liseau, R., 1994 A&A 292, 631
Heathcote, S., *et al.*, 1996, AJ 112, 1141
Herbig, G.H., 1974 Lick Obs. Bull., 658
Holtzmann, J.A., *et al.*, 1995, PASP, 107, 1065
Kenyon, S.J., Gomez, M., Marzke, R.O., Hartmann, L., 1994 AJ 108, 251
Knapp, G.R., Kuiper, T.B.H., Knapp, S.L., Brown, R.L., 1976 ApJ 206, 713
Luyten, W.J., 1963 Harvard Annu. Card No. 1589
Luyten, W.J., 1971 *The Hyades*, Univ. Minnesota P., Minneapolis
Lynds, B.T., 1962 ApJS 7, 1
Mundt, R., Fried, J.W., 1983, ApJ, 274, L83
Mundt, R., Stocke, J., Strom, S.E., Strom, K.M., Anderson, E.R., 1985 ApJ 297, L41
Neckel, Th., Staude, H.J., 1987 ApJ 322, L27
Reipurth, B., 1994, *A general catalogue of Herbig-Haro objects*, electronically published via anon. ftp to ftp.hq.eso.org, directory /pub/Catalogs/Herbig-Haro.
Sandqvist, Aa., Bernes, C., 1980, A&A 89, 187
Sharpless, S., 1959 ApJS 4, 257
Snell, R.L., Loren, R.B., Plambeck, R.L., 1980 ApJ 239, L17
Staude, H.J., Elsässer, H., 1993 A&A Review. 5, 165
Strom, S.E., Grasdalen, G.L., Strom, K.M., 1974, ApJ, 191, 111
Strom, S.E., Strom, K.M., 1974, Vrba, F.J., 1976, AJ, 81, 320
Trauger, J.T., *et al.*, 1994, ApJ 435, L3

GIANT HERBIG-HARO FLOWS

JOHN BALLY AND DAVID DEVINE
Center for Astrophysics and Space Astronomy,
University of Colorado, Boulder CO 80309, USA

Abstract. Recent observations with wide field-of-view CCDs have shown that over 20 Herbig-Haro flows extend for more than one parsec from their driving sources. We review the observed properties of these giant HH flows and discuss the physical consequences for star formation, and for the physics and chemistry of the surrounding interstellar medium.

1. Introduction

The advent of large format (2048 × 2048 pixels or larger) CCD detectors have for the first time provided both high sensitivity (orders of magnitude greater than photographic emulsions) *and* a wide field-of-view (FOV) approaching one degree. The combination of these two parameters has opened opportunities for deep narrow-band imaging of entire star forming clouds. Such observations have shown that many Herbig-Haro (HH) objects trace parse-scale outflows from young stars.

The first parsec-scale Herbig-Haro flow was recognized in 1993 (Bally & Devine 1994) with the 23′ FOV provided by a Tektronix 2k CCD on the KPNO 0.9-m telescope. The HH34 jet (Reipurth et al. 1986) in the Orion A molecular cloud was found to power a 20′ long chain of HH objects with a projected length of 3 parsecs (assuming a distance of 500 pc). This system, containing the previously discovered objects HH33/40, HH34, H34N, HH85/86/87, HH126, and a new object HH173, is powered by a less than $45L_{\odot}$ young stellar object (YSO), HH34 IRS. High quality images of the entire system, obtained with the ESO NTT in January 1994, were compared with images taken in the 1980s with a variety of telescopes to determine proper motions of various knots. These measurements together with R = 6000 resolution long-slit spectra demonstrate that all HH objects north of HH34 IRS move north and are redshifted while all HH objects to the south

B. Reipurth and C. Bertout (eds.), Herbig–Haro Flows and the Birth of Low Mass Stars, 29–38.

of HH34 IRS move south and are blueshifted (see images in Devine 1997; Devine et al. 1997). Radial velocities and proper motions show systematic decrease in velocity with increasing distance from the source. The flow morphology and kinematics exhibits S-shaped point symmetry indicating possible long-term changes in the ejection direction of the bipolar jet from HH34 IRS. High velocity CO emission is confined to the inner 0.1 pc portion of the parsec-scale optical flow. However, shock excited H_2 emission has been detected near the flow ends 1.5 pc from the source (HH85 and HH33/40 in the north; HH86/87/88 in the south), but not in the internal working surfaces closer to the HH34 IRS (see Devine et al. 1997 for details).

Wide FOV imaging of other star forming clouds have demonstrated that parsec-scale HH flows are common with lengths ranging from 1 to over 10 parsecs. Some of the best known and most studied molecular outflows and HH objects contain shocks lying more than 1 pc from their sources. Ogura (1995) reported a parsec-scale flow associated with HH1/2 that contains a 0.5 pc diameter shock close to the HH1/2 outflow axis, HH401, and a counter-shock HH 402 to the southeast of HH1/2 directly opposite HH401. This flow has a projected end-to-end length of about 6 parsecs. Barnard 5 IRS1, the source of a well known molecular outflow, powers a spectacular pair of HH objects (HH366) lying beyond the projected edges of the B5 cloud about 10′ on either side of the source (Bally, Devine, & Alten 1996). Re-analysis of existing millimeter wavelength data shows that the CO flow is co-extensive with the optical flow and much larger than previously thought. The HH111 flow contains HH311 and HH113 (Figure 1), which lie more than 3.8 pc on either side of the VLA source (Reipurth, Bally, Devine 1997). In this case, the optical flow is almost 10 times larger than the associated molecular flow, which is confined to the inner 1 parsec portion of this system (cf. Cernicharo & Reipurth 1996 and references therein). The T-Tauri system is the source of a giant bow shock HH355N which subtends more than 0.2 square parsecs and lies nearly one parsec from the source, and a similar counter bow shock HH 355S on the opposite side of T Tauri; both are seen in Figure 2 (Reipurth, Bally, Devine 1997). The counter-jet in the HH83 system appears to power several shocks lying between it and HH 84, the apparent terminal shock of this flow located more than 3 parsecs from the source at the western edge of the Orion A molecular cloud. Some bow shocks powered by YSOs in L1551 lie well beyond the projected extent of detectable CO emission in regions with such low extinction that galaxies can be seen in the background. Most (but not all) parsec-scale flows discovered to date are powered by low mass YSOs.

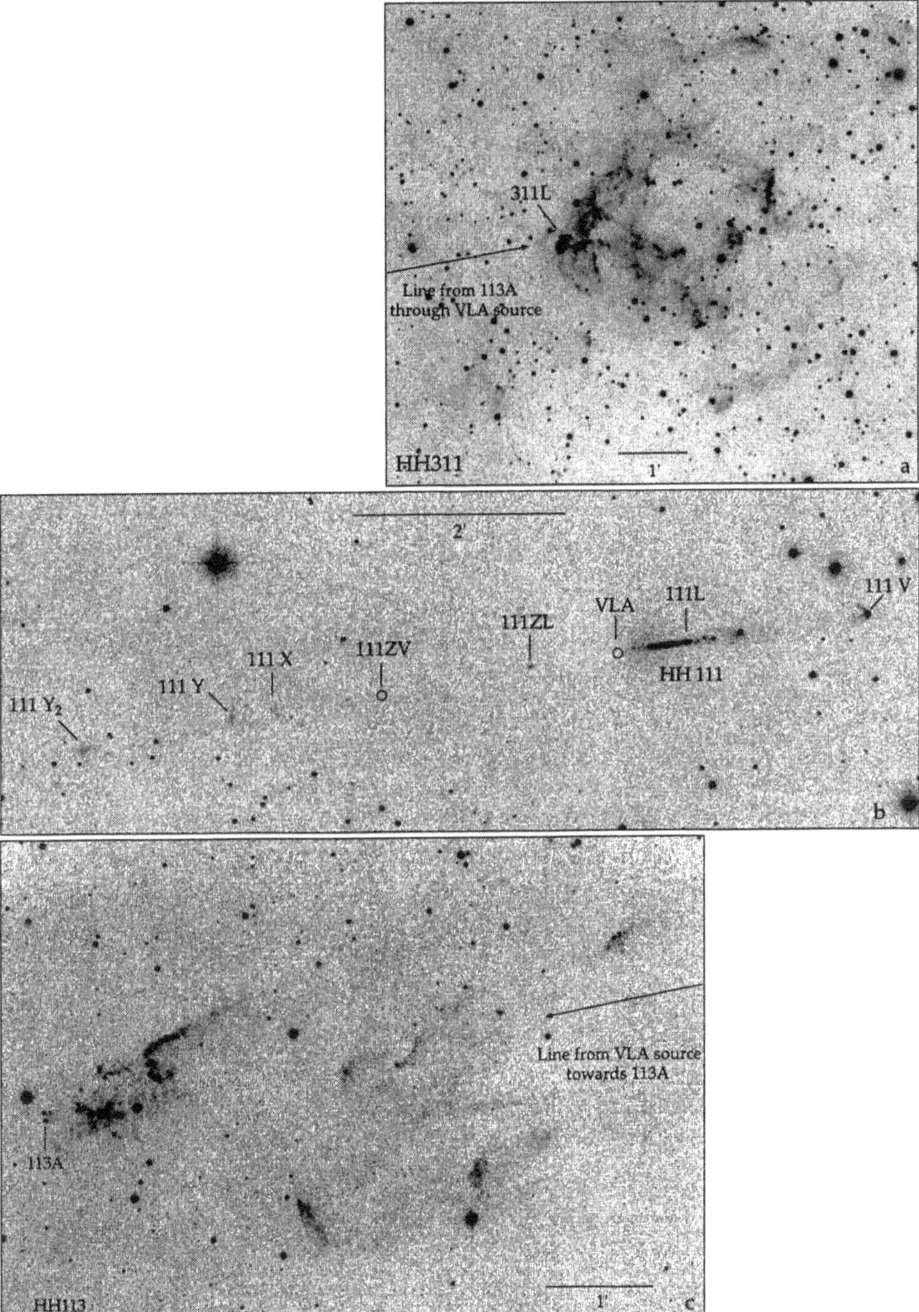

Figure 1. Close-up views of the three major components of the 7 pc long HH111 flow. *Upper Right:* NTT Hα + [S II] image showing HH311 located 3.8 pc west of HH111. *Center:* KPNO 0.9 m [S II] image of the HH111 jet and counter jet. *Bottom Left:* NTT Hα + [S II] image showing HH113 located 3.5 pc east of HH111. Proper motions show that these HH objects are moving away from the source of HH111. The eastern object, HH113, is redshifted and the western object, HH311, is blueshifted. From Reipurth, Bally, Devine (1997).

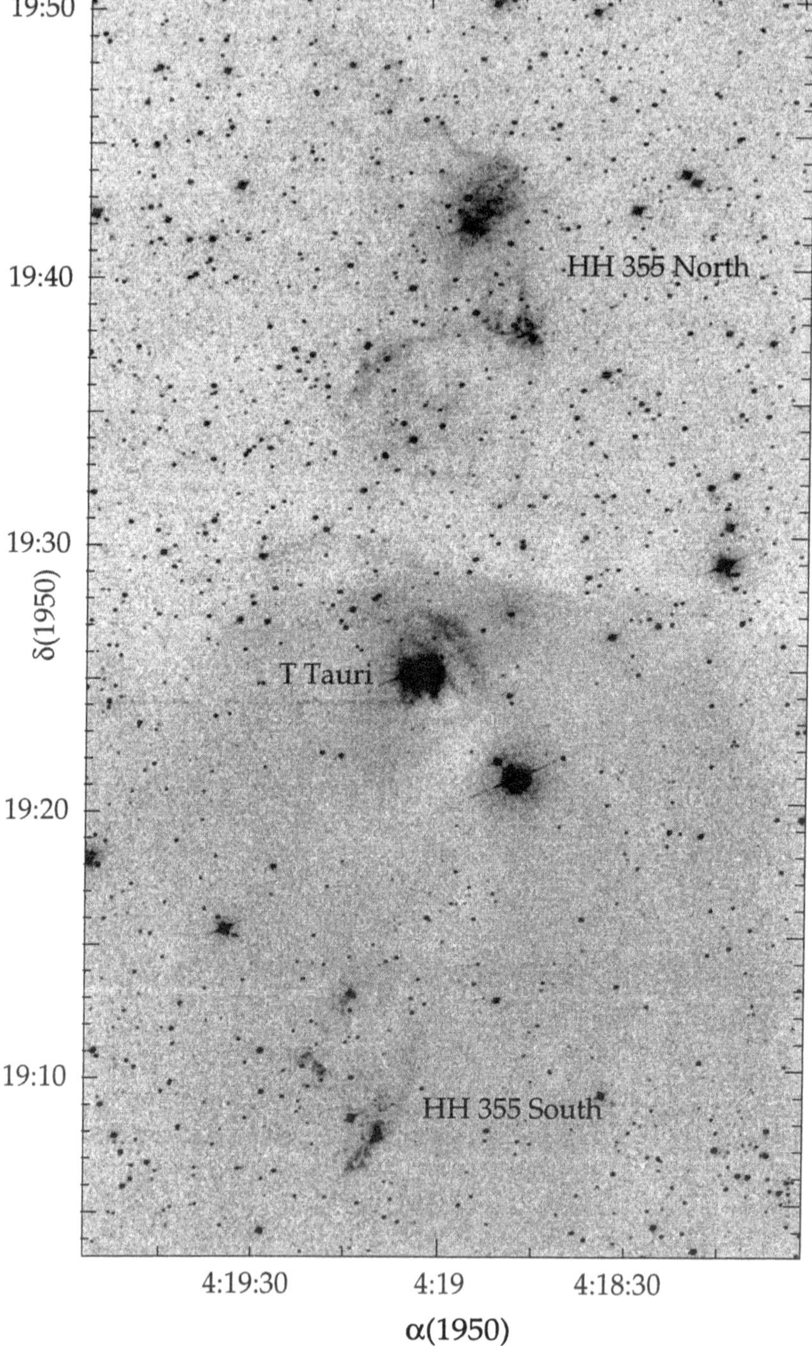

Figure 2. The T-Tauri parsec-scale Herbig-Haro flow, HH355, showing a giant bow shock extending 20′ to the north and a chain of HH objects extending 20′ to the south. From Reipurth, Bally, Devine (1997).

2. Observed Properties

Table 1 lists some giant HH flows and their sizes. In this section, we describe the observed properties of these flows.

Large Dynamic Ages: The dynamical ages of parsec-scale outflows are about an order of magnitude larger than the ages of the previously recognized Herbig-Haro flows. A typical dynamical age for a shock in a parsec scale Herbig-Haro flow is $\tau_{dyn} = 10^4 \ d_{pc}/v_{100}$ years, where d_{pc} is the flow length in parsecs and v_{100} is the apparent velocity in units of 100 km s^{-1}. For the flows discussed here, the dynamical ages of the outermost visible components range from 10^4 to 10^5 years. The longer time scale is comparable to the estimated duration of both the stellar accretion phase and the lifetimes of outflows estimated from CO observations.

Multiple Internal Working Surfaces: Most giant HH flows contain multiple groups of HH objects along the outflow axis, providing evidence for time-variable ejection velocity, ejection direction, and possibly mass-loss rate and degree of collimation. The spacing of shocks tends to increase with increasing distance from the source. Some HH flows contain continuous jets within 0.1 pc of the source. Others have closely spaced shocks separated by several arc minutes in the inner parsec of the flow. Most have large gaps between the outermost shocks that can be up to $10'$ long. The ratio of the observed proper motions divided by the shock velocity deduced from the emission lines is much larger than unity, indicating that most HH objects are shocks formed by faster fluid elements overtaking moving fluid with a slightly lower velocity and *not* stationary gas. Thus, most HH objects are internal shocks that trace interactions between plugs of material ejected at different times with different velocities from a YSO.

S-Shaped Point Symmetry: Many parsec-scale flows have S-shaped point symmetry about the source. Either precession or irregular source orientation changes can lead to S-shaped symmetry. Examples include HH34 and PV-Ceph (Figure 3). At least one source (B5; Bally, Devine, & Alten 1996) has C-shaped symmetry indicating either flow deflection or relative motion of the source and the medium into which the flow is propagating.

Velocities, Sizes, Morphologies: There is a systematic decrease in the mean fluid velocities as determined from both proper motion and radial velocity measurements (cf. HH34; Devine 1997). There is an increase in the linear dimensions of HH objects with increasing distance from the source. Furthermore, shocks look increasingly complex and chaotic with increasing distance from the source.

Blow-Out: Parsec-scale flows have sizes about an order of magnitude larger than the cloud cores from which they originate. Many flows have punched completely out of their parent molecular cloud cores and are in-

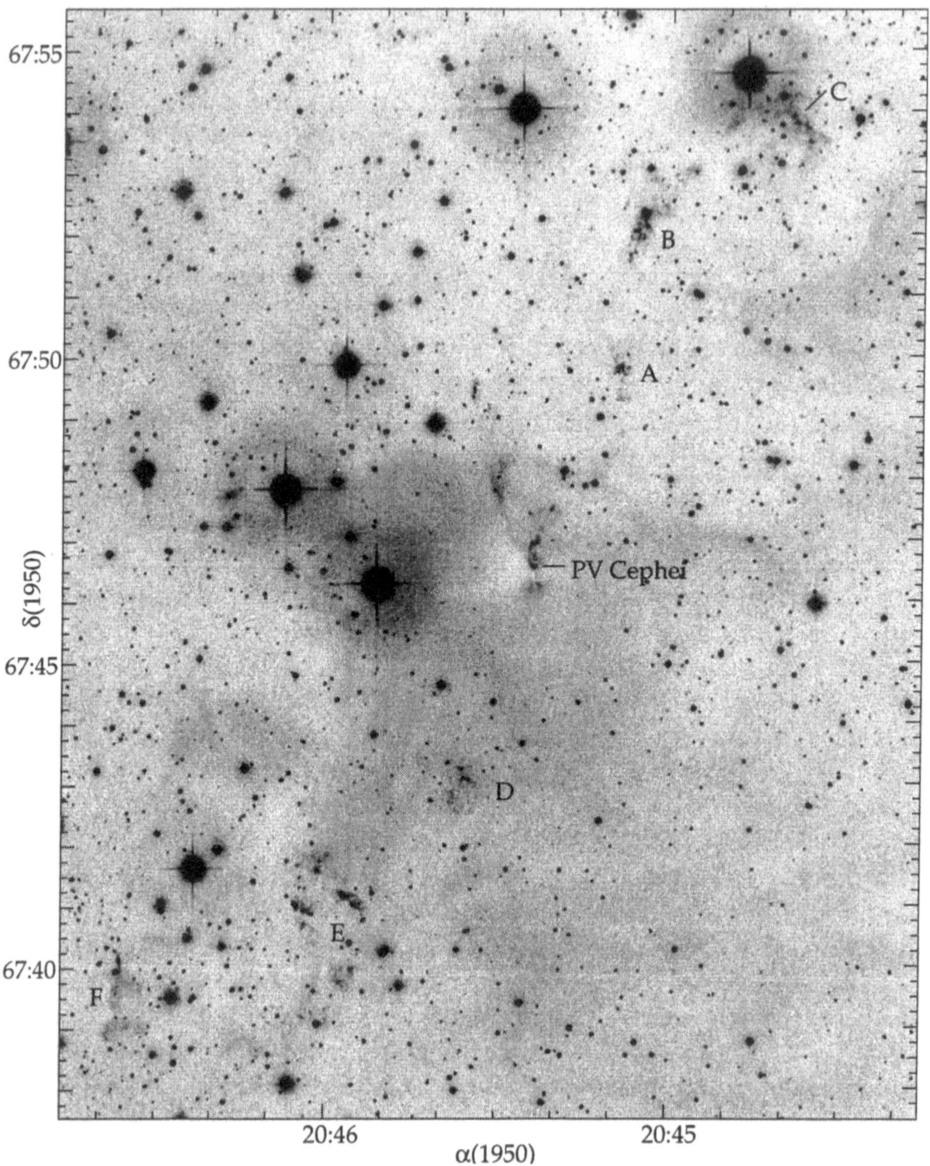

Figure 3. A KPNO 0.9 meter image of the 20′ PV-Ceph parsec-scale HH315 flow (chain of nebulae extending along a northwest – southeast axis) showing S-shaped symmetry about the source. A second flow may emerge from the cloud core towards the northeast. From Reipurth, Bally, Devine (1997).

jecting mass and energy into the surrounding atomic or ionized interstellar medium. For example, the northern shock in the T Tauri system, the north-eastern shocks in the L1551 IRS5 system, and HH311, the western terminus of the HH111 system, lie along a lines-of-sight devoid of CO emission that contain a rich distribution of background stars and galaxies.

Clustering: Most giant HH flows listed in Table 1 were found in regions of relatively isolated star formation such as L1551 which contain about a dozen or fewer YSOs. Even in these regions, there is evidence for multiple outflows that frequently overlap in the plane of the sky. However, many (possibly most) YSOs are found in highly clustered environments where many dozens to hundreds of stars have formed within a parsec-scale region. In such environments, it is hard or impossible to make specific associations between individual HH objects and sources. Images of the Perseus cloud complex reveal many HH objects located more than a parsec from the nearest known YSO or cloud core. Owing to confusion, many such HH objects have not yet been linked to specific flows or sources (cf. NGC1333; Bally, Devine, & Reipurth 1996: L1448 and L1455; Bally et al. 1997).

3. Consequences of Parsec-Scale Herbig-Haro Flows

Our observations demonstrate that HH objects are frequently found parsecs from their driving sources. Therefore, outflows traced by HH objects have physical dimensions at least an order of magnitude larger than previously thought. The large sizes of outflows from low mass stars has a number of implications.

Implications for Outflow Models: The fast component of outflows that excite HH objects can be traced at least as far from their driving sources as associated molecular outflows. In some cases, the optical outflow is an order of magnitude larger than the associated CO flow. Such observations provide support for "unified" outflow models in which a fast, collimated, time-dependent, and bipolar wind (or jet) from a YSO that excites HH objects is also the primary engine that powers the lower velocity lobes traced by CO (Raga & Cabrit 1993; Masson & Chernin 1993, 1994; Chernin et al. 1994). In this model, molecular gas is entrained by the jet from the environment. Collimated and time-variable jets can produce poorly collimated lower velocity molecular outflows by the combined action of jet orientation variations (precession) and velocity variations. As jets change their orientation, they constantly encounter fresh material which is entrained and accelerated. An outflow cavity with a relatively wide opening angle can be produced. Ejection velocity variations produce multiple internal shocks where fast fluid elements overtake slower moving ejecta (e.g. Biro & Raga 1994; Raga & Kofman 1992). The resulting 'splash' of postshock debris spreads orthogonal to the mean flow axis and can contribute to the widening of the flow cavity and the entrainment and acceleration of displaced gas.

Mass-Loss History of YSOs: The morphology and kinematics of parsec-scale flows can be used to constrain the mass loss history of the source YSOs and to probe the evolution of HH objects. Shocks in parsec-scale outflows trace ejecta that are progressively older with increasing distance from the source. The apparent deceleration of the distant ejecta may be a result of interactions with slower moving (or stationary) material in the outflow cavity. Alternatively, older, more distant material may have been ejected at a lower velocity than more recent ejecta. If the latter interpretation is correct, then the mean jet ejection velocity must have changed significantly in the lifetimes of the visible outflow components. With detailed radial velocity, proper motion, and excitation data, it may be possible to partially reconstruct the mass ejection history of the source YSO, and by inference, its mass accretion history. The distribution of shocks (continuous jets or many closely spaced low excitation shocks near the source and widely spaced higher excitation ones farther away) implies that flow velocity variations with low amplitudes occur frequently while large amplitude variations are relatively rare. The inverse correlation between ejection velocity jumps and their frequency is reminiscent of a $1/f$ process.

Probing the ICM and ISM with HH Objects: Most of the volume inside giant molecular clouds is filled with voids that do not emit strongly in the CO lines. The nature of this inter-clump medium (ICM) remains mysterious. The terminal working surfaces of giant HH flows can be used to probe the nature of the ICM. Analysis of the shock properties can provide information about the ionization state, density, and other physical characteristics of the ICM. For instance, detection of shock excited H_2 can be used to infer that the pre-shock medium is predominantly molecular; the presence of Balmer filaments would be an indication that the shock is moving into a neutral atomic medium.

Chemical Rejuvenation: Fast shocks associated with the terminal working surfaces of parsec-scale flows may dissociate molecules, resulting in the 'chemical rejuvenation' of star forming molecular clouds. Dissociation of molecules effectively re-sets the conditions of the effected region to its 'initial' atomic state. Furthermore, shocks may contribute to the production of the large abundance of species such as C I and C II found in the ICM even if UV radiation from O and B stars is absent. Chemical rejuvenation must be an important process in the vicinity of young clusters such as NGC1333 where dozens of active flows can simultaneously churn the surrounding medium and where the mean time between the passage of dissociating shocks over random parcels of gas can be short compared to the dynamic evolutionary time scale of the cloud.

Origin of Turbulence and Self-Regulation of Star Formation: Parsec-scale outflows may be the primary agents for the generation of turbulence in molecular clouds and the surrounding ISM in the absence of massive stars. They may contribute to the self-regulation of star formation. Giant flows sweep-up and expel gas from the parent cloud core, inject it into the surrounding ICM, and, in cases where the source lies near the edge of a molecular cloud, into the surrounding interstellar medium. If the ambient medium is predominantly atomic, the swept up gas will not be traceable by molecular emission lines, and the HH flow may extend much farther than any associated molecular outflow. However, if the ambient medium is predominantly molecular, an extended molecular outflow is produced. The classical molecular outflows are likely to be accelerated during the initial stages of this process and much lower velocity large scale flows formed during the latter stages. Eventually these mass motions degrade in momentum conserving interactions with their surroundings and eventually blend with the surrounding medium. The non-linear growth of instabilities during the earlier radiative phases will likely make the final state chaotic and turbulent.

Prospects for X-Ray and UV-Observations: The 'naked' portions of giant flows, where galaxies can be seen in the background, may be observable at X-ray and UV wavelengths. Emission produced by fast shocks or absorption lines originating from high ionization states of various elements may be observable. Shock-excited emission may make a substantial contribution to X-ray and UV backgrounds of the Galaxy and provide a new probe of the star formation environment. Since velocities in the ambient medium are typically only a few kms^{-1}, the shock velocity in terminal working surfaces reflects the full flow velocity. Assuming that thermal and ionization equilibrium is reached and that the fluid velocity has declined by a factor of f in propagating through the outflow cavity, the immediate post-shock temperature in the adiabatic layer that forms behind either the bow or reverse shock (depending on the density contrast between the ambient and moving fluids) will attain a value $T_{ps} > 0.08 \mu m_H f^2 v_0^2 / k \approx 3 \times 10^5 V_{300}^2 f_{0.5}^2$ Kelvin. For the faster flows (such as HH 111), this implies terminal post-shock temperatures of at least 10^6K. Since the density in the inter-clump medium of a GMC, or in the ISM that surrounds it, can be as low as 0.1 cm^{-3}, the cooling time for the hot component may approach the dynamical time scales of the flow. Conditions in the terminal shocks of parsec scale flows that punch into relatively low density environments may resemble those found in moderately old supernova remnants and the same shock diagnostics may be applied.

TABLE 1. Some Parsec-Scale Herbig-Haro Flows

Flow	Coordinates (B1950.0)	Angular Size (′)	Projected Size (parsecs)
L1448 (IRS2)	$03^h\ 22^m$, +30° 35′	~20	~1.8 × d_{300}
L1448 (IRS3)	$03^h\ 23^m$, +30° 35′	~30	~2.7 × d_{300}
B5 IRS1 (HH366)	$03^h\ 44^m$, +32° 43′	22	1.9 × d_{300}
IRAS04239+2436 (HH300)	$04^h\ 24^m$, +24° 36′	> 29	>1.1 × d_{150}
T-Tauri	$04^h\ 19^m$, +19° 25′	40	1.8 × d_{150}
L1551 IRS5	$04^h\ 29^m$, +18° 02′	32	1.4× d_{150}
HH114/115	$05^h\ 15^m$, +07° 07′	19	2.6 × d_{500}
HH243 (RNO43)	$05^h\ 30^m$, +12° 48′	26	3.8 × d_{500}
HH83/84	$05^h\ 32^m$, -06° 31′	>20	>2.9 × d_{500}
HH34	$05^h\ 33^m$, -06° 29′	20	2.9 × d_{500}
HH1/2 + HH401/402	$05^h\ 34^m$, -06° 48′	43	6.2 × d_{500}
HH111	$05^h\ 49^m$, +02° 48′	57	7.7 × d_{500}
Z CMa	$07^h\ 01^m$, -11° 28′	10.7	3.6 × d_{1150}
HH80/81	$18^h\ 16^m$, -20° 49′	>12	> 5.3 × d_{1500}
PV-Ceph	$20^h\ 45^m$, +67° 46′	19	2.6 × d_{500}
L1228 (HH199)	$20^h\ 58^m$, +77° 24′	20	1.8 × d_{300}
L1228 (HH200)	$20^h\ 58^m$, +77° 25′	18	1.6 × d_{300}

Acknowledgements: We thank Bo Reipurth for his collaboration on this project. This research was partially funded by NASA grants NASA grant NAGW-4590 (Origins) and NASA grant NAGW-3192 (LTSA).

References

Bally, J., & Devine, D. 1994, ApJ, 428, L65
Bally, J., Devine, D., Alten, V., & Sutherland, R. S. 1997, ApJ, 478, 603
Bally, J., Devine, D., & Alten, V. 1996, ApJ, 473, 921
Bally, J., Devine. D., & Reipurth, B. 1996, ApJ, 473, L49
Biro, S., & Raga, A. C. 1994, ApJ, 434, 221
Cernicharo, J., & Reipurth, B. 1996, ApJ, 460, 57
Chernin, L. M., Masson, C. R., Pino, E. M. G. D., & Benz, W. 1994, ApJ, 426, 204
Devine, D., Bally, J., Reipurth, B. & Heathcote, S. 1997 AJ (in press)
Devine, D. 1997, in *Low Mass Star Formation - from Infall to Outflow: Poster Proceedings of IAU Symposium No. 182*, ed. F. Malbet, A. Castets, p. 95
Masson, C. R., & Chernin, L. M. 1993, ApJ, 414, 230
Masson, C. R., & Chernin, L. M. 1994, 1994, Ap&SS, 216, 113
Ogura, K. 1995, ApJ, 450, 23
Raga, A. C., & Kofman, L. 1992 ApJ 386, 222
Raga, A. C., & Cabrit, S. 1993 A&A, 278, 267
Reipurth, B, Bally, J., Graham, J. A., Lane, A. P., & Zealey, W. J. 1986, A&A, 164, 51
Reipurth, B., Bally, J., Devine, D. 1997, AJ, in press

HERBIG-HARO OBJECTS IN THE ORION NEBULA REGION

C. R. O'DELL
Max Planck Institute for Astronomy
Heidelberg, D-69117 GERMANY

Abstract. The Orion Nebula Region has two different systems of objects classified as HH objects. The North System is associated with the H_2 fingers seen in the infrared and is probably the result of Rayleigh-Taylor instabilities in shocked material moving into the near side of the giant molecular cloud OMC-1. The South System is associated with source(s) within the Trapezium cluster, with the shocked HH objects occuring where jets from pre-main sequence stars impinge on the neutral lid of material that lies across the front of the Orion Nebula. Such jets are different from those driving other HH objects in that these are passing through photoionized material and two of the Orion jets may have been detected.

1. Introduction

The Orion Nebula region is fundamentally different from most of the regions where Herbig Haro (HH) objects are studied. This is because it contains a very high density star cluster, the Trapezium Cluster (Prosser et al. 1994) and this cluster contains many hot, high mass stars. The objects now called HH 203 and HH 204 (Reipurth 1994) are obvious on all well exposed ground images (Münch & Wilson 1962) and were succeeded in discovery by HH 201 (Münch & Taylor 1974), the complex source HH 202 (Cantó et al. 1980), and then numerous sources (HH 205-210) were found in surveys in the [O I] emission line (Axon & Taylor 1984). A determination of the proper motion of HH 201, 205-6 (Jones & Walker 1985) indicated that these objects appear to be emanating from a common center. That result was emphasized by the determination by Allen & Burton (1993) that these last HH objects are at the tips of radial spokes of molecular hydrogen emission and that there seems to be a common region of geometric origin near the intense infrared source IRc2 (Schild et al. 1997, McCaughrean & Mac Low 1997). IRc2 seems to be a luminous hot star imbedded within the OMC-1

B. Reipurth and C. Bertout (eds.), Herbig–Haro Flows and the Birth of Low Mass Stars, 39–46.

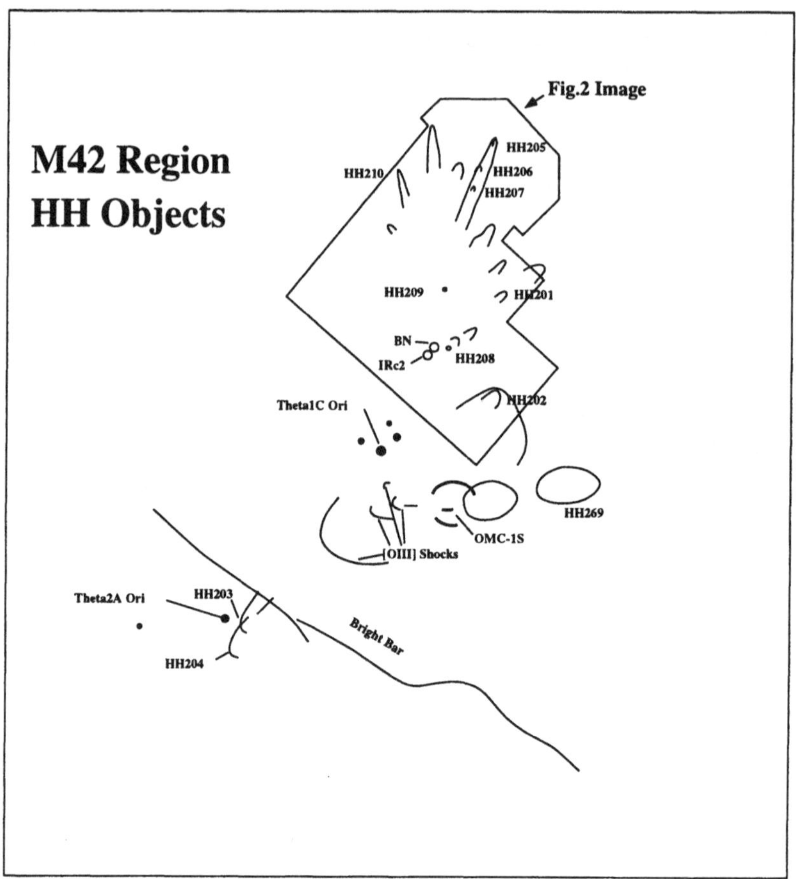

Figure 1. The main features of the Orion Nebula Region are shown, with the brightest stars and infrared and radio sources marked. The angular distance between θ^1C Ori and θ^2A Ori is 135", corresponding to 0.28 pc. The cataloged HH objects are numbered and the brightest additional shock forms are also given. The outline shows the region imaged in Figure 2.

molecular cloud. A final entry in the current catalog of HH objects was added with the determination that HH 269 possesses many of the features of HH objects, although its morphological form is quite different from all of the others (Walter et al. 1995). Hubble Space Telescope (HST) images of this region (O'Dell & Wong 1996) have revealed many features that appear to be shocked gas and a special study of the northern part of the nebula has shown that there is a myriad of shock structures. The line drawing shown as Figure 1 illustrates the location of the major features discussed in this paper.

2. Structure of the Orion Nebula Region

An appreciation of the HH objects we see in the Orion Nebula requires a basic understanding of its three dimensional structure. In the background lies the giant molecular cloud OMC-1, which has imbedded within it the luminous objects producing the IRc2 and Becklin-Neugebauer (BN) infrared sources in the north and the OMC-1S radio source in the south. The Trapezium Cluster lies in the foreground and its hottest member θ^1C Ori lies at a distance of only 0.2 pc (corresponding to 95" at the distance of 430 pc) in front of the cloud surface (O'Dell 1994). The Orion Nebula (M42, NGC 1976) is a thin (0.05 pc) zone of ionized material on the front of OMC-1 (Wen & O'Dell). This "blister" of material is where most of the material of the optical nebula is contained. The recently ionized gas finds itself in an over-pressure situation and freely expands away from the ionization front, so that there is a velocity gradient of about 10 km/s as the gas is accelerated away from the ionization front. Although all stages of low through intermediate ionzation are found along the line of sight from θ^1C Ori to the ionization front, most of the material in the visible nebula is of low ionization and high density, while the material around the young cluster member stars is largely of higher ionization and low density. This is the important difference from other regions of star formation with HH objects, in that the stars that potentially create the HH objects exist in an already photoionized gas, which may cause them to appear quite different from most members of this class. In front of the cluster lies a lid of neutral gas and dust, whose presence is revealed by its absorption of 21-cm emission and the formation of absorption lines in the cluster stars (van der Werf & Goss 1989, O'Dell et al. 1993a). The details of this three dimensional model of the region are summarized in recent articles (Wen & O'Dell 1995, O'Dell 1994), although the basic model has been progressively refined by many authors since the blister nature of the nebula was first demonstrated over 20 years ago (Zuckerman 1973).

3. The Two HH Object Systems in this Region

Study of the various HH objects here indicate the common property that they all have blueshifted emission lines, although there is a division of the objects in the north and the south, the latter having much lower velocities. In terms of size and form, there also seems to be a division between north and south objects. This leads me to divide the objects into North and South Systems.

3.1. THE NORTH SYSTEM

The North System objects are numerous and relatively homogeneous. The defining feature is their size of a few arcseconds, high blue shifts (several hundred km/s, Axon & Taylor 1984, Taylor et al. 1986, Hu 1996), and symmetry of placement around the BN-IRc2 location. HH 201, 205-7, and 210 share all these features, and all of them except HH 201 lie at the tip of one of the H_2 fingers. Recent HST images made in [O I] and [S II] show (O'Dell et al. 1997) that each of the H_2 fingers has associated low ionization optical features and that there are multiple shock form structures in the inner region near IRc2-BN, (Figure 2). HH 208 and HH 209 have the shared features of high blueshifts and small sizes, but do not have the symmetry of a bow shock viewed from an angle. HH 208 has numerous associated shocks out in front of it, and that combined geometry, like HH 201, indicates that the picture is not a simple one of a point of common origin in the IRc2-BN location.

The North System objects probably represent a new subclass of HH objects, recognizing that most HH objects known today are shocks driven by jets arising from pre-main sequence stars. The North System complex spans such a wide range of angles but similar dynamic ages that it must be driven by some large scale force. The most attractive mechanism to explain this is that of an intense stellar wind arising in the O star exciting IRc2, as advocated by Stone (1995). However, in its original formulation, Stone argued that the instabilities that produce the fingers and HH objects originated from a second fast wind overtaking a previous wind. This would produce fingers and HH objects in all angles along the line of sight, whereas the fingers all have optical counterparts at their tips, which indicates that they are preferentially located on the near side of OMC-1. The mechanism for explaining this is also in the Stone paper, in that the authors point out that a single stellar wind producing a shock moving into a region of decreasing density would also become Rayleigh-Taylor unstable. Such a density decrease is exactly what one would expect on the near side of OMC-1, so that instabilities in a single wind shock are the most likely origin. The lack of exact orientation with IRc2 is easily accounted for by shaping of the general flow as it passes through inhomogenous regions prior to reaching the observer's side of OMC-1. The North System objects probably should remain in the HH category, since they still arise from outflow from a young star, and we should be alert that similar objects may be found near other imbedded massive stars.

Figure 2. The region recently imaged by HST which includes all of the North System HH objects plus HH 202. The irregular outline is due to the image being a combination of two different pointings. The image is the sum of [O I]+[S II] minus Hα+[N II], thus subtracting the bright nebular background and enhancing the regions of low excitation. The latter are primarily formed at the shock excited HH objects and the main ionization front of the nebula. The brightest portions of HH 202 and 205-210 are all saturated in this depiction, in order to display the new faint optical features.

3.2. THE SOUTH SYSTEM

The South System objects include HH 202-4 and HH 269 and their locations are sketched in Figure 1. Again, all of the objects are blueshifted with respect to OMC-1 (O'Dell et al. 1991, O'Dell et al. 1993b, Walter et al. 1995), but are significantly larger and have much lower blueshifts. All probably arise from interactions of jets from one or more pre-main sequence stars with the neutral lid that covers the front of the Orion Nebula, a region that must be much lower density than the OMC-1 surface regions where the North System objects are found. The nature of HH 202 remains uncertain. It has two regions of very high surface brightness placed on a small concave form and is itself a smaller feature enclosed within a much larger concave high ionization structure. HH 203-4 has previously been studied in detail (Hu 1996a,1996b), with HH 203 probably being the result of a jet causing a shock in a cloud with a strong density gradient, i.e. the jet is striking the edge of a cloud. HH 204 is more symmetric and has much fine detail in its tip. All of HH 202-4 have the characteristic of extended [O III] within the concave envelope defining their boundaries. This is very different from what is is expected if only a jet is shocking a neutral cloud. In that case the [O III] emission should be concentrated at the tip, where the least cooling has occured. A natural explanation of this is that the material behind the shock is being photoionized by θ^1C Ori (O'Dell et al. 1997).

HH 269 (Walter et al. 1995) is different in form from all other HH objects, being an ellipse oriented almost exactly east-west, and having bright knots at its two tips. The most reasonable explanation is that it is the result of a jet having passed completely through one of the concentrations in the lid, leaving behind only the ring formed well behind the tip of the shock. HST images (O'Dell & Wen 1994, O'Dell & Wong 1996) indicate that there is a very similar object to its east. That object in turn has to its east a dark incomplete oval of similar form. The bipolar outflow imbedded source OMC-1S lies within the boundaries of this object (Schmid-Burgk et al. 1990, Rodriguez-Franco et al. 1992, Bachiller 1996). It is completely uncertain if this dark object is related to HH 269 and the object in between. A possible clue to the driving source of HH 269 may lie with recent low resolution Fabry-Perot observations (Hartigan et al. 1997) that show there to be a linear "jet" of emission exactly in alignment with the centers of HH 269 and its eastward companion. On the eastward side of this jet feature are numerous high ionization bow shocks (O'Dell et al. 1993b) whose center of symmetry falls near the west end of the new jet that we see. The symmetry axes of HH 202, 203-4, and 269 intersect in this same region, indicating that this region contains one or more sources of jets.

4. Discussion

The one jet that we do see (driving HH 269) has high ionization. There is a second possible jet extending to the northwest from HH 203 which is similarly ionized. If these are the driving jets, then they are very dissimilar from those previously found. This is not surprising, since any jets would be passing through material that is already ionized. The detection of similar objects in Orion is probably rendered difficult by the strong nebular background. The remaining major question is "Why don't we see HH objects in the South System formed by the blueward jets?" Such objects would arise where those jets impinge on the concentration of material near the ionization front. Because of the high density there the cooling lengths would be short and the objects correspondingly small. The density increase as the jet encounters first the free flow away from the main ionization front and then the photodissociation region means that the mass loading would be very high. Since momentum would be conserved, the corresponding velocity of the shocked material would be little different from its original value, making the shocked material's detection against the nebula very difficult.

In conclusion we can say that the Orion Nebula region is very rich in HH objects. However, in neither the North or South Systems are we dealing with ordinary HH objects. The North System objects are probably the result of instabilities in a wind-driven shock reaching the decreasing density gradient on the near surface of OMC-1, while the South System objects are the results of jets having passed through photoionized material and finally producing low ionization shocks when they strike the foreground lid that covers the Orion Nebula.

Acknowledgements: This work was supported in part by NASA grant NAG 5-1626. The author was supported in part by the Alexander von Humboldt Foundation while at the Max Planck Institute for Astronomy while on sabbatical leave from Rice University in Houston, Texas. The observations used in this discussion were primarily obtained under STScI program 5976 and the full results of that program are published under the names of all of the team members (O'Dell et al. 1997). Particular thanks are due to team members Patrick Hartigan and Michael G. Burton.

References

Allen, D. A. & Burton, M. G. 1993, Nature, 363, 54
Axon, D. J. & Taylor, K 1984, MNRAS, 207, 241
Bachiller, R. 1996, ARAA, 34 111
Cantó, J., Goudis, C., Johnson, P. G., & Meaburn, J. 1980, A&A, 85, 128
Hartigan, P., Morse, J. A., & O'Dell, C. R. 1997, private communication
Hu, X. 1996a, PhD Thesis, Rice University, Houston, Texas
Hu, X. 1996b, AJ, 112, 2712
Jones, B. R. & Walker, M. F. 1985, AJ, 90, 1320

McCaughrean, M. J. & Mac Low, Mordecai-Mark 1997, AJ, 113, 391
Münch, G. & Taylor, K. 1974, ApJ, 192, L93
Münch, Guido & Wilson, O. C. 1962, ZAp, 56, 127
O'Dell, C. R. 1994, ApSS, 216, 267
O'Dell, C. R. & Wen, Z. 1994, ApJ, 436, 194
O'Dell, C. R. & Wong, S. K. 1996, AJ, 111, 846
O'Dell, C. R., Wen, Z., & Hester, J. J. 1991, PASP, 103, 82
O'Dell, C. R., Walter, D. K., & Dufour, R. J. 1992, ApJ, 399, L67
O'Dell, C. R., Valk, J. H., Wen, Z., & Meyer, D. M. 1993a, ApJ, 403, 678
O'Dell, C. R., Hartigan, P., W. M. Lane, Wong, S. K., Burton, M. G., Raymond, J., & Axon, D. J., 1997, AJ, submitted
Prosser, C. F., Stauffer, J. R., Hartmann, L., Soderblom, D. R., Jones, B. F., Werner, M. W., & McCaughrean, M. J. 1994, ApJ, 421, 517
Reipurth, B. 1994, A General Catalogue of Herbig-Haro Objects, available by electronic transfer through ftp to hq.eso.org in the directory /pub/Catalogs/Herbig-Haro
Rodriguez-Franco, A., Martin-Pintado, J., Gomez-González, J., & Planesas, P. 1992, A&A, 264, 592
Schild, H., Miller, S., & Tennyson, J. 1997, A&A, 318, 608
Schmid-Burgk, J., Güsten, R., Mauersberger, R., Schulz, A., & Wilson, T. L. 1990, ApJ, 362, L25
Stone, J. M., Xu. J., & Mundy. L. G. 1995, Nature, 377, 315
van der Werf, P. P. & Goss, W. M. 1989, A&A, 224, 209
Walter, D. K., O'Dell, C. R., Hu, X., & Dufour, R. J. 1995, PASP, 107, 686
Wen, Zheng & O'Dell, C. R. 1995, ApJ, 438, 784
Zuckerman, B. 1973 ApJ, 183, 863

SPECTROSCOPIC PROPERTIES OF HERBIG-HARO FLOWS

KARL-HEINZ BÖHM
Department of Astronomy, University of Washington
Seattle, Washington 98195, USA

AND

ANTHONY P. GOODSON
Department of Physics, University of Washington
Seattle, Washington 98195, USA

Abstract. While recent studies of Herbig-Haro (HH) objects have focused less on the details of their spectra than on the hydrodynamics of jets and their working surfaces, many open questions concerning these spectra remain. Attempts to quantitatively explain a wide range of lines for many HH objects point to discrepancies between theory and observation. Some lines (specifically [S II](6716+6731)) are much stronger than predicted by simple plane-shock and bow-shock models, while in general high ionization lines (e.g. lines of [O III], [Ne III] and [S III] in the optical and the [C IV] and [N V] in the ultra-violet) are much weaker than expected, pointing to difficulties with current models. On the other hand, examination of these lines has lent new insight into both the quality of our predictions and the nature of HH outflows. Examination of many Fe lines have demonstrated that our ability to estimate abundances from faint lines is surprisingly good (or surprisingly consistent). Position velocity diagrams have also been constructed (using forbidden emission lines), allowing outflows to be mapped to within 0."3 arcseconds of the source star.

1. Introduction

The description and interpretation of spectra originally played a decisive role in the early analysis of Herbig-Haro (HH) objects (Herbig 1951; see also Böhm 1956; Osterbrock 1958 for early attempts to give a quantitative explanation.) Later, the recognition that most HH spectra are generated

B. Reipurth and C. Bertout (eds.), Herbig–Haro Flows and the Birth of Low Mass Stars, 47–62.

by shocks (Schwartz 1975) offered an opportunity to explain their spectra using shock wave models (see e.g. Dopita 1978, Raymond 1979.) The further discovery that the optical spectra of many HH objects are generated in bow-shock like structures (Hartmann & Raymond 1984) encouraged the use of bow shock models to obtain a more detailed understanding of the spectra (see e.g. Hartmann & Raymond 1984; Raga & Böhm 1985, 1986, 1987; Hartigan, Raymond & Hartmann 1987; Raymond, Hartigan & Hartmann 1988.) More recent studies have focused less on the detailed explanation of the spectrum than on the hydrodynamics of the jets and their working surfaces and, if possible, on a hydrodynamical explanation of the images in specific strong lines. There have been relatively few attempts to explain quantitatively a wide range of lines. This is especially true of the many new line identifications which have been made in the last decade. (See e.g. Solf, Böhm & Raga 1988 who found 189 lines for the 3720 Å $< \lambda <$ 10830 Å in HH1.)

This paper addresses our current ability to quantitatively interpret HH spectra, and highlights some of the open questions regarding this interpretation which are applicable to HH objects in general. All discrepancies which are restricted to one or a very few objects are assumed to be related to the specific properties of those particular objects and are of less interest. Our question is: *Do we really understand HH spectra?*

Our main interest is the optical spectra (we define their range as 3700-10830 Å). Because of the close connections with optical spectra we shall also briefly discuss the UV spectra (for HH objects usually measured with IUE) and the near IR H_2 rotation-vibration spectra.

We want to study and (if possible) interpret:

1) The spectrophotometric flux of the lines.

2) The line profiles or the position-velocity diagrams (derived from long-slit spectra).

3) The full appearance (in 2 spatial dimensions) of line profiles as seen e.g. with a Fabry-Perot interferometer.

From our point of view the observations described in 1) are the most important ones. (For example, if the spectrophotometry of a given line disagrees with the models for all HH objects, there is likely something wrong with the basic interpretation.) Observations described in 2) and especially in 3) are more strongly dependent on the geometry of the shock and often require a specific interpretation of the individual object.

The following theoretical models can be and have been used for the interpretation of HH spectra:

1) Plane shocks (useful, easily understood, although in most cases not very realistic.) May be acceptable in some cases with high spatial resolution.

2) "1.5-Dimensional Bow Shock" (a bow shock composed of the superposition of oblique plane shocks.) (Hartmann & Raymond 1984; Raga & Böhm 1985, 1986; Hartigan, Raymond & Hartmann 1987.)

3) 2-Dimensional Bow Shocks (see e.g. Raga & Böhm 1987, Raga *et al.* 1988), asymmetric bow shocks (Henney 1996.)

4) Models (simulations) of working surfaces of jets (including bow shocks and jet shocks) (e.g. Blondin, Königl & Fryxell 1989; Stone & Norman 1993; de Gouveia dal Pino & Benz 1993; Raga 1995.)

5) Internal working surfaces of the jet due to the variability of the sources (Biro & Raga 1994)

6) Other important mechanisms: a) Turbulent mixing (entrainment) layers (Cantó & Raga 1991), especially important for H_2 emission. b) Emission of C-bow shocks or mixed J-C shocks (e.g. Fernandes, Brand & Burton 1995; Davis, Eislöffel & Smith 1996)

2. Some Unsolved Problems of Herbig-Haro Spectra

Any appraisal of the quality of our understanding of HH spectra requires a model which has a wide application, at least in a qualitative context. As stated above, we search not for discrepancies between observation and theory in individual objects (which may be due to using an inappropriate detailed model) but rather for discrepancies for certain lines which apply to all observed objects.

The models described in 1) and 2), which are stationary and require only the knowledge of shock velocity and pre-shock density, are used for the comparison. They permit the prediction of the whole line spectrum without difficulty. Other models are hydrodynamically much more sophisticated (especially 4) and 5)) but have difficulty predicting the full spectrum. A considerable fraction of the comparison with models 1) has been carried out by Raga, Böhm & Cantó (1996).

In the following we discuss unexpected results for the spectrophotometric flux of the lines in the optical spectrum. In our discussion we follow the distinction between high excitation objects (HEO) and low excitation objects (LEO) (Böhm, Brugel & Mannery 1980.) We show in tables 1 and 2 information about the spectrum of HH1 (high excitation object, Solf *et al.* 1988) and HH7 (low excitation object, Böhm & Solf 1990.) We list all detected ions and atoms and indicate the flux of the strongest line of the ion or atom in question. (The fluxes F are listed on the standard relative scale where $F(H_\beta) = 100$.)

The tables show the well-known facts about HH-spectra. In order to show the difference between high and low excitation objects more clearly we have also listed in table 2 (with a minus sign) ions which were not

TABLE 1. HH1: a High Excitation Object

Optical Lines		UV Lines	
Ion(atom)	F($F_{H_\beta}=100$)	Ion	F($F_{H_\beta}=100$)
H	311		
HeI	131	CII	186
$HeII$	2	$CIII$	330
$[CI]$	7	CIV	416
$[NI]$	16	$NIII$	125
$[NII]$	151	$OIII$	150
$[OI]$	159	$MgII$	163
$[OII]$	135	$SiII$	71
$[OIII]$	57	$SiIII$	70
$[NeIII]$	15	$SiIV$	200
NaI	0.4		
$MgI, [MgI]$	7, 1.2		
$[SII]$	105		
$[SIII]$	6		
$[ClII]$	1.5		
$[ArIII]$	2.4		
$[ArIV]$	0.5		
$CaII, [CaII]$	33, 45		
$[CrII]$	2.3		
$[FeII]$	31		
$[FeIII]$	4		
$[NiII]$	16		

detected in HH7 but are seen in HH1. We indicate some ions or atoms which are not so well-known, like Ar IV, Cl II, and Na I. The presence of Na I in the HH spectrum is surprising and begs the questions: In which region of the shock wave is it formed and is it excited by electron collisions (in spite of the very low ionization energy of Na I)?

An interesting problem is the unusual flux of [C I] 9849 Å in HH7 which is about 150 times as strong as the same line in HH1, a result which is not easily explained in terms of a shock wave model. However, the total line fluxes in low excitation objects are (surprisingly) well explained with low velocity J shocks.

Is the simple separation of HH objects in high and low excitation sufficient? The introduction of a new category, "intermediate excitation objects," is under consideration. Raga *et al.* (1996) suggested this because they found a reasonably large number of HH objects (13 out of 45) which

TABLE 2. HH7: a Low Excitation Object

	Optical Lines		Fluorescent H_2 Lines (not in HH7, but in HH43, HH47)
Ion(atom)	F(F_{H_β} = 100)	Ion	F(F_{H_β} = 100)
H	435		H_2
HeI	–		λ
HeII	–		1271
[*CI*]	1059		1431
[*NI*]	286		1461
[*NII*]	34		1489
[*OI*]	577		1505
[*OII*]	–		1547
[*OIII*]	–		1562
[*NeIII*]	–		
NaI	–		
MgI	–		
[*SII*]	754		
[*SIII*]	–		
[*ClII*]	–		
[*ArIII*]	–		
[*ArIV*]	–		
CaII, [*CaII*]	<14, 86		
[*CrII*]	<3		
[*FeII*]	52		
[*FeIII*]	–		
[*NiII*]	27		

seemed to show "high excitation" in the H_α/[S II] (6716+6731) ratio and low excitation in the [O III](5007)/H_β ratio.

There are several features of the spectrophotometry which are not understood for a broad range of HH objects. The first result of this type concerns the overly strong ratio of [S II](6716+6731)/H_α which cannot be understood for the majority of high and intermediate excitation objects in terms of either a plane shock or a 1.5-D bow shock model (Raga *et al.* 1996.) The authors studied 45 HH condensations. In figure 1 we compare the shock velocity determined from different line ratios for plane shock waves for the observed high excitation objects. For most line ratios the resulting shock velocities cluster at $\sim$ 100 km/sec or somewhat below. However, the ratio of [S II](6716+6731)/H_α indicates a shock velocity of between 180 and 400

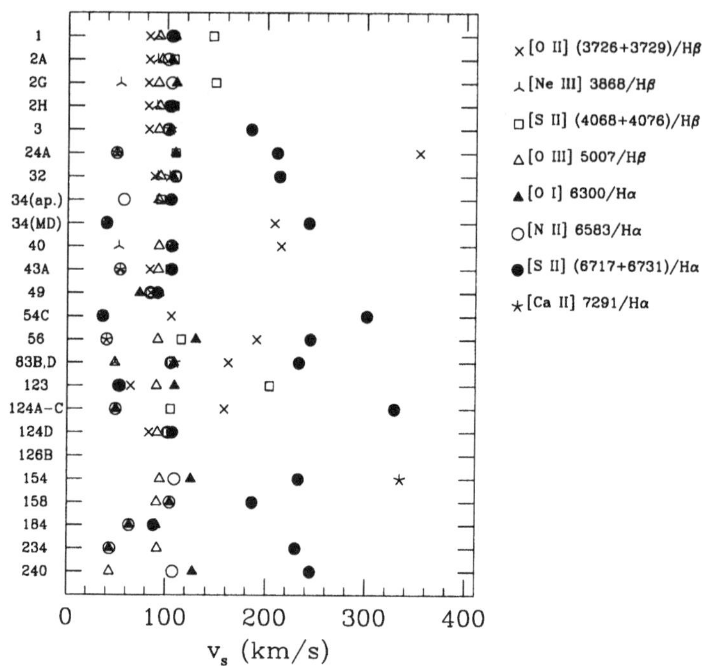

Figure 1. Shock velocities V_s of HEOs as determined from individual line ratios using a plane shock model. Practically all line ratios (and especially [O III]5007/H_β) indicate for all objects $V_s \leq 100$ km/sec. The only major exception is [S II](6717+6731)/H_α which shows for most HEOs a much higher shock velocity. This interpretation persists if bow shock models are used (see fig. 2.) (The high V_s estimate of [O II](3726+3729)/H_β is attributed to contamination of [O II] from background H II regions.) From Raga, Böhm & Cantò (1996.) Courtesy Rev Mex AA.

km/sec. This discrepancy is independent of the type of shock model (plane shock or bow shock) model used, as demonstrated by figure 2.

Most of the observed HEOs show a much larger [S II](6716+6731)/H_α ratio than predicted for these models for the range of shock velocities V_s 50–200 km/sec, which more than spans the range of shock velocities indicated by the other lines (V_s 50 – 150 km/sec.) The prediction for LEO (e.g. HH11, HH10, HH7, HH34 (jet)) does not lead to comparable contradictions. They are compatible with bow shocks in the range 20-35 km/sec. It is interesting that the discrepancy occurs in the flux ratio [S II](6716+6731)/H_α, which is considered as fundamental for the identification of HH objects.The theory confirms qualitatively the strength of the [S II](6716+6731) in shocks but

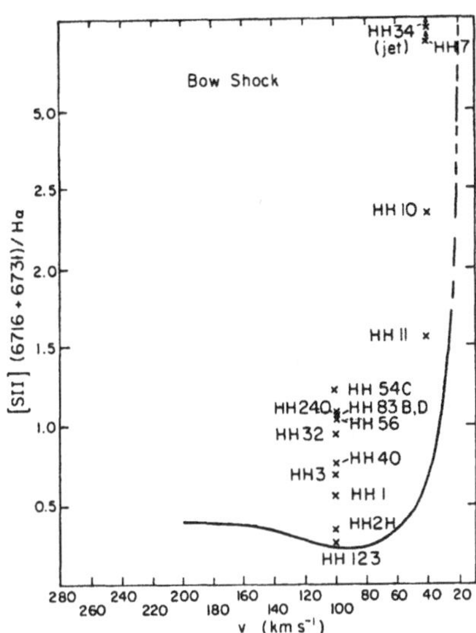

Figure 2. Prediction of the line ratio [SII](6716+6731)/H_α for 1.5-D bow shocks in the range 30 km/sec $\leq V_s \leq$ 200 km/sec (extrapolated to 20 km/sec.) Interpretation of more well-behaved line ratios (not shown) and estimates of total flux require the observational results for the HEOs from HH123 to HH54C to be fitted on the theoretical curve for velocities $\geq$ 100 km/sec. This is obviously impossible. On the other hand the observed LEOs (HH11, HH10, HH7 and HH34 jet) can be fitted easily on the steeply rising part of the theoretical curve in the range 20 km/sec $\leq V_s \leq$ 40 km/sec.

quantitatively the [S II] are considerably stronger than the simple theory predicts. It can be shown, however, that other important low ionization lines (e.g. [O I], measured by the [O I](6300+6363)/H_α ratio) agree very well with the theoretical predictions even in HEOs (Raga *et al.* 1996.)

The next observational result which seems to contradict theoretical expectations is the line ratio [O III](5007)/H_β for HEO. A preliminary discussion of this result was given by Böhm (1995) and Raga *et al.* (1996). Raga *et al.* (1996) plotted the ratio [O III](5007)/H_β as a function of [N I](5200)/H_β. (The latter ratio measures the excitation of the HH object.) In figure 3 we have compared the plot for the observed objects to the theoretical prediction for plane shock waves. The agreement appears reasonable at first sight. However, there is a surprising feature in this diagram. Even the HEOs have an [O III](5007)/H_β ratio which clusters between 0.0 and 0.7, with a definite upper limit among the objects studied so far of [O III](5007)/H_β=1.09. Table 3 shows how surprising this result is from a theoretical viewpoint. The plane and the bow shock models do not at all predict a limit at 1.09. An increase beyond a shock velocity of $\sim$ 120 km/sec would lead to a drastic increase of this ratio, which is seen in no

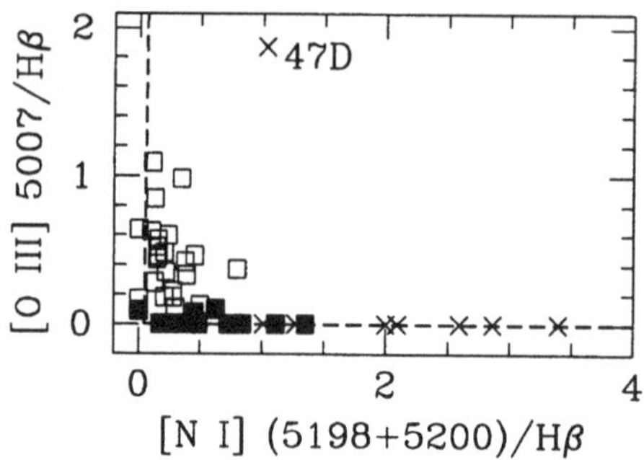

Figure 3. The observed ratio [O III]5007/H_β as a function of excitation, as measured by [N I](5198+5200)/H_β (open squares: HEOs, filled squares: intermediate excitation objects, crosses: LEOs, model prediction: dotted curve.) The line ratio [O III]5007/H_β for the HEOs cluster in the range 0-0.7 and do not become larger than 1.09, in total contradiction with theory (see table 3.) A high value occurs strangely for the LEO(!) HH47D. We follow Dopita *et al.* (1982) and consider this [O III] line strongly influenced by the surrounding H II region. From Raga, Böhm & Cantó (1996). Courtesy Rev Mex AA.

object. Observations show that this effect not only influences [O III]/H_β but also the ratios [Ne III]/H_β and [S III]/H_α. In other words, this effect does not seem to have anything to do with the special properties of the O^{++} ion but rather seems to be a (not yet understood) limitation of high ionization lines in HH objects in general. The result can also be seen in figure 1, where the shock velocities determined from the [O III](5007)/H_β ratio stay at 100 km/sec or below for all HEO.

TABLE 3. Flux ratio $R = [OIII]5007/H_\beta$ for plane and bow shock models

Plane Models V_s (km/sec)	R	Bow Shock Models V_s (km/sec)	R
90	0.07	100	0.43
110	4.17*	150	1.78
		200	2.06

An interesting result which is also visible in the compilation by Raga *et al.* (1996) may throw some additional light on the empirical distinction between HEO and LEO. Raga *et al.* (1996) plotted the average electron density (determined from the ratio (6731)/(6716) of [S II]) as a function of the [N I]5200/H_β ratio and compared this diagram to theoretical plane shock predictions. The theoretical average electron density has been cal-

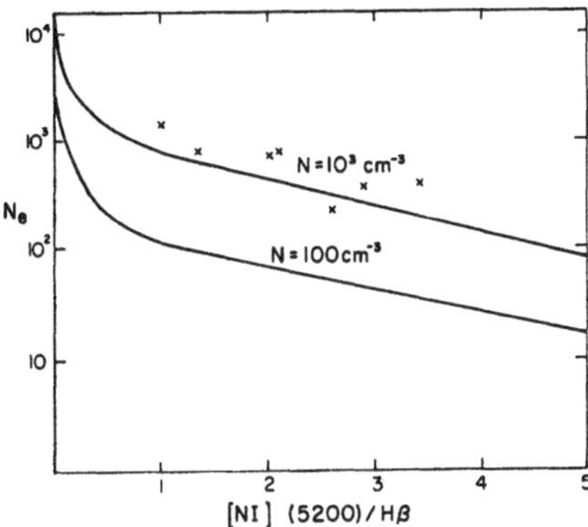

Figure 4. The observed relation between N_e (based on the flux ratio [S II]6731/[S II]6716) and [N I](5198+5200)/H_β for the LEO only, compared to model predictions for a pre-shock density of $N = 100cm^{-3}$ and $N = 1000cm^{-3}$. The extension of the theoretical curves to very high shock velocities (see Raga *et al.* 1996) is not shown (not relevant in the present context.) The figure shows that LEOs have an approximate pre-shock density of $10^3 cm^{-3}$, whereas Raga *et al.* (1996) showed that the HEO and intermediate excitation objects have approximately a pre-shock $N \sim 10^2 cm^{-3}$ (their figure 4b.)

culated from the model results of the (6731)/(6716) ratio and does not use any theoretical input parameters directly. Their result (their figure 4b) shows that the observational results for HEO and most of the intermediate excitation objects are centered around a theoretical curve which is based on a pre-shock density of N=10^2 cm^{-3}. It is interesting that all LEO lie considerably above this curve. In figure 4 we show that all LEO cluster around the curve for a pre-shock N=10^3 cm^{-3}. To first order, this can be treated as an observational selection effect. For a given pre-shock density the (post-shock) N_e declines very rapidly with declining shock velocity. That is especially true for the shock velocity range around 40 km/sec. For the LEO objects we expect (from theoretical models) an N_e of 10-100 cm^{-3} for a pre-shock density of N=10^2 cm^{-3}, meaning that these objects will have a very low surface brightness and are difficult to observe (in comparison to the high excitation objects.) It is therefore not surprising that we discover LEO which have on the average a higher pre-shock density than the HEO. However, there remain two open questions: Why do we not discover HEO which have also a higher pre-shock density ($\sim 10^3$ cm^{-3}), and why do the LEO seem to show an approximately constant pre-shock density of 10^3 cm^{-3}? The answer to these two questions may just be insufficient statistics, although the present results look surprising.

3. The reliability of quantitative spectrophotometry in HH objects and the gas-phase abundances of metals

The study of gas-phase abundances of Fe and Ti in HH-objects (Beck-Winchatz, Böhm & Noriega-Crespo 1996, see below) permits us to make a judgement about the presently achieved accuracy of spectrophotometric data on relatively faint lines in HH objects and the accuracy of abundances derived therefrom. The motivation for this work was as follows: The pre-shock gas-phase metal abundances are likely reduced by binding of the metals in dust grains. The passage of these grains through the HH shock, however, could destroy the dust grains (mostly through sputtering, see e.g. Seab 1987; Draine 1995.) It is to be expected that HEOs, with their high shock velocities, would destroy the dust grains more efficiently than LEOs. This scenario can be checked by determining and comparing the gas-phase metal abundances in LEOs and HEOs. This program has been carried out by Beck-Winchatz *et al.* (1996) (whose results will be quoted below) for a few HH objects.

Our main emphasis here will be on the fact that such observations show us to which degree the spectrophotometry of different moderate faint lines are consistent with each other and what they tell us about the reliability of HH object spectrophotometry and its interpretation. Carrying out this program relies on the comprehensive calculations of the collision strength, especially for Fe^+, by Zhang & Pradhan (1995) (see also Bautista & Pradhan 1995.) Fe^{++} are also included, but they are less important and we rely on older data (see Garstang, Robb & Roundtree 1978.) The internal consistency of spectrophotometric data and interpretation are checked in the following way: If the spectrophotometry of the lines is correct, if there are no errors in the collision strengths and transition probabilities, and if the statistical equilibrium and the model of the HH object are strictly correct then we should get exactly the same abundance from any Fe line in a given HH object.

A comparison of the determined Fe abundance from different lines provides an estimate of the magnitude of spectrophotometric and "theoretical" (interpretive) errors, and demonstrates the presently achieved consistency of the spectrophotometric and derived data. In table 4 we show all line fluxes of [Fe II] in HH1 which are larger than 0.02 of the H_β flux and we list the resulting abundance ratio $N(Fe^+)/N(S^+)$. We list the lines which deviate in their abundance ratio from the average by less than 20 % with an asterisk (*) and the ones which show a deviation of 20-30% with a cross. There are relatively few lines (4) which show larger deviations and among these are 3 which are relatively high, which might mean that an unrecognized blend is present.

The results of the gas-phase abundance determination of Fe for the

TABLE 4. $N(Fe^+)/N(S^+)$ for HH1 as derived from various lines (natural value *not* logarithm)

Line	F	$N(Fe^+)/N(S^+)$
4276.8	3.6	2.8*
4352.8	2.1	3.43*
4458.0	2.6	2.52+
4814.6	5.5	2.06
4889.6	4.1	2.41+
4905.4	2.0	2.81*
5111.6	3.3	3.8*
5158.8	22.1	2.91*
5261.6	8.9	3.40*
5333.7	4.7	5.13
7637.5	6.8	5.21
8617.0	30.4	3.36*
8891.9	10.0	3.48*
9033.5	2.7	4.40+
9051.9	6.5	3.16*
9267.5	3.9	5.04
	average	3.47
	solar value	2.95

HEOs HH1, HH43A, HH255 and the LEO HH7 and HH11 show unexpected results. (See table 5.) First there are no objects (among these) which show a strongly depleted gas-phase Fe abundance. So, a considerable part of Fe bound up in dust grains must have gone back to the gas phase in all of them. Moreover in our (very restricted) sample there is no fundamental difference between the HEOs and LEOs.

4. Attempts to explain UV and the 2μm IR spectra of H_2

In HEOs more energy is often emitted in the UV spectrum (1200 Å $< \lambda <$ 3000 Å) than in the optical (3700 Å $< \lambda <$ 10830 Å) (Böhm, Böhm-Vitense & Brugel 1980.) Although at first the ionization in the UV seemed to be higher than that in the optical part (Böhm 1983) this does not agree with our present conclusions (see the predictions by Hartigan *et al.* 1987.) We see a similar unexpected behavior of the high ionization lines as was found in the optical. Just as in HEO the [O III] never shows a shock velocity essentially more than $\sim$ 100 km/sec, the same statement applies to the

TABLE 5. $Fe^+/S^+ \sim Fe/S$ in HH Shocks (direct abundance ratio, *not* logarithm)

Sun	2.95
HH1	3.47
HH7*	2.74
HH11*	3.99
HH43A	1.41
HH255 (Burnham's Nebula)	1.05

[C IV] line in the UV. Related to this is the problem that the [N V] line at 1240 Å has never been seen in HH objects (although the prediction for a 160 km/sec shock says that its flux should be 718, compared to 100 for H_β.) It is of great importance but not easy (because of the reddening of HH objects) to obtain improved and reasonably high resolution UV spectra of Herbig-Haro objects.

The UV spectrophotometry of LEO is limited to HH7, HH11, HH43 and HH47 (Cameron & Liseau 1990; Schwartz 1983; Böhm, Scott & Solf 1991.) So far the fluorescent H_2 (pumped by Lyman α) lines (listed in table 2) have been detected in HH43 and HH47. Considerable progress in the interpretation of these interesting lines has been made by Wolfire & Königl (1991.)

The rotation-vibration lines of H_2 in the 2μm region (at least in some cases) are reasonably well predicted by the same models which predict the optical lines. In this context specifically HH32 (Davis, Eislöffel & Smith 1996) and HH1 agree with expectations (Davis, Eislöffel & Ray 1994), because they show that a considerable portion of the H_2 line emission originates in the "wings" of the bow shock. The lack of H_2 emission in the western "wing" of HH1 is understood by the asymmetry of the HH1 bow shock (Solf *et al.* 1991 and especially Henney 1996.) In other cases (e.g. in HH110) a mixing layer (entrainment) can best explain the H_2 emission. (Raga 1995; Noriega-Crespa *et al.* 1996)

5. The explanation of Line Profiles and Position-Velocity Diagrams of HH objects

Considerable progress was achieved in the study of line profiles by Hartigan *et al.* (1987) who pointed out that the bow shock velocity is equal to the measured FWZI (full width at zero intensity) and discussed a number of objects from this point of view. A study of direct and dust-scattered emission line profiles in the HH1/HH2 system was carried out by Solf & Böhm (1991) who showed that very useful information about the state of motion of these HH objects can be derived from the dust-scattered line profiles originating in front of or behind the HH object. (A preliminary explanation of this result is given by Noriega-Crespo, Calvet & Böhm 1991 and a very detailed and useful theory by Henney *et al.* 1994.)

We have made a considerable effort to explain the observed "position-velocity diagrams" (the spatially dependent line profiles which are obtained from a long slit spectrum, usually but not necessarily along the axis of the HH object, see Raga & Böhm 1985, 1996 and Indebetouw & Noriega-Crespo 1995.) Since the explanations are based on an approximate and geometrically simple bow shock model the agreement is usually acceptable but not ideal. But there are some cases (e.g. HH32, see Solf, Böhm & Raga 1986) where the agreement is impressive. Recently there has been an attempt also to explain the ultraviolet (I.U.E.) "long slit" spectra in a similar way (Moro-Martin *et al.* 1996.)

Following a suggestion of Solf (1989) it has recently become possible to generate position-velocity diagrams of HH jet outflows in the immediate neighborhood of the source stars. (Solf 1989, 1993; Solf & Böhm 1993; Hirth, Mundt & Solf 1994; Hirth 1994; Böhm & Solf 1994; Staude & Elsässer 1993.) The method is based on using long slit spectra taken under different position angles centered on the source star (usually a *T Tauri* star.) The photospheric contribution to the spectrum is eliminated by using the (adjusted) radiation of a star of equal spectral type as a template in the spectral range of the studied HH line. This technique works especially well for forbidden lines; position-velocity diagrams are obtained for these lines of the stellar jet very close to the star. Radial velocities are given with respect to the photospheric velocity, the position is expressed in distances to the source star. From these diagrams we can sometimes obtain useful information about the outflow close to the star (up to a fraction of an arcsecond.) In figure 5 we show as an example the [S II]6731 position-velocity diagram covering the complex outflow environment of *T Tau* for the position angles 0 and 90 degrees. In the 90° diagram we can clearly recognize a peak in intensity at ~ -130 km/sec which is extended in the negative direction (270°) and increases in velocity up to -175 km/sec and is visible up to 3."5 from the star. The fact that the velocity dispersion for this component (B and B') is considerably smaller than the centroid radial velocity shows that this is a strongly collimated flow which is visible only 0."3 from *T Tau*. (Fig 5b.) This is the origin of the outflow which continues as HH 155 to at least 32" west of *T Tau* (Schwartz 1975, Bührke, Brugel & Mundt 1986, see also Reipurth 1994.) The position-velocity diagram for 0° also shows a

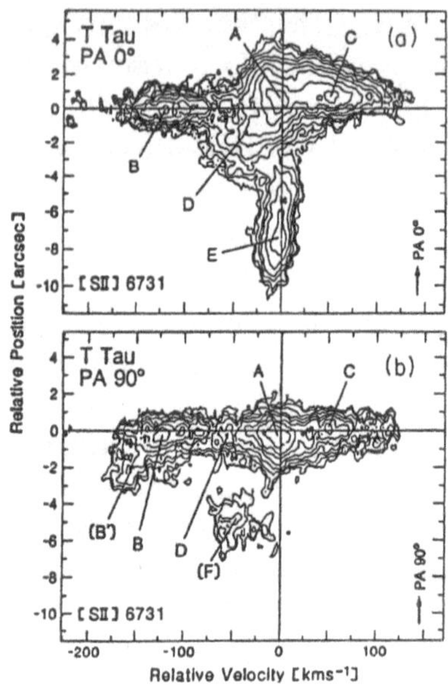

Figure 5. The position-velocity diagram for [S II]6731, indicating the kinematics of the different outflow systems in the immediate environment of T Tau using Solf's method. The upper diagram refers to a position angle 0°, The lower to a position angle of 90°. In the upper diagram we see the weakly collimated bipolar wind C, D. In the lower diagram we see the highly collimated jet (velocity dispersion $\ll$ centroid velocity) starting with v=−130 km/sec at 0."3 from the star and pointing in the direction 270° (B and B'). It is visible at large distance as HH155. The enigmatic main part (E) of Burnham's nebula has very low velocity and low velocity dispersion although it has a HEO shock wave spectrum. From Böhm and Solf's (1994) data.

(much less collimated) bipolar outflow in the directions $345^o/165^o$ (Böhm & Solf 1994.) The enigmatic properties of component E (the main part of Burnham's nebula – HH255) have been discussed by Böhm & Solf (1997).

6. Future Studies of HH Spectra

In the future it will be important to clarify the remaining general contradictions between spectrophotometric observations of certain lines in a large number of HH objects and the predictions of simple bow shock (or plane shock) models.

1) These include the unpredicted strength of the [S II] 6716,6731 lines for all high and intermediate excitation objects (as indicated by the [S II] 6716,6731/$H\alpha$ ratio.)

2) Another surprising feature is that the observed line ratios of high

ionization lines, like [O III]5007/H_β and [NeIII]3869/H_β in all high excitation objects require a shock wave velocity not essentially larger than 120 km/sec which leads to the strange conclusion that there are many HH objects with shock velocities up to (say) 120 km/sec but that there are hardly any with shock velocities above this value. This effect is confirmed by line ratios observed in the UV.

It is obvious from 1) and 2) that there is a fundamental problem with the present bow shock model.

3) The unusually high flux in the C I 9850 line and the unexpected spatial distribution of this line in some low excitation objects should be studied and explained.

4) The apparently different pre-shock density for High Excitation Objects (mostly $N \sim 10^2/cm^3$) and Low Excitation Objects (mostly $N \sim 10^3/cm^3$) should be further studied. One should attempt to find HEO's with N comparable to those of LEO's.

5) It has been shown that the spectrophotometry and its interpretation in terms of abundances in HH objects can be carried out with considerable accuracy. The gas-phase metal abundance in HH shock waves should now be determined for fairly large numbers of HH objects to see whether there are objects with drastically reduced abundance due to grain formation.

6) With regard to the study of position-velocity diagrams we expect especially interesting results from a further application of Solf's method (1989) to the jet and shock wave structure in the immediate vicinity of the source star. While the obvious observations of this type have been made (see e.g. Hirth 1994), a more detailed interpretation may be extracted from these observations, such as establishing the correlation between kinematics and ionization, which has given useful insight in the case of *T Tau* (Böhm & Solf 1997; Hartigan, Edwards & Ghandour 1995.)

References

Bautista, M.A. & Pradhan, A.K.: 1995, ApJ 442, L65
Beck-Winchatz, B., Böhm, K.H. & Noriega-Crespo, A.: 1996, AJ 111,346
Biro, S. & Raga, A.C.: 1994, ApJ 434, 221
Blondin, J.M., Königl, A. & Fryxell, B.A.:1989, ApJ 337, L37
Böhm, K.H.: 1956, ApJ 123, 379
Böhm, K.H.: 1983, Rev Mex AA 7, 55
Böhm, K.H.: 1995 in Shocks in Astrophysics, ed. T. Millar & A. Raga (Dordrecht: Kluwer), p.11
Böhm, K.H., Böhm-Vitense, E. & Brugel, E.: 1981, ApJ 245, L113
Böhm, K.H., Brugel, E. & Mannery, E.: 1980, ApJ 235, L137
Böhm, K.H. & Solf, J.: 1990, ApJ 348, 297
Böhm, K.H. & Solf, J.: 1994, ApJ 430, 277
Böhm, K.H. & Solf, J.: 1997, A&A 318, 565
Bührke, T., Brugel, E.W. & Mundt, R.: 1986, A&A 164, 83

Cameron, M. & Liseau, R.: 1990, A&A 240, 409
Cantó, J. & Raga, A.C.: 1991, ApJ 372, 646
Davis, C.J., Eisloffel, J. & Ray, T.P.: 1994, ApJ 426, L93
Davis, C.J., Eisloffel, J. & Smith, M.D.: 1996, ApJ 463, 246
De Gouveia Dal Pino, E.M. & Benz, W.: 1993, ApJ 410, 686
Draine, B.: in Shocks in Astrophysics, ed. T. Millar & A. Raga (Dordrecht: Kluwer), p. 111
Dopita, M.: 1978, ApJS 37, 117
Dopita, M., Schwartz, R.D. & Evans, I.: ApJ 263, L73
Fernandes, A. Brand, P. & Burton, M.: 1995, in Shocks in Astrophysics, ed. T. Millar & A. Raga (Dordrecht: Kluwer) p. 45
Garstang, R.H., Robb, W.D. & Roundtree, S.P.: 1978, ApJ 222, 384
Hartigan, P., Raymond, J. & Hartmann, L.: 1987, ApJ 316, 323
Hartigan, P., Edwards, S. & Ghandour, L.: 1995, ApJ 452, 736
Hartmann, L. & Raymond, J.: 1984, ApJ 276, 560
Henney, W.J.: 1996, Rev Mex AA 32, 3
Henney, W.J., Raga, A.C. & Axon, D.J.: 1994, ApJ 427, 305
Herbig, G.H. 1951: ApJ 113, 697
Hirth, G.A. 1994: Ph.D. Thesis, University of Heidelberg
Hirth, G.A., Mundt, R. & Solf, J.: 1994, A&A 285, 233
Indebetouw, R. & Noriega-Crespo, A.: 1995, AJ 109, 752
Moro-Martin, A., Noriega-Crespo, A., Böhm, K.H. & Raga, A.C.: 1996, Rev Mex AA 32, 75
Noriega-Crespo, A., Calvet, N. & Böhm, K.H.: 1991, ApJ 379, 676
Noriega-Crespo, A., Garnavich, P.M., Raga, A.C., Cantó, J. & Böhm, K.H.: 1996, ApJ, 462, 804
Osterbrock, D.E.: 1958, PASP, 70, 399
Raga, A.C.: 1995, Rev Mex AASC 1, 103
Raga, A.C. & Böhm, K.H.: 1985, ApJS 58, 201
Raga, A.C. & Böhm, K.H.: 1986, ApJ 308, 829
Raga, A.C. & Böhm, K.H.: 1987, ApJ 323, 193
Raga, A.C., Böhm, K.H. & Cantó, J.: 1996, Rev Mex AA 32, 161
Raga, A.C., Mateo, M., Böhm, K.H. & Solf, J.: 1988, AJ 95, 1783
Raymond, J.: 1979, ApJS 39, 1
Raymond, J., Hartigan, P. & Hartmann, L.: 1988, ApJ 326, 323
Reipurth, B.: 1994, A general catalogue of Herbig-Haro objects, electronically published via ftp to hq.eso.org in directory /pub/catalogs/Herbig-Haro
Reipurth, B. & Cernicharo, J.: 1995, Rev Mex AASC 1, 43
Schwartz, R.D.: 1975, ApJ 195, 631
Schwartz, R.D.: 1983, ApJ 268, L87
Seab, C.G.: 1987, in Interstellar Processes, ed. D. Hollenbach & H. Thronson (Dordrecht: Reidel), p. 491
Solf, J.: 1989, in Low Mass Star Formation and Pre-Main Sequence Objects, ed. B. Reipurth (Garching: ESO Conf. Series 33), 399
Solf, J.: 1993, in Stellar and Circumstellar Astrophysics, ed. G. Wallerstein & A. Noriega-Crespo (San Francisco: ASP Conf. Series), p. 22
Solf, J. & Böhm, K.H. 1991: ApJ 375, 618
Solf, J. & Böhm, K.H. 1993: ApJ 410, L31
Solf, J., Böhm, K.H. & Raga, A.C.: 1986, ApJ 305, 795
Solf, J., Böhm, K.H. & Raga, A.C.: 1988, ApJ 334, 229
Solf, J., Raga, A.C., Böhm, K.H. & Noriega-Crespo, A.: 1991, AJ 102, 1147
Staude, H.J. & Elsasser, H.: 1993, A&AR 5, 165
Stone, J.M. & Norman, M.L.: 1993, ApJ 413, 198
Wolfire, M.G. & Königl, A.: 1991, ApJ 383, 205
Zhang, H.L. & Pradhan, A.K.: 1995, A&A 293, 95

SPECTROSCOPIC SIGNATURES OF MICROJETS

J. SOLF
Thüringer Landessternwarte Tautenburg,
Sternwarte 5, D-77778 Tautenburg, FRG

Abstract. High-resolution long-slit spectroscopy of forbidden emission lines is used to investigate on a sub-arcsecond scale the spatial and kinematic properties and the physical conditions of the mass outflows from T Tauri stars in the immediate vicinity of the outflow source (microjets). Special attention is given to the case of DG Tau. The data permit us to distinguish physically different outflow components: (1) a high-velocity component (HVC) attributed to a fast jet, (2) a low-velocity component (LVC) attributed to gas entrained by the jet, and (3) a near-rest-velocity component (NRVC) attributed to a slow disk wind and/or disk corona.

1. Introduction

Forbidden emission lines (FELs) observed in the mass outflows from young stellar objects can be used as powerful diagnostic probes to study the morphology and kinematics as well as the physical conditions of the various manifestations of outflows, such as (bipolar) jets or Herbig-Haro (HH) objects. At optical wavelengths the spectra of the outflows are dominated by the lines of [OI], [NII] and [SII], indicating shocks as the main source of excitation in the outflowing gas. Classical T Tauri stars (TTSs) exhibit FELs superimposed upon the photospheric continuum spectrum. Since these lines are observed near the stellar position, their analysis has provided useful information on the properties of the outflows in the vicinity of their (stellar) outflow source (Appenzeller, Jankovics and Östreicher 1984; Edwards et al. 1987; Hamann 1994; Hartigan, Edwards & Ghandour 1995).

FELs observed in TTSs are more often blueshifted (with respect to the stellar velocity). This has been explained as due to a bipolar outflow of which the (redshifted) receding portion is hidden by an opaque circumstellar disk. Double-peaked FEL profiles are quite common consisting of a

B. Reipurth and C. Bertout (eds.), Herbig–Haro Flows and the Birth of Low Mass Stars, 63–72.

"low-velocity component" (LVC) and a "high-velocity component" (HVC). Several attempts have been made to understand the nature of the LVC and HVC. In most cases, a single outflow component has been assumed and it has been tried to interpret the double-peaked line profiles as due to geometric projection effects (e.g. Edwards et al. 1987; Hartmann & Raymond 1989; Gómez de Castro & Pudritz 1993; Ouyed & Pudritz 1993). In contrast, Kwan & Tademaru (1988; 1995) have noted that the [OI] line characteristics derived for the LVC and HVC are not readily explained by a single outflow component and proposed the existence of two physically distinct gas components in the outflows. According to their model the HVC is attributed to a high-speed gas in a jet-like flow due to a stellar wind that is collimated by the magnetic field in the circumstellar disk, whereas the LVC is attributed to an intrinsically low-speed gas due to a wind from the circumstellar disk and/or a disk corona.

As shown in previous work (Solf 1989), high-resolution long-slit spectroscopy of the FELs in TTSs can provide detailed information on a sub-arcsecond scale about both the spatial and the kinematic properties of the various outflow components in the immediate vicinity of the outflow source. Since in many cases the length of the outflow features observed in the vicinity of the star is quite small (less than a few arcseconds) these outflows have been nicknamed "microjets". Several long-slit spectroscopic studies of the FELs in TTSs (Solf & Böhm 1993; Böhm & Solf 1994; Hirth, Mundt & Solf 1994a, 1997; Hirth et al. 1994b) have been carried out in order to investigate the spatial and spectral properties of the microjets. The results of these studies support the interpretation of the HVC and LVC according to the Kwan & Tademaru model.

In this contribution, after a brief account on the applied observational and data reduction techniques, I will summarize the general characteristics of the spectroscopic signatures of microjets. Special attention will be given to the case of DG Tau which may be considered as a prototype for these outflows and for which more detailed observations from a period of serveral years are available.

2. Observations and Data Reduction Techniques

Using the Coudé spectrograph of the 2.2-m telescope at the Calar Alto Observatory (Spain), for each object several long-slit spectra have been obtained with the entrance slit centered on the star and oriented at various position angles (PA), preferentially parallel and perpendicular to the main outflow direction. A typical spectral resolution (FWHM) of $\sim$10 km s^{-1} and a spatial resolution (FWHM) of $0.''7 - 1.''5$ along the slit direction have been achieved. From each spectrogram a position velocity (PV) diagram of the (spatially and spectrally resolved) FEL regions covered by the slit has been deduced. A more detailed description of the observational and data reduction techniques has been given elsewhere (Solf 1989; 1994).

Perhaps the most important aspect in the data reduction and analysis of the FELs on long-slit spectrograms of TTSs comes from the presence of the underlying stellar continuum on the same spectrogram. *Firstly,* the continuum in the spectral range of the FELs has to be removed. This can be done effectively by evaluating the actual amount to be subtracted from an interpolation of the heights of the continuum on either side of the FEL. In order to account for absorption lines in the stellar spectrum underneath (and hidden by) the FEL, a template stellar spectrum (with similar absorption line characteristics but without the FEL) is used in the subtraction procedure. *Secondly,* the spatial distribution of the continuum (along the slit direction) can be used to determine accurately both the stellar position and the actual spatial resolution (FWHM). This permits us to evaluate with high accuracy (1) the position of the centroid of each FEL component relative to the stellar position and (2) its spatial extent (FWHM) corrected for the actual resolution. The strictly differential method permits us to measure the position of FEL components with an accuracy of $\sim 0.''02$ and to determine their spatial extent (corrected FWHM) down to $\sim 0.''2$.

3. Results and Discussion

3.1. SPECTROSCOPIC SIGNATURES OF MICROJETS

As already noted, the FELs observed in TTSs are more often blueshifted. If redshifted FEL features are present as well, these are usually fainter than the blueshifted ones. In many cases, the FEL profiles are rather complex presenting, e. g., double peaks, although single-peaked profiles occur quite often. In the latter case, the emission peak is usually quite narrow and only slightly blueshifted, typically at -5 km s^{-1} (Hamann 1994; Hartigan et al. 1995). This "near-rest-velocity component" (NRVC) is sometimes present in the more complex FEL profiles as well and sometimes confused by other line components. As noted by Hamann (1994), the NRVC has to be distinguished from the LVC (peaking at $\leq \pm 50$ km s^{-1}), since these two components may represent distinct emitting gas regions. In some cases, the HVC (usually observed at $\geq \pm 100$ km s^{-1}) is not well separated from the LVC in a particular FEL. In these cases, the distinction between the various components is based on their relative strengths in different FELs (e. g. [NII] and [SII]). The deduced spatio-kinematic properties provide additional criteria for distinguishing the various FEL components.

Although there may be many particularities varying from object to object, the more general characteristics of the spatial and kinematic properties

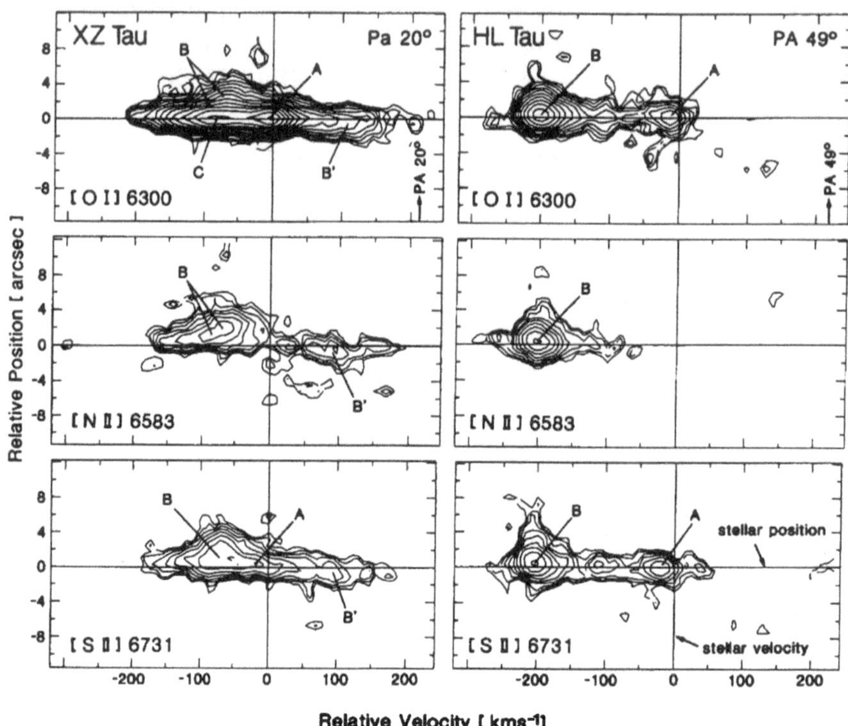

Figure 1. PV diagrams of the [OI]λ6300, [NII]λ6583 and [SII]λ6731 lines in XZ Tau and HL Tau derived from long-slit spectrograms obtained with the slit parallel to the outflow directions. The stellar continuum has been removed. Contours represent a factor of $2^{1/2}$ in the intensity. Velocities and positions are quoted with respect to the stellar velocity and stellar position, respectively.

of the outflows in the vicinity of the outflow source (microjets) are well illustrated by the observations of the FEL regions in XZ Tau and HL Tau. Fig. 1 shows the PV diagrams of the lines of [OI]λ6300, [NII]λ6583, and [SII]λ6731 obtained with the slit parallel to the bipolar outflow directions. XZ Tau presents both blueshifted and (fainter) redshifted FEL features located on opposite sides with respect to the central star as expected for a bipolar outflow of which the receding portion is partially obscured by a circumstellar disk. As can be noted on Fig. 1, in both stars the HVC (marked B) is present in all lines observed, whereas the LVC (marked A) is not present in [NII]. Compared to the LVC, the HVC is broader in velocity, spatially more extended along the outflow direction, and its centroid is separated from the star by a larger offset. Although the overall extent of the HVC is comparable in all three lines, the [OI]λ6300 emission is much more concentrated close to the central star. Nevertheless, the [OI] emission centroid of the HVC (marked C in the case of XZ Tau) is still clearly offset from the stellar position.

In a long-slit spectroscopic survey of 38 TTSs Hirth, Mundt & Solf (1997) have studied the spatial and kinematic properties of the FEL regions of [OI]$\lambda\lambda$6300,6363, [NII]$\lambda\lambda$6548,6583, and [SII]$\lambda\lambda$6716,6731 and derived the following main results: Generally, the HVC is spatially more extended than the LVC. In the [SII] lines, the emission centroid of the HVC is separated from the stellar position by a typical offset of 0.″6; the typical offset of the LVC is more than a factor 3 smaller. In the [NII] lines, detected in the HVC only, both the spatial extent and offset from the star are usually somewhat larger compared to the other FELs observed. In the [OI] lines, the offset of the emission centroid of the HVC is much smaller than in the other lines (typically ~0.″2). The LVC generally presents the highest compactness and the smallest offset of its emission centroid from the star (typically ~0.″1 in [OI] and ~0.″2 in [SII]).

3.2. THE CASE OF DG TAU

3.2.1. *General remarks*

DG Tau is a source of collimated mass outflows observed in (blueshifted) FEL condensations extending up to ~10″ towards PA 225° (Mundt & Fried 1983; Mundt, Brugel & Bührke 1987). On direct images of DG Tau in the light of FELs (see, e.g., Solf & Böhm 1993), one can recognize an inner region of the outflow (up to ~4″ from the star), containing the "microjet", and an outer region (from ~5″ to ~10″), containing several HH condensations. Adopting a distance of 140 pc, an inclination angle (with respect to the plane of the sky) of ~51° has been deduced for the outflow direction using the proper motions and radial velocities derived for the HH condensations (Eislöffel 1992).

On direct images of DG Tau, the microjet region is heavily confused by the presence of the much stronger stellar continuum. On long-slit spectrograms of DG Tau, however, the underlying continuum spectrum can be effectively removed. Fig. 2 presents PV diagrams of the lines of [SII]λ6731, [NII]λ6583, [OI]λ6300, and[OI]λ5577 deduced from long-slit spectrograms obtained with the slit parallel to the outflow direction (PA 45°). The PV diagrams permit us to distinguish various line features (designated A, A', B and C) on the basis of their kinematic and spatial characteristics.

Features B and C show both high velocities. B appears in the lines of [SII]λ6731, [NII]λ6583, and [OI]λ6300, whereas C is present in [OI]λ6300 only (and marginally detected in [OI]λ5577). B is spatially more extended than C. As already pointed out by Solf & Böhm (1993), both the (mean)

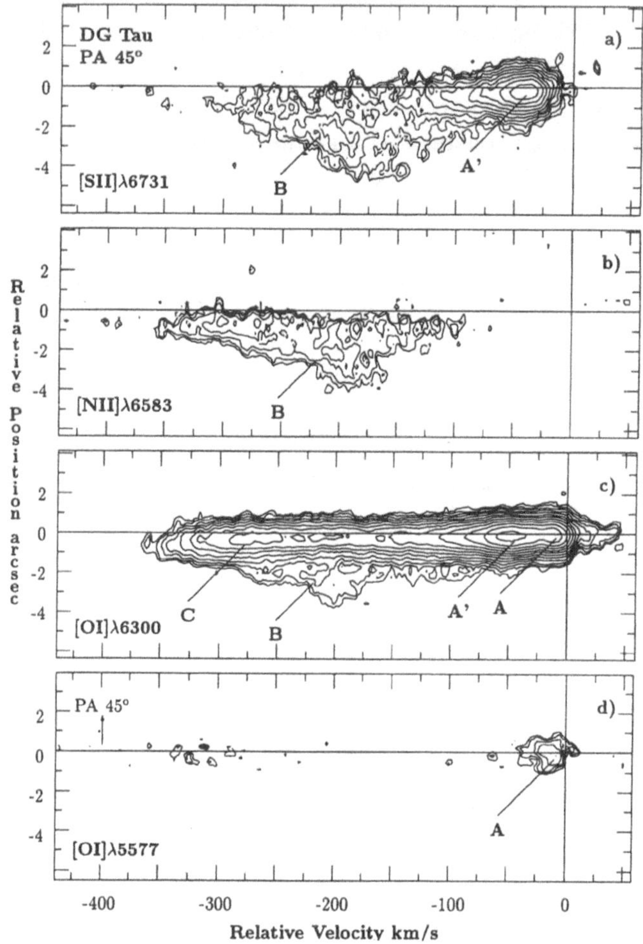

Figure 2. PV diagrams of the line emission of [SII]λ6731, [NII]λ6583, [OI]λ6300, and [OI]λ5577 in DG Tau. The diagrams have been deduced from PA 45° long-slit spectrograms obtained in 1992 (slit parallel to the outflow direction). Contours, velocities and positions as in Fig. 1

velocity and the electron density deduced within feature B are strongly increasing with decreasing distance from the central star and that trend continues within feature C. Therefore, B and C, rather than representing two separate outflow components, are considered to trace the outer (lower-density) and the inner (higher-density) regions, respectively, of one and the same "high-velocity" gas component (HVC).

Features A and A' show both low velocities. The velocity of A, lower than that of A', is near the (stellar) rest velocity. Both features are present

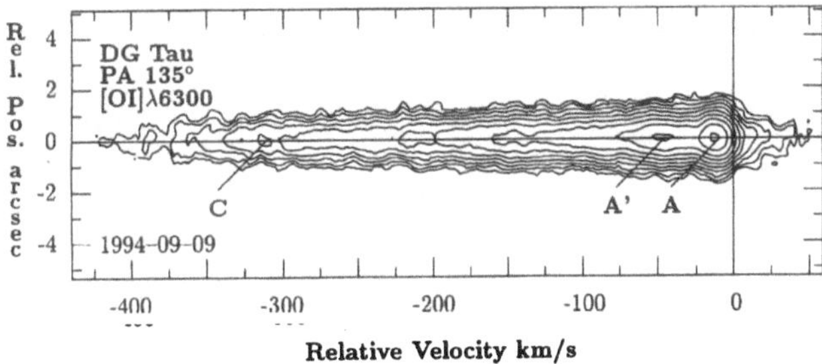

Figure 3. PV diagram of the [OI]λ6300 line in DG Tau at PA 135° (slit perpendicular to the outflow direction). Contours, velocities and position as in Fig. 1

in [OI]λ6300 and both absent in [NII]λ6583. The distinction between A and A' as two separate components is based on the fact that A is observed in [OI]λ6300 and [OI]λ5577, but not observed in [SII]λ6731, whereas A' is observed in [SII]λ6731 and [OI]λ6300, but not observed in [OI]λ5577. These findings indicate that the electron density is much higher in A ($\geq 10^7$ cm^{-3}) than in A' ($\geq 10^5$ cm^{-3}). Adopting the denominations proposed by Hamann (1994) we attribute feature A' to a "low-velocity" gas component (LVC), and feature A to a "near-rest-velocity" gas component (NRVC).

3.2.2. *Time variability*

The destinction between A and A' as two separate gas components in the outflows is strongly supported by a time series (from 1992 to 1994) of observations of DG Tau. As can be noted on the PV diagrams of the [OI]λ6300 line (Fig. 4) and the corresponding intensity profiles (Fig. 5), the relative line strengths of A and A' exhibit dramatic changes during that time period. Feature A, much weaker than A' in 1992, became the dominant feature in 1994. Furthermore, the time series reveals remarkable changes in the HVC as well. The "terminal" velocity ("blue edge") of feature C increased from ~ -350 km s^{-1} in 1992 to ~ -420 km s^{-1} in 1994.

3.2.3. *The high-velocity gas component (HVC)*

As already pointed out, the HVC is attributed to the high-speed gas of a jet-like outflow. Features B and C (see Figs. 2, 4) represent the outer and inner regions of the microjet, respectively. The relatively narrow outer tip of B (at $\sim 4''$ offset) shows a velocity of about -200 km s^{-1}, only slightly larger than that of the more distant HH condensations (-160 km s^{-1}). Both the mean velocity and velocity width deduced for the jet flow increase towards the central star. Similarily, an increase of the electron density in the jet of up to $>10^4$ cm^{-3} has been deduced from the [SII] line ratio. Due to

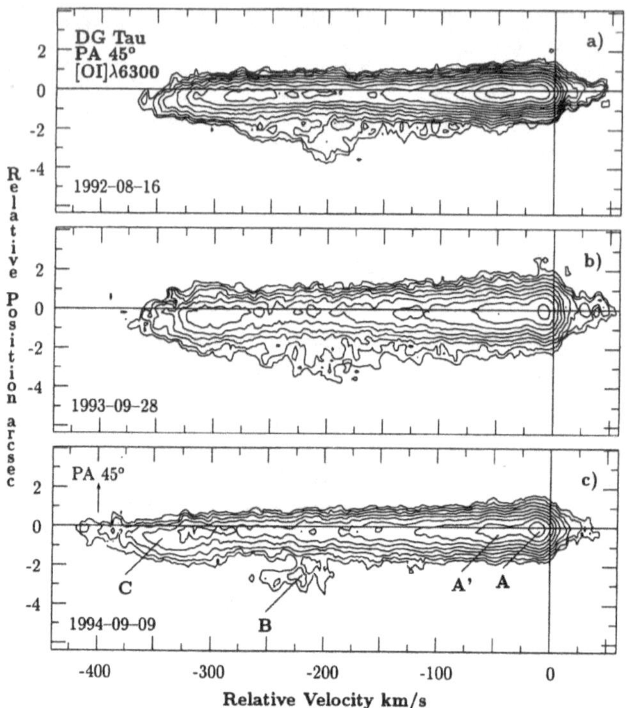

Figure 4. Time series from 1992 to 1994 of the PV diagrams of the [OI]λ6300 line in DG Tau. Position angle, contours, velocities and positions as in Fig. 2

even higher densities within 0.5″ from the star significant line quenching of [SII] occurs such that only [OI] is observable. From an analysis of the higher-density region of the HVC (feature C) on the PV diagrams at PA 45° and PA 135° (Figs. 2 and 3) one derives a spatial extent (FWHM) of ~0.″7 (~100 AU) and <0.″2 (<30 AU) for the microjet parallel and perpendicular to the outflow direction, respectively. The derived spatial offset of the [OI]λ6300 emission centroid of the HVC increases from 0.″2 to 0.″6 as a function of increasing velocity. A mean velocity of -270 km s^{-1} and a total velocity width of >250 km s^{-1} have been deduced for the microjet from the [OI]λ6300 emission.

The observed "terminal" velocity of the "blue wing" of the [OI]λ6300 line has increased significantly during 1992–1994. This suggests that probably a new mass ejection event has occurred duing that period. Interestingly, the gas in the more recent flow is faster than that in the precursor. This implies that the gas of the new ejection will eventually overtake that of the previous ejection and hence produce internal shocks in the jet flow.

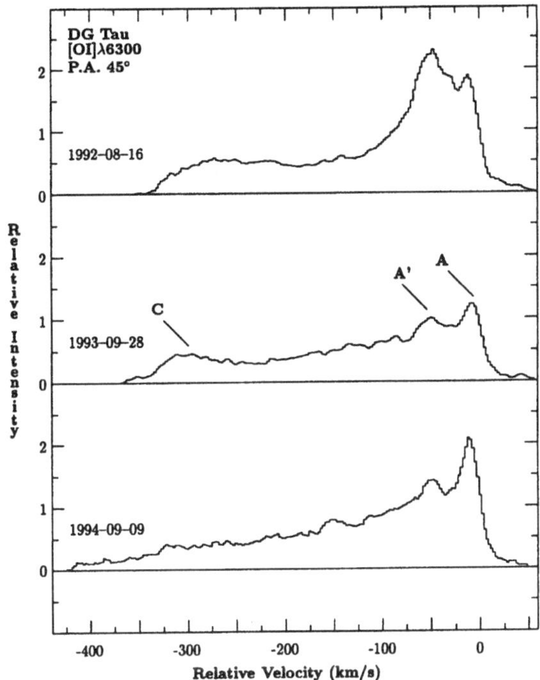

Figure 5. Time series of the [OI]λ6300 line intensity profiles in DG Tau (see Fig. 4)

3.2.4. *The low-velocity gas component (LVC)*

The LVC, represented by feature A' (see Figs. 2–4), is detected in [OI]λ6300 and [SII]λ6731. It is not detected in [OI]λ5577 and [NII]λ6583. The latter indicates a lower excitation in the LVC compared to the HVC. The derived electron density of the LVC is comparable to that of the HVC, but significantly lower than that of the NRVC. A centroid velocity of -43 km s^{-1} and velocity width (FWHM) of 39 km s^{-1} have been measured. The emission centroid of the LVC is offset from the stellar position by 0.″11 (15 AU) and 0.″25 (35 AU) in [OI]λ6300 and [SII]λ6731, respectively. The spatial width (corrected FWHM) of the LVC deduced for [OI]λ6300 is ~0.″3 (~40 AU) in both directions, parallel and perpendicular to the flow. The velocity profile of the LVC is highly asymmetric presenting an extended blueshifted wing. The spatial offset of the emission in that wing becomes progressively larger if one proceeds to larger (negative) velocities. Remarkably, the high-velocity wing of the LVC extends into the domain of the HVC (feature B) and is confused by the emission of the HVC such that no clear distinction can be made between the LVC and HVC in that velocity range. This indicates that the HVC and LVC may be physically related to each other in the sense that a transition zone exists between the associated two gas components. Because of these findings we suggest that the LVC is attributed to a slower gas in the environment of the jet which is entrained by the faster jet gas. This interpretation is well compatible with the different spatio-kinematic properties and excitation conditions deduced for the LVC and HVC.

3.2.5. *The near-rest-velocity gas component (NRVC)*

The NRVC, represented by feature A (Fig. 2–5), is detected in [OI]λ6300 and [OI]λ5577, but not detected in [NII]λ6583 and [SII]λ6731, indicating both a lower excitation and a higher density compared to the HVC and LVC. Remarkably, the [OI]λ6300 profile, peaking at -15 km s^{-1}, shows an extended redshifted wing reaching up to $\sim +50$ km s^{-1}. The offset from the stellar position deduced for the emission centroid of the NRVC is $0.''11$ (15 AU) and $\leq 0.''05$ ($\leq$7 AU) for [OI]λ6300 and [OI]λ5577, respectively. The spatial extent (corrected FWHM) derived for [OI]λ6300 is $\sim 0.''3 \times 0.''4$ ($\sim$ 42 AU $\times$ 56 AU), parallel and perpendicular to the flow, respectively. These findings indicate that the NRVC represents a compact region of an intrinsically slow, high-density gas concentrated in the immediate vicinity of the outflow source. This result lends strong support to the model of Kwan & Tademaru (1988; 1995) in which the NRVC is attributed to a slow disk wind and/or disk corona. The redshifted line wing is probably due to rotation as expected for a line forming region corotating with the disk.

References

Appenzeller, I., Jankovics, I. & Östreicher, R. 1984, A&A, 141, 108
Böhm, K.-H. & Solf, J. 1994, ApJ, 430, 277
Edwards, S., Cabrit, S., Strom, S.E., Heyer, I., Strom, K.M. & Anderson, E. 1987, ApJ, 321, 473
Eislöffel, J. 1992, Ph.D. thesis, Univ. Heidelberg
Gómez de Castro, A.I. & Pudritz, R.E. 1993, ApJ, 409, 748
Hamann, F. 1994, ApJS, 93, 485
Hartigan, P., Edwards, S. & Ghandour, I. 1995, ApJ, 452, 736
Hartmann, L.W. & Raymond, J.C. 1989, ApJ, 337, 903
Hirth, G.A., Mundt, R. & Solf, J. 1994a, A&A, 285, 929
Hirth, G.A., Mundt, R., Solf, J. & Ray, T.P. 1994b, ApJ, 427, L99
Hirth, G.A., Mundt, R. & Solf, J. 1997, A&A, submitted
Kwan, J. & Tademaru, E. 1988, ApJ, 332, L41
Kwan, J. & Tademaru, E. 1995, ApJ, 454, 382
Mundt, R., Brugel, E.W. & Bührke, T. 1987, ApJ, 319, 275
Mundt, R. & Fried, J. 1983, ApJ, 274, L83
Ouyed, R. & Pudritz, R.E. 1993, ApJ, 419, 255
Solf, J. 1989, in: ESO Conf. Proc. 33, Low Mass Star Formation and Pre-Main Sequence Objects, Ed. B. Reipurth, (Garching b. München: ESO), p. 399
Solf, J. 1994, in: ASP Conf. Ser. 57, Stellar and Circumstellar Astrophysics, ed. G. Wallerstein & A. Noriega-Crespo (San Francisco: ASP), p. 399
Solf, J. & Böhm, K.-H. 1993, ApJ, 410, L31

THE IONIZATION STATE ALONG THE BEAM OF HERBIG-HARO JETS

FRANCESCA BACCIOTTI
Dipartimento di Astronomia e Scienza dello Spazio,
Università di Firenze,
Largo E. Fermi 5, 50125 Firenze, Italy

Abstract. An original spectroscopic diagnostic technique has been recently developed that allows to estimate in a model-independent way the ionization fraction x_e and the average excitation temperature in the beam of Herbig-Haro jets (Bacciotti, Chiuderi and Oliva 1995). The procedure is based on the fact that in the low excitation conditions present in this region the ionization state of oxygen and nitrogen can be assumed to be regulated by charge exchange with atomic hydrogen. The application of this technique to long-slit spectra of several well-known stellar jets indicates that the hydrogen ionization fraction x_e starts from 0.2–0.4 at the beginning of the flow and gently decreases along the whole jet or along sections of it as a result of time-dependent recombination. The average temperature stays almost constant, ranging from 4500 to 7000 K. The momentum transfer rates evaluated with the derived total number densities ($n_H \sim 10^3 - 10^4$ cm^{-3}) give support to the picture in which the jet is responsible for the acceleration of a surrounding molecular outflow.

1. Introduction

Highly collimated, supersonic Herbig-Haro (HH) jets emanating from young stellar objects have been the subject of intense investigations in the recent past. Many of their features, however, remain largely unexplained. Still under debate are, for example, the jet acceleration mechanism (see, e.g., Königl and Ruden 1993 for a review), the excitation of the gas in the beam (Raga and Kofman 1992, Stone and Norman 1993, Bodo et al. 1994, Ray et al. 1996), and the capability of the narrow central jet to drive a coaxial molecular outflow (see, e.g., Chernin and Masson 1995, Cabrit, this volume)

B. Reipurth and C. Bertout (eds.), Herbig–Haro Flows and the Birth of Low Mass Stars, 73–82.

A crucial physical parameter for any jet model is the *total density* of the flow which, however, is poorly known for HH jets. In fact, contrary to the electron density, which is easily found from the ratio of the [SII] lines at 6716 Å and [SII] 6731 Å (see, e.g. Osterbrock 1989), no standard procedure exists to deduce directly from observations the total hydrogen density. Its estimate has been most often derived from shock models (see, e.g., Hartigan, Morse and Raymond 1994). The results obtained this way are, however, largely model-dependent, being greatly influenced by different pre-shock conditions, by (unknown) magnetic fields, and by the assumed shape of the shock front. Alternatively, the average total density can be estimated from the luminosity in one forbidden line (Edwards et al. 1987, Cabrit et al. 1990), but this kind of analysis is severely hampered by the limited knowledge of the ionization and excitation state of the emitting species and by strong dust absorption.

Recently Bacciotti, Chiuderi and Oliva (1995, hereafter BCO95) developed an original spectroscopic diagnostic technique that allows to find in a model-independent way the ionization fraction $x_e = n_{H^+}/n_H \approx n_e/n_H$ of the jet gas, and hence the total density, together with the average excitation temperature. The main features of the procedure, which uses ratios of the most frequently observed forbidden lines together with Hα, will be briefly recalled in the following. The results obtained from the application of the technique to several well-known HH jets will be reviewed, including results from spatially resolved analyses. Some interesting physical implications will be discussed.

2. The diagnostic procedure

The most often observed lines in the beams of HH jets are Hα, the [SII] doublet at 6716, 6731 Å, the [OI]$\lambda\lambda$6300, 6363 and the [NII] $\lambda\lambda$6548, 6584 lines. Here, as in all low-excitation nebulae, sulphur can be assumed to be all singly ionized. In addition, the lack of any observational evidence for high electron temperature and of any local or nearby source of energetic radiation suggests that charge exchange may be the dominant process controlling the ionization of O and N. Checking this hypothesis against competing effects, as, for example, radiative plus dielectronic recombination, one finds that for oxygen charge exchange dominates unless the plasma is very close to complete ionization, while for nitrogen the charge exchange hypothesis is valid as long as x_e is lower than about 0.5. Regions of higher ionization are identified by the technique, but can only be described on a qualitative basis. The charge exchange hypothesis allows to express the populations of neutral O and singly ionized N as a function of x_e. Once a set of relative abundances is assumed, and the electron density is determined from standard methods

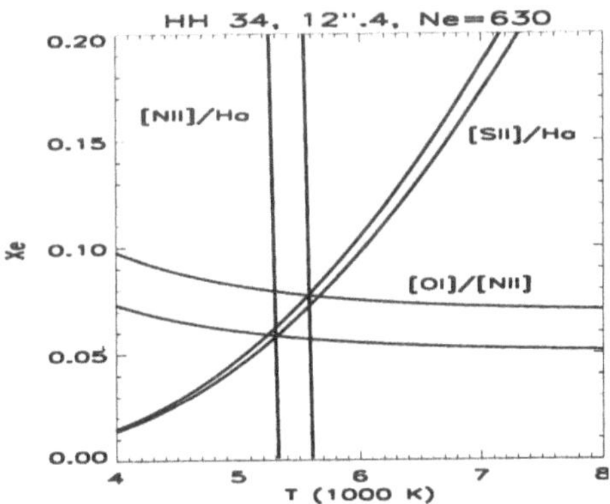

Figure 1. Example of a diagnostic diagram calculated for the position at 12.″4 in the HH 34 jet. The intersection of the crossing stripes provides $T = 5450 \pm 150$ K and $x_e = 0.067 \pm 0.008$.

any line ratio can be regarded as a known function of x_e, T, and can be calculated numerically (for more details see BCO95). A diagnostic (x_e, T) diagram like the one in Fig. 1 provides then the required values: the strips are the loci of the (x_e, T) values for which the predicted line ratios equal the observed ones within the measurement errors, and their intersections define the local (x_e, T) and their uncertainty.

3. First inspection of the HH 34 and HH 111 jets

As a first application BCO95 analyzed the physical conditions in the beams of the HH 34 and HH 111 jets, using spectra integrated over the most luminous knots (Reipurth et al. 1986, Reipurth 1989). Both jets turned out to be remarkably neutral: the diagnostic provided $x_e = 0.067 \pm 0.01$ in the HH 34 beam and $x_e = 0.11 \pm 0.02$ for the HH 111 jet. The electron density and excitation temperatures were found to be $n_e = 720$ cm^{-3} and $T_e = 5800 \pm 200$ K in the HH 34 jet, and $n_e = 900$ cm^{-3} and $T_e = 5700 \pm 200$ K in HH 111. Combining n_e and x_e an average total hydrogen density of about 10^4 cm^{-3} was found for both flows, from which the authors derived a momentum transfer rate $\dot{P} \approx 6.7\ 10^{-5}$ M$_\odot$ yr^{-1} km s^{-1} for the HH 34 jet, and $\dot{P} = 1.3\ 10^{-4}$ M$_\odot$ yr^{-1} km s^{-1} for HH 111.

The origin of the observed ionization degree is uncertain, however, since the low excitation of these jets can only arise from low velocity shocks ($v_s = 25 - 35$ km s^{-1}) that are not capable of producing *in situ* an ionization degree larger than a few per cent. On the other hand BCO95 pointed out that the typical recombination time of the jet gas is of the same order of the travel time of the bright visible section. Therefore the observed ioniza-

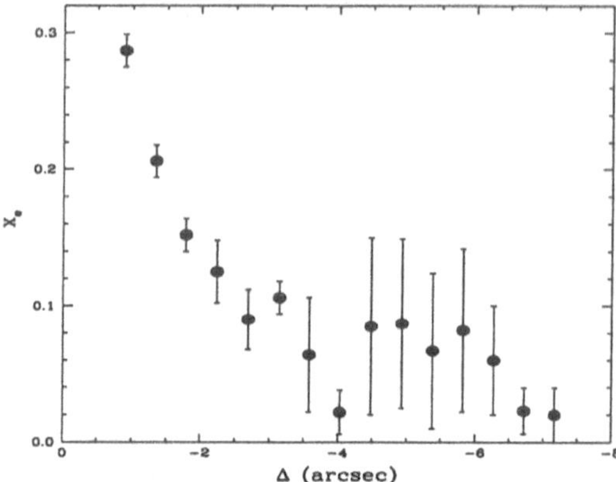

Figure 2. Ionization fraction along the axis of the optical outflow of RW Aurigae (red lobe) as a function of the distance from the star

tion state may be a remnant of the initial heating and the excitation occured in the acceleration region (for example by means of a strong steady shock heavily shielded from view by a dense circumstellar environment). Traveling away from the star the flow is subject to a sudden expansion (Mundt, Ray and Raga 1991), that 'freezes' the ionization of the gas; the jet then recollimates, but recombination is sufficiently slow to leave the gas considerably ionized also at large distances from the source. According to this scenario, x_e should decrease along the jet axis on spatial scales determined by the product of the recombination time and the flow velocity. This prediction is not affected by the presence of weak shocks that eventually form in the beam, if these are so weak to poorly contribute to ionization.

4. Analyses from long-slit spectra

The observational confirmation of the picture outlined in the previous section can come from the application of the BCO95 procedure to spectra spatially resolved along the jet axis.

The optical jet of RW Aur. As a first attempt in this direction Bacciotti, Hirth & Natta (1996) have examined the optical bipolar outflow associated with the T Tauri star RW Aurigae, for which long-slit spectra were obtained in November 1993 with the Cassegrain twin spectrograph at the 3.5-m telescope of the Calar Alto Observatory in Spain. The results for the red lobe of the flow, in which the lines were sufficiently intense to allow a

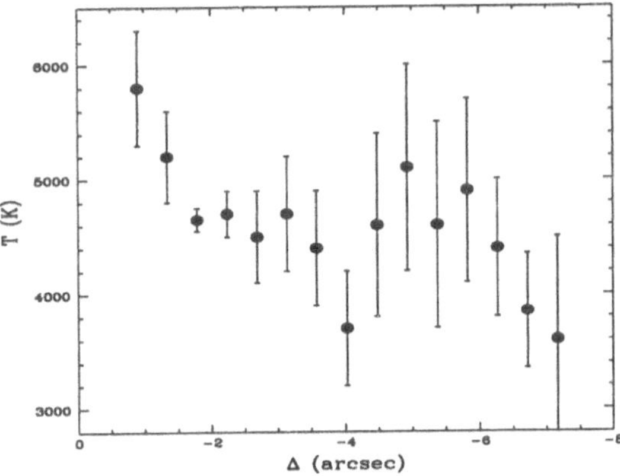

Figure 3. Average excitation temperature along the axis of the optical outflow of RW Aurigae (red lobe) as a function of the distance from the star

realistic determination, nicely confirm the suggestion offered in BCO95. As shown in Fig. 2, the ionization fraction decreases along the beam from $x_e \sim 0.28$ near the star to about 0.02 at a distance of 6–7″ ($\sim$ 1000 AU), while the temperature shows a slight decline with typical values of 4500 K (Fig. 3). The electron density (not shown) also decreases on average along the flow, following a well-defined recombination law. The derived total density is again of the order of 10^4 cm^{-3} which yields a mass loss rate $\dot{M} \sim 5\ 10^{-8}$ M$_\odot$ yr^{-1} and an average momentum transfer rate $\dot{P} = 6.5\ 10^{-6}$ M$_\odot$ yr^{-1} km s^{-1}.

The short RW Aur outflow, however, belongs to a system more evolved than the sources of "classical" Herbig-Haro jets. It is therefore of great interest to further investigate the cases of prototypical jets in order to obtain a model-independent determination of the fundamental physical parameters for each of them, and to establish if the agreement between the BCO95 model and the observed trends is confirmed on a large statistical basis. Recent analyses of long-slit spectra of a sample of well-known jets are illustrated in Bacciotti and Eislöffel (1997, hereafter BE97). Here we report on the results obtained for the HH 34 and the HH 24C/E jets as representative cases (for a detailed description of the objects see, e.g., Eislöffel and Mundt 1992, Mundt, Ray and Raga 1991). The long-slit spectra were taken in February 1987 at the ESO/MPI 2.2m-telescope on La Silla using the Boller & Chivens spectrograph. The spectral resolution was 1.6Å per pixel and the spatial resolution 1.″78 per pixel along the slit. The results are given in Figures 4, 5 and 6 in the form of columns of graphs, illus-

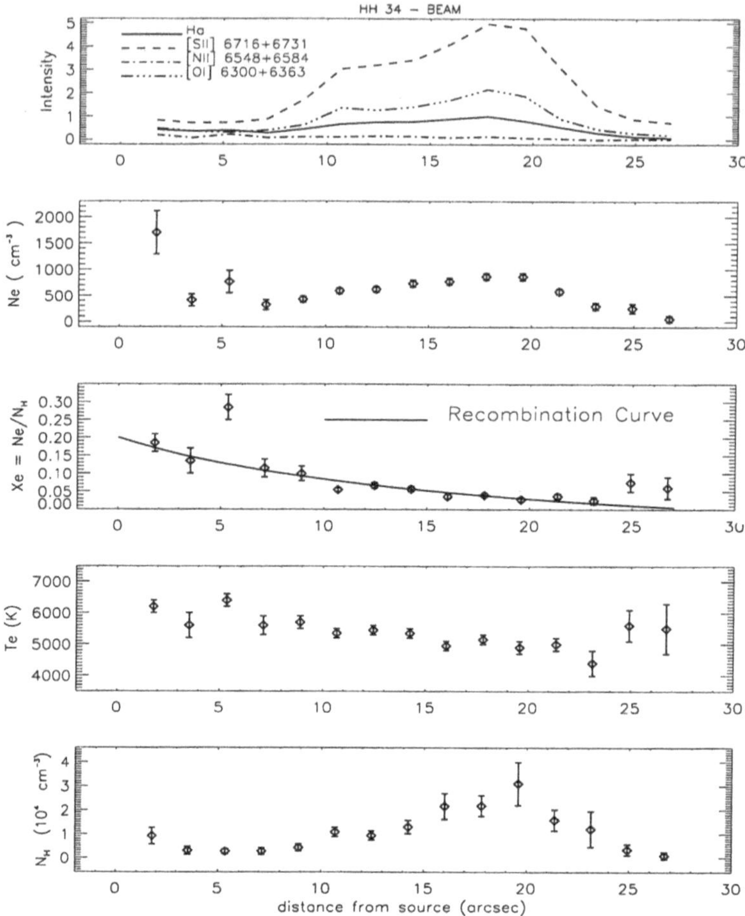

Figure 4. Physical conditions along the beam of the HH 34 jet. From top to bottom: a) intensity tracings normalized to Hα in the various lines b) electron density derived form the [SII] lines ratio c) hydrogen ionization fraction x_e and superimposed recombination curve (see text) d) average excitation temperature e) total hydrogen density $n_H = n_e/x_e$.

trating a) relative intensity tracings of the various lines integrated over the line widths, b) the electron density derived form the [SII] lines ratio, c) the hydrogen ionization fraction x_e, d) the average excitation temperature and e) the total hydrogen density n_H.

HH 34. Here the electron density is observed to increase progressively to knot J, after which it falls again (see Fig. 4). On the contrary, the ionization fraction decreases steadily from $x_e = 0.2$ to about 0.03 from the beginning to the end of the visible section. The jump in x_e at position $z = 5.''3$ is probably affected by a large measurement error due to the faintness of the lines. Disregarding this point, the ionization nicely follows a recombination curve calculated assuming that on average the gas parcels flow in a set of nested cones, whose shape is specified by the initial jet radius and by a constant opening angle (see BE97 for details). Our best-fit curve turns out to have a negative opening angle $\theta = -1^\circ\!.4$, i.e., the jet is slightly converging, in accordance with the behavior of n_e. The average excitation temperature mildly decreases along the flow from about 6000 K to about 4300 K. Finally, the total density profile shows a behavior similar to that of the electron density: after a slight decline prior to $z = 5''$ the total density increases from 4–500 cm^{-3} up to 3 10^4 cm^{-3} near knot J, then it decreases again. From the total density a mean value of $n_H \sim 1\ 10^4\ cm^{-3}$ along the jet can be deduced, in accord with BCO95.

HH 24C. In this jet n_e decreases on average along the flow, but shows two peaks at the position of the two brightest knots, with $n_e \sim 800\ cm^{-3}$ and $\sim$400 cm^{-3} respectively (see Fig. 5). The ionization fraction oscillates around a mean value of $\sim$0.33, with a slight increase in the fainter section at the middle of the flow, where the gas is apparently re-ionized: contrary to HH 34 the ionization data points are better fit by two different recombination curves, the first one appropriate for a slightly converging jet section described by an angle -2°, while the second one corresponds to flow surfaces opening by a small angle 1°. The average excitation temperature varies around 5500 to 6000 K, increasing slightly toward the end of the bright beam section. The average total density is about 1 10^3 cm^{-3}. Assuming a jet velocity of 400 km s^{-1} the average mass loss rate obtained for this jet is $\dot{M} = 6.8\ 10^{-8}\ M_\odot\ yr^{-1}$ while the momentum transfer rate is $\dot{P} = 2.7\ 10^{-5}$ $M_\odot\ yr^{-1}\ km\ s^{-1}$.

HH 24E. The presence of re-ionization episodes is even more evident in the HH 24E counterjet, where x_e shows two pronounced jumps at positions $z = -10.''7$ and $z = -21.''3$, i.e. at the beginning of knot E and of the bright terminal condensation HH 24A, respectively. The ionization fraction decreases smoothly downstream of each jump, albeit along different recombination curves, that correspond to opening angles of 2°, 1°, and 6°, respectively (with an assumed space velocity of 350 km s^{-1}). On average, x_e becomes progressively higher from one section to the next. The average excitation temperature also increases suddenly from 5100 to 6800 K at position $z = -10.''7$, while near knot A the increase is smoother and only of 1000 K. The electron density is observed to markedly increase in knot A, while no apparent enhancement is seen near the first jump at $z = -10.''7$. The derived total density is higher in the first section of the flow. As a mean

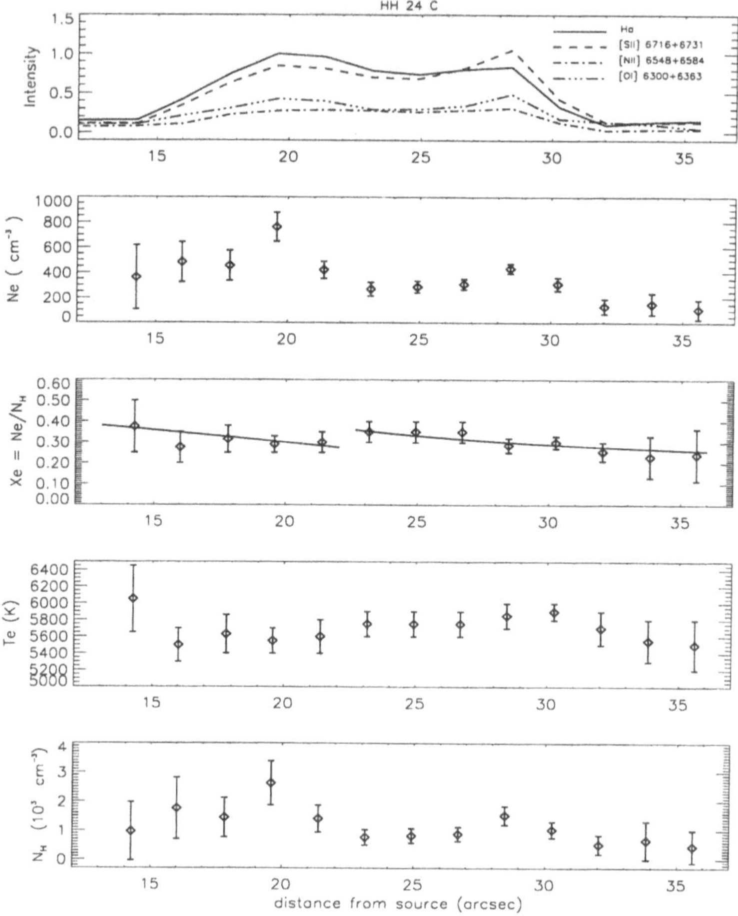

Figure 5. Physical conditions along the beam of the HH 24C jet. Panel description as in Fig. 4.

along the jet one can take $n_H = 2.5\ 10^3\ \mathrm{cm}^{-3}$, which yields $\dot{M} = 6.1\ 10^{-8}$ $\mathrm{M}_\odot\ \mathrm{yr}^{-1}$ and $\dot{P} \approx 2.1\ 10^{-5}\ \mathrm{M}_\odot\ \mathrm{yr}^{-1}\ \mathrm{km\ s}^{-1}$.

5. Conclusions

The BCO technique has been demonstrated to be a powerful tool for a model-independent investigation of the physical conditions in the beams of HH jets. A summary of the results is presented in the following.

– The gas in the beam is *partially ionized*; x_e ranges between 0.05 and 0.4, being higher for more excited and lighter jets. Both high temperatures and low densities, in fact, disfavour recombination.

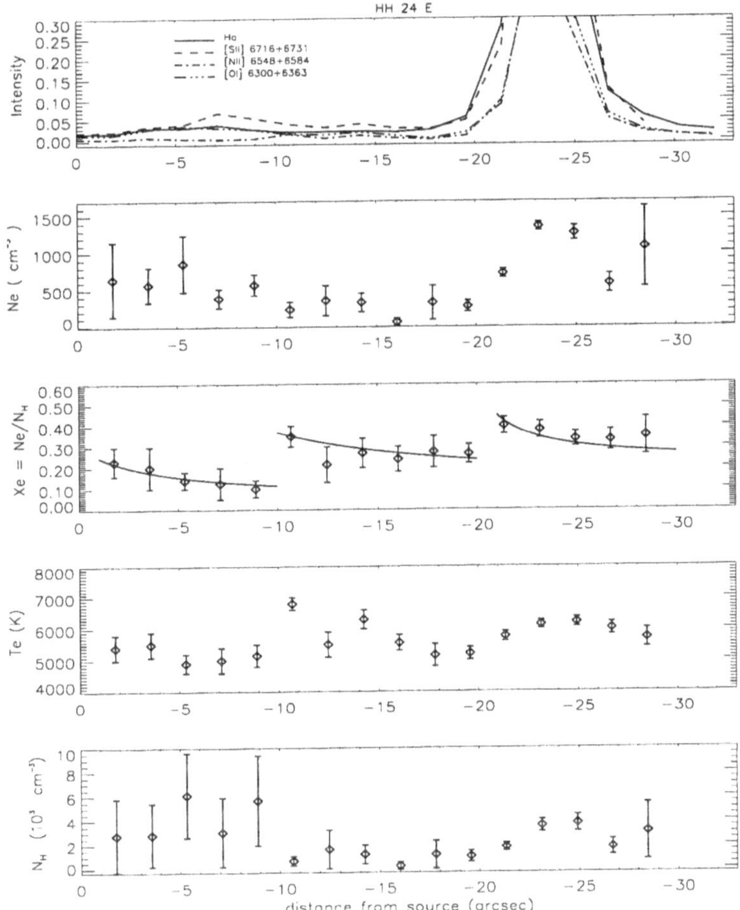

Figure 6. Physical conditions along the beam of the HH 24E jet. Panel description as in Fig. 4.

– *The ionization fraction slowly decreases along the jet or along sections of it.* As it was originally suggested in BCO95, the ionization state of the jet gas is probably produced in the acceleration region; then the gas slowly recombines traveling away from the source. If the shocks shaping the beam have $v_s < 30 - 35$ km s^{-1}, they can only add a negligible contribution to the ambient ionization, as it apparently occurs in the optical outflow of RW Aur and in the HH 34 jet. The jumps in x_e followed by slow decay observed in the HH 24C/E flow are probably due to stronger internal shocks, that substantially re-ionize the gas.

– The excitation temperature varies between 4500 and 7000 K. The limited spatial resolution of the investigated spectra does not allow to study in detail the post-shock cooling region and therefore the obtained values are averages over a large parcel of gas. Taking into account this constraint, the diagnostic results are not in contradiction with the prediction of shock models (see Hartigan et al. 1994, BE97).

– The average total hydrogen density $n_H = n_e/x_e$ ranges between about 10^3 and a few 10^4 cm^{-3}. Average mass loss and momentum transfer rates calculated under the assumption of constant density over the jet section are of the order required to drive surrounding molecular outflows.

– Partial ionization should be taken into account in any jet model. Reliable calculations for beam shocks should consider that the fronts advance in a substantially ionized medium. In addition, partial ionization may introduce important differences in the modeling of magnetized jets, due to the effects introduced by collisions between charged particles and neutrals (see Bacciotti, Chiuderi and Pouquet 1997 for a discussion).

Acknowledgements: It is a pleasure to thank Alex Raga, Jochen Eislöffel and Tom Ray, without the help of whom the work presented in this paper would not have been possible.

References

Bacciotti, F., Chiuderi, C., and Oliva, E., 1995, A&A, 296, 185 (BCO95)
Bacciotti, F., Hirth, G. and Natta, A., 1996, A&A, 310, 309
Bacciotti, F., Chiuderi, C. and Pouquet, A., 1997, Ap J, 478, 594
Bacciotti, F., and Eislöffel, J., 1997, A&A, submitted (BE97)
Bodo, G., Massaglia, S., Ferrari, A., and Trussoni, E., 1994, A&A, 283, 655
Cabrit, S., Edwards, S., Strom, S., and Strom, K., 1990, ApJ 354, 687
Chernin, L., and Masson, C., 1995, ApJ, 455, 182
Edwards, S., Cabrit, S., Strom, S., Heyer, I., Strom, K., and Anderson, E., 1987, ApJ 321, 473
Eislöffel, J., and Mundt, R., 1992, A&A 263, 292
Hartigan, P., Morse, J., and Raymond, J., 1994, ApJ, 436, 125
Königl, A., and Ruden, S.P., 1993, in Protostar and Planets III, Levy G. and Lumine, J. eds. Tucson: University of Arizona press, 641.
Mundt, R., Ray, T.P., and Raga, A. C., 1991, A&A 252, 740.
Osterbrock, D.E., 1989, *Astrophysics of Gaseous Nebulae and Active Galactic Nuclei*, University Science Books, Mill Valley, CA.
Raga, A.C., 1992, private communication.
Raga, A.C., and Kofman, L., 1992, ApJ 386, 222
Ray, T.P., Mundt, R., Dyson, J.E., Falle, S.A.E.G. and Raga, A.C., 1996, ApJ, 468, L103
Reipurth, B, 1989, Nature, 340, 42
Reipurth, B., Bally, J., Graham, J.A., Lane, A.P., and Zealey, W.J., 1986, A&A 164, 51
Stone, J.M., and Norman, M.L., 1993, ApJ 413, 198

THERMAL RADIO JETS

LUIS F. RODRIGUEZ
Instituto de Astronomía, UNAM
Apdo. Postal 70-264
México, DF, 04510 México

Abstract. Since thermal radio jets can be observed with subarsecond angular resolution and are unaffected by dust absorption, they provide a useful tool to study collimated outflows very close to the young stars that produce them. Here, I review recent results on this area, giving emphasis to the study of thermal jets in the case of the quadrupolar outflows in L723 and HH 111.

1. Introduction

Young stars are known to possess powerful collimated winds that interact with their surrounding gaseous medium producing the Herbig-Haro objects and the molecular bipolar outflows. As demonstrated by these Proceedings, great advance has been made in our understanding of these large scale (tenths of pc or more) manifestations of the mass loss activity in young stars. Our understanding of the "engine" that powers these outflows is, however, still limited. There is a degree of consensus in the sense that the acceleration and collimation of these winds involves magnetohydrodynamic processes in an accretion disk, but many basic questions remain. What is the collimation scale? Why are the terminal velocities of these winds comparable with the escape velocity of the central object? What is the ratio between accreted and ejected material? Are these winds the mechanism by which the forming star gets rid of excess angular momentum and magnetic flux?

Certainly, observations with the highest angular resolution possible are required to address these important issues. In this paper, I review recent results in the field of thermal jets. At centimeter wavelengths the continuum emission from young stellar objects is dominated by free-free

B. Reipurth and C. Bertout (eds.), Herbig–Haro Flows and the Birth of Low Mass Stars, 83–92.

(bremsstrahlung) emission from ionized, collimated outflows (thermal jets), that can be observed using radio interferometers with angular resolutions in the range of 0''.1. This field was pioneered by the studies made toward T Tauri stars with the Very Large Array by Cohen *et al.* (1982), and this radio interferometer is still the instrument of choice given its great sensitivity. Recent reviews concerning thermal jets have been given by Anglada (1995; 1996), and Rodríguez (1995; 1996).

2. Thermal Jets

2.1. HOW ARE THERMAL RADIO JETS IDENTIFIED?

Since the thermal jets are detected via their centimeter continuum emission, it could be easy to erroneously identify another type of source as one of them. There are, however, a number of criteria that allow the reliable identification of a radio source as a thermal jet.

1. One expects them to be located at the centroid of the outflow region. A calculation of whether or not the source detected in the solid angle considered could be an unrelated background source is also valuable. This test can be made using background source counts. For example, extrapolating from the results of Condon (1984) to 3.6-cm (the most sensitive wavelength at the VLA) one expects $\sim 0.01\ S^{-0.8}$ background sources per square arc min above a flux density of S (in mJy). Then, if for example other observations of the region suggest that the exciting source should be located in a region of solid angle of 0.5 square arc min and a source of 0.5 mJy is detected at 3.6-cm in this solid angle, most likely this source is associated with the region considered since the *a priori* possibility of the source being a background object is only $\sim$0.01.

2. The thermal jets are expected to be elongated along the large-scale outflow axis, as traced by the Herbig-Haro objects and/or the bipolar outflow.

3. Thermal jets should show characteristic dependences with frequency for the total flux density, S_ν, and the deconvolved major axis, θ_{maj}. The theoretical radio continuum spectra of a confined thermal jet has been calculated by Reynolds (1986). For a collimated wind of constant temperature, velocity, and ionization fraction, the flux density and angular dimension of the source depend on frequency as $S_\nu \propto \nu^{1.3-0.7/\epsilon}$ and $\theta_{maj} \propto \nu^{-0.7/\epsilon}$, respectively, where ϵ is the power law index that describes the dependence of the jet half-width, w, (perpendicular to the jet axis) with the distance to the jet origin ($w \propto r^\epsilon$). For the simplest case of a biconical jet (that is, a jet with constant opening angle) one has $\epsilon = 1$ and an expected behavior of $S_\nu \propto \nu^{0.6}$ and $\theta_{maj} \propto \nu^{-0.7}$. Anglada (1996) presents a list of sources where these frequency dependences have been measured.

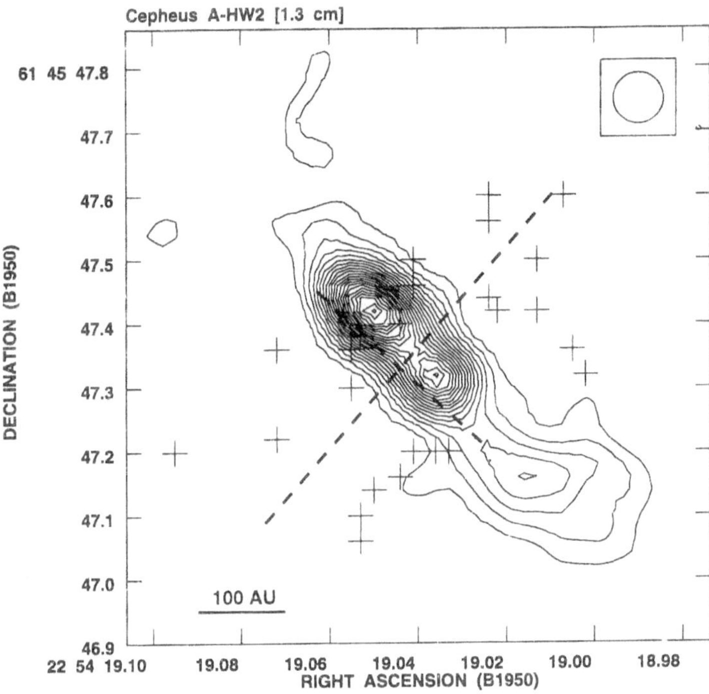

Figure 1. VLA-A continuum map of the Cepheus A HW2 thermal radio jet at 1.3 cm. Contours are -6, -3, 3, 6, 9, 12, 15, etc ... times 0.12 mJy $beam^{-1}$, the rms noise of the map. The beam is shown in the top right-hand corner. Crosses indicate the position of the H_2O maser spots detected in a region of about 0.8 arcsec around HW2. Dashed lines indicate the major and minor axes of the disk that is proposed to be traced by the maser spots.

4. The gas detected in the radio continuum has left the star in the last months or years. One then expects to be able to measure variations and/or proper motions of condensations in the jet on timescales of years or even less. The case of HH 80-81 is well studied by Marti *et al.* (1993; 1995), who have been able to detect proper motions in several of the condensations in the jet. The corresponding velocities of the condensations are in the range 600-1400 km s^{-1}, confirming that we are dealing with a high-mass object (since a correlation is expected between outflow velocity and the mass of the star).

5. Thermal jets are intimately associated with warm, dense molecular gas (as traced by ammonia: see Torrelles *et al.* 1992, 1993a; Gómez *et al.* 1994; or CS: see Yang *et al.* 1997). Finally, it has become evident recently

that at least in some cases the thermal jets can be associated with H_2O masers (Gómez *et al.*, 1995). The relation between the thermal jets and the H_2O masers is unclear: while in Cep A HW2 the masers seem to be tracing a circumstellar disk perpendicular to the jet (see Figure 1 and Torrelles *et al.* 1996), in L1448C the masers appear to be related morphologically and kinematically with the jet (Chernin, 1995).

2.2. ADVANTAGES AND DISADVANTAGES

The high angular resolution and accurate positional accuracy provided by the radio observations, together with the fact that they are practically unaffected by dust obscuration have allowed important new advances in our understanding of the collimated outflow phenomenon:

1. Since the exciting sources of the outflows are usually deeply embedded objects, in several cases sensitive centimeter observations have been the first to discover these sources (e.g., HH 1-2: Pravdo *et al.* 1985; L1448C: Curiel *et al.* 1990; NGC 2264G: Gómez *et al.* 1994; HH 24: Bontemps *et al.* 1995, 1996) with posterior observations at other wavelengths confirming these detections.

2. In many cases, the centimeter observations have provided a significant improvement in the positional accuracy (by more than two orders of magnitude in some cases) of the outflow sources. In general, positions from IRAS and other FIR observations have been notably improved (e.g., RNO 43, B 335: Anglada *et al.* 1992; HH 114, HH 199: Rodríguez and Reipurth 1996).

3. These radio continuum observations have permitted, in a number of cases, to distinguish and to discriminate among several candidates for the outflow excitation and even to propose alternative candidates. A recent example of this use of thermal jets is the finding by Rodríguez et al. (1997) of a radio source previously undetected at other wavelengths at the core of the HH 7-11 outflow (see Figure 2). These authors favor this new object as the most likely exciting source of this classic HH system, although noting that high-angular resolution molecular observations of the region are required to test more conclusively if the powering source of the outflow is this new radio object or the infrared/optical source SVS 13. Useful additional criteria that can be used to favor between candidates are the proximity of the source to a high density and/or a temperature molecular peak, the association with high velocity molecular gas, and the jet-like morphology of the radio source.

4. Given the high angular resolution of the interferometers (tenths of arc sec), the mapping of thermal jets provides direct evidence that collimation is already present very close to the star (tens of AUs or even less).

5. In the case of low luminosity objects ($L_\star \leq 300\ L_\odot$) it has been pos-

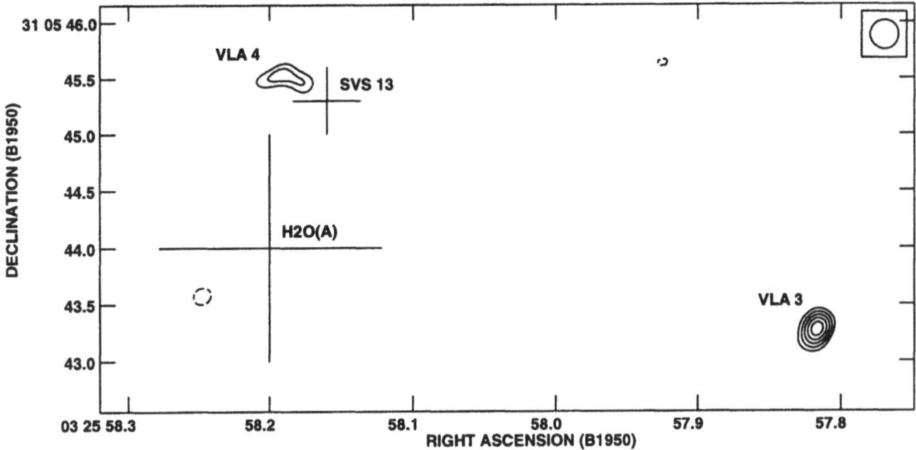

Figure 2. Natural-weight VLA map at 3.6 cm wavelength made in the A configuration of the sources VLA 3 and VLA 4 at the central region of the HH 7-11 outflow. The positions of the near-infrared/optical source SVS 13 and of the water maser H_2O(A) are indicated with crosses. The source VLA 4 is associated with the infrared/optical source SVS 13. The source VLA 3 is a newly detected object that has been proposed by Rodríguez *et al.* (1997a) as a candidate to excite the HH 7-11 outflow. The half power contour of the beam is shown in the top right corner. Contour levels are −3, 3, 4, 5, 6, and 7 times the rms noise of 20 μJy beam^{-1}.

sible to establish (Anglada, 1996) that a correlation exists between the centimeter radio continuum luminosity and the momentum rate in the outflow (as measured by the high-velocity CO emission). This correlation strongly supports the notion that the large scale phenomena (molecular outflows) are being produced by the small scale jets.

A major disadvantage with the subarcsecond observations of thermal processes is the modest flux density contained in the very small synthesized beam. When observing a source with brightness temperature T_B, at a frequency ν, with a circular beam with angular diameter θ_B, the flux density inside the beam will be given by

$$\left[\frac{S_\nu}{mJy}\right] = 0.40 \left[\frac{\nu}{8.4\ GHz}\right]^2 \left[\frac{T_B}{10^3\ K}\right] \left[\frac{\theta_B}{0.1\ arc\ sec}\right]^2. \quad (1)$$

This equation implies that very sensitive instruments are required for the study of thermal processes with high angular resolution. For free-free emission in typical astrophysical conditions, it is expected that $T_B \leq 10^4\ K$. The relatively low brightness temperature of thermal processes limits at present the use of very long baseline interferometry, that could provide milliarcsecond angular resolution. In addition to the relative weakness of the sources, since we are dealing with a continuum emission process we do not have direct radial velocity information. This limitation can be alleviated if proper motions can be measured in the future in the condensations of thermal jets (Marti *et al.*, 1995).

2.3. ARE THERMAL JETS ALWAYS PRESENT IN OUTFLOW REGIONS?

This is a difficult question to answer, since it seems to depend on how deep one is able to integrate on a given region. Rodríguez and Reipurth (1996; 1997a) have made relatively deep integrations (one to two hours) with the VLA at 3.6-cm, detecting the possible exciting source in 11 out of 14 regions studied. Many of these sources had been previously observed unsuccessfully with the VLA in the snapshot mode (integrations with 10 to 20 minutes of duration). The sources detected include HH 34, B5, HH 83, and IRAS 04368+2557 (= L1527). Most well-studied outflow sources have thermal jets at their centers. One notable exception is HH 212 (Zinnecker *et al.*, 1997).

3. Quadrupolar Outflows

One example of the possibilities provided by the study of thermal jets is given by the case of the quadrupolar outflows. It has become increasingly evident that in some regions of star formation quadrupolar outflows, that is, flows that appear to be the close superposition in the sky of two distinct bipolar outflows, are observed. There are, however, several explanations proposed to explain this peculiar four-lobed morphology: (1) Each pair of red and blue lobes could be the limb-brightened walls of the evacuated cavities of a single bipolar outflow (Avery *et al.*, 1990), (2) A single outflow lobe could be split into two lobes as a result of the interaction with a high density molecular clump (Mizuno *et al.* 1990; Torrelles *et al.* 1993b), (3) Multiple episodes of outflow activity, with precession of the outflow axis, could produce a complex lobe morphology (e.g., Fukue and Yokoo 1986; Narayanan and Walker 1996), and (4) The four-lobed structure could be produced by two independent bipolar outflows driven by two stars or even by a binary system (e.g., Anglada *et al.* 1991; Garay *et al.* 1996).

The known cases of outflows with a quadrupolar structure are L723 (Goldsmith *et al.*, 1984), IRAS 16293−2422 (Walker *et al.* 1988; Mizuno *et al.* 1990), Cepheus A (Bally and Lane, 1991), IRAS 21334+5039 (Smith and Fischer, 1992), IRAS 20050+2720 (Bachiller *et al.*, 1995), and HH 111 (Cernicharo and Reipurth, 1996). Here we will discuss recent results related to L723 and HH 111.

3.1. L723

L723 is an isolated molecular cloud located at a distance of 300 ± 150 pc (Goldsmith *et al.*, 1984). A peculiar quadrupolar molecular outflow has been observed in this region (Goldsmith *et al.* 1984; Moriarty-Schieven and Snell 1989; Avery *et al.* 1990; Hayashi *et al.* 1991). The morphology of this outflow, that is particularly evident in the CO maps of Avery *et al.* (1990), consists of two pairs of lobes with a common center. The larger pair of lobes extends along a direction with a position angle PA $\simeq 110°$, while the smaller pair extends along a direction with PA $\simeq 30°$.

Two radio continuum sources, VLA 1 and VLA 2, were found at the center of this outflow through Very Large Array (VLA) observations at 3.6 cm (Anglada *et al.*, 1991). The two radio continuum sources are separated by $15''$ (4500 AU in projection), and both lie within the error ellipsoid of IRAS 19156+1906. The recent high angular resolution study of Anglada *et al.* (1996) reveals that while the source VLA 1 appears unresolved at their angular resolution of $\sim 0''.3$, the source VLA 2 appears as clearly elongated approximately along the direction of the larger pair of lobes of the molecular outflow (see Figure 3). This alignment and the flux density and deconvolved angular size dependences with frequency observed between 3.6 and 6 cm are consistent with VLA 2 being a thermal radio jet, and suggest that this source is related to the excitation of the larger pair of outflow lobes. Additional evidence in support of this interpretation comes from the VLA ammonia study of Girart *et al.* (1997), who find heating and line broadening toward VLA 2, while no emission is detected at the position of the source VLA 1. The exciting source of the second, more compact lobe pair is still to be determined.

3.2. HH 111

HH 111 is a spectacular HH jet located in L1617 in the Orion B cloud complex (Reipurth, 1989). The central jet complex stretches over $6'$, which at the distance of 460 pc corresponds to 0.8 pc. It consists of a bright highly collimated jet, a small faint counterjet, and at least four bow shocks, two on each side of the source. The exciting source was detected in the radio continuum at 2 and 3.6-cm by Rodríguez and Reipurth (1994). This thermal jet appeared to align very well with the optical flow (at a position angle of about 277°). Recently, Gredel and Reipurth (1993; 1994) and Cernicharo and Reipurth (1996) found a second outflow in the infrared and in high-velocity CO. Then, HH 111 is a quadrupolar outflow.

Motivated by this result, Rodríguez and Reipurth (1997b) undertook a deeper integration with the VLA toward the core of HH 111. The resulting map (see Figure 4) reveals that, in addition to the previously known elonga-

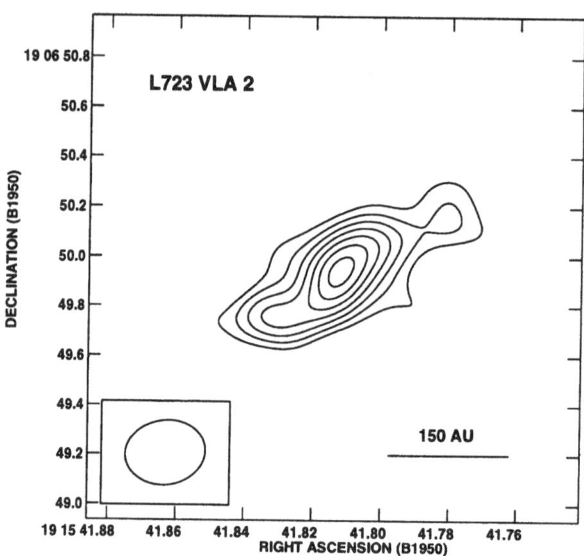

Figure 3. VLA map at 3.6 cm wavelength of the source VLA 2 at the core of the quadrupolar outflow in L723. Contours are −3, 3, 4, 5, 6, 7, 8, and 9 times the rms noise of 11 μJy beam^{-1}. The half power contour of the synthesized beam is also shown. The major axis of VLA 2 aligns with the axis of the large CO lobe pair that has a position angle PA $\simeq$ 110°.

tion along the optical HH flow, the radio source shows weaker elongations approximately in the north-south direction, *closely aligned with the axis of the second outflow* . Then, the HH 111 thermal jet is also quadrupolar. Is this the first detection of a close binary radio jet? The centroids of the two jets appear to coincide in projection within $0\rlap{.}''1$, that is, 50 AU at the distance of the source. This superposition could be just a chance alignment, with the two sources actually being much more separated physically. However, if the system is indeed a binary jet, it could be used to test models of the formation of collimated outflows (in the same spirit that twins are used to test hypothesis in genetics). Most models for collimation of jets require organized magnetic fields over scales larger than the apparent separation of the components of this binary jet. Again, if the association is real, the period of the binary would be in the range of 10^2 years, while the age of the optical jet is more in the range of 10^3-10^4 years. How is it possible that the binary has completed many orbits without wrapping significantly the magnetic fields? Is the collimation produced very close to the star, so that

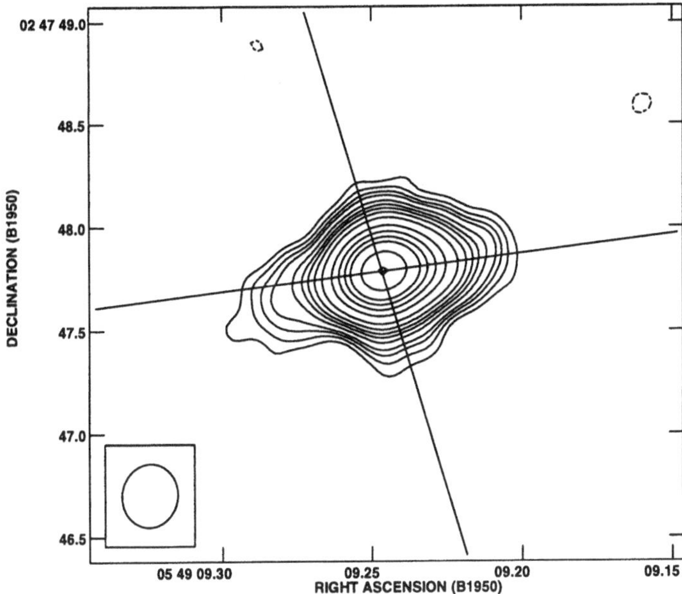

Figure 4. VLA map at 3.6 cm wavelength of the exciting source at the core of the quadrupolar outflow in HH 111. Contours are −3, 3, 4, 5, 6, 8, 10, 12, 15, 20, 30, 40, 50, 60, 80, and 100 times the rms noise of 5.4 μJy beam^{-1}. The half power contour of the synthesized beam is also shown. The straight lines give the position angles of the two flows in the region.

at 50 AU we have mostly ballistic motions of the gas? The study of close binary jets should help us set constraints in the modeling of jets from young stars.

Acknowledgements: The author gratefully acknowledges comments from G. Anglada, J. Martí, and J. M. Torrelles and the continued support of DGAPA, UNAM and CONACyT, México.

References

Anglada, G. 1995, in Rev. Mexicana Astron. Astrofís. Ser. Conf. 1, Circumstellar Disks, Outflows and Star Formation, ed. S. Lizano & J. M. Torrelles (México, D.F.: Inst. Astron., UNAM), 67

Anglada, G. 1996, in ASP Conf. Ser. 93, Radio Emission from the Stars and the Sun, eds. A.R. Taylor & J.M. Paredes, (San Francisco: ASP), 3

Anglada, G., Estalella, R., Rodríguez, L.F., Torrelles, J.M., López, R, and Cantó, J. 1991, ApJ 376, 615

Anglada, G., Rodríguez, L.F., Cantó, J., Estalella, R., and Torrelles, J.M. 1992, ApJ 395, 494

Anglada, G., Rodríguez, L.F. and Torrelles, J.M. 1996, ApJ 473, L123

Avery, L.W., Hayashi, S.S., and White, G.J. 1990, ApJ 357, 524

Bachiller, R., Fuente, A. and Tafalla, M. 1995, ApJ 445, L51
Bally, J., & Lane, A.P. 1991, in Astrophysics with Infrared Arrays, ed. R. Elston, ASP. Conf. Ser. 14, 273
Bontemps, S., André, P., and Ward-Thompson, D. 1995, A&A 297, 98
Bontemps, S., Ward-Thompson, D., and André, P. 1996, A&A 314, 477
Cernicharo, J. and Reipurth, B. 1996, ApJ 460, L57
Chernin, L.M. 1995, ApJ 440, L97
Cohen, M., Bieging, J.H. and Schwartz, P.R. 1984, ApJ 253, 707
Condon, J.J. 1984, ApJ 287, 461
Curiel, S., Raymond, J., Rodriguez, L.F., Canto, J., and Moran, J.M. 1990, ApJ 365, L85
Fukue, J. and Yokoo, T. 1986, Nature 321, 841
Garay, G., Ramírez, S., Rodríguez, L.F., Curiel, S., and Torrelles, J.M. 1996, ApJ 459, 193
Girart, J.M., Estalella, R., Anglada, G., Torrelles, J.M., Ho, P.T.P., and Rodríguez, L.F. 1997, submitted to ApJ
Goldsmith, P.F., Snell, R.L., Hemeon-Heyer, M. and Langer W.D. 1984, ApJ 286, 599
Gómez, J.F., Curiel, S., Torrelles, J.M., Rodríguez, L.F., Anglada, G., and Girart, J.M. 1994, ApJ 436, 749
Gómez, Y., Rodríguez, L.F., and Marti, J. 1995, ApJ 453, 727
Gredel, R. and Reipurth, B. 1993, ApJ 407, L29
Gredel, R. and Reipurth, B. 1994, A&A 289, L19
Hayashi, S.S., Hasegawa, T. and Kaifu, N. 1991, ApJ 377, 492
Marti, J., Rodríguez, L.F., and Reipurth, B. 1993, ApJ 416, 208
Marti, J., Rodríguez, L.F., and Reipurth, B. 1995, ApJ 449, 268
Mizuno, A., Fukui, Y, Iwata, T., Nozawa, S., and Takano, T. 1990, ApJ 356, 184
Moriarty-Schieven, G.H. and Snell, R.L. 1989, ApJ 338, 952
Narayanan, N. and Walker, C.K. 1996, ApJ 466, 844
Pravdo, S.H., Rodríguez, L.F., Curiel, S., Cantó, J., Torrelles, J.M., Becker, R.H., and Sellgren, K. 1985, ApJ 293, L35
Reipurth, B. 1989, Nature 340, 42
Reynolds, S.P. 1986, ApJ 304, 713
Rodríguez, L.F. 1995, in Rev. Mexicana Astron. Astrofís. Ser. Conf. 1, Circumstellar Disks, Outflows and Star Formation, ed. S. Lizano & J. M. Torrelles (México, D.F.: Inst. Astron., UNAM), 1
Rodríguez, L.F. 1996, in Rev. Mexicana Astron. Astrofís. Ser. Conf. 4, VIII Reunión Regional Latinoamericana de Astronomía de la IAU, ed. E. Falco, J.A. Fernández, & R.F. Ferrero (México, D.F.: Inst. Astron., UNAM), 7
Rodríguez, L.F. and Reipurth, B. 1994, A&A 281, 882
Rodríguez, L.F. and Reipurth, B. 1996, Rev. Mexicana Astron. Astrofís. 32, 27
Rodríguez, L.F., Anglada, G., and Curiel, S. 1997, to appear in ApJL
Rodríguez, L.F. and Reipurth, B. 1997a, in preparation
Rodríguez, L.F. and Reipurth, B. 1997b, in preparation
Smith, H.A. and Fischer, J. 1992, ApJ 398, L99
Torrelles, J.M., Gómez, J.F., Curiel, S., Ho, P.T.P., Eiroa, C., and Rodríguez, L.F. 1992, ApJ 384, L59
Torrelles, J.M., Rodríguez, L.F., Cantó, J., and Ho, P.T.P. 1993a, ApJ 404, L75
Torrelles, J.M., Verdes-Montenegro, L., Ho, P.T.P., Rodíguez, L.F., and Cantó, J. 1993b, ApJ 410, 202
Torrelles, J.M., Gómez, J.F., Rodríguez, L.F., Curiel, S., Ho, P.T.P., and Garay, G. 1996, ApJ 457, L107
Walker, C.K., Lada, C.J., Young, E. T. and Margulis, M. 1988, ApJ 332, 335
Yang, J., Ohashi, N., Yan, J., Liu, C., Kaifu, N., and Kimura, H. 1997, ApJ 465, 683
Zinnecker, H., McCaughrean, M., and Rayner, J. 1997, in preparation

MOLECULAR HYDROGEN EMISSION IN EMBEDDED FLOWS

JOCHEN EISLÖFFEL
Institute for Astronomy, University of Hawaii,
2680 Woodlawn Drive, Honolulu, HI 96822, U.S.A.
Max-Planck-Institut für Astronomie,
Königstuhl 17, D-69117 Heidelberg, Germany
Thüringer Landessternwarte Tautenburg,
Sternwarte 5, D-07778 Tautenburg, Germany

Abstract. This overview of recent observations of molecular hydrogen (H_2) in outflows from young stars discusses the morphology, the excitation and the kinematics of the H_2 gas in these flows. A comparison between the H_2 and CO outflows will lead us to the conclusion that highly-collimated jets may drive the latter. We also discuss the mechanisms which can entrain ambient molecular gas into the jet flow.

1. Introduction

Bipolar outflows from young stars are among the most spectacular manifestations of star formation. We now recognize that they are one of the fundamental astrophysical processes that regulate star formation both in single objects and on scales of molecular clouds. Outflows are observed in the emission lines of many atoms and molecules in various excitational states. It is widely believed that we see radiatively cooling gas, which has been excited by shocks in the supersonic flows or by their interaction with the ambient medium.

In regions of relatively low extinction optical emission, e.g. in the Hα and [SII]$\lambda\lambda$6716,6731 lines, typical of Herbig-Haro objects, can be seen. More deeply embedded flows are more accessible in tracers at longer wavelengths: molecular hydrogen (H_2), the most abundant molecule in the Universe, is an important tracer of low-velocity shocks at near- to mid-infrared wave-

B. Reipurth and C. Bertout (eds.), Herbig–Haro Flows and the Birth of Low Mass Stars, 93–102.

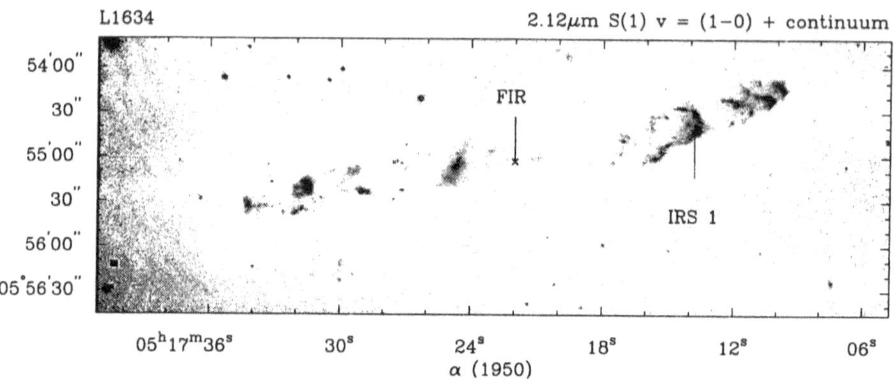

Figure 1. The L1634 outflow in the 1-0 S(1) line of H_2 at 2.12μm (from Hodapp & Ladd 1995).

lengths. Because of its short cooling time (of the order of a year) it illustrates regions where interactions are currently taking place: the H_2 emission is like the sparks that fly when outflowing gas hits something in its way. On the other hand, CO, which is excited in cool gas ($T_{ex} \geq 5\,K$), shows us the coadded history of the interactions at submm- and mm-wavelengths. All of this information is vital if we are to understand how winds from young stars interact with the surrounding molecular gas, how the wind energy and momentum is dissipated, if and how ambient gas is entrained, and consequently if and how winds can drive massive molecular outflows.

Here I will give an overview of the morphology, excitation, and kinematics of H_2 in embedded outflows. Then, I will compare H_2 and CO in outflows, and finally I will describe some recent results on entrainment of ambient gas into the flows.

2. Morphology of H_2 in outflows

Molecular hydrogen has been observed in optically visible Herbig-Haro (HH) objects both by direct imaging (see, e.g. Hartigan, Curiel & Raymond 1989; Stapelfeldt et al. 1991; Eislöffel et al. 1994; Davis, Mundt & Eislöffel 1994; Noriega-Crespo et al. 1996) and by spectroscopy (see, e.g., Gredel 1994, 1996; Fernandes & Brand 1995). However, not all HH objects are bright in H_2 emission, most likely because many are already rather evolved and have broken out of their molecular cloud cores. Deeply embedded outflows from Class 0 or Class I sources are probably much younger and are still ploughing their way through a denser molecular environment (see Fig. 1). Such flows usually show one or several bright and bow-shaped knots in H_2 (see, e.g., Bally, Lada & Lane 1993; Hodapp 1994; Davis & Smith 1996a; Eislöffel et al. 1996; Ladd & Hodapp 1997). These bright knots are thought to be the working surfaces of the flows. In addition, the flows often show faint and more diffuse H_2 emission stretching from the bows back towards the source along both sides of the flow axis (see, e.g., Davis & Eislöffel 1995; Dent, Matthews & Walter 1995; Hodapp & Ladd 1995). Such condensations may result from interaction of the flow with the ambient medium in a shear layer along the flow channel walls.

3. Excitation of H_2 in outflows

Various types of shock structures may be present in the shock waves that form in the flows or in their violent interaction with the ambient medium. If the shock speed is sufficiently high a "J-shock" may arise, in which a jump in velocity, density, temperature, pressure and entropy takes place, so that the shock front appears as a discontinuity. On the contrary a "C-shock", in which the shock energy is dissipated via ambipolar diffusion between ions and neutrals, will show continous changes of quantities within the shock structure (e.g. Draine & McKee 1993). To understand the shock structures better, a detailed knowledge of the excitation state of the gas, as measured by the column density of H_2 in each energy level, is clearly necessary.

Several outflows have been studied in detail so far, with differing results: While observations of OMC-1 peak 1 (Nadeau & Geballe 1979; Moorhouse et al. 1990) could be well fitted by C-shock models (Smith, Brand & Moorhouse 1991; Smith 1991), J-shocks seem to be present in the bow shock of HH 91 (Smith 1994). Observations of HH 1 by Noriega-Crespo & Garnavich (1994) were consistent with either J-type or C-type shocks, while Gredel (1994, 1996) could fit his data on various outflows with models of slabs of gas at a single temperature. On the other hand, Fernandes & Brand (1995) required a combination of C-type bow shocks and fluorescence to explain the excitation in HH 7. If we assume that similar processes are at work in all objects, these confusing results may follow from the fact that all the spectroscopic observations covered only a limited range in upper level energy of H_2, or that they lacked spatial resolution (or both). Recently, Eislöffel et al. (1996) tried to overcome these limitations in a different observational approach: the Cep E outflow was imaged at high S/N with narrow-band filters in the three near-infrared H_2 lines 1-0 S(1), 2-1 S(1), and 3-2 S(3) (and in the nearby continuum), lines which cover a wide range in upper level energy (6950 K to 19080 K). These lines were found to show relative brightnesses close to 1 : 0.1 : 0.01, the line ratios being constant to within 1% everywhere.

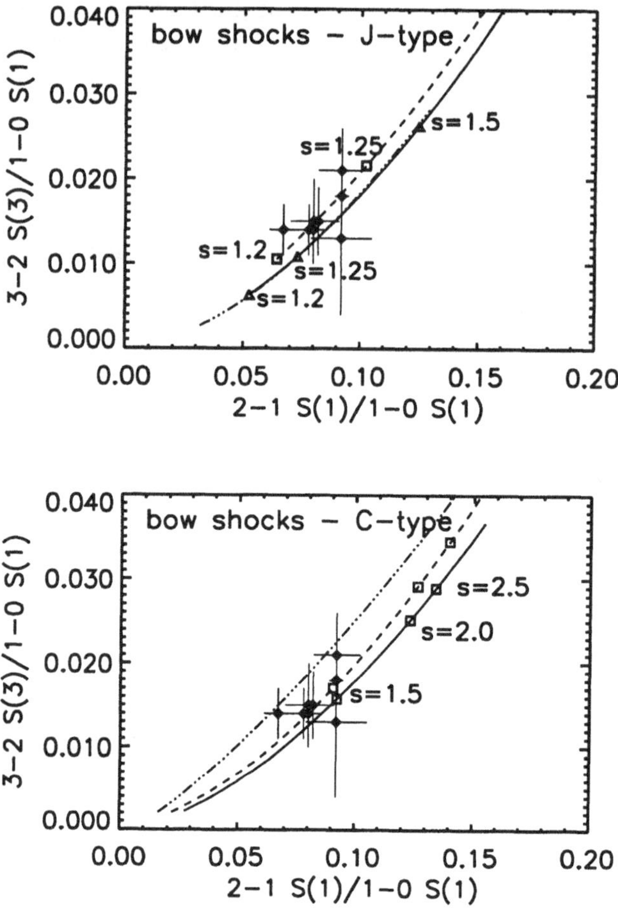

Figure 2. upper panel: J-type bow shock models superimposed on Cep E line ratios. lower panel: C-type bow shock models superimposed on the same data (from Eislöffel et al. 1996).

Comparison of the data with models of slabs of gas at one or two fixed temperatures cannot account for the observed line ratios, nor can fluorescence nor planar J- or C-type shocks. J-type bow shocks (Fig. 2, upper panel) cannot be excluded althought most data points fall above the theoretical curve, and the model bows are very wide with a shape parameter s = 1.2 – 1.3 (where the bow shape is defined as $z = r^s$). Numerical simulations by Suttner et al. (1996) produce shape parameters of s = 1.5 – 1.7. C-type bow shocks seem to be consistent with the observations of a large range of densities (Fig. 2 lower panel). They require a shape parameter of about s

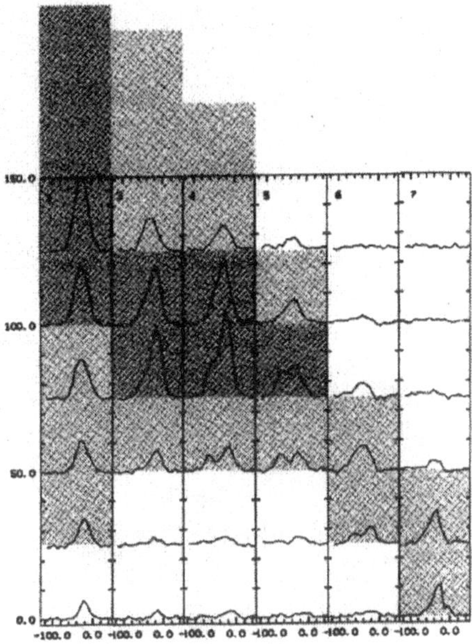

Figure 3. H_2 1-0 S(1) spectra from regions around knot A in L1448. The spectra are superimposed onto a grey-scale image of the bow shock A. Double-peaked line profiles, typical of bow shocks, are clearly indicated (from Davis & Smith 1996a).

= 1.5, in agreement with the numerical simulations. However, the rather constant line ratios observed across the entire outflow region are not easily understood. These would not be expected for a resolved bow shock. It has therefore been suggested that the large bows in Cep E may have broken up due to instabilities, and may now consist of unresolved mini-bow shocks which move at more or less the same speed.

Various accepted programmes for ISO will do imaging and spectroscopy of purely rotationally excited lines of H_2 with upper level energies from 1680 K to 4600 K in the thermal infrared. This will extend the total observed range of upper level energy to more than a factor of 10 and will constrain the excitation models further than is possible at present.

4. Kinematics of H_2 in outflows

The kinematics of the H_2 gas studied with high-resolution spectroscopy provides another invaluable means to understand the shock structures: wide, double-peaked profiles are expected in bow shocks, whereas wide, but single-peaked Gaussian profiles are expected in turbulent boundary layers, and narrower, symmetric lines are expected in planar shocks (Smith &

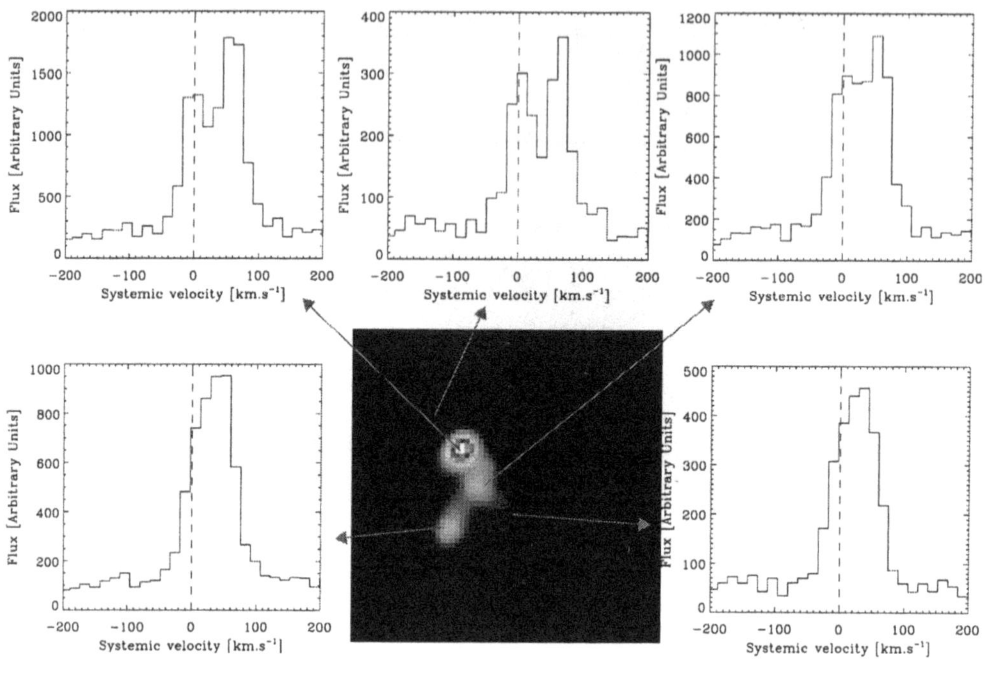

Figure 4. Spectro-image of the northern bow N1 in Cep E in the H_2 1-0 S(1) line, taken with BEAR at CFHT. Several of the line profiles extracted at various positions are double-peaked, typical of bow shocks (Eislöffel et al., in prep.).

Brand 1990a, 1990b). Centroid velocities and observed asymmetries in the line profiles may be used to predict shock speeds and outflow orientations (see e.g. Hartmann & Raymond 1984). This second parameter is of vital importance to outflow studies at all wavelengths (optical through to radio), since outflow energetics may only be estimated (from radial velocities) if one knows the inclination angle of the flow.

So far, only a few outflows have been studied in the H_2 1-0 S(1) line at high spectral resolution, thus resolving the line profiles. They include HH 1/2 and HH 7-11 (Zinnecker et al. 1989), L1448 (see Fig. 3, and Davis & Smith 1996a), DR21 (Davis & Smith 1996b), HH32 (Davis, Eislöffel & Smith 1996) and Cep E (see Fig. 4, Eislöffel et al., in prep.).

Evidence for turbulent shear layers between the flow and its environment is seen in the spectra of DR21 and parts of L1448, which show Gaussian line profiles. On the other hand, the bright condensations in L1448 and Cep E show wide, double-peaked profiles, typical of bow shocks. Also the profiles in HH 32 are consistent with a high-velocity bow shock pointed away from

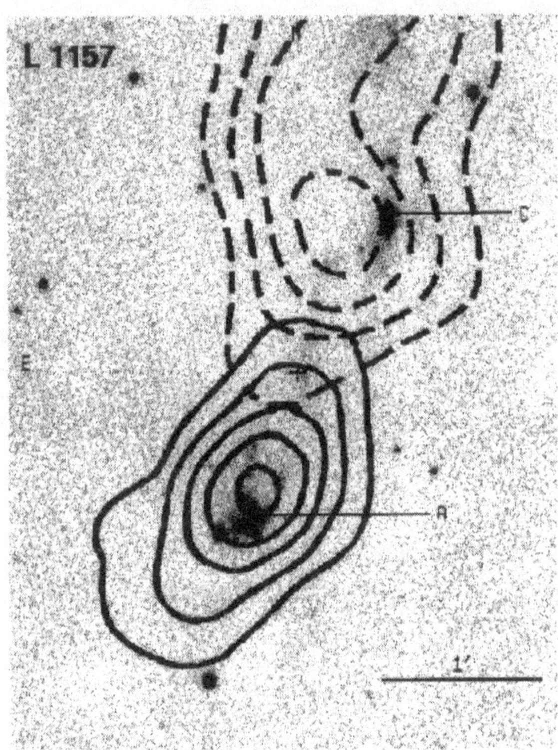

Figure 5. H_2 1-0 S(1) image of the L1157 outflow region with contours of the CO 1-0 outflow. The H_2 regions A and C are clearly associated with the peaks in the CO outflow lobes (from Davis & Eislöffel 1995).

us.

These observations bring up the question whether the H_2 emission in these outflows derives from internal or "working-surface" bow shocks, or whether supersonic turbulence in a "mixing layer" around the flows plays a dominant role. A comparison of the H_2 and CO in the outflows may help to answer this question.

5. Comparison of H_2 and CO outflows

Recent observations have shown that at least some CO outflows are quite well-collimated (see the review by Bachiller 1996 and references therein). When overlaying CO maps onto H_2 images it becomes evident that CO and H_2 gas are intimately related to each other (see Fig. 5 and other examples in

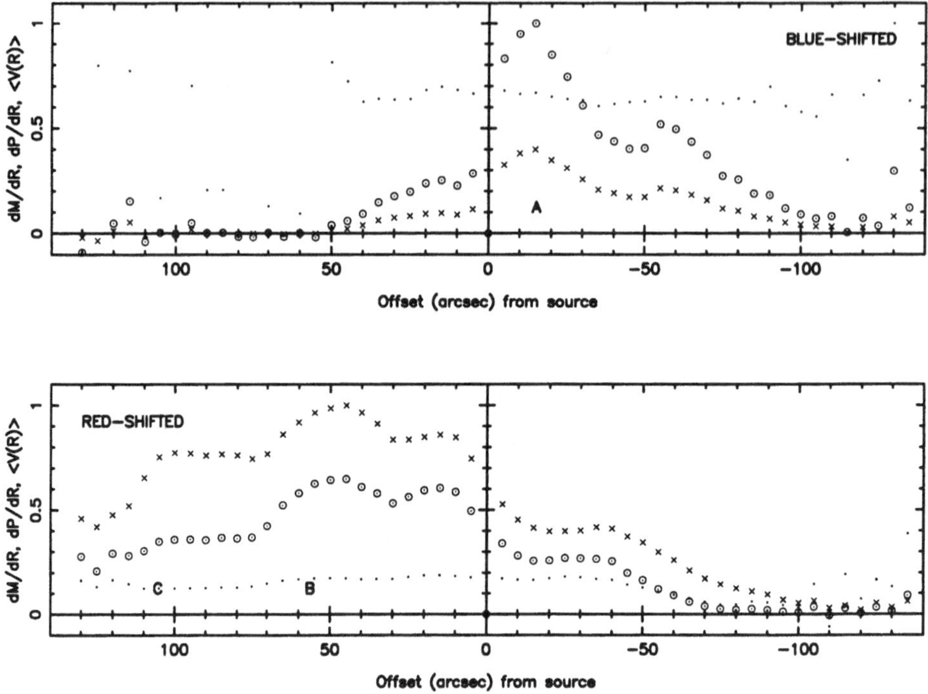

Figure 6. Distribution of mass (integral T_b.dv – crosses), momentum (integral T_b.v.dv – dots+circles), and mean velocity (dots) along theRNO 15-FIR outflow axis. Profiles for the blue-shifted (v < 2.0 km/s) and red-shifted (v > 7.4 km/s) high-velocity gas are plotted, measured at 5" intervals, from the sum of spectra in strips perpendicular to, though centred on, the outflow axis. dM/dR, dP/dR and <V(R)> are normalised to the maximum measured values. The offsets are in arcseconds from RNO 15-FIR (positive 'x' is towards the northeast). The positions of the molecular shocks along the outflow axis are labelled A, B, and C.

Davis & Eislöffel 1995). In almost all outflows the CO peaks just upstream of the bright H_2 bows. The other H_2 condensations which are seen between the bright H_2 bows and the outflow source turn out to line up along the sides of the CO outflow lobes. These observations suggest that the CO outflows consist of ambient gas that has been entrained by a highly-collimated "jet".

6. Entrainment mechanisms

How can ambient molecular gas be entrained by collimated jets to form massive, molecular (CO) outflows? Theoretical studies suggest that ambient gas may be accelerated, either in a turbulent mixing layer along the length of the jet or a jet segment (Cantó & Raga 1991; Masson & Chernin 1992), or through the bow shock at the head of the jet (Masson & Chernin 1993; Raga & Cabrit 1993). In the turbulent model, a shear layer develops along the interface between the supersonic jet and its surroundings which thickens with distance from the source, as jet material and ambient gas enter this layer. Conversely, entrainment in the bow shock or "prompt" model is quite localised. Here, the ambient, molecular gas swept up through each bow shock cools rapidly, and although some of the gas may spill down the sides of the jet, a high-velocity "clump" will develop just behind each shock front.

In the turbulent entrainment model, the momentum per unit length should increase with distance from the source, as more of the jet momentum is transfered to the swept up molecular gas (Chernin & Masson 1995). In the prompt entrainment model, however, the momentum should decrease with distance from the source if the flow "accelerates" out of a dense protostellar core into lower density gas.

Recent CO data taken by Davis et al. (1997a) allow them to distinguish between entrainment models in the RNO15-FIR outflow. This seemingly rather simple outflow consists of a sequence of compact line emission features, which imply the presence of a highly-collimated, bipolar outflow driven by RNO15-FIR (Davis et al. 1997b). The spatial coincidence between three peaks in the CO map and three corresponding H_2 shocks suggests that (a) these H_2 shocks are those which entrain much of the ambient gas to form the CO outflow, and that (b) the prompt entrainment mechanism dominates over turbulent entrainment. Fig. 6 shows the mass per unit length (dM/dR), momentum per unit length (dP/dR), and mean velocity ($<V(R)> = [dP/dR]/[dM/dR]$) along the outflow axis, integrated across the width of the flow and over the high-velocity blue-shifted and red-shifted line wings. In both the northeastern, red-shifted flow lobe, and the southwestern, blue-shifted flow lobe, the mass and momentum decrease with distance from the source. This decrease is most dramatic in the blue-shifted lobe (at negative offsets in Fig. 6). Also, the molecular (H_2) shocks observed along the flow lobe coincide with peaks in these distributions. These H_2 features therefore probably represent the entraining bow shocks. Indeed, it seems very likely that prompt or "bow shock" entrainment dominates in this particular outflow.

7. Conclusions

In this overview of molecular hydrogen in outflows from young stars we have seen that the H_2 is observed as bright bow shocks at the ends of the flows (or flow segments) and as fainter emission along the flow axis. The excitation mechanism of this H_2 gas is not yet clear. A comparison of the H_2 and CO maps shows that the CO is probably entrained by a highly-collimated jet.

High-resolution spectroscopic studies of the kinematics of both H_2 and CO indicate that this happens mostly via prompt entrainment in the leading bows, and to a minor extent also via turbulent entrainment in shear layers along the flow axis.

Acknowledgements: I would like to thank Chris Davis very much for our pleasant collaboration on molecular hydrogen and embedded outflows. My thanks also go to Mike Smith, Jérôme Bouvier, Reinhard Mundt, Tom Ray and François Rigaut for their help and participation in many projects. Finally, I wish to thank the organizers for their support.

References

Bachiller, R. 1996, Ann. Rev. Astron. Astrophys. 34, 111
Bally, J., Lada, E.A., Lane, A.P. 1993, ApJ 418, 322
Cantó, J., Raga, A.C. 1991, ApJ 372, 646
Chernin, L.M., Masson C.R. 1995, ApJ 455, 182
Davis, C.J., Eislöffel, J. 1995, A&A 300, 851
Davis, C.J., Eislöffel, J., Ray, T.P., Jenness, T. 1997a, A&A, in press
Davis, C.J., Eislöffel, J., Smith, M.D. 1996, ApJ 463, 246
Davis, C.J., Mundt, R., Eislöffel, J. 1994, ApJ 437, L55
Davis, C.J., Smith, M.D. 1996a, A&A 309, 929
Davis, C.J., Smith, M.D. 1996b, A&A 310, 961
Davis, C.J., Ray, T.P., Eislöffel, J., Corcoran, D. 1997b, A&A, in press
Dent, W.R.F., Matthews, H.E., Walther, D.M. 1995, MNRAS 277, 193
Draine, B.T., McKee, C.F. 1993, Ann. Rev. Astron. Astrophys. 31, 373)
Eislöffel, J., Davis, C.J., Ray, T.P., Mundt, R. 1994, ApJ 422, L91
Eislöffel, J., Smith, M.D., Davis, C.J., Ray, T.P. 1996, AJ 112, 2087
Fernandes, A.J.L, Brand, P.W.J.L. 1995, MNRAS 274, 639
Gredel, R. 1994, A&A 292, 580
Gredel, R. 1996, A&A 305, 582
Hartigan, P., Curiel, S., Raymond, J. 1989, ApJ 347, L31
Hartmann, L.W., Raymond, J. 1984, ApJ 276, 560
Hodapp, K.-W. 1994, ApJS 94, 615
Hodapp, K.-W., Ladd, E.F. 1995, ApJ 453, 715
Ladd, E.F., Hodapp, K.-W. 1997, ApJ 474, 749
Masson C.R., Chernin, L.M. 1992, ApJ 387, L47
Masson C.R., Chernin, L.M. 1993, ApJ 414, 230
Moorhouse, A., Brand, P.W.J.L., Geballe, T.R., Burton, M.G. 1990, MNRAS 242, 88
Nadeau, D., Geballe, T.R. 1979, ApJ 230, L169
Noriega-Crespo, A., Garnavich, P.M. 1994, Rev. Mex. AA 28, 173
Noriega-Crespo, A., Garnavich, P.M., Raga, A.C., Cantó, J., Böhm, K.-H. 1996, ApJ 462, 804
Raga, A.C., Cabrit, S. 1993, A&A 278, 267
Smith, M.D. 1991, MNRAS 253, 175
Smith, M.D. 1994, A&A 289, 256
Smith, M.D., Brand, P.W.J.L. 1990a, MNRAS 243, 498
Smith, M.D., Brand, P.W.J.L. 1990b, MNRAS 245, 108
Smith, M.D., Brand, P.W.J.L., Moorhouse, A. 1991, MNRAS 248, 730
Stapelfeldt, K.R., Scoville, N.Z., Beichman, C.A., Hester, J.J., Gautier, T.N. 1991, ApJ 371, 226
Suttner, G., Smith, M.D., Yorke, H.W., Zinnecker, H. 1996, A&A 318, 595
Zinnecker, H., Mundt, R., Geballe, T.R., Zealey, W.J. 1989, ApJ 342, 337

TURBULENT MIXING LAYERS IN HERBIG-HARO AND EMBEDDED H_2 JETS

ALBERTO NORIEGA-CRESPO
Infrared Processing and Analysis Center
California Institute of Technology 100 - 22
Pasadena, California, 91125, USA

Abstract. Outflows arising from young stellar objects interact with their surrounding medium through different mechanisms, such as shocks, turbulence and /or entrainment. These two last mechanisms have recently begun to be fully developed in an effort to understand the coupling of the stellar jet gas (mostly atomic) with the molecular environment. Observationally a number of objects are being studied in the near infrared and millimeter wavelengths to map the warm (H_2) and cold (CO) molecular gas, respectively. In this paper we discuss some of the properties in the near infrared of various embedded jets in the context of turbulent entrainment.

1. Introduction

A large fraction of the energy released by stellar jets in their surrounding medium is produced by the collisional excitation of the gas. We associate this excitation process with shocks, but turbulence and entrainment are processes which dissipate energy in a similar way (Taylor & Raga 1995). Most non-adiabatic 2-D hydrodynamical simulations of jets show how the stellar jet engulfs and pushes the surrounding molecular gas at the front and the sides, i. e. entraining it. The gas in a stellar jet is typically at a temperature of 10^4 K, while the molecular gas can be at a temperature of 100 K or lower. In the interaction of the hot atomic gas with the cold molecular gas, there must be a region with "intermediate" properties in chemical composition, temperature and velocity, i. e. a mixing layer where the atomic and molecular gasses meet.

In some optical Herbig-Haro (HH) jets (e.g. HH 1-2, HH 7-11) the morphology is very similar in the atomic lines (Hα and [S II] $\lambda\lambda$6717/31) and

B. Reipurth and C. Bertout (eds.), Herbig–Haro Flows and the Birth of Low Mass Stars, 103–110.

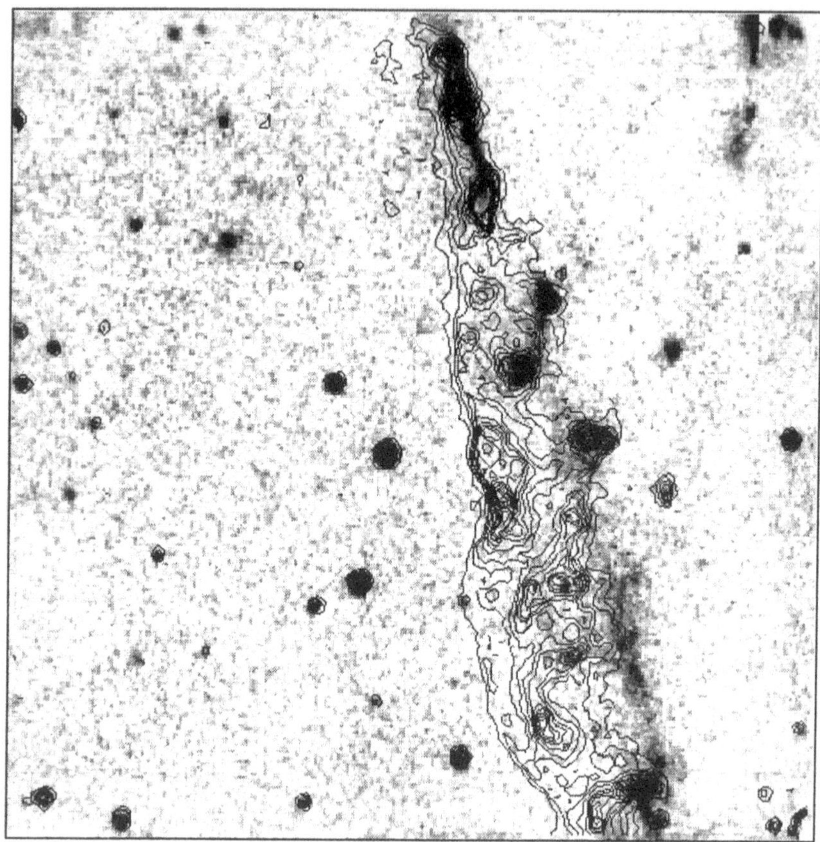

Figure 1. The HH 110 jet in Hα (contour) and H_2 2.121 μm emission (grayscale) in a 2′ field.

the molecular hydrogen lines in the near infrared (e.g. 1-0 S(1) 2.121 μm). Furthermore, there is a growing number of embedded jets without optical counterparts, which look very well collimated and closer in shape to the optical jets than to the CO outflows. The morphological similitude prompts various questions, for instance: is the H_2 emission observed in HH jets produced in a mixing layer? Are the highly supersonic optical jets coupled with the CO molecular outflows through turbulent entrainment? What are the physical properties and emitting characteristics of a mixing layer? We do not have yet definitive answer for these questions, but see e. g. Chandler & Richer (1997) and Raga (1995) for some observational and theoretical arguments in favor of this picture.

The numerical and analytical models for mixing layers along the jet beam show that the layer is thin, long and with small column densities

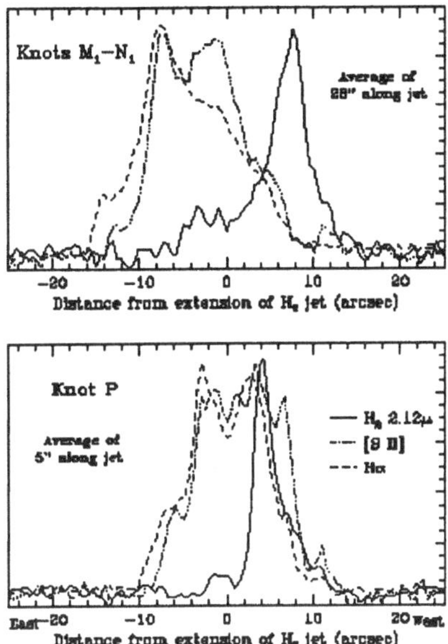

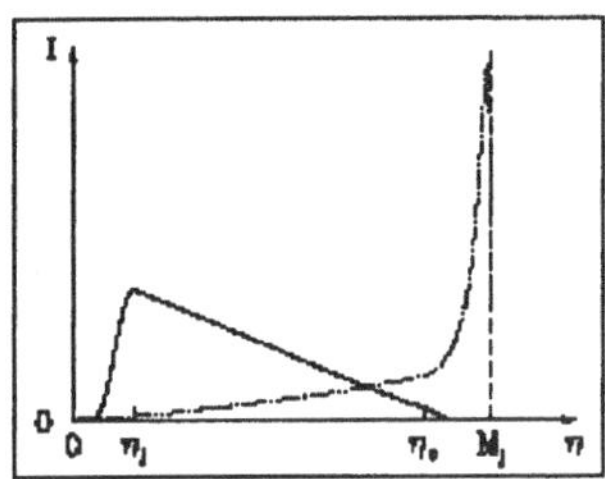

Figure 2. The observed (left) and predicted (right) spatial intensity distribution across two knots (M-N & P) of the HH 110 jet in Hα, [S II] and 2.121 μm(see for details Noriega-Crespo et al. 1996).

(Taylor & Raga 1995), which perhaps is not the best general model for a large CO outflow, but similar to some of the observed molecular hydrogen jets. In this paper we look at the HH 110, IRAS 05487+0255 and Cep E jets in the near infrared (spectroscopy and images), in order to gather more data to enrich and better constrain the theoretical models of turbulent mixing layers.

2. The HH 110 Jet

The HH 110 jet has some characteristics that suggest a good starting point to search for the effects of entrainment and turbulence, since the jet has a turbulent optical morphology (Reipurth & Olberg 1991, Reipurth et al. 1996). The lack of an obvious source along its axis may be explained as if the jet originates in a grazing jet-dense cloud core collision (Raga & Cantó 1995). Even more remarkable is the fact that its H_2 emission (Davis et al. 1994) is narrow and "straight", and shifted towards the west (see fig 1) with respect to the optical Hα and [S II] emission (Noriega-Crespo et al. 1996).

Furthermore the ratio of the (1,0) 2.121 to (2,1) 2.248 μm lines, which is a measure of the excitation of the gas, does not remain constant along the jet condensations, i. e. the excitation is not uniform as could be the case for shocks with a similar velocity.

It is possible to study the spatial intensity distribution of the optical and H_2 emission across and along the jet, and to compare them with simple analytical models of the cross section of a two-dimensional mixing layer. For instance, the fraction of H_2 across the mixing layer should reach a maximum value near the molecular cloud ($\sim$ 1) and it should decrease linearly to a minimum value closer to the atomic jet ($\sim$ 0) (Cantó & Raga 1991). The emission from this layer then should reflect directly this hydrogen fraction, since its structure across is essentially isothermal and in transverse pressure equilibrium (Cantó & Raga 1991). This means that the Hα emission peaks close to the edge of the jet beam and that the $H_2\,(1,0)$ S(1) peaks close to the interface between the mixing layer and the undisturbed molecular environment (Noriega-Crespo et al. 1996). This is the tendency observed in knots M-N and in a less pronounced way in knot P (see fig 2).

3. The IRAS 05487+0255 Jet

Fifty arcseconds west from HH 110 there is a remarkable object that we have named the IRAS 05487+0255 jet because it is near that infrared source. The jet is not visible at optical wavelengths, but is clearly detected in the near infrared. At 2.121 μm the jet (and counter-jet) are very well collimated with a length-to-width ratio $\sim 10-20$ (see fig 3). The spectra of the jet and counter-jet in the K-band show a few H_2 emission lines. Perhaps the most noticeable property of the spectra is the shift in radial velocity of the 1-0 S(1) 2.121 μm line (see fig 4). The radial velocities based on this line for the jet and counter-jet are $\sim -275 \pm 50$ km s^{-1} and $\sim 180{\pm}50$ km s^{-1} respectively. These velocities are comparable to the radial and flow velocities measured in other *optical* jets using the atomic emission lines. This result suggests that at least in IRAS 05487+0255 the molecular hydrogen is moving at a velocity comparable to that of the atomic jet gas.

The H_2 emission of the entire jet extends for at least 40″ or $\sim$ 0.1 pc at the distance of Orion. If the flow velocity is comparable to that of the radial velocities, then the dynamical age of the system is quite short ($\sim$ 500 yrs), consistent with a young jet arising from an embedded source. The high degree of collimation of the molecular gas, the short dynamical age and the magnitude of the radial velocities are consistent with the properties expected in a turbulent mixing layer (Garnavich et al. 1997).

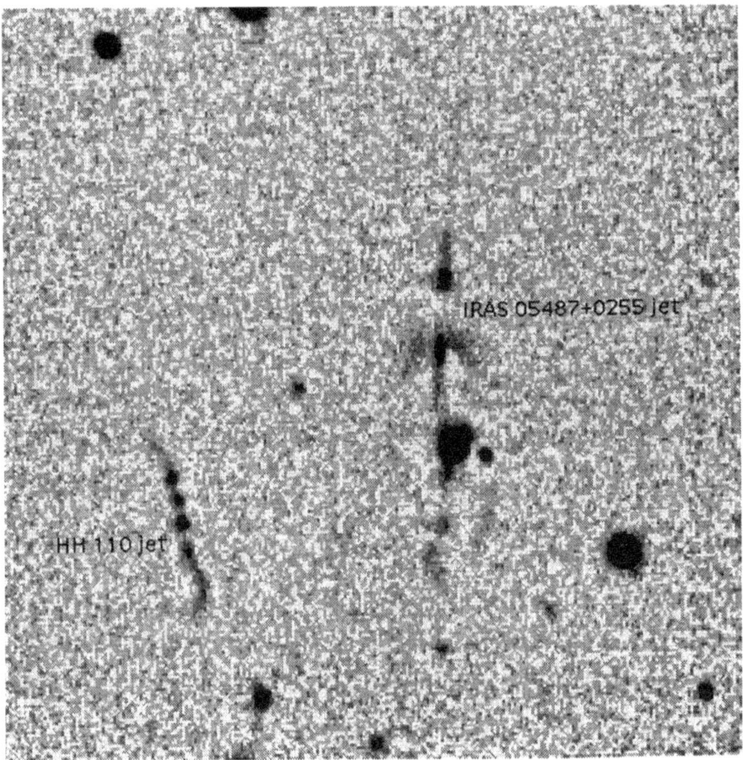

Figure 3. The IRAS 05487+0255 jet (center) in the H_2 2.121 μm line; the HH 110 jet is east (left). The field is 2′.

4. The Cepheus E Outflow

The two previous outflows seem to be particular examples where it is possible to infer the presence of entrainment in a mixing layer by observing the molecular hydrogen gas. One would like to compare these tendencies with those from some other embedded outflows. In this sense the Cepheus E bipolar outflow is a good candidate, since at visible wavelengths, e. g. Hα and [S II], only a bright south knot is detected (now called HH 377 - Devine et al. 1997), while in the near infrared (Hodapp 1994; Ladd & Hodapp 1997; Eislöffel et al. 1996) there is a well defined bipolar flow (see fig 5).

The images taken at the different molecular hydrogen lines show a complex structure (Eislöffel et al. 1996), and the K-band spectra indicates a higher excitation than that expected in C-type shocks, at least in the best fitted model (Ladd & Hodapp 1997). We obtained H and K-band spectra of both the South and North outflow components to try to detect any possible

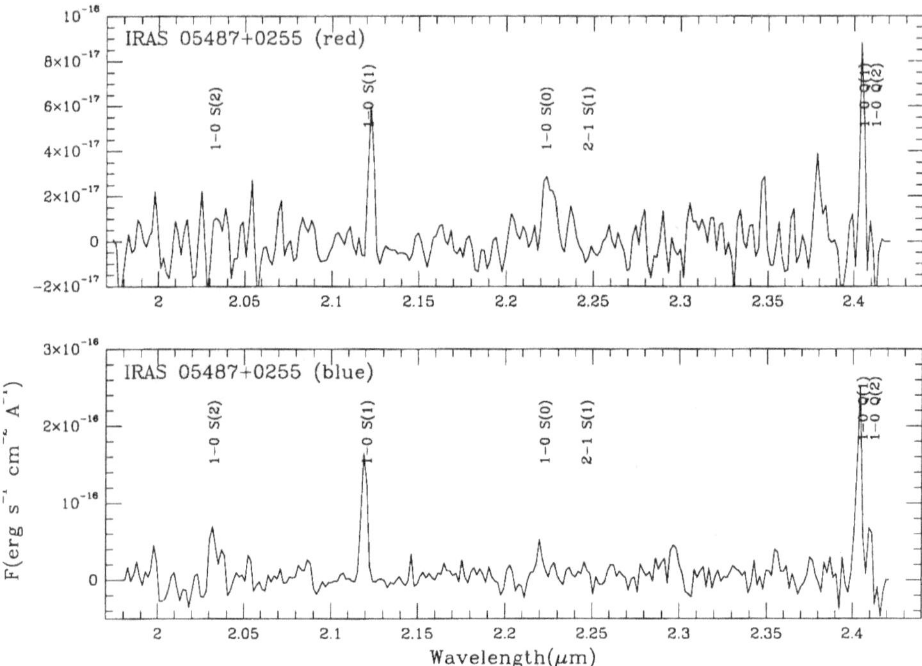

Figure 4. The IRAS 05487+0255 jet & counter-jet K-band spectra.

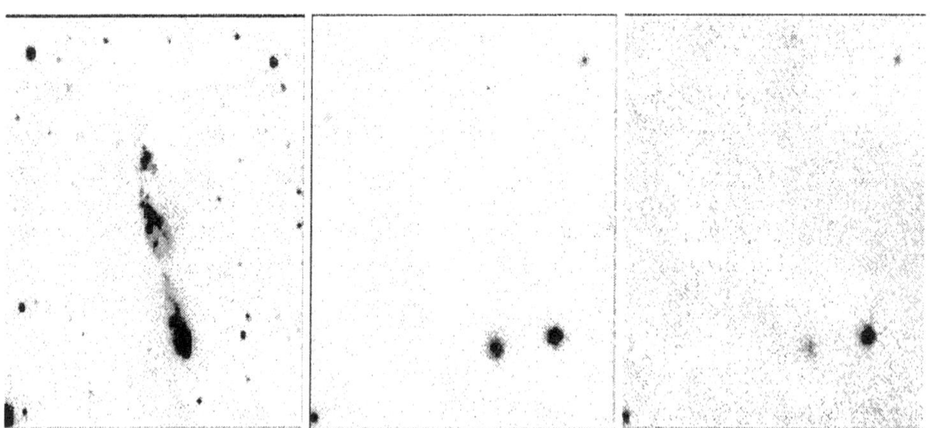

Figure 5. Cepheus E in the NIR at 2.212 μm (left), and at optical wavelengths, [SII] $\lambda\lambda$6717/31 (center) and Hα (right). The field is $\sim 2'$.

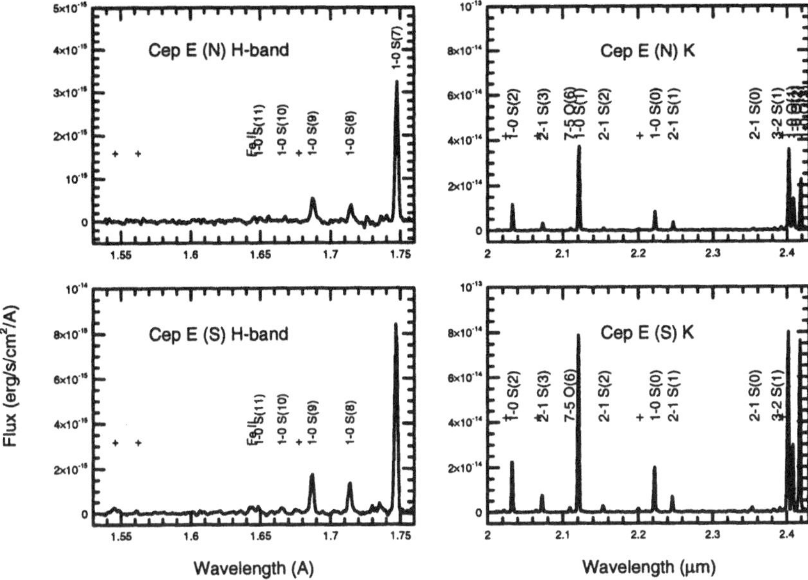

Figure 6. The spectra of the north and south Cepheus E outflow components in the H and K bands.

shift of the brightest lines.

The spectra were obtained with the same set-up and resolution as for the IRAS 05487+0255 jet, and were calibrated in the same way. The spectra (see fig 6) also have a higher signal-to-noise, so in principle it should be possible to detect shifts ≥ 50 km s^{-1}, which corresponds to a measurement of $\sim 1/3$ of a pixel. In the case of Cep E, it was difficult to detect a radial velocity shift for the individual north and south components. Nevertheless it is possible to measure a relative shift between the two outflow components of ~ 110 km s^{-1}. From the morphology the flow is not far from the plane of the sky, and at a first approximation, the relative radial velocity is of the order of the relative flow velocity. Essentially the molecular hydrogen gas in Cep E seems to be moving at a velocity of ~ 50 km s^{-1}, i.e. higher than the $30 - 40$ km s^{-1} dissociation shock velocity. If shocks are not driving the bulk of the motion of the molecular hydrogen gas in Cep E, then the alternative is that some of it is being excited and entrained by turbulence at the mixing layer.

5. Summary

We have presented observations in the near infrared of three different jets which have different morphological characteristics. For the HH 110 jet the analysis of the spatial intensity distribution of the Hα, [S II] and H_2 emission can be explained in the context of a turbulent mixing layer. For the IRAS 05487+0255 jet the narrowly collimated H_2 emission and the large radial velocities of their beams ($\sim$ 100 km s^{-1} suggest again the presence of a turbulent mixing layer. And finally in the Cep E outflow the H_2 could be moving at velocities of $\sim$ 50 km s^{-1}, which are difficult to explain by shock acceleration, but more easy to understand as the result of entrainment by the fast moving gas beam of the jet.

Acknowledgements: The research that has led to this manuscript has been done in collaboration with some excellent persons. I would like to thank particularly Karl-Heinz Böhm, Peter Garnavich & Alejandro Raga. Bo Reipurth kindly provided me with the optical images of HH 110, and Nichole King helped me with the NIR observations. I thank Salvador Curiel, Jorge Cantó, Adam Frank, Rosario López, and Dave Van Buren for very useful conversations, and to Amaya Moro-Martin for her careful reading of the manuscript. The generous access to the IAUNAM computer facilities to complete this paper is gratefully acknowledged. My research has been supported by the NASA Long Term Astrophysics program through a contract with the Jet Propulsion Laboratory (JPL) of the California Institute of Technology.

References

Cantó, J., & Raga, A. C. 1991, ApJ, 372, 646
Chandler, C. J., & Richer, J. S. 1997 in "Low Mass Star Formation from Infall to Outflow", ed. F. Malbet & A. Castets, p76
Davis, C.J., Mundt, R., & Eislöffel, J. 1994, ApJ, 437, L55
Devine, D., Reipurth, B., Bally, J.: 1997, in *Low Mass Star Formation - from Infall to Outflow*, poster book for IAU Symp. No. 182, eds. F. Malbet & A. Castets, Observ. de Grenoble, p. 91
Eislöffel, J., Smith, M.D., Davis, C. J., Ray, T. P. 1996, AJ, 112, 2086
Garnavich, P. M., Noriega-Crespo, A., Raga, A. C., & Böhm, K. H. 1997, ApJ (in press).
Hodapp, K. W. 1994, ApJS, 94, 615
Ladd, E.F., & Hodapp, K. W. 1997, ApJ, 474, 749
Noriega-Crespo, A., Garnavich, P.M., Raga, A.C., Cantó, J., & Böhm, K.H. 1996, ApJ, 462, 804
Raga, A. C. 1995, RMxAA, 1 (Cnf Sr), 103
Raga, A. C., and Cantó, J. 1995, RMxAA, 35, 51
Reipurth, B., & Olberg, M. 1991, A&A, 246, 535
Reipurth, B., Raga, A.C., Heathcote, S.: 1996, A&A, 311, 989
Taylor, S. D., & Raga, A.C. 1995, A&A, 296, 823

FAR-IR SPECTROPHOTOMETRY OF HH FLOWS WITH THE ISO LONG-WAVELENGTH SPECTROMETER

R. LISEAU
Stockholm Observatory
S–133 36 Saltsjöbaden, Sweden
e-mail: rene@astro.su.se

T. GIANNINI, B. NISINI, P. SARACENO AND L. SPINOGLIO
CNR-Istituto di Fisica dello Spazio Interplanetario
I-000 44 Frascati, Italy

B. LARSSON
Stockholm Observatory
S–133 36 Saltsjöbaden, Sweden

D. LORENZETTI
Osservatorio Astronomico di Roma
I-000 40 Monte Porzio, Italy

AND

E. TOMMASI
LWS *Instrument Dedicated Team*
ISO *Science Operations Centre*
Villafranca de Castillo, Spain

Abstract. Full ISO-LWS spectral scans between about 45 to 190 μm of 17 individual HH objects in 7 star forming regions have revealed essentially only [O I] 63 μm line emission, implying that the FIR cooling of these objects is totally dominated by this line alone. In this case, J-shock models can be used to determine the mass loss rates of the HH exciting sources. These mass loss rates are in reasonably good agreement with those estimated for the accompanying CO flows, providing first observational evidence that HH and molecular flows are driven by the same agent. The $L_{\rm mech} - L_{\rm bol}$ relation, based on our results with the LWS, implies that young stellar objects of lower mass are loosing mass at relatively higher rates than their more massive counterparts.

B. Reipurth and C. Bertout (eds.), Herbig–Haro Flows and the Birth of Low Mass Stars, 111–120.

1. Introduction

Herbig-Haro objects (HH) have commonly been identified on the basis of observations in the *visual* spectral region. As such, the known objects are, when viewed against dark clouds, surface phenomena in regions of stellar formation. Recent advances in detector array development have opened the near infrared (NIR) for the observational study of HH flows and a growing body of 2 μm-data is now becoming available also for cloud embedded flows (cf. article by J. Eislöffel). At NIR-wavelengths, extinction of radiation by dust can still present a problem, particularly in regions of high column density ($A_K \sim 0.1\, A_V$), and HH flows deeply embedded or on the far side of the clouds might not be readily accessible to NIR observations. On the other hand, at far infrared (FIR) wavelengths, extinction is generally negligible ($A_{100\,\mu m} \lesssim 10^{-3}\, A_V$), except perhaps on the very smallest spatial (subarcsec) scales.

First FIR spectroscopic observations of HH flows have been presented by Cohen et al. (1988). These data were obtained from airplane altitude (KAO) and were restricted, therefore, to the few existing spectral windows of diminished telluric opacity. The space borne LWS, in contrast, is free of atmospheric limitations and has observed, for the first time, fully covering FIR-spectra of HH flows.

1.1. PRIMARY OBJECTIVES OF THE LWS-HH PROGRAMME

The LWS (Long-Wavelength Spectrometer: Clegg et al. 1996, Swinyard et al. 1996) aboard ISO (Infrared Space Observatory: Kessler et al. 1996) operates in the wavelength region $\sim$40–200 μm and FIR-spectrophotometry of unextinguished HH flows offered thus the opportunity to address:

- the spatial distribution (extent) and morphology (degree of symmetry) of HH flows;
- the appearance of FIR-spectra and the importance of FIR line cooling (atomic and molecular) for the overall energy balance of HH flows;
- the nature of the HH-excitation in molecular clouds, i.e. to what degree the shocks are predominantly J- or C-type;
- the physics of mass loss from young stellar objects and the interrelation of molecular ('CO ouflows') and HH flows.

In addition, since HH objects were expected to emit little, if any, continuous radiation at long wavelengths, LWS observations should also be suitable to obtain the FIR energy distributions of the associated clouds at high spectral resolution. This should permit the study of the dust opacity in dark, molecular clouds and the search for solid state spectral features in a previously unexplored wavelength regime.

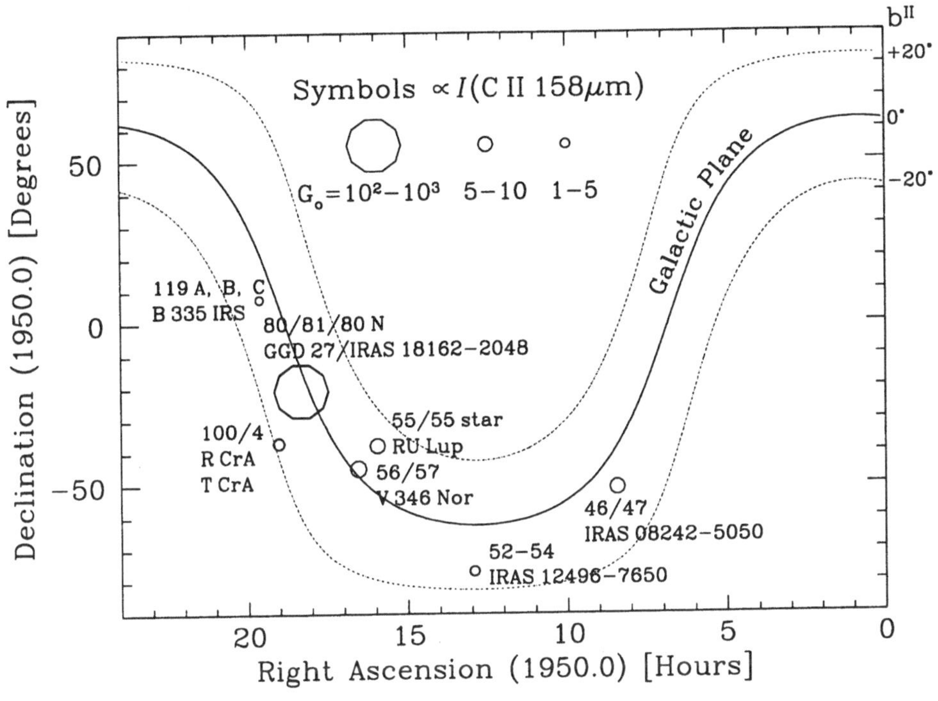

Figure 1. The sky distribution, in equatorial coordinates, of 17 HH flows in 7 star forming regions observed with the LWS is shown together with the driving stellar sources. Symbol sizes correspond roughly to the value of the interstellar radiation field, expressed in terms of G_0 (see text). For convenience, the run of the galactic plane and of the galactic latitudes $\pm 20°$ are also shown

In this contribution, we concentrate mainly on the aspects of shock-type and mass loss during the early phases of stellar evolution.

2. HH observations with the LWS

The data presented in this contribution have been obtained during Guaranteed Time observations of the LWS-team. In 7 different regions of star formation, the number of individual HH objects observed was 17 and their distribution in the sky is shown in Fig. 1. The lack of objects in the constellations of, e.g., Perseus, Taurus-Auriga and Orion reflects the visibility restrictions of ISO. HH flows in these regions are expected to be observed during the extended ISO-mission in 1997/98.

Full LWS-grating scans (43–193 μm) have been obtained for all objects, at the spectral resolution $R_\lambda \sim 150$–300. The observations with the $80''$ circular beam were made both in pointed and in mapping mode, i.e. point-

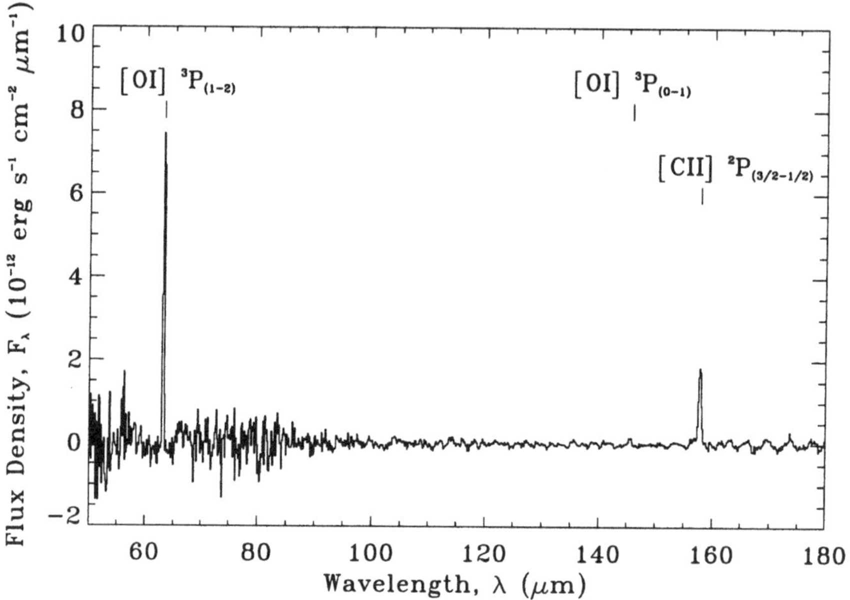

Figure 2. Typical FIR line spectrum of HH objects

ing the telescope towards optically identified HH positions as well as obtaining unbiased raster scan maps of the flows. The absolute flux calibration is believed to be better than about 30%. Within integration times of about 700 to 4 500 s per position a continuum sensitivity ($1\,\sigma$) of ~ 1 Jy at 60 and 100 μm was achieved, comparable to that of IRAS at much lower spectral resolution. The sensitivity ($1\,\sigma$) to spectral line flux was about $2\,10^{-13}\,\mathrm{erg\,cm^{-2}\,s^{-1}}$ at 60 μm and $3\,10^{-14}\,\mathrm{erg\,cm^{-2}\,s^{-1}}$ at 160 μm respectively.

3. Results and discussion

3.1. EXTENDED [C II] 158 μm EMISSION AND MOLECULAR CLOUD PDR

First results have been published by Liseau et al. (1996) and by Nisini et al. (1996). These regarded our observations of HH 52–54 and revealed the remarkably rich line spectrum of HH 54, displaying, in addition to atomic fine structure lines, rotational molecular lines of CO, OH and H_2O. These latter lines were not seen towards either HH 52, HH 53 nor in the mapped part of the molecular outflow and it was reasonable to assume that the non-detections were due to insufficient signal-to-noise in these somewhat weaker spectra. The LWS observations of the significantly increased sample of HH flows presented here show, however, that molecular line emission is rarely present in the far infrared spectra of these objects. The median spectrum of the flows (Fig. 2) emphasizes this point: the only lines present at any appreciable strength are due to the fine structure transitions [O I] 63 μm and [C II] 158 μm.

From the mapping observations it became evident that the [C II] 158 μm emission is spatially extended and relatively uniform in intensity, contrary to what is observed for [O I] 63 μm being localized and of variable strength.

The relative constancy of the extended [C II] 158 μm emission leads us to suspect that this line is not dominated by the HH shocks, but rather originates in the PDR (Photon Dominated Region) of the molecular host clouds. This suspicion is supported by the theoretical predictions for [O I] 63 μm and [C II] 158 μm of PDR models (Wolfire et al 1990; Hollenbach et al. 1991) and those of J-shock models (Hollenbach & McKee 1989). As long as the interstellar UV-field, expressed in terms of the local value of the solar neighbourhood, G_0, does not greatly exceed 10^3, the intensity ratio [C II] 158 μm/[O I] 63 μm $\gg 1$ in a low density PDR, whereas this ratio is inverted behind fast dissociative J-shocks. From the measured [C II] 158 μm intensities towards the HH flows and the PDR models, we have estimated the values of G_0 for the different environs. With the exception of HH 80–81, these values are generally in the range 1–10 (cf. Fig. 1).

3.2. [O I] 63 μm LINE LUMINOSITY AND STELLAR MASS LOSS RATE

The localized character of the [O I] 63 μm emission certainly associates it with the HH objects. As evidenced in Fig. 2, the level of molecular line emission (in particularly CO and H_2O) is very low *compared to that in* [O I] 63 μm. From comparison with the theoretical models of C-shocks by Draine et al. (1983) we infer that HH shocks appear generally not to be dominated by the magnetically supported, non-dissociative C-type. Rather, our LWS observations demonstrate that the FIR cooling of these flows is completely dominated by [O I] 63 μm emission. For the HH excitation, our data are in favour, therefore, of dissociative J-shock models (Hollenbach & McKee 1989).

It was realized by Hollenbach (1985; see also: Hollenbach & McKee 1989) that, in certain conditions, the [O I] 63 μm line luminosity is a direct measure of the rate at which the central stellar source looses mass. Specifically, for a dissociative 'wind shock' with particle flux, Φ_H, expressed in terms of the pre-shock particle density n_0 and the shock velocity v_s, such that

$$\left[\frac{\Phi_H}{\mathrm{cm^{-2}\,s^{-1}}}\right] \lesssim 1.0 \left[\frac{n_0}{10^5\,\mathrm{cm^{-3}}}\right]\left[\frac{v_s}{100\,\mathrm{km\,s^{-1}}}\right],$$

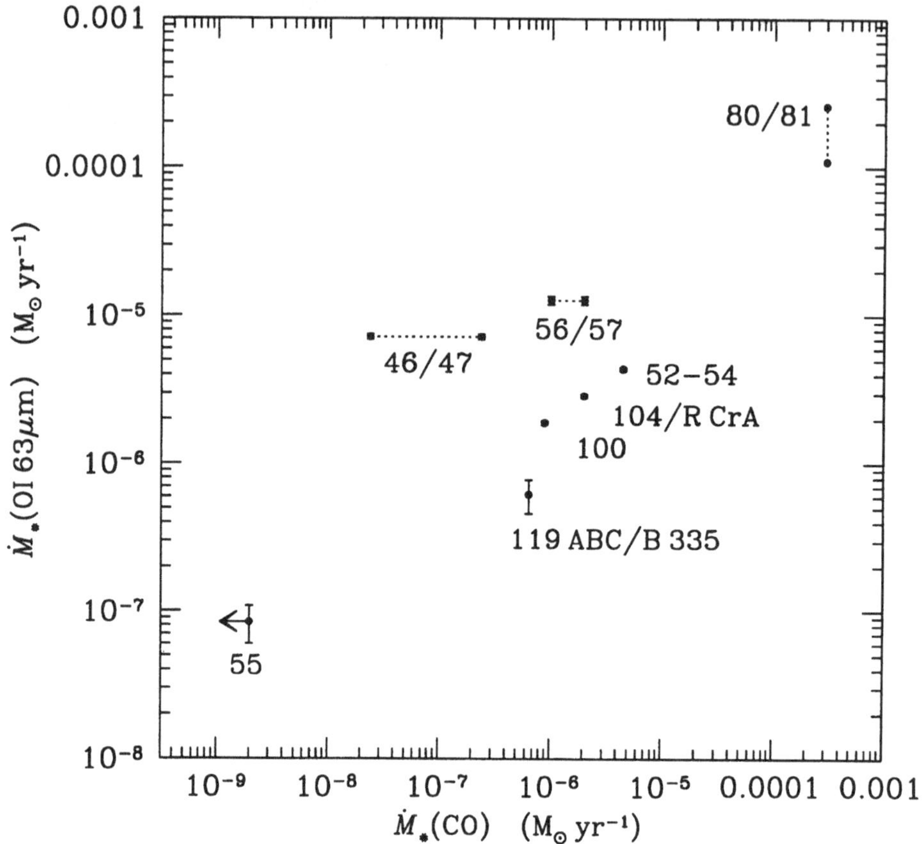

Figure 3. Stellar mass loss rates, based on observations of [O I] 63 μm and on CO observations taken from the literature, for the HH flows observed with the LWS

the [O I] 63 μm luminosity is directly proportional to the stellar mass loss rate, viz.

$$\dot{M}_* = 10^{-6} \left[\frac{L(\mathrm{O\,I\,63\,\mu m})}{10^{-2}\,\mathrm{L}_\odot}\right] \quad (\mathrm{M}_\odot\,\mathrm{yr}^{-1})\,.$$

Similarly, from spatial maps of the wing intensity of mm-wave CO lines, estimates of $\dot{M}_*$ can be deduced, commonly by assuming (for low luminosity sources) momentum to be conserved, viz. $\dot{P}$ = constant, so that

$$\dot{M}_* = 10^{-6} \left[\frac{M}{0.1\,\mathrm{M}_\odot}\right] \left[\frac{\Delta v}{10\,\mathrm{km\,s}^{-1}}\right] \left[\frac{\Delta t}{10^4\,\mathrm{yr}}\right]^{-1} \left[\frac{v_\infty}{100\,\mathrm{km\,s}^{-1}}\right]^{-1} \quad (\mathrm{M}_\odot\,\mathrm{yr}^{-1})\,,$$

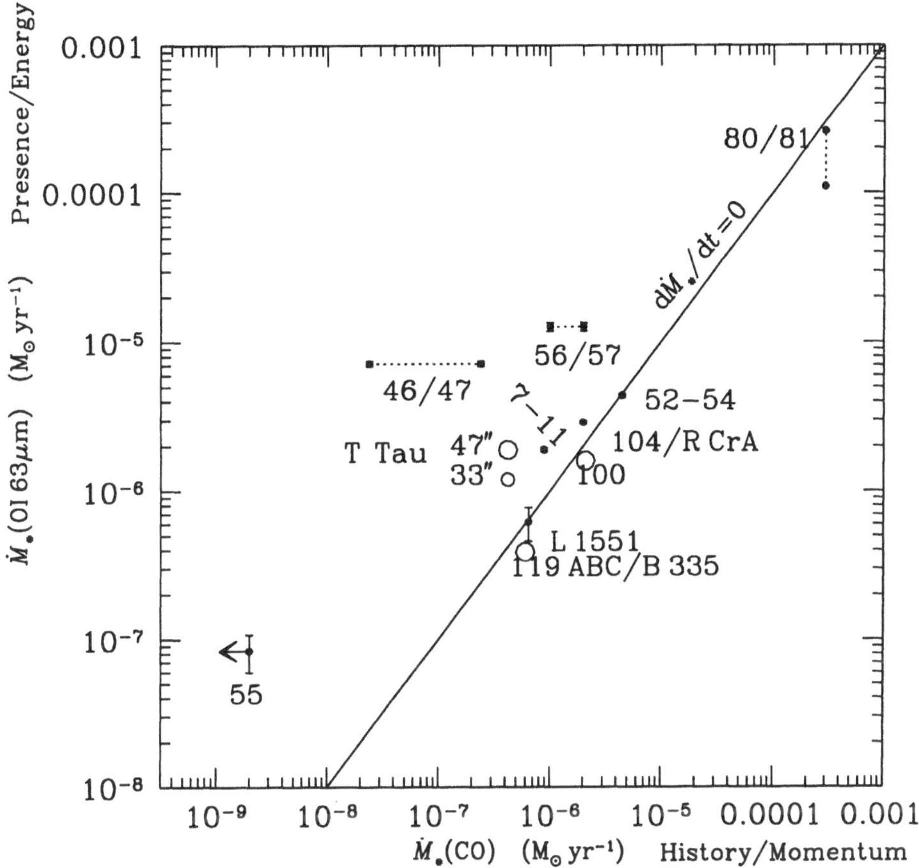

Figure 4. Same as Fig. 3 but data for HH 7–11, L 1551 IRS 5 and T Tau (Cohen et al. 1988) are also shown (open symbols). In addition, a line of unit slope is shown for reference

where M and Δv are the mass and the velocity of the molecular gas, respectively, Δt is an estimate of the CO flow time scale and v_∞ is the terminal velocity of the stellar wind which has accelerated the molecular gas. Neglecting contributions from the CO line core, not accounting for opacity and/or projection effects will imply uncertainties of the mass loss rates as large as more than *one order of magnitude* (e.g. Cabrit & Bertout 1992). Further, what regards distance uncertainties, the dependencies for these two methods are different, viz. $\dot{M}_*([\mathrm{O\,I}]\,63\,\mu\mathrm{m}) \propto D^2$, whereas $\dot{M}_*(\mathrm{CO}) \propto D$.

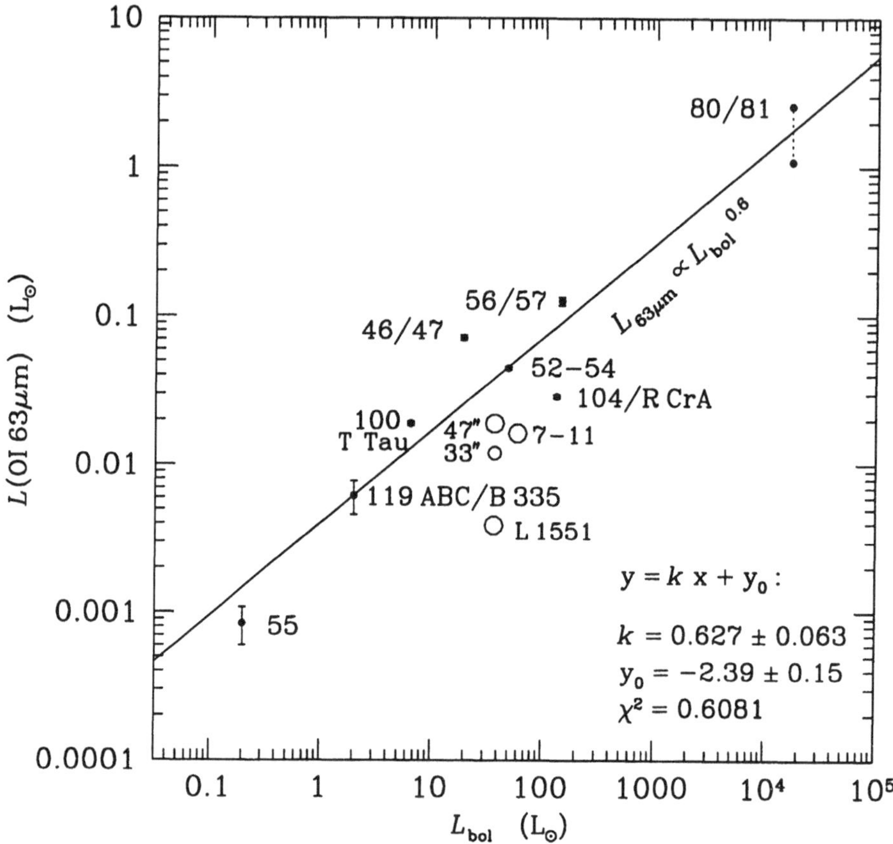

Figure 5. Luminosity in [O I] 63 μm versus stellar $L_{\rm bol}$. The former is directly proportional to the stellar mass loss rate, $\dot{M}$. Filled points are LWS data, whereas open symbols refer to data by Cohen et al., shown for comparison. The least squares fit, shown by the straight line, refers to LWS data only

Given these sources of large uncertainty we would expect considerable scatter in a $\dot{M}_*$([O I] 63 μm)–$\dot{M}_*$(CO) plot. However, the contrary is revealed by Fig. 3 showing these quantities to correlate. This trend is clearly supported by data from Cohen et al. (1988), which are shown superposed in Fig. 4.

Also shown in Fig. 4, for reference, is a straight line of unit slope, designating the locus of complete congruence between $\dot{M}_*$([O I] 63 μm) and $\dot{M}_*$(CO) (note that this is *not* a fit to the data points).

These two determinations of $\dot{M}_*$ refer to two different aspects of the mass loss, though. In the [O I] 63 μm emission, the instantaneous energy dissipation is observed, whereas the CO flows measure the momentum transfer over time scales, Δt, much longer than the cooling time of [O I] 63 μm. One interpretation of the straight line would be, then, that of temporal constancy of the mass loss from the central stellar source. According to the figure, for most sources $\dot{M}_*$ seems not to vary much. Interestingly, however, the position of HH 46/47 appears most disparate: for this system independent evidence for time variable flows does exist (e.g. Hartigan et al. 1993).

3.3. [O I] 63 μm LINE AND STELLAR BOLOMETRIC LUMINOSITY

One 'prediction' of the correlation of Figs. 3 and 4 is that there should exist a Lada-relation for outflows (Lada 1985), i.e. for the mechanical luminosity (the [O I] 63 μm luminosity) and the bolometric luminosity of the mass loosing stellar source. Such dependence is indeed implied by the data displayed in Fig. 5. Formally, our LWS data yield (with luminosities in $L_\odot$)

$$L_{63\mu\mathrm{m}} \sim 0.004\, L_{\mathrm{bol}}^{0.6} \quad .$$

This power law exponent, 0.6, is identical to that found by Edwards et al. (1993) for stellar jets, and similar to that (0.8) determined by Cabrit & Bertout (1992) for CO outflows, but contrasting to earlier findings, according to which exponents $\gtrsim 1$ have commonly been inferred (see, e.g., review by Bally & Lane 1991). These works were generally biased towards high luminosity objects. Consequently, the more recent findings suggest that low luminosity (low mass) young stellar objects are relatively more efficiently 'generating' mass loss than their high mass counterparts. However, whether this could mean that the mass loss processes (the 'drivings') are different in this stellar mass dichotomy, is not possible to tell on the basis of the presented observational evidence.

4. Conclusions

ISO-LWS observations of 17 HH Objects in 7 regions of stellar formation lead us to conclude:

- The FIR spectra of HH flows show a general lack of prominent molecular line emission, indicating that magnetically supported shocks do not play a dominant rôle in HH excitation.
- The uniformly extended [C II] 158 μm line emission from (low luminosity) HH regions can generally be attributed to weak PDRs at the cloud surface ($G_0 \sim 1-10$).
- In mapped flows, the [O I] 63 μm line displays strong spatial variations, apparently related with optically identified HH emission.

- Comparison of stellar mass loss rates estimated using the [O I] luminosity with those determined from mm-wave CO observations results in remarkable agreement, providing empirical proof that HH flows and molecular flows ('CO outflows') are physically associated.
- These data provide also a demonstration of the potential of the [O I] 63 μm line as a mass loss diagnostic.
- The [O I] luminosity, L([O I] 63 μm), being a measure of the stellar mass loss rate, correlates with the bolometric luminosity, L_{bol}, of the central driving source.
- From this relation it is inferred that low mass young stellar objects are more efficient 'mass loosers' than are young stars of high mass.

References

Bally J., Lane A.P., 1991, in: Lada C.J. & Kylafis N.D. (eds.) The physics of star formation and early stellar evolution, Kluwer, p. 471

Cabrit S., Bertout C., 1992, A&A 261, 274

Clegg P.E. et al., 1996, A&A 315, L 38

Cohen M., Hollenbach D.J., Haas M.R., Erickson E.F., 1988, ApJ 329, 863

Draine B.T., Roberge W.G., Dalgarno A., 1983, ApJ 264, 485

Edwards S., Ray T., Mundt R., 1993, in: E.H. Levy & Lunine J.I. (eds.), Protostars and Planets III, U.Arizona, p. 567

Hartigan P., Morse J.A., Heathcote S., Gerald C., 1993, ApJ 414, L 121

Hollenbach D.J., 1985, Icarus 61, 36

Hollenbach D., McKee C.F., 1989, ApJ 342, 306

Hollenbach D.J., Takahashi T., Tielens A.G.G.M., 1991, ApJ 377, 192

Kessler M. et al., 1996, A&A 315, L 27

Lada C.J., 1985, ARA&A 23, 267

Liseau R. et al., 1996, A&A 315, L 181

Nisini B. et al., 1996, A&A 315, L 321

Swinyard B.M. et al., 1996, A&A 315, L 43

Wolfire M.G., Tielens A.G.G.M., Hollenbach D., 1990, ApJ 358, 116

II. The Physics and Chemistry of Molecular Outflows

OBSERVATIONAL PROPERTIES OF MOLECULAR OUTFLOWS

RACHAEL PADMAN, STEPHEN BENCE AND JOHN RICHER
Mullard Radio Astronomy Observatory
Cavendish Laboratory, Madingley Rd., Cambridge, CB3 0HE, ENGLAND

Abstract. Molecular outflows are intimately related to the highly collimated Herbig–Haro jets emanating from young stars. In consequence, the usual dynamical timescale significantly underestimates the true age of an outflow. If we correct for this factor, and assume an intrinsic outflow speed similar to that of the underlying jet, we predict that molecular outflows should have an overall extent of several parsecs, in accordance with recent results. It seems likely therefore that outflows are a major source of interstellar turbulence, and have a profound impact on the process of star formation.

Whilst interpretation of jet-like outflows is relatively straightforward, the origins of shell-like outflows, such as that from L 1551–IRS5, are less obvious. We discuss the current observational status of both types of flow, and hypothesize an evolutionary connection between them. A large and well-defined outflow sample is urgently required, to permit the establishment of an age-sequence; such a sample would also provide the basis for a proper investigation of outflow energetics and interaction with the ISM.

1. Introduction

What is a molecular outflow? Until moderately recently the words suggested sources like the L 1551–IRS5 outflow, in which CO and other low-excitation species are observed in a pair of oppositely-directed and poorly-collimated "lobes", symmetrically located about an embedded YSO. It is now clear that molecular outflows are usually accompanied by many other phenomena, such as Herbig-Haro jets, radio free-free jets and even H_2O masers. In this review we restrict attention primarily to the phenomenon as origi-

B. Reipurth and C. Bertout (eds.), Herbig–Haro Flows and the Birth of Low Mass Stars, 123–140.

nally understood. For the most part we will discuss the kinematics of the low-excitation molecular component of the gas, as traced by single-dish CO observations in the mm and submm wavebands. Fortunately, the interstellar chemistry of CO is quite simple, and the CO/H_2 abundance can thus be assumed to be constant for all values of extinction greater than about unity. On the other hand, instrumental resolution is generally not better than about 1000 AU, and we are perforce restricted to relatively large scale phenomena. Interferometer observations, such as those from the Plateau de Bure, are discussed by Guilloteau et al. elsewhere in this volume.

We also need to decide what is meant by "observational properties", and in particular whether to concentrate on individual outflows or on ensemble properties, such as the statistics of lobe length, maximum observed velocity, mechanical luminosity and thrust. Our approach here is to use individual outflows as examples, but to emphasize the role of ensemble properties in constraining the theoretical models.

It is not possible in a paper of this length to give a comprehensive review of all of the relevant observations. We hope we might be forgiven for illustrating our main points with examples drawn from our own observations and those of our collaborators. For an extremely comprehensive and up to date review of the field the reader is referred to the 1996 Annual Reviews article by Bachiller [1].

2. Properties of individual outflows

2.1. L1551–IRS5 AS AN ARCHETYPE?

As already noted, the "archetypal" outflow is that from L 1551–IRS5 [2]. The CO in this source is poorly collimated (the length to width ratio is only about 3). A Herbig–Haro jet emanates from the source (see Fridlund, this volume), and there are several Herbig–Haro objects, amongst them HH 28 and HH 29. The source itself is a 30 $L_\odot$ YSO [3], which is moderately deeply embedded: it is surrounded by a dust core which is just optically thick in $C^{17}O$ $2 \rightarrow 1$ [4].

At higher resolutions the CO gas is observed to lie in an open shell, with the source, IRS5, lying just at the apex [5] [6]. There is little evidence for a more collimated component to the outflow, although a number of "hotspots" suggest a current interaction with an invisible jet (which would have to be somewhat misaligned with the current axis of the outflow, as given by the Herbig–Haro jet from IRS5).

In about 1990 it was discovered that several outflow sources showed distinctly jet-like structures (particularly at high radial velocity offsets). Classic examples of these molecular jets can be seen in NGC2024 [7][8], L1448 [9], Orion South [10] and VLA 1623 [11]. Several lines of evidence

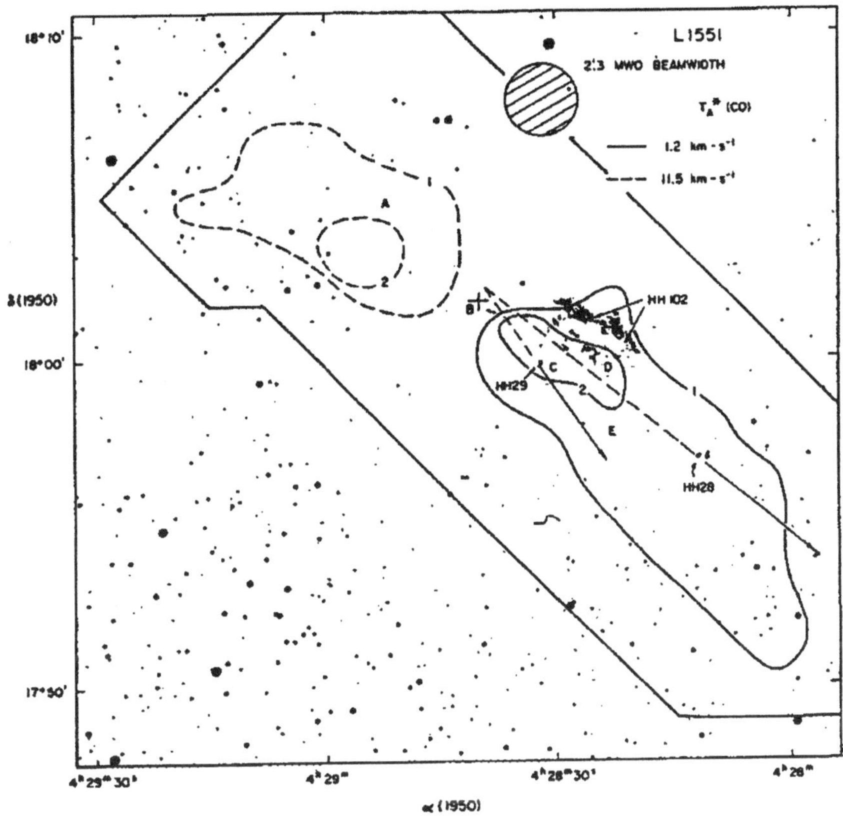

Figure 1. CO 1–0 in L 1551 (from Snell *et al.* 1980 [2])

suggest that the CO emitting gas in these sources consists of ambient cloud material swept-up by an underlying, and mostly invisible, jet:

- The collimation increases monotonically with radial velocity offset, whereas the apparent mass *decreases*, suggestive of an entrainment process.
- Most such sources show associated H_2 v=1-0 S(1) jets and bowshocks coincident with the most highly collimated CO (see, e.g. [8][12][13]).
- A few of these sources also show strong SiO emission, which is thought to be due to grain sputtering in shocks [14].
- Characteristic bow shock structures are associated with the ends of the jets, as deduced from the 3-dimensional spatial-velocity structure [15].
- The mass of the outflowing molecular gas is comparable with that deduced for the lobes in the absence of outflow, and is much greater

than could have been transported in the jet (assuming canonical mass loss rates and lifetimes).

Although interaction with an underlying high-velocity jet appears to be a good working hypothesis for the origin of the high-velocity CO emission in these sources, there are still many unresolved issues. For example, there is evidence that the jet varies both in strength and in direction (see, *e.g.* [15][16]). While this is undoubtedly reflected in the structures visible in CO, it does not seem that CO observations can elucidate this behaviour except indirectly. The CO observations are of much more direct relevance to the issue of how the ambient gas is entrained in the jet. It is certainly not clear yet (to observers!) whether this takes place at the head of the jet ('prompt' entrainment, [17]), along the sides of the jet ('lateral' or 'steady-state' entrainment, [18][19]) or at internal working surfaces [20]. Probably all three operate, and we would like to understand when each is important.

2.2. RNO 43: A JET-LIKE OUTFLOW

The outflow associated with RNO 43 [15] illustrates many of these phenomena. A very highly collimated CO 'jet' is associated with the inner parsec of the outflow, and is also seen in H_2 S(1) emission [21]. The overall extent of RNO 43 is of order 5 parsecs, and characteristic bow shock signatures are seen at each of the high-integrated-intensity "blobs" near the extremities of the source. The location and radial velocities of these blobs can be fitted to the surface of a cone of opening angle approximately 15°. It appears that the CO observations can be well fitted by assuming the CO to be entrained in an episodic and wandering underlying jet.

The axis of the RNO 43 outflow lies very close to the plane of the sky, and motions along the jet therefore have only a small radial component of velocity. Using simple symmetry properties, we can thus determine the 'longitudinal' and 'transverse' velocity distributions independently. The relative magnitudes of forward and transverse momenta provide useful constraints on models of outflow acceleration [25][26][27][28][29] (see also paper by Wilkin in this volume).

Of course, not all outflows are jet-like. Ironically, the intense interest in molecular jets has meant that the rest have been somewhat overlooked. Outflows like that from L 1551–IRS5 itself [2], and from the RNO 91 source in the L43 dark cloud [30], are completely different in structure. The molecular gas in both sources appears to lie in an open shell, with the source at the apex. It is probable that this is an evolutionary effect, with the shell-like outflows being older than the jet-like sources. But there has been no serious statistical study of this possibility, and it is also possible that simple environmental effects are responsible for the differences.

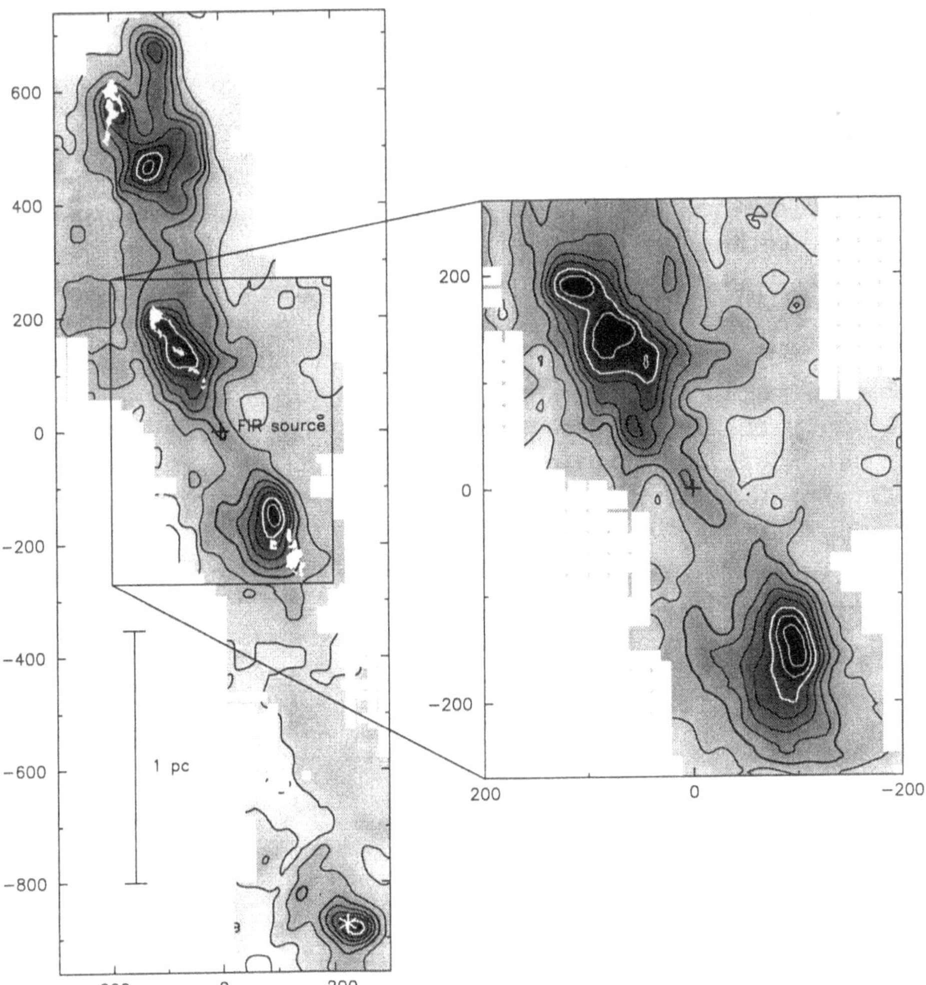

Figure 2. CO 2–1 integrated intensity in the RNO 43 outflow (adapted from Bence *et al.* 1996). The optical H–H objects observed by Mundt *et al.* (1987; [22]) and Reipurth (1991; [23]) are overlaid on the CO map, and the position of HH 179 (Reipurth, private communication; see also Reipurth *et al.* [24]) is indicated with a star. Offsets in both R.A. and Dec. are in arcseconds from the FIR source at $\alpha = 05^h29^m30^s.6$, $\delta = +12°47'35''$

3. Properties of the ensemble

3.1. STATISTICS

Lada [31] used early observations to plot the correlation between the luminosity of the central sources and the thrust and luminosity in the outflowing molecular gas. The correlation has now been re-examined and applied to a larger sample of sources [32][33], using the total source luminosity $L_{\rm bol}$

instead of L_* and applying a consistent correction for the inclination factor. The newer results confirm two important points:

- The momentum supply rate to the CO-emitting gas (the thrust, F_{CO}) is typically 100 times *larger* than that available in radiation from the central source (L_*/c).
- The mechanical luminosity in the CO-emitting gas, L_{CO} is typically 10 to 100 times *less* than the stellar luminosity (L_*).

Given the very large thrust, it is clear that the outflowing molecular gas can not be accelerated directly by radiation pressure — the opacity required is very much higher than we observe. But since much of the stellar luminosity is probably due to accretion, and knowing that the typical velocity of a jet is about equal to the escape velocity from the (proto)stellar surface, these correlations are strongly suggestive that the outflow is driven by some (magneto)hydrodynamic mechanism operating deep in the star's potential well.

There is, however, one major caveat to this interpretation. Calculations of luminosity and thrust have traditionally used the dynamical timescale, τ_{dyn} as an estimate of the age when converting observed energy and momentum to supply rates:

$$L_{CO} = \frac{1/2 \int_{-\infty}^{\infty} v^2 M(v)\, dv}{\tau_{dyn}};$$

$$F_{CO} = \frac{\int_{-\infty}^{\infty} |v| M(v)\, dv}{\tau_{dyn}};$$

where $M(v)$ is the total gas mass per unit velocity interval, and

$$\tau_{dyn} = \frac{R_{max}}{V_{max}}, \tag{1}$$

R_{max} is the maximum observed extent of the outflow, and V_{max} is its maximum observed velocity (in CO). Typically, τ_{dyn} turns out to have a value less than a few $\times 10^4$ years.

On the other hand, Parker *et al.* [34] and Fukui *et al.* [35] both deduced statistical lifetimes for outflows a factor of ten higher than the dynamical times. Parker *et al.* found that over 75% of IRAS sources embedded in Lynds class VI dark clouds had associated outflows. Assuming that these IRAS sources would go on to become T-Tauri stars, and would appear as visible objects at an age of around 200 000 years, then *both* (a) more than 75% of embedded sources have outflows at some time, *and* (b), in those, outflow persists for more than 75% of the embedded lifetime. Fukui *et al.* used similar arguments to deduce the lifetime of outflows associated with visible T-Tauri stars, assuming an age of about 3 million years for the stars. The average *age* of these sources should then be just half their statistical lifetime, which for most sources is still much greater than the (observed) dynamical time.

Parker *et al.* did not correct for background sources, which were estimated to be up to 50% of the total near the Galactic plane, and the actual fraction of embedded sources with outflows must therefore be higher than 75%. Bontemps *et al.* [33] detected CO outflow in 80% of 45 class 0 and class I sources in a more heterogeneous sample. Outflow from YSOs therefore *must* persist for almost the entire duration of the embedded phase (although these statistics say nothing about whether it carries on into the class II and class III stages also).

3.2. IMPLICATIONS OF JET-DRIVEN MODELS

Arguments such as these show that the dynamical lifetime is most definitely not the *age* of the source. What is it then? Is it useful at all?

As defined, $\tau_{\rm dyn}$ depends both on the extent of the outflow and on the observed velocity. Both are subject to quite severe selection effects, depending on the sensitivity of the available instruments. Although we tend preferentially to see outflows with short dynamical times (i.e. those with the highest velocities), this cannot explain the discrepancy between dynamical time and statistical age in samples selected for some property *not* involving outflow.

In fact, in the "swept-up cloud" model, there is no reason to suppose that the molecular material should be moving at anything like the speed of the underlying jet. In the early stages of the outflow, where the jet is confined within the dense inner regions of the cloud core, we expect it to be underdense with respect to the quiescent cloud material. The rate of advance of the head of the jet, and also the highest longitudinal velocities in the swept-up gas, will therefore be limited by the momentum supply rate to the displaced quiescent material, and will be much less than the actual jet speed. In the outer regions of the cloud it is more likely that the jet will be ballistic, and both the rate of advance of the bow shock and the highest velocities in the molecular gas will approximate the jet speed. The dynamical time clearly depends on the time history of both these quantities, and therefore on the total run of densities in the cloud between the source and the present bow shock position [8].

It is interesting to note that neither observers nor theorists have got into the same tangle with optical jets. The reason appears to be that the dynamical timescale of those sources is *so* short (given the observed velocities of several hundred km/s and the relatively short jets) that it has never been remotely feasible that τ_{dyn} could represent the age [22]. There just aren't enough T-Tauri stars in the sky! Clearly in this case τ_{dyn} is just the time taken for the material in the jet to propagate from the source to the observation position. On the other hand, this might have been used to *predict* that optical jets should have very much larger extents than they were thought to have, until very recently. The discovery in 1994 of parsec scale jets, by Bally & Devine [16], should not have been a surprise....

The same argument applies to molecular outflows. A jet moving at 200 km/s will travel 20 pc in 10^5 years. It would thus not be surprising to see outflows on this scale. In a series of recent observations using the QUARRY array receiver on the FCRAO 14-m telescope, Bence *et al.* (in preparation) have identified at least two very large scale outflows (Figures 3–6), from sources in the Parker *et al.* sample. Thus, of the handful of sources from the Parker *et al.* sample examined closely so far, both the L 1262 and L 588 outflows extend to more than 2 parsecs. RNO 43 (already mentioned) is at least 5 parsecs long. Large scale outflows are not rare.

3.3. IMPLICATIONS OF LARGE SCALE OUTFLOWS

The predicted scale size for outflows (up to 20 parsecs) is very much larger than that of the typical dark clouds from which they emanate. Even if the average scale size is only a few parsecs, the outflows still extend well beyond the visible cloud boundaries. In practical terms, if the opening angle of outflows is not too small, then a star-forming molecular cloud will feel the effects of the outflows originating within it throughout most of its volume.

One long-standing problem in star-formation and ISM theory is the origin of turbulence. When observed in CO and other abundant species, most molecular clouds have highly superthermal line widths. Since most clouds are also in virial equilibrium [37], this implies that turbulent motions (whether purely hydrodynamic or Alfvenic in character) dominate the cloud energetics. This turbulence must in turn largely regulate the star formation rate and efficiency. While there have been several large scale observational studies aimed at characterizing the turbulence (see, e.g. [38]), and a great deal of related theoretical work (see, e.g. [39]), the actual energy sources have not yet clearly been identified.

Norman & Silk [40] envisaged that winds from T-Tauri stars were responsible for the turbulence observed in molecular clouds, but this idea lost popularity with the discovery of molecular outflows. Outflows themselves have been seen as a possible source of turbulence ever since their discovery [41][42][43], but the interpretation has been complicated by the relatively small scale of outflows, apparently measuring a few tenths of a parsec at most. It was not clear how they could then affect more diffuse clouds on

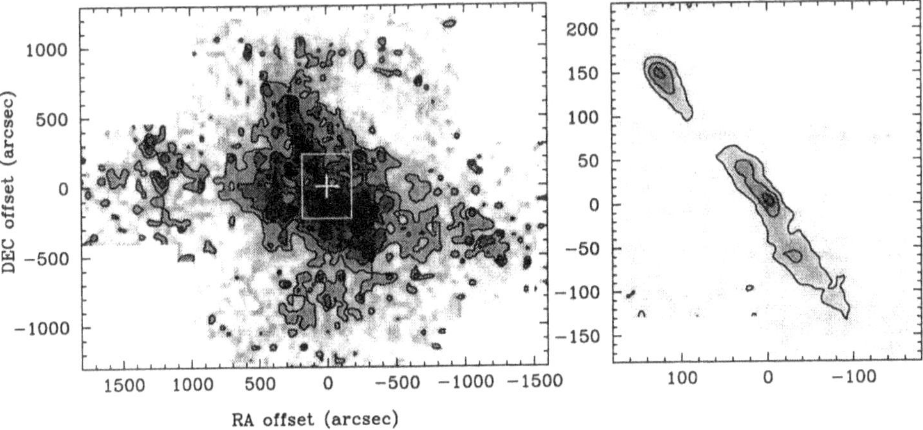

Figure 3. Integrated intensity maps of the L 1262 molecular cloud, in CO 1–0 (left; FCRAO) and CO 3-2 (right; JCMT). Channel maps of the CO 1–0 emission (not shown here) clearly show that most of the integrated intensity is due to local line broadening associated with outflowing gas, albeit with relatively small velocity offsets.

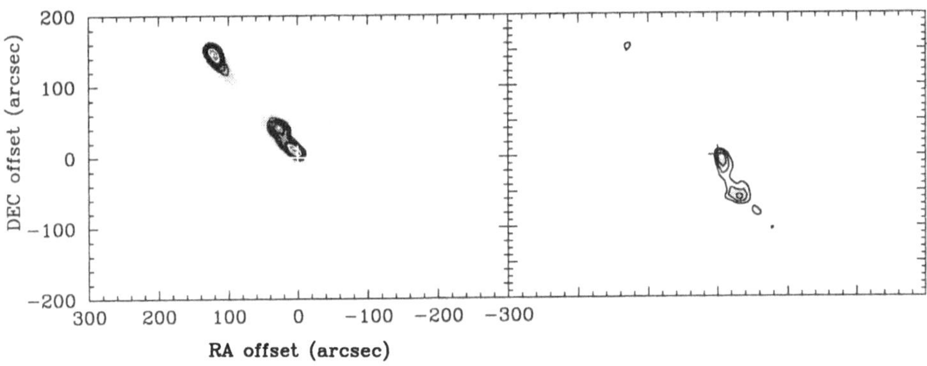

Figure 4. CO 3–2 blue- (left) and red- (right) shifted emission in L 1262

much larger scales.

This was partially resolved in 1994 by Chernin & Masson [46]. They noted that in HH 34 and HH 1-2 the momentum available in the optical jet far exceeded that which could be traced in the molecular medium, and deduced that the jet was depositing much of its momentum into the more diffuse ISM. Very strong support for this idea comes from the beautiful observations of very large scale optical jets — see the review by Bally & Devine, in this volume. Our new observations confirm that *molecular* outflows may well also inject significant turbulence into the more diffuse

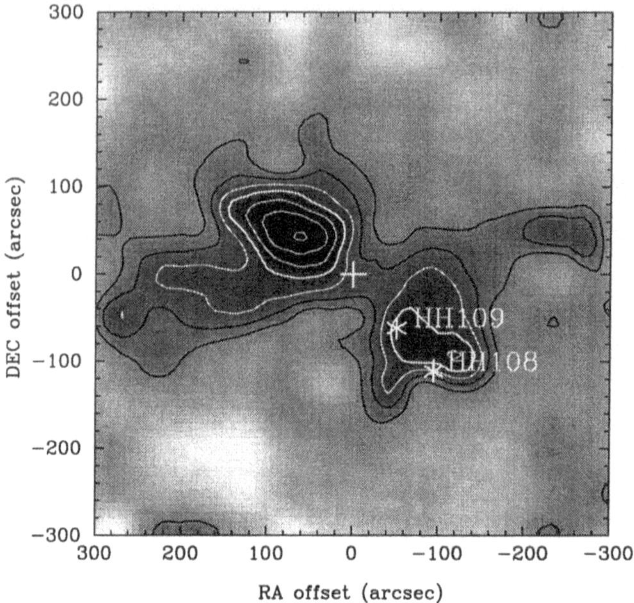

Figure 5. CO 1–0 integrated intensity in the inner regions of L 588, as observed at FCRAO using QUARRY. The line wings are very broad in this source, so that the integrated intensity is completely dominated by outflow. Two Herbig–Haro objects [36] are shown in the blue-shifted lobe of the CO outflow; the position of the IRAS source 18331-0035 is indicated with a cross.

ISM, although we have not yet managed to produce good numbers for the energy and mass loss rates at large scales — this is complicated by considerations of UV dissociation (see below).

In the cases we have observed in detail, we see that the orderly structure of the inner, collimated, outflow apparently breaks down in the outer regions, perhaps as a result of Rayleigh-Taylor and other instabilities affecting an overdense jet propagating in the diffuse ISM [44]. Furthermore, the size scales here are much larger than the typical separation of YSOs in an active star forming region such as Taurus-Auriga. Even with modest collimation, the "volume-filling factor" may well approach unity or greater, in which case we would expect frequent cases of outflows criss-crossing and interacting with each other, as observed for optical jets in NGC 1333 [45]. The situation is thus not so different from that suggested by Norman & Silk, with the only real departure being the substitution of collimated outflows for isotropic winds.

The second issue raised by the discovery of large-scale outflows is the survival of the molecular gas at very low visual extinctions. Again, this should come as no surprise — the L 1551 discovery paper of Snell *et al.* [2]

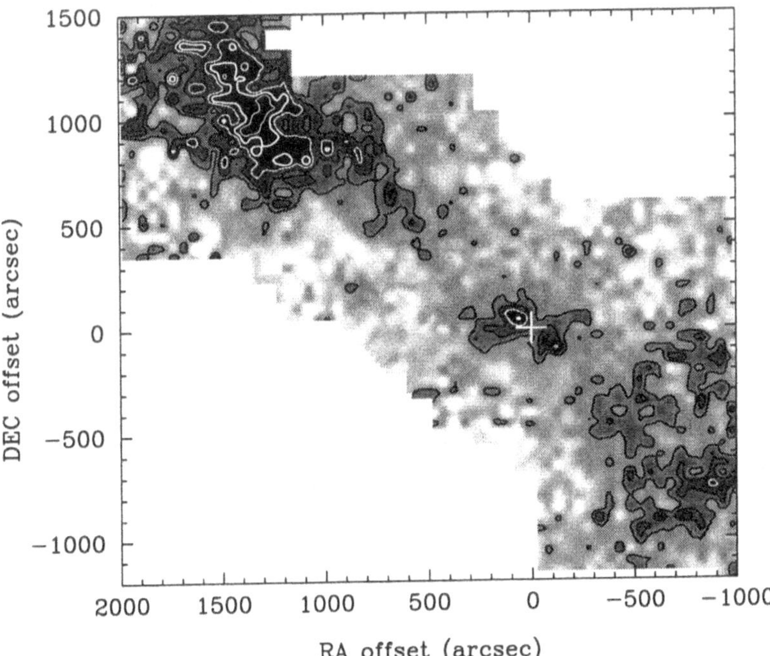

Figure 6. CO 1-0 integrated intensity in the environs of L 588. Note the fragmented emission to the NE and SW of the source, aligned with the inner outflow. Spectra taken in these two regions have the same radial velocity offset as those in the lobes of the inner outflow. The map is approximately 1 degree square.

presents an overlay of the CO outflow on the POSS plate, and background stars are clearly visible in the outer regions (see Figure 1). The L 1551–IRS5 outflow extends well beyond the L 1551 dark cloud itself. Standard theory (e.g. van Dishoeck & Black [47]) suggests that CO will be photo-dissociated on a time scale of order 50 000 years in a standard interstellar UV field ($G/G_0 = 1$), and proportionately faster as the UV intensity increases. We would therefore expect [CO]/([H] + [H2]) to be substantially reduced in the outer regions of the outflow. This may help to explain why large-scale outflows have not been discovered previously, and why many of the largest outflows already known are strongly asymmetric (see, e.g., the L 43–RNO 91 outflow presented by Bence *et al.* [48].) The *state* of the extended CO is not yet well understood, and observations of higher-J CO transitions are required so that we can estimate the temperature and the density.

Of course, the outflows do eject dust from the cloud core, as well as gas. The sky survey plates show dusty filaments around the CO shell in L 43, and in several other sources too, and the presence of hot dust is confirmed by the IRAS 60 and 100 μm images. But as the swept up material expands away from the cloud the dust rapidly becomes too diffuse to be visible, or to provide any shielding for the CO.

The discovery of large scale outflows, and the doubt this casts on the use of traditional dynamical timescale as a measure of age, implies that we should now be very curious about the origin of the correlation between $F_{\rm CO}$ and L_* or $L_{\rm bol}$. If the dynamical timescale really bears no resemblance to the true age, why are these correlations so good?

A partial answer has been provided by Masson & Chernin [26], who replotted Cabrit & Bertouts' 6 cm fluxes for outflow sources [32] as a function of $P_{\rm CO}$ rather than as $F_{\rm CO}$ — which is effectively the same as assuming a constant age for all sources. The correlation is noticeably improved. Of course, this would not have been possible had it already been very good — the improvement shows that the correlation holds *despite* an erroneous attribution of source age, not *because* of it.[1]

Finally, we note that we have said nothing at all yet about the supposed "$v \propto r$" law for outflows. It has been suggested by several authors [29][49] that outflows obey a Hubble-like law, with the highest radial velocities being observed furthest from the source. Such a correlation implies velocity sorting associated with an explosive, or very rapid, origin for outflows or outflow episodes. This is hard to understand in the context of a jet-driven model, where the observed material has been accelerated *in situ*, rather than originating at the source.

Our own observations, and our interpretation of others', suggest that for many sources such a law does apply *locally*, at particular bow shocks. That is, as predicted by most theoretical models of bow shocks, the velocity dispersion peaks at the bow shock and falls behind it (towards the source) until it has the same value as that of the quiescent cloud. For small or incompletely-mapped sources, where only the most recent outflow episode is visible, this may well masquerade as a more global phenomenon. But there is little evidence in our data for any global "Hubble-law" in sources such as NGC 2024 [8]. This source appears already to have undergone multiple outflow episodes, as evidenced by the presence of several bow shock like features in the position-velocity diagram; the maximum velocity is essentially constant over at least 75% of the outflow extent.

[1]Insofar as it now compares a time-integrated property ($P_{\rm CO}$) with a current value ($S_{6\,\rm cm}$), the new plot is less aesthetically pleasing, and it would clearly be desirable to find a better measure of $F_{\rm CO}$, either by using more accurate values of $\tau_{\rm dyn}$ or, preferably, by using an instantaneous measure such as the estimate developed in [33].

4. Outflow evolution: where did L 43 come from?

It is often said that "collimation decreases with age". Given the uncertainty attached to the use of τ_{dyn} as an age indicator, this has to be taken with a pinch of salt. The cynic might well rewrite the original statement as "the highest CO velocities are observed in the most collimated sources", which of itself would not be too surprising. In fact the personal bias of at least one of the current authors is that the best indicator of age is *size* — which is the same as saying that we should assume that all outflows have the same maximum velocity. At least in most models size increases monotonically with age, whereas τ_{dyn} itself may well decrease in periods of activity.

At any rate, there is little doubt that the recently discovered (small) outflows in L 1527 [50] and TMC 1/TMC 1A [51] are quite young, as are those presented by Ohashi *et al.* [52]. Yet all of these sources have a characteristic butterfly shape, closely related to the ogive-shaped shells at the base of the outflow in L 1448, as revealed by IRAM Plateau de Bure Interferometer maps [53]. Shepherd & Watsons' poster paper [54] shows a similar CO structure around a massive star, as observed at OVRO. So even sources which show highly-collimated CO at *high* velocities, show cavity structure at *low* velocities. Collimation is thus dependent on the velocity range chosen for the observation, and presumably on the sensitivity of the observation also.

The view which emerges is much less black and white than the one we are used to. If most sources show jet and shell structures simultaneously, then the perceived collimation may be largely a function of the relative intensity of the two structures. Sources such as L 1448 and NGC 2024, in which the jet is very bright at high velocities, appear to be highly collimated; as the jet weakens the limb-brightened cavity walls start to dominate, as in L 1551–IRS5 and L 43–RNO 91.

Still, the CO jets in L 1551 and L 43 are *very* weak (if they even exist), and the cavities relatively large. We hypothesize that both of these are relatively evolved sources, in which the outflow is perhaps a few hundred thousand years old. Then most of the material originally within the cavity would have been dispersed into the diffuse ISM, and any residual jet would propagate freely, without interacting with the medium, and hence without radiating significantly. Some support for this hypothesis is obtained from the recent observations of the extended optical jet in L 1551 (Bally & Devine, this volume), while L 43 shows evidence for a weak CO jet roughly aligned with the middle of the cavity.

Detailed observations of the L 43 outflow ([30][48]) show that the cavity is expanding slowly, with a speed of order 1–2 km/s. It is not clear whether this is just a coasting shell created during the jet's passage, or if the motion is sustained by the pressure of a high-pressure low-emissivity medium within the cavity (or by the ram pressure of a low-emissivity wide-angled wind from the star). To complicate matters further, this velocity is also close to the terminal velocity for a bubble of size equal to the scale height rising under buoyancy. Further observations will be required to distinguish between these several possibilities.

As noted by both Cabrit and Shu in comments during this meeting, if the outflows from L 43–RNO 91 and L 1551–IRS5 actually extend much further than shown in conventional CO maps, then the actual collimation of these sources is greater than thought up until now. In effect, the relatively large cavities may be just the base of a very large structure which scales with time in a self-similar way. One very nice observation which seems to support this idea is that of the jet and outflow in the southern dark cloud, Sandqvist 136. Bourke's CO map [55] and I-band image [56] appear to show a source intermediate in appearance between the jet-like sources and the cavity sources.

5. Conclusion

It is hard to draw many general conclusions beyond those already discussed. As *the* conclusion to the review, and as a way of bringing together a few nice observational results not dealt with in the rest of our review, we briefly discuss both the outstanding problems (from an observational perspective) and what we see as the major areas requiring further work.

5.1. OUTSTANDING PROBLEMS

Single-dish CO observations are unlikely to have much direct impact on our understanding of the origins and properties of the jets themselves. They are of much more use, however, when considering the interactions of the jets with the surrounding, initially quiescent, ISM. It seems to us that there are (at least) two areas which have not yet been sufficiently explored, and which may cast a good deal of light on this process.

First, a very large fraction of outflows show overlapping red- and blue-shifted emission on both sides of the source. For a simple source model, in which the CO-emitting gas flows radially out from the source within a cone of constant opening angle, this is generally taken as an indication that the outflow axis lies within the cone half-angle of the plane of the sky. For example, for a half-angle of 30°, then we would expect 50% of sources to show such overlapping red and blue emission from a single lobe, and another 13% to appear 'pole-on' [57].

In fact, most outflows have opening angles much less than 60° — the average is perhaps only half this — and we would thus expect fewer than

30% of sources to have overlapping red and blue lobes. Projection factors mean that the apparent opening angle is always more than the true angle, which again decreases the fraction expected to show this phenomenon. This can be summed up, somewhat tongue-in-cheek, as "All outflows are in the plane of the sky". [2] It may be that in fact this is just telling us not to ignore the transverse velocity dispersion in the outflow. In any case, this observation must be an important constraint on the jet-driven model.

Our second observation is that "All outflows are self-absorbed". Even before bipolar molecular outflow was first identified, it was known that most of the sources were deeply self-reversed in lines of CO [58][59][60]. At the time this was interpreted as arising from *inflow* onto a hotter core. It is now clear that self-absorption is a widespread phenomenon, and Narayanan & Walker [61] have presented some lovely work showing how this can be used to discriminate between outflow episodes in Cep A. Much more needs to be done to establish this very promising tool as a more general diagnostic of outflow structure.

5.2. FURTHER WORK

First, there is a pressing need for a complete, high-sensitivity, *sample*[3] with well-understood selection effects. There have been two serious attempts to date. Parker *et al.* [34] defined a complete sample of about 25 sources, but due to instrumental limitations were able to observe only half of these in CO (and those not completely, as the more recent discovery of very large scale structure quite clearly shows). More importantly, it was not possible to map even that many sources in submm continuum, so the class of most of the YSOs is not yet known. Bontemps *et al.* [33] observed a more heterogeneous sample of class 0 and class I sources, and found similar statistics. Both samples are fundamentally limited by being IRAS-selected, which means that many class 0 sources may have been missed. Fortunately, instrumental throughput is improving very rapidly, especially with the advent of array receivers, and unbiased surveys of useful samples should soon be feasible.

The sample needs to be large enough to yield the distribution of objects with central-source luminosity, and to permit the establishment of an *age-sequence*. It may be possible to use the properties of the central object as an independent dating mechanism, but one should, however, beware the risks of circular reference....

[2]It is not easy to deduce the inclination by independent means, and almost the only sources for which it can be done at all are those for which we have both proper motions and radial velocities for the associated optical jets.

[3]There are lots of catalogues — these are not the same thing at all.

It is strange, but true, that after many years the state of the outflowing gas (density, temperature, filling factor) is still not well known [15][62][63], although it is clear that much of it is warm (more than about 30 Kelvin), dense (more than about $10^9/m^3$) and clumpy (of unknown filling factor). Partly this is because it has been too difficult to observe both the high-frequency high-J CO lines and the lower-frequency low-brightness isotopomeric lines with adequate sensitivity and at the same resolution in all lines. Again, thanks to instrumental advances, observations can be made orders of magnitude more quickly than hitherto. We really need to know how the CO excitation varies near bow shocks, along and away from the jet, and within the cavities of the more evolved sources. Recent ISO results [64] indicate the great potential of high-J CO observations in particular.

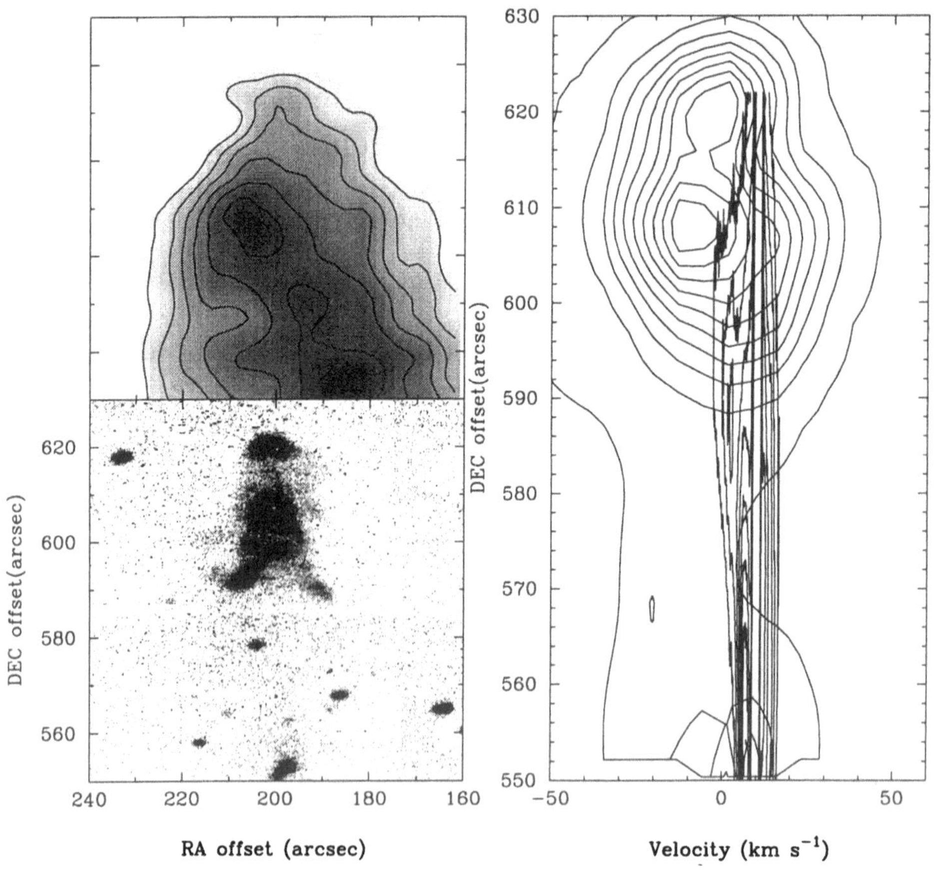

Figure 7. Images of the H_2 S(1) and CO 2–1 emission in RNO 43 (from Bence *et al.* [21]). We also show the overlaid position-velocity diagrams, where both datasets have been convolved to the same spatial resolution. The offset between the maximum velocities of the peaks in each species can be clearly seen.

Finally, we emphasize that molecular outflows are *not* distinct from the other outflow phenomena which together constitute the subject of this meeting. CO traces the low-excitation component of the gas merely, and any real understanding must involve a synthesis of observations of species with a wide range of excitation conditions. As a parting shot, we illustrate our point with an overlay of the position velocity diagrams of H_2 v=1-0 S(1) and CO emission in the N4 region of RNO 43 [21] (Figure 7). The emission from the two species is clearly different, but complementary. Presumably that from [SII], say, will be different again. Understanding the relationship between these plots is the key to an understanding of the outflow phenomenon in general.

Acknowledgements: Thanks to all our MRAO collaborators - particularly Adrian Russell, Nick Parker, Paul Scott, Richard Hills, Claire Chandler, Kate Isaak, Martina Wiedner and Richard Blundell. RP also thanks Colin Masson, for many scientific discussions, for his insight into the problems raised by molecular outflows, and especially for his generous hospitality during several visits to Pasadena and CfA.

References

1. Bachiller, R. 1996. Ann. Rev. Astron. Astrophys. 34, 111
2. Snell, R. L., Loren, R. B. & Plambeck, R. L., 1980, ApJ 239, L17
3. Cohen, M,. Harvey, P.M., Schwartz, R.D. & Wilking, B.A. 1984. ApJ 278, 671
4. Fuller, G.A., Ladd, E.F., Padman, R., Myers, P.C.,& Adams, F.C. 1995. ApJ 454, 862
5. Moriarty-Schieven, G. H., Snell, R. L., Strom, S. E., Schloerb, F. P., Strom, K. M. & Grasdalen, G. L., 1987, ApJ 319, 742
6. Moriarty-Schieven, G.H. & Snell, R.L. 1988. ApJ 332, 364
7. Richer, J. S., Hills, R. E., Padman, R. & Russell, A. P. G., 1989, MNRAS 241, 231
8. Richer, J. S., Hills, R. E. & Padman, R., 1992. MNRAS 254, 525
9. Bachiller, R., Cernicharo, J., Martin-Pintado, J., Tafalla, M. & Lazareff, B., 1990, A&A 231, 174
10. Schmid-Burgk, J., Gusten, R., Mauersberger, R., Schulz, A. & Wilson, T. L., 1990. ApJ 362, L25
11. Andre, Ph., Martin-Pintado, J., Despois, D. & Montmerle, T., 1990, A&A 236, 180
12. Davis, C.J., Dent, W.R.F., Matthews, H.E., Aspin, C. & Lightfoot, J.F. 1994. MNRAS 266, 933
13. Bally, J., Lada, C.J. & Lane, A.P. 1993. ApJ 418, 322
14. Bachiller, R., Martin-Pintado, J. & Fuente, A. 1991. A&A 243, L21
15. Bence, S.J., Richer, J.S. & Padman, R. 1996. MNRAS 279, 866
16. Bally, J. & Devine, D. 1994, ApJ 428, L65
17. De Young, D.S., 1986, ApJ 307, 62
18. Stahler, S. 1994. ApJ 442, 616
19. Raga, A.C., Canto, J., Calvet, N., Rodriguez, L.F. & Torrelles, J.M. 1993 A&A 276, 539
20. Raga, A.C. & Cabrit, S. 1993. A&A 278, 267
21. Bence, S.J., Richer, J.S. & Wright, G.S. 1997. MNRAS (submitted)
22. Mundt, R., Brugel, E. W. & Buhrke T., 1987. ApJ 319, 275
23. Reipurth, B. 1991. in *The Physics of Star Formation and Early Stellar Evolution.*,

ed C.J. Lada & N.D. Kylafis. NATO ASI Series C vol 342, 497.
24. Reipurth, B., Bally, J. & Devine, D. 1997. Astron. J. (in press)
25. Chernin, L.M., Masson. C., Gouveia Dal Pino, E.M. & Benz, W. 1994. ApJ 426, 204
26. Masson, C. R. & Chernin, L. M., 1994. in *Clouds, Cores and Low Mass Stars*. ed D.P. Clemens & R. Barvainis. ASP Conf. Series. 65, 350
27. Chernin, L.M. & Masson, C.R. 1995. ApJ 455, 182
28. Wilkin, F.P. 1996. ApJ 459, L31
29. Lada, C.J & Fich, M. 1996. ApJ 459, 638
30. Bence, S.J., Padman, R., Isaak, K., Wiedner, M.C. & Wright, G.S. 1997. MNRAS (submitted)
31. Lada, C.J. 1985. Ann. Rev. Astron. Astrophys. 23, 267
32. Cabrit, S. & Bertout, C., 1992. A&A 261, 274
33. Bontemps, S., Andre, Ph., Terebey, S. & Cabrit, S. 1996. A&A 311, 858
34. Parker, N. D., Padman, R. & Scott, P. F., 1991, MNRAS 252, 442
35. Fukui, Y., Iwata, T., Mizuno, A., Bally, J. & Lane, A. P., 1993, in *Protostars and Planets III*, ed E.H.Levy & J.I.Lunine. Univ.Arizona Press, Tucson. p603
36. Reipurth, B., & Eiroa, C. 1992. A&A 256, L1
37. Larson, R.B. 1981. MNRAS 194, 809
38. Falgarone, E., Puget, J.-L. & Perault, M. 1992. A&A 257, 715
39. Falgarone, E. & Phillips, T.G. 1990. ApJ 359, 344
40. Norman, C. & Silk, J., 1980. ApJ 238, 158
41. Solomon, P.M., Huguenin, G.R. & Scoville, N.Z. 1981. ApJ 245, L19
42. Margulis, M. & Lada, C.J. 1986. ApJ 309, L87
43. Fukui, Y., Sugitani, K., Takaba, H., Iwata, T., Mizuno, A., Ogawa, H. & Kawabata, K. 1986. ApJ 311, L85
44. Stone, J.M., Xu, J. & Mundy, L.G. 1995. Nature, 377, 315
45. Bally, J., Devine, D. & Reipurth, B. 1996. ApJ 473, L48
46. Chernin, L.M. & Masson, C.R. 1995. ApJ 443, 181
47. van Dishoeck, E.F. & Black, J.H. 1988. ApJ 334, 771
48. Bence, S., Padman, R. & Isaak, K. 1997. in *Low Mass Star Formation from Infall to Outflow*, ed. F. Malbet & A. Castets, Poster Proc. IAU 182, 57
49. Wolf, G.A., Lada, C.J. & Bally, J. 1990. AJ 100, 1892
50. Zhou, S., Evans, N.J. & Wang, Y. 1996. ApJ 466, 296
51. Chandler, C.J., Terebey, S., Barsony, M., Moore, T.J. & Gautier, T.N. 1996. ApJ 471, 308
52. Ohashi, N., Hayashi, M., Kawabe, R. & Ishiguro, M. 1996. ApJ 466, 317.
53. Bachiller, R., Guilloteau, S., Dutrey, A., Planesas, P. & Martin-Pintado, J. 1995. A&A 299, 857
54. Shepherd, D. & Watson, A.M. 1997. in *Low Mass Star Formation from Infall to Outflow*, ed. F. Malbet & A. Castets, Poster Proc. IAU 182, 175
55. Bourke, T.L., Garay, G., Lehtinen, K.K., Kohnenkamp, I., Launhardt, R., Nyman, L.-A., May, J., Robinson, G. & Hyland, A.R. 1997. ApJ 476, 781
56. Bourke, T.L., Hyland, A.R., Robinson, G. & James, S.D. 1993. Proc. Astron. Soc. Australia 10, 236
57. Cabrit, S. & Bertout, C., 1990. ApJ 307, 313
58. Phillips, T.G., Huggins, P.J., Wannier, P.G. & Scoville, N.Z. 1979. ApJ 231, 720
59. Phillips, T.G., Knapp, G.R., Huggins, P.J., Werner, M.W., Wannier, P.G., Neugebauer, G. & Ennis, D. 1981. ApJ 245, 512
60. Loren, R.B., Plambeck, R.L., Davis, J.H. Snell, R.L. 1981. ApJ 245, 495
61. Narayanan, G. & Walker, C.K. 1996. ApJ 466, 844
62. Plambeck, R.L., Snell, R.L., & Loren, R.B. 1983. ApJ 266, 321
63. Wootten, A., Loren, R.B., Sandqvist, A., Friberg, P. & Hjalmarson, A. 1984. ApJ 279, 633
64. Liseau, R. *et al.* 1996. A&A 315, L181

THE MOLECULAR OUTFLOW AND CO BULLETS IN HH111

J. CERNICHARO
Instituto de Estructura de la Materia, Dpto Física Molecular
Serrano 123, 28006 Madrid, Spain

R. NERI
IRAM, Domaine Universitaire, 300 rue de la Piscine
38406 St Martin d'Hères, France

AND

BO REIPURTH
European Southern Observatory
Casilla 19001, Santiago 19, Chile

Abstract. We present high angular resolution observations of the molecular outflow associated with the optical jet and HH objects of the HH111 system. Interferometric observations in the CO J=2–1 and J=1–0 lines of the high velocity bullets associated with HH111 are presented for the first time. The molecular gas in these high velocity clumps has a moderate kinetic temperature and a mass of a few 10^{-4} $M_{\odot}$ per bullet. We favor the view that HH jets and CO bullets, which represent different manifestations of the same physical phenomena, are driving the low-velocity molecular outflow.

1. Introduction

The often deeply embedded stars which drive Herbig-Haro (HH) flows are, as a group, considered among the youngest stars known. The supersonic shocked jets which emanate from these stars testify to the current release of large amounts of energy in or near the nascent stars. In the first steps of star formation, orbital angular momentum of the collapsing material has to be evacuated so that matter continues to be accreted on to the protostar. This can happen by ejecting a fraction of the accreting material - which takes away most of the angular momentum - in a wind of magnetic

B. Reipurth and C. Bertout (eds.), Herbig–Haro Flows and the Birth of Low Mass Stars, 141–152.

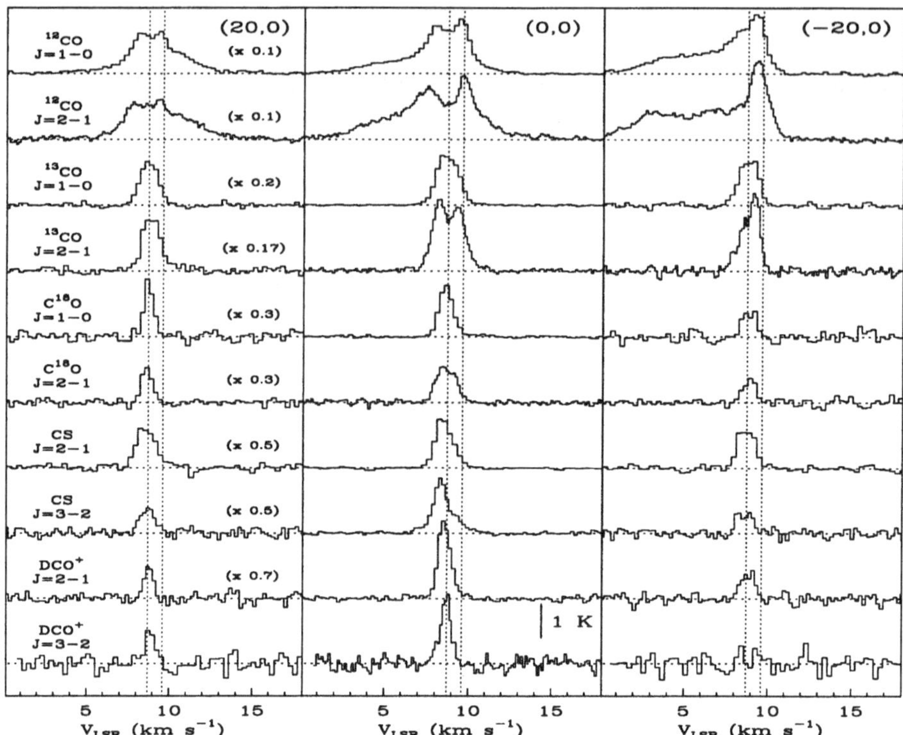

Figure 1. Molecular line emission of CO, ^{13}CO, C^{18}O, CS and DCO$^+$ toward selected positions along the optical jet of HH111 (indicated in arcsec at the top-right of each upper panel). The central position corresponds to the VLA source (α_{1950}=05^h 49^m 9.3^s, $\delta_{1950)}$=2° 47'48"). The high velocity gas is particularly prominent over the blue part of the jet.

origin (Shu *et al* 1987; Pelletier & Pudritz 1992; see also the papers by Shu and Pudritz in this volume) with velocities as high as 10^3 km s^{-1}. The wind interacts with the ambient gas producing shocks that are detected as optical jets and HH objects (Reipurth 1989; Reipurth and Cernicharo 1995; Ray *et al* 1996; see also the reviews by Reipurth and Bally in this volume).

The HH jets seem to drive the bipolar molecular outflows (Cernicharo & Reipurth 1996; hereafter CR96) detected around protostars, a second mass-loss driven phenomenon taking place during the earliest evolutionary stages of the star formation process (Bachiller, 1996). A particular nice example of optical and molecular jets can be found in HH111. This HH system, with its high velocity CO bullets and its highly collimated molecular outflow, is the ideal object to test the different models proposed to explain how molecular outflows are driven (CR96). We present in this paper a description of the molecular gas at low and extremely high velocity gas around the HH111 system.

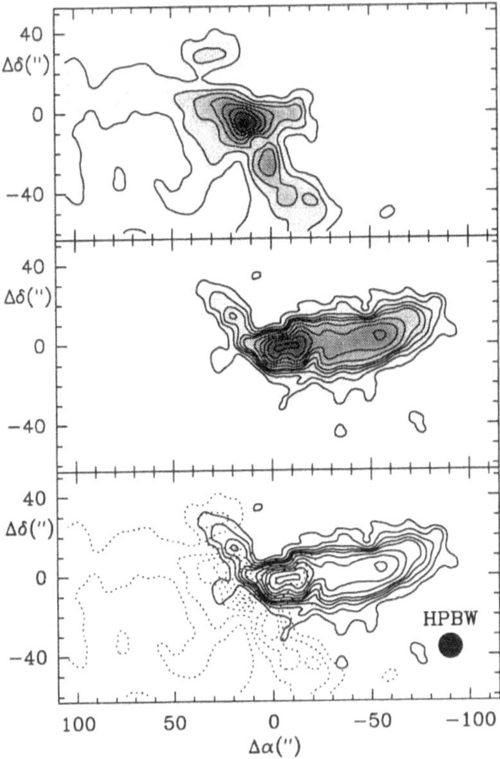

Figure 2. The two upper panels show integrated CO intensity for velocities between -5 and +7 kms^{-1} and between +10.5 and +15 kms^{-1}, respectively. The beam-width is 12" and the spacing of the data over the full region is 7". The third panel combines the blue and redshifted lobes. An optical picture of HH111 can be seen in Reipurth & Heathcote (this volume).

2. The HH111 system

The HH111 optical jet is located in the L1617 cloud in the Orion B cloud complex and was discovered by Reipurth (1989). The HH complex stretches over 367", corresponding, for a distance distance of 460 pc, to 0.82 pc. It consists of a bright highly collimated jet, a small faint counter-jet and at least 4 bow shocks, two on each side of the source. Additional very distant bow shocks on either side of the central source have also been detected by Reipurth, Bally & Devine (1997) (see also the review by Bally & Devine in this volume). The proper motions along the jet are large, between 300 and 600 kms^{-1} and directed away from the driving source, IRAS 0591+0247, which is a deeply embedded 25 $L_{\odot}$ source (Reipurth *et al*, 1992). The source has been detected at 2.0 and 3.6 cm at the VLA (Rodriguez & Reipurth, 1994). It has also been detected in the continuum at 3 mm by Stapelfeldt

& Scoville (1993) and Yang et al (1997). We have recently detected the continuum emission from the central source, together with extended emission from the surrounding core, at 1.3 mm (Cernicharo *et al* 1997). The observed flux ($\simeq$450 mJy) strongly suggests that the central source could be a class 0 source. The HH flow lies almost in the plane of the sky, with an inclination of only 10°. The energy source of HH 111 is embedded in a 30 $M_\odot$ cloud core. A molecular outflow has been detected along the axis of the HH complex (Reipurth & Olberg, 1991). The red (eastern) lobe has an extension of 0.7 pc, which is twice as large as the blue lobe. This difference probably originates because the blue lobe moves into almost empty space, whereas the red lobe is interacting with the cloud (Reipurth & Olberg (1991); CR96).

The ^{13}CO interferometric map of HH111 by Stapelfeldt and Scoville indicates a North-South velocity gradient in the direction of the core where the powering source of HH111 has been formed. The recent CS map of Yang et al (1997) taken with the Nobeyama interferometer also suggests such a velocity gradient. It has been interpreted by Yang et al (1997) as infalling gas around the exciting source of HH111. However, the fact that the emission from the CS lines show wings associated with the molecular outflow (see Figure 1) renders difficult the interpretation of the data. The presence of a well defined cavity in the ^{12}CO J=2-1 emission as observed with the 30-m IRAM telescope (CR96) and in the ^{12}CO J=1-0 interferometric maps of Cernicharo *et al* (1997; see below) and Nagar *et al* (1997) (see also Reipurth & Cernicharo 1995), indicates that the kinematics of the gas around the central source is affected by several different physical processes.

The emission of ^{12}CO, ^{13}CO and C^{18}O in the J=2-1 and J=1-0 lines, together with that of the J=2-1 and J=3-2 lines of CS and DCO$^+$, is shown in Figure 1 at selected positions along the optical jet of HH111 (CR96, and unpublished data). The line wings are detected even in the emission of the rotational transitions J=2-1 and J=3-2 of CS. The molecular hydrogen density in the core is as large as 10^5 cm^{-3}. The ^{12}CO J=2-1 and J=1-0 emission reaches its maximum at a velocity of 9.7 kms^{-1}. The gas at this velocity extends over the blue part of the jet. It has its emission maxima at positions -10 to -20 arcsec from the exciting source of the HH111 system (Fig.1) and has been interpreted by CR96 as the result of the interaction of the jet with the ambient gas, constituting the rear part of an expanding cavity around the optical jet. The gas kinematics is similar to that proposed by Martín-Pintado & Cernicharo (1987) for HH12 (see also Cernicharo 1991).

In the direction of the core some molecular lines show signs of self-absorption (CS J=2-1 and J=3-2, ^{13}CO J=2-1 and C^{18}O J=2-1). However, this double peaked line profile (see Figure 1) is probably due to the kine-

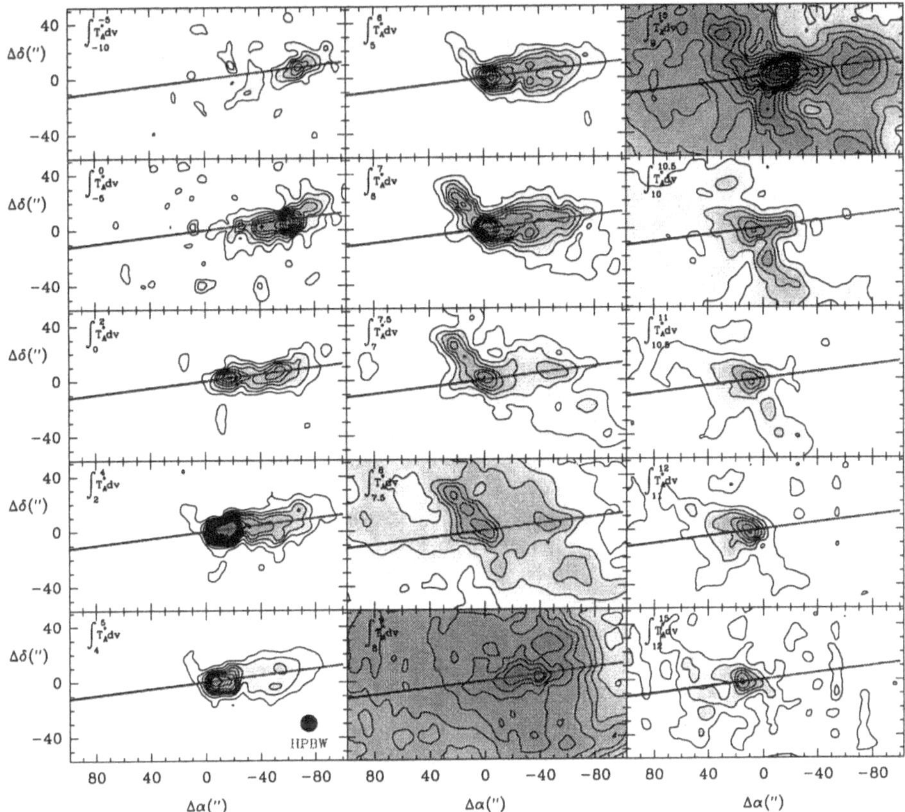

Figure 3. Integrated CO J=2-1 intensity for the velocity intervals indicated at the top-left of each panel. The thick line crossing the different panels indicates the axis of the HH flow. The first feature that appears in this figure is the long molecular jet coinciding with the HH flow, see panels corresponding to velocities lower than +2 kms^{-1}. At velocities between +5 and +7.5 kms^{-1} a second jet is clearly visible running from the VLA source towards the Northeast. This jet has its red counterpart running from the VLA source towards the Southwest at velocities between +10 and +12 kms^{-1}. The red counterpart of the main blue molecular jet also appears at these velocities but it is much less collimated and is oriented in the direction traced by the optical jet. The ambient gas appears at velocities between +7.5 and +10 kms^{-1} (adapted from data shown in CR96)

matics of the gas in the expanding cavity mentioned above.

3. The low velocity outflow

The short blue lobe found by Reipurth & Olberg (1991) is seen in the higher resolution CO J=2-1 data of CR96 to be a highly collimated molecular jet about 90 arcsec long with a collimation factor of about 9 (see Figures 2, 3, and 4 and CR96). The blue CO gas lobe ends at the bow shock P and the optical knots Q,R,S (see Reipurth 1989). The CO blue lobe is shorter than the HH lobe, in which the most distant bow shock V is located almost 150 arcsec from the VLA source. At these larger distances the quiescent gas of the cloud has a low visual extinction and volume density (CR96). Hence, the HH jet is emerging into a low density medium and escaping the cloud. CR96 suggest that the molecular outflow is formed by material dragged along from inside the globule rather than gas that has been swept up from the thin ambient medium and compressed. As shown in the infrared images of Gredel & Reipurth (1993,1994) there is a second bipolar jet, HH121, emanating from the VLA source. In Figure 3 the quadrupole flow is particularly apparent (see also Figure 2 and CR96). The presence of two outflows with nearly spatially coinciding powering sources makes it difficult to develop any simple model of the kinematics of the gas around the VLA source (see also above).

Although the blue and red integrated intensity maps shown in Figure 2 seem to indicate a rather smooth appearance for the HH111 molecular jet, the velocity interval maps of Figure 3 and the velocity-position diagram shown by CR96 reveal that there are several clumps in the outflow. In particular, the spatial distribution of the gas at velocities between +4 and +6 kms^{-1}, and at distances larger than 20″ West, seem to delineate the walls of an expanding cavity or tube surrounding the optical jet. The front and the rear of such a cavity could be traced by the gas at velocities between +7 and +10 kms^{-1} (rear) and between 0 and +4 kms^{-1} (front), see also Cabrit & Bertout (1990). The same spatial structure, i.e. an expanding tube or cavity, could also fit the kinematics of the gas near the VLA source as indicated in Figures 3 and 4.

The maps in Figure 2 and 3, and those presented in CR96, were taken with a full sampling in the J=2-1 line of ^{12}CO (HPBW=12"). Recently, Nagar et al (1997) have shown an interferometric map of the emission of the J=1-0 line of ^{12}CO. The angular resolution of their observations (5.5"x6") allow them to resolve the structure of the low velocity gas around the optical jet. They found, like CR96, a hollow tubular structure around the jet. CR96 also reported the presence of different clumps associated with the cavity. Some prelimary results of our PdBI CO J=1-0 interferometric maps were presented by Reipurth & Cernicharo (1995). Figure 4 shows our final interferometric map from a mosaic covering 5 fields along the blue part of the jet (Cernicharo *et al* 1997). The uv coverage, sensitivity and spatial resolution (3"x3.5") are better than those of Nagar *et al* (1997) and, in addition to the cavity reported by CR96 and Nagar *et al*, confirm the presence of several clumps in the emission of CO at low velocity. They are located along the optical jet and are associated with some of the HH objects. In figure 4 we can easily see the rear (v=9 to 10 kms^{-1}) and front (v$\simeq$3-5 kms^{-1}) parts of the cavity. The walls of the cavity are seen at velocities between 5.8 and 7.2 kms^{-1}, being particularly prominent at 6.4 kms^{-1}. The red counterpart of the HH111 molecular outflow appears, like in the single dish maps of Figure 3, much less collimated. In addition, the CO blue emission shows two clumps (bow shocks ?) at positions 25"-35" West of the VLA source (v=7.2-8.1 kms^{-1} in Figure 4) and 60" West of the VLA source (v=9.3-9.5 and 6.4 kms^{-1}). These clumps are also detected in the single dish maps of CR96 (see their Figure 3). The HH121 flow is also seen in Figure 4.

4. The high velocity gas along the Herbig-Haro jet

CR96 obtained a CO spectrum of the inner part of the HH jet by averaging all the ^{12}CO J=2-1 data between the VLA source and position 60" West along the jet. The spectrum consists of a broad and weak emission feature centered at -70 kms^{-1}. This weak emission could be associated to the neutral counterpart of the inner HH jet. **Why is this high velocity CO emission so weak ?** One possibility is that the presence of shocks in the HH jet has dissociated the CO in the jet stream. Alternatively, as suggested by CR96, any CO gas in the body of the jet may have so high a kinetic temperature that the opacity in the J=1-0 and J=2-1 line of CO drops, resulting in the CO emission being weak. Liseau *et al* (1996) have detected emission from high-J lines of CO in HH54. They have derived a kinetic temperature for the shocked gas in this HH object of 330 K. With such a high value for T_K, and taking into account the low CO column densities involved in the high velocity molecular jet, the expected intensities for the J=2-1 and J=1-0 lines of CO could be very low. By assuming a kinetic temperature of 100 K, a CO abundance of 10^{-4}, a distance of 460 pc, thermalization for the J=2-1 CO line, and the CO/H_2 conversion factor of Cernicharo & Guélin (1987), CR96 derived a mass for this high velocity neutral gas of 0.0007 $M_\odot$; the associated momentum is $\simeq$ 0.3 $M_\odot$ kms^{-1}.

5. The high velocity gas at the end of the Herbig-Haro jet

The gas associated with the low velocity molecular outflow in the HH111 system covers the region where the optical jet has been detected (see figure 2 and CR96) and ranges from -10 to $+15$ kms^{-1}. However, just beyond bow shock P and optical knots Q, R and S (see Reipurth 1989), where no more outflow emission could be expected, CR96 found strong CO emission at extremely high velocity (EHV).

Beyond the optical bow shock P, which is at position 75" from the VLA source, the CO emission appears at velocities between -40 and -80 kms^{-1} (see figure 5 of CR96). Assuming an angle of 10° to the plane of the sky for the HH 111 complex as derived from radial velocity and proper

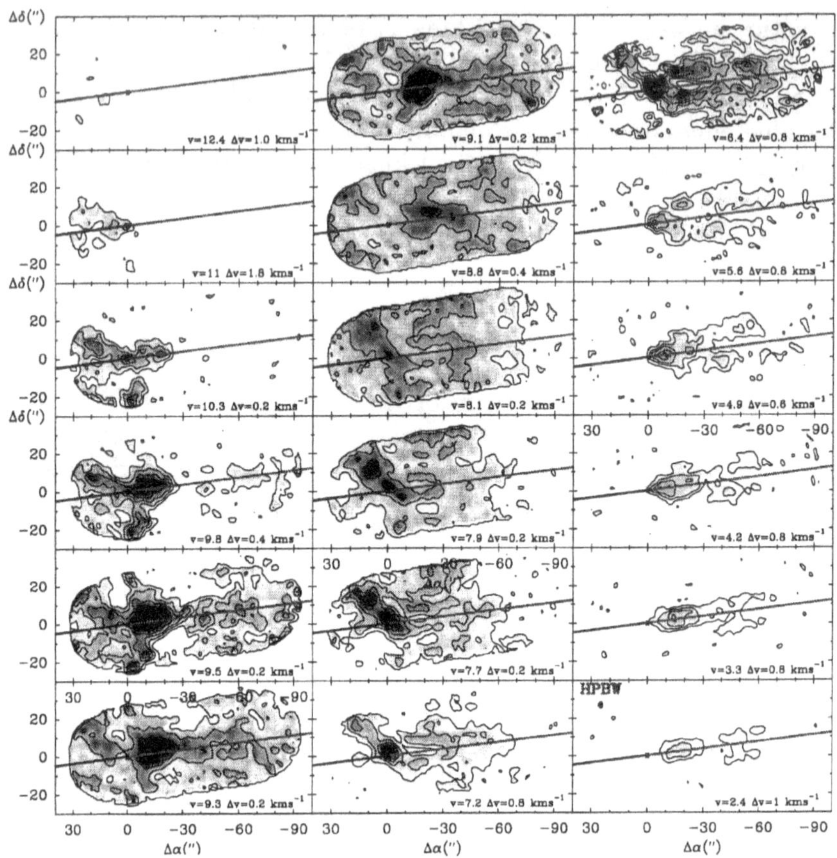

Figure 4. Interferometric maps of the J=1-0 CO emission taken with the PdBI. The maps result from a mosaic of 5 overlapping fields covering 120". The HPBW is 2.5"x3.5" and is indicated in the lower-right panel. Zero spacing data from the 30-m has been merged with the PdBI data. The thick line in each panel indicates the orientation of the HH111 optical jet.

motion measurements of the HH jet (Reipurth *et al*, 1992), the true space velocity of the high velocity CO emission is between 250 and 500 kms^{-1}. This gas is detected extending from 80" to 150" West of the VLA source. The first EHV CO emission appears just after knot P at a LSR velocity of -60 kms^{-1} which is identical to the LSR velocity of the ionized gas in this knot (Reipurth, 1989). The EHV CO emission consists of four bullets at positions 80" (knot P), 110" , 125" and 150" (knot V) West (see Figures 5 and 6 of CR96). Like for the low velocity gas at these positions, the velocity dispersion in the EHV bullets is very large as it could be expected from the interaction of the EHV gas with the ambient gas (see above). The EHV

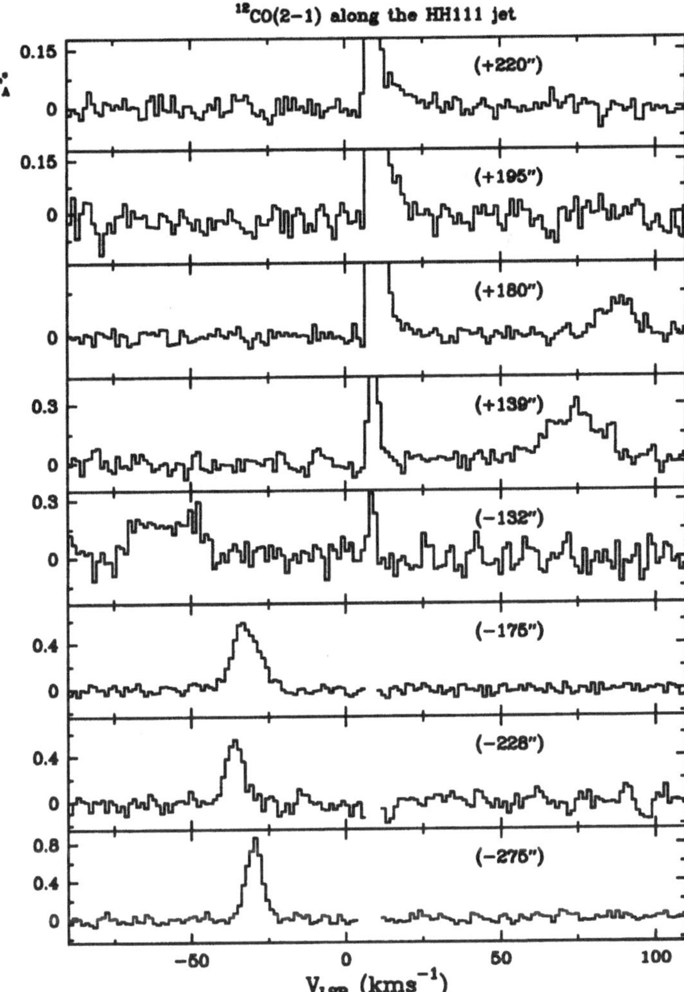

Figure 5. ^{12}CO J=2–1 spectra at selected positions along the axis of the HH111 jet. The CO bullets and the CO high velocity gas were discovered first in the blue part of the jet (Cernicharo & Reipurth, 1996). The redshifted high velocity gas (upper 4 panels) has also been observed at positions nearly symmetric to that of the blue high velocity CO gas, lower 3 panels (Cernicharo et al, 1997)

CO gas disappears beyond knot V. This optical knot has an LSR velocity of −70 kms^{-1} (Reipurth 1989), which is also identical to the velocity of the neutral gas at this position.

The red counterpart of the blue high velocity gas has been recently detected by Cernicharo *et al* (1997) -see Figure 5. The red high velocity gas appears at positions nearly symmetric to those of the blue components. It is barely resolved in the direction of the jet at position +139" (symmetric to knot V) and also shows a very broad velocity dispersion. Further away, at +180" from the VLA source, another red high velocity emission is detected. But this time, like for the high velocity blue bullets of CR96 (see below), the emission is unresolved by the 12" beam of the 30-m IRAM telescope and could be the red counterpart of Bullet 1 described below.

6. The high velocity gas further away : the CO bullets

Beyond the optical bow shock V, CR96 discovered three CO bullets (hereafter refered to as Bullet 1, 2 and 3, in order of increasing distance from the source) with separations of $\simeq$55". No optical emission is associated with these bullets. However, they are perfectly aligned along the direction defined by the optical jet. The Bullet 3 appears at 0.63 pc from the VLA source (position 280" West). The de-projected velocity of the three bullets is about 240 kms^{-1}. This velocity implies an episodic ejection of matter every $\simeq$500 years. The outermost bullet could be emitted from the central source some 2600 years ago. Note that the distance between Bullet 1 and that associated with the optical bow shock P is also about 55". The bright bow shock V, however, is located midway between knot P and Bullet 1. From the maps shown in Figure 5 of CR96 the bullets appear to be barely resolved with the 12" beam at 230 GHz. Recently, Cernicharo *et al* (1997) have made high angular resolution maps with the PdBI interferometer in the J=1–0 and J=2–1 lines of CO. Figure 6 shows the resulting spatial distribution of the CO J=1–0 emission. The J=2–1 line looks very similar and the emission of both lines is resolved by the resulting beams. The CO J=2–1/J=1–0 intensity ratio in the three bullets is about 1.5 (see also CR96). In the optically thin case and assuming equal excitation conditions for the J=1-0 and J=2-1 lines, this ratio could correspond to excitation temperatures of about 12-20 K. Assuming optically thin emission and T_{ex}=15 K, then the masses of the bullets are a few times 10^{-4} solar masses, i.e. similar to that of the bullets in other young low mass star forming regions (see Bachiller et al. 1990; Bachiller & Cernicharo 1990).

7. Molecular outflows and Herbig-Haro jets

Since HH jets are episodic, with dynamical ages of the order of 10^3 yr, the longer lasting molecular outflows would be the time averaged result of the momentum transferred from the many pulses of HH jets passing through their ambient medium.

Based on line ratios and shock models, Hartigan, Morse, Raymond (1995) derive a mass loss rate of 1.8×10^{-7} $M_\odot$ yr^{-1} for the HH 111 jet. CR96 derived a time averaged mass transport rate for the CO bullets of 4×10^{-7} $M_\odot$ yr^{-1} Given the uncertainties involved, the mass loss de-

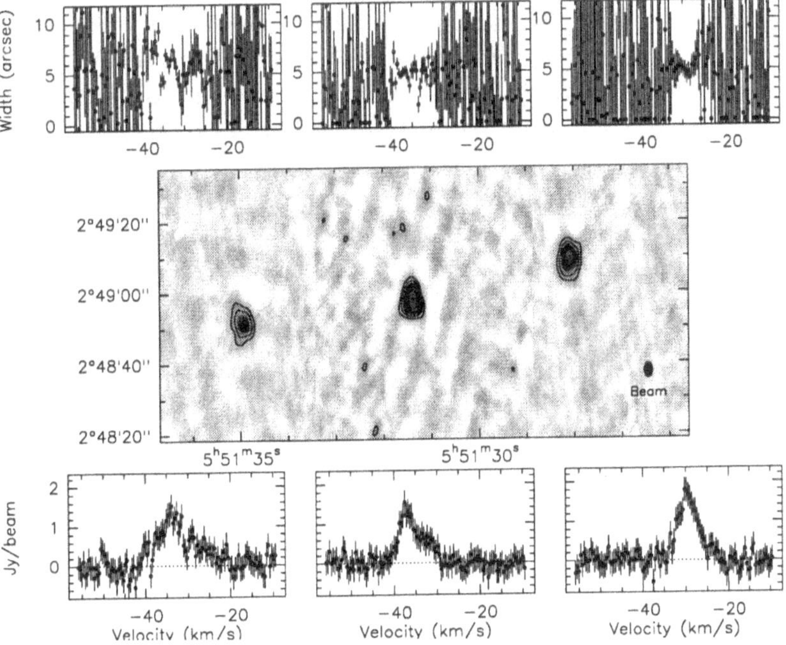

Figure 6. Maps of ^{12}CO J=1–0 taken with the IRAM PdBI intereferometer of the three bullets discovered by CR96. The central map show the integrated intensity distribution of the gas in each bullet. The three lower panels show the average CO J=1–0 spectrum in each bullet. The top panels show the spatial size for each velocity channel. Note that the three bullets are resolved with the PdBI interferometer and that they have a typical size of 5 arcsec. There is considerable velocity structure inside each CO bullet. From Cernicharo *et al* (1997)

rived from optical and radio observations are identical. It would seem that the similar mass transport rates in the HH jet and the CO bullets provide strong support for the view that these CO bullets are not made of ambient material that have been accelerated to these very high velocities, but are mostly composed of material ejected from the central engine (note also the excellent agreement in velocities at optical and radio wavelengths found by CR96). The mass loss rate quoted above corresponds to a momentum rate of 1.2×10^{-4} $M_{\odot}$ yr^{-1} kms^{-1}. The momentum in the molecular gas at intermediate velocities is about 1 $M_{\odot}$ kms^{-1}. If the jet momentum were transfered to the ambient gas with 100% efficiency, the jet would have to maintain on average the same level of activity for 8×10^3 yr. In this time span, the VLA source would have roughly 20 ejections of matter, if the frequency is one every 500 yr. The momentum in each bullet is typically 0.05 $M_{\odot}$ kms^{-1}. Consequently, this number of bullets could account for the momentum in the intermediate velocity gas if it was transfered with 100% efficiency. The fact that the outermost bullets have very high velocities tells us that the momentum transfer cannot be 100% efficient, and consequently the HH jet could be driving the observed molecular outflow only if the dynamical age of the complex is larger than we derive. As mentioned earlier, new observations by Reipurth, Bally & Devine (1997, see also the review by Bally & Devine in this volume) have shown that the HH 111 jet complex is much larger than hitherto thought, so it appears that the HH jet may indeed be able to drive the molecular outflow, even with a much lower momentum transfer efficiency.

Acknowledgements: J. Cernicharo gratefully acknowledges the support of the Spanish DGES for this research under grant number APC96-0168 and ESP96-2529-E.

References

Bachiller, R., 1996, Annu. Rev. Astron. Astrophys., 34, 111
Bachiller R., Cernicharo J., 1990, A&A, 239, 276
Bachiller R., Cernicharo J., MartinPintado J., Tafalla M., Lazareff B., 1990, A&A, 231, 174
Cernicharo, J., 1991, in *The Physics of Star Formation and Early Stellar Evolution*, , pag. 287, Eds. C. Lada & N.D.Kylafis, NATO ASI Series, Kluwer Academic Publishers
Cernicharo J., Guélin M., 1987, A&A, 176, 299
Cernicharo, J., Reipurth, B., 1996, ApJ, 460, L57 (CR96)
Cernicharo, J., Neri, R., Reipurth, B., 1997, in preparation
Gredel, R., Reipurth, B., 1993, ApJ, 407, L29
Gredel, R., Reipurth, B., 1994, A&A, 289, L19
Liseau, R., Ceccarelli, C., Larsson, B., et al, 1996, A&A, 315, L181
Martín-Pintado, J., Cernicharo, J., 1987, A&A, 176, L27
Nagar, N.M., Vogel, S.N., Stone, J.M., Ostriker E., 1997, ApJ, in press (see also the poster-book of this conference)
Pelletier G., Pudritz R.E., 1992, Astrophys. J., 394, 117
Ray, T.P., Mundt, R., Dyson J., Falle, S.A.E.G., Raga, A., 1996, ApJ, 468, L103-L106
Reipurth, B., Nature, 1989, 340, 42
Reipurth, B., Olberg, M., 1991, A&A, 246, 535
Reipurth, B., Raga, A.C., Heathcote, S., 1992, ApJ, 392, 145
Reipurth, B., Cernicharo, J., 1995, Rev. Mex. Astron. Astrophys., (Serie de Conferencias), 1, 43, 1995
Reipurth, B., Bally, J., Devine, D., 1997, AJ, *submitted*
Rodriguez, L.F., Reipurth, B., 1994, A&A, 281, 882
Shu F. Adams, F.C., Lizano, S., 1987, Annu. Rev. Astron. Astrophys., 25, 23
Stapelfeldt, K.R., Scoville, N.Z., 1993, ApJ, 408, 239
Yang J., Ohashi, N., Yan, J., Liu, C., Kaifu, N., Kimura, H., 1997, ApJ, 475, 683

SHOCK CHEMISTRY IN BIPOLAR MOLECULAR OUTFLOWS

R. BACHILLER & M. PÉREZ GUTIÉRREZ
IGN Observatorio Astronómico Nacional
Apartado 1143, E-28800 Alcalá de Henares, Spain

Abstract. Chemical studies have a great potential to study the structure and evolution of the bipolar molecular outflows driven by young stellar objects. In this paper, we discuss some very recent mm-wave studies of L 1157, a bipolar molecular outflow driven by a Class 0 protostar. These observations are very useful to illustrate the chemical alterations produced by a violent highly-collimated outflow. Different molecular lines are observed to trace different components of the gas. Some molecules are abundant in the quiescent medium but are not observed in the shock (e.g. C_3H_2, N_2H^+, $H^{13}CO^+$, DCO^+), whereas some otherwise rare molecules are very enhanced at the shocked region (e.g. SiO, CH_3OH, H_2CO, HCN, CN, SO, SO_2). In addition, we have observed strong gradients in the chemical composition across the outflow blue lobe. We briefly discuss the chemistry of the most important molecules, devoting special attention to the species which are thought to be abundant in interstellar ice mantles.

1. Introduction: chemistry and star-forming regions

One of the main contributions of radioastronomy to astrophysics is the discovery that there is a very rich chemistry operating in the interstellar medium (ISM). Up to 120 molecular species are presently known in the ISM, and this number increases at a rate of 4 new molecules identified per year in average. Interstellar chemistry involves processes in gas phase and on the surfaces of dust grains, and it is very different from the chemistry on Earth (including laboratory chemistry). Besides the extreme conditions found in the ISM, the time scales are so long that the chemical equilibrium can be attained in many situations. However, transient phenomena should

B. Reipurth and C. Bertout (eds.), Herbig–Haro Flows and the Birth of Low Mass Stars, 153–162.

also be present, in particular in regions subjected to the passage of shock waves.

Radio observations of different chemical species have been used for a long time to diagnose the physical conditions in star forming regions. Different species can probe different cloud regions, and help to provide a complete description of the whole cloud. Essentially for the same reasons, chemical studies have a great potential to understand the structure and evolution of the bipolar outflows driven by young stellar objects (YSOs).

2. Outflows from Class 0 protostars

Bipolar molecular outflows are observed in virtually all YSOs (e.g. Bachiller 1996), even around the youngest protostars already identified (the so-called "Class 0" objects, André et al. 1993). In fact, outflows from Class 0 sources are particularly violent (Bontemps et al. 1996): the mechanical power can reach 30 % of the bolometric luminosity of the central engine instead of the 0.001–0.01 % observed in most standard outflows. Such violent outflows accompanying the birth of a new star impact on the ambient cloud dispersing the surrounding molecular material, and determining the evolution of the dense core where the star was born. Outflows drive shock waves that compress and heat the gas triggering important chemical processes that do not operate in quiescent environments, producing dramatic effects on the structure and composition of the surrounding molecular cloud. In addition, outflows from Class 0 sources are highly collimated (Bachiller 1996), so the region where the shock occurs is well defined spatially, and well separated from the vicinity of the driving source which keeps the properties of the quiescent ambient gas. Processes such as endothermic reactions, sublimation of ices, and disruption of dust grain cores lead to a "shock-chemistry" which is radically different from that operating in the quiescent medium. Some of these processes are fast, and the cooling times of the shocked region are short, making the chemistry strongly time-dependent. Thus, the study of different chemical species can in principle be used as an indication of the state of evolution of the bipolar outflow and of its driving YSO.

Recent observations with high angular resolution and sensitivity are providing important insights in these transient phenomena. For instance, in the case of the outflow around IRAS 03282+3035, the SiO emission is only detected at the end of the outflow lobe, where there is the maximum interaction of the outflow with the ambient gas (Bachiller et al. 1994). In the case of L1448, the SiO emission is associated with the high-velocity molecular "bullets" at the main outflow axis (Bachiller et al. 1991), probably tracing shocks produced by the central jet (Dutrey et al. 1997). In both cases, no SiO emission is observed at ambient velocities.

To summarize, shocks have detectable, and even spectacular, manifestations in the chemical composition of the gas. Outflows from Class 0 protostars are of the highest interest to study the chemical alterations caused by shocks. The observations of different chemical species can provide the physical conditions in different regions of the shock. Due to its time-dependence, shock-chemistry has a great potential to be used as a clock to study the evolution of outflows.

3. L 1157-mm as an example of Class 0 source

In the following, we will highlight one of the most illustrative cases of outflows from Class 0 sources: the bipolar molecular outflow in the dark cloud L 1157. Due to its proximity (440 pc), high collimation, favorable orientation in the sky, and high column densities in the shocked region, this is one of the best objects to study the rich phenomenology associated with young bipolar outflows, in particular the effects of the shock-chemistry.

The highly collimated molecular outflow in L 1157 was discovered by Umemoto et al. (1992). The outflow is driven by L 1157-mm (IRAS 20386 +6751), a very cold source of about 11 $L_{\odot}$, embedded within a dense NH_3 core (Bachiller et al. 1993). The central source presents all the attributes of Class 0 sources, including a relatively strong flux density at millimeter wavelengths due to thermal dust emission. Such high flux density makes it possible to obtain detailed images of the protostellar vicinity with mm-wave interferometers. A recent high-angular resolution study by Gueth et al. (1997) shows that the continuum emission near λ 2.7 mm consists of a 35 mJy source of size $\leq 1''$, surrounded by spatially-extended low-level emission. The extended emission seems to arise from the heated edges of the cavity excavated by the high-velocity bipolar outflow. High-angular resolution images of the source in the J=1–0 line of CO, ^{13}CO, and $C^{18}O$ lead to a picture in which the $C^{18}O$ emission is associated with the protostellar condensation, whereas the ^{13}CO line emission mostly originates from a more extended envelope and the limb-brightened edges of the outflow. The CO line wings trace the bulk of the outflowing material. Possible gravitational infall motions are recognized by comparing the self-absorbed ^{13}CO profiles with the narrow $C^{18}O$ lines.

4. The CO bipolar outflow in L 1157

The bipolar outflow in L 1157 has a size of about 0.6 pc, and it is highly collimated but asymmetric (Umemoto et al. 1992). Figure 1 shows a CO 2$\rightarrow$1 integrated intensity map of the whole outflow which was recently obtained at the IRAM 30-m telescope. It appears that the southern blue-shifted lobe is more compact and brighter than the northern lobe, suggesting that

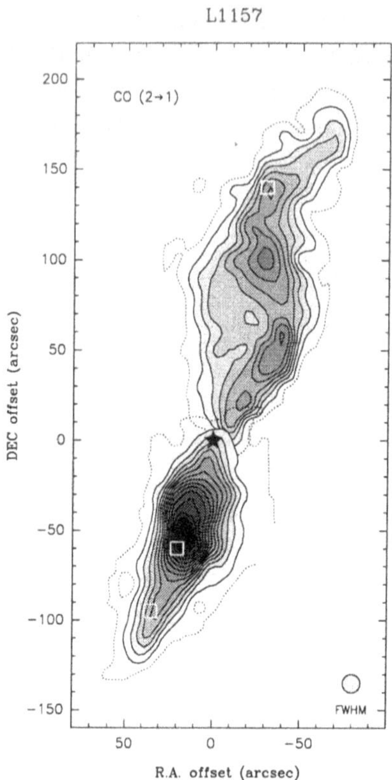

Figure 1. CO 2→1 emission from the L 1157 outflow integrated over velocity intervals from −20 to 3 km/s (blueshifted gas at the South of the driving source) and from 3 to 30 km/s (redshifted gas toward the North of the source). Position offsets are in arcsec with respect L 1157-mm, the outflow exciting source. First contour and step are 11 K km/s. The squares mark the positions in which we carried out a survey of molecular lines (see text).

the blue-shifted gas is shocking against some kind of obstacle, whereas the northern lobe is flowing more freely. The brightness of the blue-shifted lobe makes it one of the best places to study the structure of the shock. Gueth et al. (1996) obtained an interferometric high-resolution CO 1→0 image of the blueshifted lobe. Such image reveals at least two prominent limb-brightened cavities which seem to be created by the propagation of large bow shocks. Interestingly, the two cavities are not well aligned on a single line passing through the exciting source, L 1157-mm, as if the axis of the underlying jet had precessed from the first ejection event to the second one. A simple spatio-kinematic model in which the jet precesses on a narrow cone (of opening angle close to 6°) provides an accurate description of the observations.

Indeed the sharp increase of the CO brightness temperature towards the blueshifted lobe in L 1157 could be due to the gas heating produced by the shock. Multiline observations of ammonia, the best interstellar thermometer, confirm that this is the case, and provide a good estimate of the gas kinetic temperature in the shock, which results to be in the range 60–100 K (Bachiller et al. 1993). The bow shocks at the head of the cavities are particularly well observed in NH_3 (3,3) emission, and VLA images (Tafalla & Bachiller 1995, 1997) reveal the hot regions of the bow shock structures seen in the high-velocity CO. The ammonia abundance can be estimated in both the quiescent gas and in the high-velocity outflow, and it results that this abundance is enhanced by at least a factor of 10 in the shocked region.

5. A chemical survey of the L 1157 outflow

In addition to ammonia, some other molecular abundances have been long suspected to be enhanced as a result of the action of bipolar outflows on the surrounding gas. One of the most extreme examples is SiO, whose gas phase abundance is known to be enhanced by several orders of magnitude at the heads and along the axes of some molecular outflows (Bachiller et al. 1991; Martín-Pintado et al. 1992; McMullin et al. 1994). The case of L 1157 is not less spectacular, and a wealth of observations (Mikami et al. 1992, Zhang et al. 1995, Avery & Chiao 1996) indicate that SiO is enhanced by a factor of $\sim 10^6$ in the shocked region.

Spectral line surveys at mm wavelengths are recognized to be the best tool to explore the chemical composition of the vicinity of low-mass YSOs. Recent studies have considered two important Class 0 sources: NGC1333 IRAS4 (Blake et al. 1995) and IRAS16293 (Blake et al. 1994; van Dishoeck et al. 1995). Very recently, we have carried out a survey of the blueshifted lobe in L 1157 to study the chemical composition of the gas affected by the passage of the shock (Bachiller & Pérez Gutiérrez 1997). We observed about 50 different lines of 27 molecules (including rare isotopic species) toward the position of the central source (0,0) and several positions at the shocked regions (three of which are indicated as white squares in Fig. 1), in particular the two prominent NH_3 peaks of the blueshifted lobe: B1 at position (20″, –60″), and B2 at (35″, –95″). Some of the profiles are shown in Fig. 2. Indeed, the narrow line profiles observed toward the position of the source arise from cold quiescent gas, whereas toward the bow shock region the profiles are dominated by the broad lines associated with the shock.

One of the first results emerging from these observations is that some molecular lines such as those of C_3H_2, N_2H^+, $H^{13}CO^+$, and DCO^+ are

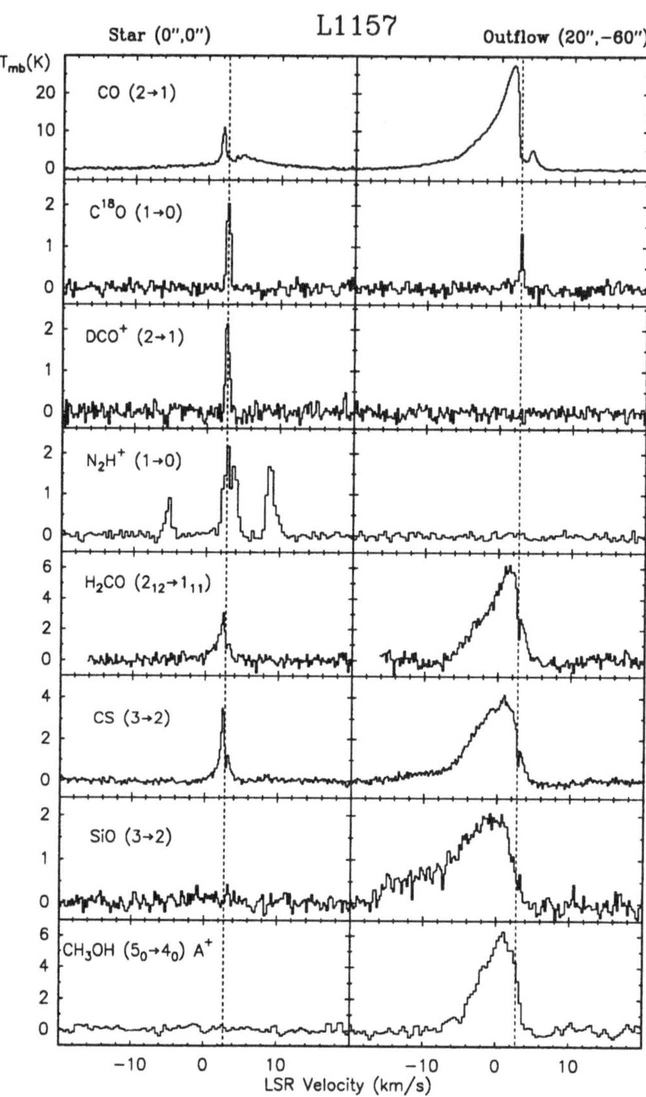

Figure 2. Molecular line profiles observed towards two representative positions in the L 1157 outflow: the (0,0) position on the central star and position (20″, -60″) on one of the bow shocks in the blueshifted lobe. The lines toward the star positions are narrow, since they arise from quiescent material associated with the dense core. The lines observed toward the outflow are very broad and blueshifted with respect to the ambient gas velocity (marked with a dashed line).

only observed toward the cold gas condensation around the exciting source, whereas some other molecules (such as SiO) only trace the warm gas in the shock. This result is important for studies of the structure of the central infalling condensation, since one can avoid the outflow contribution to the line profiles by observing the source vicinity in lines of C_3H_2, N_2H^+, $H^{13}CO^+$, and DCO^+. This does not mean that the outflowing gas is fully free of these molecules, but the column densities are insufficient to produce significant wings in the profiles.

All the other detected molecules (SiO, CH_3OH, H_2CO, HCO^+, HCN, HNC, CN, SO, SO_2, CS, OCS, H_2S, H_2CS) present emission in the outflow.

6. Spatial distribution of different molecular species in L 1157

To better characterize those gradients in the chemical composition of the gas, we have recently mapped the most interesting molecular lines with the IRAM 30-m telescope. Some preliminary maps are presented in Fig. 3. Note the large differences observed in the distribution of the different molecules. In particular the emission of N_2H^+ is well concentrated around the exciting object, allowing the study of the protostellar vicinity without any contamination from outflow emission.

A first interesting result from the mapping observations concerns the relative intensity of the lines at the two positions B1 and B2, in the blueshifted lobe. Both positions are almost equally strong in the CO isotopes, SiO and CH_3OH. However, the peak B1 is more prominent in H_2CO, HCO^+, and CN, whereas the B2 peak is somewhat brighter (in peak intensity) in most sulphur-bearing molecules like SO and SO_2. On the other hand, the behavior of CS is different from the other sulphuretted molecules, and seems more similar to SiO or H_2CO. All these striking differences cannot be explained as a mere result of possible differences in the excitation. Important chemical differences do exist between positions B1 and B2.

The abundances of the different molecules have been estimated from the ratios of the column densities to the CO column density, and by assuming a standard CO/H_2 ratio. The CO column densities in the quiescent gas and in the outflow have been derived from the $C^{18}O$ lines and the CO line wings, respectively, by assuming optically thin emission and thermalization. For the other molecules we used a Large-Velocity-Gradient (LVG) code, or we did similar assumptions as in the case of CO. The enhancement factors of the different species in the shock are thus estimated as the ratios of the abundances in the shock to those estimated in the quiescent gas. We next discuss some particular molecules, devoting special attention to the species that are thought to be abundant in the dust grain mantles, since such molecules are expected to be desorbed from grains by the shocks.

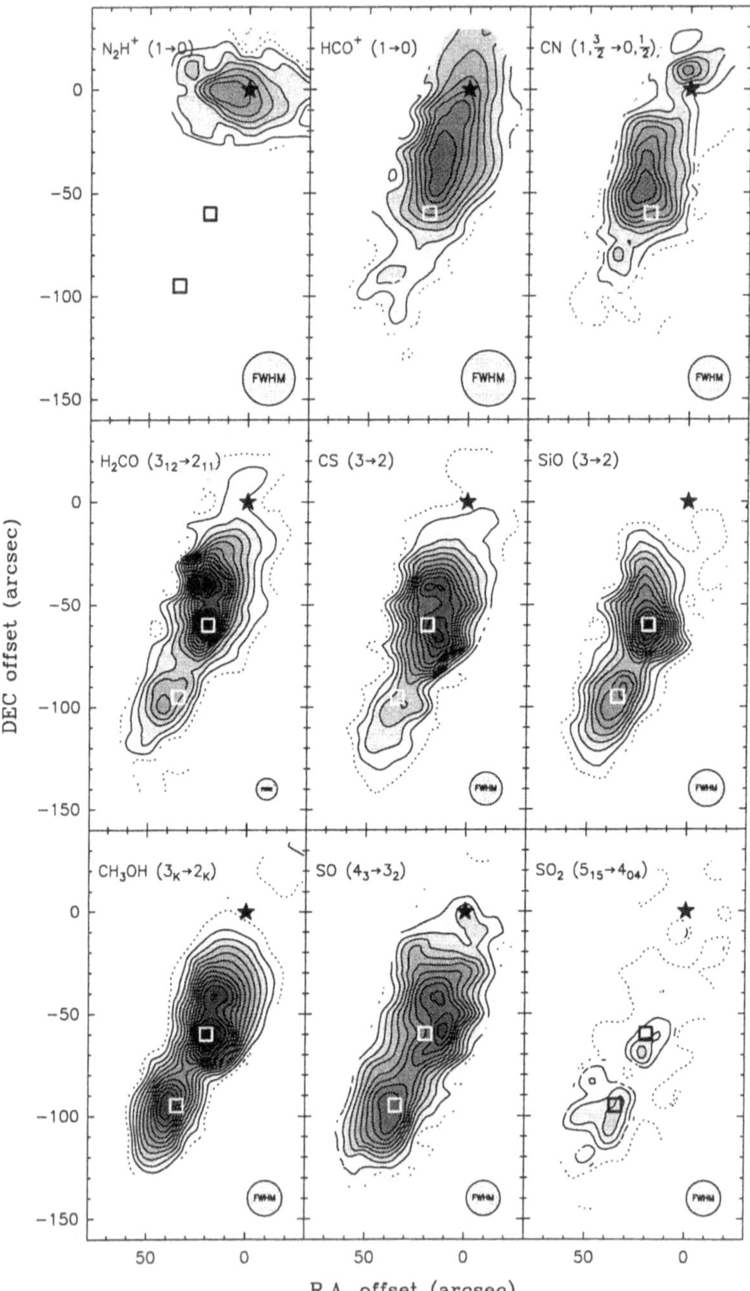

Figure 3. Map of integrated intensities of different molecular lines in the L 1157 outflow obtained at the IRAM 30-m telescope. The white squares mark the positions B1 and B2 where an initial molecular line survey was carried out (see text).

7. Chemical processes

The molecular species detected in solid phase in star forming regions include H_2O, CH_3OH, CO, CO_2, CH_4, and OCS. Methanol (CH_3OH), with an abundance of a few percent, is thought to be the most abundant molecule in ices, after water (Allamandola et al. 1992). Indeed, the millimeter-wave observations of L 1157 (Bachiller et al. 1995) show that the abundance of methanol is enhanced by a factor of 400 in the shock (see also Avery & Chiao 1996). Such methanol enhancements have been observed in some other outflows (Sandell et al. 1994; Bachiller et al. 1995; Garay et al. 1996).

There is increasing evidence that high abundances of methanol in gas phase are able to drive a chemistry completely different from the usual chemistry operating in dark clouds. In particular, laboratory experiments show that a significant fraction of the methanol released to gas phase could produce H_2CO in a relatively short time (e.g. Bernstein et al. 1995). Solid formaldehyde (H_2CO) has also been recently discovered in the vicinity of protostars (Schutte et al. 1996). Our observations of L 1157 indicate that H_2CO in enhanced by a factor close to 100 in the shocked region. The chemistry of formaldehyde is probably closely related to that of methanol in both solid and gas phases.

The chemistry of sulphur is also severely affected by shocks (e.g. Pineau des Forêts et al. 1993), and SO has been found to be enhanced in several molecular outflows (Martín-Pintado et al. 1992; Schmid-Burgk & Muders 1995; Chernin et al. 1994). The observations of L 1157 reveal significant enhancements of SO, SO_2, OCS, H_2S, and H_2CS in the shocked region, with the peculiarity, as mentioned above, that the B2 peak is richer in most sulphur-bearing molecules. It is interesting to note that the SO/H_2S and SO_2/H_2S ratios change significantly from position B1 to B2. H_2S is probably formed by grain-surface reactions (Duley et al. 1980) and could be directly released from the mantles. Once released into the gas phase, H_2S is expected to produce SO and SO_2 very quickly (in a few 10^3 yr) via reactions with H, OH, and O_2 (Millar 1993). Thus, the differences in the SO, SO_2, and H_2S abundances between B1 and B2 could be related to their different ages. It is indeed expected that B2 is older than B1, since B2 is more distant to the central star. Another attempt to use the sulphur chemistry as a rough clock in star forming regions has recently been done by Charnley (1997) to explain observations of the Orion Hot Core.

Molecular spectroscopy is also a powerful tool to investigate the physical conditions in the shocked gas. As discussed above, NH_3 provides good estimates of the kinetic temperature. Once the temperature is known, multiline studies of SiO and CH_3OH provide reliable estimates of the volume densities. In the case of the L 1157 bow shock, the density is found to be in excess of 10^6 cm^{-3} (Bachiller et al. 1993, 1995). The region emitting the radio lines is significantly colder ($T_K \sim 10^2$ K) than the gas traced by the near infrared lines of H_2 (in which $T_K \sim$ a few 10^3 K). However, the images of the L 1157 outflow in the H_2 lines near 2.1 μm (Davis & Eisloeffel 1995) also reveal the bow shock structures similar to those seen in CO and NH_3. It seems that sharp gradients in the kinetic temperature are present across the bow shocks.

References

Allamandola, L.J., Sandford, S.A., Tielens, A.G.G.M., Herbst T. 1992, ApJ 399, 134
Avery, L. W., Chiao, M. 1996, ApJ 463, 642
Bachiller, R. 1996, ARAA 34, 111
Bachiller, R., Liechti, S., Walmsley, C.M., Colomer, F. 1995, A&A 295, L51
Bachiller, R., Martín-Pintado, J., Fuente, A. 1991, A&A 243, L21
Bachiller, R., Martín-Pintado, J., Fuente, A. 1993, ApJ 417, L45
Bachiller, R., Pérez Gutiérrez, M. 1997, in preparation
Bernstein, M.P., Sandford, S.A., Allamandola, L.J., Chang, S., Scharberg, M.A. 1995, ApJ 454, 327
Blake, G.A., van Dishoeck, E.F., Jansen, D.J., Groesbeck, T.D., Mundy, L.G. 1994, ApJ 428, 680
Blake, G.A., Sandell, G., van Dishoeck E.F., Groesbeck, T.D., Mundy, L.G., Aspin, C. 1995, ApJ 441, 689
Bontemps, S., André, P., Terebey, S., Cabrit, S. 1996, A&A 311, 858
Charnley S.B. 1997, ApJ, in press
Chernin, L.M., Masson, C.R., Fuller, G. 1994, ApJ 436, 741
Davis, J.J., Eisloeffel, J. 1995, A&A 300, 851
Duley, W.W., Millar, T.J., Williams, D.A. 1980, MNRAS 192, 945
Dutrey, A., Guilloteau, S., Bachiller, R. 1997, A&A, in press
Garay, G., Koehnenkamp, I., Rodríguez, L.F. 1996, The ESO Messenger 84, 31
Gueth, F., Guilloteau, S., Bachiller, R. 1996, A&A 307, 891
Gueth, F., Guilloteau, S., Dutrey, A., Bachiller, R. 1997, A&A, in press
Martín-Pintado, J., Bachiller, R., Fuente, A. 1992, A&A 254, 315
McMullin, J.P., Mundy, L.G., Blake, G.A. 1994, ApJ 437, 305
Millar, T.J. 1993, in *Dust and Chemistry in Astronomy*, eds. T.J. Millar & D.A. Williams (Kluwer, Dordrecht)
Mikami, H., Tomofumi, U., Yamamoto, S., Saito, S. 1992, ApJ 392, L87
Pineau des Forêts, G., Roueff, E., Schilke, P., Flower, D. 1993, MNRAS 262, 915
Sandell, G., Knee, L.B.G., Aspin, C., Robson, I.E., Russell, A.P.G. 1994, A&A 285, L1
Schmid-Burgk, J., Muders, D. 1995, in *Stellar and Circumstellar Astrophysics*, eds. G. Wallerstein, A. Noriega-Crespo (ASP Conf. Ser., San Francisco)
Schutte, W.A., Gerakines, P.A., Geballe, T.R., van Dishoeck, E.F., Greenberg, J.M. 1996, A&A 309, 633
Tafalla, M., Bachiller, R. 1995, ApJ 443, L37
Tafalla, M., Bachiller, R. 1997, unpublished observations
Umemoto, T., Iwata, T., Fukui, Y., Mikami, H., Yamamoto, S., Kamaya, O., Hirano, N. 1992, ApJ 392, L83
van Dishoeck, E.F., Blake, G.A., Jansen, D.J., Groesbeck, T.D. 1995, ApJ 447, 760
Zhang, Q., Ho, P.T.P., Wright, M.C.H., Wilner, D.J. 1995, ApJ 451, L71

MODELS OF BIPOLAR MOLECULAR OUTFLOWS

SYLVIE CABRIT
DEMIRM, Observatoire de Paris
61 Avenue de l'Observatoire
75014 Paris, France

ALEX RAGA
Instituto de Astronomía, UNAM, Ap. 70-264
04510 México D. F., México

AND

FREDERIC GUETH
IRAM, 300 rue de la Piscine
38406 Saint Martin d'Hères, France

Abstract. This paper reviews various hydrodynamical models proposed for explaining bipolar molecular outflows from young stars. The different possibilities discussed are : wind-driven molecular shells, steady state filled flows, and jet-driven bow shocks. Particular emphasis is placed on comparisons between model predictions and observations of molecular flows. While it appears that jet-driven bow shocks are at the present time the most popular mechanism for explaining molecular outflows, other models have interesting characteristics that might be relevant for at least some objects.

1. Introduction

Bipolar molecular outflows from young stellar objects (YSOs) are a crucial diagnostic of mass-loss properties during the earliest stages of stellar evolution. They contain unique information about a number of fundamental parameters of the primary wind that created them, including not only its global energetics (Lada 1985; André, this volume), but also its velocity, intrinsic collimation, and time variability (in both velocity and direction), which could set strong constraints on the wind ejection mechanism. In ad-

B. Reipurth and C. Bertout (eds.), Herbig–Haro Flows and the Birth of Low Mass Stars, 163–180.

dition, molecular outflows appear to have a large impact on the chemistry and turbulence of the interstellar medium. They might also play a role in dispersing infalling envelopes around YSOs. Addressing these questions requires modelling the acceleration process and long term dynamical evolution of the flows.

A number of observational properties seem to characterize molecular outflows and are useful to constrain models. Recent summaries are given e.g. in Masson & Chernin (1993), Bachiller (1996), and Lada & Fich (1996). They include: (1) average line shape indicating a power-law mass distribution with velocity $m(v) \propto v^{-\gamma}$ with $\gamma \sim 1.3$–2.1 (Rodríguez et al. 1982; Masson & Chernin 1992; Stahler 1994), (2) moderate collimation with initial opening angles $\sim 20^\circ - 90^\circ$ and apparent length-to-width ratios of 3–10; (3) limb-brightened cavities at low-velocity, and higher collimation at high velocity; (4) apparent linear acceleration ('Hubble-law') in flows with well-separated outflow lobes; (5) rare occurence of overlapping blueshifted and redshifted emission within the same flow lobe, implying only a small amount of transverse motion (Cabrit et al. 1988; Meyers-Rice & Lada 1991); (6) in some cases, extremely high-velocity (EHV) features reaching 150 km/s, seen along the flow axis in CO, vibrationally excited H_2, and shock-enhanced species like SiO or CH_3OH (cf. Bachiller & Pérez Gutiérrez, this volume).

Proposed dynamical models can be separated into three broad classes: (1) molecular shells driven by wide-angle winds, (2) steady-state filled flows with internal stratification, (3) jet-driven molecular bow shocks. The present review will summarize their strong and weak points, and mention further developments that would help to distinguish between them.

2. Molecular shells driven by wide-angle winds

The first hypothesis for the generation of bipolar molecular outflows involved an initially isotropic wind blowing into a stratified molecular environment, resulting in the formation of bipolar swept-up shells (Snell et al. 1980). This was supported by the typically low momentum flux in ionized components of YSO jets, requiring an additional neutral component to drive molecular flows (Mundt et al. 1987; Lizano et al. 1988). While there is now clear evidence for collimated, high-velocity neutral gas that could solve the "jet momentum problem", wind-driven shells remain attractive to explain flows of large lateral extents and low collimation.

The expansion of wind-driven shells has been the subject of detailed analytical studies (e.g. Koo & McKee 1992 and refs. therein). In the case of molecular flows from YSOs, any mixing with ambient molecular gas should strongly enhance wind cooling, and the shells are often assumed to be in the highly radiative "momentum-driven" regime (even though the nominal criterion may not always hold; Frank & Mellema 1996). Hot wind bubbles seem excluded as they would predict too much sideways expansion (Masson & Chernin 1993). The next sections describe two analytical models of momentum-conserving wind-driven outflows, and recent numerical results.

2.1. RADIALLY EXPANDING MOMENTUM-DRIVEN SHELLS

Moriarty-Schieven & Snell (1988) and Meyers-Rice & Lada (1991) showed that properties of the L1551 and Mon R2 flows can be empirically reproduced by a thin shell with higher velocity toward the more distant parts. Shu et al. (1991) proposed a simple, self-similar dynamical model relying on radial expansion of the swept-up shell (i.e., assuming that shocked wind and ambient material instantaneously mix inside the shell, resulting in a net outward momentum in the same radial direction as the wind). Ambient density varies with radius as $1/r^2$, cancelling the decrease in wind ram pressure due to radial expansion, so that the shell velocity is constant in time (for fixed polar angle θ), but variations of density and wind thrust as a function of θ produce faster expansion along the axis and formation of bipolar lobes. At any given time t, $V(\theta) = t \times r(\theta)$ so that more distant parts of the shell also have a larger velocity, creating the observed 'Hubble law'. The cavity maintains the same shape and degree of collimation as its expands.

Masson & Chernin (1992) pointed out a potential problem of too much mass at high velocity. They found that the steep observed mass spectrum $m(v) \propto v^{-1.8}$ in molecular flows required loosely collimated winds and nearly evacuated polar regions, contrary to usual expectations. Modelling of the CO intensity in a cavity structure in the Mon R2 outflow yields the same result (Xie et al. 1993). However, Li & Shu (1996) noted that such thrust and density distributions are naturally encountered in models of MHD winds from the stellar/disk magnetosphere (Shu et al. 1995) and of self-similar critical magnetized cores. The global cavity shape and average $m(v)$ are then consistent with observations. Furthermore, the uncomfortable need for two distinct mass-loss components (a wind and a jet) can be avoided by interpreting the jet as a mere axial density enhancement in the MHD wind (see also Shu & Shang, this volume).

This simple model still meets some limitations. To reproduce the largest flows, e.g. Mon R2, the appropriate density law would have to hold over distances reaching a few pc. Another problem with radially expanding thin shells is the predicted velocity range at low inclination (cf. Cabrit & Bertout 1986;1990). As an illustration, in Figure 1 we show position-velocity (P-V) diagrams calculated for a wind thrust per solid angle $F_w \propto 1/\sin^2\theta$ and ambient density $\rho_e \propto \sin^2\theta(1 + 1.14\sin^2\theta)/r^2$ (close to the $n = 2$ solution

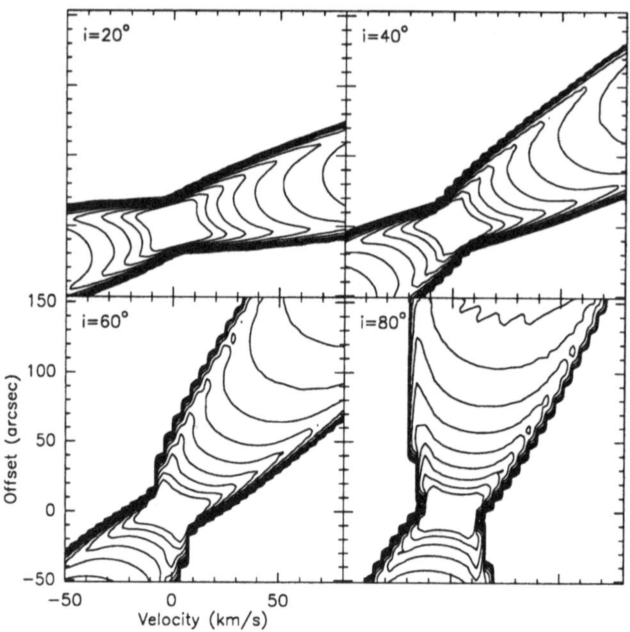

Figure 1. Synthetic position-velocity diagrams at various angles i to the line of sight for a radial wind-driven shell model similar to that of Li & Shu (1996), assuming isothermal optically thin CO emission. Contours increase by factors of 2.

from Li & Shu 1996). The two ridges trace the front and back side of the shell. At large angles to the line of sight $i > 70°$, the back side is projected close to zero velocity, and the diagram agrees with that seen in many flows such as NGC2264-G (Lada & Fich 1996) or HH111 (modelled as a radially wind-driven shell by Nagar et al. 1997). Unfortunately, it also resembles the diagram predicted for bow shocks (see Figure 3) so that there is no unambiguous interpretation.

At lower inclination $i < 70°$, most of the shell becomes blueshifted and has no more low-velocity emission (while it is still present in a bow shock, cf. Fig. 3). Such line profiles 'detached' from cloud emission are not frequently reported. Examples are Mon R2 (Meyers-Rice & Lada 1991) and TMC1 and TMC1A (Chandler et al. 1996). However, the latter may suffer from subtraction of low-velocity CO emission from the off-position. Shock-enhanced molecules, being free from confusion by ambient gas but affected by velocity-dependent chemistry, provide complementary information on the velocity range in shocked regions of outflows. In L1448, redshifted SiO emission extends all the way down to zero velocity, apparently more compatible with a bow shock (Guilloteau et al. 1992; Dutrey et al. 1997). Alternatively, wind-driven shells might have more low-velocity emission if the shell is thick and turbulent. More data on flows at low inclination would be useful to further assess this problem.

Finally, the shell in the Li & Shu model does not close back on axis. The polar regions would be dominated by a broad 'cap' of wind material, possibly shocking against itself to produce the multiple EHV features. It is not yet known whether it could reproduce the geometry and kinematics seen e.g. in H_2 images of young flows (e.g. McCaughrean et al. 1994).

2.2. STEADY WIND-BLOWN CAVITIES

Steady bipolar configurations may exist where the shocked wind is in pressure balance with the ambient medium. In the adiabatic case, transonic de Laval nozzles are established (Königl 1982; Smith et al. 1983). In the strongly cooling case, perhaps more relevant for YSOs, the wind refracts through an oblique shock and slides supersonically along the cavity walls, exerting a centrifugal pressure that contributes to balance the ambient pressure (e.g. Cantó 1980). Bipolar cavities with acute tips can be formed for a broad range of ambient pressure laws $P \propto r^{-n}$, with $0 \leq n < 2$ (Barral & Cantó 1981; Smith 1986). In particular, the Barral & Cantó model shows surprisingly good agreement with ovoid cavities observed at high resolution in CO and H_2 in the L1157 and HH46/47 outflows (Gueth et al. 1996; Eislöffel et al. 1994), including the presence of terminal shocks.

Steady state cavities would alleviate wind momentum requirements, as the flow could be older than its dynamical time (Dyson et al. 1988). However, they encounter several problems: (1) as material flows out parallel to the surface and refocusses toward the axis, projection of the front and back sides of the shell creates an X-shaped pattern in P-V diagrams which is not consistent with observations; (2) The presence of cavities at different orientations in the same flow, like in L1157, is not easily explained (it is in fact a problem for all wind-driven shells); (3) finally, shocked wind material that converged at the tip of the cavity should form a dense high-velocity jet (Cantó et al. 1988) which would presumably create a separate, non-stationary, jet-driven outflow.

2.3. NUMERICAL SIMULATIONS

The above wind-driven shell models easily reproduce outflow *shapes*, but the predicted kinematics are not fully satisfying. This may be due in part to the simplifying extreme assumptions: full mixing in the radially expanding model, full steady-state, instant cooling, infinitely thin shell.

Detailed numerical simulations have only been performed quite recently, and they show that the actual situation could be more complex than anticipated. Adiabatic simulations with an ambient stratification similar to that of Li & Shu (1996), although with density decreasing as $\rho \propto r^{-1.5}$, reveal a mixture of bubble and jet behavior, as the hot wind is focussed through a non-spherical inner shock (Frank & Mellema 1996; see also this volume). The strongly cooling case produces strong convergence toward the axis, as in the steady-state solution of Barral & Cantó (1981), but also complex vortical motions at the wind/ambient interface (Mellema & Frank 1995). Such studies must be pursued to allow realistic comparisons between the dynamics of wind-driven cavities and molecular outflow observations.

3. Steady-state filled flows

Low-velocity 'cavities' could readily be obtained at all view angles if the outflow is not restricted to a thin shell but fills the lobe volume and has a velocity decreasing toward the flow edges (Levreault 1988; Cabrit & Bertout 1990). Several dynamical models predict such an internal structure.

3.1. TURBULENT MIXING-LAYERS

Kelvin-Helmholtz instabilities along the jet/environment boundary lead to the formation of a turbulent viscous mixing layer that grows both into the environment and into the jet. Eventually the whole flow becomes turbulent (recent numerical simulations for cooling jets are presented by e.g. Massaglia et al. and Stone, this volume, and Downes, 1997). The possibility that molecular outflows are turbulent layers entrained around YSO jets has been investigated in several analytical studies.

Stahler (1994) considered a fully turbulent flow of arbitrary structure. He pointed out that the mass distribution with velocity $m \propto v^{-\gamma}$ could result from a velocity cross-section of the form $v \propto r^{-2/(\gamma-1)}$, while the observed acceleration would be an artifact of growing flow thickness at larger distances; however there was no self-consistent dynamical justification for the transverse profiles and flow sizes.

Cantó & Raga (1991) derived a "mixing-length" parameterization of the turbulent viscosity, calibrated with laboratory experiments for jet Mach numbers M_j =1–20, that allowed them to derive the growth of the mixing-layer width with distance. However, for the isothermal case relevant for radiative YSO jets, the layer is too narrow, producing outflows that have length-to-width ratios q of $\sim$ 20, approximately 5–10 times larger than the ones observed in molecular outflows. Different parameterizations of the turbulence were proposed by Lizano & Giovanardi (1995) and Dyson et al. (1995), but as no calibrations were carried out for these parameterizations, it is not possible to use them for determining the flow opening angles.

Raga et al. (1993) showed that turbulent mixing-layers could be broadened by sideways ejection of overpressured jet material at velocity discontinuities along the jet. The overall properties agree much better with molecular flows. However, q would still be very high, ≥ 10, compatible only with the most collimated outflows known to date. This problem might be alleviated by the presence of a strong pressure gradient in the surrounding environment (Richer et al. 1992). Another drawback is the predicted velocity gradient: after providing strong acceleration where the first working surface forms, the ejected jet momentum is progressively shared among larger and larger amounts of ambient gas, and the mean velocity in the turbulent envelope rapidly decreases away from the star. In contrast, most molecular outflows show velocities linearly increasing with distance from the star, followed by sudden deceleration (e.g. L1448, Bachiller et al. 1990).

Therefore, turbulent entrainment as currently modelled does not explain all properties of molecular outflows. It is not clear whether models with less drastic simplifying assumptions might show better agreement with observations. Nevertheless, chemical and thermal balance calculations show that molecules could survive in jet mixing layers, and produce collimated emission at high velocities (Taylor & Raga 1995). H_2 images suggest this process is active, at least close to the jet injection point (e.g. Gredel & Reipurth 1993; Noriega-Crespo, this volume). It would also be an attractive explanation for the EHV CO jets observed in some flows such as Orion B and HH211 (Richer et al. 1992; Guilloteau et al., this volume), the alternative explanation being that the jet is intrinsically molecular.

3.2. GLOBAL INFALL-OUTFLOW PATTERN

A new approach was recently proposed by Henriksen & Valls-Gabaud (1994) and fully developed by Fiege & Henriksen (1996a): the bipolar outflow is not entrained by an underlying wind or jet, but it traces infalling matter that has been deflected toward the poles in a central torus of high MHD pressure, and accelerated above escape speeds by local heating. The magnetic/stream lines are quadrupolar. The self-similar outflow solution has decreasing speed and increasing mass at larger angles from the axis up to $\theta_{max} \sim 50°$, nicely reproducing the observed conical cavities, increasing collimation at higher velocity, decreasing line wings, and apparent acceleration with distance (Fiege & Henriksen 1996b). However, the model has several limitations: radiative acceleration requires unrealistically high central luminosities $\sim 10^5 L_{\odot}$, suggesting that Poynting flux driving or local dissipative heating should be included. The fastest axial regions are not part of the self-similar solution, and will require special treatment.

Independently of further refinements, general model features may be accessible to observational tests. Sub-mm and mid-IR polarimetry could set constraints on a global quadrupolar magnetic configuration (e.g. Holland et al. 1996). More important, mass-flux through the outflow should be entirely provided by infall and directly related to the mass in the thick torus. Relevant data are becoming available for such a comparison (e.g. Fuller et al. 1995; Bontemps et al. 1996).

4. Jet-driven bow shocks

Recent observational findings have made this model increasingly attractive, in particular (1) the low ionized fraction in optical jets (e.g., Hartigan et al. 1994), and the highly collimated EHV atomic and molecular features in some flows (e.g. Bachiller et al. 1990; Masson et al. 1990; Richer et al. 1992), both indicate that jets may have enough momentum for driving outflows; (2) the close spatial association of curved shocks seen in H_2 and shock-enhanced species (SiO, NH_3) with local acceleration in the swept-up CO (e.g. Davis & Eislöffel 1995), and with the tip of CO cavities in interferometric maps (e.g. L1157, Gueth et al. 1996), is strongly suggestive of jet working surfaces.

4.1. GENERAL PROPERTIES OF BOW SHOCKS

Bow Shocks have been extensively studied to model optical properties of Herbig-Haro objects (see Raga 1995, and references therein). The structure of a bow shock at the head of a jet is illustrated in Figure 2. Across the jet, a working surface is formed where jet and ambient gas are arrested in a two-shock layer. High pressure gas between the shocks is ejected sideways out of the jet beam; it then interacts with unperturbed ambient gas (flowing back at $-V_{BS}$ in the frame of the shock) through a broader shock surface (the bow shock); the swept-up ambient gas is shocked at an oblique angle and streams back parallel to the bow surface in a dense layer. Some postshock gas reexpands in the low-pressure region behind the bow shock, creating a low-density cocoon with mixed ambient and jet material. Bow Shocks are appealing, as they readily explain several observed properties of molecular flows (see e.g. the analytical studies of Masson & Chernin 1993; Raga & Cabrit 1993):

(1) they naturally create an apparent acceleration: As one moves from the broader wings to the narrow apex of the bow shock the shock becomes less oblique, and the net forward velocity increases (up to V_{BS} at the bow shock apex). (2) There is more swept-up mass at low velocities: the intercepted mass flux is determined by the bow shock cross-section, which steadily grows in the bow wings while velocity decreases.

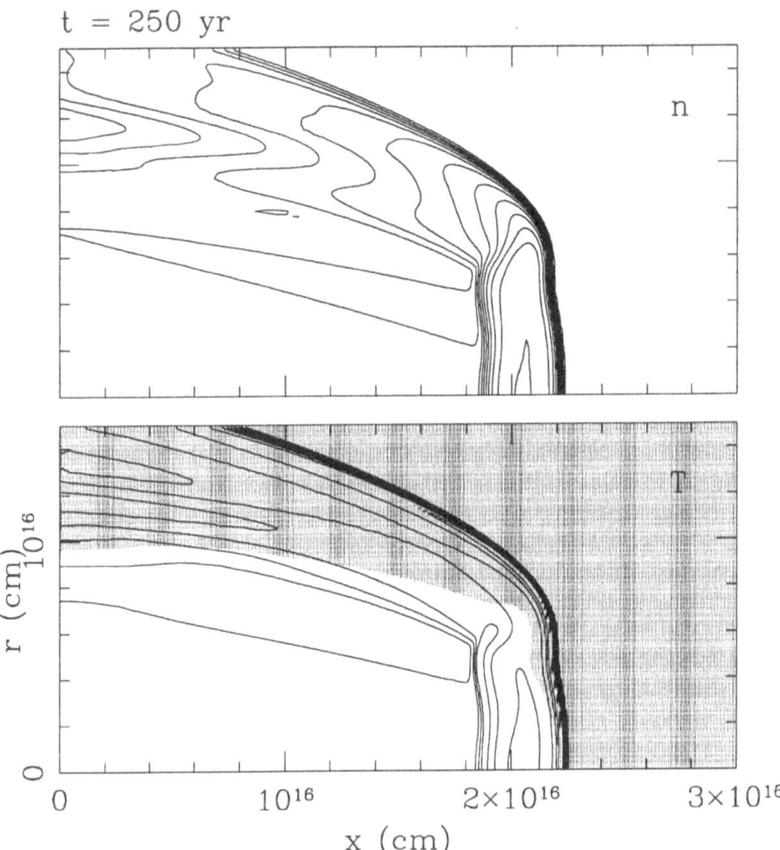

Figure 2. High-resolution 2D numerical simulation of the bow shock at the head of a jet. Top panel shows density and bottom panel temperature. Contours increase by factors of 2. The shaded area shows the locus of ambient gas. From Raga & Cantó (1997)

Recently, efforts have been made to simulate jet bow shocks with conditions more appropriate for molecular outflows (SPH with atomic cooling: Chernin et al. 1994; 2D high resolution grid with H_2 destruction and cooling: Raga et al. 1995; 2D slab jet with H_2: Downes 1996). An important result is that H_2 and CO molecules can survive dissociation in the bow shock (cf. Smith et al. 1991; Wolfire & Königl 1991), producing molecular emission at velocities up to 30-60 km/s (e.g. Raga al. 1995). The most complete exploration so far is that of Suttner et al. (1997), who present simulations of the whole flow (from 1D to 3D) including H_2 chemistry and cooling, and consider high densities $\sim 10^4 - 10^5$ cm^{-3} as well as the effect of jet pulsing and precession. The H_2 emission maps, as well as the CO line profile shape, P-V diagram, and channel maps show remarkable agreement with observations (Smith et al. 1997).

Despite their successful aspects, Lada & Fich (1996) argued that bow shocks may meet two problems: they would be expected to produce too much sideways expansion, and lobes that are too narrow. We discuss solutions to these two important issues in the following sections, and in particular investigate the effects of jet time variability on the flow structure.

4.2. SIDEWAYS MOTIONS IN BOW SHOCKS

A bow shock pushes ambient material into a motion normal to the surface of the shock. In the oblique bow shock wings, the postshock velocity vector makes a large angle to the jet axis, producing overlapping blueshifted and redshifted emission over a broad range of view angles. While this description can fit channel maps of flows almost in the plane of the sky, such as RNO 43 (Padman & Richer 1994), it is not consistent with the rarity of significant blue/red overlap among molecular flows (e.g. Cabrit & Bertout 1986).

The situation is not so clear cut in hydrodynamical simulations: P-V diagrams calculated by Raga et al. (1995) do show substantial overlap of blueshifted and redshifted emission at an inclination of 45°. On the other hand, the lower resolution 3D simulations of Chernin et al. (1994) and Suttner et al. (1997) show a much better alignment of the velocity of emitting material with the symmetry axis of the flow. Smith et al. (1997) show (through a simple, analytic model) that the higher forward velocities of emitting gas could arise from turbulent mixing inside the narrow, cold swept-up layer. Emission then includes contribution of material that went through the bow shock closer to the jet head, and has a larger forward momentum. The result is somewhat equivalent to the intuitive global momentum conservation argument used by Masson & Chernin (1993). Higher forward velocities could also be obtained through mixing of postshock ambient gas with jet gas, a process investigated under simplifying assumptions by Raga, Cabrit & Cantó (1995).

Interesting in this context is the stellar wind bow shock model of Wilkin (1996; see also Wilkin et al., this volume) which assumes full mixing between postshock ambient *and* wind gas, and solves exactly for the bow shock shape and velocity. This flow is qualitatively similar to the flow found around an internal working surface (see Raga & Cabrit 1993). In the model of Wilkin (1996) the layer velocity depends on an extra parameter $\alpha = V_{BS}/V_w$ where V_{BS} is the bow shock velocity and V_w the wind velocity (i.e. in the present context the velocity of the material ejected from the working surface, which is approximated as an isotropic wind source). The final opening angle of ve-

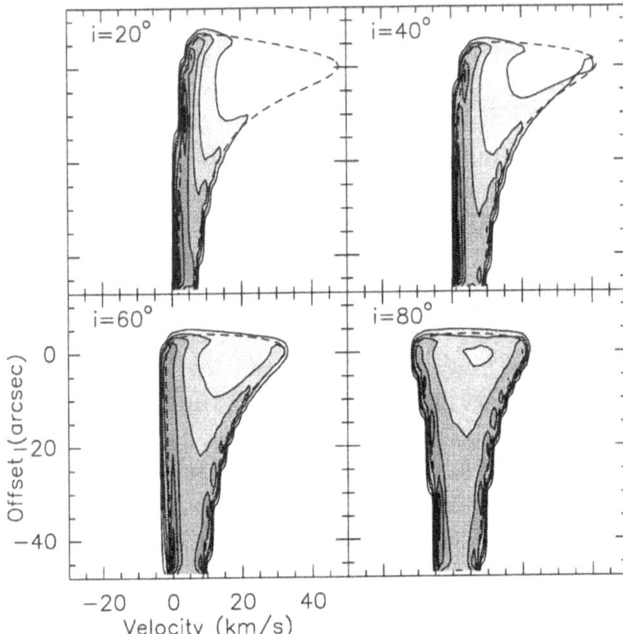

Figure 3. Synthetic position-velocity diagrams at various inclinations i to the line of sight for the analytical bow shock model of Wilkin (1996), assuming similar bow shock and wind velocities $V_{BS} = V_w = 50$ km/s, and isothermal optically thin CO emission.

locity vectors in the bow shock wings θ_{max} decreases for larger α (Wilkin et al. 1997). P-V diagrams for $\alpha = 1$ ($\theta_{max} \sim 40°$) are shown in Figure 3. Negative velocities are essentially absent until $i \geq 50°$. However, low-velocity emission remains present at all view angles, in contrast with P-V diagrams of wind-driven shells in Fig. 1.

One must mention another situation leading to more forward-directed velocities, namely when the bow shock propagates into material already moving away from the source, as expected for internal working surfaces produced in a jet from a variable source (cf. Sect. 4.4.1). In this case, the post-bow shock velocity vector is no longer perpendicular to the surface, and is oriented closer to the outflow axis (see e. g. Hartigan et al. 1987).

4.3. WIDTH AND COLLIMATION OF JET-DRIVEN BOW SHOCKS

Bow shocks may produce conical flow lobes toward the star, both because of the initial transient after jet injection (while the bow shock grows to its full size), and because of considerably reduced expansion in the much denser environment around the protostar (Masson & Chernin 1993; Raga

& Cabrit 1993; cf. Fig. 4). On the other hand, the low collimation and large flow radii do not seem as straightforward to explain.

Hydrodynamical simulations of radiative bow shocks yield maximum lobe radii of $R_c \sim 3 - 5r_j$ (where r_j is the jet radius) for lobe lengths of $L \sim 25 - 80r_j$, hence $q \sim 4 - 8$. This agrees well with the collimation of outflows observed at sufficiently high resolution. However the inferred jet radii would be very large, e.g. $r_j \sim R_c/4 \sim 7 \times 10^{16}$cm in NGC2264G (Lada & Fich 1996). Furthermore, as bow shocks grow more slowly in the transverse direction than in the forward one, collimation should increase over time.

In particular, Masson & Chernin (1993) and Wilkin (1996) noted that for strong cooling, material in the bow shock wings can be described as a shell undergoing a momentum-conserving expansion. In a uniform medium, a cubic asymptotic shape is obtained : $R_c^3 \sim 3\pi L R_o^2$ where $L = tV_{BS}$ is the distance from the apex, and R_o is the bow shock characteristic radius — for the head of a jet, the above hydrodynamical simulations would suggest $R_o \sim 0.4r_j$. Hence collimation should increase as $L^{2/3}$ and width as $L^{1/3}$ only. Several possible solutions may be invoked:

1- Some molecular lobes may trace the 'truncated' inner regions of a much longer jet. HH111 is a possible example: while the CO cavity is only 80″ long, the full jet stretches over $L \sim 367''$. For the measured jet radius $r_j \sim 1.5''$, the above cubic approximation would predict $R_c \sim 11''$, as observed (Cernicharo & Reipurth 1996). The increasing number of parsec-scale Herbig-Haro flows, and of faint molecular lobes beyond previously mapped areas (see contributions by Bally & Devine and by Padman et al., this volume) further supports this possibility. However, broader flows with $R_c \sim 2.5 \times 10^{17}$cm (NGC2264G, L1551) would require very long jets (>7 pc!) if $r_j \leq 2 \times 10^{16}$cm.

2- Time variability in jet velocity could produce new, shorter cavities inside the old one, and possibly increase the outflow width by sideways ejection of material from the jet (Raga & Cabrit 1993).

3- Jets may have an 'effective radius' that grows with distance roughly as $L/20 - L/40$, maintaining a low bow shock collimation $q \sim 3 - 5$. This could happen through jet wandering in direction (Masson & Chernin 1993), and/or through growth of a viscous mixing-layer (cf. Sect. 3.1).

Observations indeed bring growing evidence for time variability in flows, and we explore its effect in more detail in the next section.

4.4. EFFECT OF JET TIME VARIABILITY

Herbig-Haro jets often contain multiple bow shocks; similarly, molecular flows often present successive EHV peaks roughly equally spaced along the flow axis, with point symmetry between blue and red lobes suggesting that they originate from intrinsic variability at the source (e.g. L1448, Bachiller et al. 1990; RNO 43, Bence et al. 1996). Recent interferometric imaging confirms the presence of multiple cavities, sometimes with differing orientations (Gueth et al. 1996; Ladd & Hodapp 1997). Hence, time variability in velocity and/or direction is an important ingredient in the formation of molecular flows.

4.4.1. *Velocity Variability*

The effect of velocity variability of supersonic amplitude in YSO jets has been the object of numerous studies (see e.g. Raga 1993 for a review). As faster jet material catches up with slower, previously ejected gas, a two-shock structure develops (internal working surface), which propagates along the jet and drives a secondary bow shock into the low-density cocoon. Raga & Cabrit (1993) showed analytically that its transverse size could be substantially larger than the jet radius and suggested that it could explain the low collimation of molecular flows. In particular, if the cocoon density drops with distance as $1/d^2$, the secondary bow shock radius would grow linearly with d, reproducing the observed conical outflow shapes. At the same time, secondary bow shocks would produce local re-acceleration at positions of EHV features (which would trace the internal working surfaces in the jet). However, their simple analytical model did not include the confining effect of the main shell created by the leading jet head.

A 2D numerical simulation of a variable velocity jet in a molecular medium of decreasing density is shown in Fig. 4. H_2 destruction and cooling are included (cf. Raga et al. 1995). While local forward reacceleration of the cocoon clearly occurs at each internal working surface, the sideways expansion of secondary bow shocks is indeed severely limited by the dense layer created by the leading jet head, and at this stage the flow is not much broader than without time variability (a similar result is found for a constant ambient density by Suttner et al. 1997). However, the decreasing environmental density gradient broadens the overall cavity at its end compared with the case of a uniform ambient medium (see also Gouveia dal Pino & Birkinshaw 1996). The bow shock shape and its ratio to jet size are in striking agreement with e.g. the HH 211 flow (Guilloteau et al., this volume). Furthermore, the full opening angle at the origin increases with time, reaching 20° in the last panel of Fig. 4. Therefore, it is not excluded that the bow shock could eventually reach the full opening angles of 30–45° typically observed in flows (Bachiller et al. 1995; Chandler et al. 1996 and references therein). In particular, the broadening effect of internal working surfaces may reveal itself on longer time scales than considered in current simulations, and needs further investigation.

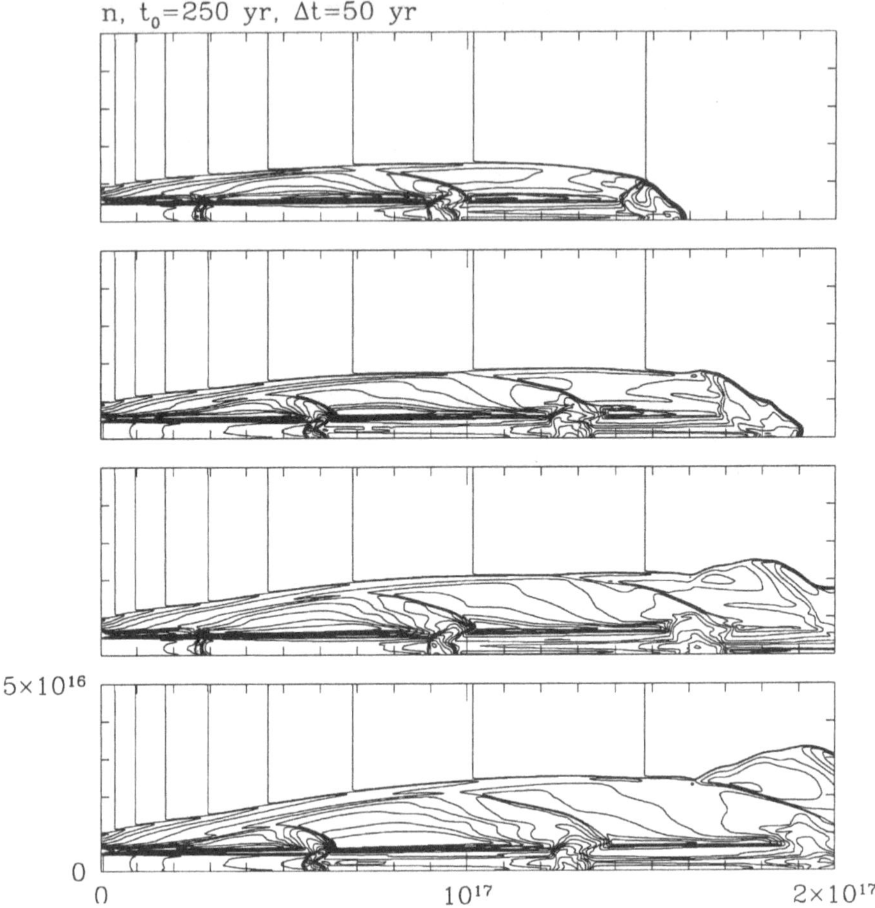

Figure 4. Total density at four time steps in a molecular jet with time-variable ejection velocity ($V_j(km/s) = 150 + 50\sin(2\pi t/100yrs)$), from Cabrit & Raga (1997). Each internal working surface ejects material sideways, driving a new bow shock into the ambient cocoon. Molecular cloud density decreases outward as $\rho_e = \rho_j \times (1 + x/10^{16}\mathrm{cm})^{-2}$.

Another effect is noteworthy: at a given distance from the star, secondary bow shocks grow broader at each successive passage, as the main cavity continues expanding in the intervening time. If CO emission became dominated by one of the 'new' working surfaces, and not by the leading jet head, the observed collimation would be much lower than the true collimation of the full cavity. This is another illustration of the 'truncation' effect invoked above for explaining flow widths.

4.4.2. *Directional Variability*

The propagation of a jet with continuously variable ejection direction was investigated analytically by Raga, Cantó & Biro (1993). At growing distance from the source, the jet beam becomes increasingly misaligned with the direction of propagation. A sideways bow shock is driven into the environment and a jet shock into the inclined jet beam. In 2D, a series of alternating one-sided bow shocks is formed (Biro et al. 1995); non-axisymmetric K-H instabilities could produce similar structures (Stone et al. 1997). In precessing 3D jets, the spiral geometry is more complex (Cliffe et al. 1996). In either case, the jet 'effective radius' of interaction is increased, enhancing entrainment of surrounding gas, and the overall cavity could be substantially broadened for large precession angles (Cliffe et al. 1996).

Fully decoupled cavities at different angles, as observed in L1157 and Cep E (Gueth et al. 1996; Ladd & Hodapp 1997), could be obtained with a combination of both velocity and directional variability. As shown analytically and numerically by Raga & Biro (1993), the jet beam eventually breaks at the positions of the internal working surfaces, for some parameters giving rise to independent jet-bullets travelling in different directions, which interact with the environment through their own individual bow shock. Bullet deceleration is then expected, and may be seen in e.g. L1448 (Bachiller et al. 1990) and HH34 (Devine, 1997).

5. Conclusions

We have compared various models with the observed properties of outflows. Detailed morphologies and PV diagrams appear to provide stronger constraints than the global mass distribution with velocity. In particular, turbulent mixing layers along jet walls appear too narrow and/or decelerated to explain most outflows. The 'infall-driven' model of Fiege & Henriksen (1996a,b) is promising, but needs inclusion of non-radiative acceleration and demonstration of similar infall and outflow mass flux rates. Wind-driven shells and jet-driven bow shocks remain so far the most appealing explanations for molecular outflows. Both models, though, currently have weak points that preclude a universal interpretation of observations. Possibly, each of these two models may apply to a different stage of YSO evolution or to a different luminosity range. Several important areas for future modelling work can be identified.

Wind-driven flows have so far been studied mostly through analytical work. Simple models are successful at reproducing the shape and integrated line profile of observed flows, but they meet several limitations. For example, radially expanding shells require nearly evacuated polar regions over large distances, and lack low-velocity emission at small view angles. However, these predictions rely on extreme simplifying assumptions. Realistic numerical simulations would be useful to (1) clarify the actual constraints on wind and ambient density stratification; (2) see whether the shape, kinematics, and excitation of shock-heated molecular features at cavity ends can be reproduced; (3) investigate the effect of time variability for comparison with the multiple cavities now found in interferometric maps. In particular, cavities at different angles may be challenging to wind-driven shell models, which involve external collimation.

Jet-driven molecular bow shocks have been modelled numerically in increasing detail. They are very successful at explaining the velocity field of molecular outflows, and the associated shock-excited EHV features. Predicted H_2 and CO emission maps are in striking agreement with the young flows discovered recently. However, bow shocks might produce too much transverse velocities, and lobes that remain too narrow (for typical jet radii). Both problems have possible solutions that need further theoretical backing: (1) the small amount of overlapping blue/red emission could be explained by very efficient turbulent mixing in the bow shock. It would be desirable to further study this effect; (2) broader, moderate collimation outflows might be obtained through a combination of time variability in velocity and direction (increasing the effective jet radius), and outflow 'truncation' (the outer parts leave the cloud and become unobservable, e.g. HH111; Cernicharo & Reipurth 1996). These effects, and the long term evolution of bow shocks, need to be better understood and quantified with realistic parameters.

Finally, the wind-driven and bow shock models discussed here ignore magnetic fields, although they could have a substantial effect on the shock structure and emission properties (cf. Hollenbach; Pineau des Forêts & Flower, this volume). This conference has seen several explorations of jet propagation in the presence of a magnetic field, and our understanding should progress substantially in the near future.

Future observations should also bring crucial informations on the formation processes of molecular outflows. More statistics on the true velocity range in flows, in particular at low inclinations, would provide interesting constraints. For such studies, shock-enhanced species (e.g. SiO) will prove very useful. Unbiased large-scale mapping beyond known outflow limits, such as presented by Padman et al. (this volume), are also necessary to determine true flow extents and ages, and build an evolutionary sequence. Most importantly, interferometric high-resolution mapping is now revealing the inner structure (e.g. the multiple cavities and their actual morphologies) of a growing number of flows, and bring in some cases unique information on jet sizes and molecular content (Guilloteau et al., this volume). More observations with such high angular resolution are needed, to allow direct comparisons with numerical simulations.

Clearly, many issues regarding the formation and propagation of molecular outflows are still not settled, However, present ideas on this subject are quite promising. It will be necessary to focus on carrying out close comparisons between theoretical models and observations of selected flows, in order to narrow down the still quite wide range of possible mechanisms for explaining molecular outflows.

References

Bachiller, R. 1996, ARAA, 34, 111
Bachiller, R., Martín-Pintado, J., Tafalla, M., et al., 1990, A&A, 231, 174
Bachiller, R., Guilloteau, S., Dutrey, A., Planesas, P., Martín-Pintado, J., 1995, A&A, 299, 857
Bence, S.J., Richer, J.S., Padman, R., 1996, MNRAS, 279, 866
Barral, J.F., Cantó, J., 1981, RevMexAA, 5, 101
Biro, S., Raga, A.C., Cantó, J., 1995, MNRAS, 275, 557
Bontemps, S., André, P., Terebey, S., Cabrit, S. 1996, A&A, 311, 858
Cabrit, S., Bertout, C., 1986, ApJ, 307, 313
Cabrit, S., Bertout, C., 1990, ApJ, 348, 530
Cabrit, S., Goldsmith, P.F., Snell, R.L., 1988, ApJ, 334, 196
Cabrit, S., Raga, A.C., 1997, in preparation
Cantó, J., 1980, A&A, 86, 327
Cantó, J., Raga, A.C., 1991, ApJ, 372, 646
Cantó, J., Tenorio-Tagle, G., Rózyczka, M., 1988, A&A, 192, 287
Cernicharo, J., Reipurth, B. 1996, ApJ, 460, L57
Chandler, C., Terebey, S., Barsony, M., Moore, T.J.T., Gautier, T.N, 1996, ApJ, 471, 308
Chernin, L.M., Masson, C.R., Gouveia dal Pino, E.M., Benz, W., 1994, ApJ, 426, 204
Cliffe, J.A., Franck, A., Jones, T.W., 1996, MNRAS, 282, 1114
Davis, C.J., Eislöffel, J., 1995, A&A, 300, 851
Devine, D. 1997, in Low Mass Star Formation - from Infall to Outflow, eds. F. Malbet and A. Castets, p.95
Downes, T.P. 1996, in Solar and Astrophysical MHD flows, ed. K. Tsinganos (Kluwer)
Downes, T.P., Ray, T. Drury, L. 1997, in Low Mass Star Formation - from Infall to Outflow, eds. F. Malbet and A. Castets, p. 98
Dutrey, A., Guilloteau, S., Bachiller, R., 1997, A&A, in press
Dyson, J.E., Cantó, J., Rodriguez, L.F., 1988, in Mass Outflows from Stars and Galactic Nuclei, L. Bianchi and R. Gilmozzi, ed(s)., Kluwer, Dordrecht, p. 299
Dyson, J. E., Hartquist, T. W., Malone, M. T., and Taylor, S. D. 1995, RevMexAA Conf. Series, 1, p. 119.
Eislöffel, J., Davis, C.J., Ray, T.P., Mundt, R., 1994, ApJ, 422, L91
Fiege, J.D., Henriksen, R.N., 1996a, MNRAS, 281, 1038
Fiege, J.D., Henriksen, R.N., 1996b, MNRAS, 281, 1055
Frank, A., Mellema, G., 1996, ApJ, 472, 684
Fuller, G.A., Ladd, E.F., Padman, R., Myers, P.C., Adams, F. C. 1995, ApJ, 454, 862
Gouveia dal Pino, E.M., Birkinshaw, M., 1996, ApJ, 471, 832
Gredel, R., Reipurth, B., 1993, ApJ, 407, L29
Gueth, F., Guilloteau, S., Bachiller R., 1996, A&A, 307, 891
Guilloteau, S., Bachiller, R., Fuente, A., Lucas, R., 1992, A&A, 265, L49
Hartigan, P., Patrick, R.J., Hartmann, L., 1987, ApJ, 316, 323
Hartigan, P., Morse, J., Raymond, J., 1994, ApJ, 436, 125

Henriksen, R.N., Valls-Gabaud D., 1994, MNRAS, 266, 681
Holland, W.S., Greaves, J.S., Ward-Thompson, D., André, P., 1996, A&A, 309, 267
Königl, A., 1982, ApJ, 261, 115
Koo, B.C., McKee C.F., 1992, ApJ, 388, 93
Ladd, E.F., Hodapp K.W., 1997, ApJ, 474, 749
Lada, C.J., 1985, ARAA, 23, 267
Lada, C.J., Fich M., 1996, ApJ, 459, 638
Levreault, R.M., 1988, ApJ, 330, 897
Li, Z.Y., Shu F.H., 1996, ApJ, 472, 211
Lizano, S., Heiles, C., Rodriguez L.F., et al., 1988, ApJ, 328, 763
Lizano, S., Giovanardi, C., 1995, ApJ, 447, 742.
Masson, C.R., Chernin, L.M., 1992, ApJ, 387, L47
Masson, C.R., Chernin, L.M., 1993, ApJ, 414, 230
Masson, C.R., Mundy, L.G., Keene, J., 1990, ApJ, 357, L25
McCaughrean, M.J., Rayner, J.T., Zinnecker, H., 1994, ApJ, 436, L189
Mellema, G., Frank, A., 1995, ApSS 233, 145
Meyers-Rice, B.A., Lada, C.J., 1991, ApJ, 368, 445
Moriarty-Schieven, G.H., Snell, R.L., 1988, ApJ, 332, 364
Mundt, R., Brugel, E.W., Bührke, E., 1987, ApJ, 319, 275
Nagar N.M., Vogel, S.N., Stone, J.M., Ostriker, E.C., 1997, ApJL, in press
Padman, R., Richer, J.S., 1994, ApSS, 216, 129
Raga, A.C., 1993, ApSS, 208, 163
Raga, A.C., 1995, RevMexAA Conf. Series, 1, p. 103
Raga, A.C., Cantó, J. 1997, in Open problems on astrophysical jets, eds. S. Massaglia and G. Bodo (Gordon and Breach, in press).
Raga, A.C., Biro, S., 1993, MNRAS, 264, 758
Raga, A.C., Cabrit, S., 1993, A&A, 278, 267
Raga, A.C., Cantó, J., Biro, S., 1993, MNRAS, 260, 163
Raga, A.C., Cantó, J., Calvet., N., Rodriguez, L.F., Torrelles., J.M., 1993, A&A, 276, 539
Raga, A.C., Cabrit, S., Cantó, J., 1995, MNRAS, 273, 422
Raga, A.C., Taylor, S.D., Cabrit, S., Biro, S., 1995, A&A, 296, 833
Richer, J.E., Hills, R.E., Padman, R., 1992, MNRAS, 254, 525
Rodríguez, L.F., Carral, P., Moran, J.M., Ho, P.T.P. 1982, ApJ, 260, 635
Shu, F.H., Ruden, S.P., Lada, C.J., Lizano, S., 1991, ApJ, 370, L31
Shu, F.H., Najita, J., Ostriker, E.C., Shang, H, 1995, ApJ, 455, L155
Smith, M.D, 1986, MNRAS, 223, 57
Smith, M.D., Brand, P., Moorhouse, A., 1991, MNRAS, 248, 451
Smith, M.D., Suttner, G., Yorke, H.W., 1997, A&A, in press
Smith, M.D., Smarr, L., Norman, M.L., Wilson, J.R. 1983, ApJ, 264, 432
Snell, R.L., Loren, R.B., Plambeck, R.L, 1980, ApJ, 239, 17
Stahler, S., 1994, ApJ, 422, 616
Stone, J.M., Xu, J., Hardee, P.E. 1997, ApJ, in press
Suttner G., Smith, M.D., Yorke, H.W., Zinnecker, H., 1997, A&A, in press
Taylor, S., Raga, A.C., 1995, A&A, 296, 823
Wilkin, F.P., 1996, ApJ, 459, 31
Wilkin, F.P., et al. 1997, in preparation
Wolfire, M.G., Königl, A., 1991, ApJ, 383, 205
Xie, T., Goldsmith, P.F., Patel, N, 1993, ApJ, 419, 33

THE PHYSICS OF MOLECULAR SHOCKS IN YSO OUTFLOWS

DAVID HOLLENBACH
MS 245-3, NASA Ames Research Center
Moffett Field, CA, 94035-1000, USA

Abstract. Shock waves light up the jets, winds and outflows around YSOs and diagnose the physical conditions and processes resident in these regions. This paper discusses the differences between the jet/wind shock and the ambient shock, between C shocks and J shocks, and between the shocks produced by pure jets and by collimated wide angle winds. Basic shock physics is briefly reviewed, with a special focus on the temperature structure in shocks and the Wardle instability of C shocks. Application is made to the origin of shocked H_2 emission and to H_2O masers.

1. Introduction

Protostellar birth is marked by the powerful ejection of a significant fraction of the accreting material by the rotating, magnetized, luminous protostar. These outflows or protostellar winds are extremely supersonic; typically, their speeds ($\gtrsim 100$ km s^{-1}) are of order the escape speed from the stellar surface or from the inner disk, whereas the internal sound speed or the sound speed in the ambient medium surrounding the protostar is $\lesssim 10$ km s^{-1}. Therefore, strong shocks form both in the wind and where the wind impacts the circumstellar material.

A working description of a shock wave is a region where a supersonic, coherent flow of gas is suddenly decelerated, heated and compressed. From the reference frame of the shock wave, the ordered supersonic flow of particles into the shock is suddenly largely randomized, leading to an irreversible increase of entropy and a rise in temperature. This process is often collisional and may be visualized as a flow of gas particles striking a "wall" of relatively stationary particles, thereby randomizing the directions of their motion. At high shock speeds ($\gtrsim 100$ km s^{-1}) the increase in entropy may

B. Reipurth and C. Bertout (eds.), Herbig–Haro Flows and the Birth of Low Mass Stars, 181–198.

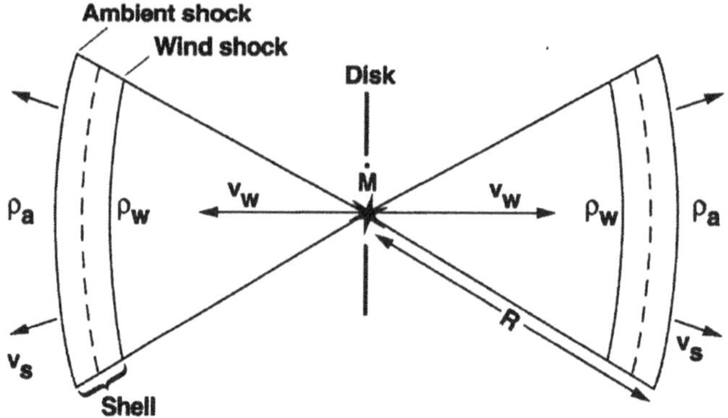

Figure 1. The "two-shock" structure characteristic of winds or jets driving outflows

be produced by collisionless processes, such as the generation and dissipation of plasma turbulence (see references in Draine & McKee 1993).

Figure 1 schematically illustrates a steady, diverging bipolar flow emanating from near a protostar and producing two shocks as it drives a shell into the ambient medium. The inner shock, sometimes called the "wind" or "jet" shock, decelerates the wind/jet to the shell speed. The outer (downwind) shock, or "ambient" shock, accelerates ambient medium to the shell speed. If the wind is sufficiently collimated to be called a "jet", the jet shock is sometimes called the "Mach disk" and the whole structure including the shell and the two shocks is called the "working surface."

Winds and jets from Young Stellar Objects (YSOs) are sufficiently complex to create shocks in more diverse ways than illustrated in Figure 1. For example, their collimation by magnetic fields and the ambient medium may lead to converging flows towards the polar axis, creating internal shocks. Shocked jet gas can also expand sideways, perpendicular to the jet, leading to a second shock as it encounters ambient medium (see §3.2). In addition, the jets and winds are time dependent, so that the jet/wind or shells may supersonically overtake and shock previously ejected wind or shells. Nevertheless, we shall use the simple case illustrated in Figure 1 to help us understand the basic physics of shocks and the nature of the observed shocks in outflow regions.

There are a number of motivations for studying shocks around YSOs. Shock waves compress, heat and accelerate the preshock medium. The compression forms clumps and may trigger local star formation. On the other hand, the acceleration increases turbulent energy in the ambient medium which may lead to the support or even dispersal of gravitationally bound clouds, and the suppression of more global star formation. The heating alters the chemistry and dissipates the total kinetic energy of the system. The heating and compression may lead to H_2O masers, as we will discuss in §3.3. Perhaps more importantly, the heating leads to characteristic shock spectra (often dominated by infrared emission for molecular shocks) which diagnose the presence of activity, the source of the activity, the nature of the outflow, and the characteristics of the ambient preshock medium. For example, the shocks diagnose the wind/jet origin, morphology, mass-loss rate, velocity and mechanical luminosity. They further diagnose the mass, velocity and pressure of swept-up outflow material, and the density structure and chemical abundances in the preshock gas.

A number of papers in this volume have described and reviewed observations of shocks in YSO outflows. Eislöffel, and Noriega-Crespo discuss the H_2 $2\mu m$ vibrational emission which delineates shocks in jets and outflows around low mass YSOs. Claussen et al present recent work on H_2O masers near low mass protostars, and Bachiller & Gutiérrez reviews shock chemistry. Similarly, shocks have been observed in the outflows around young high mass stars (e.g., the spectacular shocks around θ^1C in Orion, Allen & Burton 1993). Bachiller (1996) presents a comprehensive recent review of molecular outflows. Theoretical studies of molecular shocks predict signatures including rotational transitions of H_2, CO, OH, and H_2O, fine structure emission from [OI], [SI], [SiII], and [FeII], and the chemical enhancement of a number of molecular species (e.g., ; Draine et al 1983; Draine & Roberge 1984; Flower et al 1985, 1989; Pineau des Forêts et al 1986a,b, 1987, 1989; Hollenbach et al 1989; Hollenbach & McKee 1989; Neufeld & Dalgarno 1989a,b; recent review of Draine & McKee 1993).

2. Molecular Shock Physics

In this brief review we shall not discuss the atomic physics of the chemistry or cooling in molecular shock waves (see, e.g., Hollenbach & McKee 1979, McKee & Hollenbach 1980, Draine & McKee 1993), but shall concentrate on shock structure (§2.1), a magnetic instability in C shocks (§2.2), and the application of molecular shock models to outflows (§3.1-3.2) and H_2O masers (§3.3). However, to understand shock waves one should keep in mind a few facts about molecular shock chemistry and cooling. Above about 300 K, atomic oxygen O reacts with H_2 to form OH which reacts with H_2 to form H_2O. This reaction tends to drive all oxygen not incorporated in CO into H_2O in hot postshock gas with high H_2 abundances. Further, in postshock gas which is largely molecular, shock cooling is dominated by H_2 and CO at low density ($\lesssim 10^5 - 10^6$ cm^{-3}) and by H_2O and gas-grain collisions at higher density (Kaufman & Neufeld 1995).

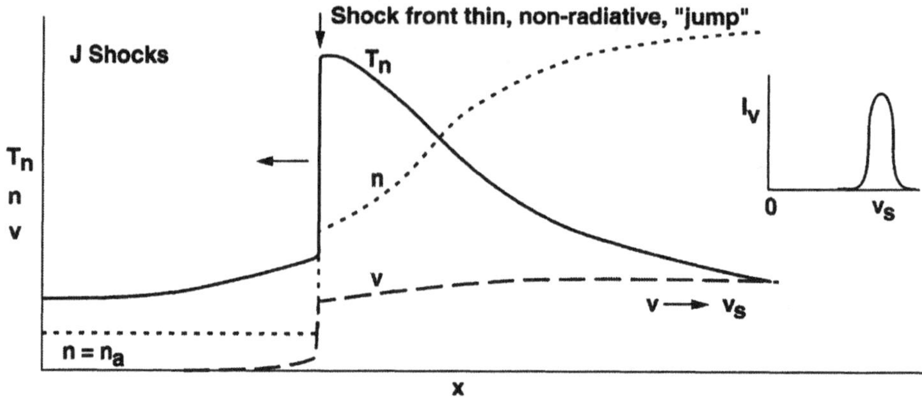

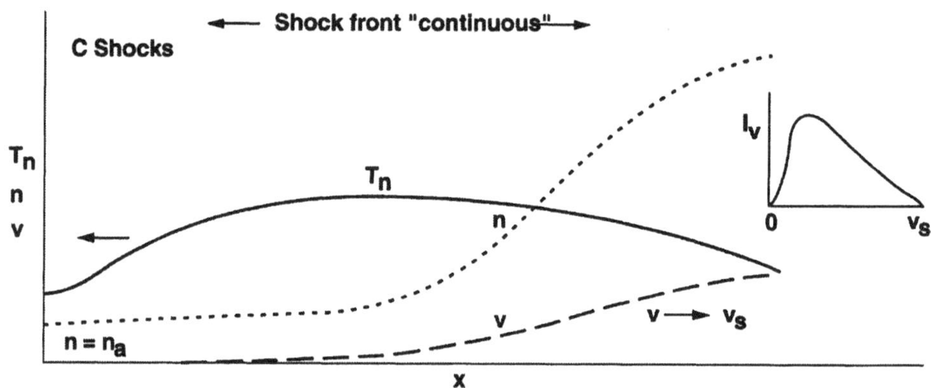

Figure 2. The structure of J shocks and C shocks

2.1. SHOCK STRUCTURE: J SHOCKS AND C SHOCKS

Figure 2 shows the shock structure and a schematic of the line profile for two types of shock waves: J shocks and C shocks. If we define the "shock front" as the region where the bulk flow is transformed into the random thermal motion of the particles (i.e., where the heat is deposited), then a J shock can be described as a shock where the heat deposition length is short compared to the cooling length. The shock front is then non-radiative, and one can use the three Rankine-Hugoniot Jump (hence "J") conditions to determine the conditions immediately behind (postshock or "ps") the shock front, which is treated as being infinitely thin.

$$n_{ps} = 4n_a, \tag{1}$$

$$v_{ps} = (1/4)v_s, \tag{2}$$

$$T_{ps} = 3.18 \times 10^5 v_{s7}^2 x_{ts}^{-1} \text{ K}, \tag{3}$$

where v is measured in the frame of the shock front, v_s is the shock velocity (flow velocity into shock front), $v_{s7} = v_s/100$ km s^{-1}, n is the hydrogen nucleus density, n_a is the preshock ambient density, and x_{ts} is the number of gas particles per hydrogen nucleus. Note that in Figure 2 we have plotted v in the frame of the ambient gas – often the observer's frame. The emission from the region behind the shock front may be calculated by modeling the subsequent cooling and compression of the postshock gas. In the ambient frame all of the emission occurs between velocities of $(3/4)v_s$ and v_s. Therefore, an observer in the ambient frame of an approaching or receding shock would see a relatively narrow line shifted by roughly v_s from line center.

Magnetic fields can limit the postshock compression of the cooling gas. If we parameterize the preshock magnetic field component perpendicular to v_s as

$$B_{0\perp} = bn_a^{1/2}\ \mu\text{G}, \tag{4}$$

where n_a is in cm^{-3}, then the maximum postshock density is given

$$n_m = 77 n_a v_{s7} b^{-1}. \tag{5}$$

Note that the Alfvén speed is $v_A = 1.8b$ km s^{-1}, and that $b \sim 0.1 - 3$ under typical interstellar conditions. For shocks which radiate most of the shocked flow energy, the total intensity in all of the cooling radiation is given

$$I_T = 9 \times 10^{-5} n_a v_{s7}^3\ \text{erg cm}^{-2}\ \text{s}^{-1}\ \text{sr}^{-1}. \tag{6}$$

Equations (5 and 6), which follow from general conservation equations, are valid for both J and C shocks. In J shocks we treat the gas as a single fluid, so that the ions and grains are assumed to be completely frozen to the neutral component. As we shall discuss further below, dissociative shocks ($v_s \gtrsim 30 - 50$ km s^{-1}) are J shocks, as are lower velocity shocks incident upon interstellar gas with relatively high levels of ionization ($x_{ia} >> 10^{-6}$).

On the other hand, a different type of shock, the C shock, occurs at relatively low velocities and moderate to low ionizations and depends upon the presence of a magnetic field (Mullan 1971, Draine 1980). When the cold gas contains a well coupled magnetic field, disturbances propagate at the bulk Alfvén speed

$$v_A = \left(\frac{B^2}{4\pi\rho}\right)^{1/2}. \tag{7}$$

Shocks are possible if $v_s > v_A$, otherwise only damped waves exist. In the partially ionized ISM the bulk medium includes ions which respond nearly instantaneously to changes in the magnetic field and neutrals which couple to the ions by collisions. Here we use "ions" in a general sense to include atomic and molecular ions and charged dust grains. Since such collisions are infrequent, it is possible to transmit Alfvén disturbances in just the ion plus field components. These disturbances damp as they propagate, but they travel at the *ion* Alfvén speed, $v_{iA} = (\rho/\rho_i)^{1/2} v_A$ (ρ_i is the ion mass density), which can be very large for small ion fraction. In C shocks $v_s < v_{iA}$, and the magnetic field and ion number density vary continuously, moving as a subsonic wave to transmit information faster than the (neutral) shock speed. If the neutrals as well as the ions vary Continuously across the shock, it is a "C" shock (Draine 1980). The oncoming shock sends a message to the upstream gas via the ions and magnetic field, warning of the approach of compressed, heated and accelerated neutral gas. In the upstream gas, the ions begin to compress and accelerate, so that they drift with respect to the neutrals, gradually heating and accelerating the cold gas via collisions. The neutral flow is continuous when the radiation from the shock front is significant, suppressing the increase in the gas temperature. The thickness or column density of the heated C shock wave is proportional to ρ_i^{-1}, or the length scale for each hydrogen molecule to be struck by an ion. On the other hand, the ambipolar heating rate is proportional to ρ_i. Thus, with increasing ionization fraction, smaller columns of hotter gas are produced until a transition to J shocks occurs. The columns and temperatures of the C shocked gas also depend on the preshock magnetic field. The neutrals move through ions which have already been compressed, and their compression is limited by the magnetic field or b (see Eq.(4)). Thus, higher magnetic fields lead to lower ion densities and greater columns of cooler shocked neutral gas.

Nearly the entire flux of a C shock is produced as infrared emission. In the C shock the radiation is emitted as the gas is being heated in the shock front; in the J shock it is given off after the impulsive heating event. To predict the emission from a C shock we must compute the heating and cooling rates throughout the flow, and we must calculate the continuous structure of the shock front.

These C shocks are generally non-dissociative (in order that molecular cooling is available), depend on low ionization fractions (so that the ambipolar heating does not drive the molecules to dissociate or collisionally ionize), and form in shocks of relatively low velocity. C shocks occur for $v_s \lesssim 40 - 50$ km s^{-1} if $x_{ia} \lesssim 10^{-6}$. The peak temperature rarely exceeds $\sim 3000 - 5000$ K (to avoid molecular dissociation). The neutral gas radiates copiously before significant acceleration or neutral compression occurs. An observer (at rest with respect to the preshock gas) of an approaching or receding C shock sees lines broadened by the velocity range of emission with a significant contribution from velocities $<< v_s$ (see Figure 2).

The temperature structure in J and C shocks are shown in Figures 3 and 4 for two illustrative models. Figure 3 shows a $v_s = 80$ km s^{-1} dissociative

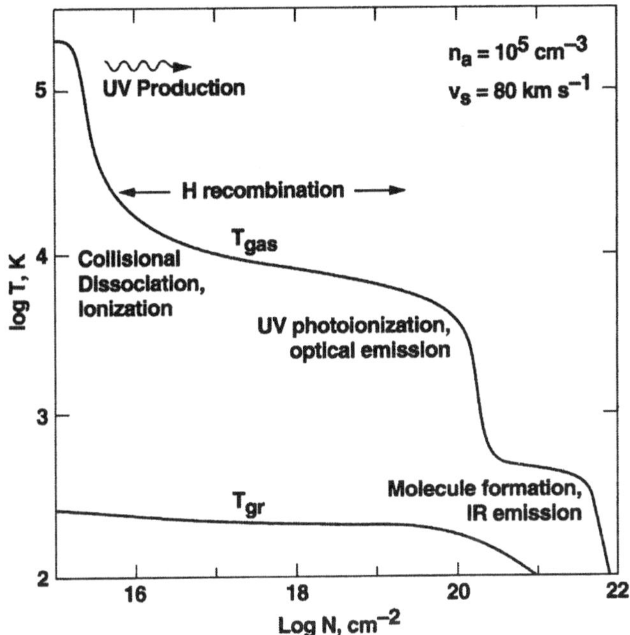

Figure 3. J shock model (Hollenbach & McKee 1989)

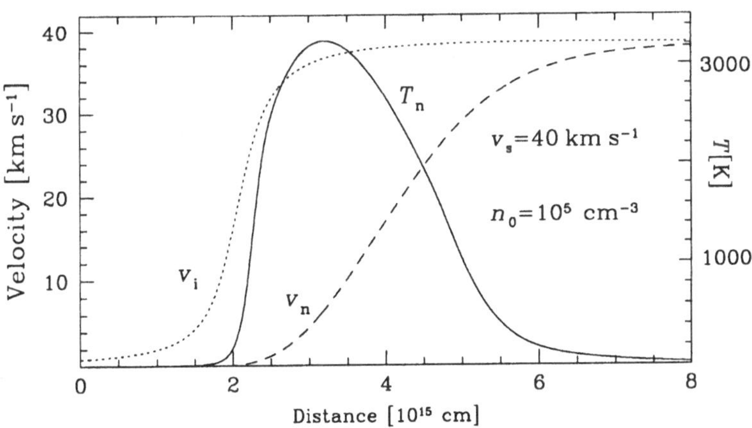

Figure 4. C shock model (Kaufman & Neufeld 1996a)

J shock incident upon ambient gas of density $n_a = 10^5$ cm^{-3}. Figure 4 shows a $v_s = 40$ km s^{-1} C shock with the same n_a. Immediately evident is that the peak gas temperature (see Eq. (3)) in the J shock, where the heat is added impulsively, is much higher than the peak temperature in the C shock, even when the difference in v_s is taken into account. Figure 3 also plots the dust temperature T_{gr}; in both C and J shocks the dust temperature is much less than the gas temperature because of the weak gas-grain coupling and the efficient radiative cooling of dust.

The C shock thickness is of order 4×10^{15} cm, whereas the thickness of the $T > 100$ K J-shocked material is $\lesssim 10^{15}$ cm. The J shock would be even thinner if the heating due to the reformation of molecular hydrogen did not maintain a ~ 400 K temperature plateau to $N \sim 4 \times 10^{21}$ cm^{-2} (see also Neufeld & Dalgarno 1989). This effect only occurs for $n_a \gtrsim 10^4 - 10^5$ cm^{-3}, so that the newly-formed, vibrationally-excited H_2 can be collisionally de-excited before it radiates away its vibrational energy. Thus, J shocks tend to have smaller columns of hotter (and more atomic) gas than C shocks.

In the 80 km s^{-1} dissociative J shock, the gas is first heated to $T \gtrsim 10^5$ K where it collisionally dissociates and ionizes, and where it emits considerable ultraviolet radiation. The gas cools to 10^4 K, where the Lyman continuum radiation maintains a temperature plateau through photoionization heating of H. Once these UV photons are absorbed, at $N \sim 10^{20}$ cm^{-2} in this model, hydrogen atoms predominate and the gas cools by atomic fine structure emission, especially [OI] 63 μm. Hollenbach (1985) shows how the [OI] 63 μm luminosity from a dissociative J shock is proportional to the mass flux into the shock; thus, [OI] 63 μm from wind shocks (Figure 1), which are generally radiative J shocks for low mass protostars, can measure the wind or jet mass loss rate (Cohen et al 1988, Ceccarelli et al 1997).

Molecular hydrogen does not significantly reform behind dissociative J shocks until $T \lesssim 500$ K. Therefore, collisional excitation of the 2 μm H_2 vibrational lines is insignificant, and only a relatively weak emission from newly-formed H_2 is predicted (Hollenbach & McKee 1989). This weak emission has never been definitively detected. However, the heating due to H_2 reformation when $n_a \gtrsim 10^4 - 10^5$ cm^{-3} leads to a relatively large column $N \sim 10^{22}$ cm^{-2} of ~ 400 K gas which is increasingly molecular downstream. H_2O is formed in this temperature plateau, and the conditions are ripe for H_2O maser production when $n_a \gtrsim 10^6$ cm^{-3} (see §3.3). Dissociative J shocks also radiate copiously in rotational OH, SiO, and CO lines and in the fine structure lines of [OI], [SiII], [FeII], [NeII], and [SI] (Hollenbach & McKee 1989, Neufeld & Dalgarno 1989, Haas et al 1991).

Figure 4 demonstrates that C shock temperatures are never large enough to dissociate H_2; however, large columns of $T \sim 1000 - 3000$ K H_2 can be produced, and H_2 2 μm emission can be very luminous. The hot molecular

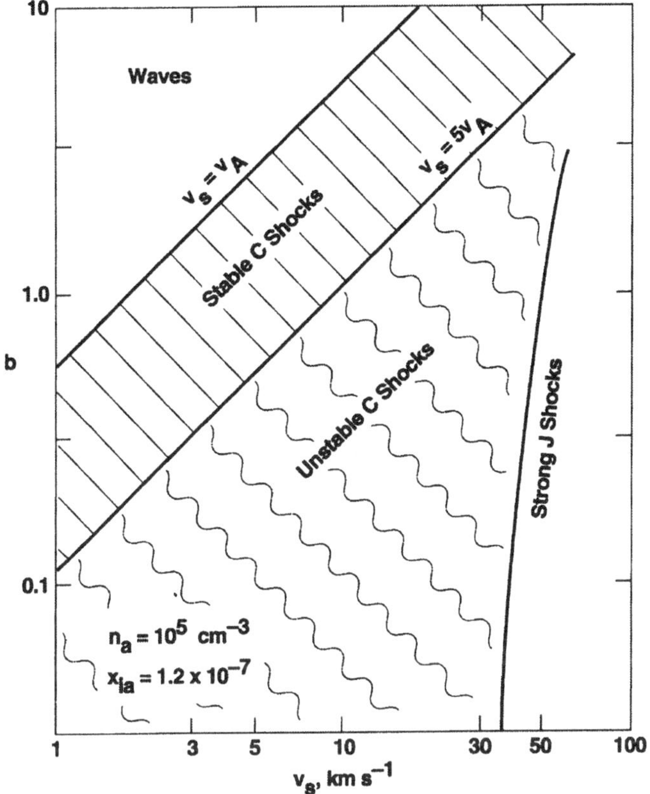

Figure 5. C and J shock parameter space (Hollenbach et al 1989)

gas also rapidly transforms all oxygen not in CO into H_2O. Hence, intense infrared H_2O emission is predicted, along with even warmer masers than in J shocks (Neufeld & Melnick 1991, Melnick et al 1993, Kaufman & Neufeld 1996b). Figure 4 compares ion and neutral flow velocities measured in the preshock frame, and demonstrates that ions are accelerated first so that they can then drag the neutrals to the shock speed. Pineau des Forêts et al. (this volume) discuss other interesting aspects of C shocks.

Figure 5 illustrates the parameter space occupied by C and J shocks as a function of the shock velocity and preshock magnetic field (b is defined in Eq. (4)). We have chosen representative values for n_a and x_{ia}; J shocks become more prevalent if $x_{ia} >> 10^{-6}$ (e.g., Hollenbach et al 1989, Smith & Brand 1990). In the upper left corner $v_s > v_A$ and no shocks forms. C shocks form between the solid lines. J shocks form when the shock velocities are high enough ($\gtrsim 40$ km s^{-1}) to dissociate H_2, which reduces the cooling and leads to impulsive shock heating.

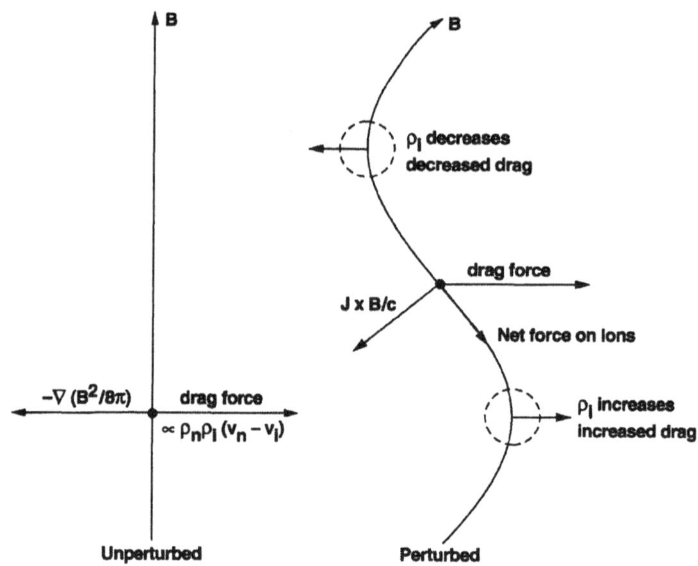

Figure 6. The Wardle Instability (adapted from Draine & McKee 1993)

2.2. THE WARDLE INSTABILITY OF C SHOCKS

Wardle (1990, 1991a,b) described an important dynamical instability in C shocks involving deformation of the magnetic field. Figure 6 schematically illustrates the Wardle instability. Consider a plane parallel shock with B perpendicular to v_s. The ions and (B field) move ahead of the compressed neutral gas, driven by the $\nabla(B^2/8\pi)$ force until a steady state wave forms, retarded by the drag force of the neutrals on the ions. Consider a perturbation of the magnetic field as in Figure 6. The drag force now has a component parallel to the local B field that cannot be balanced by the $\mathbf{J} \times \mathbf{B}$ force, and ions will therefore be accelerated along field lines to collect in magnetic "valleys". As a consequence, ρ_i will increase in the valleys, the drag force will increase there, and the field lines will be further distorted. Wardle found C shocks to be unstable for $M_A \gtrsim 5$ (see Figure 5).

This significant discovery left shock modelers in a bad state. Until the non-linear development of the clumps and spectra produced by these C shocks could be followed by numerical MHD codes, there was no consistent way to compare observations with shock models, since a large portion of C shock parameter space (Figure 5) was now uncertain. The Chamonix meeting therefore marked an historic development in interstellar shock modeling as two teams (MacLow & Smith 1997; Stone 1997, Neufeld & Stone 1997) reported results from 2D MHD codes which treated the weakly coupled ions and neutrals separately and followed the Wardle instability to saturation.

The surprising and interesting (though possibly disappointing to the code developers) results reported by Neufeld & Stone is that the C shock spectra from Wardle unstable shocks do not appreciably deviate from the spectra found by the older 1D steady state C shock codes which suppressed the instability. The neutral gas emits before it is appreciably clumped. Thus, the older model spectra (e.g., Draine et al 1983, Draine & Roberge 1984) are still applicable. MacLow & Smith emphasize that the weaker, high excitation lines are affected more, and that the instability may lead to measurable spectral variations with time.

3. Applications

3.1. ASPECTS OF THE TWO-SHOCK PICTURE

A useful approximation to the two-shock model illustrated in Figure 1 is that often a pseudo "steady state" is established where the pressure in the shell is equal to the ram pressure of the supersonic gas hitting it from both sides. Therefore the ram pressure of the ambient gas equals the ram pressure of the overtaking wind

$$\rho_a v_s^2 \simeq \rho_w (v_w - v_s)^2. \tag{8}$$

If $\rho_a >> \rho_w$ (a "light" jet/wind), then $v_s \simeq (\rho_w/\rho_a)^{1/2} v_w$, the shell moves much slower than the wind. If the wind shock is radiative, the total luminosity of the wind shock is greater than the ambient shock by a factor (v_w/v_s). However, the wind shock will likely be a dissociative J shock radiating much of its luminosity in the UV, and the ambient shock may dominate the H_2 emission. The mass in the shell is dominated by the flux of material from the ambient shock by a factor (v_w/v_s).

On the other hand, if $\rho_a << \rho_w$ (sometimes called a "heavy" jet/wind), then the shell propagates at nearly the wind speed, $v_s \sim v_w$, and the wind shock is a slow shock while the ambient shock may be a dissociative J shock. This occurs early in the evolution of a jet/wind, or possibly when the wind breaks out of the core into a low density interclump medium. In this case, the ambient shock dominates the luminosity whereas the mass in the shell is mostly injected by the wind.

We note that Eq.(8) can only be strictly valid when the ambient gas and wind strike the shock fronts perpendicularly. Wilkin (1996, this volume) has found analytic solutions to more general cases, and applied these solutions to bow and outflow shocks near protostars.

Davis and Eislöffel (1995) noted and studied an interesting predicted correlation between the ambient shock luminosity L_{sa} and the mechanical luminosity L_{CO} of the swept material between the two shocks (Figure 1).

L_{sa} is given

$$L_{sa} = \frac{1}{2}\rho_a v_s^3 A_{sa}, \tag{9}$$

where A_{sa} is the area of the shock. L_{CO} is derived from CO measurements. The earliest and most prevalent signature of protostellar outflows has been CO rotational line emission detected in line wings, at $v \sim 5-30$ km s^{-1} from line center, indicating material that was not gravitationally bound to the star-forming clump and was therefore an "outflow". The CO measurements indicated hydrogen masses $M_{CO} \lesssim 200$ M$_\odot$; such large masses indicate that this outflow is not the wind from the protostar/disk, but must be swept-up ambient material (e.g., Lada 1985, Fukui et al 1993, Bachiller 1996). L_{CO} is defined

$$L_{CO} = \frac{\frac{1}{2}M_{CO}v_s^2}{\tau}, \tag{10}$$

where $\tau = R/v_s$, the flow "lifetime". Assuming any ambient density distribution less steep than r^{-3} and that most of the shell mass is swept-up ambient gas and not shocked wind (light wind, $v_s << v_w$), $M_{CO} \simeq A_{sa}R\rho_a$. Substitution into Eq. (10) leads to

$$L_{CO} \simeq L_{sa} \tag{11}$$

The current ambient shock luminosity is equal to the shell kinetic energy divided by its age. Davis & Eislöffel (1995) do not actually observe this correlation in mapping 5 sources, but they attribute the discrepancies to difficulties in determining v_s and τ from CO observations, the extinction at 2 μm to correct for the intrinsic L_{sa}, and the possibility of a heavy wind.

3.2. JETS AND WIDE ANGLE WINDS

Observations of protostellar outflows have demonstrated two types of phenomena whose relationship is not yet clear. On the one hand, optical observations of relatively unobscured regions of the outflow reveal tightly collimated ($\lesssim$ 100 AU) jets which extend to large distances ($\gtrsim$ 0.1 pc) from the protostar (e.g., Mundt & Fried 1983, Bachiller 1996). These jets are generally observed in Hα, SII(6730 Å), and other low excitation optical emission lines, presumably excited in the $T \sim 10^4$ K ionized gas behind dissociative J shocks (Figure 3) formed either internal to the jet or at the jet/ambient gas interface. On the other hand, the CO outflows, especially those in more obscured regions, generally show a much less collimated structure (references in Bachiller 1996; Cernicharo et al, Padman et al, this volume).

Two classes of theories have arisen to reconcile or unify these seemingly contradictory or disparate phenomena. On the one hand are theories that assume that the basic stellar or disk wind is very tightly collimated into a

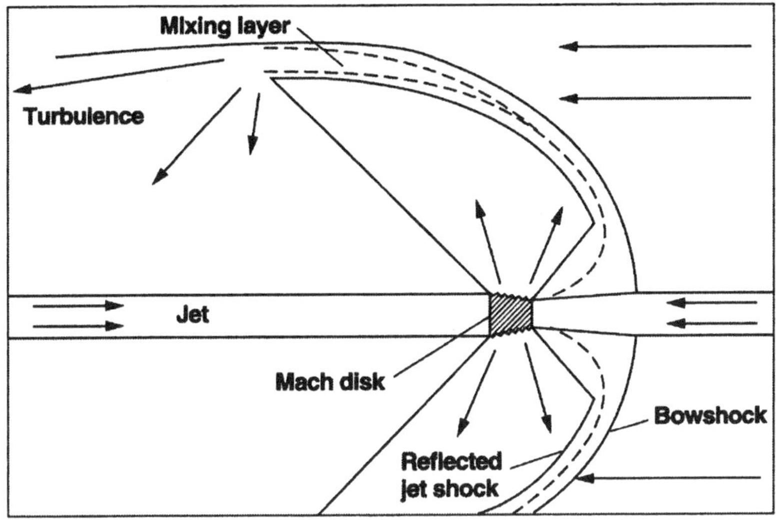

Figure 7. A jet creating a bow shock (adapted from Raga & Cabrit 1993)

jet by an unspecified mechanism. However, the interaction of this jet with the ambient medium leads to a bow shock and turbulent entrainment of ambient material (the CO outflow) that extends far from the jet axis (e.g., Raga & Cabrit 1993, Masson & Chernin 1993, Stahler 1994, Chernin & Masson 1995). On the other hand, Shu et al (1995) (see also Shu & Shang, this volume) have derived the streamlines of the MHD wind driven from the inner accretion disk of a protostar by the interaction with the magnetic dipole of a rotating star. They find that the pinching effect of the toroidal component of the magnetic field set up by this rotating configuration leads to a natural collimation of the wind. The density in the wind is higher along the pole than away from the cylindrical axis, and thus the emission measure of the optical lines strongly peak along the axis, giving the appearance of a jet. However, the streamlines are actually quite radial, and there is still considerable mass loss, $\sim$ 0.5 of the total, which is ejected at relatively wide angles to the axis. The wind is really a "collimated wide angle wind." This second class of model, therefore, relies on the *direct* impact of the wide angle component to drive the less collimated CO outflows.

Both of these classes of models can be understood by extending the simple two shock picture of Figure 1. Here, we shall only focus on the shocks and the shock physics of these models. Figure 7 shows a schematic figure of one of the first class of models, where a solitary jet sets up a bow shock. In this model the pressurized shell between the jet shock and the ambient shock can "squirt" gas out sideways at high speed. This ejection from the

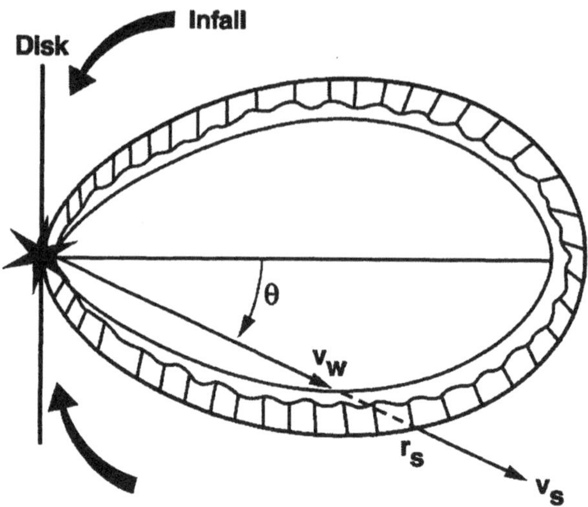

Figure 8. The wide angle collimated wind (adapted from Shu et al 1991)

working surface acts then like a much less collimated wind, which interacts with the ambient gas to form an inner shock of the deflected jet with an outer ambient bow shock (Figure 7). Material coming off the distant reaches of the bow, far from the axis, form vortices and turbulence downstream which extends along the jet towards the source. The bow and turbulence entrain the more extended, less collimated CO outflow. A variation of this class of models is that the turbulence set up by the shear of the jet through the ambient gas creates a thick turbulent layer, and the CO outflow is entrained in this turbulent layer (Stahler 1994).

Figure 8 shows a schematic of the collimated wide angle wind model. The shape of the outflow shell depends on the intrinsic collimation of the wind $\rho_w(\theta)$ and the density distribution $\rho_a(\theta)$. In both Figures 7 and 8 we have only illustrated one lobe of the bipolar flow, and the accretion disk is viewed edge-on and vertical. Li & Shu (1996) show that using the Shu et al (1995) wind solution for $\rho_w(\theta)$ and the density contours of flattened magnetized cores for $\rho_a(\theta)$, one can produce in the shell the observed mass in the CO outflow as a function of velocity (Masson & Chernin 1992).

Hartigan et al (1996) present a good review of the various possibilities for the origin of the observed H_2 emission in bow shocks. We highlight here a few important observations and suggestions. The 2μm vibrational lines lie 6000 - 18,000 K above ground, and the line ratios suggest temperatures ~ 2000 K. Brand et al (1988, 1989) has shown that observations of different regions often show remarkably similar line ratios, suggestive of an effective thermostating mechanism. Smith et al (1991) model H_2 spectra from bow shocks, and show that the ambient shock gets progressively more oblique further back on the bow. The head of the bow, therefore, has the highest ambient shock velocity which in many cases may be a dissociative J shock, weak in H_2 2μm emission. Further back on the bow, the shock velocity (the perpendicular component of the ambient flow velocity striking the bow shell) drops below the critical velocity v_{cr} for the J/C transition, and a strong (oblique) C shock is produced, intense in H_2 2μm emission. As long as the bow speed v_b into the ambient gas exceeds v_{cr}, a range of C shock velocities from the dissociation limit down to very weak shocks are therefore produced. If the beam encompasses the entire bow structure, the resultant H_2 spectra are insensitive to v_b for $v_b > v_{cr}$.

We further note that *global* bow shocks and shells are created by the jets and collimated winds discussed above, but that much smaller bow shocks may be produced if the ambient gas has small clumps which are overtaken by the shell. In this case, a small bow forms around the clump, with the head of the bow pointing toward the outflow source. The global bow points away. Such clumps might be produced by episodic mass loss, which sends previous generations of global bow shocks through the ambient gas. Wardle instabilities in these shocks produce clumps of size scale the order of the thickness of the C shocks, which may be of order $10^{16} - 10^{17}$ cm for $n \sim 10^5 - 10^4$ cm^{-3} (Figure 4).

3.3. H_2O MASERS ASSOCIATED WITH YSOS

Interstellar H_2O masers at 22 GHz often appear to be individual clumps, streaming away from some center of activity at velocities up to 200 km s^{-1}. Individual features have apparent sizes of 10^{13} cm and brightness temperatures usually in the range $T_b \sim 10^{11} - 10^{14}$ K (Genzel 1986). The isotropic luminosity of individual maser spots range from $\lesssim 10^{-6}$ to 0.08 $L_\odot$ in the Galaxy (Walker et al 1982). Pumping by an external source of radiation is ruled out by observations (e.g., Genzel 1986), and an internal source of pump energy, such as produced in a shock, seems required. The development of powerful shocks in maser regions is inevitable in light of the high velocities observed in the sources, and the H_2O maser luminosity correlates with the mechanical luminosity in the observed outflows (Felli et al 1992), as would be expected in a shock model.

Litvak (1969), Strelnitski & Sunyaev (1973), Schmeld et al (1976), Hollenbach et al (1987), Elitzur et al (1989), Kaufman & Neufeld (1996b), and Hollenbach et al (1997) propose that H_2O masers originate in the warm molecular gas behind shock waves driven by YSO winds. C shocks simply heat the molecular gas; fast J shocks dissociate the molecular gas, but as H_2 reforms in the postshock gas, it delivers part of its formation energy as heat

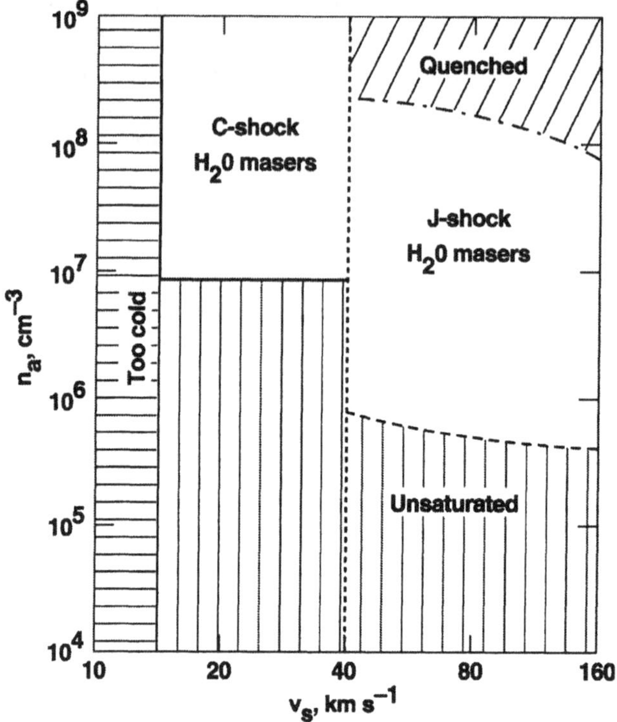

Figure 9. The shock parameter space for strong 22 GHz H_2O masers, assuming $v_A \sim 1$ km s^{-1} and $x_{ia} \lesssim 10^{-6}$ (Hollenbach et al 1997)

which maintains a large postshock molecular column at $T \sim 400$K (Figure 3). In either case, the warm $T \gtrsim 300$ K postshock gas drives all oxygen not locked in CO to form H_2O, and collisionally excites the maser levels which typically lie $\gtrsim 600$ K above the ground state. Shock compression (especially in J shocks) helps the gas attain sufficiently high densities to produce strong masers. The sheet-like geometry of the shocked gas allows infrared photons to escape normal to the sheet, thereby enhancing non-LTE level populations necessary for the maser inversion. It also provides long coherence paths in the plane of the shock for the maser lines.

Figure 9 summarizes the shock parameter space which produces 22 GHz H_2O masers. The vertical dashed line at $v_s \sim 40$ km s^{-1} marks the approximate boundary between C and J shocks. Above the dash-dot line, the maser is quenched in J shocks. Below the dashed and solid line, the maser is unsaturated and weak. J shocks in the range ($v_s \gtrsim 40$ km s^{-1}, 10^6 cm^{-3} $\lesssim n_a \lesssim 10^8$ cm^{-3}) and C shocks in the range (15 km s^{-1} $\lesssim v_s \lesssim 40$ km s^{-1}, 10^7 cm^{-3} $\lesssim n_a \lesssim 10^9$ cm^{-3}) produce strong, saturated, beamed 22

GHz H_2O masers.

Genzel (1986) reviews the observations of H_2O masers and we summarize here the shock model explanation of the data. Their space velocities are often $\gtrsim 30$ km s^{-1}; this is the velocity required to produce strong C or J shocks. Their densities are inferred to be $\sim 10^9$ cm^{-3}; a J shock compresses gas of density $\sim 10^7$ cm^{-3} to this density, while a C shock requires higher preshock densities. Their spot sizes are about 10^{13} cm; the shock thickness and unsaturated core size are of this order in shock models. Their observed isotropic luminosities range from $10^{-7} - 10^{-1}$ $L_\odot$ and their brightness temperatures T_b range from $10^{11} - 10^{14}$ K; the shock models in the parameter range shown in Figure 9 produce these numbers with coherence lengths of 10^{14-15} cm. Lower radial velocity masers tend to have higher T_b; in shocks, lower radial velocites mean that the line of sight is close to the shock plane, resulting in larger coherence lengths and T_b. The H_2O maser luminosity correlates with the mechanical luminosity seen the the CO outflow; the mass loss produces the shocks which, in turn, produce the masers (see also Eq. 11). Millimeter observations, the detection of other H_2O maser lines (e.g., Melnick et al 1993), and the observation that there are not enough external photons to pump the maser all point to warm $\gtrsim 300$ K gas where collisions pump the maser; J shocks produce large columns of ~ 400 K H_2O and C shocks can produce even warmer H_2O. Finally, Fiebig & Güsten (1989) have observed the Zeeman splitting of the 22 GHz H_2O maser in W49 and estimated the component of the B field along the line of sight to be about 100 mG; the shock models predict exactly this order of field in the masing region, relatively independent of the preshock field (Elitzur et al 1989).

Acknowledgements: I gratefully acknowledge the support of the NASA Astrophysical Theory Program, which funds the Center for Star Formation Studies, a consortium of researchers from NASA Ames, University of California at Santa Cruz, and University of California at Berkeley.

References

Allen, D.A., Burton, M.G. 1993, Nature 363, 54
Bachiller, R. 1996, ARAA 34, 111
Brand, P.W.J.L., Toner, M.P., Geballe, T.R., Webster, A.S. 1989, MNRAS 237, 1009
Brand, P.W.J.L., Moorhouse, A., Burton, M.G., Geballe, T.R., Bird, M., Wade, R. 1988, ApJ Lett. 334, L103
Ceccarelli, C., Haas, M.R., Hollenbach, D.J., Rudolph, A.L. 1997, ApJ 476, 771
Chernin, L., Masson, C. 1995, ApJ 443, 181
Cohen, M., Hollenbach, D.J., Haas, M.R., Erickson, E.F. 1988, ApJ 329, 863
Davis, C.J., Eislöffel, J. 1995, A&A 300, 851
Draine, B.T. 1980, ApJ 241, 1021 (erratum: 246, 1045)
Draine, B.T., McKee, C.F. 1993, ARAA 31, 373
Draine, B.T., Roberge, W.G. 1984, ApJ 282, 491
Draine, B.T., Roberge, W.G., Dalgarno, A. 1983, ApJ 264, 485

Elitzur, M., Hollenbach, D.J., McKee, C.F. 1989, ApJ 346, 983
Felli, M., Palagi, F., Tofani, G. 1992, A&A 255, 293
Fiebig, D., Güsten, R. 1989, A&A Lett. 214, 333
Flower, D.R., Pineau des Forêts, G., Hartquist, T.W. 1985, MNRAS 216, 775
Flower, D.R., Heck, L., Pineau des Forêts, G. 1989, MNRAS 239, 741
Fukui, Y., Iwata, T., Mizuno, A., Ball, J., Lane, A.P. 1993, in *Protostars and Planets III*, ed E.H. Levy, J.I. Lunine (Tucson: Univ. Ariz. Press), p603
Genzel, R. 1986, in *Masers, Molecules and Mass Outflows in Star Forming Regions*, ed. A.D. Haschick (Westform, MA: Haystack Observatory), p233.
Haas, M.R., Hollenbach, D.J., Erickson, E.F. 1991, ApJ 374, 555
Hartigan, P., Carpenter, J.M., Dougados, C., Skrutskie, M.F. 1996, AJ 111, 1278
Hollenbach, D.J. 1985, Icarus 61, 40
Hollenbach, D.J., Chernoff, D., McKee, C.F. 1989, in *Infrared Spectroscopy in Astronomy*, ed. B. Kaldeich, ESA SP-290, p245
Hollenbach, D.J., Elitzur, M., McKee, C.F. 1997, ApJ, in preparation
Hollenbach, D.J, McKee, C.F. 1979, ApJ Suppl. 41, 555
Hollenbach, D.J., McKee, C.F. 1989, ApJ 342, 306
Hollenbach, D.J., McKee, C.F., Chernoff, D. 1978, in it Star Forming Regions, ed. M. Peimbert & J. Jugaku (Dordrecht: Reidel), p334
Kaufman, M.J., Neufeld, D.A. 1995, ApJ 418, 263
Kaufman, M.J., Neufeld, D.A. 1996a, ApJ 456, 611
Kaufman, M.J., Neufeld, D.A. 1996b, ApJ 456, 250
Lada, C. 1985, ARAA 23, 267
Litvak, M.M. 1969, Science 165, 855
MacLow, M.M., Smith, M.D. 1997, ApJ, submitted
Masson, C.R., Chernin, L.M. 1992, ApJ Lett. 387, L47
Masson, C.R., Chernin, L.M. 1993, ApJ 414, 230
McKee, C.F., Hollenbach, D.J. 1980, ARAA 18, 219
Melnick, G.J., Menton, K.M., Phillips, T.G., Hunter, T. 1993, ApJ 416, L37
Mullan, D.J. 1971, MNRAS 153, 145
Mundt, R., Fried, J.W. 1983, ApJ Lett. 274, L83
Neufeld, D.A., Dalgarno, A. 1989a, ApJ 340, 869
Neufeld, D.A., Dalgarno, A. 1989b, ApJ 344, 251
Neufeld, D.A., Melnick, G.J. 1991, ApJ 368, 215
Neufeld, D.A., Stone, J.M. 1997, ApJ, submitted
Pineau des Forêts, G., Flower, D., Hartquist, T., Dalgarno, A. 1986a, MNRAS 220, 801
Pineau des Forêts, G., Flower, D., Hartquist, T., Millar, T. 1987, MNRAS 227, 993
Pineau des Forêts, G., Roueff, E., Flower, D.R. 1986b, MNRAS 223, 743
Pineau des Forêts, G., Roueff, E., Flower, D.R. 1989, JChemSocFaradayTrans 85, 1665
Raga, A., Cabrit, S. 1993, A&A 278, 267
Schmeld, I.K., Strelnitski, V.S., Muzylev, V.V. 1976, SovAstron 20, 4111
Shu, F.H., Najita, J., Ostriker, E., Shang, S. 1995, ApJ Lett. 455, L155
Smith, M.D., Brand, P.W.J.L. 1990, MNRAS 242, 495
Smith, M.D., Brand, P.W.J.L., Moorhouse, A. 1991, MNRAS 248, 451
Stahler, S. 1994, ApJ 422, 616
Stone, J.M 1997, ApJ, submitted
Strelnitski, V.S., Sunyaev, R.A. 1973, SovAstron 16, 579
Walker, R.C., Matsakis, D.N., Garcia-Barreto, J.A. 1982, ApJ 255, 128
Wardle, M. 1990, MNRAS 246, 98
Wardle, M. 1991a, MNRAS 250, 523
Wardle, M. 1991b, MNRAS 251, 119
Wilkin, F.P. 1996, ApJ Lett. 459, L31

THE PHYSICAL AND CHEMICAL EFFECTS OF C-SHOCKS IN MOLECULAR OUTFLOWS

G. PINEAU DES FORÊTS
DAEC, Observatoire de Paris
F-92195 Meudon Principal Cedex
France

D.R. FLOWER
Department of Physics, University of Durham
Durham DH1 3LE
UK

AND

J.-P. CHIÈZE
DSM/DAPNIA/Service d'Astrophysique
CE-Saclay
F-91191 Gif-sur-Yvette Cedex
France

Abstract. The bow shocks which are observed in jets associated with low-mass star formation probably have C-shock characteristics over at least part of their extent. We show that ion-neutral streaming in C-type shocks can give rise to the release of significant amounts of silicon from the refractory cores of dust grains and hence to the formation of SiO in the gas phase. The computed profiles of SiO emission lines are found to be qualitatively similar to those observed in molecular outflow regions such as L1448. The question of the validity of the stationary-state assumption, adopted in all previous studies of C-shocks in the interstellar medium, is addressed. Time-dependent calculations of the evolution of a C-type shock propagating in low-density gas suggests that steady state is unlikely to be attained under the dynamical conditions in outflows.

B. Reipurth and C. Bertout (eds.), Herbig–Haro Flows and the Birth of Low Mass Stars, 199–212.

1. Introduction

Observations of the jets associated with low-mass star formation, discussed in this volume, provide abundant evidence for the presence of bow shocks. The latter are believed to have J-shock characteristics at the apex and C-shock characteristics in the wake (Smith and Brand 1990). At the apex, molecular hydrogen is collisionally dissociated, whereas, in the wake, H_2 is rovibrationally excited (Davis and Eisloeffel 1995; Gredel 1994, 1996).

Further evidence of the effects of shocks comes from observations of molecular lines in outflows, particularly of emission from rotationally excited states of SiO (Bachiller et al. 1991; Bachiller and Gomez-Gonzalez 1992; Martin-Pintado et al. 1992; Zhang et al. 1995). The SiO emission is found at velocities and positions which differ from those characterizing the ambient medium. In addition, the SiO:H_2 abundance ratio in the high velocity gas (10^{-7} to 10^{-6}) is much higher than in the ambient medium (less than 10^{-11}). Thus, the SiO emission appears to be intimately associated with dynamical activity and almost certainly with the presence of shocks.

The fraction of elemental silicon which is deduced from observations of the gas phase varies greatly amongst the different components of the interstellar medium. In diffuse clouds, from a few per cent to 70 per cent of the Si is observed to be in the gas phase (Sofia et al. 1994; Fitzpatrick 1996). In dark clouds, on the other hand, where most of the gas-phase silicon is expected to be in the form of SiO (Herbst et al. 1989), very low upper limits ($< 3 \times 10^{-12}$) have been established to the SiO:H_2 abundance ratio (Ziurys et al. 1989; Martin-Pintado et al. 1992). Indeed, SiO emission is observed only from dynamically active regions: molecular outflows, supernovae remnants and the Galactic centre. In quiescent clouds, silicon must be present in the solid phase, most probably in the form of refractory silicate grains or, possibly, in the ice mantles of the dust grains.

In this review, we shall address the questions of how the silicon is released from the solid phase and what is its subsequent fate. The discussion will centre around the role played by C-type shocks. In view of the short timescales associated with dynamical events in outflows, we also present the results of calculations of the temporal evolution of C-type shocks; all previous work has been based on the assumption of steady state.

2. Non-thermal sputtering of grains by shocks in the interstellar medium

In regions of high fractional ionization (the 'warm' or 'diffuse' interstellar medium), the neutral fluid is well coupled to the ions and thereby to the magnetic field. Under these conditions, shocks tend to be J-type and betatron acceleration of the (charged) grains occurs, owing to the compression of the gas and hence of the magnetic field. The charged grains orbit the magnetic field lines and acquire velocities, relative to neutral species, which are sufficient to give rise to sputtering (Draine and Salpeter 1979; Seab and Shull 1983; McKee et al. 1987; Jones et al. 1994, 1996). Shock speeds in the range $50 \lesssim v_s \lesssim 200$ km s^{-1} yield impact energies, for helium atoms, for example, between about 50 eV and 1 keV, which are sufficient to erode and even to destroy refractory grains. Grain-grain collisions are also important.

In regions of low fractional ionization, on the other hand, the collisional coupling between the ionized and neutral fluids is insufficient to maintain a common flow speed, and C-type shocks tend to develop. In 'low' velocity shocks ($10 \lesssim v_s \lesssim 50$ km s^{-1}), the differential flow velocity gives rise to He atom impact energies, on charged grains, of less than 50 eV, sufficient to erode the grain mantles (Draine et al. 1983; Flower and Pineau des Forêts 1994). The effects of such shocks on the grain cores is more uncertain because of the lack of pertinent measurements of the sputtering yields in the low-energy domain: few experimental data are available below about 150 eV (cf. Draine 1995).

2.1. COMPUTED SPUTTERING YIELDS

We have undertaken calculations of the sputtering yields for refractory materials, such as amorphous silica, SiO_2 (Flower et al. 1996; Field et al. 1997), and, more recently, forsterite, Mg_2SiO_4. In Fig. 1 are presented the results of calculations of the sputtering yield of Si from amorphous SiO_2 under impact by He^+ and heavier ions. The distinction between ionic and atomic impact is not significant in the context of the following discussion, because ions capture electrons as they approach the solid surface. The sputtering yield, Y, is defined as the ratio of the number of sputtered atoms of a given type to the total number of impacts and depends on both the velocity (and hence energy) and the angle, θ, of impact. The results in Fig. 1 were obtained by means of the classical trajectory Monte Carlo code TRIM; for a description of the method, see Eckstein (1991). Up to 10^7 trajectories were run for each value of θ and of the energy, E, of impact. The results were then averaged over the angle of incidence, relative to the normal to the surface, using the relation

$$Y(E) = \int_0^{\pi/2} Y(E,\theta) \sin\theta \, d\theta$$

to obtain the results plotted in the Figure.

The very sharp decrease in the sputtering yields with decreasing impact velocity is evident from Fig. 1. We shall see below that the impact velocity is comparable with the shock velocity, v_s. It follows that, for $v_s \lesssim 50$

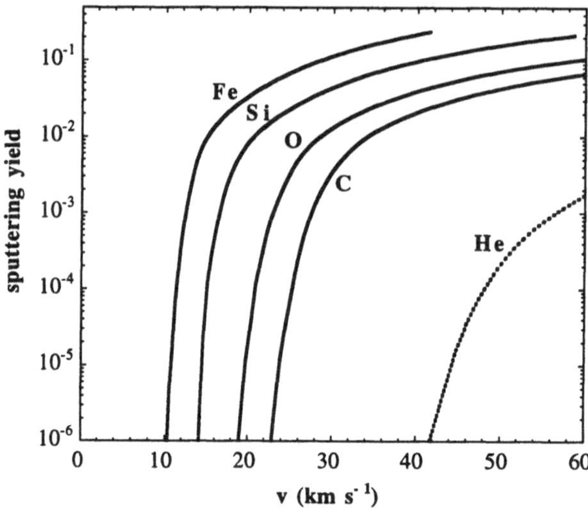

Figure 1. Sputtering yields of Si from amorphous SiO_2, computed as functions of the velocity of incidence of helium and heavier projectile ions. The yields have been averaged over the angle of incidence.

km s^{-1}, the sputtering yield for He impact becomes much smaller than the yields for impact by heavier species. Indeed, the ratio of the sputtering yields more than compensates for the lower abundances of the heavy species, which, consequently, predominate at low shock speeds. Accordingly, in the calculations reported below, we have incorporated sputtering by the more abundant atomic and molecular species that are heavier than helium, adopting the sputtering yield calculated for the ion of nearest mass. This procedure receives its justification from the calculated sputtering yields, which show the impact energy (and hence the mass of the incident particle, at a given impact velocity) to be the determining factor.

2.2. C-TYPE SHOCKS IN DARK MOLECULAR CLOUDS

In dark clouds, the fractional ionization of the medium is so low that the collisional coupling between the neutral and the charged fluids is insufficient to prevent substantial differences developing in their flow speeds and kinetic temperatures. It is the charged fluid which couples directly to the magnetic field and receives 'advance notice' of the arrival of the shock wave if the Alfvén speed in the ionized fluid exceeds the shock speed. The charged fluid is then heated and accelerated prior to the neutral fluid (Mullan 1971; Draine 1980). In a C-type shock, the neutral fluid always flows supersonically relative to the shock front: radiative cooling prevents the sound speed,

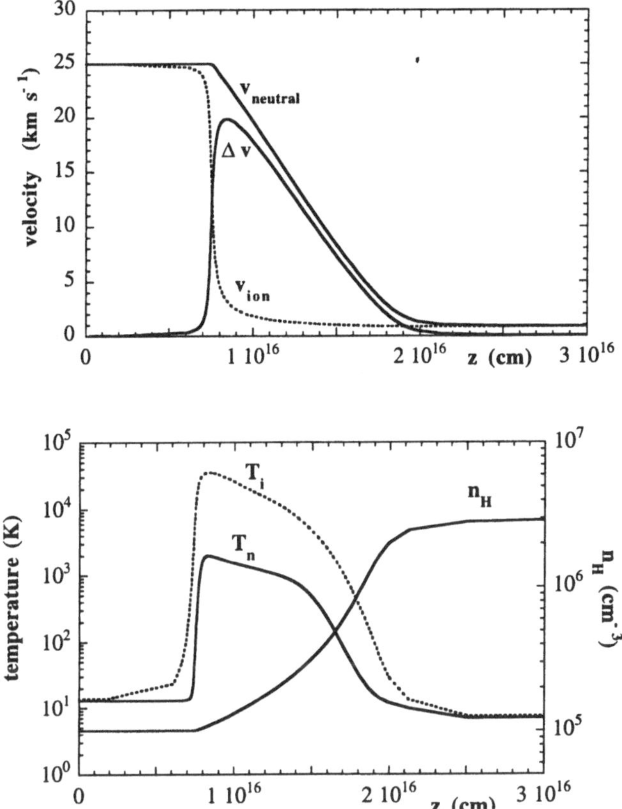

Figure 2. (a) Velocity and (b) temperature profiles for a shock wave of velocity $v_s = 25$ km s^{-1} propagating into a medium in which $n_H = 10^5$ cm^{-3} and $B_0 = 200\mu$G. The ion-neutral drift speed is $\Delta v = v_n - v_i$ in the shock frame, in which the preshock gas flows at the shock speed, v_s. 'Preshock' is to the left and 'postshock' to the right in this Figure.

c_s, from attaining the local flow speed, v_n. (The appearance of a sonic point, $v_n = c_s$, in the flow is generally associated with a discontinuity.)

In Fig. 2, we show the stationary-state velocity and temperature profiles of a C-type shock propagating at 25 km s^{-1} into a medium of density $n_H = n(H) + 2n(H_2) = 10^5$ cm^{-3} where the magnetic induction is $B_0 = 200\mu$G, transverse to the flow direction. Compression of the neutral gas occurs over an extended distance, of the order of 10^{16} cm. The difference Δv between the flow speeds of the charged and neutral fluids may be seen to peak at a value not much smaller than the shock speed. The velocity of impact of the neutral particles on the charged grains is equal to the ion-neutral velocity difference. Each grain collides with many times its own mass of gas as it passes through the shock wave.

In the calculations reported below, the size distribution of the grains has been taken from Mathis et al. (1977). Individually, the grains are much more massive than the atomic and molecular ions, and, in a medium of low fractional ionization, their contribution to momentum transfer to the neutrals must be taken into account (Draine 1995; Flower and Pineau des Forêts 1995). In this context, it is necessary to establish the degree to which the grains are coupled to the magnetic field and hence flow at the same speed as the other charged particles.

The gyromagnetic frequency of a grain of mass m_g and charge Ze in a magnetic induction B is given by

$$\nu = \frac{BZe}{2\pi m_g},$$

whereas the time, τ, required for the grain to collide with its own mass of neutral gas is determined by

$$\tau^{-1} = \frac{\rho_n}{m_g}\pi a^2 \, |v_i - v_n|,$$

where ρ_n is the mass density of the gas and a is the radius of the grain. The grains are coupled to the magnetic field if $\nu > \tau^{-1}$, whence

$$a(\mu m) < 0.02(Zb)^{\frac{1}{2}} \left[\frac{|v_i - v_n|}{10\,\mathrm{km\,s^{-1}}}\right]^{-\frac{1}{2}} \left[\frac{n_H}{10^4\,\mathrm{cm^{-3}}}\right]^{-\frac{1}{4}}$$

and we have assumed that $B(\mu\mathrm{G}) = bn_H^{\frac{1}{2}}$. In the preshock gas, $b \approx 1$, but the differential compression of the ions ensures that $b \gg 1$ in the shock. Furthermore, the magnitude Z of the grain charge may be much larger than 1 in the shock (Caselli et al. 1997). It follows that most of the grains in the population that we adopt are well coupled to the magnetic field for the range of physical conditions that will be considered.

2.3. NUMERICAL SIMULATIONS

We shall present illustrative results for a model with the parameters specified above ($v_s = 25$ km s^{-1}, $B_0 = 200\mu$G, $n_H = 10^5$ cm^{-3}), incorporating about 100 gas-phase species and 800 chemical reactions. The grains are treated as being composed of a nucleus of amorphous silica surrounded by a mantle of ice containing up to 12 species, of which the most abundant are saturated hydrogenated molecules such as H_2O. More detailed information on the model is to be found in Schilke et al. (1997).

It is essential to treat the chemistry, particularly the ion-neutral chemistry, in parallel with the dynamics. The gas is partially neutralized in the shock wave, owing to the enhanced rates of certain ion-neutral reactions which have activation energies. As a consequence, the dynamical structure of the shock wave is profoundly modified. Calculations reported in Section 3 below will serve to emphasize this point.

The mantles of the grains are rapidly removed in the shock wave, exposing the grain core. In the model presented here, a few per cent of the elemental silicon is then released from the grain nucleus by sputtering. This process is also rapid, occurring in the vicinity of the maximum of $\Delta v = | v_i - v_n |$. Once in the gas phase, the atomic silicon is oxidized, first to SiO and subsequently to SiO_2, in the reactions

$$\begin{aligned} \mathrm{Si} + \mathrm{O_2} &\longrightarrow \mathrm{SiO} + \mathrm{O} \\ \mathrm{Si} + \mathrm{OH} &\longrightarrow \mathrm{SiO} + \mathrm{H} \\ \mathrm{SiO} + \mathrm{OH} &\longrightarrow \mathrm{SiO_2} + \mathrm{H} \end{aligned}$$

The second stage of the oxidation process (formation of SiO_2 from SiO) is slower than the first but, nonetheless, it is effective owing to the enhanced abundance of OH in the shock-heated gas. SiO_2 is the most abundant silicon-bearing species for a considerable distance into the postshock gas but is invisible, as it does not possess a permanent dipole moment. Eventually, SiO_2 is removed from the gas phase by accretion on to the residual grain cores.

The flow time, t, of the neutrals is related to the distance, z, by the relation

$$t = \int \frac{1}{v_n} dz.$$

Fig. 3 compares fractional abundances, expressed as functions of the flow time, calculated including and neglecting the re-accretion of gas-phase species on to grains. In the latter case, SiO_2 is the most abundant Si-bearing species for times comparable with the estimated lifetimes of molecular clouds (10^5 yr), when the fractional abundance of SiO is only of the order of 10^{-11}. SiO_2 is ultimately destroyed in reactions with ions, yielding SiO. However, in the molecular outflows associated with low-mass star formation, the dynamical timescales ($10^3 - 10^4$ yr) are too short for the oxidation of SiO to SiO_2 to be completed, and so Si and SiO are likely to be the most abundant silicon-bearing species in the gas phase.

Once the physical and chemical structure of the shock wave has been determined, the profiles of the SiO emission lines can be calculated (Schilke et al. 1997). In Fig. 4, we compare the profiles of SiO rotational lines observed in the outflow L1448 (Bachiller et al. 1991) with the computations,

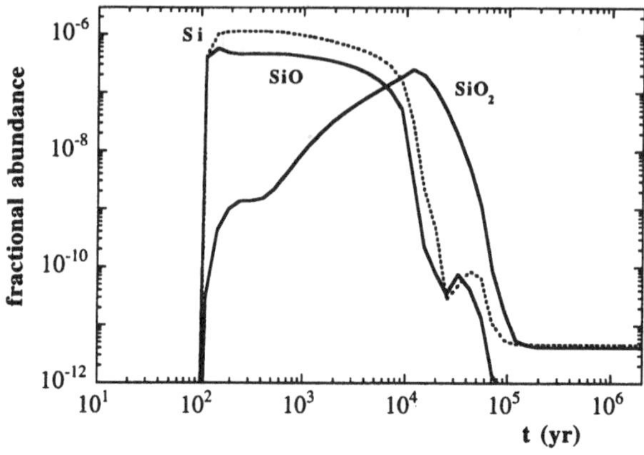

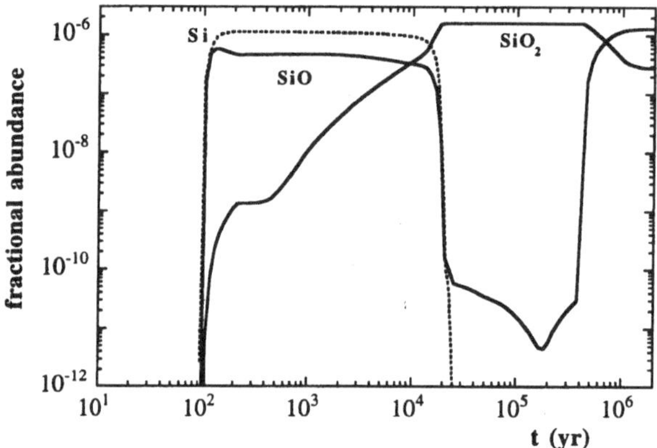

Figure 3. (a) Illustrating the release of Si into the gas phase and its subsequent oxidation to SiO and SiO_2 prior to re-accretion by the grains. The independent variable is the flow time, t. (b) Neglecting re-accretion on to the grains. In this case, the decrease in the fractional abundance of SiO_2 at large times is due to reactions with ions, which yield SiO.

which relate to the blue-shifted components. It may be seen that the characteristic shapes of the lines are well reproduced by the model. Corrections for the beam filling factors have not been made, and so the absolute values of the observed and calculated brightness temperatures are not comparable. Analogous calculations are planned incorporating sputtering yields for more realistic grain materials (e.g. forsterite). Other species observed in outflows, such as CS and SO (see the review of Bachiller and Pérez Gutiérrez in this volume), will also be studied. Furthermore, we are currently investigating the rovibrational transitions of H_2 arising from such shock waves.

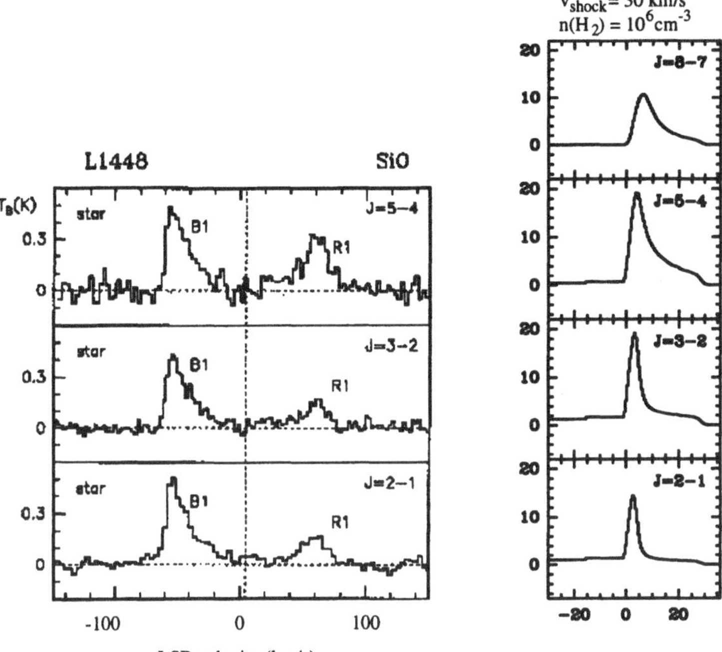

Figure 4. Observed and computed profiles of rotational transitions of SiO; the calculations refer to the blue (B1) component. Corrections for beam filling factors have not been made, and so the absolute values of the brightness temperature, T_B, are not comparable.

3. The temporal evolution from J- to C-type shocks

Molecular outflows are examples of regions where major dynamical changes in the flow are likely to occur on timescales shorter than is required for the flow to reach a stationary state. Under these circumstances, a study of the propagation of shock waves should be based on the partial differential equations which describe the variation of the fluxes of number density, mass, momentum and energy of the charged and neutral fluids with position and time.

We have already mentioned in Section 2 the dynamical importance of the chemistry, specifically the ion-molecule chemistry; this is illustrated by the calculations in Fig. 5, which compares results obtained, at steady state, for a model in which $v_s = 10$ km s^{-1}, $B_0 = 25\mu$G and $n_H = 10^3$ cm^{-3}. In one calculation, the effects of chemical reactions were included, in the other, they were neglected. In the latter case, the fractional ionization of the gas is modified only by the differential compression of the ionized and neutral fluids by the shock wave and has the same value in the preshock and the

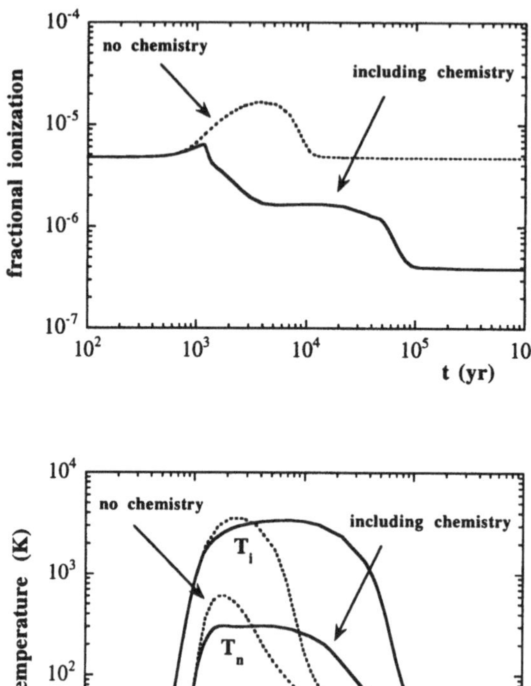

Figure 5. (a) Fractional ionization predicted by a model in which $v_s = 10$ km s^{-1}, $n_H = 10^3$ cm^{-3} and $B_0 = 25\mu$G; results are compared including and neglecting chemical reactions. (b) Corresponding temperature profiles of the ionized and neutral fluids.

postshock gas. The much lower degree of ionization, when chemical reactions are included, leads to weaker ion-neutral coupling, thereby enhancing the width of the shock wave and the time required to re-establish equilibrium, in the postshock gas. We note that the time required to attain the new equilibrium state is greater than or equal to the flow time through the shock wave, as is apparent from this Figure. The maximum value attained by the kinetic temperature of the neutral fluid is lower when chemical reactions are included. It is clear from Fig. 5 that the dynamics and the chemistry are intimately linked and have to be treated in parallel; this has been done in all the other calculations reported in this review.

A potential difficulty in treating the temporal evolution of shocks is the occurrence of sonic point(s) in the flow. The 'pseudo-viscosity' method of von Neumann and Richtmyer (see Richtmyer 1957) has been extensively used to address this problem. The technique involves artificially broadening the discontinuity so that it may be passed in a few integration steps without the necessity of impractically small step lengths. However, precautions have to be taken when collisional processes, in addition to those giving rise to viscosity, are involved. In the calculations reported below, this has been achieved by modifying the rate coefficients of all collisional processes (inelastic collisions, chemical reactions) by the ratio of the true to the pseudo-viscosity coefficient of the gas. A subsequent scaling procedure restores the correct time and distance scales to the computed variations in the extensive physical parameters; for further information, see Chièze *et al.* (1997).

Solving the time-dependent hydrodynamical equations is computer time consuming, and so the extent of the chemical network and number of chemical species must be limited. In the calculations that we shall now discuss, a network of 130 chemical reactions, involving 32 chemical species comprising the elements H, He, C, O and Fe (Fe^+ is a representative heavy ion), was included alongside the hydrodynamical equations. There is then a total of 40 dependent variables per z-mesh which must be determined as functions of the independent variable, t. The step size used for the independent variable is adjusted automatically. Effects associated with the presence of dust grains have been neglected in these calculations.

In the model being considered, it has been assumed that a 'piston' acts continuously on the medium, driving a shock wave into the gas and imposing the postshock velocity. The shock wave, which is initially J-type, advances into the gas at a velocity which slightly exceeds that of the piston. As time progresses, the mass of the postshock material gradually increases until a steady state is finally attained. A comparison may then be made with the results of a stationary state calculation.

Fig. 6 shows the evolution of the temperature and density profiles of a shock wave whose parameters are those specified above. Even at an early time ($t = 10^2$ yr), the perturbation has travelled much further in the ionized fluid (at the ion Alfvén speed) than in the neutral fluid (at the shock speed). A discontinuity in the flow variables of the neutral gas clearly remains at $t = 10^3$ yr. It is only much later ($t = 5\,10^5$ yr), at a time which is comparable with the cooling time in the postshock gas (Fig. 5), that steady state is attained. At this point, the time-dependent profiles become indistinguishable from those generated by an independent, stationary-state calculation, affording at least a partial check on the validity of the results. The discontinuity in the neutral flow becomes weaker as time progresses and disappears completely when the width of the shock wave becomes equal to the distance required for the velocities of the charged and neutral fluids to recouple.

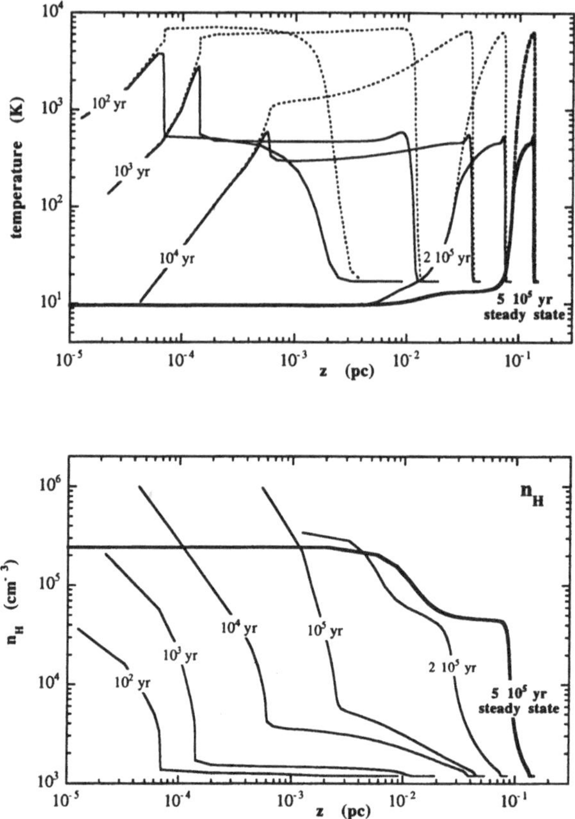

Figure 6. (a) Temperature and (b) density profiles as functions of position and time for the same model as in Fig. 5; chemical reactions are included. The shock wave advances from left to right, where a stationary state is finally attained.

In the context of the jets and outflows associated with low-mass star formation, an important point which emerges from Fig. 6 is that the time required to reach a stationary state is much greater than the dynamical timescales associated with such regions. Time-dependent calculations of, for example, the intensities of the emission lines of H_2 and the process of grain erosion, then become necessary. However, we note that the preshock gas density in this model ($n_H = 10^3$ cm^{-3}) is smaller than the densities that are believed to prevail in outflows, where the cooling times will be correspondingly shorter. Calculations for higher densities are planned but will prove very computer time consuming.

4. Concluding remarks

We have shown that C-type shocks, with velocities in the range $20 \lesssim v_s \lesssim 50$ km s^{-1}, give rise to complete sputtering of the mantles and partial sputtering of the refractory cores of interstellar grains by heavy particle impact. Release of silicon into the gas phase is followed by its rapid oxidation to SiO, which is observed in molecular outflows, and then more slowly to SiO_2 (which is not observed: SiO_2 has no permanent dipole moment). The computed profiles of the SiO emission lines are found to be qualitatively similar to those which are observed in outflows such as L1448.

Time-dependent calculations of shock propagation in a low-density ($n_H = 10^3$ cm^{-3}) molecular medium suggest that, in outflow regions, steady state may not be attained between the arrival of successive 'bullets' of outflowing material. At times significantly shorter than required to reach steady state, even a planar shock wave will have both C- and J-type characteristics, i.e. there will be an extended precursor region, followed by a quasi-discontinuous change in the flow variables. Calculations for higher preshock gas densities are necessary to confirm this result.

Acknowledgements:
We are grateful to the British Council and the Ministère des Affaires Etrangères for support under the 'Alliance' programme, contract no. 96026.

References

Bachiller, R., Martin-Pintado, J. and Fuente, A. (1991) *A&A*, 243, L21.
Bachiller, R. and Gomez-Gonzalez, J. (1992) *A&A Rev.*, 3, 257.
Caselli, P., Hartquist, T.W. and Havnes, O. (1997), *A&A*, in press.
Chièze, J.-P., Pineau des Forêts, G. and Flower, D.R. (1997) *MNRAS*, to be submitted.
Davis, C.J. and Eisloeffel, J. (1995) *A&A* 300, 851.
Draine, B.T. (1980) *Ap.J.* 241, 1021.
Draine, B.T. (1995) *Ap. Sp. Sci.* 233, 111.
Draine, B.T. and Salpeter, E.E. (1979) *Ap.J.* 231, 438.
Draine, B.T., Roberge, W.G. and Dalgarno, A. (1983) *Ap.J.* 264, 485.
Eckstein, W. (1991) *Computer Simulation of Ion-Solid Interactions*, Springer-Verlag, Berlin.
Field, D., May, P.W., Pineau des Forêts, G. and Flower, D.R. (1997) *MNRAS*, 285, 839.
Fitzpatrick, E.L. (1996) *Ap.J.* 473, L55.
Flower, D.R. and Pineau des Forêts, G. (1994) *MNRAS* 268, 724.
Flower, D.R. and Pineau des Forêts, G. (1995) *MNRAS* 275, 1049.
Flower, D.R., Pineau des Forêts, G., Field, D. and May, P.W. (1996) *MNRAS* 280, 447.
Gredel, R. (1994) *A&A* 292, 580.
Gredel, R. (1996) *A&A* 305, 582.
Herbst, E., Millar, T.J., Wlodek, S. And Bohme, D.K. (1989) *A&A* 222, 205.
Jones, A.P., Tielens, A.G.G.M., Hollenbach, D.J. and McKee, C.F. (1994) *Ap.J.* 433, 797.
Jones, A.P., Tielens, A.G.G.M. and Hollenbach, D.J. (1996) *Ap.J.* 469, 740.
Martin-Pintado, J., Bachiller, R. and Fuente, A. (1992) *A&A* 254, 315.
Mathis, J.S., Rumpl, W. And Nordsieck, K.H. (1977) *Ap.J.* 217, 425.

McKee, C.F., Hollenbach, D.J., Seab, C.G. and Tielens, A.G.G.M. (1987) *Ap.J.* 318, 674.
Mullan, D.J. (1971) *MNRAS* 153, 145.
Richtmyer, R.D. (1957) *Difference methods for initial-value problems*, Interscience, New York.
Schilke, P., Walmsley, C.M., Pineau des Forêts, G. and Flower, D.R. (1997) *A& A*, in press.
Seab, C.G. and Shull, J.M. (1983) *Ap.J.* 275, 652.
Smith, M.D. and Brand, P.W.J.L. (1990) *MNRAS* 245, 108.
Sofia, U.J., Cardelli, J.A. and Savage, B.D. (1994) *Ap.J.* 430, 650.
Zhang, Q., Ho, P.T.P., Wright, M.C.H. and Wilner, D.J. (1995) *Ap. J.* 451, L71.
Ziurys, L.M., Friberg, P. and Irvine, W.M. (1989) *Ap.J.* 343, 201.

HERBIG-HARO OBJECTS AS SEARCHLIGHTS FOR DENSE CLOUD CHEMISTRY

S. D. TAYLOR
University College London, Gower St., London WC1E 6BT, England

1. Introduction

Stars form from dense pockets of gas that have undergone gravitational collapse, and young stars are still embedded in this molecular material. There is no lack of evidence that the outflows from these stars, manifested in the form of Herbig-Haro objects and jets, interact dynamically with the dense gas either in the form of shock excitation or the imparting of momentum; molecular outflows, H_2 emission, and perhaps also CO and SiO bullets and jets, are tracers of this phenomenon. However there are a set of radio observations of molecules showing emission near HH objects that are remarkable for their apparent lack of signatures of such dynamic interaction. A likely cause of this 'quiescent' emission is a radiative interaction between the HH shock system emission and ambient gas.

The 'cores' from which stars form are localised ($\sim$ 0.1pc) density enhancements within the molecular clouds traced by the CO molecule, and there is a great deal of interest in their chemistry since many molecules, some rather complex, have been observed in them (Williams 1994). These molecules are able to form because they are shielded from the ambient interstellar radiation field, but a number of uncertainties remain, amongst them the effect of accretion of gas molecules onto the surfaces of dust grains to form ice mantles, and the possibility of a subsequent surface chemistry. These ices are observed in infrared absorption in a number of molecules (H_2O, CO, CH_3OH, CH_4, CO_2; Whittet *et al.* 1996) when suitable background sources are present, but must exist to some extent in all gas exposed to a greatly attenuated UV flux.

In this contribution I will describe a model in which HH jets propagate into a clumpy medium, and each of these clumps contains dust that can accrete ice mantles. If a clump finds itself in the vicinity of the jet it may be

B. Reipurth and C. Bertout (eds.), Herbig–Haro Flows and the Birth of Low Mass Stars, 213–222.

exposed to radiation from one of its working surfaces, and if this can remove the mantles the ensuing gas phase chemistry can produce large abundance enhancements in many molecules, perhaps rendering them observable. I will then discuss the nature of these clumps, and how the illumination of the clumps by the jet 'searchlight' can help us to understand the dust:gas interaction in dense regions.

2. Observations of Quiescent Emission

Cores are detected by means of molecular transitions that trace gas denser than that of the ambient cloud, ($n_H \geq 10^4$ cm^{-3} against $n_H \sim 10^3$ cm^{-3}, n_H is the number of hydrogen nuclei per unit volume). In order to study dense gas around HH objects it is also necessary to have high angular resolution to determine the position relative to the shocks, and so such observations have tended to be made using interferometers, although submm single dish observations can also achieve high resolution. For these reasons the HCO^+ (J=1–0) and NH_3 (1,1) and (2,2) inversion transitions have mostly been used. Rudolph and Welch (1988, 1992) mapped HH7-11 and HH34 in HCO^+, and Rudolph (1992) also looked at the jet in L1551 in the same molecule, all using the Hat Creek interferometer. HH1 and HH2 were imaged by Torrelles *et al.* (1992,1993) in ammonia using the VLA, as was HH80 (North) by Girart *et al.* (1994). In addition Davis, Dent and Bell Burnell (1990) observed HCO^+ (J=4–3) near HH1 and HH2 using JCMT and Davis & Dent (1993) found NH_3 in HH34 with the Effelsberg telescope. All of these observations share the following characteristics;

(i) The emission is confined to patches of emission close to, but downwind of, the optical shock emission. These localised areas of emission, which we term *clumps* to distinguish from the *core* emission that surrounds the source of the outflow, have sizes of $\sim 10'' - 20''$, which is typically $\sim$ 0.02pc.

(ii) The clumps appear dynamically unaffected, in the sense that the lines have widths of < 1 km s^{-1}, and centres within 1–2 km s^{-1}of the core emission surrounding the outflow source.

At this conference, Dent (1997) has presented further observations around HH2 using JCMT, and detected HCO^+ (J=3–2 and J=4–3), but also found weak emission from the new molecules H_2CO (5_{15}–4_{14}), N_2H^+ (J=4–3), and HCS^+ (J=8–7), together with the non-detection of transitions from several other molecules. Once again the HCO^+ shares the above characteristics.

3. Possible Explanations

It is necessary to explain not only why these clumps appear ahead of HH objects, but also why they do not appear elsewhere. There are three ways in which this may be possible;

(i) **Coincidental alignment.** It could be reasoned that with a small sample set it may be that clumps dense enough to emit are found by chance projected where they are. The apparently high detection rate combined with the general absence of emission elsewhere in the observed fields would seem to argue against this, as does the presence of clumps ahead of more than one shock system in the same source. For example this emission is found ahead of each of the optical knots HH1 and HH2, and ahead of each of the knots in HH7-11 except for HH7.

(ii) **Shocked cloudlet.** These observations initially seem to favour the 'shocked cloudlet' hypothesis for the origin of HH objects themselves, whereby a stellar wind strikes a density enhancement in the cloud causing a bow shock around it and a shock within it. Reasons given in the past for why this model is not generally applicable are given added weight with the addition of this radio emission, even though in theory the presence of a shock could lead to extra molecular emission compared with other ambient clumps either by chemical enhancement or an increase in density on cooling. To explain the generally high proper motions of HH objects (e.g. HH1-2) the cloudlet must be shocked and accelerated without being destroyed, but even if this is so the clumps seen in molecular emission have ambient cloud velocities. Whilst it is possible to explain the proper motion by invoking the fragmentation of the cloudlet, the radio emission is more difficult to explain since rotational temperatures derived from the ammonia transitions suggest the gas is cold, and if this is cooled postshock gas the shock velocity in the cloudlet must be very small for the gas to have a line centre so close to that of the ambient gas. Furthermore, although in HH7-11 the HCO^+ emission is very close to the optical emission, in other objects it is clearly spatially separated, and in the case of HH34 the optical arc structure points in the wrong direction and appears morphologically distinct from the radio.

(iii) **Irradiated clump.** Given the quiescent nature of the emission, an attractive alternative is that of an ambient clump that has been unaffected dynamically by the outflow, but happens to lie sufficiently close to the shock system that it is subject to the radiation field it generates, which somehow enhances the emission from certain molecules. This scenario was proposed by Wolfire and Königl (1993, WK93) specifically to model the HCO^+ observations. In that paper the HCO^+ (J=1–0) line is enhanced both by an increase in fractional abundance of the molecule $x(HCO^+)$ $(= n(HCO^+)/n_H)$

due to chemical effects, and an increase in excitation because of electron collisions in the irradiated gas which has a relatively high fractional ionisation. Both effects are the consequence of the impact of high energy photons (X-rays and EUV) from a strong shock on a thin layer at the edge of the clump (visual extinction $A_V < 0.1$ magnitudes). Photoionisation of H_2 (which is by far the dominant gas phase species) heats and partially ionises this layer and allows molecules like OH and H_2O to form quickly by the endothermic neutral-neutral reactions that are well known from work on shock chemistry, or by the ion-molecule route thought to occur in low temperature interstellar clouds but enhanced by the high abundance of H^+. Ordinarily any molecular species would be quickly destroyed in this harsh environment, but the emission from a shock is comprised mostly of line photons that do not overlap the dissociation bands of H_2 and CO, and as a consequence they retain their 'pre-shock radiation' high abundances for some time after that radiation is switched on and can contribute to the chemistry. C^+ still becomes very abundant as a result of the dissociation of CO molecules, and it is its reaction with OH and H_2O that produces HCO^+,

$$C^+ + H_2O \rightarrow HCO^+ + H \tag{1}$$

$$C^+ + OH \rightarrow CO^+ \xrightarrow{H_2} HCO^+. \tag{2}$$

Whilst this model is of relevance to objects like HH1-2 where there are clearly very fast shocks capable of producing high energy radiation, it is questionable how valid it is for 'low excitation objects' (LEO's; Böhm & Solf 1992) such as HH7-11, where the shock velocities appear to be no more than 30–40 km s^{-1}, although the linewidths are appreciably larger. HH34 appears to be an intermediate case, requiring shock velocities of $\sim$ 120–170 km s^{-1}. It is also not clear how the subsequent observations of ammonia can be similarly explained, since an enhancement over other ambient clump abundances is required. The temperatures even in the very edge of the clump are unlikely to be high enough to drive neutral-neutral formation due to the high barrier ($\Delta E = 1.6 \times 10^4$K) of the initiating reaction

$$N + H_2 \rightarrow NH + H, \tag{3}$$

and formation via nitrogen ions is unlikely to produce more NH_3 than in cold gas since the barriers in the reaction chain do not seriously affect the final abundance of NH_3.

We have considered an alternative model, which is in fact simply a variation on the WK93 model, but where the shock radiation field is considered to have negligible strength above 13.6eV. The key addition in this model is the inclusion of grain surface processes.

4. Alternative Model

The processes that result in the collapse of a cloud to form a star-forming core appear likely to result in fragmentation, and the pieces not massive enough to go on to form stars would seem to be the most likely source of the clumps presumed here to be exposed to the shock radiation. It is the presence of clumps such as these that we have invoked previously as the reason why CS (J=1–0) emission is seen quite generally to have a larger spatial extent than NH_3 (1,1) in star-forming regions, despite the CS transition having a higher critical density and therefore supposedly tracing denser gas (Taylor, Morata and Williams 1996). This is based on the different formation timescales for the two molecules as the clumps form from more diffuse gas. These clumps would exist within a relatively tenuous interclump medium that surrounds the star-forming core, and through which the stellar outflows are propagating (in this picture young outflows such as HH211 seen only in the infrared would be expected to show molecular emission at their heads since they are still embedded in the high density core).

The interclump medium will, to some extent, allow through the interstellar radiation field, and this means that the clumps are exposed to a UV radiation field at some level even in the absence of shocks. However, beyond a certain value of visual extinction the dust in the interior will accrete ice mantles, like those seen in absorption in the infrared towards suitable background sources. The mantles are likely to be hydrogenated molecules of abundant atomic species, such as H_2O, NH_3, CH_4.., and although species such as methanol and carbon dioxide have also been seen it is possible that they only occur in the densest cores where accretion of volatile CO can occur followed by surface reaction or irradiation by a protostar. If the HH shock radiation can remove these mantles then a chemistry can follow that will enhance the abundance of many molecules, HCO^+ and NH_3 included.

To this end, we have modelled such a clump as a 1D slab of constant density and temperature exposed to a reduced flux of UV photons χ_0 in units of the Draine interstellar radiation field (Draine 1978). This is then similar to a model of a photodissociation region (e.g Hollenbach, Takahashi,& Tielens 1991). To obtain the clump initial chemical conditions just prior to the arrival of the HH object we use a network of chemical equations and associated reaction rates and numerically determine the fractional abundance of each species as a function of time, starting with everything atomic apart from molecular hydrogen (consistent with collapse from a diffuse state). Freeze-out of species onto dust surfaces is included, the only surface chemistry presumed being the hydrogenation of atoms to saturation. The chemistry is halted before complete freeze-out occurs, in line with observations that suggest that less than half of the observed gas phase abundance is seen in solid form (Chiar *et al.* 1995). The time is reset to $t = 0$, and the shock radiation is switched on, and at the same time complete mantle release is assumed. The strength of this radiation field depends on the shock velocity and the preshock (interclump) density and the nature of it was examined in some detail by Wolfire and Königl (1991). They converted the spectrum into an integrated UV flux in units of the interstellar flux (here χ_1), only a component of which is continuum that can dissociate H_2 or CO (χ_c). In our models we take values of χ_1 and χ_c similar to those of WK93, for the sake of comparison. We then once again follow the time-dependent chemistry.

The results for which the clump has a density of $n_H = 5 \times 10^4 \mathrm{cm}^{-3}$, temperature T=10K, and $\chi_1 = 20, \chi_c = 2$, are shown in Figure 1. Here the fractional abundances of a number of chemical species are plotted as a function of visual extinction for four different times after the shock radiation is switched on. As the mantles are released the observed molecules (dashed lines) briefly attain high abundances before a dissociation wave traverses the clump. The largest possible abundances in the absence of additional radiation are achieved in the absence of freezeout, and for NH_3 and HCO^+ these peak at 1×10^{-8} and 2×10^{-10} respectively. There is a clear enhancement in the model abundances by orders of magnitude, that might explain the observed emission. It is obvious that NH_3 is produced directly by mantle desorption, remaining at its initial fraction until destroyed by UV photons. HCO^+ is produced by reaction (1), however now the H_2O is provided by mantle chemistry rather than gas phase reactions. Since for $A_V \gtrsim 1$ and t $\lesssim$ 20 yr the C^+ is formed mostly by the preshock χ_0 UV field a clump more shielded than we have assumed will have a lower HCO^+ abundance at these points. Models with C^+ formed entirely from CH_4 released from mantles (i.e due to the χ_1 UV field) show HCO^+ down by factors of 2-3. For $A_V \lesssim 1$ or t $\gtrsim$20 yr there is little difference.

Except at the very edge of the clump the HCO^+ is also higher than that produced in the WK models, and this difference increases if the temperature in the clump is increased (see Taylor & Williams 1996) since the destruction of HCO^+ by dissociative recombination has a steep inverse temperature dependence,

$$HCO^+ + e^- \rightarrow CO + H, \qquad k \propto T^{-1}. \tag{4}$$

A simple comparison of the heating rate due to photoelectric ejection from dust (Bakes & Tielens 1994) with cooling due to CO, CII and OI (Hollenbach & McKee 1979), however, suggests that T > 30K is unlikely except in

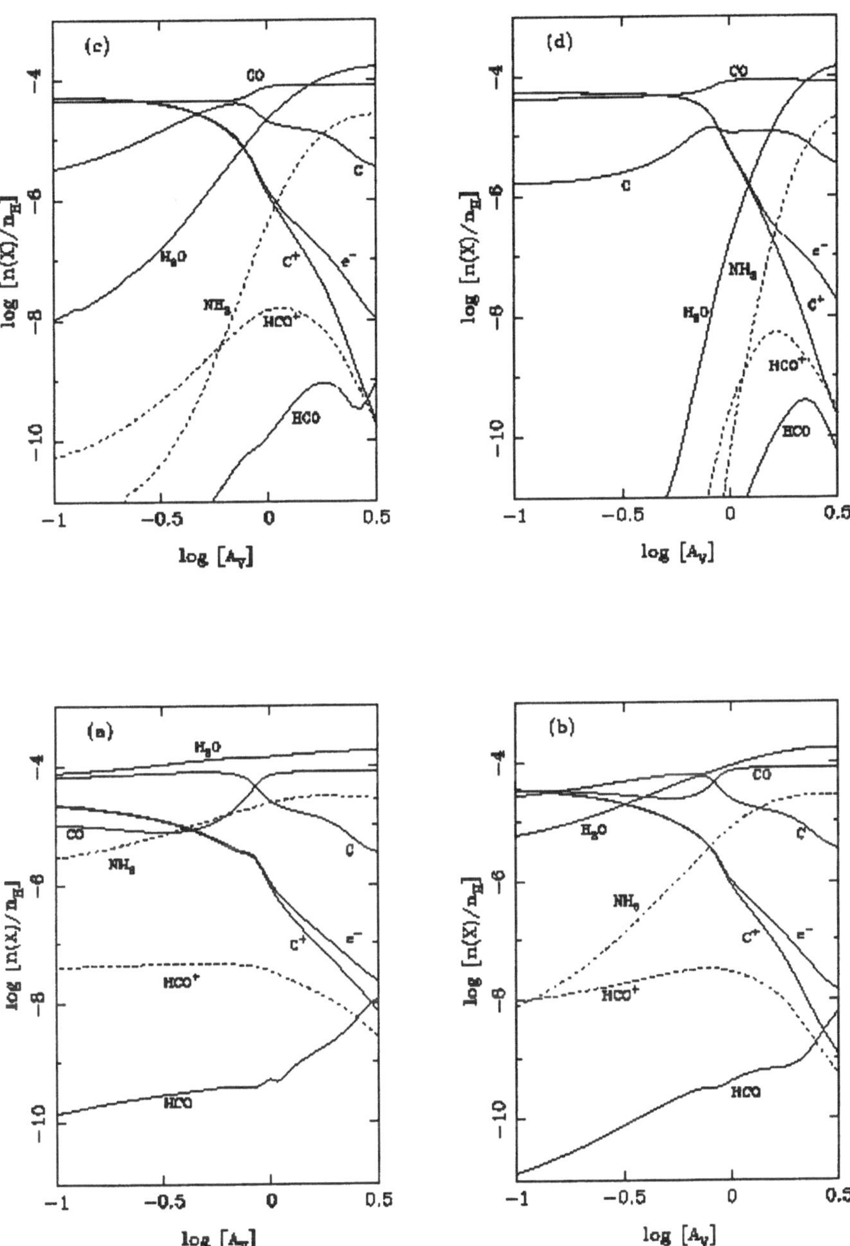

Figure 1. Fractional abundances (relative to hydrogen nuclei) of various species as a function of visual extinction (magnitudes) into a clump, for times 3, 10, 30 and 100 (a, b, c, d) years after the shock radiation is turned on.

the clump edge, as can be inferred from the thermal balance calculations in WK93.

The UV radiation field destroys molecules in cold gas and makes the chemistry very time-dependent. Lower radiation fields have the effect simply of extending the chemical timescales, hence these results will continue to hold for LEO's. The fact that the molecules shown in Fig.1 are destroyed on timescales of tens of years would imply that this chemistry may be progressing on observable timescales, a unique feature of these objects.

Other molecules are also enhanced in these models over those that would be obtained in a quiescent clump. The molecules we include in the chemistry are fairly simple since we have included relatively few species ($\sim$ 100) composed of the elements H,N,C,O,S, and Mg, and more complicated species would have to be included to properly analyse what the observational signatures might be. Photodissociation products of desorbed molecules have high abundances, such as CH_2, CH, OH, NH_2, but only the last two seem to have the potential to be observed in radio emission at reasonable angular resolution. Other potential species are derived from these products, for example CN and NO. Sulphur containing species, such as HCS^+ which is formed in a similar way to HCO^+ (C^+ reacting with H_2S) may also be abundant if the depletion of sulphur into refractory grain material is not too high. It may be that grain surface reactions involving hydrogenation of accreted CO also occur, in which case formaldehyde (tentatively detected by Dent) and methanol may also be observable.

5. Discussion Points

5.1. CLUMPS IN MOLECULAR OUTFLOWS

Dynamically unaffected clumps are an attractive proposition if they lie ahead of the leading jet bowshock, however this is clearly not the case in HH7-11, and what is more it appears that the shocks at HH1, HH2 or HH34 also do not represent the end of the jet (Reipurth & Heathcote 1992, Ogura 1996). HH objects embedded in the flow from an earlier jet episode may partly account for the fact that shock velocities inferred by excitation analysis are often much lower than those from linewidths or proper motion measurements. It seems likely that the clumps, whilst not lying in the path of the jet as would be the case in the shocked cloudlet hypothesis, are at least embedded in a lower velocity molecular outflow caused by the interaction of the jet with the ambient medium. The molecular outflows in HH1-2 and HH34 are weak (Chernin & Masson 1995), but a molecular outflow is observed in HH7-11 even at velocities $>$ 100 km s^{-1}, though this material sits on the optical axis and the clumps are more likely to find themselves in outflows with more standard velocities. Since this outflow has been traced in CO (J=1-0) (Liseau *et al.* 1988), which emits at low gas densities, and since the other outflows are so weak, we favour the view that this material is of low enough density to leave the clumps essentially unaffected, provided that entrainment of clump material in the boundary layers along the sides of the clump does not destroy the clumps. On the other hand clumps less stable will be shocked and swept up, and may be partly responsible for the HCO^+ (J=4-3) observations of HH7-11 by Dent *et al.* (1993).

5.2. MANTLE ACCRETION EFFICIENCY

Grain surfaces accrete gas phase species at the same efficiency regardless of depth in our models, which is clearly not applicable as A_V decreases, since solid state IR absorption features are not seen below a (region dependent) critical extinction. In fact this is possibly a consequence of the general interstellar IR radiation field preventing the formation of a surface monolayer (Hartquist, Williams & Whittet 1992). We have also assumed that the shock radiation field can instantly remove these mantles, which probably requires non-thermal photodesorption since thermal evaporation requires significant grain heating, and the grain temperature has only a weak dependence on the UV flux (Hollenbach *et al.* 1991). The sensitivity of the surface abundances to the accretion and desorption rates is an issue that requires further investigation.

An objection may be raised that we have postulated the release of ammonia from ice mantles, yet to date no ammonia has been observed in the solid phase. Whilst the reason why the hydrogenation of nitrogen atoms on the surface isn't as efficient as might be expected is puzzling, the current upper limits of a few percent of solid water still allow for a surface fractional abundance orders of magnitude greater than that formed from purely gas phase processes, hence our results can still be consistent with the observations.

5.3. HOT CORE ANALOGY

The irradiation model presented here bears many similarities with models of hot cores in high mass star forming regions. There the onset of stellar radiation following the birth of a new star is thought to remove mantles and start a chemistry that actually produces some of the most complex species known in space (Millar 1993). It is an intriguing possibility that some of these molecules may be observable in the low mass star forming regions where most of these HH objects are formed. A note of caution must be added though, for although the hot cores and the clumps are comparable in size they are probably considerably denser ($n_H \sim 10^7 cm^{-3}$) so that the column densities of a given species are likely to be much greater, and the surface chemistry may be more complex. It may also be that some of the chemistry that occurs in the hot cores is a consequence of the inferred gas temperature of $\sim$ 200K, which is greater than that possible in clump interiors.

Deuteration fractionation is possible in these cold clumps through purely gas phase chemistry, but Brown & Millar (1989) showed that hydrogenation or deuteration on grains can lead to much larger abundances on the surface. Since a major assumption in our model is that large quantities of (undetectable) water are released into the gas phase, a good test could be a search for HDO, which does have detectable transitions in cold clouds.

References

Bakes,E.L.O,Tielens,A.G.G.M.,1994,*ApJ*,**427**,822

Böhm,K.H,Solf,J.,1992,*AJ*,**104**,1193

Brown,P.D.,Millar,T.J.,1989,*MNRAS*,**237**,661

Chiar,J.E.,Adamson,A.J.,Kerr,T.H.,Whittet,D.C.B.,1995,*ApJ*,**455**,234

Chernin,L.M.,Masson,C.R.,1995,*ApJ*,**443**,181

Davis,C.J.,Dent,W.R.F.,1993,*MNRAS*,**261**,371

Davis,C.J.,Dent,W.R.F.,Bell Burnell,S.J.,1990,*MNRAS*,**244**,173

Dent, W.R.F.,1997, in *Low Mass Star Formation - from Infall to Outflow*, poster proceedings of IAU Symp. No. 182, ed. F. Malbet & A. Castets, Grenoble, p.88

Dent,W.R.F.,*et al.* ,1993,*MNRAS*,**262**,L13

Draine,B.T.,1978,*ApJS*,**36**,595

Girart,J.M.,*et al.* ,1994,*ApJ*,**435**,L145

Hartquist, Williams & Whittet 1992 MNRAS,258,599

Hollenbach,D.J.,McKee,C.F.,1979,*ApJS*,**41**,555

Hollenbach,D.J.,Takahashi,T.,Tielens,A.G.G.M.,1991,*ApJ*,**377**,192

Liseau,R.,Sandell,G.,Knee,L.B.G.,1988,*A&A*,**192**,153

Millar,T.J.,1993,*Dust and Chemistry in Astronomy.* Millar,T.J.,Williams,D.A. (eds.),IOP Publishing Ltd., Bristol.

Ogura,K.,1996,*ApJ*,**450**,L23

Reipurth,B.,Heathcote,S.,1992,*A&A*,**257**,693

Rudolph,A.,1992,*ApJ*,**397**,L111

Rudolph,A.,W.J.,Welch,1988,*ApJ*,**326**,L31

Rudolph,A.,W.J.,Welch,1992,*ApJ*,**395**,488

Taylor,S.D.,Morata,O.,Williams,D.A.,*A&A*,**313**,269

Taylor,S.D.,Williams,D.A.,1996,*MNRAS*,**282**,1343

Torrelles,J.M.,*et al.* ,1992,*ApJ*,**396**,L95

Torrelles,J.M.,*et al.* ,1993,*ApJ*,**417**,655

Whittet,D.C.B.,*et al.* ,1996,*A&A*,**315**,L357

Williams,D.A.,1994,*Contemporary Physics*,**35**,269

Wolfire,M.G.,Königl,A.,1991,*ApJ*,**383**,205

Wolfire,M.G.,Königl,A.,1993,*ApJ*,**415**,204 (WK93)

III. Theoretical Models

PROTOSTELLAR X-RAYS, JETS, AND BIPOLAR OUTFLOWS

FRANK H. SHU AND HSIEN SHANG
Astronomy Department, University of California
Berkeley, CA 94720-3411, USA

Abstract. We review the theory of x-winds in young stellar objects (YSOs). In particular, we consider how a model where the central star does not corotate with the inner edge of the accretion disk may help to explain the enhanced emission of X-rays from embedded protostars. We argue, however, that the departure from corotation is not large, so a mathematical formulation that treats the long-term average state as steady and axisymmetric represents a useful approximation. Magnetocentrifugally driven x-winds of this description collimate into jets, and their interactions with the surrounding molecular cloud cores of YSOs yield bipolar molecular outflows.

1. Introduction

Beginning with the notion that rapid rotation coupled with strong magnetic fields can considerably enhance the mass loss $\dot{M}_{\mathrm{w}}$ of thermally driven stellar winds (Mestel, 1968), a community effort has gone into the development of the set of ideas comprising the theory of x-winds in YSOs. Even if the resultant flow is quite cold, Hartmann and MacGregor (1982) demonstrated that $\dot{M}_{\mathrm{w}}$ could have almost arbitrarily large values, with the flow velocity at infinity dependent only on the ratio of the azimuthal and radial field strengths at the position where the gas is injected onto open field lines near the equator of a protostar that rotates near breakup. Shu *et al.* (1988) assigned the cause for the protostar to spin at breakup to a circumstellar disk that abuts against the surface of the central object and accretes onto it at a high rate $\dot{M}_D$. They also replaced Hartmann and MacGregor's arbitrary choices for the injection density and angle of magnetic field direction with the requirement that in steady state, the wind mass-loss rate $\dot{M}_{\mathrm{w}}$ and average terminal velcocity $\bar{v}_{\mathrm{w}}$ must be a definite fraction f and multiple

B. Reipurth and C. Bertout (eds.), Herbig–Haro Flows and the Birth of Low Mass Stars, 225–239.

$(2\bar{J}_w - 3)^{1/2}$, respectively, of the adjoining disk's accretion rate $\dot{M}_D$ and rotation velocity $\Omega_x R_x$ (see below).

In the interim, Blandford and Payne (1982) advanced their influential self-similar model of centrifugally driven winds from the surfaces of magnetized accretion disks. Pudritz and Norman (1983) applied these pure disk-wind models to bipolar outflows, and Königl (1989) investigated how the wind might smoothly join a pattern of accretion flow inside the disk. Heyvaerts and Norman (1989) studied how the winds might collimate asymptotically into jets, while Uchida and Shibata (1985) and Lovelace *et al.* (1991) advocated alternative driving mechanisms where magnetic pressure gradients play a bigger role in the acceleration of a disk-blown wind.

Motivated by the problem of binary X-ray sources, a parallel line of research developed concerning how magnetized stars accrete from surrounding disks. Ghosh and Lamb (1978) used order-of-magnitude arguments to show that a strongly magnetized star would truncate the surrounding accretion disk at a larger radius than the stellar radius R_* and divert the equatorial flow along closed field-line funnels toward the polar caps. Although Ghosh and Lamb thought that this inflow would spin up the central object faster than if the accretion disk had extended right up to the stellar surface, Königl (1991) made the surprising and insightful suggestion that the process might torque down the star and account for the relatively slow rate of spin of observed T Tauri stars. Observational support for magnetospheric accretion was subsequently marshalled by Edwards *et al.* (1993) and Hartmann *et al.* (1994).

In a modification of Ghosh and Lamb's ideas based on unpublished work by Arons, McKee, and Pudritz, Arons (1986) proposed that a centrifugally-driven outflow accompanies the funnel inflow [see also Camenzind (1990)]. Independently, Basri (1989, unpublished) had arrived observationally at the same suggestion by extending the synthesis work by Bertout *et al.* (1988) on ultraviolet excesses in T Tauri stars. Shu *et al.* (1994a) put these ideas together into a concrete proposal that generalized the earlier x-wind model of Shu *et al.* (1988).

2. Generalized X-wind Model

In the generalized x-wind model, if the star has mass M_* and magnetic dipole moment μ_*, the gas disk is truncated at an inner radius,

$$R_x = \Phi_{dx}^{-4/7} \left(\frac{\mu_*^4}{G M_* \dot{M}_D^2} \right)^{1/7} . \tag{1}$$

A disk of solids may extend inward of R_x to the evaporation radius of calcium-aluminum silicates and oxides [see Shang *et al.* (1997) and Meyer *et al.* (1997)]. In equation (1), Φ_{dx} is a dimensionless number of order unity that measures the amount of magnetic dipole flux that has been pushed by the disk accretion flow to the inner edge of the disk. In the preferred model of Najita and Shu (1994) and Ostriker and Shu (1995), $\Phi_{dx} = 1.15$, and the actual magnetic flux trapped in a small neighborhood of R_x is 1.5 times larger than the pure dipole value. In other words, magnetic flux is swept toward R_x from both larger and smaller radii (see below).

For a Keplerian disk, the inner edge of the disk rotates at angular speed

$$\Omega_x = \left(\frac{GM_*}{R_x^3}\right)^{1/2}. \tag{2}$$

To satisfy mass and angular momentum balance, the disk accretion divides at R_x into a wind fraction, $\dot{M}_w = f\dot{M}_D$, and a funnel-flow fraction, $\dot{M}_* = (1-f)\dot{M}_D$, where

$$f = \frac{1 - \bar{J}_* - \tau}{\bar{J}_w - \bar{J}_*}. \tag{3}$$

In equation (3) $\bar{J}_*$ and $\bar{J}_w$ are, respectively, specific angular momenta nondimensionalized in units of $R_x^2\Omega_x$ and averaged over funnel and wind streamlines, and τ is the negative of the viscous torque of the disk, $-\mathcal{T}$, acting on its inner edge and measured in units of $\dot{M}_D R_x^2 \Omega_x$. For typical application (see below), $\bar{J}_* \approx 0 \approx \tau$ and $\bar{J}_w \approx 3$ or 4; thus, $f \approx 1/3$ or $1/4$. For every gram of material that comes through the disk, the x-wind model requires that a significant fraction gets thrown back out, carrying away the excess angular momentum that the star finds difficult to accept directly.

In steady state, the star is regulated to corotate with the inner disk edge,

$$\Omega_* = \Omega_x. \tag{4}$$

If corotation did not hold, the system would react to reduce the discrepancy between Ω_* and Ω_x. For example, suppose the star turns faster than the inner edge of the disk, $\Omega_* > \Omega_x$. The field lines attached to both would then continuously wrap into ever tighter trailing spirals, with the field lines adjacent to the star tugging it backward in the sense of rotation. This tug decreases the star's angular rate of rotation Ω_* to more nearly equal the rate Ω_x. Conversely, imagine that $\Omega_* < \Omega_x$. With the star turning slower than the inner edge of the disk, the field lines attached to both would continuously wrap into ever tighter leading spirals, with the field lines adjacent to the star tugging it forward in the sense of rotation. This tug increases the star's angular rate of rotation Ω_*, again to more nearly equal the rate Ω_x. In true steady state, $\Omega_* = \Omega_x$, and the funnel-flow field lines acquire just enough of a trailing spiral pattern (but without continuously wrapping up) so that the excess of material angular momentum brought

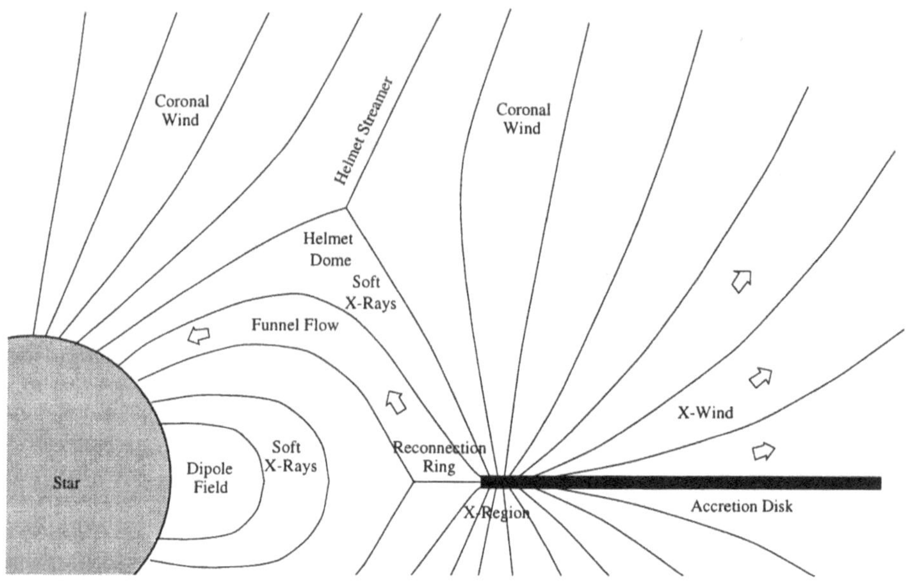

Figure 1. Schematic drawing of the x-wind model. Coronal winds from the star and the disk may help the x-wind to open field lines surrounding the helmet streamer, but this aspect of the configuration is not central to the model.

toward the star by the inflowing gas is transferred outward by magnetic torques to the footpoints of the magnetic field in the disk (Shu *et al.*, 1994a).

The x-wind gas gains angular momentum and the funnel gas loses angular momentum at the expense of the matter at the footpoint of the field in, respectively, the outer and inner parts of the X-region. As a consequence, this matter and the field lines across which it is diffusing pinch toward the middle of the X-region. In reality, Shu *et al.* (1997) point out that the idealized steady state of exact corotation probably cannot be maintained because of dissipative effects. Two surfaces of null poloidal field lines – labeled as "helmet streamer" and "reconnection ring" in Figure 1 – mediate the topological behavior of dipole-like field lines of the star, opened field lines of the x-wind, and trapped field lines of the funnel inflow emanating from the x-region. Across each null surface, which begin or end on "Y-points" [called "kink points" by Ostriker and Shu (1995)], the poloidal magnetic field suffers a sharp reversal of direction. By Ampére's law, large electric currents must flow out of the plane of the Figure along the null surfaces. Nonzero electrical resistivity would lead to the dissipation of these currents and to the reconnection of the oppositely directed field lines [see, *e.g.*, Biskamp (1993)]. The resultant reduction of the trapped magnetic flux in the x-region as the fan of field lines presses into the evacuated region of the annihilated fields would change the numerical value of the coefficient Φ_{dx} in equation (1).

When R_x changes, the angular speed Ω_x of the footpoint of magnetic field lines in the x-region will vary according to equation (2). However, the considerable inertia of the star prevents its angular velocity Ω_* from changing on the time scale of magnetic reconnection at the null surfaces of the magnetosphere. The resultant shear when $\Omega_* \neq \Omega_x$ will stretch and amplify field lines attached to both the star and the disk. The poloidal field will bulge outward from the increased magnetic pressure, inserting more magnetic flux into the fan of field lines emanating from the x-region [see the simulations of Linker and Mikić (1994) and Hayashi *et al.* (1996)]. Dynamo action inside the star would presumably replace the upward rising dipole-like poloidal fields. This dynamo action would be enhanced by the wrapping of field lines between the star and disk. Averaged over long times, we envisage a secular balance, with enough dynamo-generated poloidal field being inserted into the x-region to balance the rate of field dissipation at the null surfaces.

Johns and Basri (1995) and Johns-Krull and Hatzes (1997) have found two sources in which the hydrogen lines that diagnose for inflow and outflow exhibit a periodicity equal to the rotation period of the central T Tauri star. If the funnel flow and wind emanate in these objects from the surrounding accretion disks, then the source regions for the flowing gas corotate with the central stars – in agreement with the prediction that $\Omega_* = \Omega_x$. However, the majority of objects studied by Johns-Krull and Basri (1997) do not show such periodicity, indicating that even if Ω_* does equal Ω_x as a long-term average, Ω_* may not generally equal Ω_x instantaneously.

3. Evidence from Protostellar X-rays

Recent X-ray observations of YSOs may yield important constraints on the degree by which Ω_* does not equal Ω_x. Comparing data from *ROSAT* and *ASCA*, Carkner *et al.* (1996) find that the X-ray spectra of classical T Tauri stars (CTTSs) are harder than weak-lined T Tauri stars (WTTSs) (Feigelson *et al.*, 1987), but not as hard as embedded protostars (Koyama *et al.*, 1996). The luminosity of low-mass protostars at all X-ray energies is typically [see Grosso *et al.* (1997) and the discussion in note 11 of Shu *et al.* (1997)]:

$$L_x^{tot} \sim 1 \times 10^{32} \text{ erg s}^{-1}, \tag{5}$$

which is about 2 orders of magnitude larger than the corresponding values for WTTSs and CTTSs. Since a main difference among these objects is the rate of disk accretion, it is tempting to attribute the enhanced X-ray emission in protostars to the strong interaction between the stellar magnetosphere and the surrounding accretion disk [see also Hayashi *et al.* (1996) and Goodson *et al.* (1997)].

When $\Omega_* \neq \Omega_x$, Shu *et al.* (1997) use dimensional analysis to estimate the rate of increase of magnetic energy in field lines attached to both the star and the x-region:

$$\frac{dE_{\rm mag}}{dt} = \alpha|\Omega_* - \Omega_x|\frac{\mu_*^2}{R_x^3}, \tag{6}$$

where α is a numerical coefficient of order unity. Numerical calculations suggest that coronal mass ejections occur when the field energy increases to a point where erupting magnetized plasmoids can open the field lines under the helmet dome [see the simulation of solar flares in Figures 1 and 2 of Linker and Mikić (1994)]. As previewed in the previous section, the analogous process in the reconnection ring will sporadically inject fresh poloidal fields to offset the reconnection loss of trapped magnetic flux in the x-region.

We write $|\Omega_* - \Omega_x|$ as a fraction S of Ω_x. Using equations (1) and (2) to eliminate μ_* and Ω_x, we then get

$$\frac{dE_{\rm mag}}{dt} = (\alpha S \Phi_{\rm dx}^2)\frac{GM_*\dot{M}_D}{R_x}. \tag{7}$$

For low-mass protostars, the quantity $GM_*\dot{M}_D/R_x$ has a typical order of magnitude of 3×10^{33} erg s^{-1}. Thus, if the time-averaged X-ray luminosity $L_x^{\rm tot}$ in equation (5) is ultimately supplied by the time-averaged release of magnetic energy $dE_{\rm mag}/dt$, and if α and $\Phi_{\rm dx} \sim 1$, then equation (7) allows us to estimate that $S \sim 0.03$. This represents a small average departure from corotation. While such instantaneous departures may be important in understanding the short-term variability and eruptive magnetic activity, we henceforth ignore them for the long-term behavior of YSOs.

4. Outline of Mathematical Formulation

A mathematical formulation of the steady-state problem when $\Omega_* = \Omega_x$ was given in outline by Shu *et al.* (1988) and in detail by Shu *et al.* (1994b), who nondimensionalized the governing equations by introducing R_x, Ω_x^{-1}, and $\dot{M}_w/4\pi R_x^3\Omega_x$, respectively, as the units of length, time, and density. Assuming axial symmetry and time-independence in a frame that corotates with Ω_x, we may then introduce cylindrical coordinates (ϖ, φ, z) and a stream-function $\psi(\varpi, z)$ that allows the satisfaction of the equation of continuity,

$\nabla \cdot (\rho \mathbf{u}) = 0$, in the meridional plane:

$$\rho u_\varpi = \frac{1}{\varpi}\frac{\partial \psi}{\partial z}, \qquad \rho u_z = -\frac{1}{\varpi}\frac{\partial \psi}{\partial \varpi}. \tag{8}$$

Field freezing in the corotating frame, $\mathbf{B} \times \mathbf{u} = 0$, implies that the magnetic field is proportional to the mass flux, $\mathbf{B} = \beta \rho \mathbf{u}$, where β is a scalar. The condition of no magnetic monopoles, $\nabla \cdot \mathbf{B} = 0$, now requires that β be conserved on streamlines, *i.e.*,

$$\beta = \beta(\psi). \tag{9}$$

Similarly, the conservation of total specific angular momentum requires that the amount carried by matter in the laboratory frame, $\varpi(\varpi\Omega_\mathrm{x} + u_\varphi)$ where Ω_x is replaced by 1 in dimensionless equations, plus the amount carried by Maxwell torques, $-\varpi B_\varphi \mathbf{B}$, per unit mass flux, $\rho\mathbf{u}$, equals a function of ψ alone:

$$\varpi[(\varpi + u_\varphi) - \beta^2 \rho u_\varphi] = J(\psi). \tag{10}$$

Finally, conservation of specific "energy" in the corotating frame for a gas with dimensionless isothermal sound speed ϵ results in Bernoulli's theorem:

$$\frac{1}{2}|\mathbf{u}|^2 + \mathcal{V}_\mathrm{eff} + \epsilon^2 \ln \rho = H(\psi). \tag{11}$$

Up to an arbitrary constant, defined so that $\mathcal{V}_\mathrm{eff} = 0$ at the x-point $\varpi = 1$ and $z = 0$, $\mathcal{V}_\mathrm{eff}$ is the dimensionless gravitational potential plus centrifugal potential (measured in units of $GM_*/R_\mathrm{x} = \Omega_\mathrm{x}^2 R_\mathrm{x}^2$):

$$\mathcal{V}_\mathrm{eff} = \frac{3}{2} - \frac{1}{(\varpi^2 + z^2)^{1/2}} - \frac{\varpi^2}{2}. \tag{12}$$

Equations (9) - (11), with β, J, and H arbitrary functions of ψ, represent formal integrations of the governing set of equations. The remaining equation for the transfield momentum balance – the so-called Grad-Shafranov equation – cannot be integrated analytically and reads

$$\nabla \cdot (\mathcal{A} \nabla \psi) = \mathcal{Q}, \tag{13}$$

where $\mathcal{A}$ is the Alfvén discriminant,

$$\mathcal{A} \equiv \frac{\beta^2 \rho - 1}{\varpi^2 \rho}, \tag{14}$$

and $\mathcal{Q}$ is a source function for the internal collimation (or decollimation) of the flow:

$$\mathcal{Q} = \rho \left[\frac{u_\varphi}{\varpi} J'(\psi) + \rho |\mathbf{u}|^2 \beta\beta'(\psi) - H'(\psi) \right], \tag{15}$$

with primes denoting differentiation with respect to the argument ψ.

The quantity $\mathcal{A} = 0$, *i.e.*, $\beta^2\rho = 1$, when the square of the flow speed $|\mathbf{u}|^2$ equals the square of the Alfvén speed, $|\mathbf{B}|^2/\rho$. For sub-Alfvénic flow, $\mathcal{A} > 0$; for super-Alfvénic flow, $\mathcal{A} < 0$. If $\mathcal{A}$ were freely specifiable (which it is not), equation (13) would resemble the time-independent heat-conduction equation. What is spreading in the meridional plane of our problem, however, is not heat, but streamlines.

5. Fixing the Free Functions

When combined with Bernoulli's equation (11), the Grad-Shafranov equation (13) has three possible critical surfaces associated with it, corresponding to slow MHD, Alfvén, and fast MHD crossings (Weber and Davis, 1967). Thus, when applied to the problem of the x-wind, equation (13) is a second-order partial differential equation of elliptic type interior to the fast surface and hyperbolic type exterior to this surface [see Heinemann and Olbert (1978) and Sakurai (1985)]. The unknown functions $\beta(\psi)$, $J(\psi)$, and $H(\psi)$ are to be determined self-consistently so that the three crossings of the critical surfaces are made smoothly. This does not fix all three functions uniquely since the loci of the critical surfaces in (ϖ, z)-space are not known in advance. It turns out that we can choose one of the loci freely. Alternatively, we can freely specify one of the functions β, J, or H. The method of Najita and Shu (1994) fixes in advance the locus of the Alfvén surface and determines all other quantities self-consistently from this parameterization. Shang and Shu (1997) have invented a simplified procedure in which the function $\beta(\psi)$ is specified in advance; then $J(\psi)$, $H(\psi)$, and the loci of the slow, Alfvén, and fast surfaces are found as part of the overall solution (including force balance with the opened field lines of the star and dead zone). We may regard the procedure of choosing $\beta(\psi)$ arbitrarily as a mathematical substitute for the physical problem, where loading of matter onto field lines occurs in the x-region with the breakdown of the ideal MHD approximation of field freezing (Shu *et al.*, 1994a).

Apart for a trivial replacement of $\dot{M}_*$ for $\dot{M}_{\rm w}$, the funnel flow behaves somewhat differently from the x-wind. The funnel flow has a slow MHD crossing but probably no Alfvén or fast MHD crossings. (In steady state the star and disk can communicate with each other along closed field lines by means of the latter two signals.) As a consequence, both $\beta(\psi)$ and $J(\psi)$ can be freely specified for the funnel flow, although $J(\psi)$ must ultimately be made self-consistent with the physical assumption that no spinup or spindown of the star occurs in steady state, *i.e.*, that Ω_* remains equal to $\Omega_{\rm x}$ as a long-term average. In these circumstances, when one has a small star, Ostriker and Shu (1995) show that $J(\psi)$ is likely to be nearly zero on every funnel-flow streamline ψ = constant. In other words, $\bar{J}_* \approx 0$ and any excess angular momentum brought to the star by the matter inflow is transferred back to the disk by the magnetic torques of a trailing spiral pattern of funnel field lines.

6. The Cold Limit

The overall problem is mathematically tractable because the parameter ϵ is much smaller than unity (typically, $\epsilon \approx 0.03$). This leads to many simplifications, in particular, to the use of matched asymptotic expansions for solving and connecting different parts of the flow. For example, we may show that the slow MHD crossing must be made by matter-carrying streamlines within a fractional distance ϵ of R_x (unity in our nondimensionalization), and that to order ϵ^2, $H(\psi)$ may be approximated as zero. In the limit $\epsilon \to 0$, the gas that becomes the x-wind (or the funnel flow) emerges with linearly increasing velocities in a fan of streamlines from the X-region as if it were a single point.

In this approximation, the function $\beta(\psi)$ cannot be chosen completely arbitrarily if the magnetic field, mass flux, and mass density do not to diverge on the uppermost streamline $\psi \to 1$ as the x-wind leaves the x-region. For modeling purposes, Shang and Shu (1997) adopt the following distribution of magnetic field to mass flux:

$$\beta(\psi) = \beta_0(1-\psi)^{-1/3}, \tag{16}$$

where β_0 is a numerical constant related to the mean value of β averaged over streamlines:

$$\bar{\beta} \equiv \int_0^1 \beta(\psi)\, d\psi = \frac{3}{2}\beta_0. \tag{17}$$

The reader should not worry that equation (16) implies $\beta \to \infty$ as $\psi \to 1$. This singular behavior merely reflects the fact that the magnetic field $\mathbf{B} = \beta\rho\mathbf{u}$ is nonzero on the last x-wind streamline where by definition ρ must become vanishingly small while $\mathbf{u}$ remains finite.

7. Asymptotic Collimation into Jets

Given the form $\beta(\psi)$ from equation (16), the loci of streamlines at large distances from the origin may be recovered in spherical polar coordinates $[r = (\varpi^2 + z^2)^{1/2}, \theta = \arctan(\varpi/z)]$ from the asymptotic analysis of Shu *et al.* (1995):

$$r = \frac{2\bar{\beta}}{C}\cosh[F(C,1)], \qquad \sin\theta = \operatorname{sech}[F(C,\psi)], \tag{18}$$

where different values of the dimensionless current C correspond to different locations on any streamline ψ = constant, and where $F(C, \psi)$ is the integral function,

$$F(C, \psi) \equiv \frac{1}{C} \int_0^{\psi} \frac{\beta(\psi)\, d\psi}{[2J(\psi) - 3 - 2C\beta(\psi)]^{1/2}}. \tag{19}$$

Notice that $r \to \infty$ when $C \to 0$ and *vice-versa.*

The factor $2\bar{\beta}$ in equation (18) applies if the field lines in the dead zone are completely closed as assumed in the calculations of Ostriker and Shu (1995) and Shu *et al.* (1995), but it becomes $6\bar{\beta}$ if they are completely opened as in Figure 1 [see the discussion of Shang and Shu (1997)]. In the former case, the hoop stresses of the toroidally wrapped fields on the uppermost streamlines of the x-wind are asymptotically balanced by the magnetic pressure of a bundle of longitudinal field lines coming from the star that carries the same dimensionless magnetic flux, $2\pi\bar{\beta}$, as their opened counterparts in the x-wind. In the latter case, there are additional longitudinal field lines roughly parallel and antiparallel to the polar axis carrying oppositely directed flux $\pm 2\pi\bar{\beta}$ from the opened field lines of the dead zone. The general case would have a coefficient in equation (18) between $2\bar{\beta}$ and $6\bar{\beta}$.

The density and velocity fields associated with equation (18) are obtained from

$$\rho = \frac{C}{\beta\varpi^2}, \qquad v_{\rm w} = (2J - 3 - 2C\beta)^{1/2}, \tag{20}$$

where $\varpi \equiv r \sin\theta$ and $v_{\rm w}$ is the wind speed in an inertial frame. The solid and short dashed lines in Figure 2 show, respectively, isodensity contours and flow streamlines for a case $\beta_0 = 1$ computed on four different scales by the simplified approximate procedure discussed by Shang and Shu (1997). The dimensional units of length and velocity, $R_{\rm x}$ and $\Omega_{\rm x} R_{\rm x}$, are ~ 0.06 AU and ~ 100 km s^{-1} in typical application. The outermost density contour in the fourth panel is $\rho = 10^{-8}$ in units of $\dot{M}_{\rm w}/4\pi\Omega_{\rm x} R_{\rm x}^3$. For the model $\bar{J}_{\rm w} = 3.73$, as can be obtained from the square of the average ϖ position of the inner set of long dashes marking the location of the Alfven surface. The Alfvén and fast surfaces formally asymptote to infinity as the upper streamline $\psi = 1$ is approached, where the density becomes vanishingly small. Because numerical computations are difficult in this limit, the actual uppermost streamline displayed is $\psi = 0.98$ rather than $\psi = 1$. Although streamlines collimate logarithmically slowly, isodensity contours become cylindrically stratified fairly quickly, roughly as $\rho \propto \varpi^{-2}$ [*cf.* equation (20)]. Since the radiative emission of forbidden lines is highly biased toward regions of moderately high density, the flow will appear more collimated than it actually is. There is a greater wide-angle component to YSO

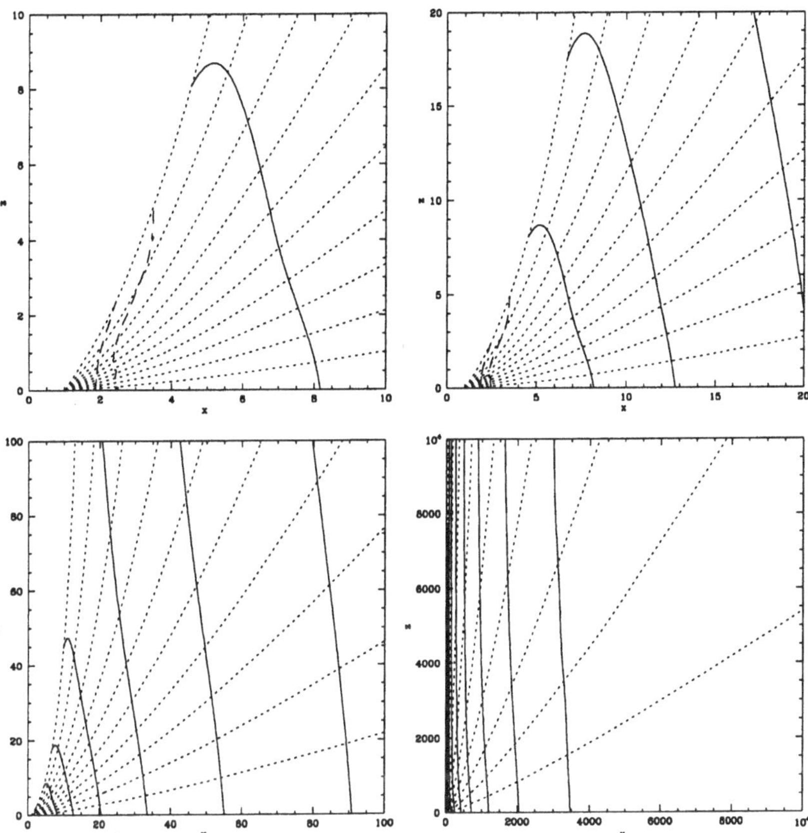

Figure 2. Isodensity contours (solid curves) and streamlines (dotted curves) for a cold x-wind with $\beta(\psi) = \beta_0(1-\psi)^{-1/3}$, where $\beta_0 = 1$. Isodensity contours are spaced logarithmically in intervals of $\Delta \log_{10} \rho = 0.5$, and streamlines are spaced so that successive dotted lines contain an additional 10% of the total mass loss in the upper hemisphere of the flow. The loci of the Alfvén and fast surfaces are marked by dashed lines. The empty space inside the uppermost streamline, $\varpi \leq \varpi_1$, is filled with open field lines from the central star that asymptotically have the field strength, $B_z = 2\bar{\beta}/\varpi_1^2$.

winds than generally appreciated. In other words, the jetlike appearance of YSO outflows is somewhat of an optical illusion.

The relative lack of streamline collimation, as opposed to density collimation, is a salient feature of the x-wind model and provides a key discriminating test via spectroscopy of forbidden line emission at high spatial and spectral resolution. The x-wind model predicts that such lines should have larger velocity widths at the base of the flow, where a typical line of sight encounters flow velocities at a variety of angles, than farther up the length of the jet, where the flow vectors are better oriented along a single direction. A hint of this behavior may already be present in the data presented by Solf at this conference.

8. Bipolar Molecular Outflows

Shu *et al.* (1991) proposed that bipolar molecular outflows are created when a YSO wind with power dependent on polar angle θ as $P(\theta)$ sweeps up static ambient material with a density distribution (generated by ambipolar diffusion) given by

$$\rho_{\mathrm{amb}}(r, \theta) = \frac{a^2 Q(\theta)}{2\pi G r^2}, \tag{21}$$

where a is the effective isothermal sound speed of the surrounding molecular cloud core and $Q(\theta)$ describes the flattening produced by magnetic forces. Li and Shu (1996) have computed the resulting model when $P(\theta) \propto 1/\sin^2\theta$ as predicted by the asymptotic analysis of the previous section and when $Q(\theta)$ of the theoretical molecular cloud core (a self-gravitating magnetized toroid) is chosen to be consistent with the observational data on core flattening (Myers *et al.*, 1991). They show that the model, which has *no* adjustable parameters in nondimensional form, gives surprisingly good recreations of the observations concerning lobe shapes, line profiles, distributions of mass with line-of-sight velocities, and distributions of momenta with bipolar axial distance (Chernin and Masson, 1995). Given the simplicity of the model, where the real magnetohydrodynamics is collapsed into a single thin-shell treatment, the agreement is quite encouraging, especially in light of the difficulty that more complex (and also more *ad hoc*) jet-driven models have traditionally had in explaining the same data (see the discussions of Cabrit *et al.* and Eislöffel in this volume).

9. Stability of Parsec Long Jets

The small amount of ambient matter swept to high velocities, which solves the original objection of Masson and Chernin (1992) to the bipolar outflow model of Shu *et al.* (1991), arises because the heaviest part of the x-wind encounters only the low-density material of the poles of the molecular toroid [where $Q(\theta)$ ideally becomes vanishingly small]. As a consequence, it is easy for the x-wind to break out of the molecular cloud core in these directions and, thus, to account for the parsec long jets presented by Bally at this conference.

It has often been asked how jets can maintain their integrity for such long paths. The glib answer provided by all proponents of magnetocentrifugal driving is that the flow is self-collimated. Self-collimation occurs by the hoop stresses of the toroidal magnetic fields that grow in relative strength to the poloidal components as the flow distance from the source increases. But such magnetic confinement schemes have long been known to the fusion community to be unstable with respect to kinking and sausaging [see Figure 10.6 in Jackson (1975); see also Eichler (1993)]. A solution to the problem has also long been known and is the principle behind working Tokamaks: the introduction of dynamically important levels of longitudinal magnetic fields along the length of the plasma column to stabilize the kinking and sausaging motions [see Figures 10.7 and 10.8 of Jackson (1975)]. Such dynamically important levels of longitudinal fields along the central flow axis are exactly what fill the hollow core of the jet in the x-wind model of Figure 2.

10. Crushing the Magnetosphere?

At this conference, Hartmann has questioned the applicability of x-wind models to YSO jets and bipolar outflows on the basis that the mass accretion rate $\dot{M}_D$ in the most active sources should be large enough to crush the magnetosphere to the stellar surface. His argument proceeds by assuming that every quantity in equation (1) is fixed at the same value in different sources except for $\dot{M}_D$. He scales according to CTTSs where $R_x \sim 5R_*$ after he has lowered $\dot{M}_D$ in CTTSs by a factor of 10 from the conventionally estimated values [see Bertout (1989) and Edwards *et al.* (1993)]. He then finds R_x to be smaller than R_* for sources in which $\dot{M}_D$ is appreciably larger than the value he adopts for CTTSs.

Clearly, more debate and re-evaluation is needed to assess the validity of the lower $\dot{M}_D$ adopted for CTTSs. Even if we accept the proposition, a weak point exists in the supposition that everything else on the right-hand side of equation (1) retains the same value in different sources. Consider the objects most likely to have crushed stellar magnetospheres – sources in FU Orionis outburst. During an FU Orionis outburst, $\dot{M}_D$ may indeed increase from the quiescent state by a few orders of magnitude. The increase in $\dot{M}_D$ should result initially in a decrease in R_x. But, in accordance with equation (6), the resulting mismatch in Ω_x and Ω_* will generate on the shear time scale a substantial increase in the magnetic energy of the configuration. This increase inserts more trapped magnetic flux into the x-region (and thereby increasing the coefficient $\Phi_{dx}^{-4/7}$). Through enhanced dynamo action, it may also increase the stellar magnetic dipole moment μ_* (thus also increasing R_x relative to Hartmann's scaling). Hence, it may not be so easy, even in the case of FU Orionis outbursts, to completely crush a stellar magnetosphere by disk accretion. A substantial portion of the outer layers of the star (the part that anchors the magnetic field) may have to be spun to breakup (so that Ω_* effectively equals Ω_x, shutting off the enhanced dynamo) before the stellar magnetosphere can be crushed. But such a situation reproduces precisely the assumption underlying the original x-wind model (Shu *et al.*, 1988), so an x-wind can still be launched in this case.

Finally, we note that a departure of Ω_* from Ω_x has no effect on the generalized theory of x-winds reviewed in this paper, since the field lines in the x-wind are disconnected from those of the star (see Figure 1). As long as the base of the x-wind rotates at or close to the Keplerian speed Ω_x given by equation (2), the footpoint of the field will have the necessary centrifugal fling to drive an x-wind. What will be affected by $\Omega_* \neq \Omega_x$ is the time-steady assumption for the funnel flow. As we have argued, the main modification may be a time-unsteady response that results in a transient amplification of the magnetic field.

Acknowledgements: This research is funded by a grant from the National Science Foundation and by the NASA Astrophysics Theory Program that supports a joint Center for Star Formation Studies at NASA/Ames Research Center, the University of California at Berkeley, and the University of California at Santa Cruz.

References

Arons, J. 1986, in Plasma Penetration into Magnetospheres, ed. N. Kylafis, J. Papamastorakis, and J. Ventura (Iraklion: Crete Univ. Press), 115

Bertout, C. 1989, ARAA 27, 351

Bertout, C., Basri, G., and Bouvier, J. 1988, ApJ 330, 350

Biskamp, D. 1993, Nonlinear Magnetohydrodynamics (Cambridge University Press)

Blandford, R.D., and Payne, D.G. 1982 MNRAS 199, 883

Camenzind, M. 1990, in Reviews in Modern Astronomy 3, ed. G. Klare (Berlin: Springer), 259

Carkner, L. Feigelson, E.D., Koyama, K., Montmerle, T, and Reid, I.N. 1996, ApJ 464, 286

Chernin, L., and Masson, C. 1995, ApJ 455, 182

Edwards, S., Strom, S. E., Herbst, W., Attridge, J., Merrill, K. M., Probst, R., and Gatley, I. 1993, AJ 106, 372.

Eichler, D. 1993, ApJ 419, 111

Feigelson, E.D., Jackson, J.M., Mathieu, R.D., Myers, P.C., and Walter, F.M. 1987, AJ 94, 1251

Ghosh, P., and Lamb, F.K. 1978, ApJ 223, L83

Goodson, A.P., Winglee, R.M., and Böhm, K.-H. 1997, ApJ, submitted

Grosso, N., Montmerle, T., Feigelson, E.D., André. P., Casanova, S., and Gregorio-Hetem, J., Nature, in press

Hartmann, L., Hewitt, R., and Calvet, N. 1994, ApJ 426, 669

Hartmann, L., and MacGregor, K.B. 1982, ApJ 259, 180.

Hayashi, M.R., Shibata, K., and Matsumoto, R. 1996, ApJ 468, L37

Heinemann, M., and Olbert, S. 1978, J. Geophys. Res. 83, 2457
Heyvaerts, J., and Norman, C. 1989, ApJ 347, 1055
Jackson, J.D. 1975, Classical Electrodynamics (New York: Wiley)
Johns, C.M., and Basri, G. 1995, ApJ 449, 341
Johns-Krull, C.M., and Basri, G. 1997, ApJ 474, 443
Johns-Krull, C.M., and Hatzes, A.P. 1997, ApJ, in press
Königl, A. 1989, ApJ 342, 208
Königl, A. 1991, ApJ 370, L39
Koyama, K., Ueno, S., Kobayashi, N., and Feigelson, E.D. 1996, PASJ 48, L87
Li, Z.Y., and Shu, F.H. 1996, ApJ 472, 211
Linker, J. A., and Mikić, Z. 1994, ApJ 430, 898
Lovelace, R.V.E, Berk, H.L., and Contopoulos, J. 1991, ApJ 379, 696
Masson, C., and Chernin, L. 1992, ApJ 387, L47
Mestel, L. 1968, MNRAS 138, 359
Meyer, M.R., Calvet, N., and Hillenbrand, L.A. 1997, AJ, in press.
Myers, P.C., Fuller, G.A., Goodman, A.A., and Benson, P.J. 1991, ApJ 376, 561
Najita, J.R., and Shu, F.H. 1994, ApJ 429, 808
Ostriker, E.C., and Shu, F.H. 1995, ApJ 447, 813
Pudritz, R.E., and Norman, C.A. 1983, ApJ 274, 677
Sakurai, T. 1985, A&A 152, 121
Shang, H., Shu, F., Lee, T., and Glassgold, A.E. 1997, in Low Mass Star Formation from Infall to Outflow, Poster Proc. IAU Symp. No. 182, ed. F. Malbet and A. Castets (Grenoble: Laboratorie d'Astrophysique), 312
Shang, H., and Shu, F.H. 1997, in preparation
Shu, F.H., Lizano, S., Ruden, S., and Najita, J. 1988, ApJ 328, L19
Shu, F.H., Ruden, S.P., Lada, C.J., and Lizano, S. 1991, ApJ 370, L31
Shu, F., Najita, J., Ostriker, E., Wilkin, F., Ruden, S., and Lizano, S. 1994a, ApJ 429, 781
Shu, F.H., Najita, J., Ruden, S.P., and Lizano, S. 1994b, ApJ 429, 797
Shu, F.H., Najita, J., Ostriker, E. C., and Shang, H. 1995, ApJ 455, L155
Shu, F.H., Shang, H., Glassgold, A.E., and Lee, T. 1997, Science, submitted.
Uchida, Y., and Shibata, K. 1985, PASJ 37, 515
Weber, E.J., and Davis, L. 1967, ApJ 148, 217

ENERGETICS, COLLIMATION AND PROPAGATION OF GALACTIC PROTOSTELLAR OUTFLOWS

Views and Perspectives

MAX CAMENZIND
Landessternwarte Königstuhl, D-69117 Heidelberg, Germany

Abstract. Formation of jets in low–mass protostellar objects and young pre–main sequence stars is ultimately related to the existence of some gaseous disk around a rapidly rotating central object. This configuration has deep parallels to extragalactic systems such as radio galaxies and quasars. Rotating black holes are still thought to be the prime–mover behind the activity detected in centers of galaxies, while, in the case of protostellar jets, rapidly rotating stars and disks are responsible for the ejection of bipolar outflows. In both cases, magnetic fields are invoked for the acceleration, the collimation and propagation of these outflows. The ultimate rooting of these fields is still under debate. We discuss models where winds injected into rapidly rotating magnetospheres of the central object drive the outflows. From these considerations it follows that the jets of young stellar objects can only be produced magnetically and that their progagation is determined by their magnetic properties. Such jets have low Mach numbers $\simeq 2$ and their instabilities are dominated by the pinch mode. Knots closest to the source are attributed to compression by the time–dependent pinches. Multiple bow shocks occur on longer time–scales (a few thousand years) and are attributed to variations in the magnetospheric structure of the star, or the disk.

1. Introduction

It is now known for a long time that young low–mass stars are capable of driving highly collimated bipolar outflows with velocities of several hundred km s^{-1}. The association of many outflows with stars deeply imbedded in their molecular cloud cores suggests that the energetic winds are intrinsic to the star formation process itself. In fact, they may provide the only means

B. Reipurth and C. Bertout (eds.), Herbig–Haro Flows and the Birth of Low Mass Stars, 241–258.

for a collapsing and rotating object to shed sufficient angular momentum (as already proposed by Hartmann & MacGregor in 1982). An additional impact of these winds consists in reversing the infall of core material and thereby determining the final mass of the star.

In addition to embedded sources, energetic winds are also observed from optically visible pre-main sequence stars with ages of the order of a few million years, or less. Unlike these classical T Tauri stars (cTTS), the embedded infrared sources (EIR) are observed over a wide range of bolometric luminosities (0.1 to $10^5\, L_\odot$), indicating that mass outflows are driven by many types of masses. The energetic outflows from the EIR are primarily traced by swept-up shells of high velocity molecular gas (tens of km s^{-1}) with structures of bipolar topology and extending up to several parsecs. In a few cases, the driving wind interior to the molecular shells has been detected as a very high speed neutral wind with velocities of the order of a few hundred km s^{-1}. The determination of the mechanical luminosities and momentum fluxes from molecular outflows has been given by Lada (1985; see also Lada & Fich, 1996).

The most spectacular and also most probably youngest outflows have been found in near-infrared surveys (McCaughrean et al., 1994; Zinnecker et al., 1996). These jets quite often show inner knots and signs of swept-up ambient molecular gas.

The most popular theoretical ideas for the ultimate source of momentum and energy in these outflow sources are based on the existence of winds driven away by magnetic fields of the surrounding accretion disk (Blandford & Payne 1982; Pelletier & Pudritz, 1992; Ferreira & Pelletier, 1995; Ouyed et al., 1997). All these models invoke a dipolar magnetic structure in the disk which is essentially advected inwards from the surrounding molecular core. Accretion disks of low-mass protostellar objects are, however, in most cases convective such that the postulated magnetic structures are unstable due to dynamo action (Camenzind, 1990). The normal modes excited by geometrically thin disks are of the quadrupolar type (Camenzind, 1990; Stepinski & Levy, 1991; Rüdiger et al., 1995; Camenzind & Lesch, 1994), and therefore unfavourable for driving efficiently disk winds.

Since the protostar itself is in a fully convective state, dynamo operation is unavoidable and leads to surface field strengths of the order of one kilo-Gauss. This seems to be quite natural for a rapidly rotating protostellar object. Rapid rotation together with unipolar induction generate a Poynting luminosity of the surface of the protostar which is in fact exactly of the same order of magnitude as the kinetic luminosity observed in protostellar outflows. This fact indicates that not the gravitational potential of the star, but the rotational energy of rapidly rotating objects is the source of the kinetic energy observed in the outflows. It is a consequence of the conservation law for the total energy consisting of magnetic energy and kinetic energy along a given flux surface of a rotating magnetosphere.

In the next Sect. we discuss possible jet formation scenarios from the point of view of rapidly rotating magnetospheres. One of the key questions for magnetically driven outflows is the origin of the magnetic fields in the protostellar environment. We discuss the dynamo equations for convective stars and disks. It turns out that the characteristic dynamo time for protostars is of the order of a few hundred years. Seed fields left over from the protostellar collapse will therefore be rapidly destroyed and restructured in a short time most probably into a dipolar structure inside the star and some complicated quadrupolar structure along the disk. Though the outcome of these dynamical systems is not known – and will remain unknown for the next years – we speculate that the stellar field is indeed a dipolar configuration.

Such a dipolar structure will strongly interact with the ambient disk. Accretion towards the stellar surface is channeled into an accretion column, and strong winds will be driven away from the polar cap regions. These magnetized winds are then collimated into a cylindrical shape at a distance of about one hundred astronomical units, corresponding to about 10000 stellar radii. In Sect. 3 we summarize the basic facts of axisymmetric MHD which show the existence of self–collimated outflows. It is interesting that such structures naturally occur when the electric fields, generated by the rapid rotation of the magnetosphere beyond the light cylinder, are taken into account in the force–balance equation. The light cylinder is in fact at a radius of 100 AU for a rotation period of 3 days. Since electric fields are neglected in the usual treatment of magnetized winds, people missed the existence of such self–collimated solutions. The resulting jet outflows have a very simple structure, as discussed in Sect. 4. The magnetic field is predominantly toroidal with a field strength of the order of a few milli–Gauss, corresponding to a fast magnetosonic speed of about 100 km/s for the typical density expected at 100 AU in the jet. This means that the sonic speed is completely irrelevant for the dynamics of the jet. In fact, these jets have a low plasma $\beta = P_{\rm gas}/P_m \simeq 10^{-2}$. Only magnetic and electric fields are responsible for the pressure–equilibrium across the jet. This also means that the gas flowing in the jet at a speed of $\simeq$ 300 km/s has a low magnetosonic Mach number of the order of 2 – 3.

Traditionally, the propagation of collimated flows through the interstellar medium has been described in terms of hydrodynamical flows (Stone & Norman, 1993; Suttner et al., 1996). This requires a high sonic Mach number $M \geq 50$ and means therefore practically no interaction with the ambient medium. The regular spacing of knots can therefore not be explained in terms of internal shocks. The fact that jets are supermagnetosonic opens up the possiblity that the interaction with the ambient medium is in fact much stronger than in hydro–jets. The most dangerous mode excited for magnetic jets is the fundamental pinch and kink mode (Appl & Camenzind, 1992). This could be responsible for mixing up with the ambient medium leading in this way e.g. to regularly spaced H_2–knots as observed quite recently. It also means that the bow shock structure is much more complicated than in simple hydro–jets. Therefore, also the interpretation of the Herbig–Haro objects has to be carefully re–analyzed.

2. Jet Formation Scenarios

Observationally, there are various constraints on the typical parameters for jet formation. HST data (Ray et al., 1996; Stapelfeldt et al., 1995) support models in which the outflow is initially poorly focused and then collimated into a cylindrical structure with diameters of at least of order of a few tens of AU. This is the most important information concerning the overall structure of young stellar jet formation. These scales correspond typically to a few thousand stellar radii. Jets are not pencil–like outflows from the inner edge of accretion disks, these are wide outflows, probably initially of spherical shape, until the flows are suddenly collimated into a cylindrical flow structure (Fig. 1). These geometric constraints are the most severe restrictions on jet formation models. This also means severe implications on the dynamic range of numerical jet formation. In the framework of magnetohydrodynamic (MHD) models this implies solar wind type outflows from the inner part of the star–disk system with a collimation mechanism operating preferentially on the scale of a few thousand stellar radii.

Rapidly rotating objects have been proposed to be the ultimate sources of collimated outflows, once they carry a sufficiently strong magnetosphere. Since Keplerian disks represent the most extreme rotational states, they are *prime candidates* for jet drivers (Pudritz & Norman, 1986; Pelletier & Pudritz, 1993; Ferreira et al., 1995). The crucial question here is whether the disk is able to maintain a *dipolar magnetic structure*.

2.1. JETS AS COLLIMATED DISK WINDS

We consider the interaction of an inward flow of material inside a disk with velocity $\vec{v}$ and the magnetic field in the disk. The disks near protostars are probably heavily convective and therefore prone to *dynamo action*. Advection and convective motion lead to a source for poloidal magnetic fields $\vec{B}_p = (\nabla\Psi \times \vec{e}_\phi)/R$ given in terms of the flux function $\Psi = RA_\phi$ (or the

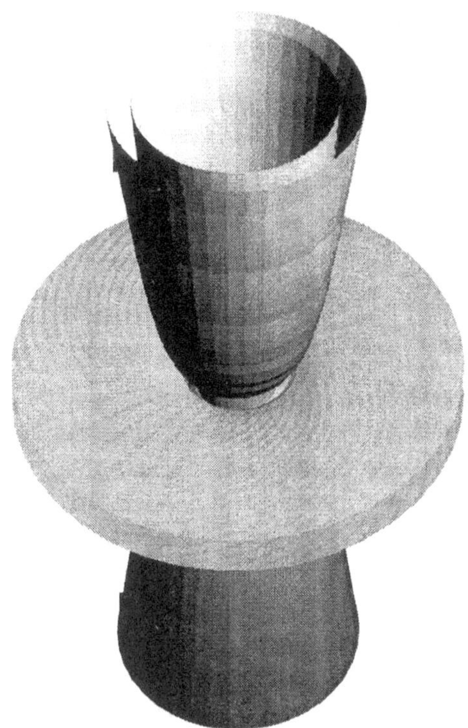

Figure 1. A schematic view on outflows driven by dipolar magnetospheres of rapidly rotating disks and stars. Rotating field lines form axisymmetric magnetic surfaces centered on the rotational axis. Plasma is streaming along these surfaces which are collimated at a certain radius into essentially a cylindrical shape. For protostellar objects, this collimation radius is observed to be about 50 AU, disks extend to a few 100 AU. The plasma is in force–balance perpendicular to these surfaces.

vector potential A_ϕ)

$$\frac{\partial \Psi}{\partial t} + \vec{v} \cdot \nabla \Psi - \eta R^2 \nabla \cdot \left[\frac{1}{R^2} \nabla \Psi \right] = \alpha \, R B_\phi \, . \tag{1}$$

Poloidal magnetic flux Ψ undergoes a time–evolution due to the advection of the flux by means of the accretion process (with velocity $\vec{v}$) and diffusion due to turbulent processes, given by the magnetic diffusivity $\eta \simeq l_c v_c/3$ with convective velocity v_c on scales $l_c \leq H$. Regeneration out of toroidal magnetic fields B_ϕ, or its current function RB_ϕ, occurs due to the dynamo effect represented by the function $\alpha \simeq v_c$. This toroidal field itself undergoes a similar time–evolution, written here for the current function $I = RB_\phi$,

$$\frac{\partial I}{\partial t} + R^2\nabla \cdot \left[\frac{\vec{v}}{R^2}I\right] - R^2\,\nabla \cdot \left[\frac{\eta}{R^2}\nabla I\right] = R^2\vec{B}_p \cdot \nabla\Omega\,. \tag{2}$$

Differential rotation in disks mainly drives the evolution of the toroidal component, winding it up to equipartition field strengths. The magnetic diffusivity η is in general R–dependent and can therefore not be taken out of the diffusive term. When dynamo action is absent, $\alpha = 0$, the poloidal field structure evolves independently of the toroidal field and is only determined by initial and outer boundary conditions. The toroidal field is essentially determined by the radial component of the poloidal magnetic field. If there is no bending of field lines in the disk, toroidal fields cannot be generated.

The above equations describing the evolution of magnetic fields inside the disk, must be supplemented by equations for the behaviour of the magnetic field outside the disk. Usually, vacuum conditions are assumed in the exterior region, i.e. no currents are flowing in the exterior region. If the magnetic fields are, however, loaded with disk plasma, the poloidal field structure has to be taken as a solution of the Grad–Shafranov equation (see later on), and currents are allowed to circulate between the disk surface and the wind region. This coupled 2D problem is quite complicated and has not yet been solved in the literature (for 1D solutions, see Ferreira, 1997).

For rapidly rotating disks, this dynamical system evolves naturally into a *quadrupolar structure* (Camenzind, 1990; Stepinski & Levi, 1991; Rüdiger et al., 1995), quite in analogy to the time evolution of galactic magnetic fields. It is quite important for the stability of the magnetic fields that accretion disks are highly diffusive. Simulations based on ideal MHD equations are not suitable to test the time evolution of magnetic fields in disks. Outflows driven by such configurations could stabilize these structures. But at the moment there is no way to couple numerically a diffusive disk with an ideal MHD flow exterior to the disk. Disks have extremely low Reynolds numbers, outflows high Reynolds numbers.

Dragging of magnetic fields in accretion disks If the outer boundary conditions are set to bring in continuously dipolar flux, the inner disk could be forced to a dipolar configuration, $\Psi = \Psi_d + \Psi_\infty$ with $\Psi_\infty = R^2 B_\infty/2$ (Reyes–Ruiz & Stepinski, 1996). This adds a source term to the diffusion equation for the poloidal flux generated in the disk

$$\frac{\partial \Psi_d}{\partial t} + \vec{v} \cdot \nabla\Psi_d - \eta R^2\,\nabla \cdot \left[\frac{1}{R^2}\nabla\Psi_d\right] = -v_R\,RB_\infty\,. \tag{3}$$

The evolution of the field in the disk is essentially determined by the magnetic Prandtl number $P_m = \eta/\nu \simeq 1$ (Reyes–Ruiz & Stepinski, 1996). Significant bending of the field lines only occurs for small Prandtl numbers, $P_m \simeq 0.01$. Only in this case the field lines are bent sufficiently to allow the development of a centrifugally driven wind (Blandford & Payne, 1982). The toroidal field in the disk depends inversely on the magnetic diffusivity. These results may change, if dynamo action is taken into account for the internal evolution.

These pure disk configurations have another drawback. Magnetic fields from the inner disk are rapidly collimated into a cylindrical shape (Ouyed et al., 1997). The resulting jet radii are generally too small.

2.2. JETS AS INDUCED STELLAR WINDS

Fields from disks can be flared up to larger radii, provided there is some extra pressure along the axis. The most natural source of such a pressure is an additional magnetosphere built up by the central star itself. Such configurations have been discussed by the present author for the extragalactic case already in 1986 (Camenzind, 1986a) and applied to protostellar outflows in 1990 (Camenzind, 1990). A somewhat similar model has been discussed by Shu et al. in 1994. Personally, I think that only the existence of such a central pressure can provide the observed large collimation radii.

The strong differential rotation of accretion disks is responsible for the excitation of quadrupolar modes in disks. Since differential rotation is probably suppressed in protostellar objects (Rüdiger, private communication), the original dipolar structure left over from the collapse process, could be stabilized over long time scales. This is still an open question, but this mechanism would provide a different scenario for jet formation (Camenzind, 1990). Initially, we thought that the interaction between the stellar magnetosphere and the surrounding disk could drive the outflows (see also Fendt et al., 1995; Goodson et al., 1997).

Such outflows could, however, also be driven by the *central object* itself. This idea has a long history and requires that the central object carries an axisymmetric magnetosphere. Jet outflows occur in magnetized astrophysical systems whenever they provide a perfect *axisymmetry*. Real three–dimensional objects such as pulsars and Weak–Line T Tauri stars never show the existence of collimated plasma flows.

Convection and shear in the protostellar cores are also driving forces for a stellar dynamo. Rapidly rotating protostars have sufficient energy to build up large–scale magnetic fields which will be strongly interacting with its surrounding accretion disk. Neglecting meridional circulations, the stellar dynamo equations have a much simpler form than the disk dynamo

$$\frac{\partial \Psi}{\partial t} - \eta_* R^2 \nabla \cdot \left\{ \frac{1}{R^2} \nabla \Psi \right\} = R_\alpha \alpha_* I \, , \tag{4}$$

$$\frac{\partial I}{\partial t} - R^2 \nabla \cdot \left\{ \frac{\bar{\eta}_*}{R^2} \nabla I \right\} = R_\Omega R^2 \vec{B}_p \cdot \nabla \bar{\Omega} - \frac{I}{\tau_{\rm loss}} f(B/B_{\rm crit}) \, . \tag{5}$$

Here $R_\alpha = \alpha_* R_*/\eta_*$, $R_\Omega = \Omega_* R_*^2/\eta_*$, and $\mathcal{P} = R_\alpha R_\Omega$ is the dynamo number. Radii are given in units of the stellar radius R_*, time in units of R_*^2/η_*, and the toroidal field B_ϕ in units of B_* which depends on the stellar density

$$B_* = \sqrt{4\pi\rho_*} \, \frac{\eta_*}{R_*} \, . \tag{6}$$

The typical scale of the α–effect is as usually given by

$$\alpha \simeq \left\{ \begin{array}{ccc} l(z)\Omega_*(z) & : & l\Omega_*/v_c \ll 1 \\ v_c(z) & : & l\Omega_*/v_c \gg 1 \end{array} \right\} \tag{7}$$

v_c is the convective velocity in the star. Convective velocity and Rossby number $R_o = (\Omega_* \tau_c)^{-1}$ follow from the energy transport in the star

$$L_* = \epsilon_c \, M_* \, v_c^3 / R_* \, , \tag{8}$$

and therefore

$$v_c \simeq 40 \, {\rm m\,s^{-1}} \, (\epsilon_c/30)^{-1/3} \, (L_*/L_\odot)^{1/3} \, (M_*/M_\odot)^{-1/3} \, (R_*/R_\odot)^{1/3} \, . \tag{9}$$

This would correspond to a mean Rossby number in the star

$$R_o = 1.5 \times 10^{-3} \, \omega^{-1} \, (L_*/4L_\odot)^{1/3} \, (\epsilon_c/30)^{2/3} \, (M_*/M_\odot)^{-5/6} \, (R_*/3R_\odot)^{5/6} \, . \tag{10}$$

The helicity α is a measure of the lack of symmetry of the convective flow. For small Rossby numbers the correct expression is given by

$$\alpha = \alpha_0 \, \bar{\alpha}(r,\theta) \quad , \quad \alpha_0 \simeq L_c \, v_c / H_P = \gamma \, v_c \quad , \quad \gamma < 1 \, . \tag{11}$$

H_P is the pressure scale height. In this kinematic dynamo scheme, magnetic fields are found to grow exponentially, or to decay exponentially. Nonlinear feedback between strong magnetic fields and turbulence suggests however the existence of a critical field strength $B_{\rm crit}$ for the onset of toroidal flux loss and quenching of the α–effect

$$\alpha = \alpha_0 \, \bar{\alpha}(r,\theta) \, q(B/B_{\rm crit}) \, , \tag{12}$$

where q is a dimensionless function which tends to zero for $B^2 > B^2_{\rm crit}$. Since this critical field strength is the natural scale for the toroidal field, the dynamo equation for the poloidal flux also provides the units Ψ_0 for Ψ

$$\Psi_0 = \frac{R^3_{\rm in}\, B_{\rm crit}\, \alpha_0}{\eta_*} = 3\, \frac{R_{\rm in}}{L_c}\, R^2_{\rm in} B_{\rm crit}\, \gamma\, R_o^{-p}\,. \tag{13}$$

$R_{\rm in}$ is the inner radius for the convective zone of the protostar.

In these natural units, the dynamo equations only contain one essential coupling constant. This is the dynamo number $\mathcal{P}$, defined by

$$\mathcal{P} = \frac{R^3_{\rm in}\, \Omega_{\rm in}\, \alpha_0}{\eta^2_*} = \frac{9}{P_m} \left(\frac{R_{\rm in}}{L_c}\right)^2 \frac{R_{\rm in}\Omega_{\rm in}}{v_c}\, \frac{\alpha_0}{v_c}\, R_o^{-2p}\,. \tag{14}$$

Using the scaling relations for rapidly rotating stars, we get

$$\mathcal{P} \simeq 4 \times 10^9\, \gamma\, \omega^3 \left(\frac{R_{\rm in}}{L_c}\right)^2 \frac{R_{\rm in}}{0.1\, R_*} \quad , \quad \omega = \frac{\Omega_*}{\Omega_K(R_*)}\,. \tag{15}$$

Here, we used the magnetic Prandtl number P_m, $\eta_* = P_m\, \nu_*$, $P_m \simeq 1$. The corresponding dynamo number is fantastically high for rapidly rotating convective stars and scales as $\mathcal{P} \propto \omega^3$ with the angular velocity. But even for moderately rotating T Tauri stars, $\omega \simeq 0.1$, we still find extremely high dynamo numbers, $\mathcal{P}_{\rm TTS} \simeq 4 \times 10^5$, while the Sun has the dynamo number $\mathcal{P}_\odot \simeq 1000 - 2000$ as obtained from other estimates. This extremely high dynamo number for protostellar objects is the second difference between solar–like stars (see e.g. Küker et al. 1996) and pre–main sequence objects. It makes any scaling between solar phenomena and protostellar properties completely obsolete. The dynamo number is responsible for the characteristic mode excited in the star. The structure which evolves under high dynamo numbers will crucially depend on the nonlinear saturation process. No models have been calculated for this domain of dynamo numbers. So we can only speculate on the resulting magnetic structure.

One of the crucial quantities is the critical field strength where saturation occurs. There are various ways to derive this quantity. In a first attempt, the critical field strength essentially follows from the condition that convection must be super–Alfvénic, i.e. the convective energy density should supercede the magnetic energy density,

$$B_{\rm crit} \simeq v_A\, \sqrt{4\pi\rho_*} = \frac{v_c}{M_T}\, \sqrt{4\pi\rho_*}\,. \tag{16}$$

In natural units we obtain

$$B_{\rm crit} \simeq 10\,{\rm kG}\, \frac{0.3}{M_T} \left(\frac{L_*}{4L_\odot}\right)^{1/3} \left(\frac{M}{M_\odot}\right)^{1/6}\,. \tag{17}$$

The typical toroidal field strength built up by dynamo action in a protostellar convection zone is of the order of 10 kilo–Gauss. This critical field strength is independent of the rotation period of the star, and is therefore also the typical field strength expected in the lower convection zone of the Sun (Küker et al., 1996). From the expression (13) we then get the poloidal magnetic flux in the interior of the star

$$\Psi_0 \simeq 3 \times 10^{26}\,\mathrm{G\,cm}^2\,\frac{\gamma\omega}{0.1}\,\frac{B_{\rm crit}}{10\,\mathrm{kG}}\,. \tag{18}$$

From this we find the typical poloidal field strength at the surface of the star

$$B_* \simeq \Psi_0/R_*^2 \simeq 6\,\mathrm{kG}\,\frac{\omega\gamma}{0.1}\,\frac{B_{\rm crit}}{10\,\mathrm{kG}}\,, \tag{19}$$

which would be about 2 kilo–Gauss for a protostar rotating with 5 days. According to this scaling law, rapidly rotating protostellar cores would have poloidal fields at the surface of the order of 10 kilo–Gauss, while T Tauri stars with $\omega \simeq 0.1$ would still have magnetic fields of the order of one kilo–Gauss at their surface.

3. Rotation and Jet Energy

Rotation is the ultimate energy source for magnetically driven winds. Though the solar wind is magnetized, it is not magnetically driven. Winds ejected from rapidly rotating stars and disks are, however, magnetically driven. When a solar type star reaches the main sequence with a rotation rate $\Omega_* \simeq 20\Omega_\odot$ at an age of 50 million years, then the thermally driven solar wind changes to a completely magnetically and rotationally driven MHD wind (MacGregor, 1996). Historically, these winds have been discussed as spherically symmetric winds in the equatorial plane of a star that rotates uniformly with angular velocity Ω_* (Weber & Davis, 1967; Belcher & MacGregor, 1976; Charbonneau, 1995). Fully 2D axisymmetric winds have been discussed so far only in the self–similar approach.

3.1. WHAT DRIVES THE ACCELERATION ?

When a rotating body carries an axisymmetric magnetosphere, rotation induces electric fields $\vec{E}_\perp$ over the unipolar induction such that the magnetic surfaces become equipotential surfaces. These electric fields together with the toroidal magnetic fields $\vec{B}_T$ give rise to a Poynting flux (Camenzind, 1986a)

$$\vec{P}_p = \frac{c}{4\pi}\,\vec{E}_\perp \wedge \vec{B}_T\,. \tag{20}$$

This can be written in the form, $I = -cRB_\phi/2$,

$$\vec{P}_p = \frac{\Omega_* I}{2\pi c}\,\vec{B}_p\,. \tag{21}$$

When integrating this Poynting flux over the entire surface of the central object, we end up with a magnetic luminosity

$$L_{\rm mag} = \frac{1}{c}\,\Omega_* I_* \Psi_*\,, \tag{22}$$

depending on the rotational state of the object, the total magnetic flux Ψ_* carried by the magnetosphere and the total current I_* driven by the rotation of the object. This magnetic luminosity is a new source of energy which is usually neglected in the energy budgets. It is however an important source of energy for strongly magnetized and rapidly rotating objects.

Young stellar objects at the age of a few hundred thousand years are observed to rotate faster than the Sun, $\Omega_* \simeq 20\,\Omega_\odot$ with $\Omega_\odot = 3 \times 10^{-6}$ rad s^{-1}, and to carry a stronger magnetosphere. Field strengths of the order of 1000 Gauss are consistent with observations. This corresponds to a magnetic luminosity of the rotating star

$$L_{\rm mag} = 0.1\,L_\odot\,\frac{\Psi_*}{10^{25}\,{\rm Gauss\,cm^2}}\,\frac{I_*}{10^{14}\,{\rm Amps}}\,\frac{\rm days}{P_*} \tag{23}$$

with a magnetic flux given in units of 10^{25} Gauss cm^2 and a current of 10^{14} Amperes. This is some fraction of the bolometric luminosity of low–mass stellar objects, but completely negligible in the case of the Sun due to its low mean magnetic field and its low rotation rate. For young stellar objects, the magnetic luminosity is close to the kinetic luminosity found in the jets of these objects

$$L_{\rm jet} = 0.1\,L_\odot\,\frac{\dot{M}_j}{10^{-8}\,M_\odot\,yr^{-1}}\left(\frac{v_j}{300\,{\rm km\,s^{-1}}}\right)^2\,. \tag{24}$$

The primary energy source driving the outflows can be either the protostellar rotation or the rotational energy stored in a surrounding disk. Protostars formed out of rotating molecular clouds will be born as rapid rotators. The rotational energy stored in the rapidly rotating protostellar object

$$E_{\rm rot} = \frac{1}{2}\,k^2 M_* R_*^2 \Omega_*^2 = 4 \times 10^{46}\,{\rm erg}\,\frac{5R_\odot}{R_*}\left(\frac{M_*}{M_\odot}\right)^2\left(\frac{P_K}{P_*}\right)^2 \tag{25}$$

is in fact of the order of the energy stored in the molecular outflows. This is a considerably energy, when it is dissipated on a time–scale of less than 100.000 years

$$L_{\rm rot} = \frac{E_{\rm rot}}{\tau_j} \simeq 1.0\, L_\odot\, \frac{10^5\,{\rm yr}}{\tau_j} \left(\frac{M_*}{M_\odot}\right)^2 \left(\frac{3P_K}{P_*}\right)^2 . \tag{26}$$

The critical period P_K for a 1 $M_\odot$ object would be about half a day. Since it scales quadratically with the mass, it can easily explain the observed mechanical luminosities (Lada, 1985). That this short time–scale is not unrealistic, follows from the braking time of a magnetized wind

$$\tau_m = \frac{J_*}{dJ_*/dt} \simeq k^2 \left(\frac{R_*}{R_A}\right)^2 \frac{M_*}{\dot{M}_j} \simeq 10^5\,{\rm yr} \tag{27}$$

for mass–loss rates of the order of $\dot{M}_j \simeq 10^{-8}\, M_\odot$ yr^{-1}. Rapidly rotating protostars have in fact sufficient energy to drive this magnetic luminosity, disk accretion is not necessarily involved.

3.2. THE LIGHT CYLINDER OF RAPIDLY ROTATING OBJECTS

In the theory of rotating magnetospheres (as for pulsars e.g.), the appearance of the light cylinder is crucial. For protostellar objects, we find a light cylinder radius $R_L = c/\Omega_*$ given by

$$R_L = \frac{c}{\Omega_*} = 100\,{\rm AU}\, \frac{P_*}{3\,{\rm days}} . \tag{28}$$

This radius is interestingly close to the observed collimation radii. Once a magnetosphere extends beyond the light cylinder, Newtonian MHD becomes obsolete and has to be replaced by the special relativistic one (Camenzind, 1986a). The deeper reason for this is the existence of electric fields which can no longer be neglected in modelling the magnetosphere (Camenzind, 1987; 1989; Appl & Camenzind, 1993a,b; Fendt et al., 1995). They contribute to the forces at the same level as magnetic fields do. Electric fields can only be neglected, if the jet radii are much smaller than one hundred AU.

3.3. PLASMA MOTION ALONG MAGNETIC FLUX–TUBES AND PLASMA DIAGNOSTICS

In ideal MHD, plasma injected at the foot points of magnetic flux tubes moves along these flux surfaces. The theory of axisymmetric and stationary plasma motion is well developed and will not be repeated here (for the Newtonian theory, see e.g. Heyvaerts & Norman, 1996; for the general theory including electric fields and gravitational forces, see Camenzind, 1996). Applications to 2D protostellar magnetospheres are discussed in Fendt & Camenzind (1996), and Paatz & Camenzind (1996a). The particular form of these outflows is used to explain the origin of the low–velocity and high–velocity components of forbidden emission lines (Paatz & Camenzind, 1996b). The differences between their profiles follow from their different critical densities and the rotational properties of the wind in the formation region.

4. Collimation of MHD Jets: Cylindrical Pinch Solutions

The collimation of magnetized winds into a conical outflow is very similar to the plasma confinement in a z–pinch. The equilibrium condition for a non–rotating axisymmetric current carrying pinch configuration has to be generalized to include various relativistic effects (Appl & Camenzind 1993a,b; Fendt et al., 1995)

$$\begin{aligned} \kappa \frac{B_p^2}{4\pi}(1-M^2-x^2) &= (1-x^2)\nabla_\perp \frac{B_p^2}{8\pi} + \nabla_\perp \frac{B_\phi^2}{8\pi} + \nabla_\perp P \\ -\frac{B_p^2\Omega_*}{4\pi c^2}\nabla_\perp(\Omega_* R^2) &+ \left(\frac{\rho j^2}{R^3} - \frac{I^2}{4\pi R^3}\right)(-\nabla_\perp R)\,. \end{aligned} \tag{29}$$

Pressure gradients acting perpendicular to the magnetic surfaces are in equilibrium with electric forces, centrifugal forces, pinch forces produced by poloidal currents, as well as with curvature forces expressed at the left hand side. This latter force is modified by the Mach number M and the light cylinder, $x = R/R_L$, and $j = R^2\Omega$.

We discuss in the following the solutions for a cylindrical pinch, where the curvature forces vanish, $\kappa = 0$. In this case, the equilibrium condition reduces to its one–dimensional form

$$(1-x^2)\frac{d}{dR}\frac{B_p^2}{8\pi} + \frac{d}{dR}\frac{B_\phi^2}{8\pi} + \frac{dP}{dR} - \frac{B_p^2 R}{2\pi R_L^2} - \left(\frac{\rho j^2}{R^3} - \frac{I^2}{4\pi R^3}\right) = 0\,. \tag{30}$$

The pressure of the toroidal field and the pinch force can be combined into one single expression. In addition, we normalize radii in units of the light cylinder R_L, $x = R/R_L$, and the poloidal magnetic field in units of its value on the central axis, B_0, $y = B_p^2/B_0^2$,

$$(1-x^2)\frac{dy}{dx} - 4xy + \frac{1}{x^2}\frac{dI^2}{dx} + \frac{8\pi}{B_0^2}\frac{dP}{dx} - \frac{8\pi\rho}{R_L^2 B_0^2}\frac{j^2}{x^3} = 0\,. \tag{31}$$

Electric fields, $E_\perp = (R/R_L)B_p$, due to the rotation of the magnetic surfaces modify the classical pinch equilibrium in two ways. First, the equation has a singular point at the light cylinder, and secondly, the additional term $4xy$ is crucial for the form of the equilibrium. In contrast to force–free models (Appl & Camenzind, 1993b), the specific angular momentum j of the plasma and the poloidal current I are not arbitrary, but follow from the MHD wind theory (Camenzind, 1996).

We only discuss solutions of the relativistic pinch equilibrium neglecting the influence of the centrifugal force. In this case the equilibrium is determined by the condition

$$(1-x^2)\,\frac{dy}{dx} - 4xy + \frac{1}{x^2}\,\frac{dI^2}{dx} + \frac{8\pi}{B_0^2}\,\frac{dP}{dx} = 0\,. \tag{32}$$

The toroidal current follows to a good approximation

$$I \simeq -\Omega_*(R^2 B_p)\,\frac{1}{\beta_p} = -cB_0 R_L\,(x^2\sqrt{y})\,\frac{1}{\beta_p} \tag{33}$$

and we prescribe the plasma pressure in terms of $P = \beta B_p^2/8\pi$ with a constant plasma beta. This leads finally to the equation, $\gamma_p = 1/\sqrt{1-\beta_p^2}$,

$$\left(1+\beta+\frac{x^2}{\gamma_p^2\beta_p^2}\right)\frac{dy}{dx} + \frac{4}{\gamma_p^2\beta_p^2}\,xy = 0\,. \tag{34}$$

This homogeneous equation has the remarkably simple solution

$$y(x) = \frac{C}{(1+\beta+x^2/x_c^2)^2} \tag{35}$$

with

$$x_c = \gamma_p\,\beta_p \simeq 10^{-3}\quad,\quad R_c = \gamma_p\beta_p\,R_L \simeq 10^{-3}\,R_L \tag{36}$$

as the *core radius* of the jet with $\beta_p \simeq 10^{-3}$ and $\gamma_p = 1$ for protostellar outflows. Inside the core, given by the light cylinder radius and the poloidal velocity β_p, the poloidal magnetic field is essentially homogeneous. It decays as a monopole field outside the core, $B_p \propto 1/R^2$ for $R \gg R_c$,

$$B_p = \frac{B_0}{1+\beta+R^2/R_c^2}\,. \tag{37}$$

This is not unexpected, since the jet is formed by a supermagnetosonic wind. The current function I increases quadratically inside the core and saturates towards a constant value outside the core, $I_\infty \simeq cB_0R_L\beta_p$. This is exactly the structure of the force–free solutions found previously (Appl & Camenzind, 1993b).

MHD jets always have a core–envelope structure, along the axis the magnetic field is predominantly longitudinal, beyond the core it becomes dominated by the toroidal component. The light cylinder $R_L \simeq 100$ AU is the basic scale for the jets, the core is extremely small, $R_c \simeq 10R_*$.

5. Propagation of MHD Jets: Knots as Internal Pinch Modes

Protostellar jets differ from extragalactic jets in that radiative cooling is dynamically important. Protostellar jets are observable because of their radiative losses. HD simulations are parametrized by three dimensionless quantities

- the sonic Mach number, $M_s \geq 20$;
- the density contrast $\eta = 1 - 10$, jet density to ambient density;
- cooling length $\chi =$ ratio between cooling length and jet radius.

An adiabatic jet has $\chi = \infty$ while for an isothermal jet $\chi = 0$, realistic values are $\chi \simeq 1 - 10$.

Many protostellar jets have impressive arrays of emission knots in the inner part of the jet. Such knots can occur from overpressured jets. *Reconfinement shocks* are formed with emission knots behind them. But, the knot spacing is too large for hydro–jets ($\simeq 10$) and the knots proper motion is predicted to be zero, whereas they are observed to move with 20 – 100 % of the jet speed.

The inclusion of magnetic fields into jet simulations has various effects. No MHD simulations have been done including radiative cooling. From adiabatic simulations it is known that magnetic fields have various effects on the flow structure and the stability of jets (Clark et al., 1990; Kössl et al., 1990; Lind et al., 1989). Compared to hydrodynamic models, the bow shock is in general further separated from the end of the jet due to magnetic pressure. This effect depends dramatically on the magnetic field structure and the magnetic diffusivity which is used in the simulation. An important effect is due to the reduction of the essential Mach number which is now the magnetosonic Mach number, $M_{\rm MS} \simeq 2 - 3$. The sonic speed is no longer the characteristic speed in magnetized jets. Third, Kelvin–Helmholtz instabilities are reduced by magnetic field tension. And finally, in 3D simulations, helical motions appear which are reminiscent of HH46/47.

Protostellar MHD jets are low β flows

$$\beta = \frac{P_{\rm gas}}{B^2/8\pi} \simeq 0.1 - 0.01\,, \tag{38}$$

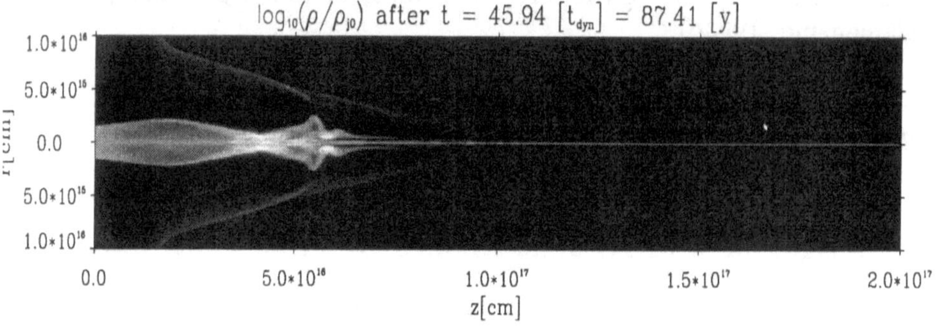

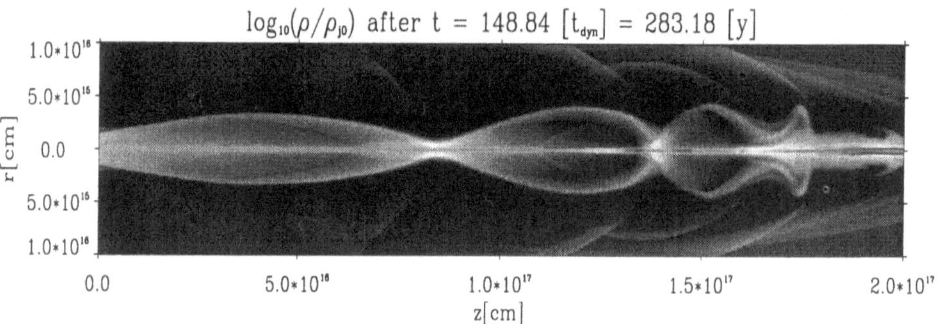

Figure 2. The time–evolution of a core–envelope jet with a monopolar poloidal magnetic field and a toroidal field initially constrained to some region (Thiele 1997, LSW Heidelberg). The beam is ten times overdense with respect to the ambient medium, the density is shown as grey scale. The inital beam radius is 100 AU, the jet has propagated for 87 years (upper panel) and for 283 years (lower panel) from left to right.

and have strong toroidal magnetic fields, $B_\phi \gg B_p$ (the observable part of the jet is formed by the envelope structure, the core is not resolvable). Under these conditions, pinch modes are excited (Fig. 2), the strong toroidal field focuses the flow towards the axis so that a quasi–periodic structure appears.

These simulations demonstrate that axisymmetric knots are produced very close to the collimation zone on time scales given by the characteristic dynamical time

$$\tau = \frac{R_j}{v_j} \simeq 0.5\,\mathrm{year}\,\frac{R_j}{50\,\mathrm{AU}}\,\frac{300\,\mathrm{km\,s^{-1}}}{v_j}\,. \tag{39}$$

They occur roughly on 10 dynamical time scales. Knot generation is therefore inherent to the dynamics of the jet itself, no episodic ejection is needed (see also Ouyed et al., 1997). These knots persist as coherent structures which propagate along the jet channel at a fraction of the jet speed. Investigations into the thermal structure of these knots will demonstrate the feasibility of this model for explaining the emission properties of the forbidden lines from these jets.

Obviously, there are multiple bow shocks on longer time scales, typically found at spacings of a few thousand years. These variations must be intrinsic to the central source or the surrounding accretion disk. These longer time scales could be related to magnetic transport in the disk, or some characteristic dynamo time scale of the central star.

6. Conclusions

MHD jet models for protostellar sources close a gap between jet formation scenarios and the propagation properties of young stellar jets. There is no doubt that protostellar outflows must be collimated by magnetic processes. One can therefore not avoid to treat the propagation of these outflows also within the framework of MHD. The main difference between hydro jets and MHD jets is the Mach number – MHD jets have low Mach numbers of order 2 – 3, since the relevant signal speed is the fast magnetosonic speed, which is of the order of 100 km/s, and no longer the sound speed. This low Mach number together with the low plasma–β makes them vulnerable to internal instabilities. Since the toroidal magnetic field is dominant in the outer parts of the jets, the pinch mode is naturally excited. These fields are extremely weak, less than a milli–Gauss. Pinch modes create oscillations that propagate as coherent structures along the jet and evolve probably into weak magnetosonic shocks. The emission knots can probably be associated with these structures.

The very origin of poloidal and toroidal magnetic fields in the jets is still under debate. Magnetized central objects are good candidates, since their energetics can explain the observed properties of the jets. The ultimate energy source behind these outflows is the rotational energy of the central star, or its surrounding disk.

Acknowledgements: Part of this work is supported by the Schwerpunktsprogramm *Physik der Sternentstehung* of the Deutsche Forschungsgemeinschaft. The time–dependent MHD simulations have been provided by Markus Thiele.

References

Appl, S., Camenzind, M.: 1992, A&A **256**, 354
Appl, S., Camenzind, M.: 1993a, A&A **270**, 71
Appl, S., Camenzind, M.: 1993b, A&A **274**, 699

Blandford, R.D., Payne, D.G.: 1982, MNRAS **199**, 883
Belcher, J.W., MacGregor, K.B.: 1976, ApJ **210**, 498
Camenzind, M.: 1986a, A&A **156**, 137
Camenzind, M.: 1986b, A&A **162**, 32
Camenzind, M.: 1987, A&A **184**, 341
Camenzind, M.: 1989, in *Accretion Disks and Magnetic Fields in Astrophysics*, ed. G. Belvedere, Kluwer (Dordrecht), p. 129
Camenzind, M.: 1990, in *Reviews of Modern Astronomy* **3**, ed. G. Klare, Springer–Verlag (Heidelberg), p. 234
Camenzind, M.: 1993, in *The Jets of Radio Galaxies*, eds. H.J. Röser & K. Meisenheimer, Lecture Notes in Physics **421**, Springer-Verlag (Heidelberg), p. 109
Camenzind, M. 1996, in *Solar and Astrophysical Magnetohydrodynamic Flows*, ed. K.C. Tsinganos, Kluwer (Dordrecht), p. 699
Camenzind, M., Lesch, H.: 1994, A&A **284**, 411
Charbonneau, P. 1995, ApJS **101**, 309
Clark, D.A., Stone, J.M., Norman, M.L.: 1990, BAAS **22**, 801
Fendt, C., Camenzind, M., Appl, S.: 1995, A&A **300**, 791
Fendt, C., Camenzind, M. 1996, A&A **313**, 591
Ferreira, J., Pelletier, G.: 1995, A&A **295**, 807
Ferreira, J.: 1997, A&A, in press
Goodson, A.P., Winglee, R.M., Böhm, K.H.: 1997, ApJ, submitted
Hartmann, L., MacGregor, K.B.: 1982, ApJ **259**, 180
Heyvaerts, J., Norman, C.A. 1996, in *Solar and Astrophysical Magnetohydrodynamic Flows*, ed. K.C. Tsinganos, Kluwer (Dordrecht), p. 459
Kössl, D., Müller, E., Hillebrandt, W.: 1990, A&A **229**, 378, 397
Küker, M., Rüdiger, G., Pipin, V.V.: 1996, A&A **312**, 615
Lada, C.J.: 1985, AAR&A **23**, 267
Lada, C.J., Fich, M.: 1996, ApJ **459**, 638
Lind, K., Payne, D.G., Meier, D.L., Blandford, R.D.: 1989, ApJ **344**, 89
MacGregor, K.B.: 1996, in *Solar and Astrophysical Magnetohydrodynamic Flows*, ed. K.C. Tsinganos, Kluwer (Dordrecht), p. 301
McCaughrean, M.J., Rayner, J.T., Zinnecker, H.: 1994, ApJL **436**, L189
Ouyed, R., Pudritz, R.E., Stone, J.M.: 1997, Nature **385**, 409
Paatz, G., Camenzind, M.: 1996a, A&A **308**, 77
Paatz, G., Camenzind, M.: 1996b, Astro. Lett. and Communications **34**, 315
Pelletier, G., Pudritz, R.E.: 1992, ApJ **394**, 117
Pudritz, R.E., Norman, C.A.: 1986, ApJ **301**, 571
Ray, T.P., Mundt, R., Dyson, J.E., Falle, S.A.E.G., Raga, A.C.: 1996, ApJL **468**, L103
Reyes–Ruiz, M., Stepinski, T.F.: 1996, ApJ **459**, 653
Rüdiger, G., Elstner, D., Stepinski, T.F.: 1995, A&A **298**, 934
Shu, F.H., Najita, J., Wilkin, F., Ruden, S.P., Lizano, S.: 1994, ApJ **429**, 781
Stapelfeldt, K.R., et al.: 1995, ApJ **449**, 888
Stepinski, T.F., Levy, E.H.: 1991, ApJ **379**, 343
Stone, J.M., Norman, M.L.: 1993, ApJ **413**, 210
Suttner, G., Smith, M.D., Yorke, H.W., Zinnecker, H.: 1997, A&A, in press
Weber, E.J., Davis, L.: 1967, ApJ **148**, 217
Zinnecker, H., McCaughrean, M.J., Rayner, J.T.: 1996, in *Disks and Outflows around Young Stars*, eds. S. Beckwith, J. Staude, A. Quetz and A. Natta, Lecture Notes in Phys. **465**, (Heidelberg, Springer–Verlag), p. 236

NUMERICAL SIMULATIONS OF JETS FROM ACCRETION DISKS

RALPH E. PUDRITZ
Canadian Institute for Theoretical Astrophysics, University of Toronto, Toronto, Ontario, Canada M5S 3H8

AND

RACHID OUYED
Department of Physics and Astronomy, McMaster University Hamilton, Ontario, Canada L8S 4M1

Abstract. Hydromagnetic disk winds have great potential as a general theory for the production and collimation of astrophysical jets in both protostellar and black hole environments. We first review the analytic stationary theory of these outflows as well as recent numerical simulations of MHD disk winds. We then focus on simulations that we have done on winds from magnetized disks using the ZEUS 2-D code of Stone and Norman. We treat the Keplerian disk as a fixed platform throughout the simulations. We show that both stationary and episodic, jet-like outflows are driven from disks depending upon their magnetic structure and mass loss rates.

1. Introduction

The nature and origin of bipolar outflows and jets in protostellar systems continues to be an important and lively area of research seventeen years after the seminal discovery paper of Snell *et al.* (1980). Many significant observational advances over this period of time leave little doubt that outflows are associated with the process of star formation itself. This connection between star formation and outflow is well seen in the recently discovered, so-called Class 0 sources which are probably the best candidates yet found for protostars (André *et al.* 1993). Class 0 sources are associated with highly collimated outflows whose mechanical luminosities are comparable to the source bolometric luminosity, $L_{mech} \simeq L_{bol}$. This suggests that the efficient

B. Reipurth and C. Bertout (eds.), Herbig–Haro Flows and the Birth of Low Mass Stars, 259–274.

conversion of gravitational binding energy, released during accretion, into the mechanical energy of the jet is occurring at this early stage in star formation. Moreover, outflows continue to play an important role during the entire collapse and accretion disk formation stage; the evidence shows that outflows persist for at least 10^5 years (Parker, Padman, and Scott, 1991). The role of outflows has yet to be fully integrated into our theories of star formation (see review; Pudritz, McLaughlin, and Ouyed 1997).

The driver for molecular outflows on larger scales is believed to be something akin to optical jets (Masson and Chernin, 1992). The HST images of the optical jet in HH30 (Stapelfeldt 1996) reveal that these jets are already well collimated on scales of 15 AU. This jet contains a series of knots whose proper motions have been measured. In fact the observations of a large sample of optical jets (eg. Reipurth 1989, Eislöffel and Mundt 1994) indicates that nature rarely, if ever, produces the stationary outflows which are the domain of much intensive theoretical work (eg. Blandford and Payne, 1982; Pudritz and Norman, 1986, Königl 1989, Heyvaerts and Norman 1989, Lovelace *et al.*, Li 1995, to cite only few). Jets observed in protostars and AGNs are highly time variable and episodic. As to the question of how astrophysical jets are driven, there is now substantial evidence that they are associated with accretion disks (eg. Cabrit and André 1991; Calvet, Hartmann, and Kenyon 1993, Edwards, Ray, and Mundt 1993) which are known to exist around most, if not all young stellar objects (eg. Beckwith *et al.* 1990). This is seen in the HST image of the jet in HH30 as an example, where a disk is clearly present in the system. In general, it appears that whenever a jet or outflow is detected in protostellar environments, there is good evidence that an accretion disk is also present. This correlation has recently been extended to the case of Class 0 sources with the detection of a disk in VLA 1623 (Pudritz *et al.* 1996).

Current models for the central engines that produce and collimate jets emphasize the importance of magnetized rotors. Such a mechanism can accelerate gas in systems whose radiation fields are known observationally to be insufficient to provide the observed thrust associated with outflows. Two good candidates for such rotors are (i) magnetized Keplerian disks, or, (ii) the interaction region between a stellar magnetosphere and a surrounding disk (the so-called X-celerator). A complete theory of outflows in protostellar systems must necessarily address both of these aspects of the problem. The most important difference between these two types of models lies in the role played by the central object. For disk winds, (eg. reviews by Königl and Ruden 1993, Pudritz, Pelletier, and Gomez de Castro 1991) only the mass of the central object is important; it dictates the depth of the gravitational potential well and hence the overall energetics of the outflow. In the second class of models on the other hand (eg. Shu *et al.* 1994), the outflow also depends upon the presence of a sufficiently strong magnetic field on the central star. Given that stars of all types are now known to be associated with bipolar outflows, this implies that a wide class of stellar types generate strong magnetic fields in their pre-main-sequence stage. Since energetic, highly collimated outflows are also associated with black holes, both massive as in AGNs (eg. Blandford 1990) as well as stellar mass holes (eg. Hjellming and Rupen, 1995), a universal theory for jet production is in principle possible in the context of disk wind models.

In this contribution, we review some of the recent results of simulations of magnetized winds from accretion disks. We then describe our own simulations of axisymmetric disk wind simulations in some detail. The general reader may consult Ouyed, Pudritz, and Stone (1997, hereafter OPS) for the basic results. We have found that both stationary (details in Ouyed and Pudritz 1997a, OPI) and episodic outflows (details in Ouyed and Pudritz 1997b, OPII) can be produced depending on the magnetic configuration threading the disk and on the rate at which mass enters the disk corona from the disk (details in Ouyed and Pudritz 1997c, OPIII).

2. Current Picture of Disk Winds

Theoretical models of hydromagnetic disk winds originated with the paper of Blandford and Payne (1982, hereafter BP) who analyzed the structure of stationary, self-similar outflows. Their theory extended the models of winds from magnetized rotating stars to the case of accretion disks. One of the simplest predictions of this model arises from Bernoulli's theorem; if magnetic field lines threading a disk make an angle $\theta_o \leq 60^o$ with respect to the surface of the disk, then no outflow should occur. If this condition is satisfied, then gas is accelerated along the field line; in effect one has a rather elegant sort of centrifuge. Gas will continue to be accelerated as long as it can be held in co-rotation with the disk. Thus, the role of the magnetic field is interesting; it is the deformable tool that extracts the gravitational potential energy of particles in a disk allowing those particles to spiral slowly inwards. No magnetic energy is actually used in this process. The Poynting flux associated with the outflow can be traced to the gravitational potential energy that is tapped by the torque exerted upon the disk by the threading magnetic field. The torque extracts angular momentum of material in the disk enabling the accretion process to occur. It is easy to show using the disk angular momentum equation that the torque exerted by the field and the wind necessarily makes the disk accretion rate proportional to the wind mass loss rate from the disk.

Consider a field line threading the disk at some radius r_o. Acceleration of the gas on any field line occurs predominantly out to the Alfvén point, at the radius r_A on that field line. The collection of all of these points for field lines threading the disk defines the *Alfvén surface* $r_A(r_o)$ of the wind. On this surface, the kinetic energy density in the poloidal motion of the outflow just equals the energy density in the poloidal component of the field, ie., the outflow speed equals the Alfvén speed. The importance of the Alfvén surface can be seen in another guise; it is directly related to the accretion rate through the disk. The disk angular momentum equation yields the relation $\dot{M}_a =< (r_A(r_o)/r_o)^2 > \dot{M}_w$, where the appropriate average of the lever arm occurs as one sums up outflow over the entire disk surface. Beyond the Alfvén surface, the field is not able to maintain the rigid co-rotation of the gas with the disk. Nonetheless, some centrifugal acceleration continues out to the point where the outflow speed reaches the fast magnetosonic speed, generally the highest signal speed that occurs in magnetized gas. The surface consisting of all these critical points in the outflow is the *fast magnetosonic* ($\equiv$ FM) *surface.* Beyond the FM surface, the flow can no longer communicate with the disk. The dynamics of the jet in this far field regime appears to be dominated by the toroidal magnetic field.

The magnetic structure of the outflow gradually changes as the Alfvén point is passed. Because of the inertia of matter, gas on field lines beyond r_A falls behind the rotation rate of the disk as mentioned, and a toroidal field is produced. This field dominates the dynamics of the flow beyond the FM surface. It provides the intrinsic collimation mechanism of outflows through the associated $\mathbf{J_p} \times \mathbf{B}_\phi$ Lorentz force which squeezes the flow radially inwards towards the outflow axis.

The discussion so far has assumed that the magnetic field is frozen into the disk plasma. In steady state models, this assumption cannot be strictly valid however. Matter in the disk must cross field lines as it moves radially inwards. If this did not happen in such models, the magnetic flux would be dragged into small disk radii, accumulate there, and eventually shut off the accretion flow itself. Thus, steady state theory involves a balancing act between the outward diffusion of the magnetic field, and the inward drag via the accretion flow (eg. Lubow, Papaloizou, and Pringle 1994). Some dissipation must occur in the magnetized accretion process. Indeed, these dissipative processes are important in determining exactly how the gas in the disk is divided between the small amount that is ejected by the wind, and the far larger portion that accretes onto the star (eg. Wardle and Königl 1993, Ferreira and Pelletier 1995).

One of the theoretical issues left unresolved by the theory is how the disk's magnetic field is distributed (ie. as a function of disk radius). This is a boundary condition for the outflow problem and in all generality, the physics of the accretion process must specify this function in some way. If one invokes some additional mathematical assumption, such as self-similarity, then this flux distribution function is fixed (eg BP). There is no physical reason why self-similarity should occur however. The production of an ordered, large scale field could proceed by dynamo action (eg Pudritz 1981, Tout and Pringle 1992, Stone *et al.* 1996), global modes in the Balbus-Hawley instability (Curry, Pudritz, and Sutherland 1994), or by dragging in a field from the external medium (eg. Lubow *et al.* 1994, Reys-Ruis and Stepinski 1996). A class of tractable non-self-similar models was found by Pelletier and Pudritz (1992) for outflows in a large variety of magnetic flux distributions.

2.1. NUMERICAL SIMULATIONS

These analytical models, while useful in revealing the physics of outflow, cannot address the observations in any direct way. The reason is that real astrophysical jets are episodic, not stationary. This strongly suggests that one needs to use time-dependent numerical simulations in order to investigate this physics. One would like to both confirm the steady state models by using numerical simulations and to explore the presumably much larger class of models that are truly time-dependent.

Not many published MHD disk wind calculations are currently available. The existing calculations divide into two general types; (i) in which the dynamics of the disk is also followed (Shibata and Uchida 1986, Stone and Norman 1994, Bell and Lucek 1995), and (ii) in which evolution of the accretion disk is ignored. In the latter class, the disk acts like a platform providing fixed boundary conditions for the problem (eg. Ustyugova *et al.* 1995, OPS). In axisymmetric simulations, the Balbus and Hawley (1991) instability leads to very rapid contraction of the disk for models of type (i). This leads to the formation of a transient mass ejection which follows from the winding up of the field lines threading the radially collapsing disk (eg. Uchida and Shibata 1985). Thus, it is not clear from such simulations that outflows would continue over the 10^5 years of outflow in protostellar systems. Models of type (ii), while limited by the neglect of disk evolution, have the virtue that one can isolate outflow physics from disk dynamics. Presumably real disks retain their Keplerian character, which makes this assumption palatable.

3. Description of our Model

Our simulations are solutions to the time-dependent MHD equations in cylindrical coordinates (r,ϕ,z) calculated with the ZEUS-2D code of Stone and Norman (1992 a,b). The model consists of a point mass representing a star or black hole, the *surface* of a surrounding accretion disk which is taken to have an inner radius r_i (this can be either the surface of the star, the magnetopause radius of a star with the disk, or the last stable Kepler-like orbit around a black hole), an initial disk corona which is in hydrostatic equilibrium within the gravitational field of the central object as well as in pressure balance with the disk below, and an initial magnetic field that threads the disk and corona (see OPS). The gas has an assumed polytropic ($\gamma = 5/3$) equation of state. The equilibrium density of the corona is then $\rho \propto (r^2 + z^2)^{-3/4}$. *This density structure has the same form as that expected for a collapsing singular isothermal sphere (Shu 1977).*

We choose two initial magnetic configurations which are current free; $\mathbf{J} = 0$. Surfaces of constant magnetic flux are specified by the level surfaces of the scalar function A_ϕ where $\mathbf{B} = \nabla \times (A_\phi \mathbf{e}_\phi)$. We used two types of initial magnetic configurations A_ϕ; (i) a "potential field" solution corresponding to a vacuum field that has a conducting plate as its boundary (eg. Cao and Spruit 1994); and (ii) a uniform vertical field.

We set the initial toroidal field component B_ϕ and gas velocity everywhere in the corona equal to zero. We avoid the use of a softening parameter for the (Newtonian) potential and take particular care to establish the disk corona in *numerical* hydrostatic equilibrium so that there are no initial motions.

The Keplerian accretion disk surface (the base of the corona at $z = 0$) provides fixed boundary conditions for the velocity at every disk radius; the rotational speed is Keplerian v_K and since any radial inflow speed is expected to be tiny compared to this, we ignore it. Matter is introduced into the corona (active zones of the grid) from the disk at very small injection speeds $v_{inj.}$; typically $v_{inj.} \simeq 10^{-3} v_K$ (one hundredth of the disk's sound speed). The disk's toroidal magnetic field is assumed to scale with radius as $B_\phi \propto 1/r$, although this is not essential (we have run models for which $B_\phi = 0$ and still observe outflows). The initial radial and vertical magnetic field components in the disk are continuous with the field in the corona. There is no disk, rotation, or toroidal field inside the radius r_i.

All length scales and speeds in the simulations are measured in units of innermost disk radius r_i and the Kepler speed at that radius, $v_{K,i}$, so that results can be scaled to a central object of any desired mass. All time scales are measured in units of $t_i = r_i / v_{K,i}$. Typical conditions in protostellar or AGN disks are, for the former, $r_i \simeq 3$ stellar radii $= 0.04$ astronomical units (AU), and for the latter, 10 black hole radii.

Our numerical simulations are governed by five parameters defined at r_i. The ratio of the thermal gas pressure to the magnetic field pressure, $\equiv \beta_i$; the square of the ratio of the Kepler speed to the sound speed, $\equiv \delta_i$; the ratio of the disk density to the coronal density, $\equiv \eta_i$; and the disk's toroidal to poloidal field ratio, $\equiv \mu_i$. For all Figures, the injection speed is $v_{inj.,i}/v_{K,i} = 10^{-3}$; and $(\beta_i, \delta_i, \eta_i, \mu_i) =$ (1.0,100.0,100.0,1.0). All the models are run at high spatial resolution of (500 x 200) zones. We use inflow boundary conditions at the disk surface and open outflow conditions (i.e., *zero normal gradient*) on the remaining boundaries (except along the axis of symmetry, coinciding with the disk axis, taken to be reflecting (i.e., *zero normal velocity*)).

4. Stationary outflows

We show a series of snapshots of the flow in Fig. 1 that illustrate the eventual realization of a stationary hydromagnetic outflow. As expected, the first stage of evolution in our simulations is a brief transient as a torsional Alfvén wave front propagates through the corona from the disk surface. Thereafter the flow either relaxes to a steady state outflow pattern, or an episodic outflow is established. In Fig. 1a, we plot surfaces of constant magnetic flux for an assumed potential field configuration and density isocontours at the initial instant. The evolution of the two-dimensional magnetic field structure is shown in Figs. 1b and 1c, at 100 inner time units, and at 400 time units by which time the working surface of the jet has long left the end of our mesh. At time 100, the region behind the working surface of the flow has already achieved a cylindrically collimated state while the region ahead of the working surface still retains the original field structure. At time 400, the entire flow in our mesh has been collimated. Cylindrical collimation is predicted for steady MHD winds (Heyvaerts and Norman 1989). In Fig. 1c; we plot the location of Alfvén surface (filled hexagons), as well as the FM surface (stars). There is good agreement between our numerical results and steady state theory for the position of the Alfvén surface as is shown in Fig. 1d. At the base of the outflow, the flux surfaces in Fig. 1c (or equivalently, the stream-lines) make an angle of $\simeq 55^o$ with respect to the disk across most of its surface. This geometry satisfies the outflow condition of steady state theory that $\theta_o \leq 60^o$.

The two-dimensional distribution of the toroidal magnetic field and vertical outflow speed as compared with the density is shown in OPS and OPI. An obvious bow shock traces the working surface of the outflow with respect to the undisturbed corona. There is also a highly collimated jet-like stream of gas that starts at the disk surface. The comparison of the jet density and toroidal magnetic field shows that the outflow density is delimited by the region of strong toroidal field. The physical reason for this is the strong radial pinch force that is exerted towards the jet axis by the magnetic force arising through the combination of a dominant toroidal field, and its associated current which flows primarily along the outflow axis. The highest speed gas is nearest the outflow axis with lower speeds at larger radii (see Fig. 2, OPS).

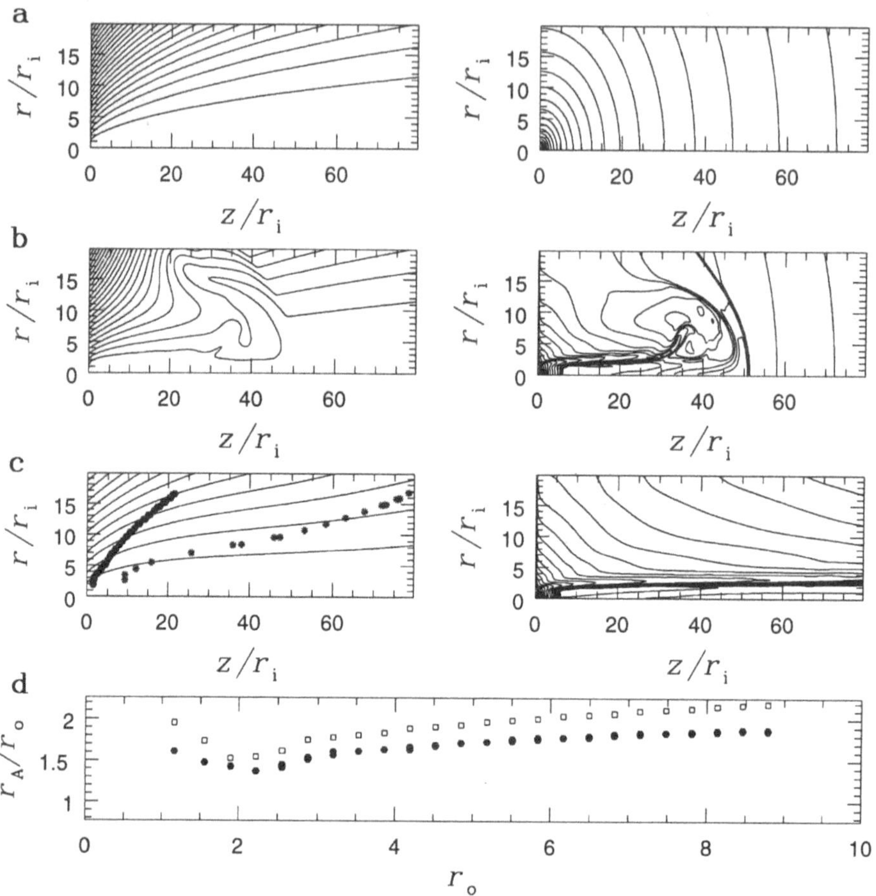

Figure 1. In all figures, the axis of symmetry is plotted horizontally and the disk surface, vertically. Frame 1a. (top) The initial magnetic configuration, shown in the left panel, is the potential field. The right panel shows the initial isodensity contours of the corona. Frames b and c; evolution of the initial magnetic and density structure (the left and right panels respectively) at 100 and 400 inner time units. Frame c; location of the Alfven critical surface (filled hexagons) and of the FM surface (stars). Frame d; comparison of the Alfvén lever arm r_A/r_o for each field line in the simulation (filled hexagons), with predictions of steady state theory (squares). (Adapted from OPS).

In order to show the details of the acceleration mechanism, we plot in Fig. 2 the values of various physical quantities at increasing distance along a fiducial field line, anchored near $r_o = r_i$. Fig. 2a shows that the Alfvén and FM Mach numbers achieve values of 5 and 1.5 respectively. The ratio of the toroidal to poloidal field strength builds from an initial value of unity to 2 (Fig. 2b). Thus, beyond the FM surface, the toroidal magnetic field

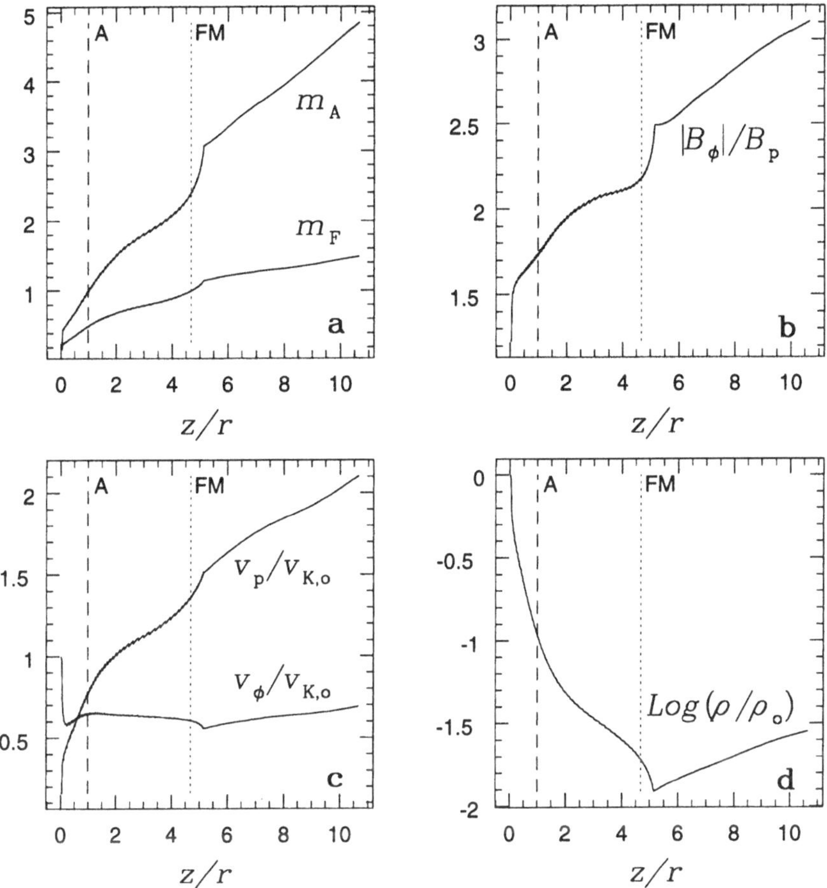

Figure 2. Physical quantities along field line anchored at r_i plotted against $z/r = cotan^{-1}(\psi)$ where ψ is the polar angle as measured from the origin of the coordinate system. Frame 2a; Alfvén Mach numbers m_A and the FM numbers m_F, Frame 2b, ratio of the toroidal to poloidal magnetic field along field line; Frame 2c, poloidal velocity v_p and toroidal velocity v_ϕ, in units of the Kepler speed at the footpoint of the field line; and Frame 2d, the density ρ in units of the density at the footpoint of the field line. The position of the Alfven (marked A) and FM points (marked FM) on the field line are also given. (Adapted from OPS)

dominates the dynamics of the jet. Fig. 2c shows that the poloidal speed of the gas along a field line starts at its input value of a thousandth of the Kepler speed at the base of that field line ($v_{K,o}$), reaches 1.5 times this value at the FM point, and continues to accelerate to 2.1 $v_{K,o}$ at the end of the grid. Thus, flow speed from our fiducial, low mass young stellar object reaches upwards of 220 km s^{-1}, comparable to observed values. The density of the flow decreases as one moves outwards along the field line because of the divergence of the flow streamlines (Fig. 1c).

The outflow achieved in this simulation has properties predicted by steady MHD disk-wind theory. The acceleration of material occurs in two separate stages. In stage 1 (the region along a field line between the disk and the FM point) the centrifugal effect dominates. In stage 2, (region beyond the FM point), the flow speed scales as $v_z \propto z$, which is theoretically expected when outflow becomes super-Alfvénic (supersonic for hydrodynamic jet models; see Raga and Kofman 1992).

This simulation produces mass, momentum and energy flux rates across the outer axial boundary of our simulation, comparable to those for AGN and protostellar outflows (see OPI). The disk accretion rate, deduced from our mass outflow rate using steady state theory, is $\dot{M}_a \simeq 6\dot{M}_w$.

5. Episodic outflow

The field lines in this simulation comprise an initially uniform and vertical magnetic field structure that penetrates the disk and overlying corona. According to steady-state models, this configuration is unfavourable for launching an outflow. In the four snapshots taken at 100, 200, 300, and 400 time units, Fig. 3 shows that outflow is nevertheless launched from the accretion disk for reasons we describe below. The density structure is shown in the left panels, and the toroidal field contours on the right. The highly collimated, jet-like flow has a density structure that is dominated by discrete knots. At time 100 and 200, the working surface of the outflow is clearly evident. The working surface in this simulation has the shape of a cone that advances into the flow with a swept-back bow shock along its flanks. Such nose cone structures have been seen in pure MHD jet simulations (Clarke *et al.* 1986, Lind *et al.* 1989). This surface leaves our grid long before 400 time units have elapsed.

The density structure in this simulation *never* achieves steady state. We have run simulations up to 1000 time units and find that the episodic nature of the flow persists unabated. The density contours show that the jet is strongly confined near the axis. We find that the associated toroidal field in the flow is tightly *anti-correlated with the knots*; high (low) density regions are associated with the lowest (highest) toroidal field strength. The average speed of material in this outflow is $0.5v_{K,i}$, with the highest outflow velocities $\simeq v_{K,i}$. Although this velocity is not as large as is sometimes observed on larger scales in real outflows, our simulations explore only a limited region of physical and parameter space. Movies of this simulation (http://www.physics.mcmaster.ca/Grad_Ouyed/ROuyed.html) show that

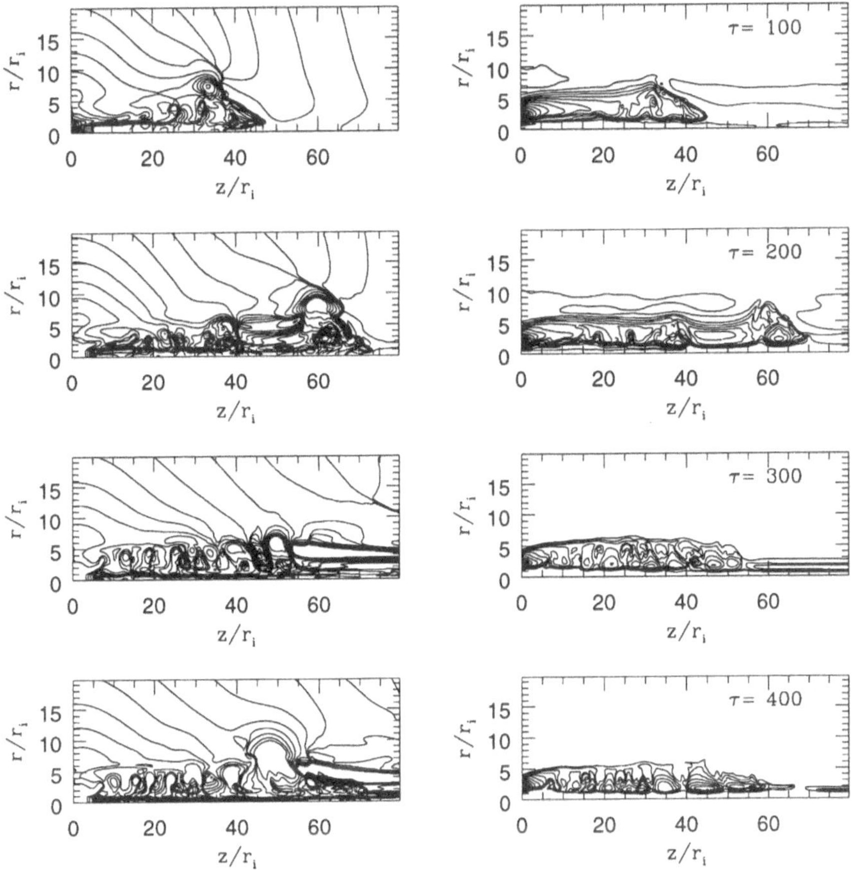

Figure 3. Jet density (left) and toroidal field (right) showing evolution of episodic outflow at 4 different times. 20 logarithmically spaced contour lines are shown for density, 20 equally spaced contours for B_ϕ. (Adapted from OPII).

the knot generator stays fixed in space and that the knots, once formed, persist as coherent structures that propagate down the length of the jet.

To explore the knot generating region further, we performed a simulation at 4 times the axial and twice the radial resolution, but otherwise identical to that shown in Fig. 3; $(10 \times 20)r_i$.

In Figure 4, we show the magnetic field configuration of the outflow at times 37.2, 42.6, and 48.0 during which the new knot (B) is formed (knot A moves away from the generating region as knot B is formed). This process takes a time $\tau_{knot} \simeq 11$ time units. The knots are produced at a distance of $z_{knot} \leq 6 - 7r_i$ from the central source. Since the variation in axial velocity associated with the knots is supersonic, they are associated with MHD shocks (see OPII) . Comparison of frames a and c shows that knots are *spatially* separated by a distance of $\delta z_{knot} \simeq 5r_i$ and thus move at a speed of $\simeq 0.5v_{K,i}$ out of this generating region.

The plots in the right hand panels of Fig. 4 reveal how the outflow and its knots are generated. A narrow radial region of field lines in the innermost parts of the disk has been opened up, making an angle of 50° with respect to the disk surface. This narrow region drives the outflow and is produced by the toroidal field which is strongest in the inner regions of the outflow where the underlying Kepler rotation is the largest. A radial, outwardly directed, toroidal magnetic pressure gradient is produced by the newly created toroidal field and this pushes open the field lines setting up a condition favourable for outflow. The knots are produced in a region of the flow beyond the Alfvén (marked on the field lines in Fig. 4) and FM surface in the outflow.

The size of the wind production region on the surface of the disk depends upon the ease with which field lines can be pushed aside by the toroidal field pressure (Ouyed 1996). The rigidity of the field is measured by the β parameter and is very different for our two cases. For the more magnetically dominated episodic case, the magnetic pressure associated with the self-generated toroidal field in the narrow band of outflowing gas has a local maximum so that the pressure force, acts both radially outwards, and *inwards towards the axis.* The accelerating outflow, on encountering the inwardly directed pressure gradient, is reflected back towards the axis. Because the outflow is rotating however, the gas spins up as it moves inwards and reflects off an inner "centrifugal barrier" when it reaches a radius comparable to its footpoint radius (see Fig. 4 in OPII). The resulting nearly harmonic oscillation in the width of the flow (Sauty and Tsinganos 1994) between these two "barriers" must, by mass conservation, lead to variations in flow velocity (much like a constricted garden hose). These variations in flow speed rapidly steepen into fast MHD shocks (Gomez de Castro and Pudritz 1993, Ouyed and Pudritz 1993, Ouyed and Pudritz 1994).

Estimating that the toroidal Alfvén speed in our simulation is $v_{A,\phi} \simeq 0.5 - 0.6v_{K,i}$, and the width of the jet to be $\delta r_j \simeq 3 - 4r_i$, the oscillation period of the jet is $t_{osc} = 2\delta r_j / v_{A,\phi} \simeq 11 - 13t_i$, which is indeed the knot production time scale. By contrast, the initial magnetic pressure in the steady flow case drops significantly as one moves outwards from the axis. The accelerating outflow does not encounter a strongly magnetically overpressured outer barrier, and continues to expand radially finally achieving a quiet, cylindrically collimated state.

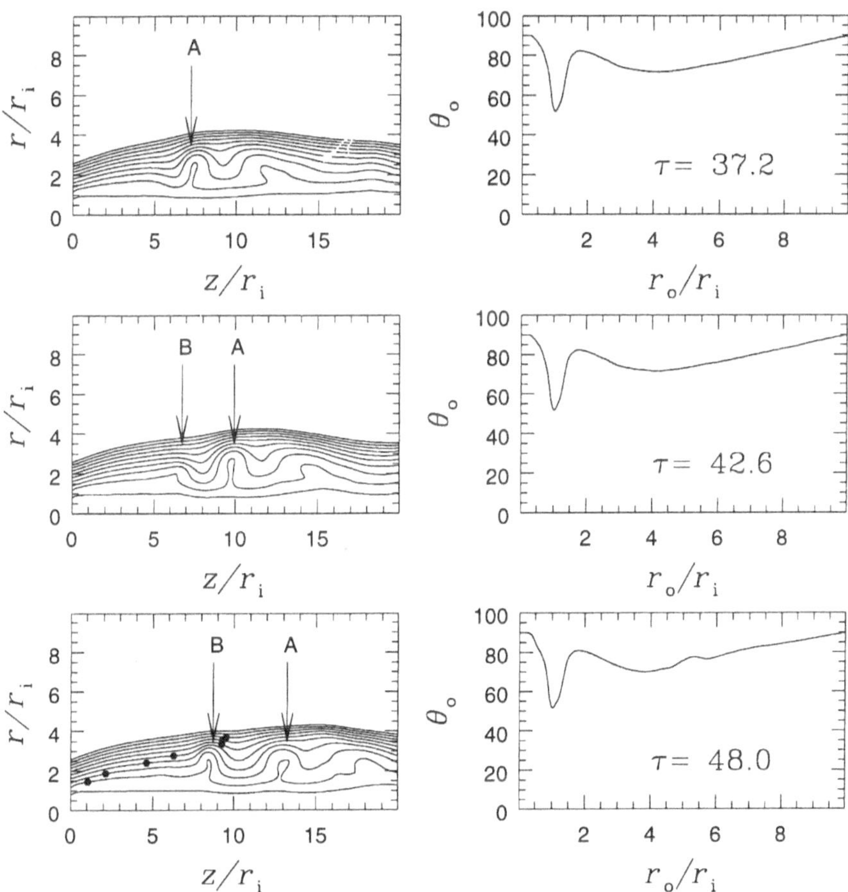

Figure 4. The left panels show the magnetic field structure of the knot generating region, at the three times; 37.2, 42.6, and 48.0 inner time units. The right panels show the angle θ_o of field lines at the base of the flow, with the disk surface, at these times. Note the narrow band of field lines which is sufficiently opened ($\theta_o \leq 60^o$) so as to drive the outflow. Only field lines involved in the knot generation process are shown; field lines at larger disk radius stay reasonably vertical as seen in the right panels. (Adapted from OPS.)

6. Transition from Episodic to Stationary Flow

In Ouyed (1996) and OPIII, we have investigated the importance of toroidal field in producing episodic jet behaviour. We found that when the number $N = B_{\phi,i}^2/8\pi\rho_i v_{inj.}^2 > 1$ (N is the ratio of the confining toroidal field pressure to the ram pressure of gas injected into the corona; this ratio is a constant for the uniform field configuration), then the outer toroidal barrier is strong and the flow is episodic. However, if $N < 1$, then the toroidal magnetic pressure is overwhelmed and a stationary flow should result. We performed a series of eight simulations identical to those shown in Figs 3 and 4, except that the injection speed in the initially uniform vertical field configuration was varied from a value of 0, to a maximum of $10^{-2} v_{K,i}$. The results of these calculations are shown in the accompanying Table (adapted from OPIII). The transition between episodic and stationary flow appears to respect this criterion very well.

TABLE 1. Effect of Varying the Injection Velocity v_{inj}^{a} (adapted from OPIII)

$\frac{v_{inj}/v_K}{10^{-3}}$	W^b	N_K^c	$\frac{z_{gen.}}{r_i}$	$\frac{\delta r_{wind}}{r_i}$	$\frac{\Delta z_{knot}}{r_i}$	$\frac{\Delta t_{knot}}{t_i}$	$\frac{\bar{v}_\infty}{v_{K,i}}$	$(\frac{B_\phi^2/8\pi}{\rho v^2})^d$
0.0	3	1	4	–	–	–	0.3	∞
0.1	2	2	5	1	7	15	0.3	3.6×10^3
0.5	2	3	$7-8$	$3-4$	$5-7$	$11-13$	0.5	1.45×10^2
1.0	2	3	$7-8$	$3-4$	$5-7$	$11-13$	0.5	3.6×10^1
5.0	1	0	–	10	–	–	0.8	1.45
10	1	0	–	10	–	–	0.9	0.36

[a]The rest of the parameters are the standard ones. Namely, $\beta_i = 1.0$, $\delta_i = 100.0$, $\eta_i = 100.0$ and $\mu_i = 1.0$.
[b]Wind properties: $1 \equiv$ steady; $2 \equiv$ episodic; $3 \equiv$ transient.
[c]Number of Knots: Total time of simulation $t_S = 50\ t_i$. The simulated region is $(r \times z) = (10 \times 20) r_i$ with a resolution of (200×500) cells.
[d]Estimated at the disk surface (injection values).

7. Conclusions

The existing numerical simulations, including our own, only explore a small part of parameter space. To date, the published simulations assume ideal MHD and some polytropic equation of state. Our own work simulates a region close to the central engine which is still a decade and a half smaller than could be resolved by HST. The two assumptions of our simulations which are most important however, are that of axisymmetry as well as the fixed nature of the disk boundary conditions.

The main effect of reducing the dimensionality of the dynamics is to eliminate some modes of MHD instabilities, such as the potentially important non-axisymmetric modes. These modes can play an important role in the dynamics of the winds. The major constraint is that the accretion disk is not allowed to respond to the changing torques exerted by the jet.

This assumption is reasonable as long as disk evolution is slower than the evolution of the jet. Given that the Alfvén crossing time in the jet is much shorter than in the disk, this is a plausible assumption. Thus, in spite of their limitations, our simulations may provide a good first step towards a detailed theory of astrophysical jets as they are observed.

Acknowledgements: The authors are grateful to the organizers for the opportunity to present this work at this most stimulating conference. We are indebted to Jim Stone, David Clarke, and Patricia Monger for stimulating conversations and assistance. REP thanks CITA for the stimulating environment and support that he enjoyed during a sabbatical leave there as this article was being written. The research of REP is supported by an operating grant from the Natural Sciences and Engineering Research Council of Canada.

References

André, P., Ward-Thompson, D., & Barsony, M. 1993, Ap. J., **406**, 122
Balbus, S. A. & Hawley, J. F. 1991, Ap. J., **376**, 214
Beckwith, S. V. W., Sargent, A. I., Chini, R. S., & Güsten, R. 1990, A. J., 99, 924
Bell, A. R. & Lucek, S. G. 1995, MNRAS, **277**, 1327
Blandford, R. D. 1990, In SAAS Fée Lectures on Advanced Astrophysics, *Active Galactic Nuclei*, eds. R. D. Blandford, H. Netzer and L. Woltjer, Berlin:Springer
Blandford, R. D & Payne, D. R 1982, MNRAS, **199**, 883 (BP)
Cabrit, S., & André, P. 1991, Ap. J., **379**, L25
Calvet. N., Hartmann, L., & Kenyon, S. J. 1993, Ap. J., **402**, 623
Cao, X. & Spruit, H. C. 1994, A&A, **287**, 80
Clarke, D. A., Norman, M. L. & Burns J. O. 1986, Ap. J., **311**, L63
Collin-Souffrin, S. 1992, In AIP Conf. No. 254, *Testing the AGN Paradigm*, eds. S. S. Holt, S. G. Neff and C. M. Urry, p. 119
Curry, C., Pudritz, R. E., & Sutherland, P. G. 1994, Ap. J., **434**, 206
Edwards, S., Ray, T. P. & Mundt R. 1993, In *Protostars and Planets III*, eds. E. Levy and J. Lunine, University of Arizona Press, p. 567
Eislöffel, J. & Mundt, R. 1994, A&A, **284**, 530
Ferreira, J. & Pelletier, G. 1995, A&A, **295**, 807
Gomez de Castro, A. I., & Pudritz, R. E. 1993, Ap. J., **409**, 748
Heyvaerts, J. & Norman, C. 1989, Ap. J., **347**, 1055
Hjellming, R. M, & Rupen, M.P. 1995, Nature, **375**, 464
Königl, A. 1989, Ap. J., **342**, 208
Königl, A., & Ruden, S. P. 1993, in *Protostars and Planets III*, E. H. Levy and J. I. Lunine eds. (Tucson: University of Arizona Press), 641.
Li, Z.-Y. 1995, Ap. J., **444**, 848
Lind, K. R., Payne, D. G., Meier, D. L. & Blandford, R. D. 1989, Ap. J., **344**, 89
Lovelace, R. V. E., Wang, J. C. L., and Sulkanen, M. E. 1987, Ap. J., **315**, 504
Lubow, S. H., Papaloizou, J. C. B., & Pringle, J. E. 1994, MNRAS, **268**, 1010
Masson, C. R. & Chernin, L. M. 1992, Ap. J., **387**, L47
Ouyed, R. 1996, Ph.D. Thesis, McMaster University
Ouyed, R., & Pudritz, R. E. 1993, Ap. J., **419**, 255
Ouyed, R., & Pudritz, R. E. 1994, Ap. J., **423**, 753
Ouyed, R., & Pudritz, R. E. 1997a, Ap. J., June 20 issue (OPI)
Ouyed, R., & Pudritz, R. E. 1997b, Ap. J., August 1 issue (OPII)

Ouyed, R., & Pudritz, R. E. 1997c, MNRAS, submitted (OPIII)
Ouyed, R., Pudritz, R. E., & Stone, J. 1997, Nature, **385**, 409 (OPS)
Parker, N. D., Padman, R., & Scott, P. F. 1991, MNRAS, **252**, 442
Pelletier, G. & Pudritz, R. E. 1992, Ap. J., **394**, 117
Pudritz, R. E. 1981, MNRAS, **195**, 881
Pudritz, R. E., McLaughlin, D. E., & Ouyed, R. 1997, in A.S.P. Conference Series: 12th Kingston Conference on Theoretical Astrophyiscs, *Computational Astrophysics*, D.A. Clarke & M.J. West eds. (San Francisco: ASP), in press.
Pudritz, R. E., Pelletier, R., & Gomez de Castro, A. I. 1991, in *Physics of Star Formation and Early Stellar Evolution*, C. Lada and N. Kylafis eds. (Dordrecht: Kluwer), 539.
Pudritz, R. E., Wilson, C. D., Carlstrom, J. E., Lay, O. P., Hills, R. E., & Ward-Thompson, D. 1996, Ap. J., **470**, L123
Raga, A. C. & Kofman, L. 1992, Ap. J., **386**, 222
Reipurth, B. 1989, In *Low Mass Star Formation and Pre-Main-Sequence Objects*, ed. Bo Reipurth, Garching: ESO, 247
Reys-Ruis, M. & Stepinski, T. 1996, Ap. J., **459**, 653
Sauty, C., & Tsinganos, K. 1994, A&A, **287**, 893
Shibata, K., & Uchida, Y. 1986, PASJ, **38**, 631
Shu, F. H. 1977, Ap. J., **214**, 488
Shu, F. H., Najita, J., Wilkin, F., Ruden, S.P., & Lizano, S. 1994, Ap. J., **429**, 781
Snell, R. L., Loren, R. B., & Plambeck, R.L. 1980, Ap. J. L., **239**, L17
Stapelfeldt, K. R. 1996, private communications.
Stone, J. M., & Norman, M. L. 1992a, Ap. J. S., **80**, 753
Stone, J. M., & Norman, M. L. 1992b, Ap. J. S., **80**, 791
Stone, J. M., & Norman, M. L. 1994, Ap. J., **420**, 237
Stone, J. M., Hawley, J. F., Gammie, C. F., & Balbus, S. A. 1996, Ap. J., **463**, 656
Tout, C. A., & Pringle, J. E. 1992, MNRAS, **259**, 604
Uchida, Y., & Shibata, K. 1985, PASJ, **37**, 515
Ustyugova, G. V., Koldoba, A. V., Romanova, M. M., Chechetkin, V. M. & Lovelace, R. V. E. 1995, Ap. J. L., **439**, L39
Wardle, M., & Königl, A. 1993, Ap. J., **410**, 218

ASYMPTOTIC STRUCTURE OF ROTATING MHD WINDS AND ITS RELATION TO WIND BOUNDARY CONDITIONS

J. HEYVAERTS
Observatoire de Strasbourg
and Université Louis Pasteur
11, rue de l'Université, 67000 Strasbourg, France

AND

C.A. NORMAN
Physics and Astronomy, Johns Hopkins University
and Space Telescope Science Institute
STScI, 3700, San Martin Drive, Baltimore, MD 21218, USA

Abstract. Approximate asymptotic solutions for rotating MHD winds are obtained analytically in terms of the first integrals of the motion. It is shown that the paraxial region of such winds is a line-shaped boundary layer which has, even at large distances, the structure of a pressure-supported current pinch. A necessary condition for cylindrically focused asymptotics to be possible is derived. A simplified model by which the asymptotic structure of such winds can be obtained in terms of general boundary conditions at the wind source is introduced. Results of semi numerical solutions of the model are reported. The model is analytically solved in the limit of very fast rotators, giving in this particular case an explicit and complete description of the wind outputs and asymptotic structure in terms of arbitrary boundary conditions at the wind source.

1. Introduction

Young Stellar Objects have collimated outflows as do other types of more exotic astrophysical objects, such as Active Galactic Nuclei. It has been suggested that the reason for the focusing of these outflows could be found in the action of the "magnetic hoop stress" which develops in MHD winds as a result of their rotation. Due to flux freezing, rotation builds an azimuthal (or "toroidal") component $\vec{B}_\theta$ of the magnetic field which is supported

B. Reipurth and C. Bertout (eds.), Herbig–Haro Flows and the Birth of Low Mass Stars, 275–290.

by a "poloidal" (i.e. in the meridional plane) electric current density $\vec{j}_P$. By Ampere's law the cross product $\vec{j}_P \times \vec{B}_\theta$ has a component, the "hoop stress" or "pinching force", which pushes the plasma to the rotation axis, while its component along the poloidal magnetic field $\vec{B}_P$ causes magnetic acceleration of the wind. The azimuthal force $\vec{j}_P \times \vec{B}_P$ exerts a torque on the wind which is ultimately transmitted to the wind source.
The pinching force competes with other forces, such as the centrifugal force, the gas pressure or the poloidal magnetic pressure and tension $\vec{j}_\theta \times \vec{B}_P$, the sum of which is generally oriented outwards in the asymptotic region of the flow. The wind will be turned into a collimated jet if the pinching force dominates over these other forces. We have shown this to be generally the case for polytropic non-relativistic winds (Heyvaerts and Norman 1989), and found that the focusing of the flow may be to a cylindrical type of geometry if the asymptotic total poloidal current per hemisphere (lim for $z \to \infty \int \vec{j}_P.d\vec{S}$) does not vanish, while it must be to a parabolic type of geometry if it does. However, these very general results have left open the question of deciding from an examination of boundary conditions on the wind source which of these possible situations was to be met, nor did they provide any explicit asymptotic solution for the wind flow nor for the global energy, mass and angular momentum outputs of such winds.
It is the purpose of this communication to report progress on these issues.

2. General Properties of Magnetized Rotating Winds

The basics of stationary, axisymmetric, rotating, perfect-MHD winds are well known from the works of Weber and Davis (1967) who described the field-aligned dynamics and of Heinemann and Olbert (1978) and Okamoto (1975) who described the cross-field force balance. A recent introduction to this subject is given by Heyvaerts (1996). The main results of this classical analysis are as follows.

2.1. REPRESENTATION OF THE MAGNETIC FIELD

Cylindrical coordinates r, θ, z and the MKSA system of units are used, μ_0 being the magnetic permeability of vacuum. The axisymmetric stationary magnetic field, as any other vector, can be split into a poloidal part and a toroidal part. It can be represented as

$$\vec{B} = \frac{\vec{\nabla} a \times \vec{e}_\theta}{r} + B_\theta \, \vec{e}_\theta \tag{1}$$

B_θ and a are functions of r and z and $\vec{e}_\theta$ is the unit azimuthal vector. Field lines of the poloidal part of magnetic field are lines of constant $a(r, z)$.

Magnetic surfaces are generated by rotation about the polar axis of such lines. They form a family of nested axisymmetric surfaces on which field lines of the total field $\vec{B}$ are drawn.

2.2. SURFACE FUNCTIONS

The toroidal component of the induction equation in the perfect MHD limit, $\vec{E} + \vec{v} \times \vec{B} = \vec{0}$, integrates as $\rho \vec{v}_P = \alpha \vec{B}_P$ and the mass conservation equation implies that α is a constant on a magnetic surface. Such functions are called "surface functions".

$$\alpha(r, z) = \alpha(a(r, z)) = \alpha(a) \tag{2}$$

The poloidal component of the induction equation implies that the electric potential ϕ is a surface function, $\phi(a)$. The angular velocity "of the field", Ω, is defined to be $\Omega(a) = d\phi/da$ and the poloidal part of the induction equation can be written in terms of it as

$$v_\theta = r\Omega(a) + \alpha(a)B_\theta/\rho \tag{3}$$

The flow then consists of a field-aligned part combined with the rotation of the field, a result known as the isorotation law. Since the plasma flows on magnetic surfaces, the function Q involved in the polytropic relation $P = Q\rho^\gamma$ is itself a surface function, $Q(a)$. The toroidal component of the equation of motion integrates into an angular momentum conservation law which states that the specific angular momentum (i.e. per unit escaping mass) is conserved following the fluid motion, defining yet another surface function $L(a)$

$$rv_\theta - rB_\theta/(\mu_0\alpha) = L(a) \tag{4}$$

which consists of a kinetic and magnetic part. The equation of motion projected on the poloidal magnetic field also integrates into an energy conservation law which states that the total specific energy, in kinetic and magnetic forms, is conserved following the fluid motion and equals another surface function $E(a)$:

$$\frac{1}{2}(v_P^2 + v_\theta^2) + \frac{\gamma}{\gamma - 1}Q(a)\rho^{\gamma-1} + G(r, z) - \frac{r\Omega(a)B_\theta}{\mu_0\alpha} = E(a) \tag{5}$$

This equation is referred to as the Bernoulli equation. $G(r, z)$ is the gravitational potential. At this stage all of the relevant equations have been integrated once but for the component of the equation of motion perpendicular to magnetic surfaces, and five surface functions, or "first-integrals" or "constants of the motion", have been introduced, namely:

$$\alpha(a), \quad \Omega(a), \quad Q(a), \quad L(a), \quad E(a) \tag{6}$$

Not all of them, however, are determined by the boundary conditions at the wind source. Only $Q(a)$ and $\Omega(a)$ can be known that way. The other three have to be determined from regularity requirements described below.

2.3. ALFVÉN DENSITY AND RADIUS

It is possible to eliminate the toroidal components by using the angular momentum conservation law and the isorotation law. They can be expressed as:

$$v_\theta = \frac{L}{r} + \frac{\rho}{r}\frac{L - r^2\Omega}{\mu_0\alpha^2 - \rho} \tag{7}$$

$$B_\theta = \frac{\mu_0\alpha\rho}{r}\frac{L - r^2\Omega}{\mu_0\alpha^2 - \rho} \tag{8}$$

For regularity it is necessary that when the density equals the Alfvén density, ρ_A, defined by $\rho_A = \mu_0\alpha^2$, the radius r equals the Alfvén radius, r_A, defined as $r_A = L/\Omega$. Both ρ_A and r_A depend on the magnetic surface. The ratio ρ_A/ρ is the square of the alfvénic Mach number of the poloidal flow relative to the poloidal Alfvén velocity. The flow then passes the Alfvén speed at this point which for this reason is called the Alfvén point.

2.4. TRANSFIELD EQUATION

The equation of motion projected perpendicular to the magnetic surface gives a non-linear partial differential equation for the function $a(r, z)$. This so-called generalized Grad-Shafranoff equation or transfield equation determines the shape of magnetic surfaces. It can be written as:

$$\frac{\alpha}{\rho r}\left[\frac{\partial}{\partial z}\frac{\alpha}{\rho r}\frac{\partial a}{\partial z} + \frac{\partial}{\partial r}\frac{\alpha}{\rho r}\frac{\partial a}{\partial r}\right] - \frac{1}{\mu_0\rho r}\left[\frac{\partial}{\partial z}\frac{1}{r}\frac{\partial a}{\partial z} + \frac{\partial}{\partial r}\frac{1}{r}\frac{\partial a}{\partial r}\right] =$$
$$E'(a) - \frac{Q'(a)\rho^{\gamma-1}}{\gamma - 1} + \frac{\alpha'}{\alpha}\frac{\mu_0\alpha^2\rho}{r^2}\left(\frac{L - r^2\Omega}{\mu_0\alpha^2 - \rho}\right)^2$$
$$-\frac{\rho}{r^2}\frac{(L' - r^2\Omega')(L - r^2\Omega)}{\mu_0\alpha^2 - \rho} - \frac{LL'}{r^2} \tag{9}$$

where a notation like L' means dL/da. The fluid density ρ is implicitly given in terms of r and a by the Bernoulli equation $\mathcal{B}(\rho, r) = E$, the Bernoulli function $\mathcal{B}$ being defined by:

$$\mathcal{B} = \frac{1}{2}\frac{\alpha(\nabla a)^2}{\rho^2 r^2} + G + \frac{\gamma Q\rho^{\gamma-1}}{\gamma - 1} - \rho\Omega^2\frac{r_A^2 - r^2}{\rho_A - \rho} + \frac{1}{2}\frac{\Omega^2 r_A^4}{r^2}\left(\frac{\rho_A r_A^2 - \rho r^2}{r_A^2(\rho_A - \rho)}\right)^2 \tag{10}$$

The problem reduces to solving the coupled system of Bernoulli and transfield equations. However, at this point, three surface functions are left undetermined.

2.5. CRITICALITY EQUATIONS

It is known that no solution of the algebraic Bernoulli equation can be found that is regular from small to large r's unless this solution passes points where the differential of the Bernoulli function vanishes. These "critical points" are located where the poloidal velocity equals the speed of either the slow or the fast magnetosonic mode for propagation along the poloidal field (Weber and Davis 1967, Heinemann and Olbert 1978; see also Heyvaerts 1996). The solution then passes in the (r-ρ) plane a "slow point" and a "fast point". The condition that the differential of the Bernoulli function vanishes at these points gives implicitly their position and density, r_s, ρ_s and r_f, ρ_f in terms of Ω, Q, α, L. Moreover, one must have

$$E = \mathcal{B}(\rho_s, r_s) = F_s(\Omega, Q, \alpha, L) \tag{11}$$

$$E = \mathcal{B}(\rho_f, r_f) = F_f(\Omega, Q, \alpha, L) \tag{12}$$

which provides two equations, the so-called criticality relations, which relate the five surface functions Ω, Q, α, L and E.

2.6. ALFVÉN REGULARITY EQUATION

All transalfvénic solutions to the Bernoulli equation pass the Alfvén point (r_A, ρ_A) so that no condition has to be imposed at that point for its solution. Not so, however, for the transfield equation. It is now well known that this equation is singular at the Alfvén surface, the locus of Alfvén points, where all of its highest order derivative terms vanish (Sakurai 1985, Heyvaerts and Norman 1989, Heyvaerts 1996). A more direct and physical approach to the singularity of the transfield equation at the Alfvén surface is as follows. The transfield equation can be given the following form:

$$\left[\frac{(\vec{\nabla} \times \vec{B}_P) \times \vec{B}_P}{\mu_0 \rho} - (\vec{\nabla} \times \vec{v}_P) \times \vec{v}_P\right] \cdot \frac{\vec{\nabla} a}{\nabla a^2} = E' - \frac{Q' \rho^{\gamma-1}}{\gamma - 1}$$
$$+ \frac{\alpha'}{\alpha} \frac{\rho \rho_A}{r^2} \left(\frac{L - r^2 \Omega}{\rho_A - \rho}\right)^2 - \frac{\rho}{r^2} \frac{(L' - r^2 \Omega')(L - r^2 \Omega)}{\rho_A - \rho} - \frac{L L'}{r^2} \tag{13}$$

The velocity is related to the field by $\rho \vec{v}_P = \alpha \vec{B}_P$. Note also that the modulus of the poloidal velocity is implicitly given by the Bernoulli equation.

So, it must be possible to turn the transfield equation into an equation for the direction of the poloidal velocity. For that we write

$$\vec{v}_P = |v_P| \ \ (cos\psi \ \vec{e}_r + sin\psi \ \vec{e}_z) \tag{14}$$

and insert this in the transfield equation, transforming it into an equation for the curvature of poloidal field lines, $d\psi/ds$, s being the curvilinear abcissa along such a field line. Let also $\vec{n} = -\vec{\nabla}a/|\nabla a|$ be the unit vector normal to magnetic surfaces. We obtain a new form of the transfield equation:

$$\frac{(\rho_A - \rho)}{\rho_A \ r \ B_P}\left(v_P^2 \frac{d\psi}{ds} - \frac{\alpha^2}{2\rho^2}\vec{n}\cdot\vec{\nabla}B_P^2\right) = \frac{B_P}{2r}\vec{n}\cdot\vec{\nabla}(\frac{\alpha^2}{\rho^2}) + E' - \frac{Q'\rho^{\gamma-1}}{\gamma-1}$$
$$+\frac{\alpha'}{\alpha}\frac{\rho\rho_A}{r^2}\left(\frac{L-r^2\Omega}{\rho_A-\rho}\right)^2 - \frac{\rho}{r^2}\frac{(L'-r^2\Omega')(L-r^2\Omega)}{\rho_A-\rho} - \frac{LL'}{r^2} \tag{15}$$

All the second order derivative terms are gathered in the l.h.s of this equation and vanish at the Alfvén surface, where $\rho = \rho_A$. The physical reason why the curvature term vanishes at the Alfvén point is that this term sums the inward poloidal magnetic curvature force, $B_P^2/\mu_0\mathcal{R}$, where $\mathcal{R}$ is the radius of curvature, with the outward centrifugal force due to the curvature of the poloidal motion $\rho v_P^2/\mathcal{R}$. At the Alfvén point these two forces balance each other exactly. This is why at this position the transfield equation degenerates into an ordinary differential equation involving functions of a, expressing cross field balance of "locally" defined forces. On the Alfvén surface the transfield equation reduces to the vanishing of the right hand side of equation (15), where the ratio $(L - r^2\Omega)/(\rho_A - \rho)$ must be understood in a limit sense. Its value can be expressed in terms of

$$q = \lim\left(\frac{r(\rho-\rho_A)}{\rho(r-r_A)}\right) = \left(\frac{dln\rho}{dlnr}\right)_{a,A} \tag{16}$$

In this expression the limit is to be taken at constant a approaching the Alfvén point, as indicated by the subscript (a, A) on the r.h.s. The value of q depends on the solution of the Bernoulli equation achieved on magnetic surface a, so it is a function of the five surface functions, i.e.:

$$q = -\frac{2\Omega r_A}{\sqrt{2E - 2G_A - 2\frac{\gamma}{\gamma-1}Q\rho_A^{\gamma-1} - \Omega^2 r_A^2 - v_{PA}^2}} \tag{17}$$

The transfield equation at the Alfvén surface becomes an ordinary differential equation which relates these five functions of a and the angle $\psi_A(a)$

that the tangent to the poloidal field line a makes with the equatorial plane at the Alfvén point. It can be written as (Heyvaerts and Norman, 1989):

$$\frac{\alpha'}{\alpha} - q\frac{r'_A}{r_A} + q\frac{sin\psi_A}{r_A|\nabla a|_A} + \frac{E'}{v_{PA}^2} - \frac{Q'\rho_A^{\gamma-1}}{(\gamma-1)v_{PA}^2}$$
$$+\frac{\Omega^2 r_A^2}{v_{PA}^2}\left(\frac{4}{q^2}\frac{\alpha'}{\alpha} + \frac{4+2q}{q}\frac{r'_A}{r_A} - \frac{\Omega'}{\Omega}\right) = 0 \tag{18}$$

Failure to satisfy it at the Alfvén surface would cause the curvature of calculated poloidal field lines to become infinite at the Alfvén point and these lines to get an unphysical sharp kink there.

This Alfvén regularity equation provides another relation among the five surface functions E, Q, L, Ω, α. Together with the criticality relations, it provides the means for determining the surface functions which are not directly given by boundary conditions. The angle ψ_A which appears in equation (18) is implicitly a function of them, resulting from the solution of the transfield equation between the wind source and the Alfvén surface.

3. The Asymptotic Transfield Equation

It is possible to derive an asymptotic form of the transfield equation in those regions of the flow where the axial distance becomes very large as compared to the Alfvén radius, r_A. We have shown that in this limit the potentially dominant terms of the transfield equation can be written as

$$v_\infty^2|\nabla a|\frac{d\psi}{ds} = \frac{\Omega}{\alpha}\vec{\nabla}a\cdot\vec{\nabla}\left(\frac{\rho r^2\Omega}{\mu_0\alpha}\right) + \frac{1}{\rho}\vec{\nabla}a\cdot\vec{\nabla}(Q\rho^\gamma) \tag{19}$$

Examination of the terms of this equation further shows that the curvature term can be neglected in a first approximation and that the pressure term must be considered only in the vicinity of the polar axis and of the equatorial plane. Indeed, the hoop stress (first term on the r.h.s) cannot dominate near the axis because the current enclosed in a circle centered on the axis vanishes with its radius. On the other hand, when symmetry is of a dipolar type, the field vanishes at the equatorial plane. There is then a rapid change, a jump, in toroidal field at its crossing, implying a sheet-like current concentration at the equator. Plasma in this region finds a sheet-pinch type of equilibrium, as indeed observed in the solar wind, the equatorial sheet being in this case warped.

3.1. ASYMPTOTIC HAMILTON-JACOBI EQUATION

So, in the lowest order approximation the transfield equation simply reduces in the "field", i.e. away from the equatorial and polar regions, to

$$\vec{\nabla} a \cdot \vec{\nabla} I = 0 \tag{20}$$

where I, defined by $I = \rho r^2 \Omega / \alpha$ is proportional to the electric current enclosed in a circle centered on the axis and passing at the point under consideration. Indeed, ρr^2 is proportional to rB_θ in this limit. This is a generalized form of Heyvaerts and Norman's (1989) solvability condition at infinity. This equation integrates as $I = I(R)$ where R is a variable that labels orthogonal trajectories to poloidal field lines in the meridional plane. We can take it as being the value of z on the axis on such a curve. Two cases must be considered according to whether the current $I(R)$ approaches zero for large R or not. According to Heyvaerts and Norman (1989) and Heyvaerts and Norman (1996) the magnetic surfaces would asymptotically approach paraboloids in the former case while in the second case they should approach, close to the axis, a cylindrical structure, possibly nested into a conical one. That this is indeed so can be seen directly from this equation, for when $I(R)$ approaches a non-vanishing constant I the asymptotic Grad Shafranoff equation integrates as

$$\frac{r|\nabla a|\Omega(a)}{v_\infty(a)} = I \tag{21}$$

which can be transformed into the the following Hamilton-Jacobi equation for the function $S(a)$ defined by:

$$\frac{\Omega|\nabla a|}{I\sqrt{2(E - I\Omega/\mu_0\alpha)}} = |\nabla S| = \frac{1}{r}. \tag{22}$$

This reduces the search for orthogonal trajectories to magnetic surfaces to a ray tracing problem in a medium with a refractive index proportional to $1/r$. The Hamilton Jacobi equation is easy to solve analytically that way in full generality. These orthogonal trajectories are found to form a family of circles centered on the rotation axis. Imposing the boundary condition that they become perpendicular to the equatorial plane at large r's further shows that they must asymptotically tend to be centered on the origin. Hence, those magnetic surfaces on which r can actually reach infinitely large values approach a family of nested cones, as expected. The question of the connection between the cylindrical and the conical part of the flow is considered in subsection 3.3.

3.2. ASYMPTOTIC PARABOLIC SOLUTION

When the poloidal current asymptotically vanishes, the transfield equation still integrates as $I = I(R)$ in regions which are neither close to the polar axis nor to the equator. It is possible to turn this into a solution valid in all the asymptotic field when $I(R)$ declines only very slowly to zero, so that a WKB method can be used. Actually, in that case, the solution is similar to one with a constant I in large regions of the asymptotic domain. In such regions orthogonal trajectories to poloidal field lines must then be close to circles centered on the origin. The distribution of flux along them can be calculated as well as the structure of the polar boundary layer at distance R from the origin. Treating locally the magnetic surfaces approximately as a set of nested cones leads us to solve along circles centered on the origin the asymptotic transfield equation

$$\frac{\Omega}{\alpha}\vec{n} \cdot \vec{\nabla}\left(\frac{\rho r^2 \Omega}{\mu_0 \alpha}\right) + \frac{1}{\rho}\vec{n} \cdot \vec{\nabla}(Q\rho^\gamma) = 0 \tag{23}$$

where $\vec{n}$ is the normal to magnetic surfaces and the curvature term has been neglected, a consistent treatment when the WKB method is indeed justified. In the "field", the pressure term is also negligible and the equation integrates into

$$\frac{r\Omega|\nabla a|}{\mu_0\sqrt{2E}} = C(R) \tag{24}$$

The fluid velocity at distance R has been taken equal to its asymptotic value. When the flux is distributed with angle as $a(\psi)$, $r|\nabla a| = sin\psi \, da/d\psi$, and ψ is found to be given in terms of a by:

$$cos\psi = tanh\left(\frac{1}{C(R)}\int_a^A \frac{\Omega(a')da'}{\mu_0\sqrt{2E(a')}}\right) \tag{25}$$

This solution is valid away from the polar region. It can be regarded as valid near the equator as well if the equatorial sheet-like current is treated as a surface singularity. Near the pole, the surface functions Ω, α and Q can be taken as almost constant and equal to their polar values Ω_0, α_0, Q_0. The transfield equation with its pressure term integrates approximately as

$$\frac{r\Omega_0|\nabla a|}{\mu_0\sqrt{2E_0}} + \frac{\gamma}{\gamma - 1}\frac{\alpha Q_0}{\Omega_0}\rho^{\gamma-1} = C(R) \tag{26}$$

with the same integration constant $C(R)$ as in eq. (24). A length scale l defined by

$$l^2 = \frac{\gamma}{\gamma - 1}\frac{Q\rho_{A0}^{\gamma-1}}{\Omega_0^2} \tag{27}$$

thus appears naturally in this equation, which can be solved to give the relation between flux and radius in the polar boundary layer parametrically in terms of $x = \rho/\rho_0(R)$, where $\rho_0(R)$ is the density on the axis at distance R:

$$sin^2\psi = \frac{\gamma}{\gamma - 1}\frac{\rho_{A0}}{\rho_0}\frac{Q_0\rho_0^{\gamma-1}}{\Omega_0^2 R^2}\left(\frac{1}{x} - \frac{1}{x^{2-\gamma}}\right) \tag{28}$$

$$a = \frac{\gamma}{2(\gamma-1)}\frac{Q_0\rho_0^{\gamma-1}\rho_{A0}\sqrt{2E_0}}{\Omega_0^2\alpha_0}\left(\ln\frac{1}{x} - \frac{2-\gamma}{\gamma-1}(1 - x^{\gamma-1})\right) \tag{29}$$

The condition that the solution in the polar boundary layer asymptotically matches the outer "field" solution can be reduced to the following equation for the ratio $n_0 = \rho_0(R)/\rho_{A0}$:

$$\frac{\gamma\, n_0^{\gamma-1}Q_0\rho_{A0}^{\gamma}}{\sqrt{2}(\gamma-1)\alpha_0\Omega_0\, \int_0^A da' \frac{\Omega(a')}{\sqrt{E(a')}}}\left(2\, log(2R/l) + (2-\gamma)\, log n_0\right) = 1 \tag{30}$$

The second term in the parenthesis on the l.h.s can be neglected for an approximate solution. The poloidal current is seen to vanish at large R's. The shape of magnetic surfaces can be calculated. In the region intermediate between the polar boundary layer and the equatorial region, they become a set of nested paraboloids of variable exponent, $m(a)$ given by

$$m(a)\int_0^A da'\Omega(a')\left(E(a')\right)^{-1/2} = \int_0^a da'\Omega(a')\left(E(a')\right)^{-1/2} \tag{31}$$

$$m(a)(r/l) = 2(z/l)^{m(a)} \tag{32}$$

In the polar boundary layer, r is found proportional to $(log\ z)^{1/2(\gamma-1)}$.

3.3. A NECESSARY CONDITION FOR CYLINDRICAL ASYMPTOTICS

In the case of cylindrical asymptotics, when the jet is unconfined by an external pressure, it is necessary that the solution fills all space for otherwise the total pressure balance could not be satisfied at the jet outer boundary. The radius $r_\infty(a)$, of the last "cylindrical" magnetic surfaces must approach infinity, for otherwise a vacuum gap would be left between the cylindrically focused region and the outer conical one, if any. The asymptotic cylindrical radius of such lines is given by the following equation, derived from the asymptotic transfield equation in the "field" zone:

$$\frac{dr_\infty}{r_\infty} = \frac{\Omega da}{\mu_0 I\sqrt{(2(E - I\Omega/\mu_0\alpha)}} \tag{33}$$

For r_∞ to diverge the quantity under the square root denominator must have a double zero at the value $a = a_\star$ which corresponds to the last cylindrical magnetic surface. For positiveness $(\alpha E/\Omega)$ must reach its minimum there. The poloidal current brought per hemisphere to infinity must be this minimum value and the amount of flux trapped in the asymptotically cylindrical region is $2\pi a_\star$. One should note that this criterium is unfortunately not expressed directly in terms of the boundary conditions at the wind source, since it requires a knowledge of the surface functions, which can be known only after solving for the flow near the source.

4. A Simplified Model

4.1. DESCRIPTION OF THE MODEL

All our results presented above have regarded the surface functions as known. In reality they have to be determined by solving the criticality relations and Alfvén regularity conditions, which implicitly requires a solution for the shape of magnetic surfaces near the source. No exact general analytical solution can be obtained. Only solutions of a self-similar or separable type are known, which cannot cope with general boundary conditions. Numerical solutions are difficult to construct. To investigate the relation between the boundary conditions at the wind source and the asymptotic structure, we therefore adopted a strategy that sacrifices exactness to allow for general boundary conditions. We study a simplified model, which makes the following simplifications and assumptions.
(a) The wind-emitting object is treated as a point source, be it a star or an accretion disk.
(b) The magnetic field configuration is regarded as a-priori known from the wind source to the fast critical surface. This is our main simplification. Such an assumption is certainly justified between the source and the Alfvén surface, where the magnetic energy dominates over the kinetic energy of the flow, but it is weaker further out. In the present state of our work, a model where the magnetic surfaces in this region are represented as nested cones has been studied. This model then consists of a family of off-equator Weber Davis flows in cross field equilibrium with each other at their Alfvén surface.
(c) Cross field balance is not imposed everywhere, but only near the wind source, at the Alfvén surface, where it takes the form of the Alfvén regularity equation, and at infinity, where it determines the asymptotic structure. For low-β conditions near the source, a split-monopole field approximately solves the transfield equation there.
(d) We impose on each magnetic surface the slow mode and fast mode criticality relations, which is easy thanks to assumption (b). We could solve for the field-aligned dynamics from the source to the fast surface, but did

not explicitly do so. This set of assumptions allows to solve for the surface functions within the framework of this model, which, we recall, is not an exact representation of the reality, though it can be improved iteratively. We checked by try and error that the fast critical point tends to be located in regions where magnetic surfaces do not grossly deviate from conical geometry, so that assumption (b) is in practice somewhat better than could anticipated.

4.2. SEMI-NUMERICAL RESULTS FOR CONFINED JETS

We solved this model for a variety of different conditions for rotation (solid body or not) and distribution of the "entropy" surface function Q, assuming the jet to be externally confined by a plasma of very low density and of a given, uniform, pressure. This extra assumption has been made both for simplicity and to incorporate a relevant physical aspect of actual jets. This uniform confinement forces a cylindrically symmetric asymptotic structure with a limited jet radius, calculated by imposing continuity of the total pressure. The methods for analytically/numerically solving the model have been presented at this conference by T. Lery, and are reported in the poster proceedings book and in Lery et al.(1997). It appeared that electric current and mass flow profiles are more concentrated for fast than for slow rotators. An important aspect of these results is that the total poloidal current brought to infinity decreases regularly when the confining pressure is decreased, with a much slower decline for faster rotators, and a very small one indeed for very fast rotators (Lery et al. 1997). This seems to indicate that fast rotators are more likely to give rise to asymptotically cylindrical structures when unconfined. This is also supported by the study of the variations of the function $\alpha E/\Omega$ given by the model. This function appears to have a more pronounced minimum for faster rotators (Norman et al. 1997).

4.3. ANALYTICAL SOLUTION FOR VERY FAST ROTATORS

We have obtained an explicit solution of the equations of the model in the limit case of a very fast rotator. In that case, it turned out that a cylindrically symmetric unconfined asymptotic solution is possible, though perhaps not unique. It is described below. A split monopole model is assumed from the source to the fast surface, the angle of the surface a with the equator being $\theta_0(a)$. The degree of rotation on a magnetic surface can be measured by the parameter $\omega = \Omega r_A/v_{PA}$ while the specific energy can be similarly normalized to the square of Alfvén velocity at the Alfvén point as $\epsilon = 2E/v_{PA}^2$. In terms of these parameters, the criticality relations can be written, in the limit of very fast rotators ($\Omega \rightarrow \infty$), as:

$$\epsilon = 3\omega^{4/3} \qquad\qquad 2\omega^2 = 3\omega^{4/3} \tag{34}$$

So, for very fast Weber-Davis types of rotators,

$$\omega = (3/2)^{3/2} \qquad\qquad \epsilon = 27/4 \tag{35}$$

Given these results and the fact that the shape of the magnetic surfaces is taken as known, the Alfvén regularity equation becomes a logarithmic differential in a which can be integrated, C being an integration constant, as:

$$\left(\frac{\alpha E}{\Omega}\right)\left(\frac{r_A}{cos\theta_0}\right)^{\frac{16}{27}\sqrt{\frac{27}{19}}} r_A^{-2(1-\sqrt{\frac{19}{27}})} = C \tag{36}$$

From these relations the unknown surface functions, E, α and the spherical distance to the Alfvén point, R_A, can be calculated and expressed in terms of a constant K which derives from the constant C above as:

$$R_A = K \ \ cos\theta_0^{\frac{\left(2\sqrt{\frac{19}{27}}-1\right)}{\left(3-2\sqrt{\frac{27}{19}}\right)}} \tag{37}$$

$$\alpha = \sqrt{\frac{27}{8}} \ \frac{A}{2\pi\mu_0 K^3 \Omega(\theta_0)} \ cos\theta_0^{-\frac{\left(\frac{60}{27}\sqrt{\frac{27}{19}}\right)}{\left(3-2\sqrt{\frac{27}{19}}\right)}} \tag{38}$$

$$E = K^2\Omega^2(\theta_0) \ \ cos\theta_0^{\frac{2\left(2-\frac{2-16}{27}\sqrt{\frac{27}{19}}\right)}{\left(3-2\sqrt{\frac{27}{19}}\right)}} \tag{39}$$

The function $\alpha E/\Omega$ is then expressed as

$$\frac{\alpha E}{\Omega} = \sqrt{\frac{27}{8}} \ \frac{A}{2\pi\mu_0 K} \ cos\theta_0^{\frac{-\frac{92}{27}\sqrt{\frac{27}{19}}-4}{3-2\sqrt{\frac{27}{19}}}} \tag{40}$$

It decreases towards the equator, as it should to allow for cylindrical asymptotics, but diverges at the polar axis. Note also that R_A approaches zero at the pole according to this expression. These pathologies at the pole are related to the fact that the wind on the polar axis cannot be in the fast rotator regime. The fast rotator approximation then breaks down in some region near the pole, which should be treated as a boundary layer. Similar calculations to those described above, but carried in the slow rotator regime, give the Alfvén regularity relation in the form:

$$R_A\left(\alpha\sqrt{E}\right)^{\frac{1}{\sqrt{2}}} = C_0 \tag{41}$$

and the criticality relations can be similarly solved to obtain E, α and R_A. The slow-rotator paraxial solution can then be matched to the fast rotator solution by demanding continuity of the Alfvén surface at the flux variable a where the regime changes. This relates the two integration constants K and C_0. As said above, the profile of $\alpha E/\Omega$ allows in this case for cylindrical asymptotics, with $a_\star = A$, the total flux. The solution is entirely cylindrically focused at infinity. Once the surface functions have been determined, it is possible to calculate the total jet outputs, i.e., mass loss, torque and thrust. We do not present these expressions explicitly here for lack of space (Heyvaerts and Norman, 1997). Each of them contains a contribution from the "field", where the fast rotator approximation applies, and a contribution from the polar region which is in the slow rotator regime. The total current emitted per hemisphere is $2\pi I/\mu_0$, and I is the minimum value of $\alpha E/\Omega$, expressed in terms of integration constant K as:

$$I = \sqrt{\frac{27}{8}\frac{A}{K}} \tag{42}$$

In the paraxial region, gas pressure must be considered, even far from the source. Regarding Ω, α and E as approximately constant in this region, a solution to the transfield equation can be found in parametric form, much as in subsection (3.2), namely:

$$r_\infty^2 = \frac{\gamma}{\gamma-1}\frac{\rho_{A0}}{\rho_0}\frac{Q_0\rho_0^{\gamma-1}}{\Omega_0^2}\left(\frac{1}{x} - \frac{1}{x^{2-\gamma}}\right) \tag{43}$$

$$a = \frac{\gamma Q_0 \rho_0^{\gamma-1}\rho_{A0}\sqrt{2(E_0 - \frac{\gamma}{\gamma-1}Q_0\rho_0^{\gamma-1})}}{2(\gamma-1)\Omega_0^2\alpha_0}\left(\ln\frac{1}{x} - \frac{2-\gamma}{\gamma-1}(1-x^{\gamma-1})\right) \tag{44}$$

In the region where gas pressure is negligible and $r_\infty \gg r_A$ the solution takes the form

$$r_\infty = r_0 \; exp\left(\int_0^a \frac{\Omega da'}{\mu_0 I\sqrt{2(E(a') - I\Omega(a')/\mu_0\alpha(a'))}}\right) \tag{45}$$

The polar boundary layer solution asymptotically matches this far field solution if the scale of the "polar pinch" is

$$r_0^2 = \frac{\gamma}{\gamma-1}\frac{\rho_{A0}}{\rho_0}\frac{Q_0\rho_0^{\gamma-1}}{\Omega_0^2} \tag{46}$$

and if

$$\frac{\gamma}{\gamma-1}Q_0\rho_0^{\gamma-1} = \frac{I\Omega_0}{\mu_0\alpha_0} \tag{47}$$

This latter relation gives the current that flows in the pinch in terms of the pressure which it confines, and is akin to relations known as Bennet pinch relations in the field of laboratory plasma confinement. The density on the axis of the jet at large distances is then a function of the current that it supports, itself related to the total mass loss rate. The latter is actually obtained from α by the quadrature:

$$\dot{M} = A \int_0^{\pi/2} cos\theta_0 \ \alpha(a(\theta_0)) \ d\theta_0 \tag{48}$$

Since α is expressed both in the fast and slow rotator region in terms of the integration constant K (or C_0 which is related to it) $\dot{M}$ is expressible in terms of K and conversely. We have seen above how the total electric current I emitted by the jet is expressed in terms of K. This relation can be turned into a (rather complicated) relation between I and $\dot{M}$. When the current I is weak we find that this relation becomes approximately

$$\dot{M} = \frac{2\pi\sqrt{8}(81 - 54\sqrt{\frac{27}{19}})\Omega_0^2 A^2}{108(168\sqrt{\frac{27}{19}} - 162)\pi^2 \mu_0 E_0^{3/2}} \left(\frac{2\sqrt{2} \ \pi \ \mu_0 \ I \ \sqrt{E_0}}{\Omega_0 A} \right)^{\frac{12 - \frac{246}{27}\sqrt{\frac{27}{19}}}{3 - 2\sqrt{\frac{27}{19}}}} \tag{49}$$

5. Winds, Breezes and Asymptotics

It is interesting here to comment on whether the flow can become faster than the fast mode speed at infinity or not. If, on a certain magnetic surface, the asymptotic poloidal wind speed v_∞ is to exceed the fast mode speed, the following inequality must be satisfied

$$\lim \left(\frac{B_\theta^2}{\mu_0 \rho} \right) < v_\infty^2 \tag{50}$$

because the fast mode speed is in the limit of zero pressure the Alfvén speed associated to the total magnetic field. For a magnetic surface which diverges at large z from the polar axis or approaches a cylindrical radius much larger than its Alfvén radius, $v_\infty^2 = 2(E - I\Omega/\mu_0\alpha)$ and ρr^2 approaches $\alpha I/\Omega$. The preceding inequality can thus be written as

$$\frac{I\Omega}{\mu_0 \alpha} < 2 \left(E - \frac{I\Omega}{\mu_0 \alpha} \right) \tag{51}$$

If the flow approaches a parabolic shape and I approaches 0 this inequality is satisfied since E is positive. However, as shown in subsection 3.3, if the flow contains an asymptotically cylindrical part, $(E - I\Omega/\mu_0\alpha)$ must vanish on the last cylindrical magnetic surface. We then meet a contradiction since a finite current current should flow in such a cylindrical region while this is not allowed by the above inequality. This shows that the flow cannot pass the fast mode speed on this "last cylindrical" surface and must therefore have the character of a "breeze" rather than of a "wind". Another way to state this is to say that the fast point must be rejected to infinity on this magnetic surface (Heyvaerts and Norman, 1989). However, if the flow becomes super-fast-mode on magnetic surfaces polewards from the last cylindrical one, so that the fast point only approaches infinity when the last cylindrical surface is approached, the above inequality turns into an equality, and I is still seen to vanish on the last cylindrical surface. It thus appears that MHD "winds" can only be cylindrically focused if bordered by a region where the wind remains strictly sub-fast-mode. The approximate analytical solution presented above for the very fast rotator has, in the adopted quasi Weber Davis description of the flow at finite distances, its fast point rejected to infinity on all magnetic surfaces which are in the fast rotator regime. This is implied by the special value of $(3/2)^{3/2}$ taken by the parameter ω (Heyvaerts, 1996). This solution would subsist in a more elaborate description of the magnetic configuration only if the flow on the outer-most magnetic surfaces would turn from asymptotically fast-critical to strictly sub-fast. In the quasi Weber-Davis representation, rejection of the fast point to infinity is possible only for infinitely fast rotators. Therefore cylindrical collimation appears in this simplified model as a limit-property, which is also consistent with the results reported in subsection 4.2.

References

Heinemann, M. and Olbert, S. (1978) Axisymmetric ideal MHD stellar wind flow, *J. Geophys. Res.*, **83**, pp. 2457–2460

Heyvaerts, J. (1996) Rotating MHD winds, in *Plasma Astrophysics*, Proceedings of the 7th EADN Astrophysics school, San Miniato (Italy), October 1994, C.Chiuderi and G. Einaudi, ed., Lecture Notes in Physics series, Springer Verlag, pp. 31–99

Heyvaerts, J. and Norman, C.A. (1989) The collimation of magnetized winds, *Ap.J.*, **347**, pp. 1055–1081

Heyvaerts, J. and Norman, C.A. (1996) Collimation of magnetized outflows, in *Solar and Astrophysical MHD flows* Proc. of the NATO Advanced Study Institute, Heraklion (Greece) June 1995, K. Tsinganos, ed., Kluwer Academic Publishers, pp. 459–474

Lery, T., Heyvaerts, J., Appl, S. and Norman, C.A. (1997), Outflows from magnetic rotators *Astron. Astrophys.*, in preparation.

Norman, C.A., Heyvaerts, J. and Lery, T. (1997) Latitudinal variations of the specific energy in radial magnetized winds. *Astron. Astrophys.*, in preparation.

Okamoto, I. (1975) Magnetic braking by a stellar wind, *Mon. Not. R. astr. Soc.*, **173**, pp. 357–379

Sakurai, T. (1985) Magnetic Stellar Winds, *Astron. Astrophys.*, **152**,pp. 121–129

Weber, E.J. and Davis, L. Jr. (1967) The angular momentum of the solar wind, *Ap.J.*, **148**, pp. 217–227

HYDRODYNAMIC COLLIMATION OF YSO JETS

ADAM FRANK
Department of Physics and Astronomy,
Bausch and Lomb Building, University of Rochester,
Rochester, NY 14627-0171, USA

AND

GARRELT MELLEMA
Stockholm Observatory, S-13336 Saltsjöbaden, Sweden

Abstract. We present the results of numerical hydrodynamic models for the collimation of outflows from young stellar objects. We show that the presence of a toroidal environment can lead to efficient formation of jets and bipolar outflows from initially uncollimated central winds. The interaction between the wind and the environment leads to two types of collimation, one which is dominated by radiative cooling effects, and one which works when cooling is less efficient. We describe the two types of jets as they appear in the simulations and we suggest a description for the long term evolution of these structures in more realistic time-dependent wind sources.

1. Introduction

The physical processes behind the origin of the subjects of this conference, HH jets, clearly still requires a lot of study. Observations are, however, taking us closer to the young stars and it seems well established that for low mass stars the only way to generate the observed momentum and energy in the jets and outflows is to have some form of MHD process accelerate material into a wind. Beyond this realization it is still not clear what type of MHD process is at work: is it the interaction of a stellar magnetosphere with the inner edge of the accretion disk (the so-called X-wind model, Shu *et al.* 1994); or a centrifugally driven wind coming off the inner parts of the accretion disk (Königl, 1989; Pudritz, 1991)? It should also be realized that for massive stars the situation is quite different; there the driving

B. Reipurth and C. Bertout (eds.), Herbig–Haro Flows and the Birth of Low Mass Stars, 291–302.

force behind the wind may be the stellar radiation. It is also not clear if all MHD models can produce well collimated jets on the observed lengthscales, something which was also recently pointed out in a numerical study by Romanova *et al.* (1996).

Based on these concerns we propose to separate the issues of wind acceleration and jet and bipolar outflow collimation. We believe such a distinction is useful as it allows us to address issues such as the relation between jets and outflows in a focused way and allows the possibility that the acceleration processes may be different in low and high mass stars.

In this work we study the collimating properties of the environment. Our approach is related to the earlier work on hydrodynamic collimation, such as DeLaval nozzles (Königl, 1982; Raga & Cantó, 1989). Such models have fallen out of favor because of length scale requirements (Königl & Ruden, 1993) and stability considerations (Koo & McKee, 1992). We have found however that these objections do not hold when considering the time-dependent evolution of these types of flow.

The background to our present study is the high degree of collimation found when studying the interaction of a stellar wind with a toroidal environment in the context of the formation of aspherical Planetary Nebulae (PNe) (Icke *et al.*, 1992). The mechanism has been called "Shock-Focused Inertial Confinement" (SFIC). In the SFIC mechanism it is the inertia of a toroidal environment rather than its thermal pressure, which produces a bipolar *wind-blown bubble.* In our studies of bipolar PNe formation we found the bubble's wind shock (which decelerates the wind) takes on an aspherical, prolate geometry (Eichler, 1982; Icke *et al.*, 1992). The radially streaming central wind strikes this prolate shock obliquely focusing it towards the polar axis and initiating jet collimation. Other effects such as instabilities along the walls of the bubble, help to maintain the collimation of the shocked wind flow.

This mechanism was first applied to the case of young stars by Frank & Noriega-Crespo (1994), and more recently by Frank & Mellema (1996) (henceforth Paper 1) and Mellema & Frank (1997) (henceforth Paper 2). A related study is the one by Peter & Eichler (1995).

2. Numerical Method

To study the collimating effects of the environment we choose the case where the central wind is maximally uncollimated, i.e. perfectly spherical. This can be considered the worst case scenario for MHD winds since most mechanisms will results in some degree of focusing towards the axis. The central protostellar wind is fixed in an inner sphere of cells on the computational grid. The relevant input parameters are simply the mass loss rate

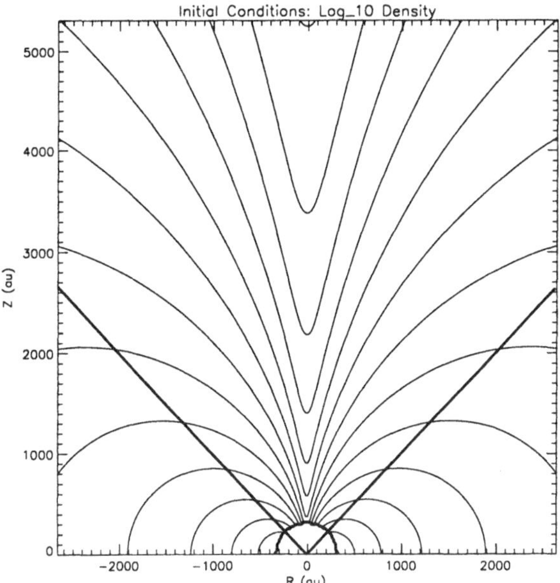

Figure 1. Initial density distribution. Shown are the $\log_{10}$ contours of density from Eq. 1 with an equator to pole contrast $q = 70$. The two solid lines show the angle at which $\rho = .5\ \rho_{\max} = .5\ \rho(90°)$. These occur at $\theta \sim 45°$ making the opening angle of the density distribution $\sim 90°$

$\dot{M}_{\rm w}$ and velocity $V_{\rm w}$ in the wind.

For the environment we choose a density distribution with a toroidal shape. Such toroidal density distributions are theoretically expected for the collapse of a rotating cloud (Terebey, Shu & Cassen, 1984), a flattened filament (Hartmann, Calvet, & Boss, 1996), or a magnetized cloud (Li & Shu, 1996). These are admittedly all axi-symmetric models, and it would be interesting to see how this holds under more general conditions. Lucas & Roche (1997) present some observational evidence for toroidal environments of proto-stars.

The actual form of the toroid we use is given by

$$\rho(R,\theta) = \frac{\dot{M}_a}{4\pi R^2}(\frac{2GM}{R})^{-\frac{1}{2}}\{1 - \frac{\zeta}{6}[13P_2(\cos(\theta)) - 1]\} \qquad (1)$$

in which R is the spherical radius (we will use r to denote the cylindrical radius). In Eq. 1 $\dot{M}_{\rm a}$ is the accretion mass loss rate and M is the mass of the star. Equation 1 is a modified form of Eq. 96 from Terebey, Shu & Cassen (1984) (originally derived by Ulrich 1976). We use it here because it produces the required toroidal geometry as well having the $R^{-\frac{3}{2}}$ radial dependence, appropriate to a freely falling envelope. The parameter ζ determines the flattening of the cloud, which we prefer to quantify with the density contrast q, the ratio between the density at the equator and at the pole. We note that this density distribution is actually very similar to the solution found by Li & Shu (1996) for a magnetized cloud. The shape of this density distribution is shown in Fig. 1, which serves to illustrate that the opening angle of the toroid is actually quite wide. A more detailed description of the initial conditions can be found in Paper 1.

The evolution of this system is followed by solving numerically for the well-known Euler equations, including a source term in the energy equation to account for radiative losses. The numerical method used to solve these equations is based on the Total Variation Diminishing (TVD) method of Harten (1983) as implemented by Ryu *et al* (1995). For cooling we used the standard coronal cooling curve of Dalgarno & McCray (1972). Details of the methods and the implementation of cooling can be found in Paper 2.

3. Results without cooling

In Paper 1 we presented the results of simulations without cooling. Although this is not a realistic assumption in most cases, we considered it to be a good first step. The fact that some of the features of the non-cooling jets also appear in the simulations with cooling (see below), justifies this approach.

Figure 2 shows a typical result from Paper 1. The outflow and environment parameters are: $\dot{M}_{\rm w} = 10^{-7}$ $M_\odot$ yr^{-1}, $V_w = 200$ km s^{-1}, $\dot{M}_{\rm a} = 10^{-5}$ $M_\odot$ yr^{-1}, $q = 70$.

The collimated flow that forms has all the usual features expected for gaseous jets: bow and jet shocks; turbulent cocoons; crossing shocks and internal Mach disks. In Paper 1 we demonstrated that the flow pattern has characteristics of both a supersonic jet and a wind-blown bubble, something which may be important for understanding the connection between jets and molecular outflows. These simulations also showed that strong collimation is principally achieved by the refraction of flow vectors across the aspherical wind shock. Even a small degree of asphericity in the wind shock (a ratio between equatorial and polar shock positions smaller than 0.8) is sufficient to produce strong flow focusing (cf. Icke 1988). In the case of stronger asphericities it is even possible to have the flow remain supersonic across the shock.

4. Results with cooling

In Paper 2 we presented the results of simulations with radiative cooling included. Although the cooling is only implemented in a simplified way (by

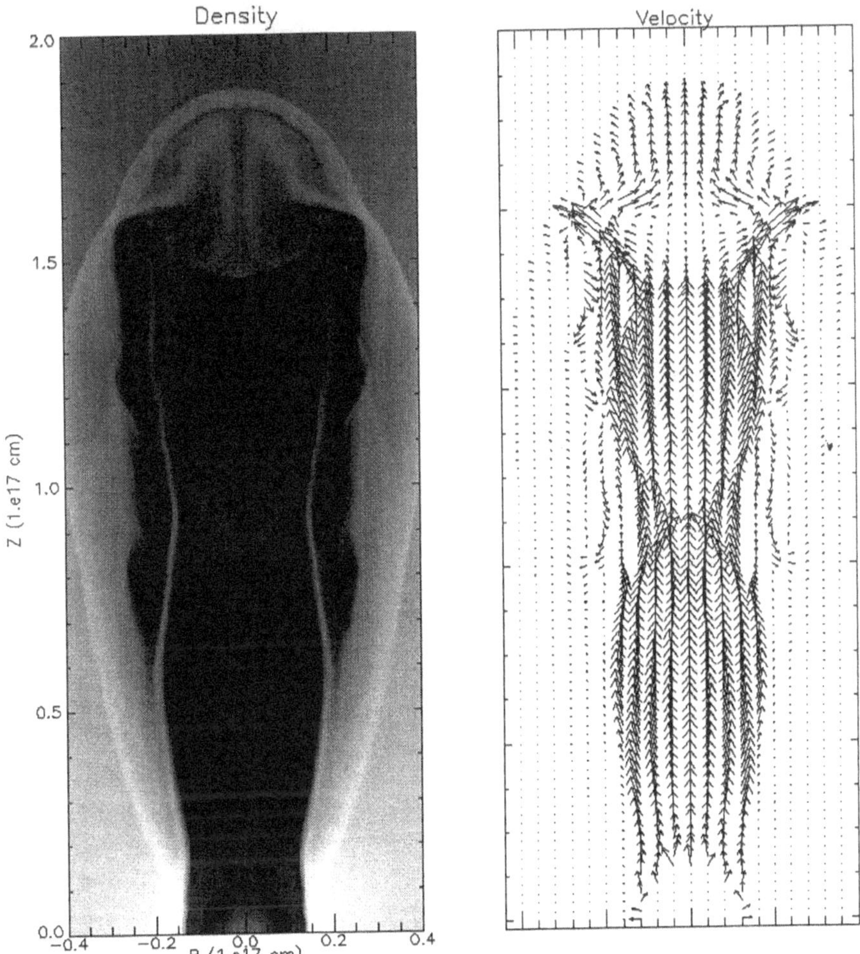

Figure 2. Density and velocity for a non-cooling model. Shown are a gray scale map of $\log_{10}(\rho)$ and a map of velocity vector field for model A after 1035 years of evolution. In the density map dark (light) shades correspond to low (high) densities. In the velocity field map vectors in the inner, freely expanding wind zone have not been plotted. Thus the first "shell" of vectors maps out the wind shock.

using a coronal cooling curve) this should still recover the most important dynamical effects of the cooling.

A typical sequence can be seen in Figs. 3 and 4. In this simulation the outflow has a velocity of 350 km s^{-1} and a mass loss rate of 10^{-7} $M_\odot$ yr^{-1}. The environment has a pole to equator contrast of 70, and an accretion

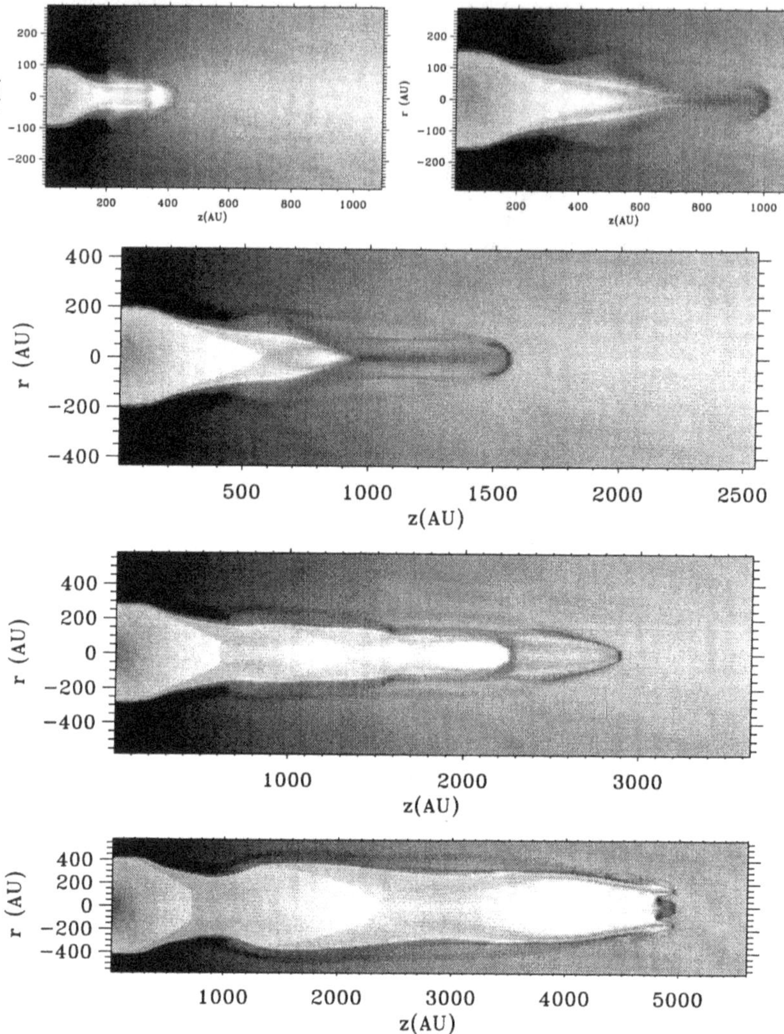

Figure 3. Density evolution for a cooling simulation. The grey scales show $\log_{10}$ contours for times t =25, 62, 87, 137, 225 years. The darker shades are higher density. One sees the initial radiative phase, the development of the 'cool jet' and the development of the 'hot jet'.

rate $\dot{M}_a = 10^{-6}$ $M_\odot$ yr^{-1}. We find that the evolution of the flow proceeds in two phases. First a dense cool jet forms ($t = 62$ years), followed at later times by a lower density, hot jet ($t > 100$ years). This second 'hot jet' is similar to the jets found in the non-cooling calculations of Paper 1. The sequence is most clearly seen in the temperature plots in Fig. 4.

The formation of this 'cool jet' early in the evolution is an interest-

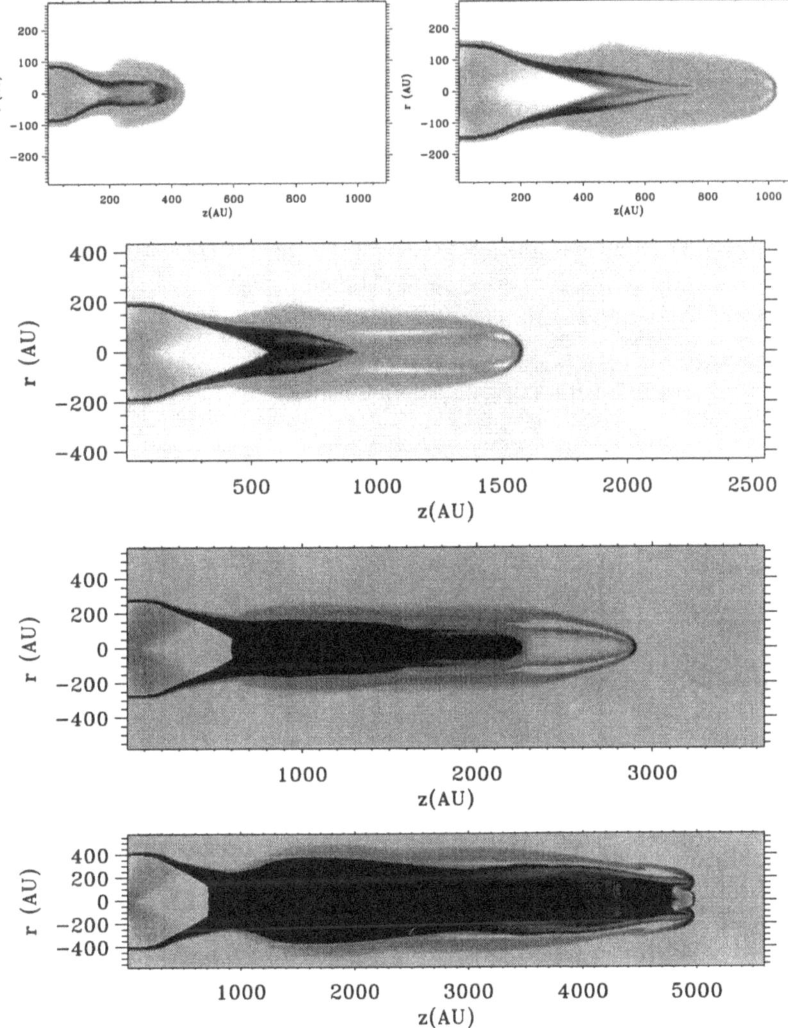

Figure 4. Temperature evolution for a cooling simulation. The grey scales show $\log_{10}$ contours for times t =25, 62, 87, 137, 225 years. The darker shades are higher temperature. One sees the initial radiative phase, the development of the 'cool jet' and the development of the 'hot jet'. Notice the shape of the inner shock.

ing new effect seen in these radiative simulations. It forms when the bubble is still mostly radiative and its formation mechanism appears to be the same as that of the "converging conical flows" originally proposed by Cantó & Rodriguez (1980), and later studied analytically and numerically by Cantó, Tenorio-Tagle & Różyczka (1988) and Tenorio-Tagle, Cantó & Różyczka (1988). When the thermal energy produced at the wind shock is completely lost to radiation the post-shock flow collapses and the only path available for the shocked wind is to slide along the contact discontinuity towards the poles of the bubble, leading to converging conical flows. The collision of fluid parcels at the poles produces a shock wave where the transverse velocity components are lost and a narrow well collimated beam or jet of dense, cool gas is formed.

In general one can say that hydrodynamic collimation is more effective when cooling is included, which given that most of the sideward expansion is due to thermal pressure is not too surprising. Another effect that helps in achieving better collimation is that due to radiative losses the wind shock is much more aspherical, leading to very efficient focusing. In fact even during the hot jet phase the focusing is so strong that at some positions the flow stays supersonic across the wind shock.

The duration of the cool jet phase depends very much on the efficiency of the cooling, and hence on the velocity of the outflow and the density of the wind and the environment. In deeply embedded objects which have higher densities than we used in the model presented here, the cool jet will last much longer, and a hot jet may never form.

5. The Effect of Wind Variability

The results from the previous section demonstrate that strong collimation can be achieved from a purely hydrodynamic interaction between winds and protostellar environments. But to apply this model to young stars we must account for the time scales inherent to YSO jets and molecular outflows (assuming that jets are connected to the outflows, Chernin *et al.* 1994). Recent deep exposure images of HH jets such as HH34 (Bally & Devine, 1994) and HH46/47 (Heathcote *et al.*, 1996) reveal multiple bow shocks that imply jet lifetimes of many thousands of years or more. In addition, the molecular outflows have dynamical lifetimes on the order of 10^3 to 10^5 years (Bachiller, 1996).

Because our bubbles are not in pressure equilibrium with the environment, they will continue to grow in the lateral direction, leading to wide jets at later times. In order to account for the observed jet widths, this expansion has to be restricted. An attractive means for stopping this equatorial growth comes directly from the observations. As was mentioned above many HH jets show clear signs of variations in velocity along the jet beams. In the most extreme case the jet appears to be temporarily shut off, which explains the multiple bow-shock structures. In the HH34 superjet Bally & Devine (1994) find a periodicity in the beam with a period between 300 and 900 years. In addition smaller scale variations with periods $\tau < 100$ years are seen in many HH jet beams (Morse, private communication). It is reasonable that the velocity variations in the jet beam are a measure of velocity variations in the source material of the jet. There might also be links with FU Orionis outbursts or related events, which occur on similar time scales as the ones mentioned above (see Bell & Chick, this volume)

One can therefore envisage a scenario where the wind-blown bubble is driven by a time-variable wind. In our simulations we did not include the effect of either gravity or the inward directed accretion flow. Both of these effects will decelerate the flow and constrain the bubble, particularly near the equator where the bubble radius is small so that the accretion velocity ($\propto R^{-0.5}$) and gravitational force density is high.

We developed a simple model for the interaction of a periodic stellar wind with an accreting environment. The model assumes spherical symmetry and strong radiative losses from the wind and ambient shocks so that we can use a thin shell approximation. The equations for mass and momentum conservation for a shell of mass $M_{\rm s}$ and radius $R_{\rm s}$ are

$$\frac{dR_{\rm s}}{dt} = V_{\rm s} \tag{2}$$

$$\frac{dM_{\rm s}}{dt} = 4\pi {R_{\rm s}}^2 \rho_{\rm w}(R_{\rm s})(V_{\rm w} - V_{\rm s}) + 4\pi {R_{\rm s}}^2 \rho_{\rm a}(R_{\rm s})(V_{\rm s} + V_{\rm a}) \tag{3}$$

$$\frac{dM_{\rm s}V_{\rm s}}{dt} = 4\pi {R_{\rm s}}^2 \rho_{\rm w}(R_{\rm s}) V_{\rm w}(V_{\rm w} - V_{\rm s}) - 4\pi {R_{\rm s}}^2 \rho_{\rm a}(R_{\rm s}) V_{\rm a}(V_{\rm s} + V_{\rm a}) - \frac{GM_* M_{\rm s}}{{R_{\rm s}}^2} \tag{4}$$

(Volk & Kwok, 1985). Here ρ_a and V_a are the radially dependent accretion flow density and velocity. These equations can be rewritten in the form of a simple set of coupled ODEs. We solved the ODEs using a 4th order Runge-Kutta method with an adaptive step-size. For the variable wind we used a sinusoidal variation with an amplitude of 100. $\dot{M}_{\rm w}$ is kept constant, but we have also tried the case in which $\dot{M}_{\rm w}$ varies in the same way as $V_{\rm w}$, which gives similar results.

The parameters for the wind and the environment are the same as for the model shown in Sect. 4. Other details of the model can be found in Paper 2. We assume that before the start of the integration, the bubble was set in motion, perhaps by an energetic episodic outburst from the protostar. This initial condition gives a total energy in the shell of 10^{40} ergs which is more than five orders of magnitude less than what is released in a typical FU Orionis outburst (Hartmann & Kenyon, 1996).

The results for variable winds with periods of $\tau_{\rm w} = 10^6$, 500, 400, 300 and 200 years are shown in Fig. 5. For the constant wind ($\tau_{\rm w} = 10^6$ years) the bubble expands monotonically although it does experience changes in velocity due to accretion ram pressure and gravitational forces. When the wind is allowed to vary, these forces produce dramatic changes in the bubble's evolution. For all four variation periods we see that the expansion

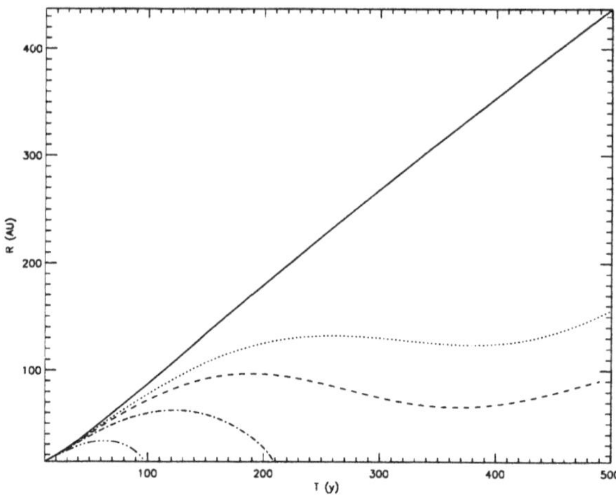

Figure 5. Evolution of spherical wind blown bubbles in accretion environment. Shown are the radius of 5 bubbles driven winds with different periods as a function of time. Solid line corresponds to a period of $P = 10^6$ years. Dotted line: $P = 500$ years. Dashed line: $P = 400$ years. Dashed-dot line: $P = 300$ years. Dashed-dot-dot line: $P = 200$ years.

of the bubble can be reversed ($V_s < 0$) for some time. For longer periods the bubble gains enough momentum before the wind enters a minimum to either continue a slow expansion ($\tau_W = 500$ years) or maintain a constant average radius ($\tau_W = 400$ years). For shorter periods the bubble is "crushed" during the wind's quiescent phase. Note that we ended the calculation if the shock radius R_s became smaller than 10^{12} cm.

From these results we conclude that, in principle, time varying winds can produce wind-blown bubbles whose size does not increase beyond some upper limit. If these bubbles produce jets through the hydrodynamic mechanisms described above, the model age and collimation scales can be made consistent with observations. We imagine that during a periodic increase in mechanical luminosity the YSO will begin driving a bubble which in turn produces collimated jets. As the wind speed varies the bubble either oscillates around some average radius (producing variations in the jet beam) or it collapses entirely. Jet production begins again when the momentum in the wind has increased enough to produce another bubble.

6. Connection to observations

The results so far can be considered as physical experiments rather than models for specific observed objects, since there are several important effects which we have left out of the numerical simulations (gravity, accretion flow, rotation). Clearly the next step should be to include these, and also to add microphysics and diagnostics to make predictions of observables, similarly to the work of Frank & Mellema (1994) and Suttner *et al* (1997). This will allow us to check whether the cool jets reproduce the observables in terms of temperature and density and in how far the hot jet can be observed.

There are however already two observational results in support of the model. Firstly the detection of non-thermal radio emission from the source W3(OH) (Reid *et al.*, 1995). These authors found strong synchrotron emission arising at the center of a linear chain of maser sources. Recent high resolution images show the synchrotron emission actually appears as a two sided jet emanating from the geometric center of the two maser flows (Wilner *et al*, 1997). Such a morphology is likely to occur from first-order Fermi acceleration at strong shocks around the central object.

Secondly a geometry very reminiscent of the converging conical flows has been observed in the proto-planetary nebulae He 3-1475 (Harrington, private communication), an object which is producing collimated outflows with velocities of several times 100 km s^{-1} (Riera *et al.*, 1995).

7. Conclusions

- The interaction of an outflow with a toroidal environment is an efficient means to collimate unfocused or wide-angle winds.
- The interaction gives rise to two jet phases, an initial 'cool jet', which form due to converging conical flows at the top of the bubble, and a later 'hot jet' which forms when cooling starts to be less efficient. The cool jet is dense and is the one which should be most easily observed.
- Both the variability and the longevity of the jets could be explained by a model in which the outflow is time-dependent, producing a series of cool jets.
- The occurrence of non-thermal emission in some radio jets can be seen as supporting our model, since its production requires strong shocks.

Acknowledgments: The authors gratefully acknowledge Dong-Su Ryu for making the TVD code available to us. We also benefited from discussions with Alberto Noriega-Crespo, Alex Raga, Mark Reid, Tom Jones. This work was supported by the Minnesota Supercomputer Institute.

References

Bachiller, R., 1996, ARAA, 34, 111
Bally, J., Devine, D., 1994, ApJ, 428, L65
Cantó, J., Rodrí guez, L.F. 1980, ApJ, 239, 982
Cantó, J., Tenorio-Tagle, G., Różyczka, M. 1988, A&A, 192, 287
Chernin, L.M. Masson, C.R., Dal Pino, E.M., Benz, W., 1994, ApJ, 426, 204

Dalgarno, A., McCray, R., 1972, ARAA, 10, 375
Eichler, D., 1982, ApJ, 263, 571
Frank, A., Noriega-Crespo, A., 1994, A&A, 290, 643
Frank, A., Mellema, G., 1994, ApJ, 430 800
Frank, A., Mellema, G., 1996, ApJ, 472, 684 (Paper 1)
Harten, A., 1983, J. Comp. Phys, 49 357
Hartmann, L., Calvet, N., Boss, A., 1996, ApJ, 464, 387
Hartmann, L., Kenyon, S., 1996, ARAA, 34, 207
Heathcote, S., Morse, J., Hartigan, P., Reipurth, B., Schwartz, R., Bally, J., Stone, J., 1996, AJ 112, 1141
Henriksen, R., Ptuskin, V., Mirabel I., 1991 A&A, 248, 221
Icke, V., 1988, A&A, 202, 177
Icke, V., Balick, B., Frank, A., 1992, A&A, 253, 224
Ip, W.H., 1995, A&A, 300, 283
Königl, A., 1982 ApJ, 261, 115
Königl, A., 1989 ApJ, 342, 208
Königl, A., Ruden, S.P. 1993, in "Protostars and Planets III", E.H. Levy, J.I. Lunine (eds). Univ. of Arizona Press. p. 641
Koo, B., McKee, C.F., 1992, ApJ, 388, 103
Li, Z., & Shu, F., 1996, ApJ, 472, 211
Lucas, P.W., Roche, P.F., 1997, MNRAS in press
Mellema, G., Eulderink, F., Icke, V., 1991, A&A, 252, 718
Mellema, G., Frank, A., 1997, submitted to MNRAS (Paper 2)
Peter, W. Eichler, D, 1995, ApJ, 438, 244
Pudritz, R.E., 1991, in "The Physics of Star Formation and Early Stellar Evolution", C.J. Lada and N.D. Kylafis (eds). NATO ASI Series, Kluwer, Dordrecht. p. 365
Raga, A.C., Cantó, J., 1989 ApJ, 344 404
Reid, M. J., Argon, A. L., Masson, C. R., Menten, K. M. Moran, J. M., 1995, ApJ, 443, 238
Riera, A., Garcia-Laro, P., Manchado, A., Pottasch, S.R., Raga, A.C., 1995, A&A, 302, 137
Romanova, M., Ustyugova, G., Koldoba, A., Chechetkin, V., Lovelace, R., 1996, preprint
Ryu, D., Brown, G.L., Ostriker, J.P., Loeb, A., ApJ, 452, 364
Shu, F., Najiata, J., Ostriker, E., Wilkin, F., Ruden, S., Lizano, S., 1994, ApJ, 429, 781
Suttner, G., Smith, M.D., Yorke, H.W., Zinnecker, H., 1997, A&A, 318, 595
Tenorio-Tagle, G., Cantó, J., Różyczka, M., 1988, A&A, 202, 256
Terebey, S., Shu, F., Cassen, P., 1984, ApJ, 286, 529
Ulrich, R.K., 1976, ApJ, 210, 377
Volk, K., Kwok, S., 1985, A&A, 153, 79
Wilner, D.J., Reid, M.J., Menten, K.M., Moran, J.M, 1997, in: Poster proceedings of IAU Symposium 182 on Herbig-Haro Objects and the Birth of Low Mass Stars, F. Malbet, A. Castets (eds.). Obs. de Grenoble. p. 193

THE CLASS 0 OUTFLOW HAMMERED OUT?

MICHAEL D. SMITH, ROLAND VÖLKER
GERHARD SUTTNER AND HAROLD W. YORKE
Astronomisches Institut der Universität Würzburg
Am Hubland, D-97074 Würzburg, Germany

Abstract. Dense molecular jets cutting through dense molecular clouds are simulated here using a ZEUS-type hydrodynamics code extended with molecular physics. H_2 'hot snapshots' and CO 'historical views' are nicely modelled with overdense uniform jets. Attention is drawn to some remaining hot problems: observed jet knots are bow shaped, bipolar outflows can be highly asymmetric, some proper motions within jets are enigmatically low and H_2 excitation can be exceedingly uniform.

We present the Hammer Jet, in which the nozzle introduces high velocity variations, as well as a strong ripping and spray action. Prominent jet bow shocks in H_2 and CO emission lines are then produced. Wide tubular CO structures with concave bases show up. Proper motions are simulated. The asymmetries are modelled by jet break-out from a molecular core into a diffuse atomic environment.

1. Introduction

The earliest outflows from protostars are well collimated and extremely long. Despite being deeply enshrouded by molecular gas, the warm shocked regions can be observed in the infrared H_2 lines, thus revealing the present impact regions. The cool accumulated material can be observed in the sub-millimetre CO lines, delineating the total swept-up and swept-out material. The constraints imposed are encapsulated in Tables 1 & 2.

The hydrodynamic jet model for molecular flows has been developed and applied by Suttner *et al.* (1997) and Smith *et al.* (1997a). Its successes and failures are denoted by ticks and crosses in the boxes. The numerical hydrocode simulations include molecular dissociation and reformation. We found that three-dimensional models are necessary although one- and two-

B. Reipurth and C. Bertout (eds.), Herbig–Haro Flows and the Birth of Low Mass Stars, 303–312.

TABLE 1. Molecular Hydrogen signatures, typical examples and the successes (ticks) and failures (crosses) of the Suttner *et al* jet model

Multiple Bows (L 1634, Davis *et al* 1997) ✔	Complex Filamentation (HH 90/91, Davis *et al* 1994) ✔
Twin/single-peaked profiles (L 1448, Davis & Smith 1996) ✔	Constant Excitation (Cepheus E, Eislöffel *et al* 1996) ✘
Pulsed jets (HH 111, Gredel & Reipurth 1993) ✔/✘	Asymmetric Lobes (L 1660, Davis et al 1997) ✔/✘

dimensional models can provide some insight. The high number of ticks is encouraging.

One major problem with our early simulations is the structure and number of the jet knots. As shown in the pulsed jet of Fig. 2, the model knots are few, weak and elliptical. Although this is not inconsistent with all the observations, the HH 111 jet possesses many knots and the HH 212 jet has inter-knot emission. Moreover, knots are usually bow shaped (Reipurth et al 1997, and these proceedings).

2. The Hammer Jet

Nozzle designs which we have recently studied are cartooned in Fig. 1. The parameters for all four models discussed here are given in Table 3. The new models include high-amplitude pulsations and are thus termed 'Hammer Jets'. They also include a 50% velocity shear, with the velocity in the jet sheath being lower than on the axis (as might be expected from disk-wind models). The result is: *prominent outward-facing bow shocks, and extended wakes* (Fig. 2). Also found are (1) wider outflows and stronger H_2 emission (due to the chosen spray and precession angles), (2) more structure in the leading bow shock, including mini-bows, (3) wave patterns on the cavity surface and (4) the highest surface brightness knots in the lobe fall well behind the leading edge.

The bows are all concave to the jet origin. In contrast, if the largest

TABLE 2. CO signatures and the successes (ticks) and failures (crosses) of the Smith *et al* jet model. NGC 6334G (Lada & Fich 1996) displays all these signatures.

Collimation increases with Velocity ✔	Red/blue lobe symmetry (high-v) ✔
Power-law line profiles (γ = -1.3 to -2.0) ✔	Hubble Law (v $\propto$ distance) ✔
High-speed CO Jets & Bullets ✔	Strong Forward Motion ✔

velocity variations are applied to the inner axial region while the jet sheath maintains smaller pulsations, concave and convex bow shocks appear. Figure 3 demonstrates this aptly, where knot collisions (boxes d-h) generate intricate structures. Jet production models can in principle be constrained. For example, de Laval nozzles should possess lower axial speeds corresponding to the higher axial pressure. Clear observed examples of reverse bow shocks, however, have not been found although HH 11 and HH 2 could be cited.

TABLE 3. Jet parameters for the models displayed. The average jet axis speed is 100 km/s, jet density is 10^5 cm^{-3}, ambient density is 10^4 cm^{-3}, both initially fully molecular.

Model Label	Shear v_o/v_i	Spray ψ	Precession P(yrs),β	Pulsation Q(yrs),A(km/s)
3D-1	-	-	100, 1°	-
3D-2	-	-	100, 1°	50, 30
MODEL 1	50%	-	46, 1°	50, 90
MODEL 2	50%	2°	36, 2°	40, 90

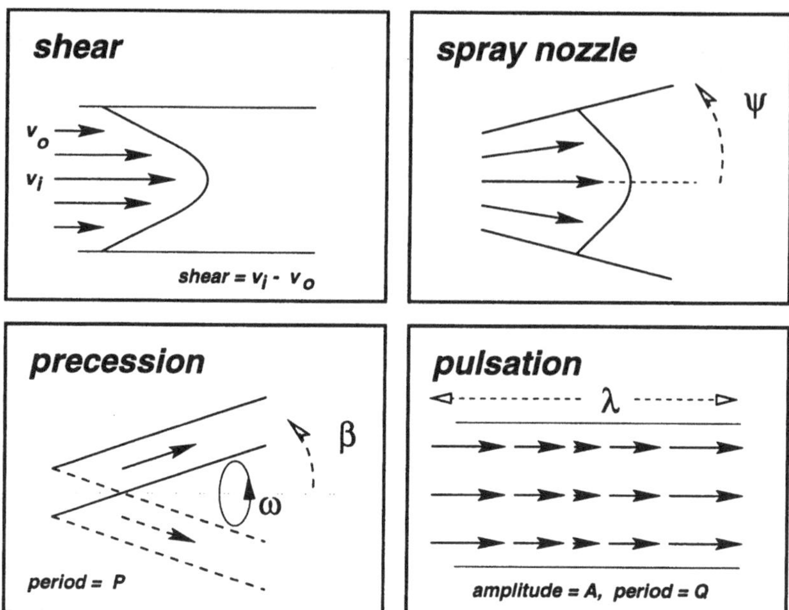

Figure 1. Nozzle geometry for the simulations

3. Lobe Asymmetries

H_2 asymmetry is not directly predicted by our one-sided models. Nevertheless, asymmetries can be incorporated either through jet-related asymmetry or environmental asymmetry.

First, we study jet asymmetry. In Fig. 4 we present MODEL 2, as it would appear in the two perpendicular planes. Also note that the axis is aligned to 60° out of the sky plane. Fine-scale asymmetries are produced. Prominent are the wave structure and sheath asymmetries, comparable to some observations (e.g. HH 211, McCaughrean et al 1994).

Large-scale asymmetries are clearly a mark of the environment. We can suppose that one jet approaches the cloud edge, and undergoes the transition from molecular cloud to atomic diffuse inter-cloud gas, as schematically shown in Figure 5. One then only sees the jet shock in H_2. The result of our 2D simulation is shown in Fig 6. The jet survives the transition but is broadened. However, 3D simulations are required before generation of H_2 distributions would be fruitful.

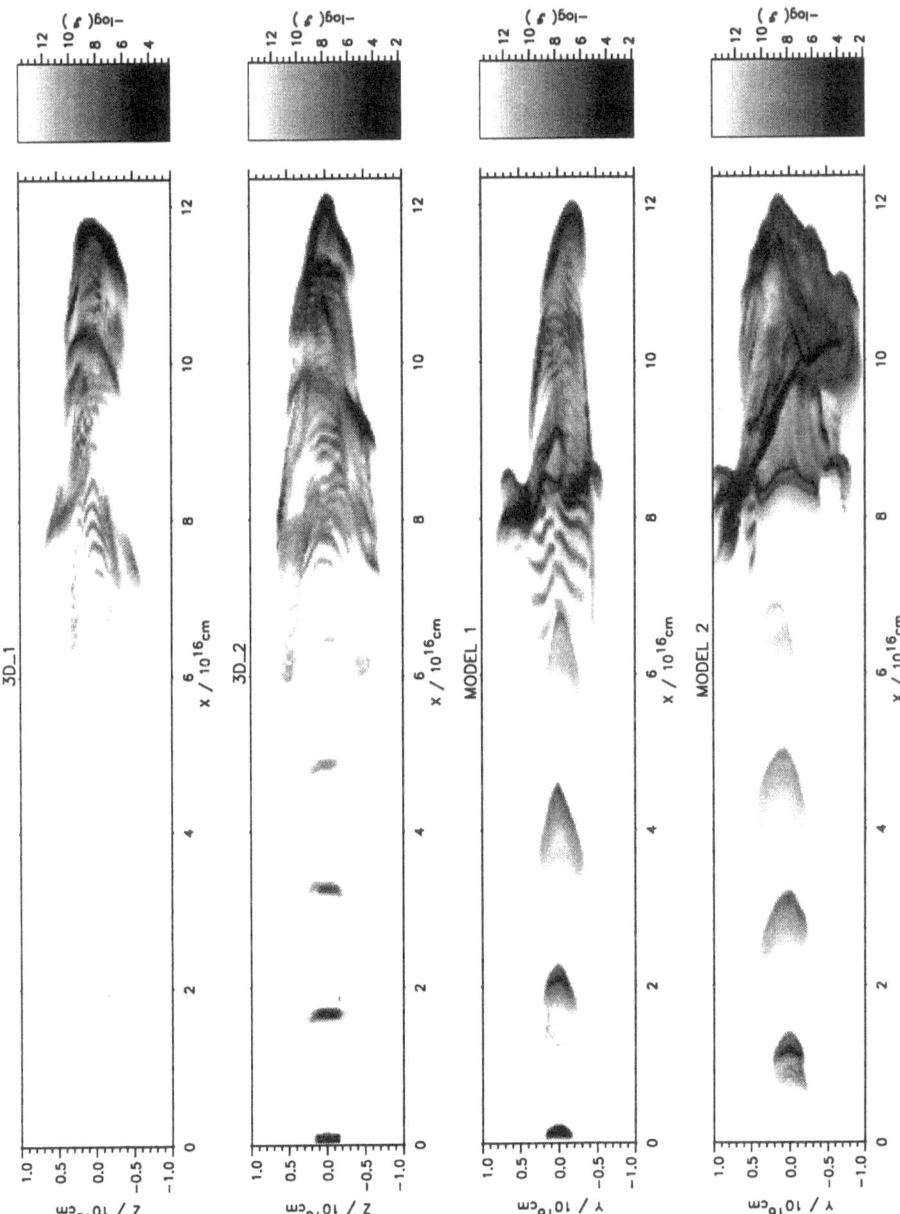

Figure 2. The four jet models at a comparable size in molecular hydrogen emission (from Suttner *et al* (1997) & Völker et al (1997)). The Cartesian grid is 500×80×80 zones, the jet radius is 1.5×10^{15} cm or 6 zones.

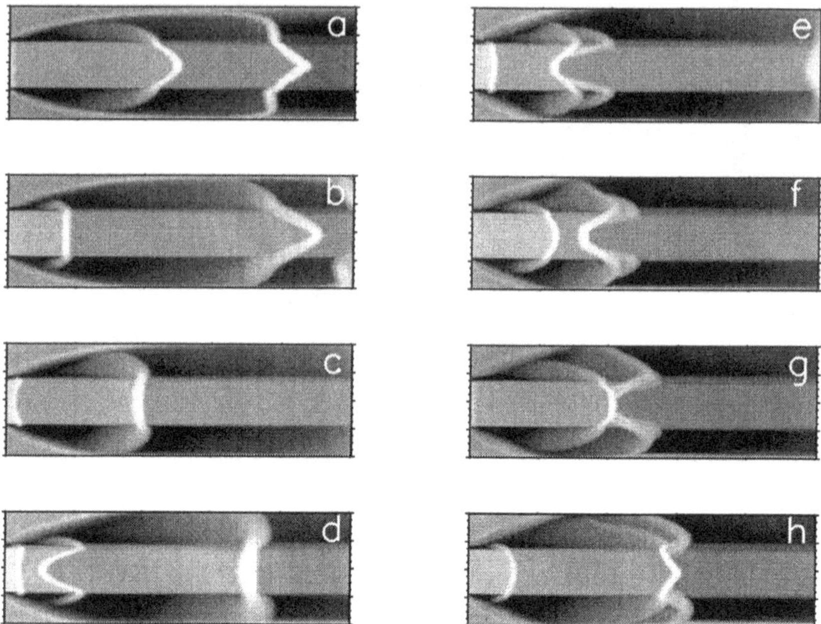

Figure 3. Variable Shear Jets. The density distributions for a sequence of 2D (Cartesian) molecular jets in which the cross-jet shear is given both positive and negative values. While the outer jet layer is injected with 90% 25 yr periodic variations about the speed of 100 km s^{-1}, the inner jet modulates this by a further 75% with a period of 61 yr. Times are, from (a) to (h), 149, 166, 183, 208, 230, 240, 248 & 258 yr.

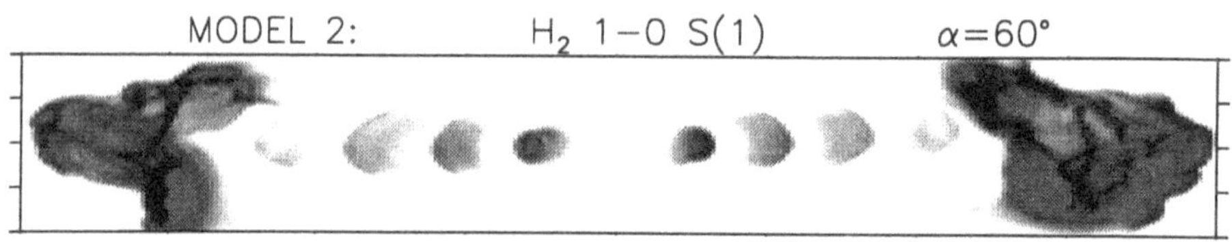

Figure 4. The twin-jet: MODEL 2 seen at 60° to the sky plane and in the two perpendicular planes (left: XZ, right: XY). See Völker et al (1997)). Note the round non-aligned bows, as well as the lobe asymmetry.

4. Tubular CO

One complaint about the jet model has been that it cannot produce wide CO structures since strong molecular cooling would shrink the cocoon. This is not true: an extremely low density and very broad *atomic gas* cocoon

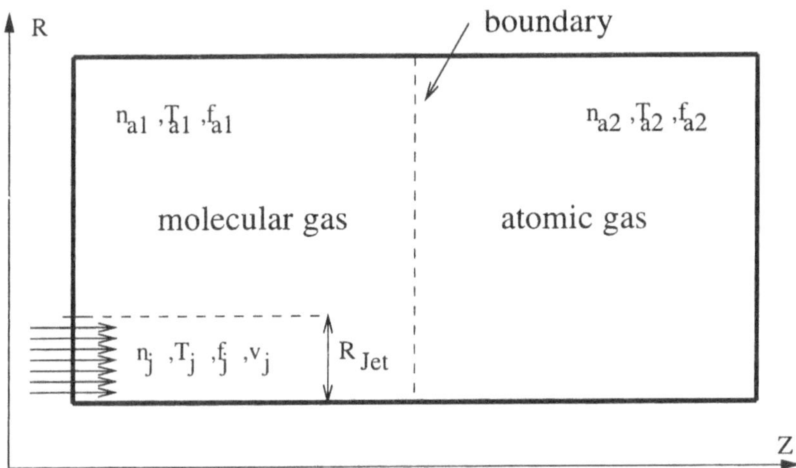

Figure 5. The initial conditions for the axially-symmetric calculation of molecular jet 'break-out' from a dense molecular core into an atomic ambient medium. Here $n_j = n_{a1} = 10^5\,cm^{-3}$ and $n_{a2} = 68.2\,cm^{-3}$. $T_{a1} = 10\,K$, $T_{a2} = 8000\,K$.

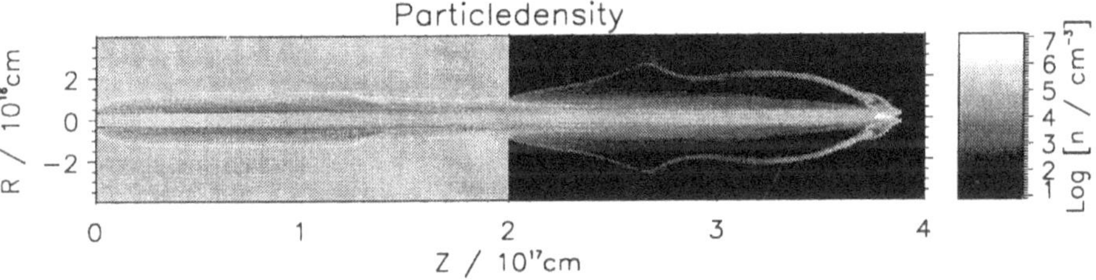

Figure 6. After 2000 years, the jet has broken well out of the core. A wide bow shock is present - note the increase in cocoon size at the boundary. The jet shock is extremely weak and the jet molecules are driven sideways into the atomic medium.

is trapped between the jet and the swept-up shell of cool gas. The jet is hypersonic (M = 200) and the swept-up shell too inertial to squeeze out the cocoon. The result is a *tubular CO structure many times wider than the jet* (Fig. 7). Note also the concave structure at the jet base. We thus find good agreement with the observations of low-velocity gas (e.g. of HH 111, Nagar et al 1997) as well as the (usually very weak) high-velocity component (Smith et al 1997). CO lobe and jet predictions from zero (Suttner *et al* 1997) to high speeds (Smith *et al* 1997) are in remarkable agreement with observations.

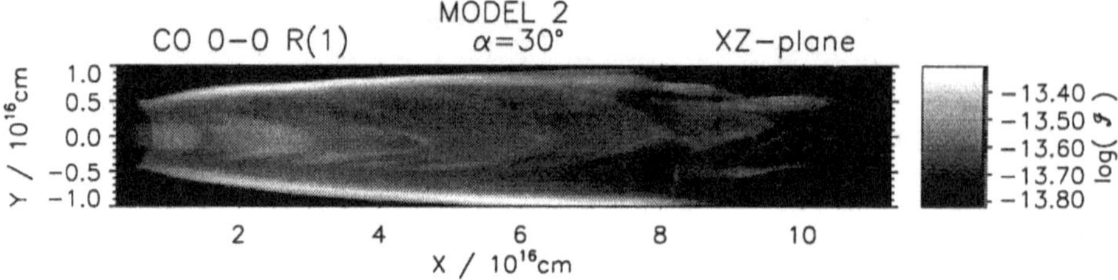

Figure 7. CO emission from the J=2-1 transition for MODEL 2. The *integrated CO* emission shown here is dominated by the low velocity tube, several times wider than the jet. Some CO bullets are also evident (the same CO/H_2 ratio in the jet and ambient gas is here assumed). Note that the *high velocity CO* distributions appear very different (Smith *et al 1997*).

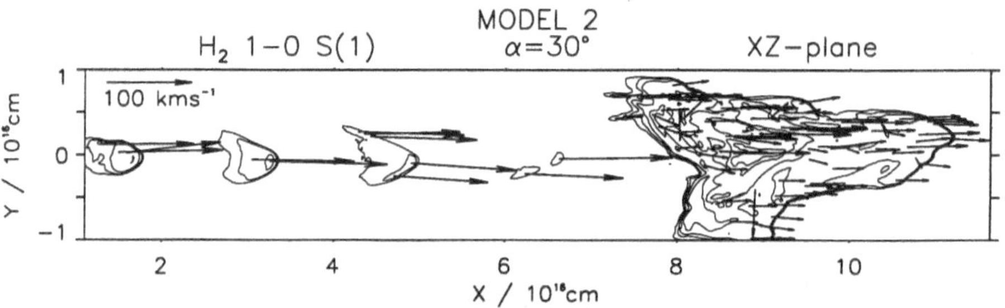

Figure 8. Proper motions in MODEL 2, demonstrating the fast motions within the jet. Proper motions were calculated from intensity peaks in consecutive simulated H_2 1-0 S(1) images.

5. Proper Motions

Individual knots in observed jets possess high proper motions. Data on the H_2 jets are lacking, but in optical jets knot pattern speeds (i.e. from the proper motions) appear to be lower than predicted by the fluid speeds (from the radial speeds) (Eislöffel 1994, Eislöffel *et al* 1994). In the HH 34 jet, the ratio of the pattern speeds of the jet bows and their fluid speeds are not too discrepant (0.7-0.9). The ratio in several jets increases with propagation distance from values as low as 0.2 towards unity. In contrast, models to date predict that the fluid and pattern speeds do not widely differ in the jet.

However, the hammer jet shown in Figures 8 and 9 possesses fluid speeds of up to 190 km s^{-1}, yet proper motions for the jet bows are 110-150 km s^{-1}. If one assumes that the (projected) knot fluid speed corresponds to the highest observed radial speed, then pattern-to-fluid speed ratios of 0.6-0.8

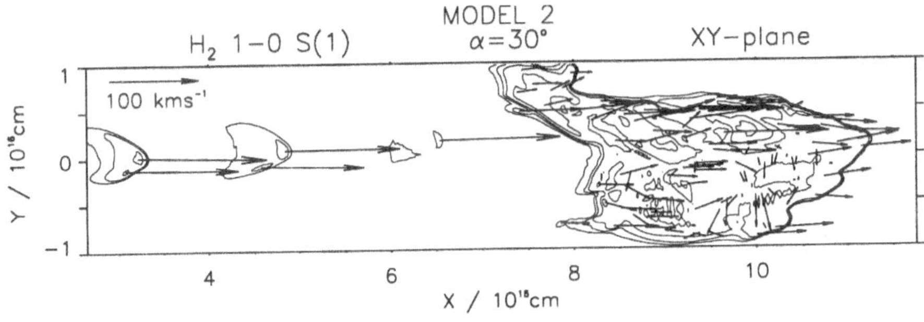

Figure 9. Proper motions in H_2 for MODEL 2, observed from along the Z-axis, otherwise as described in Fig. 8,

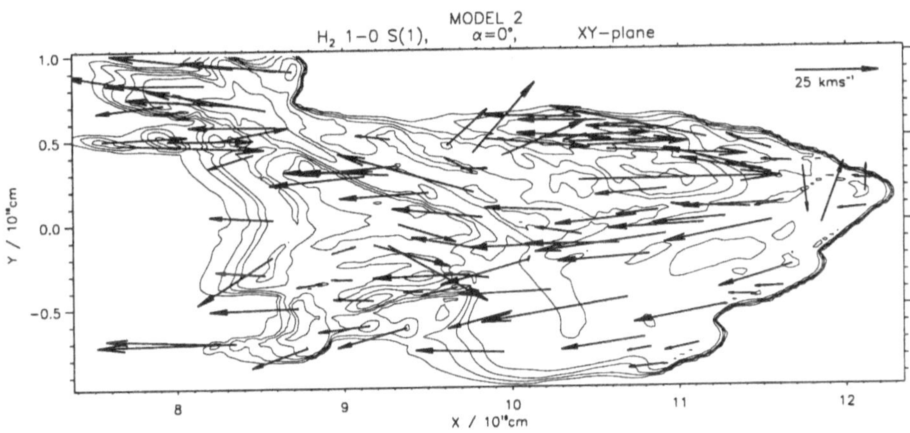

Figure 10. Proper motions in a magnified head region, demonstrating the flow away from the leading edge. That is, the average speed of the whole outflow has been subtracted. Proper motions were calculated from intensity peaks in consecutive simulated H_2 1-0 S(1) images.

would be predicted for this simulation. The high radial speeds are indeed present on the source-side of the two curved shocks (forward and reverse) which clash at each bow-shaped knot. Moreover, as the jet shocks weaken with propagation distance, pattern-to-fluid ratios should increase towards unity along the jet. We hope to confirm these predictions rigorously in the near future (Völker *et al* 1997) .

The proper-motion situation is more intricate in the lobes. Here the simulated molecular flow seen relative to the apex (Fig. 10) possesses both an acceleration away from the leading edge as well as a turbulent characteristic. These are the properties observed in the optical (Eislöffel et al 1994). HST and ground-based observations will soon provide us with the infrared data we need.

6. Conclusions

The various signatures of the collimated Class 0 outflows have here been simulated and discussed. Hammer Jets with shear provide new solutions to old problems. Remaining questions include:

- How and where do the jets form?
- How and where do the magnetic field, ambipolar diffusion and angular momentum change the picture presented above?
- Do the jets form complete with molecules? Or are the molecules entrained or formed on the jet dust?
- Can we explain the Gigantic Bipolar Outflows?
- How do outflows evolve over the Class 0 and Class I lifetimes?
- Can these outflows alter the environment sufficiently to influence the further formation of stars?

Further simulations of molecular hammers may provide some answers.

Acknowledgements: The authors gratefully acknowledge the support of the DFG.

References

Davis,C.J., Mundt, R., Eislöffel, 1994, ApJL, 437, L55.
Davis,C.J., Ray, T.P, Eislöffel, J., Corcoran, D., A&A 1997, in press.
Davis,C.J., Smith, M.D., 1996, A&A, 309,929.
Eislöffel, J., 1994, Ap&SS, 216, 135.
Eislöffel, J., Mundt, R., Böhm, K-H., 1994, AJ, 108,1042.
Eislöffel, J., Smith, M.D., Davis, C.J., Ray, T.P., 1996, AJ, 112, 2086.
Gredel, R., Reipurth, B., 1993, ApJ, 407, L29.
Lada, C.J., Fich, M., 1996, ApJ, 459, 638.
McCaughrean, M.J., Rayner, J.T., Zinnecker, H., 1994, ApJ, 436, L189.
Nagar, N.M., Vogel, S.N., Stone, J.M. & Ostriker, E. 1997, preprint (see also IAU Symp. 182, Poster Book, p 163).
Reipurth, B., Hartigan, P., Heathcote, S., Morse, J.A. & Bally, J., 1997, AJ, in press
Smith,M.D., Suttner, G., Zinnecker, H., 1997, A&A, in press.
Suttner, G., Smith, M.D., Yorke, H.W., Zinnecker, H., 1997, A&A, 318, 595

NUMERICAL SIMULATIONS OF OPTICAL KNOTS IN YSO OUTFLOWS

F. RUBINI AND S. LORUSSO
Dip. Astronomia, Univ. Florence
Largo E. Fermi 5, 50125 Florence, Italy
C. GIOVANARDI
Osservatorio Astrofisico di Arcetri
Largo E. Fermi 5, 50125 Florence, Italy
AND
F. LEEUWIN
Dip. Fisica, Univ. Bologna
Viale Berti-Pichat 2-6, 40126 Bologna, Italy

Abstract. The aim of this work is to show that, though 3D effects should be invoked to explain the less regular arc-like regions, the internal knotty emissions may be explained through a roughly axisymmetric pattern of oblique shocks.

To this purpose, numerical simulations of non-equilibrium, radiative, hydrodynamical jets in axisymmetric geometry have been performed. After showing that the internal reflecting waves, travelling in the beam of a stellar jet, steepen into oblique shocks, we verify that the optical emission arising from the shocked regions is in good agreement with observations.

Since the optical emission strongly depends on the physical and chemical parameters, (namely temperature and electron densities inside the emitting regions), special care is devoted to solving the energy transport equation.

1. Introduction

The most recent observations of HH objects show that the emission regions are, in some cases, far from being simple axially aligned structures of equally spaced knots. Instead, they reveal a morphological variety with a wide range of shapes and spatial scales.

B. Reipurth and C. Bertout (eds.), Herbig–Haro Flows and the Birth of Low Mass Stars, 313–322.

Very recent images of HH47 from HST, for instance, reveal interesting details of a quite complex structure, where the emission in [SII] seems to be decoupled from that in Hα. The former fills the whole body of the jet and, in its inner and faster part, forms a sinuous chain of small scale knots, the spacing being of 2" or 3", the latter is concentrated in wisps and filaments. The difference Hα - [SII] image clearly shows that the knots in [SII] are *internal* to the flow whereas the filaments in Hα are localized at the surface of the beam, (Heathcote et al., 1996). These authors argue that it is just the interaction between the beam and the neutral external medium that allows the formation of the observed thin Hα Balmer-dominated filaments, where sudden changes in the direction of the flow drive the formation of secondary bow-shocks. At the same time, the source of HH47 seems to undergo temporal variations at intervals of $500 - 1000$ years, creating multiple leading bow-shocks whose spacing is much larger than the one between knots, a behavior typical of many HH objects.

These observations suggest that more than one mechanism is at work to produce the optical emission, which, with respect to the spacing between two separate emissions, could be classified as large scale and small scale processes, respectively.

The arc-like Hα dominated regions, associated with the interaction between the flow and medium, extend over both large and small scales, *resp.* in the form of leading, and secondary, bow-shocks. The former would directly depend on the temporal variation of the outbursts, the latter on strong deviations of the trajectories of the particles, maybe due to the growth of kink-like instabilities, to precession of the source of the jet, or interaction with a non-homogeneous external medium. These effects could of course act separately, or together.

On the other hand, the internal, knotty emission, like the ones that are well observed in HH34, HH211, as well as in HH47, seem to be essentially small scale events, which could arise from high-frequency (Δt of a few years) perturbations of the inflow conditions rapidly steepening into shocks (Stone et al., 1993, Suttner et al., 1997), or from pinch-like instabilities originating from the toroidal component of the external magnetic field (Hardee et al., 1997).

In this work we show that it is not necessary to invoke rapid fluctuations of the inflow conditions to explain the small scale structures inside the beam of the jet. Even neglecting the presumably important magnetic effects, a pattern of emission is produced, within a purely hydrodynamic model, by internal oblique shocks.

To this purpose we have performed numerical simulations of radiative jets, fed by a stationary source, propagating into the cold, undisturbed ISM.

A recent diagnostic model allows to deduce, for the emitting regions, average values for the gas temperature, total density and ionization fraction (Bacciotti et al., 1995). In practice, we span a range of possible inflow values searching for those that lead to the right parameters for the emitting knots.

Our simulations show that internal reflecting waves can survive energy losses and steepen into oblique shocks. The compressed gas in post-shock regions is the source of optical emission, whose knotty pattern arises from the shock pattern. Emissivity in [SII] and Hα, with the proper choice of the inflow parameters, has been found to be in good agreement with observations.

Special care has been taken in treating energy losses and the cooling function. The diagnostic model of Bacciotti shows that the emitting gas average temperature is $\approx$ 7000K, which means that the cooling function should also take into account temperatures below 10^4 K, a lower bound in many numerical simulations in the literature.

2. Non-stationary dynamics of stellar jets and formation of internal shocks

Stellar jets are non-stationary and non-periodic hypersonic flows whose internal structure evolves in space and time depending on the interaction between the internal beam, the surrounding region filled with the jet particles that, at the head, have been slowed and deviated by the Mach disk, the so-called cocoon, and the shroud, the external matter entrained by the bow shock. The pressure gradient between these structures drives the formation of internal oblique shocks, and it is this mechanism which we intend to investigate as the agent responsible for optical knots.

In under-expanded jets, such as those we consider, the jet particles which leave the nozzle undergo a sudden, quasi-adiabatic expansion, which makes the trajectories diverge and the pressure decrease. Once the equilibrium pressure with the cocoon is reached, compressional waves start to travel towards the axis, and eventually steepen to form the first oblique shock, causing particles to converge into the first compression region. An expansion phase follows, after which the mechanism repeats generating a knotty structure. At its end, the particles are slowed by the Mach disk, before being deviated into the cocoon.

At the same time, particles from the external medium which have been entrained and warmed by the bow-shock, cool, forming a thick layer of fast external matter right behind the bow-shock itself. As a consequence, a shroud forms between the bow-shock and the cocoon, where, due to the low density and the large cooling length, local pressure keeps high enough to embed the jet cocoon. The relative importance of the cocoon and the shroud may be roughly measured by the relative shock number, which is the ratio between the cooling lengths of the bow-shock and the Mach disk:

$$\frac{\xi_{bs}}{\xi_{Md}} \approx \eta^3$$

(Blondin et al., 1990), where $\eta \equiv \rho_{jet}/\rho_{ISM}$ is the density ratio. This means that light jets behave as if they were adiabatic jets, with extended highly embedding cocoons flowing into an isothermal medium, whereas the shroud is dominant in heavy jets.

It follows that we should investigate jets where the inflow pressure ratio $P_r \equiv \frac{P_{jet}}{P_{ISM}}$ is as low as possible, as they should be rapidly embedded by the cocoon and develop internal shocks. In the following we shall see whether this choice is consistent with the constraints arising from observations.

As for the inflow radius r_{jet}, it is well known that the spacing between oblique shocks increases with radius. Therefore we adopt for r_{jet} a value significantly smaller than the value usually assumed for stellar outflows in numerical simulations (but consistent with the hypothesis that flows emanate from the central zone of circumstellar disks). We recall however that in under-expanded jets the effective jet radius, *i.e.* the radius of the pinching channel bounding the supersonic flow, is larger than the inflow nozzle radius, r_{jet}. Namely, it increases with the pressure ratio P_r, tending to r_{jet} only in the limit $P_r = 1$.

As a conclusion, keeping the values of both the pressure ratio $\frac{P_{jet}}{P_{ISM}}$ and the inflow radius r_{jet} as small as possible (within observational constraints) should lead to the expected pattern of small scale optical knots.

3. Jet parameters and observational constraints

Though, for given external medium parameters, internal oblique shocks exist for a wide range of the inflow parameters (r_{jet}, P_{jet}), the set leading to correct values of temperature and density in the emitting regions is a small subdomain.

The already cited diagnostic technique (Bacciotti et al., 1995) allows one to derive average values for the emitting gas temperature, total density and ionization fraction in the optical knots, starting from the observed values of emissivity ratios, namely, $[OI]/[NII]$, $[NII]/H\alpha$, $[SII]/H\alpha$. For HH34 the mean temperature in the knots is estimated to range from 5000 to 7000 K , while the electron density ranges from 100 to 1000 cm^{-3} and the ionization fraction x decreases from 0.2 to 0.032 along the flow, corresponding to a total density of about 5000 – 20000 cm^{-3} (Bacciotti & Eislöffel, 1997).

Though it would be difficult to use directly these results to infer the inflow parameters in our simulations, we can use them as a *a posteriori* check.

The total density at the inflow, ρ_{jet}, is assumed to be equal to the density in knots. The same value is taken for the undisturbed ISM ($\eta = 1$); its initial temperature is of order 100 K (with the corresponding pressure deduced from a perfect gas equation). This necessarily leads to under-expanded jets, with pressure ratios P_r greater than one; an inflow material which is too cold ($P_r \sim 1$) has no chance of reaching the correct emitting temperature, not even through compression by oblique shocks.

In fact, while the assumption of low values for P_{jet}, as already pointed out, favours the formation of oblique shocks, a lower limit for P_{jet} arises from the requirement that significant emissions in [SII] and Hα be produced, since this implies temperatures of order 7000 K.

Two regimes of pressure have been explored numerically:

- Moderately under-expanded jets:
  case A) $\rho_{jet} = 5000\ \mathrm{cm}^{-3}, x_{jet} = 0.3, T_{jet} \approx 3000$ K , $P_r = 30$;
- Strongly under-expanded jets:
  case B) $\rho_{jet} = 5000\ \mathrm{cm}^{-3}, x_{jet} = 0.3, T_{jet} \approx 6000$ K, $P_r = 60$;

The simulations show that only when the inflow temperature T_{jet} is at least ≈ 5000 K does the post-shock temperature T_{knots} reach the value of 7000 K, and the emissivity in $[SII]$ and Hα appears in agreement with the observed data.

Regarding the inflow radius, simulations with either $r_{jet} = 0.2 \cdot 10^{15} cm$ and $r_{jet} = 10^{15} cm$ have been performed.

4. The model and the numerical code

4.1. CHEMICAL EVOLUTION AND RELATED ENERGY TERMS

We consider a gas composed of molecular, atomic and ionized hydrogen plus helium and solar abundancy metals, and we describe this system via the Euler equations modified to take into account radiative losses, plus two equations that describe the non-equilibrium evolution of the electron population and of molecular hydrogen.

total density conservation + total momentum conservation +
total energy conservation $-S_{rad} + S_{ion} + S_{H2}$ (energy source) +
$\frac{\partial N_{e^-}}{\partial t} + \nabla \cdot (N_{e^-}\mathbf{u}) = N_{es} +$
$\frac{\partial H_2}{\partial t} + \nabla \cdot (H_2 \mathbf{u}) = N_{hs}$

The source term in the electron equation is given by: $N_{es} = N_{ion} - N_{rec}$, where the recombination rate N_{rec} is taken from Spitzer (1978) and the ionization rate N_{ion} is given by Lang (1975). This yields the contribution $S_{ion} = 13.6\ \mathrm{eV} \cdot N_{es}$ to the total energy variation.

The source term in the molecular hydrogen equation, N_{hs}, arises only from molecular dissociation, due to collisions with H_2 and HI (Lepp & Shull 1983), and HII and electrons (Hollenbach & McKee, 1989, Mac Low & Shull, 1986, Martin et al., 1996). The corresponding contribution to the total energy source is $S_{H2} = \Lambda_{rad} + \Lambda_{diss}$, where:

- $\Lambda_{diss} = N_{hs} \cdot 4.48$ eV results from dissociations
- Λ_{rad} is due to collisions with HI and H_2 and is taken from Lepp & Shull (1983).

The cooling function takes into account temperatures also below 10^4 K. Namely, for $T > 10^4$ K the usual cooling function by Dalgarno & Mc Cray (1972) has been used, which considers radiative losses due to de-excitation of HI energy levels.

For $T < 10^4$ K, the cooling function is extended by computing energy losses mainly due to collisions between electrons and C^+, O, N, Fe^+, O^+, S^+, or, for still lower temperatures, between HI and C^+, O, Si^+, Fe^+, (Bacciotti et al., 1995).

4.2. NUMERICAL CODE

The numerical scheme is based on a second order HLLE Riemann solver, (Einfeldt, 1991) for the spatial derivatives, combined with a second order multistage TVD-Runge Kutta time integration (Shu, 1988).

A mesh with variable spacing along the z direction was used, with at least 600×100 points, at most 2500×100 points, in R, z.

5. Numerical results

The simulations we report on differ essentially through the pressure ratio, and eventually the inflow radius. All runs assume a density ratio $\eta = 1$.

A moderately under-expanded case, with inflow parameters:
$\rho_{jet} = 5000$ cm^{-3},$T_{jet} = 3000$ K, and $r_{jet} = 10^{15}$ cm,
and with pressure ratio $P_r = 30$ (corresponding to $T_{ISM} = 100$K), showed no significant emission. Though a pattern of compression regions formed in the channel of the flow, the temperature in the knots remained indeed too low (1500K).

Fig. 1 to 3 show instead a strongly under-expanded jet $\sim$ 1300 years old.

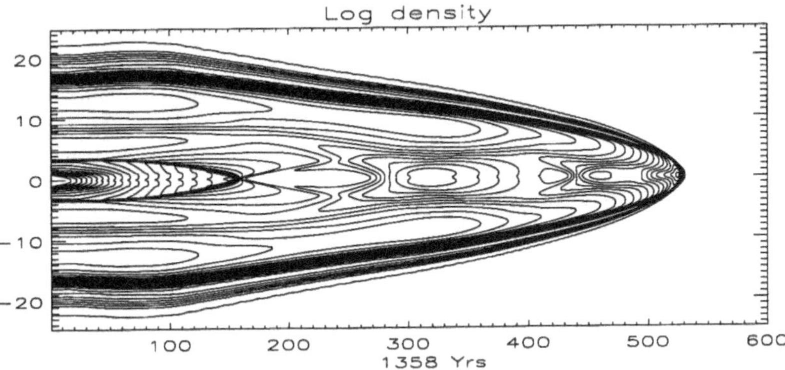

Figure 1. Density map (logarithmical) for the strongly under-expanded case with nozzle radius $r_{jet} = 10^{15}$ cm

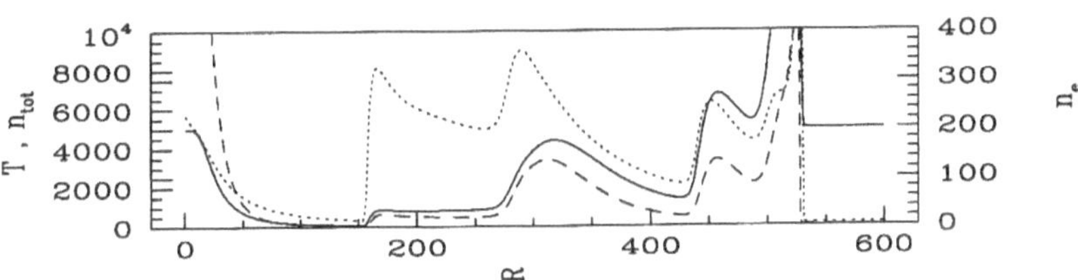

Figure 2. Profile along the jet axis of temperature (dotted line), electrons density (in cm^{-3}, dashed line) and total density (in cm^{-3}, full line).

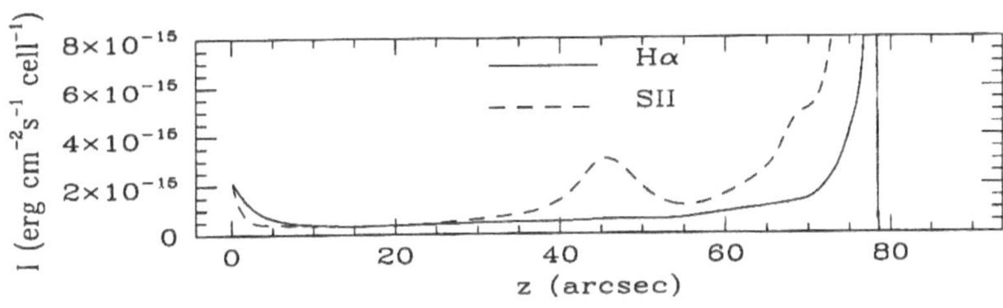

Figure 3. Profile along the jet axis of the [SII] (dashed line) and H_α (full line) fluxes, at a distance of 460 pc, from a cylindrically symmetric slice normal to the axis,with a width of 1 cell (1.2 $\times 10^{14}$ cm).

The external parameters are the same as before, whereas the jet pressure now corresponds to a pressure ratio $P_r = 60$ ($T_{jet} \approx 6000K$). The radius is again $r_{jet} = 10^{15}$ cm. In the density map displayed in Fig. 1, the expansion phase extends over $\approx 1.5 \cdot 10^{17}$ cm, and the first knot appears at a distance from the source of $\approx 3 \cdot 10^{17}$ cm. The expansion, as well as the knots, are clearly traced by the temperature and density profiles of Fig. 2. The spacing between the first and the second knot is of order $2 \cdot 10^{17}$ cm, too large with respect to the expected values. However, the emissivity results within the good range. Fig. 3 shows the emissivity in [SII] and Hα, integrated over the beam, versus the axis (the length units are given in arcsec, having assumed a distance of 460 pc). The units for the vertical axis are erg cm^{-2} s^{-1} per numerical cell, thus we need to add the contributions of all cells inside the knot to compare with observations. Since we have ≈ 50 cells inside the first knot, the numerical emissivity is of order 10^{-13} erg cm^{-2} s^{-1}, which is in good agreement with data from HH34, see Morse *et al.*, 1992.

In order to obtain a correct spatial spacing of our knots, while keeping unchanged the other parameters, we had to reduce the size of the nozzle to $2 \cdot 10^{14}$ cm. This case is shown in Fig. 4 to 6.

Fig. 4 displays the density map, after ≈ 700 years, showing a well formed pattern of internal oblique shocks. It is worth noticing that, though r_{jet} is much smaller, the effective beam radius is $\approx 2 \cdot 10^{15}$ cm.

Fig. 5 shows the total density, the electron density and the temperature measured along the axis of the jet. The adiabatic expansion, which is due to the strong pressure ratio, and causes all the values to drop, is clearly visible. The electron density, in particular, falls to zero, so that all the electrons in the emitting regions are due to post-shock ionization, independently of the inflow ionization fraction.

Four knots are clearly visible. The spacing is ≈ 5" and the emissivity, shown by Fig. 6, is again significant and in the range of observed values.

6. Conclusions

Our numerical simulations of radiative, non-equilibrium jets show that the internal oblique shocks survive the cooling losses and, with the proper choice of the inflow parameters, are responsible for the formation of a pattern of optical knots whose intensity and spacing are in good agreement with observations. We conclude that they could explain the internal knotty emissivity often observed in many HH jets, even if other effects, namely the magnetic field, $3D$ effects and low-frequency fluctuations of the inflow conditions, should be invoked to give a more general description of these objects.

Acknowledgements: The authors wish to thank Francesca Bacciotti, Tino Oliva and Claudio Chiuderi for their useful and friendly suggestions.

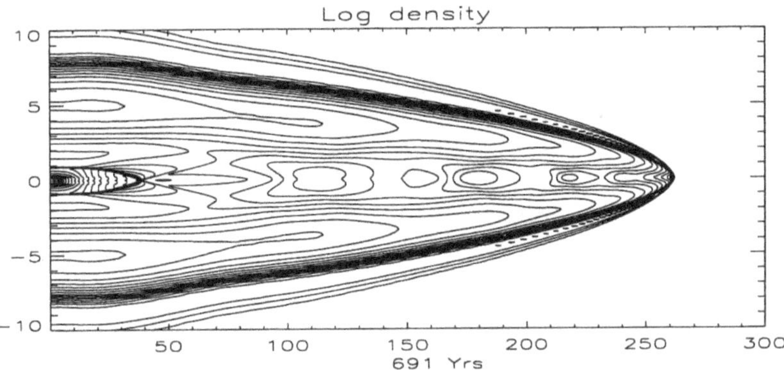

Figure 4. Density map (logarithmical) for the strongly under-expanded case with nozzle radius $r_{jet} = 0.2\ 10^{15}$ cm

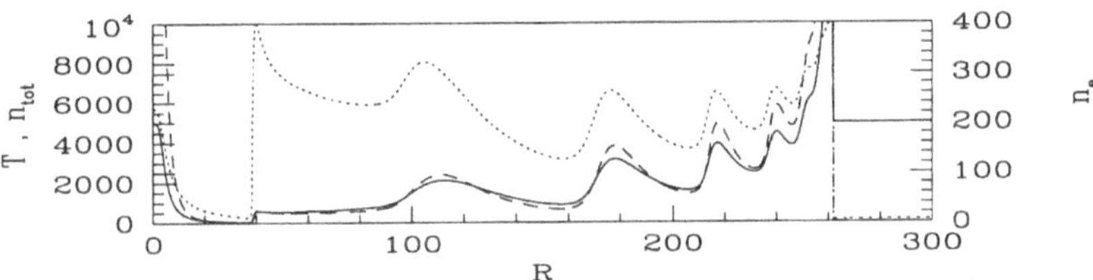

Figure 5. Profile along the jet axis of temperature (dotted line), electrons density (in cm^{-3}, dashed line) and total density (in cm^{-3}, full line).

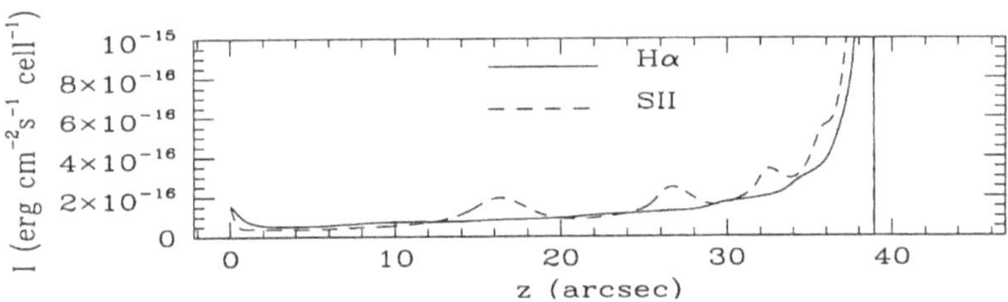

Figure 6. Profile along the jet axis of the [SII] (dashed line) and H_α (full line) fluxes, at a distance of 460 pc, from a cylindrically symmetric slice normal to the axis,with a width of 1 cell (1.2×10^{14} cm).

References

Bacciotti F., Chiuderi C., Oliva E., 1995, A& A **296**, 185
Bacciotti F., Eislöffel J., 1997, A& A, submitted.
Blondin J.M., Fryxell B.A., Königl A., 1990,ApJ **360**, 370
Cantó J., Raga A.C., Binette,L. 1989, Rev. Mex. Ast. **17**, 65
Dalgarno A., Mc Cray R.A., 1972, ARAA **10**, 375
Einfeldt B., Muntz C.D., Roe P.L., Sjogreen B., 1991, Journal of Computational Physics **92**, 273
Hardee P., Stone J., Rosen A., 1997, in 'Low Mass Star Formation from Infall to Outflow', Poster Proceedings IAU Symposium 182, ed. Malbet F., Castets A., Grenoble.
Heathcote,S. *et al.*,1996, AJ **112** ,1141
Hollenbach D., McKee C. F., 1989, ApJ **299**, 306
Lepp S., Shull J.M., 1983, ApJ **270**, 578
Lorusso, S., 1996, Tesi di laurea, Univ. Firenze
Mac Low M., Shull J.M., 1986, ApJ **302**, 585
Martin P.G., Schwarz D.H., Mandy M.E., 1996, ApJ **461**, 265
Morse J.A., Hartigan P., Cecil G., Raymond J.C., Heathcote S., 1992, ApJ **399**, 231
Reipurth B., Heathcote S., 1992, A& A **257**, 693
Sofia U.J., Cardelli J.A., Savage B.D., 1994, ApJ **430**, 650
Shu C.W., Osher S., 1988, Journal of Computational Physics 77, 439
Stone J.M., Norman M., 1993, ApJ **413**, 210
Suttner G., Smith D.M., Yorke H.,W., 1997, in 'Low Mass Star Formation from Infall to Outflow', Poster Proceedings IAU Symposium 182, ed. Malbet F., Castets A., Grenoble
Van Leer B., 1979, Journal of Computational Physics **32**, 101

ASYMMETRIC MODES OF THE KELVIN-HELMHOLTZ INSTABILITY IN PROTOSTELLAR JETS

JAMES M. STONE
Department of Astronomy, University of Maryland
College Park, MD 20742-2421 USA

Abstract. The results of a detailed analysis of the linear properties, nonlinear growth, and saturation of asymmetric modes of the Kelvin-Helmholtz instability in cooling protostellar jet beams are summarized. In the linear regime, cooling can significantly alter the growth rate and wavelength of the most unstable mode in comparison to an adiabatic jet. In the nonlinear regime, sinusoidal oscillations at the maximum growth rate lead to distortions that will be observed as 'wiggles' or 'kinks' in the jet. Strong cooling behind shocks formed in the nonlinear regime can produce emission knots and filaments. In some cases, the modes grow until the jet is disrupted. Distortions in the surface of the jet drive shock spurs into the ambient gas, resulting in longitudinal acceleration. Rapid acceleration and entrainment of ambient gas is also observed if the jet is disrupted.

1. Introduction

In the last few years, high-resolution images have been obtained for a number of protostellar jets using, for example, the HST. In fact, many of these images have been presented at this meeting (e.g. see Reipurth & Heathcote, this volume). Perhaps the most exciting aspect of these images is that the internal structure of the jet beam is resolved, in many cases for the first time, allowing the internal dynamics of the jet to be probed. In nearly every case, one finds the optical emission from the interior of the jet is highly complex, associated with knots and filaments of emission, curved 'internal bow shocks', and in some cases kinks and wiggles in the direction of propagation. As an extreme example of how complex the situation can become, I simply refer to the HST image of HH47 which graced the cover of the

B. Reipurth and C. Bertout (eds.), Herbig–Haro Flows and the Birth of Low Mass Stars, 323–333.

poster book, and which has been described in the literature by Heathcote *et al* (1996).

It seems clear that in the most complex systems, a variety of physical effects are at work to create the observed patterns of emission. For example, there is little doubt that many, if not all, protostellar jets are time-variable, and undergo periodic eruptions. Less certain is the possibility that the direction in which the jet is launched undergoes large-angle variations, perhaps due to 'precession' of the underlying accretion disk. That such 'pulsing' or changes in direction of the jet beam will lead to structure downstream is obvious: the structure is forced upon the flow by the boundary conditions at the source of the jet. However, it is also well known that the internal dynamics of a supersonic jet beam is rich enough to allow the formation of structure such as knots, filaments, and wiggles simply through dynamical instabilities associated with the jet itself. For example, detailed hydrodynamical studies of the propagation of cooling, overdense jets, beginning with Blondin *et al* (1990), has shown how Rayleigh-Taylor instabilities associated with dense sheets of gas formed by cooling fragment into dense knots (de Gouviea dal Pino & Benz 1993; Stone & Norman 1993a; 1993b; 1994). At the same time, studies of extragalactic jets have shown that various modes of the Kelvin-Helmholtz instability can produce either knots in the jet beam, or large amplitude wiggles leading to disruption at certain frequencies (e.g., see the reviews of Birkinshaw [1991], Hardee [1998]). The latter result seems especially appropriate to the observations of 'precessing' jets, however since extragalactic jets fall in a completely different region of parameter space than do protostellar jets (the former are thought to be both underdense compared to their surroundings and adiabatic, as opposed to overdense and cooling), very little of the previous work can be applied directly to protostellar systems.

Despite the obvious complexity of real systems such as HH47, it seems fruitful to consider the dynamics of an idealized supersonic jet beam in order to assess the importance of dynamical instabilities in forming structures within the beam. It is hoped that with this understanding, more complex situations can be investigated. With this in mind, my collaborators and I have recently finished a detailed analysis of the K-H instability in cooling jets in a paramenter regime appropriate to protostellar systems. In this paper, the major results and implications of this work will be summarized.

2. Problem Setup

The simplest model of an idealized jet is shown schematically in Figure 1. The jet is composed of a perfectly collimated, homogeneous stream propagating through a uniform ambient medium. Of course, in this figure the am-

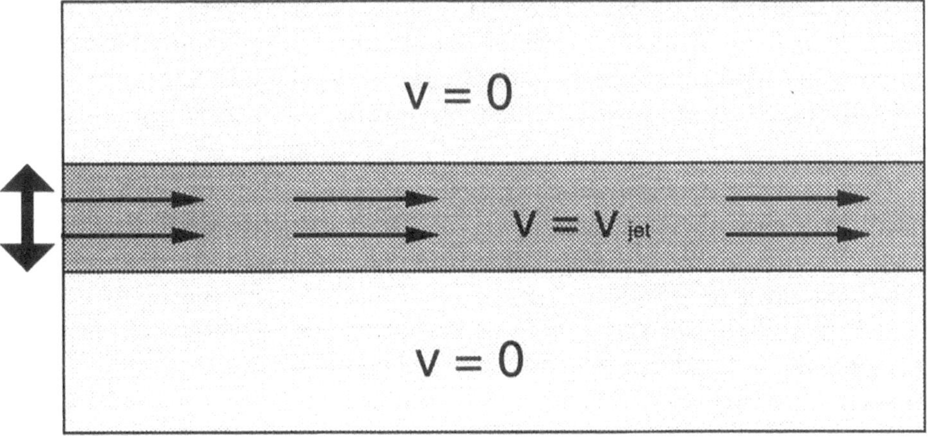

Figure 1. Schematic structure of an idealized protostellar jet beam.

bient medium does not actually represent the background fluid into which the jet propagates, rather the jet beam interacts directly only with a 'cocoon' composed of a mixture of shock processed ambient and jet gas. For this reason, the ambient medium shown in Figure 1 is never entirely uniform or quiescent in a propagating jet, nonetheless this complication is ignored in order to focus on the dynamics of the beam itself.

It is clear there will be K-H instabilities associated with the slip surface which defines the surface of the jet. However, because acoustic waves in the interior of the beam are reflected at the surfaces, the jet beam forms a resonant cavity which amplifies certain frequencies better than others. The result is that certain modes of the K-H instability will grow faster than others, and ultimately will dominate the jet. Our goal is to calculate which modes grow fastest in a protostellar jet, at what rate, and to understand what happens when the modes become nonlinear.

As mentioned in the Introduction, the analysis of the K-H instability in adiabatic jets is well developed. In cooling jets, Bührke *et al* (1988) were the first to suggest that axisymmetric pinch modes of the instability might be responsible for producing the chain of knots observed down the jet beam observed in such systems as HH34. However, the knot spacing and proper motions do not agree well with the predictions of the linear theory. Still, only recently has the evolution of the symmetric modes in cooling and overdense jets been thoroughly studied (see Massaglia *et al*, this volume). Moreover, even if the pinch modes are found not to be the primary source of symmetric knots, the asymmetric modes may still be important to produce wiggles and kinks in the beam. Finally, K-H instabilities are ultimately responsible for the production of "turbulent mixing layers" which lately have become the fashion to explain the acceleration of molecular bipolar outflows by highly collimated optical jets. Thus, if we are to understand such mixing layers we must first understand the K-H instability in cooling jets.

How does the K-H instability initially develop in the idealized jet beam shown in Figure 1? First, note that the beam is in an equilibrium state: it will not evolve unless it is perturbed. Thus, small amplitude *linear* perturbations must be introduced in the jet, usually at the nozzle, by adding a small transverse velocity (these perturbations are denoted by the thick arrow on the left hand side of Figure 1). The maximum amplitude of the transverse perturbation velocity must be much less than the sound speed in the interior of the jet in order to excite linear waves. If larger amplitude perturbations are used, they will already be of nonlinear amplitude, and will immediately produce shocks and distortions of the beam which have no relation to the resonant frequencies characteristic of the beam itself, but simply reflect the nature of the imposed variations. Fundamentally, the purpose of the perturbations is to introduce acoustic noise into the jet which can seed the instability.

Of course, our goal is to understand how the jet responds to the perturbations. If the jet is very overdense compared to its surroundings, i.e. if $\rho_{jet}/\rho_{ambient} \geq 100$, where ρ_{jet} and $\rho_{ambient}$ are the mass densities of the jet beam and ambient gas (remember this means cocoon) respectively, then the jet is effectively ballistic, and it will propagate without much notice of the ambient medium. In this case, it is clear that the maximum transverse displacement of the jet beam grows linearly with distance down the jet for a constant amplitude (transverse) perturbation. On the other hand, if $\rho_{jet}/\rho_{ambient} < 100$, the response of the jet beam to the perturbation is much more complex, and can only be computed by considering the linearized equations of hydrodynamics which describes the evolution of the perturbations. As we shall see in the next section, this leads to the maximum transverse displacement of the jet beam growing exponentially with distance down the jet. In principle, with high enough resolution, the distinction between linear and exponential growth of the amplitude of 'wiggles' in observations of the beam would indicate whether simple precession or hydrodynamical instability were responsible.

3. Linear Analysis

The evolution of small amplitude perturbations in a supersonic jet is governed by the linearized equations of hydrodynamics. If the initial equilibrium state is uniform (as it is in our case), a Fourier analysis can be used to calculate the dispersion relation for waves in the system, i.e. $k = k(\omega)$, where ω and k are the frequency and wavenumber of disturbances. Modes with complex wavenumbers then represent either exponentially growing or decaying waves, with the former resulting in instability. Generally, finding analytic solutions to the dispersion relation in closed form at arbitrary frequencies is not possible, therefore numerical root finding techniques are required to trace the solution over a broad range of perturbation frequencies.

Interested readers are referred to Hardee & Stone (1997, hereafter Paper I) for a complete discussion of the details of the analysis. The only important point to stress here is that some choice must be made for the net cooling function used in the analysis, and that the net cooling must be zero in the initial equilibrium state. In Paper I, two different cooling functions were adopted: (1) the cooling rate appropriate to photoionized gas described by MacDonald & Bailey (1981, hereafter MB), and (2) the cooling rate for interstellar gas given by Dalgarno & McCray (1972, hereafter DM). The former is appropriate for the study of Seyfert jets, the latter for the systems of interest here, namely protostellar jets. In simulations of propagating jets, Stone & Norman (1993a) found significant differences in the nonlinear evolution of the system if a more accurate non-equilibrium cooling function were adopted, however the linear analysis of K-H modes with more complex cooling functions is simply intractable. Since we are only interested in linear amplitude perturbation away from the equilibrium state, linearized cooling functions which approximate either the MB or DM cooling curves in the neighborhood of the equilibrium temperature are all that are required.

The results of the linear analysis are described in Paper I. Figure 2 shows the results for a Mach 20 jet which is 10 times denser than its surroundings (here and throughout this paper, the Mach number is defined as the ratio of the jet velocity to the sound speed in the *ambient* medium). Plotted are the dimensionless growth rate for the both the fundamental surface wave, and higher order body modes, as a function of dimensionless frequency (see paper I for a description of the classification of K-H modes into surface and body waves). The top panel plots these quantities for the adiabatic jet, the middle for a jet with DM cooling, and the bottom for a jet with MB cooling. Comparison between the panels immediately demonstrates the differences radiative cooling makes to the evolution of K-H modes in supersonic jets. For example, in the adiabatic and MB cooling jet, the growth rate of the surface wave has a maximum at some frequency (called the resonant frequency). We might expect that this resonant frequency, when it exists, will dominate the nonlinear evolution. The dots plotted on each panel are the growth rates measured from the initial linear

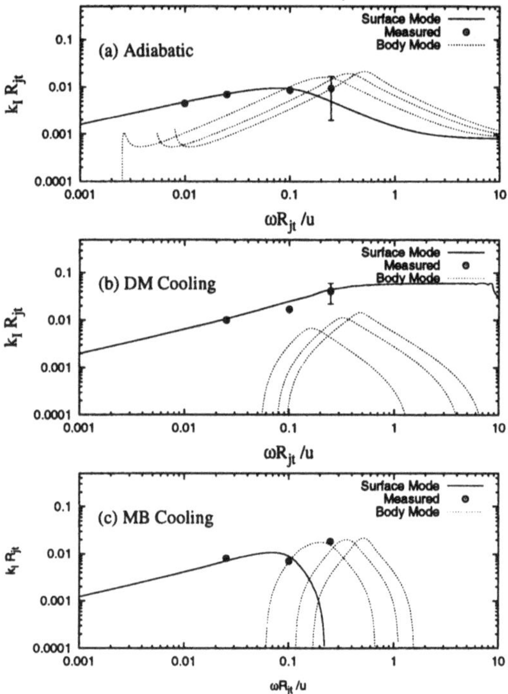

Figure 2. Growth rate of some asymmetric K-H modes in a Mach 20 protostellar jet.

evolution in a nonlinear hydrodynamical simulation of the instability - they provide both a code test and a check on the linear analysis. Generally, very good agreement is found between the two.

4. Nonlinear Hydrodynamical Simulations

In order to study the growth and saturation of K-H modes in a cooling jet in the nonlinear regime, numerical methods which solve the equations of hydrodynamics, combined with the full nonlinear cooling function over a wide range of temperatures, are required. We have used a two-dimensional (planar symmetric) hydrodynamics code based on the PPM algorithm to study the evolution of asymmetric modes in slab jets. Typically, the computational volume of size 20 R_{jt} in the transverse direction, and 400 R_{jt} in the longitudinal direction (with R_{jt} the jet radius) is divided into up to 200×800 zones. Nonuniform zones are used in the transverse direction, so that typically the numerical resolution within the jet beam is 20 zones per jet radius. The equilibrium jet is initialized across the entire mesh (with

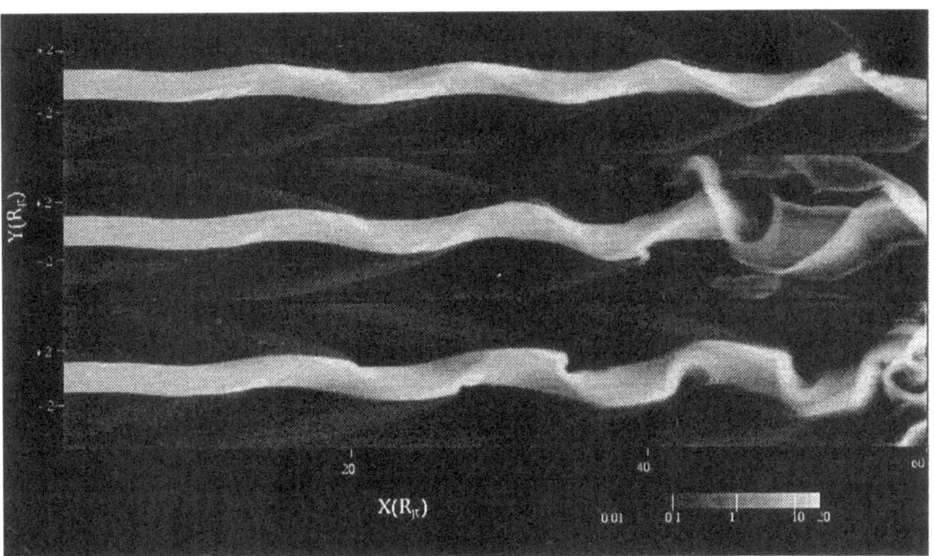

Figure 3. Log(density) for a Mach 5 adiabatic (top), DM cooling (middle) and MB cooling (bottom) jet perturbed at a frequency of $\omega R_{jt}/u = 0.4$. Note the formation of dense knots and filaments in the cooling jets.

inflow boundary condition on the left hand side of the grid, and outflow on the right), the jet is perturbed at the nozzle, and the evolution is followed for several hundred dynamical times (sound crossing times of the jet beam). The results of a series of simulations using different perturbation frequencies and cooling curves is given in Stone, Xu, & Hardee (1997, hereafter Paper II). We summarize some of the basic results here. Axisymmetric (two-dimensional) simulations of the pinch modes is described by Massaglia *et al* (this volume), and fully three-dimensional simulations of adiabatic jets are presented by Hardee *et al* (1997).

Figure 3 plots grayscale images of the logarithm of the density at late time for a Mach 5 jet computed with an adiabatic equation of state, and DM and MB cooling driven at a frequency $\omega R_{jt}/u = 0.4$ (near to the maximum growth rate of the surface wave, here u is the axial velocity of the jet) using a numerical resolution of 40 zones across a jet diameter. In each case, internal structure associated with sinusoidal oscillation grows strongly, forming a sequence of opposing shocks in the jet. Note that the DM cooling jet appears to disrupt at a lesser distance, $\approx 45R_{jt}$, than the adiabatic and MB cooling jets. In the cooling jets, the postshock gas becomes very dense. The corrugations in the edge of the jet produce oblique shock spurs in the ambient gas oriented in a sequence of opposing steps. Interestingly, the knots themselves are subject to Rayleigh-Taylor instabilities, and slowly

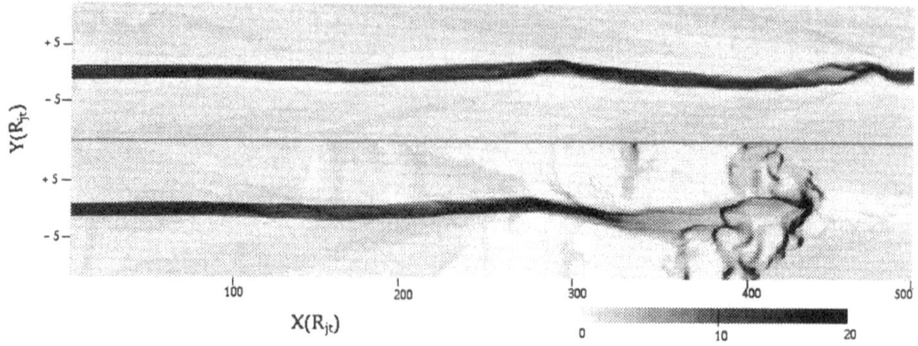

Figure 4. Log(density) for a Mach 20 adiabatic (top) and DM cooling (bottom) jet perturbed at a frequency of $\omega R_{jt}/u = 0.025$.

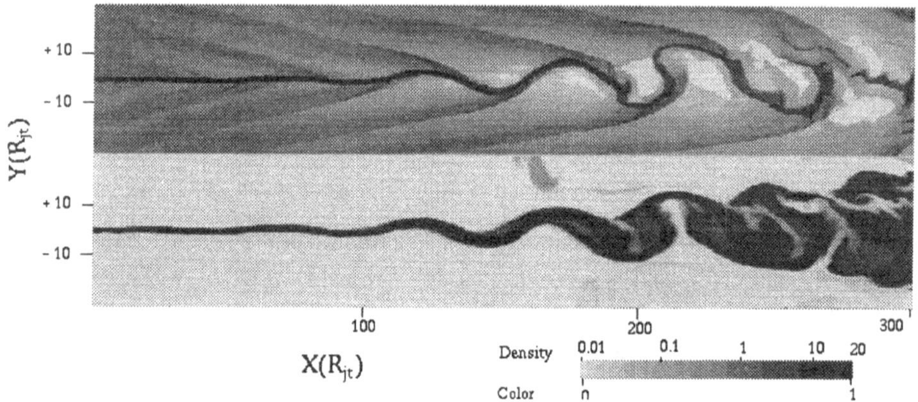

Figure 5. Log(density) (top) and 'color' (Lagrangian tracer) variable (bottom) for a Mach 5 DM cooling jet perturbed at a frequency of $\omega R_{jt}/u = 0.1$. Strong mixing is evident once the jet disrupts.

break up at late times. For comparison, in Figure 4 we plot grayscale images of the density in Mach 20 adiabatic and DM cooling jets driven at a low frequency of $\omega R_{jt}/u = 0.025$. Unlike jets perturbed at high frequency, jets perturbed at this low frequency have a long linear evolution phase. The DM cooling jet is predicted by the linear analysis to have a growth rate $\approx$ 30% larger than the adiabatic jet, and, in fact, the adiabatic jet remains relatively well collimated whereas the DM cooling jet is disrupted and forms high density knots associated with working surfaces in the ambient gas. The dense knots result from strong cooling effects.

In Figure 5 we plot grayscale images of the logarithm of the density and the color variable for simulations of a Mach 5 jet computed with the DM cooling curve. The jet is perturbed at a relatively low frequency of $\omega R_{jt}/u = 0.1$, and the result is shown after 200 dynamical times when it has reached a quasi-stationary state. This jet shows a long linear evolution phase and an overall sinusoidal pattern associated with the growth of surface waves. The surface wave grows to large amplitude at large distances from the nozzle, and disrupts the jet. Shock spurs driven into the ambient gas by the sinusoidal jet oscillations are evident. In addition to the surface wave there are oblique structures within the jet. These structures are manifested primarily as dense knots or filaments oriented on either side of the jet in an alternating pattern.

By using a Lagrangian tracer variable which can distinguish between the jet and ambient gas, we have measured the rate of mixing between these two fluids for the jet shown in Figure 5. We find that little mixing occurs near the jet nozzle, i.e., inside $x = 100R_{jt}$ where the waves are still in the linear regime. Mixing of the ambient gas starts where the perturbation reaches significant amplitude and then grows exponentially with the amplitude of the perturbation. This spatial growth of the mixing fraction appears to be in agreement with the results of Bodo et al. (1995). By tracking the momentum of the ambient gas, we also find that acceleration of the ambient gas occurs within $50R_{jt}$ of the nozzle, even before the jet disrupts and mixes significantly. Figure 5 shows that in this region strong shock spurs are driven into the ambient gas by the sinusoidal distortion of the jet. These shock spurs result in acceleration of the ambient gas. Although most previous discussions of the entrainment and acceleration of ambient gas along a jet beam have focused on turbulence either in a mixing layer, or in the turbulent "wake" of internal bow shocks in pulsed jets, our simulations show that shock trains in the ambient gas can result in acceleration of ambient gas. These shock trains may be produced by the K-H instability, as our simulations demonstrate, or they may be the wings of internal bow shocks produced by pulsing the jet.

5. Summary

Asymmetric modes of the K-H instability may be an important factor, along with other effects such as time-variability or 'precession', in producing complex structure within protostellar jets. To address this issue, detailed investigation of the properties of the instability in cooling, overdense jets, as described here, are required.

In Paper I, we presented solutions to the dispersion relation in the linear regime for K-H modes in a cooling jet over a wide range of perturbation frequencies. Our results indicated that the linear growth rates of waves in a cooling jet are substantially different from those of an adiabatic jet. Depending on the details of the cooling function adopted, the growth rate of waves could be either enhanced or reduced relative to those on the adiabatic jet.

In Paper II, we have continued this analysis by using time-dependent hydrodynamical simulations to follow the growth of unstable waves into the nonlinear regime. Growth accompanying low frequency and long wavelength perturbations can dominate the structure of the jet at late times producing large amplitude sinusoidal oscillations of the jet beam. Once the amplitude of the oscillations exceeds several jet radii, the jet is disrupted. On the other hand, higher frequency perturbations can excite body waves which result in the formation of internal sinusoidal structures. Strong internal shocks produced by these body waves can form a pattern of dense knots which are staggered asymmetrically along the length of the jet. We find that internal shocks produced by relatively short wavelength body waves can coexist on a jet with a much longer wavelength slowly growing sinusoidal oscillation of the entire jet. We also find that small amplitude oscillations of the jet associated with a growing sinusoidal surface or with a strong body wave can drive shock spurs into the ambient gas, leading to an alternating pattern of knots and shock spurs. These shock spurs can accelerate the ambient gas without mixing it with the jet fluid. Disruption of the jet by the K-H instability produces strong mixing and entrainment of ambient gas, and the formation of dense knots of shocked ambient gas swept up by jet material.

Structures observed in our simulations such as low amplitude wiggles in the jet, dense knots formed asymmetrically with respect to the center of the jet, and an alternating pattern of shock spurs driven into the ambient gas are all reminiscent of high resolution observations of protostellar jets. On the other hand, for a typical protostellar jet, dynamical periods associated with the inner ($r < 1$ AU) regions of an accretion disk are much less than the period of perturbations to which the entire jet can respond dynamically (which is about 200 years). We therefore expect protostellar jets to be dominated by the growth of body waves and internal asymmetric structures rather than the surface waves.

Of course, real protostellar jet systems will be three-dimensional. The asymmetric modes studied here are the analogue of the helical modes on a 3D jet. However, additional modes are also possible in 3D. In a companion work Hardee *et al* (1997) have presented a study of the K-H modes in a fully 3D supersonic jet including a dynamically important axial magnetic field. Actually, most models for driving jets from accreting protostars predict that the magnetic field far downstream in the outflow will be primarily toroidal – the stability properties of a cooling, toroidally magnetized jet have yet to be analyzed, although see Todo *et al* (1993) for a simulation of an adiabatic jet. Finally, it also important to remember that theoretical studies such as those described here are necessarily limited to idealized jets like that shown in Figure 1. If the jet instead has significant radial structure, e.g. it is composed of a shear layer of finite thickness, the details of the results (e.g. precise values of growth rates, etc.) will be changed, although probably the general character of unstable modes will not. It is hoped that detailed theoretical investigations of jet dynamics will ultimately enable detailed modeling of specific observations.

Acknowledgements: I thank my collaborators Phil Hardee and Jian-jun Xu for their many contributions to this work, and I gratefully acknowledge support from the NASA Astrophysics Theory Program under grant NAG-52886.

References

Birkinshaw, M. 1991, in Beams and Jets in Astrophysics, ed. P. Hughes, (Cambridge: CUP), 278
Blondin, J.M., Fryxell, B.A., & Königl, A. 1990, ApJ 360, 370
Bodo, G., Massaglia, S., Rossi, P., Rosner, R., Malagoli, A., & Ferrari, A., 1995, A& A 303, 281
Bührke, T., Mundt, R., & Ray, T.P. 1988, A&A 200, 99 (BMR)
Dalgarno, A., & McCray, R.A. 1972, ARA&A 10, 375 (DM)
de Gouviea dal Pino, E.M., & Benz, W. 1993, ApJ 410, 686
Hardee, P.E., & Stone, J.M. 1997, ApJ 483 (Paper I)
Hardee, P.E., Stone, J., Rosen, A.: 1997, in *Low Mass Star Formation - from Infall to Outflow*, poster proceedings of IAU Symp. No. 182, eds. F. Malbet & A. Castets, Observ. de Grenoble, 1997, p. 132
Hardee, P.E. 1998, in The 12th Kingston Meeting on Theoretical Astrophysics: Computational Astrophysics, eds. D. Clarke & M. West (San Francisco: ASP) in press.
Heathcote, S., Morse, J., Hartigan, P., Reipurth, B., Schwartz, R.D., Bally, J., & Stone, J.M. 1996, AJ 112, 1141
MacDonald, J., & Bailey, M.E. 1981, MNRAS 197, 995 (MB)
Massaglia, S., Trussoni, E., Bodo, G., Rossi, P., & Ferrari, A. 1992, A&A 260, 243
Stone, J.M., & Norman M.L. 1993a, ApJ 413, 198
Stone, J.M., & Norman M.L. 1993b, ApJ 413, 210
Stone, J.M., & Norman M.L. 1994, ApJ 420, 237
Stone, J.M., Xu, J., & Hardee, P. 1997, ApJ 483 (Paper II)
Todo, Y., Uchida, Y., Sato, T., & Rosner, R. 1993, ApJ 403, 164

KELVIN-HELMHOLTZ INSTABILITIES AND THE EMISSION KNOTS IN HERBIG-HARO JETS

S. MASSAGLIA, M. MICONO AND A. FERRARI
Dipartimento di Fisica Generale dell'Università,
Via Pietro Giuria 1, I-10125 Torino, Italy

AND

G. BODO AND P. ROSSI
Osservatorio Astronomico di Torino,
Strada dell'Osservatorio 20, I-10025 Pino Torinese, Italy

Abstract. We discuss the non-linear evolution of Kelvin-Helmholtz instabilities in Herbig-Haro jets performing numerical simulations by means of a PPM hydro-code modified as to include non-equilibrium, optically thin, radiation losses and heating. In this paper we discuss in particular the effects of different functional dependences of heating on density. The results obtained show a weak dependency of the instability evolution on the different forms of the heating function, that is largely unknown, therefore the simple assumption of constant heating, adopted in previous papers on this matter, does not lead to severe limitations on the general applicability of the results to the astrophysical jets and, in particular, to the origin of the emission knots.

1. Introduction

Recent wide field-of-view observations of Herbig-Haro jets (Devine 1997) have revealed that they extend for more than a few parsecs and, therefore, they last more than 10^5 yr. This implies that these jets successfully survive dynamical instabilities, the most relevant of them being the Kelvin-Helmholtz instability. Moreover, many HH jets show chains of quasi-periodically spaced, shock excited, knots (see e.g. Reipurth and Heathcote 1992; Eislöffel and Mundt 1992; Reipurth, Raga and Heathcote 1992) that emit mainly in H_α, [SII] and [OIII] optical lines.

B. Reipurth and C. Bertout (eds.), Herbig–Haro Flows and the Birth of Low Mass Stars, 335–342.

The origin of these emission knots is controversial. It has been suggested (Raga et al. 1990) that sudden variations in the mass ejection velocity from the central source, with a repetition time scale of the order of a hundred years, may give rise to working surfaces, internal to the jet, that travel as shocks at a speed very close to the jet's velocity. Alternatively, Bührke, Mundt and Ray (1988) proposed that Kelvin-Helmholtz instabilities themselves may be responsible for the knot formation, since K-H reflected modes can grow up to non-linear amplitudes leading to biconical shocks aligned along the jet's longitudinal axis.

Studies of the stability of jets against K-H instabilities has been carried out by several authors both in the linear (see Birkinshaw 1991 and references therein) and non-linear (e.g. Bodo *et al.* 1994 and 1995, Norman and Hardee 1988) regimes. In applying the K-H instability mechanism to the case of HH jets, a basic physical process must be included in the picture: radiative losses. These have been considered, in the linear limit, by Massaglia *et al.* (1992) who found that radiative losses tend to stabilize K-H reflected modes. Massaglia *et al.* (1992) adopted the Raymond-Smith (1977) cooling function with a constant heating, i.e. independent of the physical parameters. More recently, Hardee and Stone (1997) have analyzed the linear stability of a slab jet for different power-law cooling functions and heating independent of temperature and proportional to the density: they find that the growth of K-H modes is reduced for a cooling that is a steep function of temperature, while it is enhanced for a cooling that is a shallow function of temperature.

Besides linear studies of radiative K-H instabilities, the non-linear evolution has been also explored in recent years. Several authors (see Massaglia *et al.* 1996; Rossi *et al.* 1997; Micono *et al.* 1997; Stone, Xu and Hardee 1997) have tackled this problem by means of numerical hydro-codes. In general, they found that radiative losses play a significant role in the instability evolution which is crucially dependent on the assumed form of the cooling function assumed.

The instability evolution may, in general, depend crucially on the amount and density dependence of the heating term as well. The problem is that local heating is not an observable quantity and we can only argue about the relative importance of the various physical processes (ionizing radiation from the central star, turbulence, resistive dissipation of magnetic fields, etc.) that can possibly contribute to the energy input.

In this paper we analyze the spatial non-linear growth of K-H axisymmetric modes for a cylindrical jet in presence of non-equilibrium radiative losses and consider two kind of heating terms: i) a constant heating, and ii) a heating proportional to the particle density. Both kinds of heating are normalized to initially balance radiative losses. The plan of the paper is the following: in Section 2 we present the physical problem, the results obtained are discussed in Section 3 and the conclusions are drawn in Section 4.

2. The Physical Problem

We have carried out a spatial perturbation analysis by solving the hydrodynamic equations for the mass and momentum conservation for a 2-D cylindrical fluid jet, employing a two-dimensional PPM code (Colella and Woodward 1984, Bodo *et al.* 1995), modified by means of an operator splitting technique (see Rossi *et al.* 1997) to include radiative losses and heating in the energy equation. We refer to Massaglia *et al.* (1996) and Rossi *et al.* (1997) for the details of the radiation treatment and to Micono *et al.* (1997) for the characteristics of the initial configuration, perturbation, boundary conditions and the integration domain. The energy equation is written as

$$\frac{\partial E}{\partial t} + \nabla \cdot (E\vec{v}) = -p(\nabla \cdot \vec{v}) + (\mathcal{H} - \mathcal{L}),$$

where the fluid variables p, $\vec{v}$ and E are, as customary, the pressure, velocity, and thermal energy respectively; $\mathcal{H}$ and $\mathcal{L}$ represent the energy input and loss term (energy per unit volume per unit time) which includes energy lost in lines and in the ionization and recombination processes. In the equilibrium configuration, at $t = 0$, we have:

$$\mathcal{H}_0 - \mathcal{L}_0 = 0\,.$$

Our goal is to study the effects of different assumptions for the heating term, we have therefore carried out calculations setting first $\mathcal{H}$ = constant and then $\mathcal{H}$ proportional to the particle density $n_{\rm H}$. We remind that the evolution time is measured in units of $t_{\rm cr} = a/c_{\rm s}$ ($c_{\rm s}$ is the isothermal sound speed) and the velocities in units of $c_{\rm s}$.

We recall that the parameters that control the dynamics of the system are the jet Mach number M, and the infinity to on-axis density ratio $\nu \equiv n_\infty/n_0$. The radiation effects instead are fully determined setting the initial on-axis temperature T_0 and the cooling time $\tau_{\rm rad} \equiv t_{\rm rad}/t_{\rm cr}$, with $t_{\rm rad} = p/[(\Gamma-1)\mathcal{L}]$, or equivalently the column density $n_0 a$. As far as the heating time scale $\tau_{\rm heat}$ is concerned, we note that, this parameter being initially set equal to $\tau_{\rm rad}$, it is not an independent parameter. As discussed in Rossi *et al.* (1997), the initial parameters appropriate for the application to stellar jets yield $\tau_{\rm rad} \gg 1$, i.e. the initial conditions are nearly adiabatic. As unstable modes grow and shocks form, radiation effects become important locally in the compressed regions.

In the calculation we have set $M = 10$, $\nu = 1$, $T_0 = 10,000$ K and $n_0 a = 10^{18}$ cm^{-2}, which implies a jet mass flux $\approx 10^{-10} \times a_{15} M$ M$_\odot$ yr^{-1} and a momentum flux $\approx 10^{-9} \times a_{15} M^2$ M$_\odot$ km s^{-1} yr^{-1}, with $a_{15} = a/(10^{15}$ cm).

Constant heating

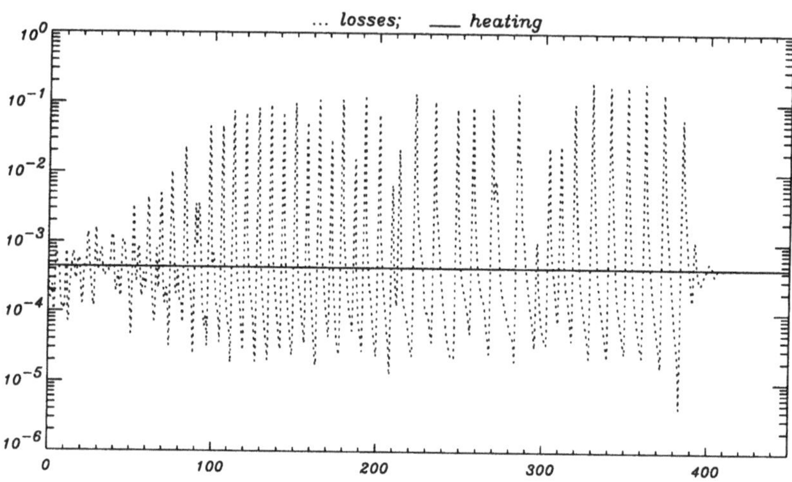

Heating proportional to density

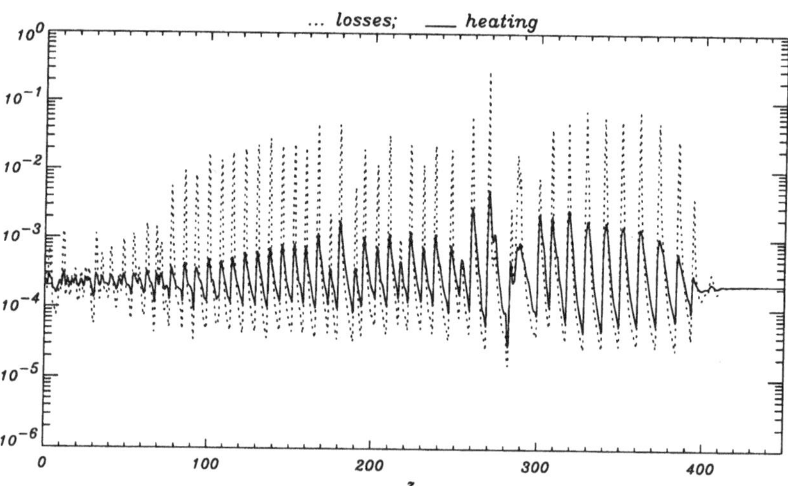

Figure 1. On-axis behavior of radiative losses (dashed line) and heating (solid line) for $\mathcal{H}$ = constant (upper panel) and for $\mathcal{H} \propto n_{\mathrm{H}}$ (lower panel) against z. Figure refers to the evolution time $\tau = 32$

In Figure 1 we plot, at the evolution time $\tau = 32$ and in arbitrary units, the on-axis behavior of radiative losses (dashed line) and heating (solid line) vs the longitudinal distance z, measured in units of a, for $\mathcal{H}$ = constant (upper panel) and for $\mathcal{H} \propto n_{\rm H}$ (lower panel). We note that a non-constant heating tends to reduce the amplitude of losses in shocks, therefore one can expect for this case radiation effects slightly reduced, i.e. larger shock temperatures and a little more vigorous mixing (see Rossi *et al.* 1997). Figure 2 is a gray-scale image of the logarithmic density distribution where we show a section of the jet going from 150 up to 400 radii and the two panels are again snapshots of the evolution as in Figure 1. We note that the two morphologies look very much qualitatively alike, albeit they present differences in the details, as shown in the magnified frames: in the non-constant heating case (bottom panel) we notice a tendency to form wider shocks, e.g. the formation at ~ 260 radii, and to send material and acoustic waves into the ambient medium in a more vigorous way.

An important tool for capturing the dynamical behavior of the system is the average momentum carried along by the jet. In Figure 3 we plot, at $\tau = 32$ (top panels) and $\tau = 38.5$ (bottom panels), the momentum carried out by the jet particles as a function of z and averaged over slices transversal to the jet axis and of longitudinal width of 50 grid points. We show, from left to right, the behavior in the adiabatic limit, the case with $\mathcal{H}$ = constant and the case with $\mathcal{H} \propto n_{\rm H}$. Examining this figure we notice that, in the adiabatic case, the jet loses almost entirely its momentum at distances $\gtrsim 230a$ due to instability-driven mixing with the ambient medium (see Bodo *et al.* 1994 and 1995). Radiation losses reduce sensibly the mixing behavior of the jet, with respect to the adiabatic case, since cooling limits the strength of the shocks that tend to inflate the jet (see discussion in Rossi *et al.* 1997 and Micono *et al.* 1997). Looking at the differences between the two different assumed $\mathcal{H}$ we see that they, qualitatively, behave in the same way, apart from some details such as a larger increase in the jet momentum at $\tau = 38.5$ and for $\mathcal{H} \propto n_{\rm H}$, at $z \sim 250a$, just before the mixing develops.

3. Conclusions

We have studied the non-linear spatial evolution of axisymmetric modes in a cylindrical jet in presence of radiative losses and two different heating terms: i) $\mathcal{H}$ = constant and ii) $\mathcal{H} \propto n_{\rm H}$. The numerical results show that the general evolution does not depend crucially on the form of $\mathcal{H}$, even though the details may do so. The physical reason for this is that the initial cooling time $\tau_{\rm rad}$ is much larger than unity, and so is the heating time $\tau_{\rm heat}$; as shocks form, losses grow as $\propto n_{\rm H}^2$ in the compression regions downstream, while heating, at most, grows as $\propto n_{\rm H}$. In the decompression

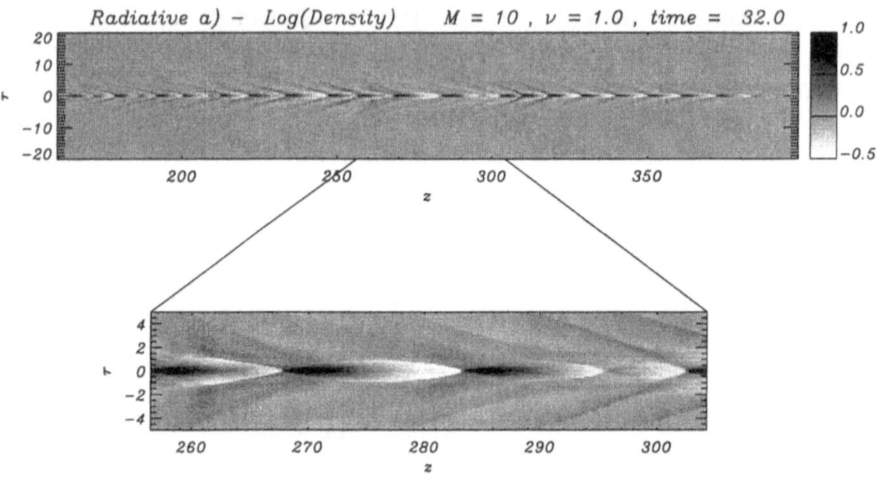

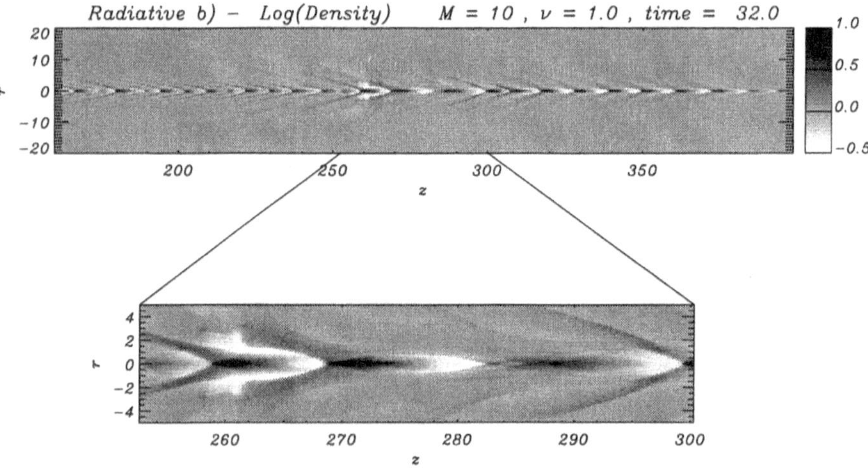

Figure 2. Gray-scale image of the logarithmic density distribution for for $\mathcal{H}$ = constant (upper panel) and for $\mathcal{H} \propto n_{\rm H}$ (lower panel). Figure refers to the evolution time $\tau = 32$

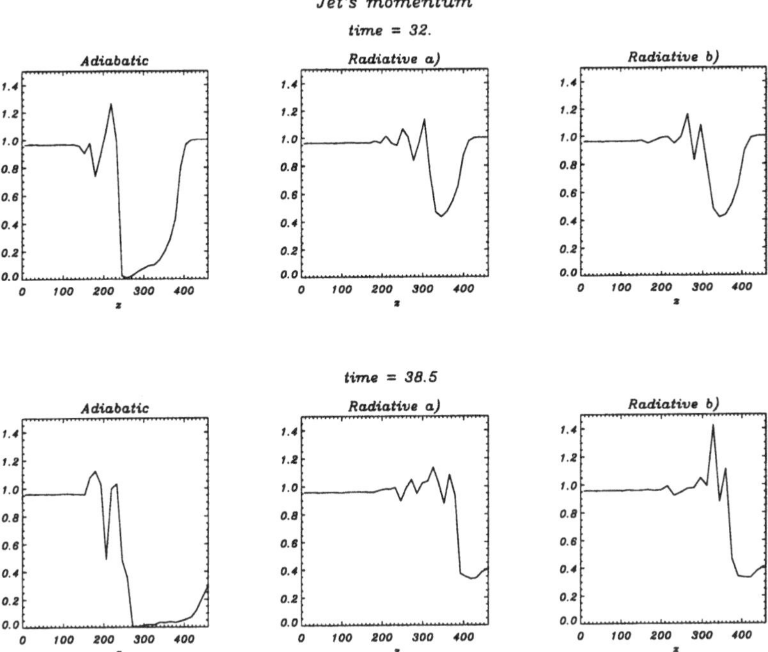

Figure 3. Average momentum carried by the jet against z. Upper panels for $\tau = 32$, lower panels for $\tau = 38.5$. From left to right: i) adiabatic limit, i) $\mathcal{H}$ = constant, $\mathcal{H} \propto n_H$

regions upstream heating dominates cooling, but anyhow the mixing history of the jet is not ruled by these regions.

These calculations show that even a very crude assumption of the (unknown) heating term does not lead to severe limits on the results.

References

Birkinshaw, M., 1991, in Beams and Jets in Astrophysics, P.A. Hughes edt, Cambridge Univ. Press (Cambridge), chap. 6

Bodo G., Massaglia S., Ferrari A., Trussoni E. 1994, A&A 283, 655

Bodo G., Massaglia S., Rossi P., Rosner R., Malagoli A., Ferrari A. 1995, A&A 303, 281

Bührke T., Mundt R., Ray T.P. 1988, A&A 200, 99

Colella P., Woodward P.R. 1984, J. Comp. Phys. 54, 174

Devine D.: 1997, in *Low Mass Star Formation - from Infall to Outflow*, Poster Proc. IAU Symp. No. 182, eds. F. Malbet & A. Castets, Observ. de Grenoble, 1997, p. 95

Eislöffel J. and Mundt R. 1992, A&A 263, 292

Hardee, P.E. and Stone J. 1997, ApJ in press.

Massaglia S., Trussoni E., Bodo G., Rossi P., Ferrari A. 1992, A&A 260, 243

Massaglia S., Rossi P., Bodo G., Ferrari A. 1996, Ap. Lett. Comm. 34, 295

Micono M., Massaglia S., Bodo G., Rossi P., Ferrari A. 1997, A&A in press.

Norman, M.L. and Hardee, P.E. 1988, ApJ, 334, 80

Raga A.C., Cantó J., Binette L., Calvet N. 1990, ApJ 364, 601
Raymond, J.C. and Smith, B.R. 1977, ApJ. Suppl. 35, 419.
Reipurth B. and Heathcote S. 1992, A&A 257, 693
Reipurth B., Raga A.C., Heathcote S.R. 1992, ApJ 392, 145
Rossi P., Bodo G., Massaglia S., Ferrari A. 1997, A&A, in press
Stone J., Xu J., Hardee, P.E. 1997, ApJ in press.

ON THE ENERGETICS AND MOMENTUM DISTRIBUTION OF BOW SHOCKS AND COLLIDING WINDS

FRANCIS P. WILKIN
Department of Astronomy, University of California
Berkeley, CA, 94720 USA

AND

JORGE CANTÓ AND ALEX C. RAGA
Instituto de Astronomía
Universidad Nacional Autónoma de México, Ap. 70-264,
04510 México, D.F., MÉXICO

Abstract. We discuss recent progress in analytic modeling of stellar wind bow shocks and colliding winds. For thin, radiative shocked layers in steady-state, the shape of the layer as well as its internal flux of mass and momentum are found from the conservation laws of mass, momentum and angular momentum. For the case that the shocked gas is well-mixed, the velocity distribution and mass column density of shocked material are also obtained. These solutions are extended to the problem of a jet bow shock, treated as a non-isotropic "wind" interacting with the ambient medium. We also examine the shell energetics for these simple analytic models. The constraint of conservation of momentum leads to an upper limit to the efficiency of thermalization and radiation of the pre-shock wind kinetic energy. Calculations are presented of this thermalization rate as a function of the input momentum rates of the pre-shock winds.

1. Introduction

Bow shocks are seen in a wide variety of circumstances in astrophysics, and many examples are known in star forming regions. This paper will primarily be concerned with analytic, dynamical models of bow shocks resulting from the collision of two supersonic flows or "winds", which may be compared with a number of different applications. While the models are necessarily

B. Reipurth and C. Bertout (eds.), Herbig–Haro Flows and the Birth of Low Mass Stars, 343–352.

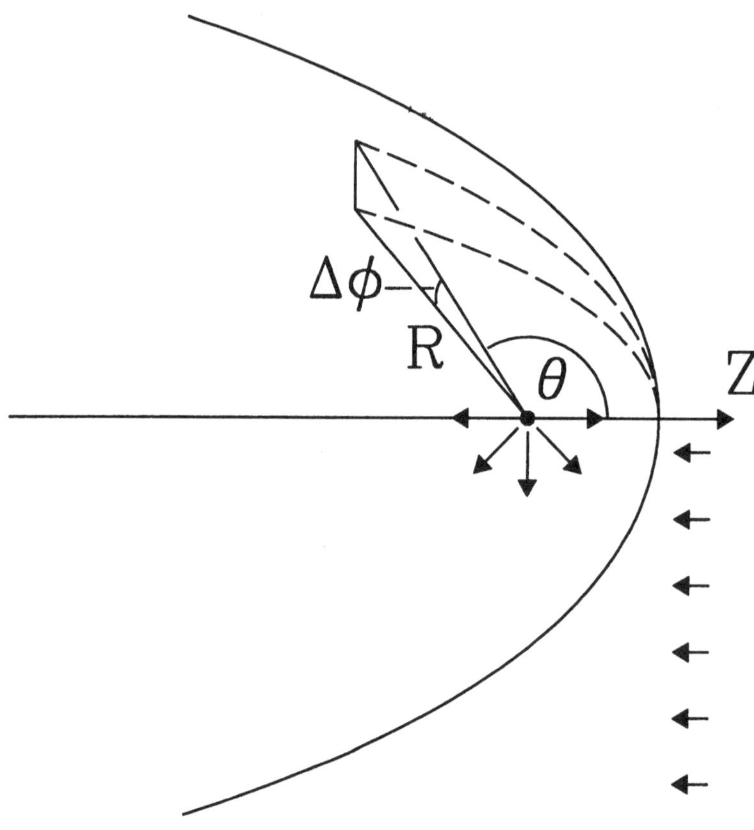

Figure 1. Geometry of the wind/ambient collision. The pre-shock flows are shown in the bottom panel, while the top panel defines the spherical coordinate system with origin at the source of the wind.

simple, in order to be analytically tractable, it is hoped that they will prove useful because they can easily be scaled and may provide insight to guide more detailed numerical studies. Because these calculations include full vector momentum conservation, they represent a step beyond previous analytic works based upon ram pressure balance arguments.

Perhaps the simplest example of a bow shock is that due to the motion of a supersonic, isotropic stellar wind, when the wind-blowing star moves supersonically with respect to the local interstellar medium. If the post-shock cooling is sufficiently rapid, the shocked fluid will lie in a thin layer that takes a steady-state, cometary form in the reference frame of the moving wind source (see Figure 1).

This model is sufficiently generic that it has been applied to many circumstances, and has been solved numerically by several authors. It was initially proposed by Baranov, Krasnobaev and Kulikovskii (1971, hereafter BKK) to describe the interaction of the solar wind with the very local interstellar medium. More recently, Van Buren *et al.* (1990) and Mac Low *et al.* (1991) have applied such a bow shock model to the confinement of cometary, ultracompact HII regions surrounding moving O stars, while Aldcroft, Romani & Cordes (1992) have applied it to pulsar wind bow shocks. The source of the wind need not be a stellar object, and similar models have been calculated, using ram pressure arguments, for a photoevaporating clump embedded in a moving medium (Dyson 1975), and for comets (Hoopis & Mendis 1980).

A new solution method for thin shell bow shocks in steady-state was given by Wilkin (1996), which yields the exact solution analytically to the above-described problem, including the shape of the bow shock and the mass column density and velocity of flow of the shocked material, assuming a mixed layer. This solution method is based upon exact, vector momentum conservation in the shocked layer, and will be described in §2. A very similar picture has been given for the bow shock due to a propagating jet, in which shocked material is imagined to be sprayed forward of the jet shock, mimicking a moving wind which interacts with the surrounding medium. The analytic solution for these bow shocks is given in §3.

A closely related problem is that of the collision of two spherical winds, which yields a family of possible bow shocks depending on the relative strengths of the two winds. An extention of the solution method (Cantó, Raga & Wilkin 1996) to consider angular momentum conservation allows the solution of this more complicated problem, discussed in §4.

The extention of these models to non-axisymmetry is obviously a necessity due to the many asymmetric observed bow shocks. Non-axisymmetric, ram pressure balance models of "proplyds", described as photoevaporating disks embedded in a stellar wind, have recently been developed by Henney *et al.* (1996); Henney (1996), see also Henney & Arthur, this volume. The extention of the analytic method to non-axisymmetric situations is given by Wilkin (1997a,b), and will not be discussed further here.

2. The Stellar Wind Bow Shock Model

The stellar wind drives a shock into the ambient medium, while the supersonic wind is abruptly decelerated, leading to two layers of shocked gas. These layers are assumed to mix, and postshock cooling is assumed to be efficient so that the dense shell has negligible thickness. The star moves with speed V_* in a uniform medium of density ρ_a. The isotropic stellar wind has mass loss rate $\dot{M}_w$ and constant speed V_w, yielding a cometary structure with the stellar trajectory as symmetry axis. The flow is assumed hypersonic, so that pressure forces are neglected. In this idealized model, the thin shell is fully described by three quantities: the shell radius $R(\theta)$, mass surface density $\sigma(\theta)$, and the tangential speed $v_t(\theta)$ of shocked material flowing along the shell.

Let the z axis be the axis of symmetry of the shell, with the stellar motion in the $\hat{\mathbf{z}}$ direction (to the right in Figure 1). In the frame of the star, the ambient medium appears as a uniform wind in the $-\hat{\mathbf{z}}$ direction. The stellar wind and the ambient medium collide head-on at $\theta = 0$, and the radius of this starting point of the shell is found by balancing the ram pressures of the wind and ambient medium, $\rho_w V_w^2 = \rho_a V_*^2$, which yields

$$R_0 = \sqrt{\frac{\dot{M}_w V_w}{4\pi \rho_a V_*^2}}. \tag{1}$$

This standoff distance sets the length scale of the shell. The shape of the shell is a universal function, which is scaled according to equation (1) to accommodate all values of the four dimensional parameters $(\dot{M}_w, V_w, \rho_a, V_*)$.

The fluxes of mass and momentum crossing an annulus of the shell, $2\pi\Phi_m(\theta)$, and $2\pi\Phi_t(\theta)$, are given, respectively, by

$$\Phi_m = \varpi\sigma v_t, \qquad \text{and} \quad \Phi_t = \varpi\sigma v_t^2, \tag{2}$$

where the cylindrical radius ϖ is $R\sin\theta$. In steady-state, the mass traversing a ring of the shell at polar angle θ from the standoff point is given by the mass flux from the stellar wind intercepted by the solid angle of the forward part of the shell plus the contribution from the ambient medium striking the circular area of the projected cross section of the shell:

$$2\pi\Phi_m = \dot{M}_w \frac{\Omega}{4\pi} + \pi\varpi^2 \rho_a V_*. \tag{3}$$

Here $\Omega = 2\pi(1 - \cos\theta)$ is the solid angle from the axis to the annulus at θ.

Following Wilkin (1996), we may calculate the rate at which vector momentum is imparted to the shell by the stellar wind, by considering a wedge of small, constant width in the azimuthal angle $\Delta\phi$ about the symmetry axis (Fig. 1). The surface integral of the wind vector momentum flux onto the shell does not depend on the detailed shape of the shell, because the coasting wind is momentum-conserving. We perform the integral over a spherical surface, using $\hat{\mathbf{r}} = \hat{\varpi}\sin\theta + \hat{\mathbf{z}}\cos\theta$:

$$\mathbf{\Phi}_w \Delta\phi = \int_{\text{wedge}} \rho_w \mathbf{V}_w (\mathbf{V}_w \cdot \hat{\mathbf{n}})\, dA$$

$$\mathbf{\Phi}_w \Delta\phi \;=\; \dot{\mathrm{M}}_w \mathrm{V}_w \, [(\theta - \sin\theta\cos\theta)\, \hat{\varpi} + \sin^2\theta \, \hat{\mathbf{z}}]\Delta\phi/8\pi. \tag{4}$$

The momentum deposited by the ambient medium is in the $-\hat{\mathbf{z}}$ direction, and depends only upon the circular cross section:

$$\mathbf{\Phi}_a \Delta\phi = -\varpi^2 \rho_a \mathrm{V}_*^2 \, \hat{\mathbf{z}} \, \Delta\phi/2. \tag{5}$$

The total momentum flux onto the $\Delta\phi$ wedge of the shell is the sum of the wind and ambient contributions. To conserve momentum in steady-state, the (tangential) momentum flux $\Phi_t \, \hat{\mathbf{t}} \Delta\phi$ traversing a $\Delta\phi$ azimuthal width of an annulus of the shell must equal the momentum flux $(\mathbf{\Phi}_w + \mathbf{\Phi}_a)\Delta\phi$ received by the shell surfaces between the standoff point and the annulus:

$$\mathbf{\Phi}_t = \frac{\dot{\mathrm{M}}_w \mathrm{V}_w}{8\pi}[(\theta - \sin\theta\cos\theta)\, \hat{\varpi} + \sin^2\theta \, \hat{\mathbf{z}}] - \frac{\varpi^2}{2}\rho_a \mathrm{V}_*^2 \, \hat{\mathbf{z}}, \tag{6}$$

where $\mathbf{\Phi}_t = \Phi_t \, \hat{\mathbf{t}}$ is the vector momentum flux in the shell, and $\hat{\mathbf{t}}$ is a tangential unit vector at constant ϕ. The momentum flux has magnitude

$$2\pi\Phi_t = \pi \mathrm{R}_0^2 \rho_a \mathrm{V}_*^2 \sqrt{(\theta - \sin\theta\cos\theta)^2 + (\tilde{\varpi}^2 - \sin^2\theta)^2}, \tag{7}$$

where a tilde indicates a length in units of R_0. We now know the *vector* momentum flux at any point in the shell. The direction of flow is that of this momentum flux, so the shell shape is given by the *trajectory equation*

$$\frac{d\tilde{z}}{d\tilde{\varpi}} = \frac{v_z}{v_\varpi} = \frac{\Phi_{t,z}}{\Phi_{t,\varpi}} = \frac{-\tilde{\varpi}^2 + \sin^2\theta}{\theta - \sin\theta\cos\theta}. \tag{8}$$

It can be shown (Wilkin 1996) that the exact integral to this equation is

$$\mathrm{R}(\theta) = \mathrm{R}_0 \csc\theta \sqrt{3(1 - \theta\cot\theta)}. \tag{9}$$

This formula for $\mathrm{R}(\theta)$, together with the momentum flux $\Phi_t(\theta)$ of equation (7), is the solution of the equations of BKK with the desired initial conditions $\mathrm{R}(0) = \mathrm{R}_0$, and $\mathrm{R}'(0) = 0$. With the previously known mass integral, we obtain the remaining shell properties. Referring to equations (2), the tangential velocity in the shell is $v_t = \Phi_t/\Phi_m$, while the mass surface density is given by $\sigma = \Phi_m^2/\varpi\Phi_t$. Both the mass surface density and the tangential velocity depend upon the nondimensional parameter $\alpha = \mathrm{V}_*/\mathrm{V}_w$.

The total amount of incident kinetic energy thermalized in the bow shock per unit time is given by the sum of the incident kinetic energy fluxes in the center of mass frame. This must be the ambient frame, since in any other frame, the ambient medium deposits an infinite amount of momentum per unit time, because the cylindrical radius of the bow shock formally diverges far back in the tail. Letting primes indicate velocities in the ambient frame, so that $\mathbf{V}'_w = V_w\,\hat{\mathbf{r}} + V_*\,\hat{\mathbf{z}}$, the rate of thermalization of incident kinetic energy is given by (Wilkin, Cantó & Raga 1997)

$$\begin{aligned} \dot{E}_{th} &= \int_{\rm shell} \frac{1}{2}\rho_w |\mathbf{V}'_w|^2 (\mathbf{V}'_w - V_*\,\hat{\mathbf{z}}) \cdot \hat{\mathbf{n}}\,{\rm dA} \\ &= \frac{1}{2}\dot{\rm M}_w (V_w^2 + V_*^2) \end{aligned} \tag{10}$$

This sets an upper limit to the rate at which energy can be radiated by the bow shock in steady-state, in the absence of other energy sources (*i.e.* radiation from the wind source). It is interesting that this very simple result was not noticed in the previous numerical calculations of the bow shock.

One would also like to describe the bow shock kinematics in the ambient, rather than stellar, frame. At any angle θ, the ratio of sideways (V'_ϖ) to forward (V'_z) velocity, and therefore sideways to forward momentum, is given by

$$\frac{V'_\varpi}{V'_z} = \frac{\theta - \sin\theta\cos\theta}{\sin^2\theta + 2\alpha(1-\cos\theta)}, \tag{11}$$

which has the limit of $\pi/4\alpha$ as $\theta \to \pi$. Clearly any ratio of sideways to forward momentum is possible, for some value of α, even for large θ, in the tail of the bow shock (See Figure 2). This may be relevant to the application of bow shock models to the ends of molecular outflows, because it is known that most of the momentum in such outflows is in the forward direction. Lada & Fich (1996) argued that this constraint excluded bow shock models, because they only considered the immediate post-shock velocity, rather than the tangential velocity in the shell after radiative relaxation and mixing with previously shocked material. We see that for values of α greater than unity, the momentum will be primarily in the forward direction.

3. Application to Jet Bow Shocks and Molecular Outflows?

The thin shell bow shock model has been applied in modified form to the problem of jet-driven molecular outflows in star forming regions. Raga & Cabrit (1993) considered internal working surfaces in a jet due to a time-dependent jet velocity, while Zhang & Zheng (1997) treated the terminal bow shock at the end of the jet. In these studies, the shocked fluid is assumed to be sprayed forward of the jet shock as a "wind" which interacts with the surrounding material. This "wind" is no longer taken to be isotropic in the reference frame of the bow shock, but is assumed to extend from the symmetry axis only to an angle $\theta_0 < \pi$ (Zhang & Zheng), or in the case of Raga & Cabrit, to extend from $\theta = \pi/4$ to $3\pi/4$. Zhang &

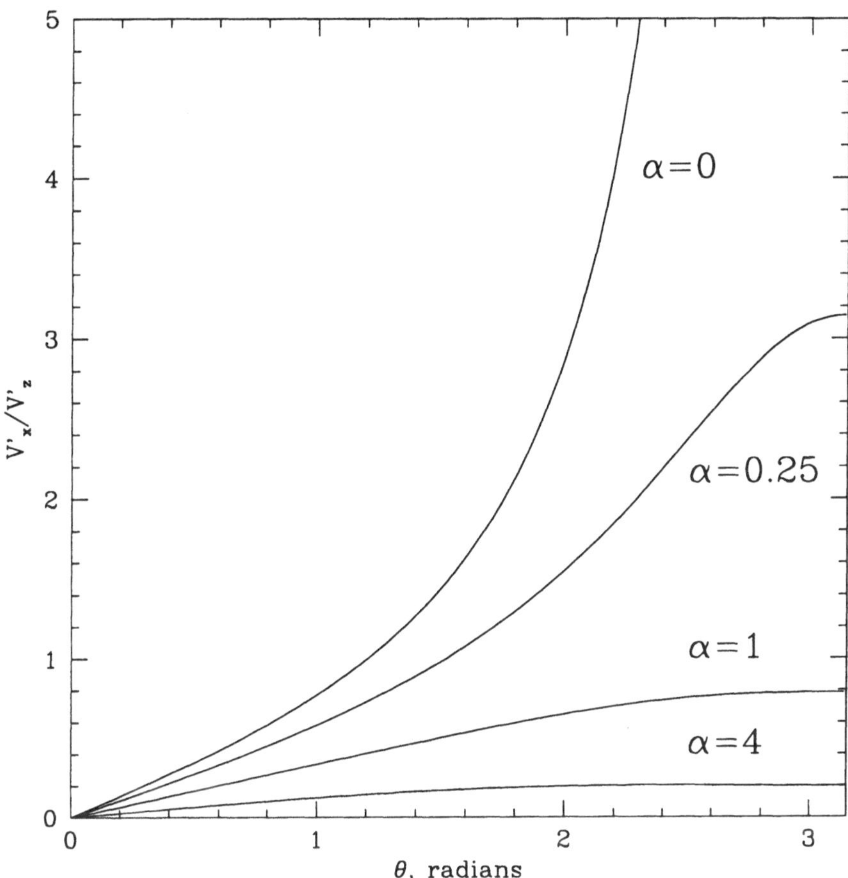

Figure 2. Ratio of sideways (V'_ϖ) to forward (V'_z) velocity for shocked gas in the ambient reference frame for the moving bow shock.

Zheng solved numerically for the shape of the bow shock and the kinematics of shocked material, pointing out that such bow shocks can have a large amount of momentum in the forward direction.

The formalism discussed in §2 can also be applied to this problem as well, although we only give the solution for the case considered by Zhang & Zheng. The solution is unchanged for $\theta \leq \theta_0$, except that $\dot{M}_w$ must be replaced by $4\pi\dot{M}_w/\Omega_0$, where $\Omega_0 = 2\pi(1 - \cos\theta_0)$ is the solid angle of the shocked jet "wind". For larger angles, the trajectory equation is the same as equation (8), except that θ is replaced by θ_0, a constant. Integration

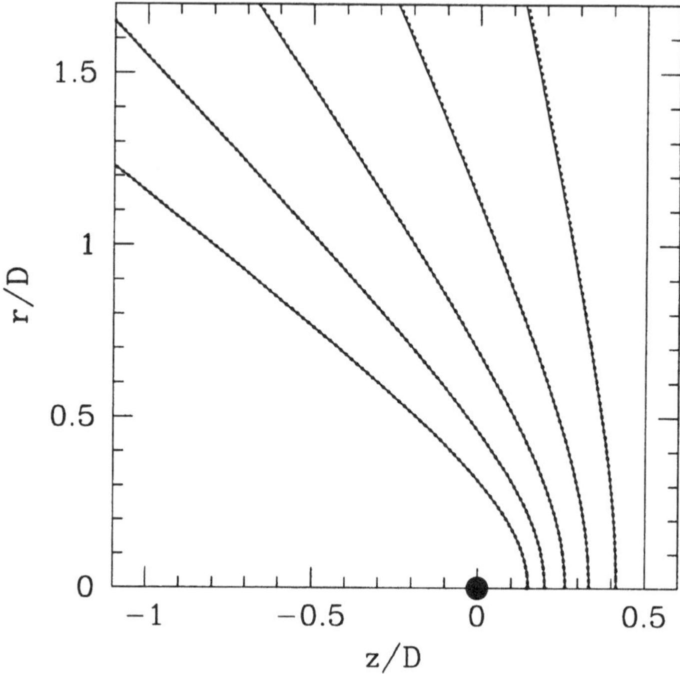

Figure 3. Solutions to the two-wind collision problem. The two wind sources are separated by a distance D, so that the weaker source is at the origin, while the stronger source is off the right edge of the graph at $z/D = 1$. Moving from right to left, the solutions correspond to $\beta = 1$, 0.5, 0.25, 0.125, 0.0625, 0.03125.

then yields the shape

$$R = R_0\sqrt{3[\sin^2\theta_0 + \cot\theta(\sin\theta_0\cos\theta_0 - \theta_0)]}. \tag{12}$$

Zhang & Zheng found that the ratio of forward to sideways velocity in the ambient frame approached a constant for large θ. Indeed, this velocity ratio is given by equation (11) with θ replaced by θ_0 if $\theta > \theta_0$. Once the shocked fluid element is beyond the maximum wind angle, its trajectory

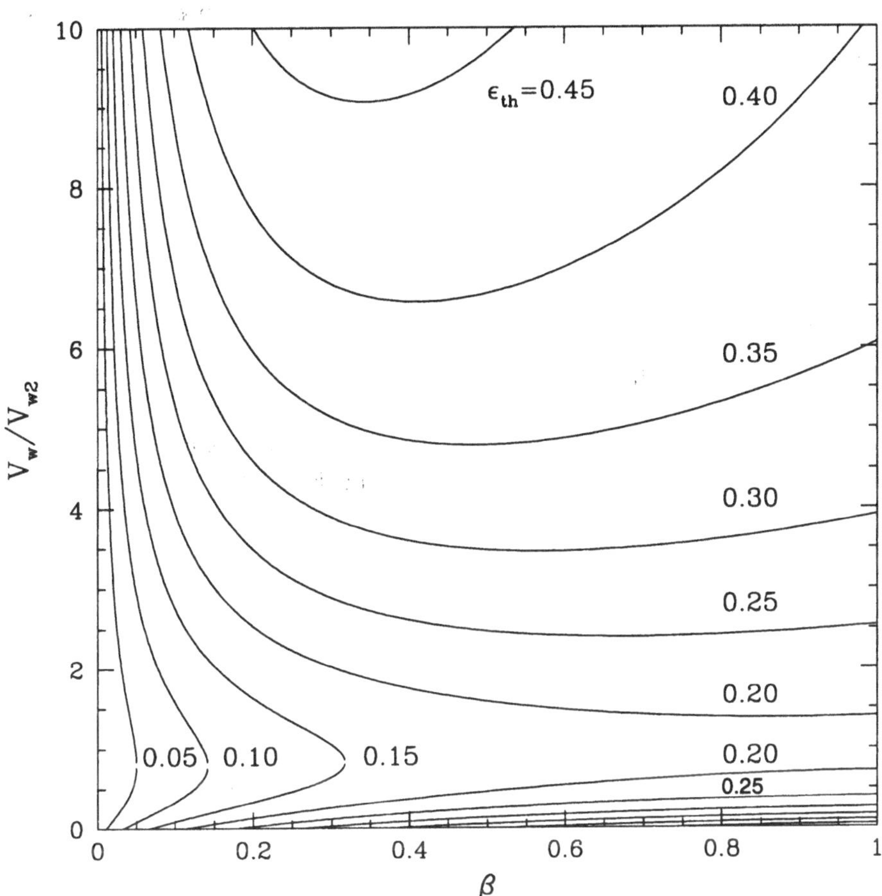

Figure 4. Efficiency of thermalization of pre-shock wind kinetic energy. Contours of equal thermalization efficiency ϵ_{th} are shown in terms of the ratio of wind momentum loss rates, β, and the ratio of wind speeds.

is a straight line, and the solution is a momentum-conserving snowplow in which the gas decelerates as it sweeps up the surrounding material.

4. Two Wind Collision in Binaries

The problem of the collision of two spherical winds has been solved by the same approach, for the case of a radiative, thin shell (Cantó, Raga, & Wilkin 1996). The ratio of momentum loss rates of the two stars is the fundamental parameter determining the shape of the bow shock, and several examples are shown in Figure 3. The solution method has been extended by including the consideration of angular momentum (about the origin). Although this may seem redundant with conservation of vector momentum, the benefit is that it provides a general, although implicit, integral to the trajectory equation (8) for more general wind collisions. If we define an angular momentum flux $\Phi_J = \Phi_m R v_\theta$, the radius of the shell is given implicitly by

$$\Phi_J = \Phi_\varpi z - \Phi_z \varpi, \tag{13}$$

where the flux functions refer to the flow in the shell, so they are determined by adding the contributions from the two winds.

Conservation of momentum implies that not all of the wind pre-shock kinetic energy can be thermalized and radiated. The maximum amount that may be thermalized and is available for radiation is given by the kinetic energy in the center of mass frame. We have calculated the rate at which energy is thermalized in the shocks and post-shock mixing. This is displayed in Figure 4, as the efficiency by which the total of the two wind kinetic energies are intercepted by the shell and thermalized. This represents an upper limit to the steady-state radiative luminosity of the bow shock.

5. Acknowledgements

F.P.W. is grateful to the organizers of this symposium for the opportunity to give this talk, and to B.Reipurth and C. Bertout and the IAU for a travel grant. This research was supported by a fellowship from the Josephine De Kármán Trust.

References

Aldcroft, T.L., Romani, R.W., & Cordes, J.M. 1992, ApJ 400, 638
Baranov, V. B., Krasnobaev, K. V., & Kulikovskii, A. G. 1971, Sov. Phys. Dokl., 15, 791
Cantó, J., Raga, A.C., & Wilkin, F.P. 1996, ApJ, 469, 729
Dyson, J. 1975, AP&SS, 35, 299
Henney, W.J., Raga, A.C., Lizano, S., & Curiel, S. 1996, ApJ 465, 216
Henney, W.J. 1996, RMxAA, 29, 192
Henney, W.J. 1997, This proceeding
Houpis, H.L.F. & Mendis, D.A. 1980, ApJ, 239, 1980
Lada, C.J., & Fich, M. 1996, ApJ 459, 638
Mac Low, M.-M., Van Buren, D., Wood, D. O. S., & Churchwell, E. 1991, ApJ, 369, 395
Raga, A. C., & Cabrit, S. 1993, A&A, 278, 267
Van Buren, D., Mac Low, M.-M., Wood, D. O. S., & Churchwell, E. 1990, ApJ, 353, 570
Wilkin, F.P. 1996, ApJ, 459, L31
Wilkin, F.P. 1997a, in *Low Mass Star Formation - from Infall to Outflow*, Poster Proceedings of the IAU Symposium No 182, Eds. F.Malbet and A.Castets, (Observatoire de Grenoble), p190
Wilkin, F.P. 1997b, (In Preparation)
Wilkin, F.P., Cantó, J. & Raga, A.C. 1997, (In Preparation)
Zhang, Q. & Zheng, X. 1997, ApJ, 474, 719

IV. Disks, Winds, and Magnetic Fields

HUBBLE SPACE TELESCOPE IMAGING OF THE DISKS AND JETS OF TAURUS YOUNG STELLAR OBJECTS

KARL STAPELFELDT
Jet Propulsion Laboratory, California Institute of Technology
4800 Oak Grove Drive, Mail Stop 183-900
Pasadena CA 91109, USA
CHRISTOPHER J. BURROWS
Space Telescope Science Institute
JOHN E. KRIST
Space Telescope Science Institute
AND
THE WFPC2 SCIENCE TEAM

Abstract. We report on Hubble Space Telescope imaging of eleven young stellar objects in the nearby Taurus molecular clouds. The high spatial resolution and stable point spread function of HST reveal important new details of the circumstellar nebulosity of these objects. Three sources (HH 30, FS Tau B, and DG Tau B) are resolved as compact bipolar nebulae without a directly visible star. In all three cases, jet widths near the sources are found to be 50 AU or less. Flattened disk structures are seen in absorption in HH 30 and FS Tau B, and in reflection about GM Aur. Extended envelope structures traced by scattered light are present in HL Tau, T Tau, DG Tau, and FS Tau. The jet in DG Tau exhibits a large opening angle and is already resolved into a bow-like structure less than 3″ from the star.

1. Introduction

The environment of a young stellar object (YSO) is a complex region where disks, infalling envelopes, outflows, and companion stars interact to define the distribution of circumstellar material. Of particular interest are the properties of protoplanetary disks believed to be present during these first stages of stellar evolution. In combination with spectral energy distributions

B. Reipurth and C. Bertout (eds.), Herbig–Haro Flows and the Birth of Low Mass Stars, 355–364.

and kinematics derived from spectroscopy, high resolution optical imaging can greatly clarify the geometry and dynamical state of material in the YSO environment. For nearby sources such as those associated with the Taurus molecular clouds, optical images with the Hubble Space Telescope / Wide Field and Planetary Camera 2 (HST/WFPC2) provide a typical linear resolution of 10 AU. This capability has led to important new discoveries since the repair of HST, and offers exciting prospects for future work. In this contribution, we review YSO imaging results obtained in Guaranteed Time Observation programs by the WFPC2 Science Team as of early 1997.

2. Observations

All the targets are low-mass pre-main sequence stars located in Taurus-Auriga at a distance of 140 pc. They were observed using broadband V, R, and I filters, usually on the Planetary Camera. Typically both short (10-30 sec) and long (200-600 sec) exposures were made in each filter, and the full sequence was completed in a single HST orbit. Emission lines from Herbig-Haro objects occur primarily in the R bandpass, and are largely absent in the I bandpass. A comparison of these two images thus allows HH objects to be distinguished from reflection nebulosity. This procedure has worked well except very close to bright objects such as DG Tau, where noise from the background continuum limits the visibility of jet features.

3. Results for Point Sources with Adjacent Nebulosity

These bright stars were deliberately saturated to provide maximum dynamic range for detection of circumstellar features. Reference point spread functions from both HST observations and TINYTIM calculations were aligned, scaled, and subtracted to reveal the circumstellar nebulosity. Imperfections in the subtraction process lead to residual artifacts along the diffraction spikes, and saturation effects prevent any study of features within a few pixels (20 AU) of the stars or where column bleeding has occurred. The fundamental contrast limit is set by scattering within the WFPC2 cameras, and improves rapidly with increasing distance from the central object. At r= $2''$, the scattered light background is already about 7 mag/arcsec^2 fainter than the integrated brightness of the central point source (Krist, 1995).

3.1. T TAURI

Only the northern component of this binary ($0.7''$/100 AU separation) was detected in the short exposures available. The optical star lies close to an arc of reflection nebulosity which encloses it on the N, S, and E. The arc shows

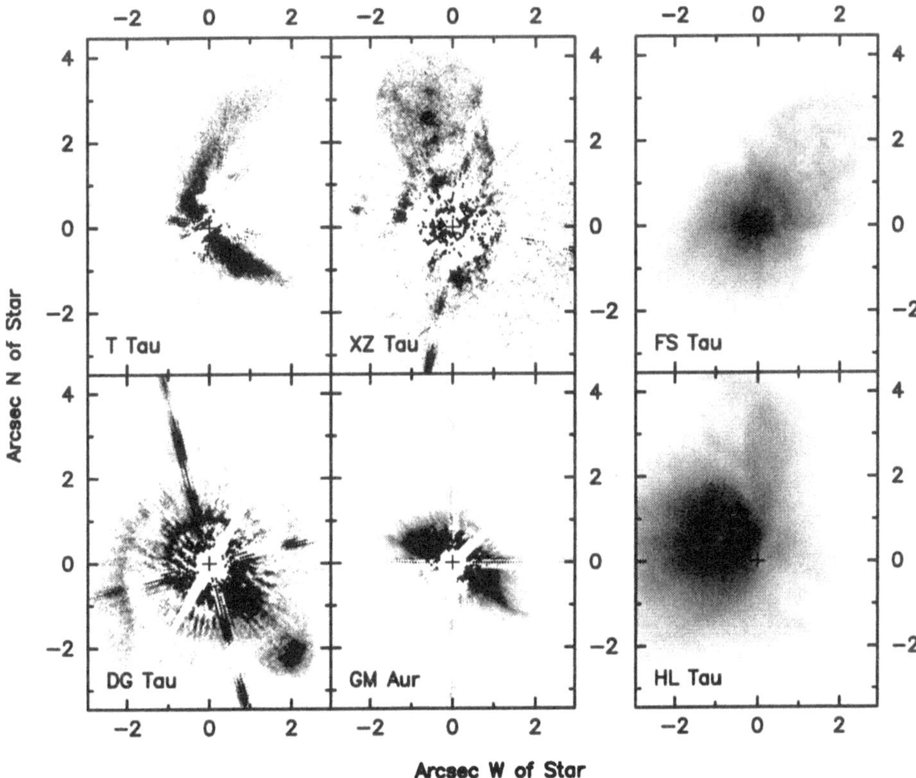

Figure 1. **Circumstellar Nebulosity of Six Young Stars.** These are PC images, with N up and E to the left. Point-spread functions were subtracted from the four R band images on the left, shown in linear stretch. I band images of FS Tau and HL Tau are shown at right in log strech.

an approximate symmetry along an axis toward the WNW, the direction of the blueshifted outflow. The arc shape appears similar to models of an illuminated outflow cavity in the circumbinary envelope. For full details, see Stapelfeldt *et al.* (1997a).

3.2. XZ TAURI

Both stars are directly seen in this close (0.3″/42 AU) binary. Little or no reflection nebulosity is present. However, an unusual "bubble" of emission nebulosity extends 4″ (550 AU) NNE of the binary. Knots within the bubble are suggestive of an enclosed jet, but do not align precisely with either star. Kinematic constraints dictate an especially small dynamical age of about 10 years for the bubble. For full details, see Krist *et al.* (1997a).

3.3. FS TAURI (HARO 6-5)

Both stars are also directly seen in this 0.24″ (34 AU) binary system. Abundant reflection nebulosity extends beyond 5″ (700 AU) from the binary, where it begins to overlap larger-scale structures associated with FS Tau B. Near the binary the nebula is wispy and complex, with only the suggestion of a symmetry axis. There is no obvious jet structure. A complete discussion will be presented in Krist *et al.* (1997b).

3.4. DG TAURI

Here the central star is directly visible adjacent to three nested arcs of reflection nebulosity. The arcs enclose DG Tau on its N and E sides, and appear offset at distances of 2″, 5″, and 10″ from the star. The innermost arc, seen for the first time in these images, is the smallest and most sharply defined of the group. All three arcs have an approximate symmetry axis aligned with the blueshifted HH jet emerging at PA 225°. This suggests that the arcs represent illuminated edges within the outflow cavity carved by the jet, similar to the situation for T Tau. The multiplicity of arcs is, however, difficult to account for in this scenario.

The DG Tau jet is clearly resolved into several broad knots within 12″ of the star. The jet has a remarkably large opening angle of 20°, and even the innermost knot shows a clear bow-like morphology at a projected distance of only 400 AU from the star (see also Lavalley, Dougados, and Cabrit 1997). The large radial velocities measured for the DG Tau jet (Solf and Böhm, 1993) suggest that the system is viewed at only a modest inclination angle from the jet axis. In this case we look "down the throat" of the outflow cavity and the jet should appear foreshortened, circumstances which seem in accord with the observed morphology.

3.5. GM AURIGAE

This star is noteworthy for its circumstellar disk mapped in CO rotational emission lines by Dutrey (1997) and Koerner, Sargent, and Beckwith (1993). The WFPC2 images reveal a flattened circumstellar reflection nebula extending symmetrically from the star to radial distances of 3″ (450 AU) along position angle 60°. The diameter and position angle of the reflection nebula agree well with values derived from the millimeter interferometer observations of the molecular gas disk. Single scattering models using circumstellar disk density distributions strongly suggest that the GM Aur reflection nebulosity arises at the illuminated upper surface of a flared, optically thick disk observed from $\approx 25°$ above its equator plane. Further details will appear in Stapelfeldt *et al.* (1997b).

3.6. SAO 76411A & HDE 283572

These two G-type weak line T Tauri stars possess no significant infrared, millimeter, or polarization excesses. The WFPC2 images show no evidence for circumstellar nebulosity. Our modeling work rules out optically thick disks for these sources unless they are viewed from high latitudes.

4. Results for Entirely Nebulous Objects

4.1. HL TAURI

It had long been assumed that this star was directly visible through an $A_V \approx 3$ mag (Beckwith *et al.* 1990). One of the first and most surprising YSO results obtained with WFPC2 was the complete absence of a visible star in HL Tauri; the "star" turned out to be the reflection nebula's compact core, only 1″ across. The actual star is located by VLA astrometry about 1″ SW of the centroid of the optical nebula, and is extincted by $A_V > 22$ mag. Most of the visible light is scattered within the cavity carved through the surrounding envelope by the blueshifted jet; the disk is only visible via the extinction edge it produces at the base of the reflection nebula. Additional details can be found in Stapelfeldt *et al.* (1995).

4.2. HH 30

HH 30 (Fig 2a.) is perhaps *the* prototype YSO accretion disk system. A flared, optically thick, circumstellar absorption disk, 450 AU in diameter, occults the central star and extends perpendicular to the highly collimated bipolar jets. By comparing the distribution of reflected light at the upper and lower surfaces of the disk to scattering models, important constraints to the disk density distribution can be derived. A detailed description of these results is presented in Burrows *et al.* (1996); a summary appears in Stapelfeldt *et al.* (1996).

HH 30 is the first YSO disk to have its vertical structure clearly resolved. Modeling of the isophotes indicates that the disk is in vertical pressure support, that it has a scale height H= 15 AU at r =100 AU, and that its scale height flares with radius more rapidly than expected for a steady-state accretion disk. Optical and mm determinations agree that the disk mass is surprisingly small, only about 10^{-3} $M_\odot$ for nominal dust properties. This result dictates that the disk should be largely optically thin to its own emergent thermal infrared radiation. The disk mass is not only much smaller than the circumstellar masses typically inferred for HH energy sources (Reipurth *et al.*, 1993), it is so small that steady accretion would consume the system in less than 10^5 years.

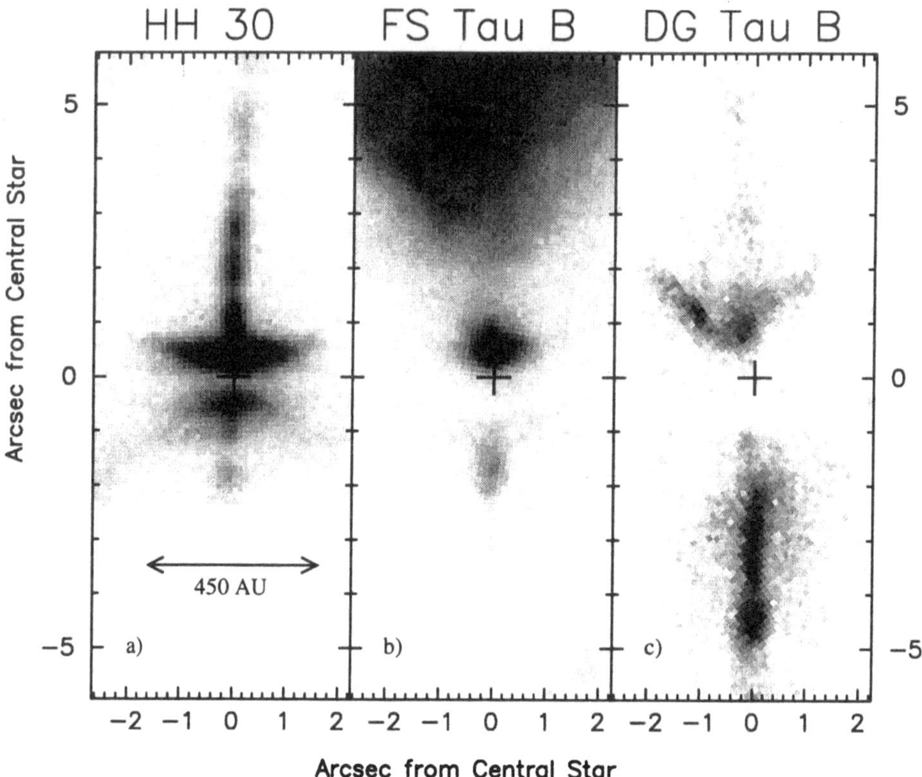

Figure 2. **Three HH Energy Sources Resolved as Compact Bipolar Nebulae.** These HST WF Camera R band images are shown in log stretch, and have been rotated such that the blueshifted jets are aligned vertically up each panel.

The jet of HH 30 is resolved into many small knots whose spacing along the flow is typically 100 AU. Combined with the measured proper motions of about 200 km s^{-1}, the implication is that new knots emerge from the source every 2-3 years. The jet is exquisitely narrow at its base, with a resolved FWHM of only 25 AU. This observation is difficult to reconcile with inertial collimation provided by the surrounding density distribution, and argues instead for intrinsic collimation very near the central star. The jet brightness fades with distance from the source in a manner consistent with cooling via free expansion. This interpretation is supported by the HST HH 30 emission line images of Ray *et al.* (1996), which show that the $H\alpha$/[S II] ratio smoothly declines with distance along the jet without peaks at the positions of bright knots. HH 30 is the only YSO jet so far where this behavior has been observed.

4.3. FS TAURI B (HARO 6-5B)

HST images of the FS Tau B source region (Fig. 2b) appear similar in several respects to those of HH 30. No stellar point source is present; a compact reflection nebula is seen where the star was presumed to be. At the base of the redshifted jet, a faint companion reflection nebulosity appears, offset from the first one along the jet axis. Both reflection nebulae extend perpendicular to the jet axis for a distance of about 350 AU. They are separated by a dark, slightly curving absorption lane which we identify as the actual circumstellar disk of FS Tau B. Significant structural differences also exist between this system and HH 30. The brightness ratio of the opposing reflection nebulae is larger in FS Tau B, indicating that the system is not so nearly edge-on as HH 30. The dark lane is wider in FS Tau B, which suggests a larger disk mass. Finally, there is an extended cone of reflected light along the blueshifted jet that has no counterpart in the HH 30 system.

The redshifted jet near the faint counter-nebula (lower half of Fig. 2b) shows 3-4 individual knots and has a FWHM of 50 AU. A significant new result is a pattern of transverse "wiggles" in the blueshifted jet at distances 20″ out from the source. This deviation from a cylindrical profile is responsible for the large apparent jet diameter found in lower resolution images of this region by Mundt, Ray, and Raga (1991). A full description of our images, including comparison of disk models and the morphology of the FS Tau B source region, will appear in Krist *et al.* (1997b).

4.4. DG TAURI B

This object lies 1′ WSW of DG Tau and drives a redshifted molecular outflow (Mitchell *et al.*, 1994). HST also resolves DG Tau B as a compact bipolar nebula, with no directly visible star. Offset astrometry places the VLA source (Rodriguez, Anglada, and Raga, 1995) within the central dark lane, and offset somewhat toward the reflection nebulosity at the base of the blueshifted jet. Here the central absorption lane is quite wide ($> 1''/140$ AU). Unlike HH 30 and FS Tau B, the scattered light nearest the source is more extended along the jet axis than perpendicular to it. This, and the lack of symmetry between the two reflection nebulae, suggests that optical light is scattered within cleared cavities in a circumstellar envelope. The disk presumably lies embedded within this envelope, and should be revealed via high resolution near-IR imaging. The bright redshifted jet has a FWHM of 50 AU that changes little along its length.

5. Discussion

Disk and jet structures close to the central object are revealed most clearly in the case of nearly edge-on systems such as HH 30. This orientation removes the glare of direct starlight through absorption in the intervening disk material, and allows the disk vertical structure to be seen in silhouette. It would be very useful for disk and jet studies to uncover other nearby YSOs viewed at similarly favorable orientations as HH 30. One way to accomplish this is suggested by the photometric properties of the three compact bipolar nebulae found so far. In Fig. 3, the K magnitude for nearby optically visible YSOs is plotted against the V-K color of each object. Photometry was taken from Kenyon and Hartmann (1995), Gauvin and Strom (1992), Hughes *et al.* (1994), Greene and Young (1992), Vrba, Rydgren, and Zak (1985), and Herbig and Bell (1988). Most of the objects fall in the upper left portion of the diagram, with a large scatter due to luminosity and spectral type variations as well as distance differences between the various star-forming regions. HH 30, FS Tau B, and DG Tau B fall in the lower right portion of this diagram, significantly offset from other nearby YSOs.

The distinctive photometric properties of the compact bipolar nebulae can be understood by considering the effects of increasing circumstellar extinction on the position of a source in this diagram. Moderate amounts of A_V (< 10 mag) will shift a source's position to the right. As the extinction is increased still further, at some point the dominant contribution to the optical light will transition from direct starlight to scattered light. When this point has been reached, the V magnitude of the system becomes much less sensitive to additional increases in A_V. Direct starlight will remain the primary contributor at 2 μm up through much larger values of A_V, however, and thus the K magnitude will continue to fade as the extinction increases. This scenario is supported by the photometry and morphology of the three compact bipolar nebulae, and can be summarized as the following hypothesis: *Systems with nearly edge-on disks are best distinguished from their local YSO population by their unusually faint K magnitudes.*

We have used this hypothesis to select additional HST targets which by this reasoning should be revealed as compact bipolar nebulae similar to HH 30. Th 28 in Lupus is an especially promising target, with bipolar jets that show small radial velocities (Krautter, 1986). Another is CoKu Tau/1, for which existing images already suggest that the reflected light is elongated perpendicular to the jet axis (Movsesyan and Magakyan, 1997). It will be interesting to see if these sources prove to be compact bipolar nebulae, and also if their reflected light shows the sharp edges and elegant symmetry of HH 30 or the more diffuse appearance of DG Tau B.

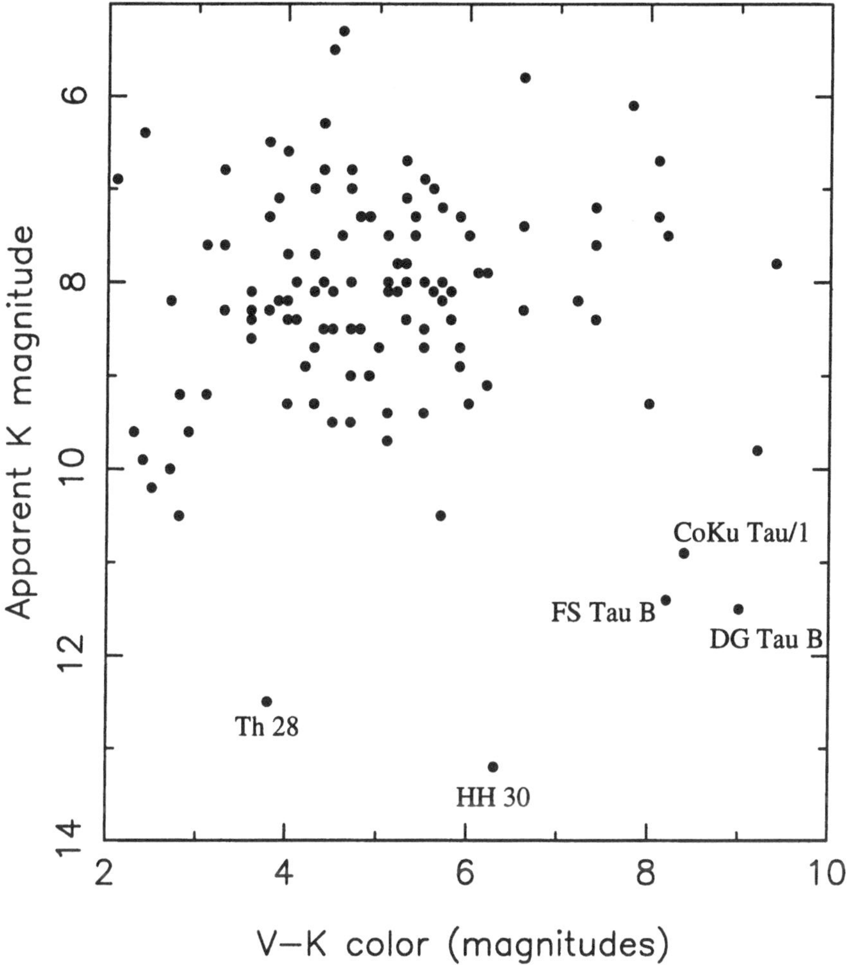

Figure 3. **Color-Magnitude Diagram of Selected Nearby YSOs.** Optically visible objects with 1.3 mm continuum emission in the Taurus, Chamaeleon, Lupus, and Ophiuchus clouds are included.

6. Future Prospects

Many more YSO imaging studies are now scheduled to occur with HST. These include follow-up observations of HH 30, HL Tau, and T Tau; the first views of the circumbinary disk in GG Tau and the expected absorption disk in Th 28; and a large snapshot survey of T Tauri stars which have millimeter continuum and/or polarization excesses. Near-infrared coronagraphic imaging with NICMOS should provide a new window into the earliest embedded phases of YSO evolution. The next few years promise to be a period of significant progress in our understanding of the structure and evolution of protoplanetary environments. Updates on our group's work will be available on the WWW at http://scivax.stsci.edu/~krist/yso.html.

Acknowledgements: This work was performed under contract to NASA as part of the WFPC2 GTO Science Program. KRS also acknowledges support from the NASA Origins of Solar Systems research program.

References

Beckwith, S.V.W., Sargent, A.I., Chini, R.S., and Gusten R. 1990, A.J. 99 924
Burrows, C.J. and the WFPC2 Science Team 1996, Ap.J. 473, 437
Dutrey, A. 1997 in preparation
Gauvin, L.S. and Strom, K.M. 1992 Ap.J. 385 217
Greene, T.P., and Young, E.T. 1992 Ap.J. 395 516
Herbig, G.H., and Bell, K.R. 1988 Lick.Obs.Bull. 1111 1
Hughes, J., Hartigan, P., Krautter, J., and Kelemen, J. 1994 A.J. 108 1071
Kenyon, S.J., and Hartmann, L. 1995 Ap.J.Suppl. 101 117
Koerner, D.W., Sargent, A.I., and Beckwith, S.V.W. 1993 Icarus 106 2
Krautter, J. 1986, A.&A. 161 195
Krist, J.E. and the WFPC2 Science Team 1997a, Ap.J., in press
Krist, J.E. and the WFPC2 Science Team 1997b, in preparation
Krist, J.E. 1995 in *Calibrating Hubble Space Telescope: Post Servicing Mission*, STScI Baltimore, p. 311
Lavalley, C., Dougados, C., and Cabrit, S. 1997 in *Low Mass Star Formation from Infall to Outflow*, Poster Proceedings of IAU Symposium 182, Grenoble, p. 147
Mitchell, G.F., Hasegawa, T.I., Dent, W.R.F, and Matthews, H.E. 1994 Ap.J. 436 L177
Movsesyan, T., and Magakyan, T. 1997 in *Low Mass Star Formation from Infall to Outflow*, Poster Proceedings of IAU Symposium 182, Grenoble, (online version only)
Mundt, R., Ray, T.P., and Raga, A.C. 1991, A.&A. 252 740
Ray, T.P., Mundt, R., Dyson, J.E., Falle, S., and Raga, A. 1996 Ap.J. 468 L103
Reipurth, B., Chini, R., Krugel, E., Kresya, E., and Seivers, A. 1993 A.&A. 232 37
Rodriguez, L.F., Anglada, G., and Raga, A. 1995 Ap.J. 454 L149
Solf, J., and Böhm, K.-H. 1993, Ap.J. 410 L31
Stapelfeldt, K.R. and the WFPC2 Science Team 1997a, submitted to Ap.J.
Stapelfeldt, K.R. and the WFPC2 Science Team 1997b, in preparation
Stapelfeldt, K.R. and the WFPC2 Science Team 1996, in *Accretion Phenomena and Related Outflows*, Proceedings of IAU Colloquium 163, in press
Stapelfeldt, K.R. and the WFPC2 Science Team 1995, Ap.J. 449 888
Vrba, F.J., Rydgren, A.E., and Zak, D.S. 1985 A.J. 90 2074

DISKS AND OUTFLOWS AS SEEN FROM THE IRAM INTERFEROMETER

S. GUILLOTEAU, A. DUTREY AND F. GUETH
IRAM, 300 rue de la piscine
F-38406 Saint Martin d'Hères Cedex, FRANCE

Abstract. The IRAM interferometer has been used to map 3 high velocity outflows from Class 0 sources in SiO and CO rotational lines. The images show that such outflows are driven by highly collimated, dense, but optically invisible jets, through large travelling bow shocks. Precession is important in explaining some morphological and kinematic aspects. A significant part of the SiO emission is directly associated with the jet. Images of the molecular core surrounding L 1157 in CO isotopic lines show direct evidence for infall in a large, flattened disk.

For more evolved objects, "surveys" of mm emission from T Tauri stars provide evidence for circumstellar dust disks with relatively flat surface density distribution and typical radii 100 to 200 AU. Much larger gas disks (radii 800 to 1000 AU) have been detected and imaged in a few objects (2 singles and 2 binaries) through CO isotopomers, but in most sources ^{13}CO is difficult to detect. The spectral line images provide unambiguous evidence for Keplerian rotation. In two cases (GG Tau and DM Tau), direct measurements of the gas mass (independently of the dust) has been made thanks to the detection of 7 molecular species. The molecules are found to be depleted by factors between 5 (for CO) to 100 (for H_2CO). These results suggest that circumstellar disks consist in a small, dense, inner disk of outer radius $\simeq$ 150AU (traced by the dust emission), surrounded by a much larger, less massive, outer disk (traced by the gas emission).

1. Introduction

Although molecular outflows have been known since 1980, they still represent an observational challenge. Many low resolution studies have been performed, but very few high angular resolution images exist because of the

B. Reipurth and C. Bertout (eds.), Herbig–Haro Flows and the Birth of Low Mass Stars, 365–380.

combination of 3 properties: long linear sizes, high collimation (implying narrow structures), and relatively low brightness for most molecular lines. With its large antennas, good site and low noise receivers, the IRAM interferometer has become the instrument of choice to study outflows since the development of new mosaicing techniques involving joint deconvolution of overlapping fields (Gueth *et al.*, 1995). Note that mosaicing is specially suited for elongated objects like outflows, while for rounder objects it would be nearly as efficient to simply use an array with smaller antennas.

When dealing with circumstellar disks around T Tauri stars, the sensitivity problem is even more critical, since the expected line widths are much smaller than for outflows, and the sizes very small. The IRAM interferometer provide unsurpassed performances at 1.3 mm, where the primary beam is well matched to the total extent of the disks. The winter conditions, with precipitable water vapor content frequently below 2 mm and average temperatures $-10°$C, provide excellent phase stability, with effective seeing below $0.3''$. With the baseline extension (Sep. 1995) up to 408 m length, observations with $0.5''$ resolution have become possible.

This paper presents an overview of research going on at IRAM on high velocity outflows from Class 0 objects and their exciting sources, and on disks around T Tauri stars.

2. Outflows from Class 0 objects

2.1. L 1448

The first highly collimated outflow studied with the IRAM interferometer is L 1448. The inner part of the flow was observed in the SiO(2-1) transition, with $2''$ resolution (Guilloteau *et al.*, 1992). These observations revealed that the high degree of collimation and the acceleration of the gas was essentially achieved within 1000 AU of the exciting star. However, SiO, because of its short lifetime and high critical density, is a very selective tracer and does not represent the bulk of the gas. Subsequent observations of the CO(1-0) line (Bachiller *et al.*, 1995) clearly revealed the outflow cavity walls. Surprisingly, the red and blue lobes present different opening angles close to the star. The picture is complicated by the possibility of a direct collision of the blue lobe with the red lobe of the outflow originating near L1448-IRS3. This collision provides a possible explanation for the strong shocks revealed in H_2 and for the general bending of the blue lobe.

To further study the molecular "bullets", which remained only marginally detected in the CO image, the complete red lobe has been mapped in SiO(2-1) using a 5-field mosaic (Dutrey *et al.*, 1997a). The SiO emission is clearly confined in a jet-like structure, well aligned with the CO cavity axis, and terminating in a strong bow shock. More detailed analysis of the

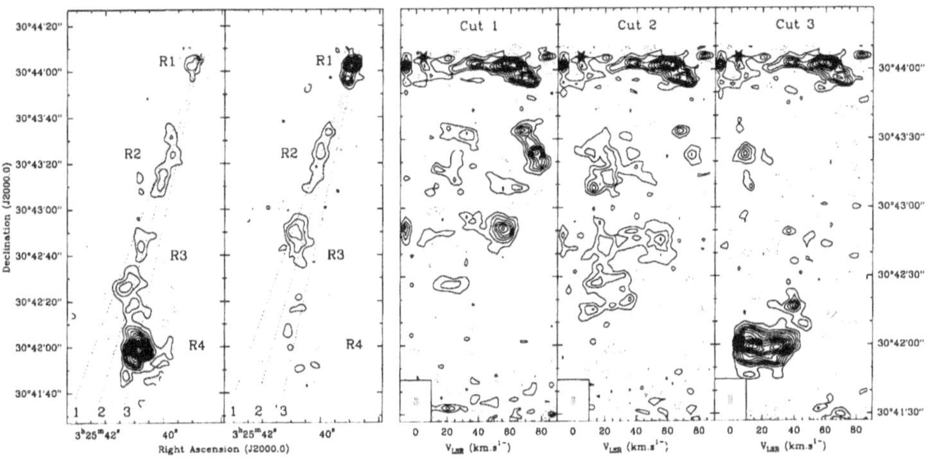

Figure 1. Left: Maps of the SiO $v = 0$ J=2-1 emission associated with the red lobe of L1448-mm. The intensity is integrated from -7 to 34 km.s^{-1} (left panel) and 35 to 80 km.s^{-1} (center panel). The beam is 4.2″ × 3.4″ at P.A. 62°. Contour step is 0.5 Jy/beam.km.s^{-1} or 0.56 K.km.s^{-1} in both maps. Right: position-velocity diagram along the lines 1-2-3. First contour and step are 20 mJy or 0.23 K. From Dutrey *et al.* 1997a

spatio-kinematic information reveals that most of the SiO emission actually originates in a few large, clumpy, bow shocks. The projected velocities of the shocks decrease with distance from the exciting source. The jet appears to wander (*"precess"*) by about 10°, but may also have been deflected against a dense molecular clump (see Fig.1).

The L 1448 example shows that jet-driven flow models may be successful in explaining molecular outflows. However, even in such a young flow, the influence of the surrounding medium has made the picture so complex that it is difficult to separate intrinsic processes (e.g. precession) from external influences (e.g. deflection). Searching for a less confused case is thus necessary.

2.2. L 1157

Compared to L 1448, L 1157 offers the double advantage of being isolated and at high declination.

The CO emission from the blue-shifted lobe of L 1157 was reconstructed from a 10-field mosaic; inclusion of short spacing information from the 30-m telescope was essential to recover the complete picture (Gueth *et al.*, 1996). The CO emission (Fig. 2) clearly delineates two well-defined cavities, with projected axes separated by about 6°. Moreover, the kinematics can be simply explained by a precessing jet model in which the "terminal" cavity C1 is nearly in the plane of the sky, while the brighter cavity C2 is about

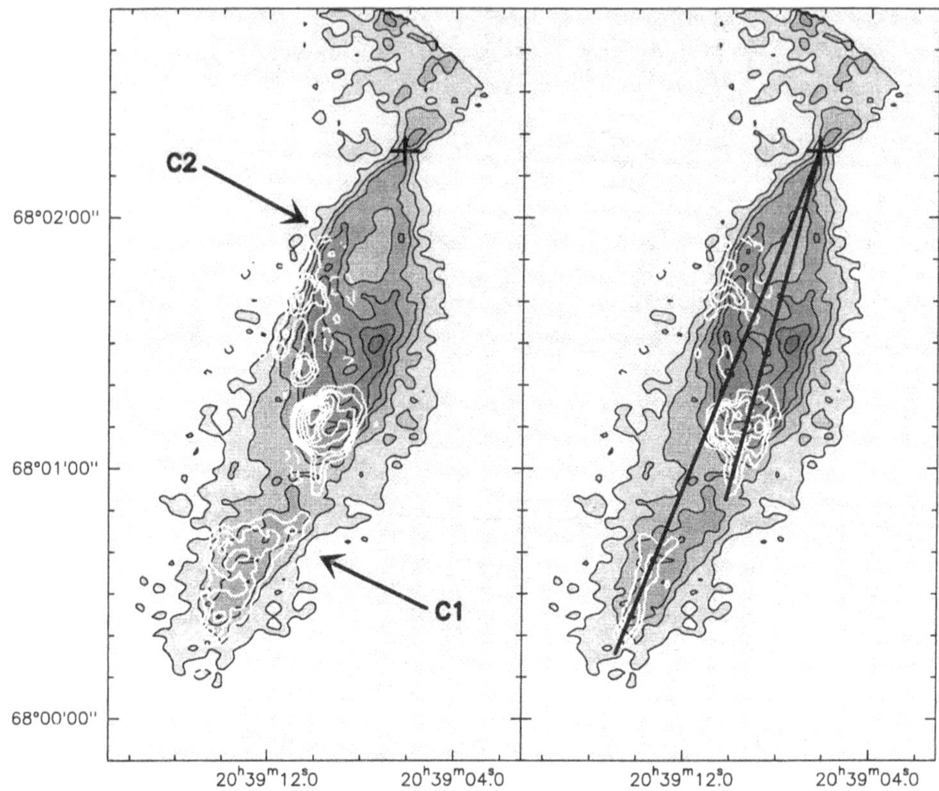

Figure 2. Left: integrated CO (greyscale) and SiO emissions (white contours) in the blue-shifted lobe of L 1157. Right: SiO emission at 2.8 km.s^{-1} superimposed on the integrated CO emission. Angular resolution is about 3″ (Gueth *et al.* 1997a)

10° from the plane of the sky towards us. The model also explains the apparent asymmetries in the brightness distribution.

The SiO image fully confirms this model, since SiO is visible along the cavity walls which are directly exposed to the precessing jet: western edge of the C1 cavity, and eastern edge of the C2 cavity (Gueth *et al.*, 1997a), see Figure 2. A very strong "shock" is visible at the apex of the C2 cavity, at a position where, according to the precession model derived from the CO observations, the driving jet crosses the wall of the C1 cavity. The CO emission of the C2 cavity is clearly in the wake of this strong SiO shock. These positional differences between SiO and CO emissions is a strong argument in favor of jet-driven flow models.

Linear features whose axes intersect exactly on the continuum emission marking the exciting source of L 1157 are well identified, but contrary to L 1448, the SiO emission does not clearly delineate a jet. Moreover, although the strong SiO emission at the C2 cavity apex has a clear bow-like appearance, its kinematics does not agree with simple bow shock models. The very well-defined linear structure just ahead of the C2 apex is intriguing: since SiO is presumably formed in the shock, it is *a priori* surprising to find emission ahead of the shock front. This structure may be a magnetic precursor or a nose cone preceeding the main shock front. Alternatively, it could be that SiO is already present in the jet, perhaps created by oblique shocks in the jet boundary layer. Note that a similar, though less clear, linear "precursor" is visible ahead of the terminal bow shock of L 1448.

2.3. HH 211

Despite their differences, the examples presented above strongly suggest that young molecular outflows may be driven by supersonic jets. However, the jets remain elusive, being revealed only by secondary indicators like SiO.

HH 211 is an IR discovered outflow (McCaughrean *et al.*, 1994), probably nearly in the plane of the sky as suggested by the small extinction difference between both lobes. It is likely to be extremely young, and may offer a unique opportunity to study the earliest phases of the outflow processes, perhaps before external influences blurr the picture.

We have performed a complete 9-field CO(2-1) mosaic of this jet, with 1.5″ angular resolution (Gueth and Guilloteau, 1997). Compared to L 1448 and L 1157, HH 211 is full of surprises (see Fig.3):

- Low velocity CO delineates the cavity walls (as expected) *but* coincides with the H_2 emission (in L 1448 and L 1157 H_2 emission is more closely related to the SiO features, although detailed comparisons are hampered by astrometric problems in the H_2 images).
- High velocity CO traces a wonderful, narrow, jet, which terminates exactly at the position of the H_2 emission (Fig.3)
- The velocity law is almost perfectly linear with distance from the star ("Hubble" law).
- Accounting for the low inclination, the deprojected velocity reaches at least 200 km.s^{-1}. Hence, the jet is very young (500 years).
- A *direct* estimate of the *mean* jet density is possible from the CO column density and apparent sizes: the CO jet is dense ($n(H_2) > 10^5 cm^{-3}$).
- The low velocity cavity shape is well reproduced by travelling bow shock models.
- There is a well resolved, flattened continuum source at the jet basis, and it is elongated nearly perpendicular to the jet.

The CO observations also suggest a preceding outflow episode whose

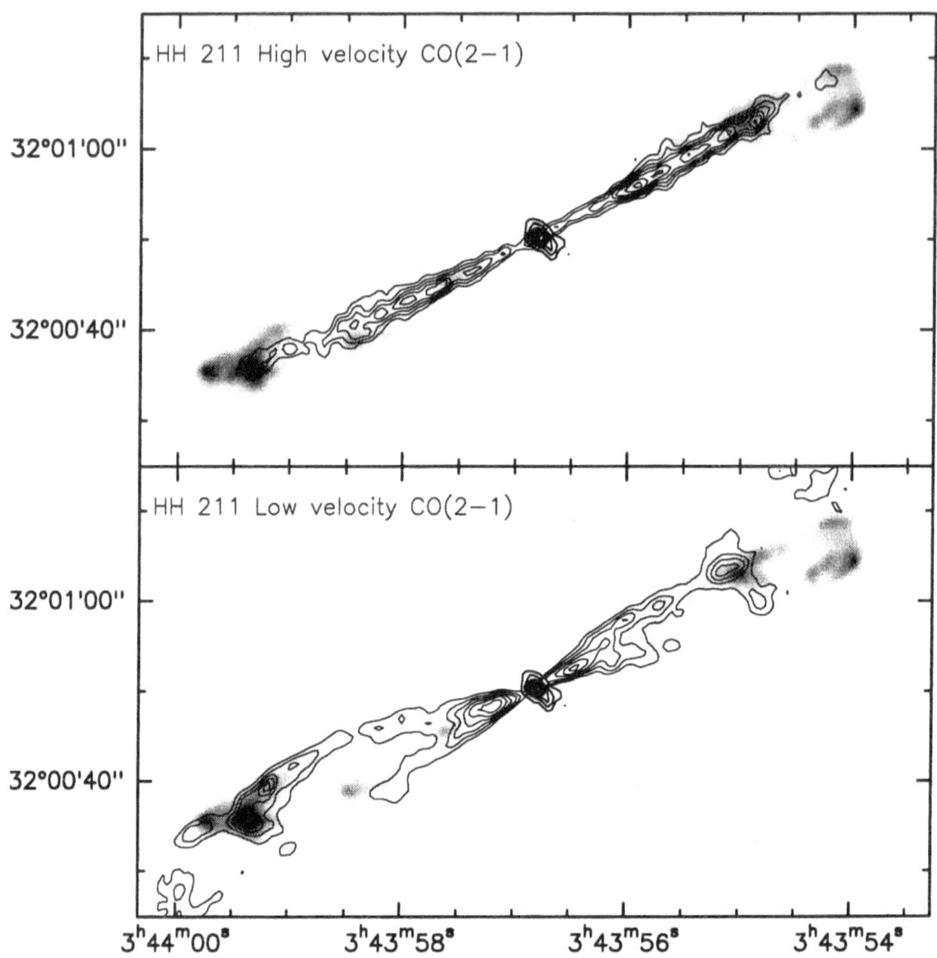

Figure 3. Contour map of the high and low velocity $^{12}CO(2\text{-}1)$ emission in HH 211, superimposed on the H_2 2.2 μm image in greyscale from McCaughrean et al. (1994). The thick contours indicate the 1.3 mm continuum emission. Angular resolution is $1.4 \times 1.1''$ at PA 83°.

axis is tilted by a few degrees from the current jet axis. HH 211 is also a strong SiO source. Interferometric observations in SiO(2-1) (Gueth *et al.*, 1997b), and SiO(1-0) (Chandler and Richer, 1997) show that SiO is *only present in the jet.* This is a striking difference with L 1448 and L 1157.

As such, HH 211 is an ideal case for very young jet-driven flows. The existence of sufficiently strong CO emission from the jet will allow much better determination of its physical parameters (density, temperature, momentum) than would be possible using any other molecular tracer, since, contrary for example to SiO or NH_3, the CO abundance is not affected by variations in the chemistry.

2.4. CONCLUSIONS

From the above examples, a few general properties of outflows from Class 0 sources can be derived. First, jets are dense, with mean densities of at least $10^5 cm^{-3}$. Second, the natural opening angles for the cavities of jet driven flows is around 30-40°. Third, precession (or jet wandering) may account for about 10° in the opening angle of the outflow cavities, but not more. Fourth, shock structure is complex, with small linear precursors ahead of the most striking bow shocks. However, the variety of characters and the importance of the surrounding medium should be stressed. Only 3 objects have been studied so far, but all present very different aspects.

The above conclusions should not be generalized to all outflows. In the case of L1551-IRS5, the outflow cavity is apparently widely opened. Whether this results from an age difference or from the different characteristics of the star formation processes in the Taurus region is entirely open. More systematic studies (e.g. small surveys) would be required to properly address this problem.

3. Infalling envelopes ?

As mentioned above, a resolved continuum source lies at the origin of the outflow in HH 211. L 1448 also presents a compact continuum source at the base of the outflow (Bachiller *et al.*, 1995). Among the sources discussed above, only L 1157 has been studied in details so far, and the situation is here somewhat more complex. Although a compact core exists, and is elongated perpendicular to the flow, a large fraction of the 2.7 mm continuum flux is spread over 10-15″ (5000 AU) (Gueth *et al.*, 1997c). This extended envelope actually contains most of the mass. In addition, the images show that the continuum emission is brighter along the edges of the outflow cavity (Fig.4), suggesting that the outflow has been energetic enough to affect a significant fraction of the envelope mass. Spectral line observations also reveal that the continuum (dust) core and $C^{18}O$(1-0) data are nearly coextensive, while ^{13}CO(1-0) is clearly visible at the edge of the outflow.

The spectra reveal a deep, *redshifted*, self-absorption feature in ^{13}CO (1-0) towards the dust core, strongly suggesting *infall motions*. Note that in general, interferometric observations can easily produce "false" self-absorption features, because an interferometer resolves out the extended structures. To avoid such a bias, it is essential to include short spacing information (see Fig.5). Because ^{13}CO has a relatively low dipole moment and is easy to excite, the deep self-absorption feature observed in L 1157 requires special conditions to be produced, namely a disk-like geometry with a large

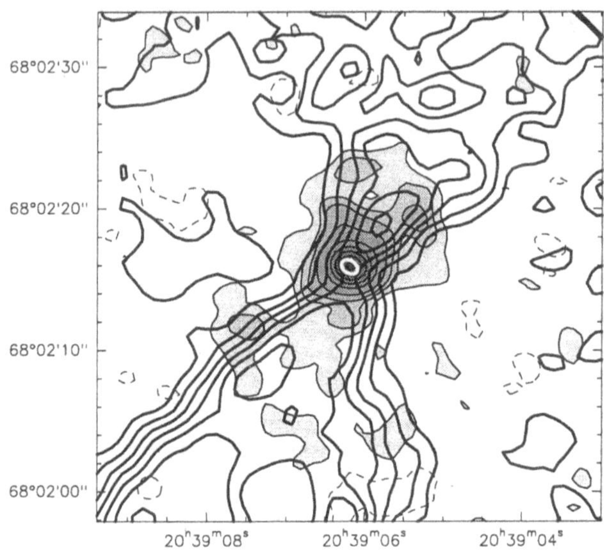

Figure 4. Overlay of the λ 2.7 mm continuum emission (greyscale) and the integrated $^{12}CO(1\text{-}0)$ emission (thick contours). The extended component of the continuum emission is clearly associated with the edges of the molecular outflow. The thick white ellipse indicates the best fit gaussian model to the compact core of the continuum emission (Gueth *et al.*, 1997c).

disk (radius 8000 AU) seen nearly edge-on. While such a situation is *a priori* unlikely, it is in agreement with the inclination derived for the outflow (Gueth *et al.*, 1997c).

A common point to all 3 sources is that the diameters of the compact cores are about 1000 AU, although except for HH 211 this is close to the resolution limit. Confirmation of these values by higher angular resolution observations would be important.

4. Disks around T Tauri stars

4.1. DUST DISKS

The diameters found for the dust cores in Class 0 objects are much larger than those found for T Tauri stars. In a survey of 2.7 mm emission in 33 T Tauri star systems, Dutrey *et al.* (1996) found that the emission from the stronger sources was marginally resolved, with angular sizes about $1''$. Interpreted in terms of disk models with power law surface density and temperature distributions, this suggests that circumstellar disks around T Tauri stars have a relatively flat density distribution, with outer radii around 150 AU. A similar conclusion was reached for HL Tau from sub-

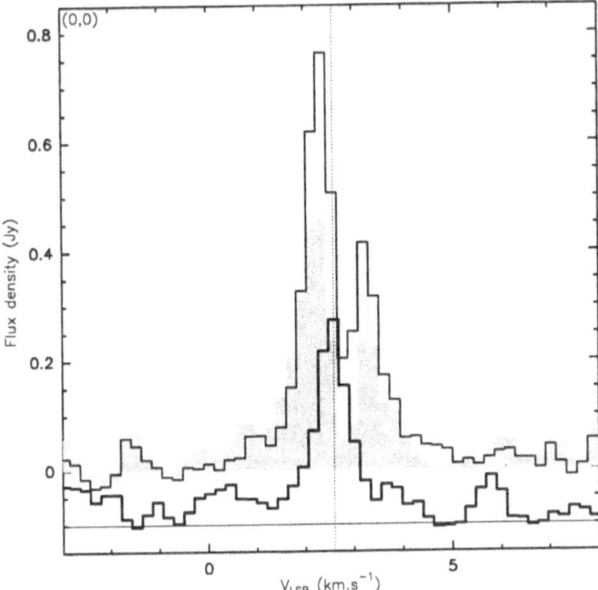

Figure 5. ^{13}CO(1-0) (thin line), C^{18}O(1-0) (thick line) towards the mm source of L 1157, showing the pronounced red-shifted self-absorption. The grey-scale histogram is the ^{13}CO(1-0) interferometric spectrum before the addition of short spacing information (Gueth *et al.*, 1997c).

arcsecond resolution images from the BIMA array (Mundy *et al.*, 1996). Comparison of the measured 2.7 mm flux densities with bolometer measurements at 1.3 mm and 0.8 mm indicate a mean dust emissivity spectral index $\beta \simeq 1.0$ (see Dutrey *et al.* 1996 and references therein).

However, interpretations based on single frequency size measurements are ambiguous, since the same angular size can be produced by various combinations of radial temperature and opacity distributions in the circumstellar disks. To remove this ambiguity, we have started a "mini-survey" of 12 T Tauri stars using the newly available dual-frequency capabilities and extended baselines (400 m) of the IRAM interferometer (Dutrey *et al.*, 1997b). These new data have sub-arcsecond angular resolution, and also allow to measure simultaneously the angular sizes at 2.7 and 1.3 mm. The preliminary results confirm the initial measurements. In particular, the sizes are identical at both frequencies, a property which is most easily explained if the brightness distributions are identical. Since the dust opacity at 1.3 mm is 2-3 times higher than at 2.7 mm, this rules out steep surface density distributions, and also implies moderate dust opacities.

Another direct result of these angular size measurements is the apparent elongation and position angles of the dust disks. From the aspect ratio, a lower limit to the disk inclination (taking into account the influence of the seeing) can be derived.

4.2. GAS DISKS KINEMATICS

Large gas disks have been detected and imaged in 4 systems: GG Tau in ^{13}CO(1-0) (Dutrey *et al.*, 1994), DM Tau in CO(1-0) (Guilloteau *et al.*, 1997), GM Aur in ^{12}CO(2-1) (Dutrey *et al.*, 1997c) and UY Aur in ^{13}CO(2-1) and (1-0) (Duvert *et al.*, 1997). All 4 sources are either isolated from molecular clouds (GG Tau, DM Tau) or in regions of low extinction near the cloud edges (GM Aur, UY Aur).

Inclination and position angles of the dust disk can be compared to the morphology observed using a spectral line tracer. This has been done for ^{12}CO(2-1) in GM Aur (Dutrey *et al.*, 1997c). A position angle of $55° \pm 5°$ and an inclination of $58° \pm 5°$ (0 meaning face on) are measured from the continuum emission. The kinematic pattern of the ^{12}CO(2-1) emission clearly indicates Keplerian rotation around the dust disk axis, and the derived position angle and inclination agree remarkably well (see Fig. 6). The quality of the ^{12}CO image is even sufficient to recover the radial temperature profile across the gaseous disk, and also the full 3-D orientation. Since the temperature decreases radially outwards, self-absorption effects are expected in an inclined flared (or geometrically thick) disk. This has been observed in GM Aur, and allows to indicate that the south-eastern part of the disk is closer to the observer than the north-western part.

Position angles, inclination and orientation measured with these 1.3mm data are indeed in excellent agreement with those derived independently from the HST optical image (Stapelfeldt *et al.*, 1997).

Keplerian rotation has also been found in GG Tau (Dutrey *et al.*, 1994), DM Tau (Guilloteau *et al.*, 1997), and UY Aur (Duvert *et al.*, 1997). In GM Aur and GG Tau, the major axis defined by the spectral line emission coincides perfectly with that defined from the dust emission. In UY Aur, however, the continuum emission is too weak to allow to define a major axis independently from the CO pattern. Determination of Keplerian rotation requires the kinematics to be sampled by a sufficiently large number of channels (10 to 20), i.e. very high spectral resolution, and thus high sensitivity. Fig.7 illustrates the accuracy reached for UY Aur. Furthermore, it should be pointed out that it is essential that the images have a circular beam (or that the modelling incorporates properly the effective synthesized beam). Otherwise, the convolution by an elliptical beam can produce an apparent skewness in the velocity distribution which can easily be mis-interpreted as a combination of infall and rotation, specially when the kinematic pattern is insufficiently resolved.

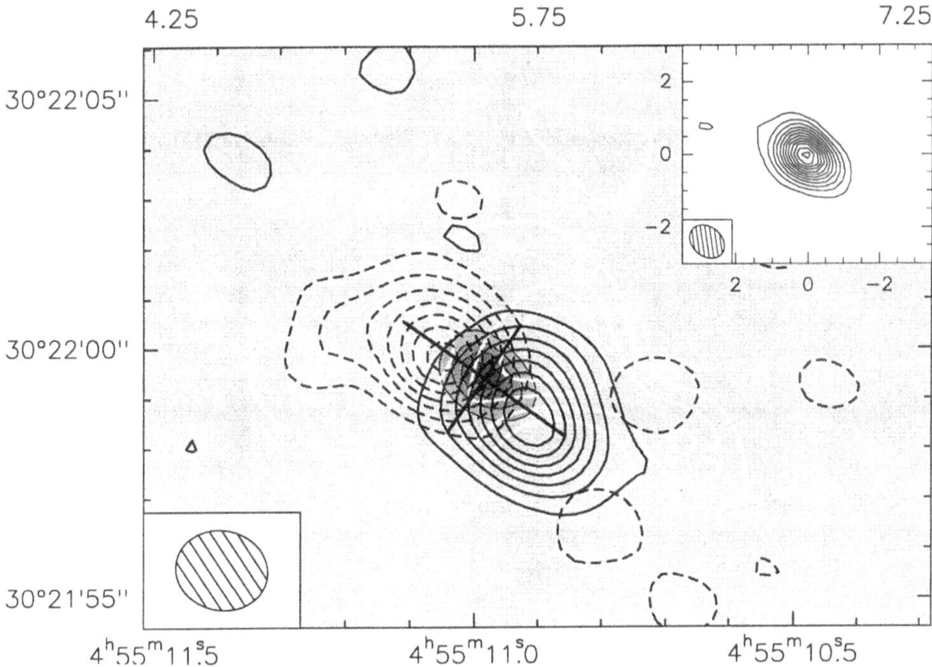

Figure 6. Contours of ^{12}CO(2-1) emission at 4.5, 5.75 (systemic), and 7.0 km/s towards GM Aur, superposed on the continuum image at 1.3 mm in grey scale. The CO contour steps are 200 mJy/beam. The cross indicates the continuum peak position, orientation and aspect ratio. Insert: higher resolution continuum image (synthesized beam $0.6'' \times 0.7''$ at PA 56°); contour step is 7.5 mJy/beam (from Dutrey *et al.* 1997c).

Both DM Tau and GM Aur are relatively old T Tauri stars ($\simeq 10^7$ years). The kinematics of gas around younger objects like HL Tau does not seem dominated by Keplerian motions: infall has been suggested for HL Tau (Hayashi *et al.*, 1993), although outflow motions make the picture ambiguous (Cabrit *et al.*, 1996). This may suggest that Keplerian rotation is established relatively late in the evolution of circumstellar disks. On the other hand, GG Tau and UY Aur are very young ($< 5\,10^5$ yr), but both are binary systems. Since binaries are more likely to form when the original specific angular momentum is large, it is perhaps not surprising to find Keplerian rotation established earlier in binary systems than in single stars. The evolutionary picture is also complicated by possible confusion with the surrounding medium (clouds, outflows). In this respect, it may be symptomatic that rotation has been unambiguously detected only in objects located in low extinction regions.

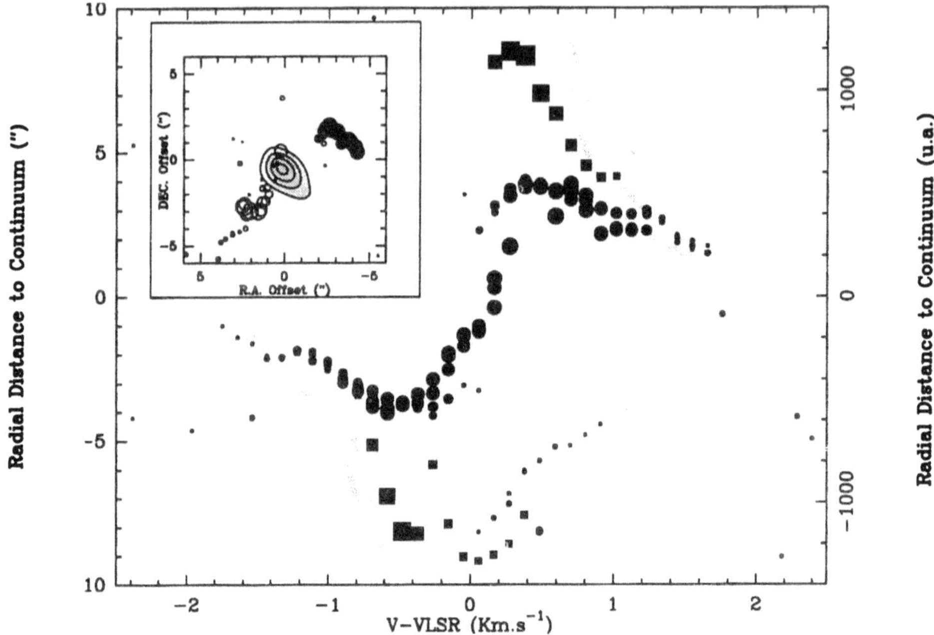

Figure 7. Velocity structure of the CO emission around the binary star UY Aur (separation 0.88″). The radial distance of the centroid of emission is plotted against velocity. Squares are for ^{13}CO(1-0), circles for ^{13}CO(2-1). The marker sizes are proportional to the emission strength. The grey lines are Keplerian rotation curves for a central mass 0.7 $M_\odot$, edge-on (dashed line) or inclined by 60° (solid line). The inset shows the loci of the emission centroid with respect to the position of the 1.3 mm continuum emission (filled contours). Filled circles are for redshifted gas, empty circles for blueshifted gas (from Duvert *et al.* 1997).

4.3. BINARIES

More than half of all stars are members of binary systems. The influence of the binarity on the circumstellar environment is important. In the 2.7 mm survey of 33 systems, Dutrey *et al.* (1996) confirmed that single stars have on average stronger emission than binaries. GG Tau, the second strongest mm source among TTauri stars in Taurus, is the most flagrant counter-example to this general rule, but provides excellent clues on the origin of the difference. The mm emission from GG Tau was first resolved by Simon and Guilloteau (1992). Subsequent detailed mapping with 2″ resolution indicated that the dust around GG Tau is actually forming a ring of inner radius 180 AU, inclined about 40° from face on (Dutrey *et al.*, 1994). The central part of this ring has presumably been cleared up by tidal interaction with the binary. Using the model from Artimowicz and Lubow (1994), Dutrey *et al.* (1994) predicted that the orbit of GG Tau was eccentric.

These results have since been completely confirmed by IR adaptive optics images (Roddier *et al.*, 1996).

If the typical radius of the circumstellar dust disk around single T Tauri stars is about 150 AU as derived from the dual frequency survey (Dutrey *et al.*, 1997b), the above mechanism provides a natural explanation for the weakness of the continuum emission in binaries of separations in the range 30 to 300 AU. Closer binaries can still maintain a circumbinary ring with radius larger than about 100 AU, while wider binaries can maintain individual circumstellar disks of radius up to 100 AU. This situation is nicely illustrated by the UZ Tau hierarchical quadruple system, where a spectroscopic binary UZ Tau-e (Matthieu *et al.*, 1996) is separated by 500 AU from the close binary UZ Tau-w (separation 50 AU). UZ Tau-e is surrounded by a disk typical of singles, while UZ Tau-w has only weak continuum emission (Dutrey *et al.*, 1996; Jensen *et al.*, 1996). This also suggests that *gaseous* disks, which extend out to 800 AU radius in single stars like DM Tau, may exist around binaries regardless of the continuum strength. This simple idea actually led to the discovery of a large disk in ^{13}CO(2-1) and (1-0) around the binary UY Aur (separation 0.9″), see Fig.7 (Duvert *et al.*, 1997).

4.4. DISK MASSES FROM DUST

The mm domain is the best frequency range to determine the total dust content of molecular disks, since the dust emission is optically thin, yet strong enough to be detectable over a useful range of masses. This was done by bolometer surveys near 1.3 mm for the Taurus region (Beckwith *et al.*, 1990), and the interferometer data confirm that the optically thin assumption is valid for most of the disks.

For more distant star forming regions, the angular distances between stars become comparable to or smaller than the typical beams of bolometer-equipped single-dish telescopes. Moreover, confusion with dust from the surrounding molecular clouds also become critical. Evidence for circumstellar disks has been demonstrated by the images of the proplyds in the Orion nebula (e.g. McCaughrean and O'Dell, 1996), but these observations provide little constraints on the average disk masses, since dust is easily opaque in the optical domain. Studies with mm interferometers are thus essential to properly measure the masses of circumstellar dust disks. However, because the flux density drops as the square of the distance, the sensitivity was a major problem, specially at 2.7 mm where most instruments operate (Mundy *et al.*, 1995). Using the new 1.3 mm capabilities of the IRAM interferometer, Lada *et al.* (1996) have been able to detect 3 out of $\simeq 15$ proplyds or cm VLA sources in a field of the Orion Trapezium cluster. The estimated disk masses of the detected disks range from 0.007 to 0.016 $M_\odot$.

4.5. DISK MASSES FROM GAS

For normal abundances and temperatures relevant to circumstellar disks (10-40 K at distances larger than 100 AU from the star), the ^{13}CO (1-0) or (2-1) lines are much more optically thick than dust: the opacity ratio is of order 1000-2000 (Dutrey *et al.*, 1996). Hence, it is surprising that only 4 systems out of 33 were detected in ^{13}CO(1-0) in the 2.7 mm survey (Dutrey *et al.*, 1996). Several possibilities can be invoked to resolve the discrepancy: 1) the gas disks are small, 2) confusion with the ambient cloud dominates, 3) circumstellar disks are dominated by a compact central core of radius 100-200 AU, 4) molecules are strongly depleted, and 5) dust absorption coefficients are incorrect.

Proposition 1) is contradicted by the existence of large gas disks around GG Tau, GM Aur, DM Tau and UY Aur. Proposition 2) could be valid, since all previous sources are free of confusion. Size measurements do suggest that dust is indeed confined within 200 AU from the star (for single stars), supporting proposition 3). Such a compact core would be optically thick in spectral lines but would escape detection because of beam dilution. A similar situation is advocated for the circumbinary ring of GG Tau (Dutrey *et al.*, 1994), where modeling suggest that 90% of the mass is in a relatively narrow (width $<$ 200AU) ring. Using a standard dust absorption coefficient, molecular depletion was also found necessary to explain the weak isotopic CO lines, indicating that propositions 4) and 5) could be valid.

If neither the dust absorption coefficient, nor the molecular content of the gas disks are known, estimating the disk masses becomes a challenge. To address this key question, Dutrey, Guilloteau and Guélin (1997, DGG) have used a completely new approach. Using the 30-m telescope, they have searched for molecular lines with varying degrees of excitation in the DM Tau and GG Tau disks, and detected for the first time CN, HCN, HNC, CS, C_2H, H_2CO and HCO^+, in addition to the ^{12}CO, ^{13}CO and $C^{18}O$ isotopes. Using the Keplerian disk model substantiated by the interferometer maps of CO or its isotopes, the higher excitation lines like CN(2-1) and HCO^+(3-2) provide minimal values for the H_2 densities, while HCN(3-2) provides upper limits. Comparing this *direct* measurement of the mean H_2 density with the measured molecular column densities, DGG were able to 1) derive *directly* the gas disk mass, and 2) evaluate the average molecular depletions. In DM Tau, CO is depleted by a factor 5 (refered to standard Taurus molecular cloud abundances), the other molecules being even more depleted (up to a factor 100 for H_2CO). The mass of the DM Tau gas disk is about 4 10^{-3} $M_{\odot}$, still several times smaller than the mass derived from dust emission. However, if the dust is confined within the inner 100 AU, molecules from this compact core would still remain undetected, despite the long integration times.

5. Conclusions

Most of the work presented above was based on 3 mm images with relatively low (2-3″) angular resolution. With the recent opening of the 1.3 mm "window" at the IRAM interferometer, angular resolutions of 1.5″ are routinely reached, and 0.5″ is possible. The spectacular images of the HH 211 outflow and of the circumstellar disk of GM Aur clearly illustrate the potential gains resulting from this improvement.

The increase in imaging speed obtained from the addition of antenna 5 in 1996 also allows more objects and topics to be studied. The results presented above offer some guidelines for future projects. Without being exhaustive, let us cite a few obvious targets: proper measurements of the angular sizes of Class 0 cores, studies of more evolved outflows (Class I, II), importance of the binarity for protostellar objects, etc....

The sub-arcsecond imaging capabilities should also enable detailed comparisons with optical and near-IR images. However, all the expected benefits of such comparisons will only be realized if the absolute astrometric accuracy of optical images matches the angular resolution.

Acknowledgements: The work presented here would not have been possible without the constant effort of the IRAM technical staff in supporting the observations. We also thank the many colleagues who participate in this work, and in particular R.Bachiller, M.Simon and G.Duvert for many fruitful discussions.

References

Artimowicz P., Lubow S.H., 1994, ApJ 421, 651
Bachiller R., Guilloteau S., Dutrey A., Planesas P., Martin-Pintado R., 1995, A&A 279 857
Beckwith S.V.W., Sargent A.I., Chini R.S., Gusten R., 1990, AJ 99, 924
Cabrit S., Guilloteau S., André P., Bertout C., Montmerle T., Schuster K., 1996, A&A 305, 527
Chandler C., Richer J., 1997, in *Low Mass Star Formation - from Infall to Outflow*, poster proceedings of IAU Symp. No. 182, eds. F. Malbet & A. Castets, Observ. de Grenoble, p.76
Dutrey A., Guilloteau S., Simon M., 1994, A&A 286, 149
Dutrey A., Guilloteau S., Duvert G., Prato L., Simon M., Schuster K., Ménard F., 1996, A&A 309, 493
Dutrey A., Guilloteau S., Bachiller R., 1997a, A&A in press
Dutrey A., Guilloteau S., Prato L., Simon M., Duvert G., Schuster K., Ménard F., 1997b, in preparation
Dutrey A., Guilloteau S., Duvert G., Simon M., Prato L., Ménard F., Schuster K., 1997c, in preparation

Dutrey A., Guilloteau S., Guélin M., 1997d, A&A 317, L55
Duvert G., Dutrey A., Guilloteau S., Ménard F., Schuster K., Prato L., Simon M., 1997, A&A in preparation
Gueth F., Guilloteau S., Viallefond F., 1995, in: Green D.A. Steffen W. (eds) YERAC95: Proc XXVIIth Young European Radio Astronomers Conf. Cambridge Univ Press.
Gueth F., Guilloteau S., Bachiller R., 1996, A&A 307, 891
Gueth F., Guilloteau S., Bachiller R., 1997a, A&A in preparation
Gueth F., et al. 1997b, in preparation
Gueth F., Guilloteau S., Dutrey A., Bachiller R. 1997c, A&A in press
Gueth F., Guilloteau S., 1997, in preparation
Guilloteau S., Bachiller R., Fuente A., Lucas R., 1992, A&A 265, L49
Guilloteau S., Dutrey A., Guélin M., 1997, in preparation
Hayashi M., Ohashi N., Miyama S.M., 1993, ApJ 418, L71
Jensen E.L.N., Koerner D.W., Matthieu R.D., 1996, AJ 111, 2431
Lada E.A., Dutrey A., Guilloteau S., Mundy L. 1996, BAAS 189, 530
Matthieu R.D., Martin E.L., Maguzzu A., 1996, BAAS 188, 6005
McCaughrean M., O'Dell, C.R., 1996, AJ 111, 1997
McCaughrean M., Rayner J.T., Zinnecker H., 1994, ApJ 436, L189
Mundy L.G., Looney L.W., Lada E.A., 1995, ApJ 452, L137
Mundy L.G., Looney L.W., Erickson W., Grossman A., Welch W.J. Forster J.R., Wright M.C.H., Plambeck R.L., Lugten J., Thornton D.D., 1996, ApJ 464, 169
Simon M., Guilloteau S., 1992, ApJ 397, L476
Roddier C., Roddier F., Northcott M.J., Graves J.E., Jim L., 1996, ApJ 463, 326
Stapelfeldt *et al.*, 1997, this volume.

NMA IMAGING OF ENVELOPES AND DISKS AROUND LOW MASS PROTOSTARS AND T TAURI STARS

YOSHIMI KITAMURA
Institute of Space and Astronautical Science
3-1-1 Yoshinodai, Sagamihara, Kanagawa 229, Japan

MASAO SAITO
Department of Astronomy, University of Tokyo
2-11-16 Yayoi, Bunkyo-ku, Tokyo 113, Japan

AND

RYOHEI KAWABE AND KAZUYOSHI SUNADA
Nobeyama Radio Observatory
Nobeyama, Minamimaki-mura, Minamisaku-gun,
Nagano 384-13, Japan

Abstract. We are intensively studying low mass star formation with the radio telescopes at Nobeyama in Japan. Using both the Nobeyama 45 m dish equipped with a 2 × 2 array receiver and the Nobeyama Millimeter Array (NMA), we can cover a very wide spatial range from overall molecular clouds down to compact protoplanetary disks. With the 45 m dish we are investigating hierarchical structures of molecular clouds including star-forming cores. With NMA we are imaging disklike structures (i.e., envelopes, accretion disks, and protoplanetary disks) around protostars and T Tauri stars. Recently, we have completed our survey for dense disklike envelopes around eleven Class 0 & I protostars by NMA. In this paper, we will present our recent results of the disklike envelopes in addition to the previous NMA results of the disks around three T Tauri stars. On the basis of the data, we will discuss the evolution of the disklike structures (dense envelopes → tenuous ones → dispersing ones → accretion disks → protoplanetary ones), and propose a new scenario for the formation of low mass stars.

B. Reipurth and C. Bertout (eds.), Herbig–Haro Flows and the Birth of Low Mass Stars, 381–390.

1. Background

In low mass star formation, there are many interesting physical and chemical processes which one should understand. We are now studying or will study some of the processes with the Nobeyama 45 m telescope and the Nobeyama Millimeter Array (NMA).

In molecular clouds (age $\sim 10^7$ yr, size $\sim 10^7$ AU), hierarchical structures generally exist (e.g., Scalo 1985), and the cloud turbulence plays an important role in the hierarchy. In order to reveal the hierarchy, Sunada *et al.* (1997) have mapped the entire region of $1° \times 1°.5$ of Heiles Cloud 2, the central region of Taurus Molecular Cloud Complex, with a spatial resolution as high as 50" by using the 45 m telescope equipped with a 2×2 array receiver (Figure 1). It is noted that our homogeneous map covers a wide spatial range from 3 to 0.03 pc, two orders of magnitude! Consequently, we have successfully revealed the cloud hierarchy: six filaments with sizes of $\sim 1 \times 0.4$ pc, five major clumps with sizes of $\sim 0.3 \times 0.1$ pc, and numerous fragments with sizes of ~ 0.03 pc. We should understand these components in connection with the fractal structure found by Larson (1995).

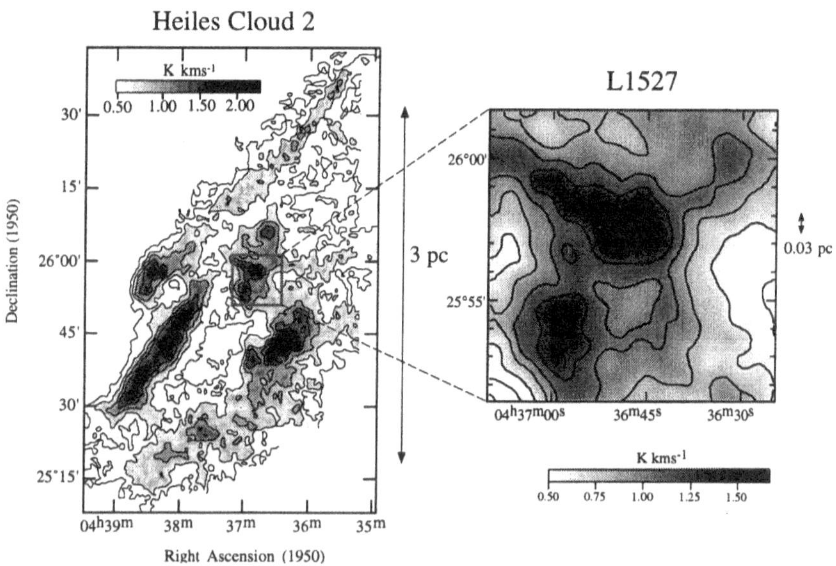

Figure 1. $C^{18}O(1\text{-}0)$ Total Map of Heiles Cloud 2

In the cloud hierarchy molecular cores (age $\sim 10^{5-6}$ yr, size $\sim 10^4$ AU) are the birthplace for new stars. One of the most important properties of the cores is their mass function, directly linked to IMF of stars (e.g., Nakano *et al.* 1995). To obtain the mass function, we need statistical studies of the core formation in the hierarchy as shown in Figure 1. Another important property is how the cores gravitationally collapse to form stars: Shu's solution (Shu 1977) or Larson's one (Larson 1969). The initial conditions of the solutions would be influenced by the turbulence and magnetic fields, which work against gravitation. Therefore, their decay processes are essential for triggering the collapse.

After the gravitational collapse of the cores begins, Class 0 protostars with ages of $\sim 10^4$ yr (e.g., B335 and L1527) are considered to appear at the center of the cores (André *et al.* 1993). The luminosity of the protostars is due to the release of the gravitational energy of infalling gas, and is controlled by the mass accretion rate. In this stage, the formation of disklike envelopes and the ignition of outflows simultaneously occur, and they essentially determine the mass accretion onto the central protostars.

When the major part of the core gas infalls or disperses, the protostars are believed to evolve from Class 0 to Class I with ages of $\sim 10^5$ yr (e.g., L1551 IRS 5 and HL Tau). In this stage the gas in the disklike envelopes is falling onto compact accretion disks and the disks are growing. The infalling motions in the envelopes around L1551 IRS 5 and HL Tau were detected by the NMA imaging observations (Saito *et al.* 1996; Hayashi *et al.* 1993). Recently, Mundy *et al.* (1996) succeeded in imaging directly the compact accretion disk with a radius of 90 - 160 AU around HL Tau.

At the transient stage between the Class I protostars and classical T Tauri stars (CTTSs) with ages of $\sim 10^{5-6}$ yr, there are several flat-spectrum T Tauri stars (e.g., DG Tau and T Tau). It is very important to observe the sources for understanding the evolution from the large envelopes into the compact accretion disks. In the disposal of the envelopes outflow as well as accretion plays an important role. In the disklike envelope around DG Tau, Kitamura *et al.* (1996a) found a dispersing motion which might be driven by the stellar wind. Momose *et al.* (1996) found that the core around T Tau is being dispersed by the outflow. These dispersing processes would determine the star formation efficiency in the cores (Nakano *et al.* 1995).

To understand how the initial conditions for planet building (Cameron 1985; Hayashi *et al.* 1985) are determined during the evolution of the disklike structures, we should observe compact disks around weak-line T Tauri stars (WTTSs) with ages of $\sim 10^{6-8}$ yr as well as CTTSs. The accretion disks around CTTSs are considered to evolve into protoplanetary disks around WTTSs. In fact, the spectral energy distributions (SEDs) towards many T Tauri stars strongly suggest the disk evolution (e.g., Strom *et al.* 1989; Moriarty-Schieven *et al.* 1994). For the two CTTSs of DM Tau and GG Tau, rotating compact disks with a few hundred AU radii have been imaged by NMA (Saito *et al.* 1995; Kawabe *et al.* 1993). However, no one has succeeded in imaging the disks around WTTSs.

2. Our Survey

We have completed our survey for the dense disklike envelopes around Class 0 & I protostars with NMA (Saito 1997). The purpose of our study is to understand the formation and evolution of the disklike envelopes around low mass protostars and to reveal the evolution of the protostars. The protostellar evolution from Class 0 to I has already been proposed by André *et al.* (1993) based on SEDs. The profiles of SEDs, however, are very sensitive to viewing angles and do not necessarily represent the evolution, as pointed out by Tamura *et al.* (1996) and Ohashi *et al.* (1997). Therefore, it is worthwhile to re-investigate the evolution from a more physical standpoint.

We have observed ten protostars in Taurus, which are listed in Table 1, in addition to B335. For our survey it is essential to select the best molecular lines, and we selected the $H^{13}CO^{+}$(1-0) line as a high-density tracer. This is because the line is expected to be optically thin (Butner *et al.* 1995), and because the 45 m observations by Mikami (1995, private communication) showed that the $H^{13}CO^{+}$ emission is distributed more compactly than ^{13}CO(1-0) and $C^{18}O$(1-0). Although the CO lines seem good, as shown by Hayashi *et al.* (1993) and Ohashi *et al.* (1996 and 1997), the lines tend to be influenced by outflowing gas because of their low critical densities.

3. Results and Discussion

3.1. DISKLIKE ENVELOPES AROUND TYPICAL PROTOSTARS, B335 AND L1551 IRS 5

First of all, we will show our results for the typical Class 0 protostar, B335, and the typical Class I protostar, L1551 IRS 5. In B335, Chandler *et al.* (1993) imaged an elongated core perpendicular to the molecular bipolar outflow along the east-west line. The protostellar collapse has been suggested by Zhou *et al.* (1993). They showed that velocity profiles of several molecular lines agree with the inside-out collapse model proposed by Shu (1977). At NMA we have imaged an infalling disklike envelope with a radius of $\sim$ 2000 AU, whose major axis is perpendicular to the outflow axis (Saito *et al.* 1997). The total flux of $H^{13}CO^{+}$ and the 87 GHz continuum flux density were 1.3 Jy km s^{-1} and 18 mJy, respectively, and the mass of the envelope is estimated to be 0.06 $M_{\odot}$. We have detected velocity shifts along both the disk minor and major axes, which can be interpreted as infalling and rotating motions, respectively. The infall velocity and the mass accretion rate are estimated to be 0.14 km s^{-1} at $r = 2200$ AU and $1.2 \times 10^{-6} M_{\odot}yr^{-1}$, respectively. The presence of the infall is supported by the small rotational velocity of 0.14 km s^{-1} at $r = 490$ AU, only one third of the Kepler velocity. The infall velocity, however, is smaller than the free-fall velocity toward the protostar with 0.1 $M_\odot$, and the calculated accretion luminosity is also smaller than the actual luminosity. Our model calculations show that the differences are partly due to a finite beam, a finite velocity resolution, and the uncertainty of the disk thickness, the disk inclination angle, and the stellar mass.

With the Nobeyama 45 m telescope, we have also investigated 10000 AU-scale structures including the 1000 AU-scale envelope in order to study the initial conditions of the B335 core before the gravitational collapse. We have obtained a density profile obeying a power law of $r^{-1.85}$, and the coefficient of the profile is slightly larger than that expected in a hydrostatic isothermal core. Furthermore, we have found a velocity gradient along the line parallel to the major axis of the envelope, suggesting rigid rotation with an angular velocity of $1.1 \times 10^{-14}\mathrm{s}^{-1}$. Consequently, it is possible that the protostar in B335 has been formed in a rigidly rotating isothermal core.

Many authors have studied L1551 IRS 5 as a prototype of protostars. Sargent *et al.* (1988) revealed a disklike envelope with 700 AU radius, which is perpendicular to the molecular outflow. Ohashi *et al.* (1996) recently suggested possible infall in a ^{13}CO disklike envelope observed by NMA. By the $H^{13}CO^+$ NMA observations, we have imaged an infalling disklike envelope with a radius of 2800 AU and a mass of 0.27 $M_\odot$, the major axis of which is perpendicular to the outflow axis (Saito *et al.* 1996). In the envelope, we have detected probable infall and slow rotation. The infall velocity and the mass accretion rate are estimated to be 0.6 km s^{-1} at $r = 800$ AU and $1.1 \times 10^{-5} M_\odot \mathrm{yr}^{-1}$, respectively. The rotational velocity of 0.23 km s^{-1} at $r = 900$ AU is smaller than the Kepler velocity, which fact is consistent with the presence of the infall. The infall velocity is smaller than the free-fall velocity toward the 1 $M_\odot$ object, which is similar to the B335 case. The accretion luminosity, however, is four times larger than the actual luminosity, that is the luminosity problem.

3.2. NEW SCENARIO FOR EVOLUTION OF DISKLIKE ENVELOPES AROUND PROTOSTARS

Our NMA imaging of the dense envelopes around the Class 0 & I protostars is shown in total maps of Figure 2 (Saito 1997). For the sources with intense $H^{13}CO^+$ emission, dense disklike envelopes are clearly seen, while for the other sources envelopes can not be well recognized. To discuss these maps together with other important properties, we summarize significant parameters including our data of the $H^{13}CO^+$ total fluxes and the 87 GHz continuum flux densities in Table 1 (Saito 1997).

Our results of the $H^{13}CO^+$ imaging can be explained in terms of the evolution of the envelopes, as follows. (1) The $H^{13}CO^+$ emission was detected for some sources and not detected for the other sources, while CO

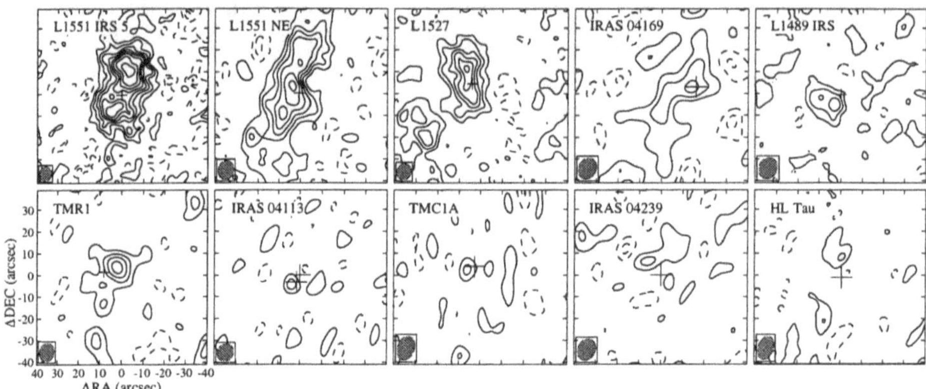

Figure 2. $H^{13}CO^+$(1-0) Total Maps of 10 Protostars in Taurus

TABLE 1. Properties of Protostars

Source Name	$H^{13}CO^+$(1-0) total flux (Jy km s^{-1})	CO disk?	CS(5-4) wing emission?	at the center of an $H^{13}CO^+$ parent core?	T_{bol} (K)	87 GHz flux density (mJy)	CLASS
L1551 IRS 5	7.3	$C^{18}O^a$	Y	Y^g	97	74	E
L1551 NE	5.7	$C^{18}O^b$	Y	---	75	63	E
L1527	3.3	$C^{18}O^c$	Y	Y^g	59	22	E
IRAS 04169	2.7	$C^{18}O^d$	Y	Y^g	170	< 26	E
L1489 IRS	1.5	$C^{18}O^d$	Y	---	238	< 18	F
TMR1	0.52	$^{13}CO^e$	Y	N^h	144	< 28	F
IRAS 04113	(0.09)	---	N	---	606	52	G
TMC1A	(0.07)	$^{13}CO^d$	N	N^h	172	< 23	G
IRAS 04239	< 0.1	---	N	---	236	< 24	G
HL Tau	< 0.1	$^{13}CO^f$	---	---	576	44	G

CO disk : a. Sargent *et al.* (1988), b. Momose (1997, private communication), c. Ohashi *et al.* (1997) d. Ohashi (1997, private communication), e. Terebey *et al.* (1990), f. Hayashi *et al.* (1993)
CS wing : Moriarty-Schieven *et al.* (1995)
$H^{13}CO^+$ parent core : g. Mizuno *et al.* (1994), h. Takakuwa (1996)
T_{bol} : Chen *et al.* (1995)

disks seem to exist in all the sources. This fact directly shows the density evolution of the envelopes. (2) The detection of the CS(5-4) wing emission is well correlated to the detection of the $H^{13}CO^+$ emission. This fact also supports the density evolution in the envelopes, because the CS wing emission is considered to represent the interaction between the dense gas of parent cores and outflows. Note that no detection of the wing emission does

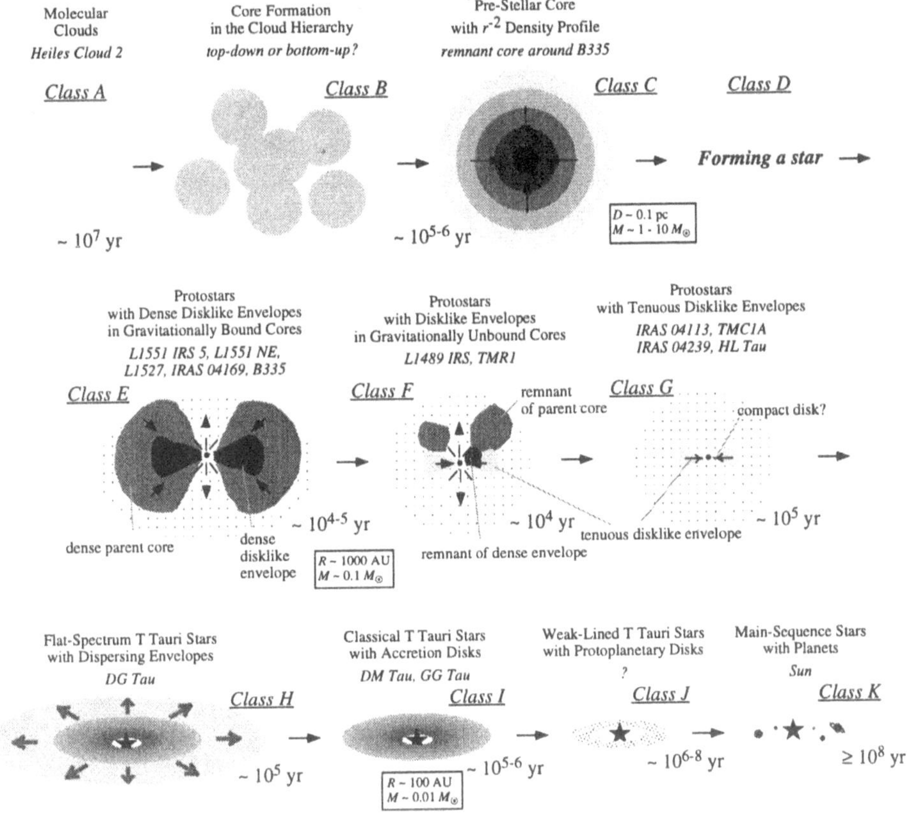

Figure 3. Our Proposed Scenario for Low Mass Star Formation

not indicate the termination of the outflows, because the outflow phenomena are seen all over the sources. (3) It is natural that the association with dense parent cores is correlated with the presence of the dense envelopes. (4) Although the bolometric temperatures estimated by Chen *et al.* (1995) tend to increase from the top to bottom in Table 1, there is no clear correlation between the $H^{13}CO^+$ intensities and the temperatures. This is because the temperatures derived from SEDs are sensitive to the geometrical factor of viewing angles rather than the protostellar evolution. (5) We can find no clear correlation between the 87 GHz continuum flux densities and the $H^{13}CO^+$ total fluxes. In addition, the submillimeter continuum flux densities at 264 GHz (Moriarty-Schieven *et al.* 1994) do not correlate with the $H^{13}CO^+$ fluxes. The continuum flux densities would not be good indicators for the protostellar evolution, because the emission comes not only from the envelopes but also from the accretion disks.

According to the above discussion, we propose a new scenario for the evolution of the disklike envelopes around protostars during low mass star formation, as schematically shown in Figure 3. We divide protostars into the three classes of *Class E*, *F*, and *G*: Dense envelopes evolve into tenuous envelopes owing to outflow and accretion processes, and simultaneously, dense parent cores accrete and disperse. Our classification is considered to be better than both the classifications of Class 0 by André *et al.* (1993) and T_{bol} by Myers *et al.* (1993). This is because the discrimination between protostars with and without gravitationally bound parent cores is physically clearer than that based on SEDs which are sensitive to viewing angles. It is very important to know whether a parent core surrounding a star is gravitationally bound or not, because the stellar mass is determined by the supply of gas from the bound core (Nakano *et al.* 1995).

3.3. COMPACT DISKS AROUND CLASSICAL T TAURI STARS

The envelope evolution newly found by us can be consistently connected with the disk evolution revealed by the previous NMA imaging of the disks around the three CTTSs of DG Tau, DM Tau, and GG Tau, as shown in Figure 3. For DG Tau, a flat-spectrum T Tauri star, we imaged a dispersing disklike envelope with a radius of 2800 AU and a mass of 0.03 $M_{\odot}$, whose major axis is perpendicular to the optical jet (Kitamura *et al.* 1996a). In the outer part of the envelope, we detected an expanding motion with a velocity of 1.5 km s^{-1}. The expansion velocity is larger than both the Kepler and free-fall velocities, and the expansion could be driven by the stellar wind from an energetic point of view. Since DG Tau is now evolving into a CTTS from a protostar, the star can be considered as *Class H* in our scenario: The envelope is disappearing and a compact accretion disk will remain. Recently, Kitamura *et al.* (1996b) have succeeded in imaging the accretion disk with a radius of $\sim$ 100 AU around DG Tau by achieving the highest spatial resolution of one arcsec in NMA. This study revealed that almost all the mm flux density at 147 GHz is coming from the compact disk and that the contribution of the envelope to the SED is negligible.

For the two CTTSs of DM Tau and GG Tau, rotating compact disks with radii of 350 and 500 AU, respectively, were imaged by NMA (Saito *et al.* 1995; Kawabe *et al.* 1993). The rotation roughly agrees with Kepler rotation. Although GG Tau is a binary system, both the stars can be regarded as *Class I* in our scenario: A compact accretion disk is rotating around a central star and will evolve into a protoplanetary disk.

In order to reveal the initial conditions for planet formation, we should accurately estimate the disk properties independently of instrumental parameters. For the disk around DM Tau, we have performed radiative transfer calculations including the observational parameters, based on a Kepler rotating disk (Kitamura *et al.* 1993), and have applied the calculations to the observed data. The fitting is quite good as shown in Figure 4. The model parameters are as follows: $\Sigma(r) = 0.02(r/R)^{-2}\mathrm{gcm}^{-2}$, $T(r) = 30(r/R)^{-0.5}\mathrm{K}$, $R = 350\mathrm{AU}$, $i = 40°$, $M_* = 0.48 M_\odot$, (beam size) = 6".5 × 4".9, and (velocity resolution) = 0.81 km s^{-1}. The disk mass is calculated to be 0.02 $M_\odot$, which supports the minimum-mass nebula proposed by Hayashi *et al.* (1985). Here we assume no condensation of CO on grains. The power law index of $\Sigma(r)$ was estimated to be 2 by Saito *et al.* (1995), which index is larger than the value of 1.5 in the Hayashi model.

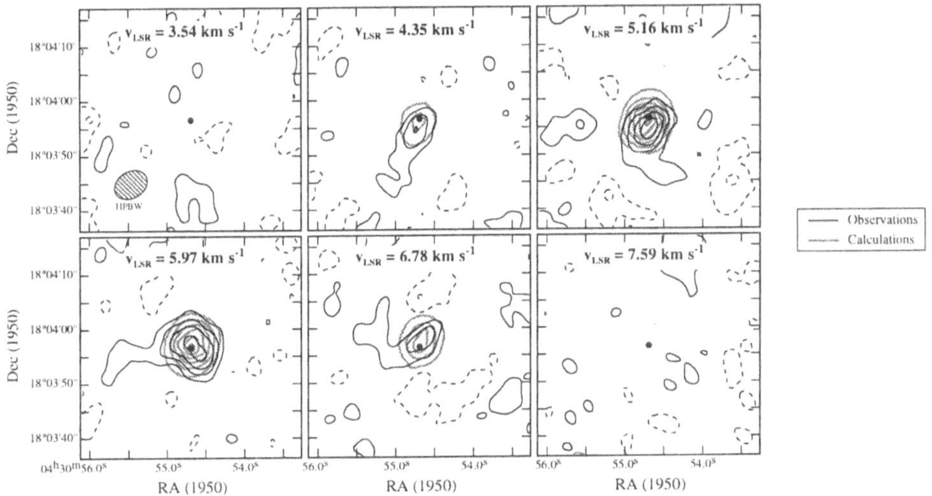

Figure 4. Model Fitting to ^{12}CO(1-0) Channel Maps of DM Tau

4. Our Future Studies

1. To understand the hierarchy of molecular clouds including star-forming cores, we will efficiently perform large-scale and high-resolution mapping for several clouds with the Nobeyama 45 m telescope equipped with a new 5 × 5 array receiver.
2. To check the generality of our proposed scenario, we should increase our sample number of NMA imaging in each evolutionary stage of low mass star formation. The sensitivity and spatial resolution of NMA will be improved by incorporating the 45 m dish into the array (*Rainbow project*) and by a phase correction method, respectively.
3. To reveal planet formation, we will directly image planet-forming regions around T Tauri stars by the Large Millimeter and Submillimeter

Array (LMSA) in Chile in 2000's. The spatial resolution of LMSA will reach 0.01 arcsec, that is 1.4 AU at the distance of Taurus!

References

André, P., Ward-Thompson, D. and Barsony, M. 1993, ApJ 406, 122

Butner, H. M., Lada, E. A. and Loren, R. B. 1995, ApJ 448, 207

Cameron, A. G. W. 1985, in *Protostars & Planets II*, ed. D. C. Black & M. S. Matthews (Tucson: Univ. of Arizona Press), 1073

Chandler, C. J. and Sargent, A. I. 1993, ApJ 414, L29

Chen, H., Myers, P. C., Ladd, E. F. and Wood, D. O. S. 1995, ApJ 445, 377

Hayashi, C., Nakazawa, K. and Nakagawa, Y. 1985, in *Protostars & Planets II*, ed. D. C. Black & M. S. Matthews (Tucson: Univ. of Arizona Press), 1100

Hayashi, M., Ohashi, N. and Miyama, S. M. 1993, ApJ 418, L71

Kawabe, R., Ishiguro, M., Omodaka, T., Kitamura, Y. and Miyama, S. M. 1993, ApJ 404, L63

Kitamura, Y., Kawabe, R. and Saito, M. 1996a, ApJ 457, 277

Kitamura, Y., Kawabe, R. and Saito, M. 1996b, ApJ 465, L137

Kitamura, Y., Omodaka, T., Kawabe, R., Yamashita, T. and Handa, T. 1993, PASJ 45, L27

Larson, R. B. 1969, MNRAS 145, 271

Larson, R. B. 1995, MNRAS 272, 213

Mizuno, A., Onishi, T., Hayashi, M., Ohashi, N., Sunada, K., Hasegawa, T. and Fukui, Y. 1994, Nature 368, 719

Momose, M., Ohashi, N., Kawabe, R., Hayashi, M. and Nakano, T. 1996, ApJ 470, 1001

Moriarty-Schieven, G. H., Wannier, P. G., Keene, J. and Tamura, M. 1994, ApJ 436, 800

Moriarty-Schieven, G. H., Wannier, P. G., Mangum, J. G., Tamura, M. and Olmsted, V. K. 1995, ApJ 455, 190

Mundy, L. G., Looney, L. W., Erickson, W., Grossman, A., Welch, W. J., Forster, J. R., Wright, M. C. H., Plambeck, R. L., Lugten, J. and Thornton, D. D. 1996, ApJ 464, L169

Myers, P. C. and Ladd, E. F. 1993, ApJ 413, L47

Nakano, T., Hasegawa, T. and Norman, C. 1995, ApJ 450, 183

Ohashi, N., Hayashi, M., Ho, P. T. P. and Momose, M. 1997, ApJ 475, 211

Ohashi, N., Hayashi, M., Ho, P. T. P., Momose, M. and Hirano, N. 1996, ApJ 466, 957

Saito, M. 1997, Ph. D. Thesis, University of Tokyo

Saito, M., Kawabe, R., Ishiguro, M., Miyama, S. M., Hayashi, M., Handa, T., Kitamura, Y. and Omodaka, T. 1995, ApJ 453, 384

Saito, M., Kawabe, R., Kitamura, Y. and Sunada, K. 1996, ApJ 473, 464

Saito, M., Sunada, K., Kawabe, R., Kitamura, Y. and Hirano, N. 1997, to be submitted to ApJ

Sargent, A. I., Beckwith, S., Keene, J. and Masson, C. 1988, ApJ 333, 936

Scalo, J. M. 1985, in *Protostars & Planets II*, ed. D. C. Black & M. S. Matthews (Tucson: Univ. of Arizona Press), 201

Shu, F. H. 1977, ApJ 214, 488

Strom, K. M., Strom, S. E., Edwards, S., Cabrit, S. and Skrutskie, M. F. 1989, AJ 97, 1451

Sunada, K. and Kitamura, Y. 1997, submitted to ApJ

Takakuwa, S. 1996, Master Thesis, University of Tokyo

Tamura, M., Ohashi, N., Hirano, N., Itoh, Y. and Moriarty-Schieven, G. H. 1996, AJ 112, 2076

Terebey, S., Beichman, C. A., Gautier, T. N. and Hester, J. J. 1990, ApJ 362, L63

Zhou, S., Evans II, N. J., Kömpe, C. and Walmsley, C. M. 1993, ApJ 404, 232

THE OBSERVATIONAL EVIDENCE FOR ACCRETION

LEE HARTMANN
Center for Astrophysics, 60 Garden St.,
Cambridge, MA 01238 USA

Abstract. Outflows from low-mass young stellar objects are thought to draw upon the energy released by accretion onto T Tauri stars. I briefly summarize the evidence for this accretion and outline present estimates of mass accretion rates. Young stars show a very large range of accretion rates, and this has important implications for both mass ejection and for the structure of stellar magnetospheres which may truncate T Tauri disks.

1. Introduction

Star formation requires the contraction of typical molecular cloud cores by factors of a million or more to reach stellar densities. The contraction process must involve both redistribution of angular momentum and release of gravitational potential energy to produce a star. The necessary angular momentum transfer seems to occur mainly in circumstellar disks, which allow accretion to a central stellar core as a smaller amount of matter is pushed out (or viscously spread) to larger radii. Much of the energy release appears to take the form of radiation; however, a substantial fraction of the energy generated by accreting material apparently is put into outflows or jets which produce Herbig-Haro objects (see Calvet, this volume). The rapidly-rotating disk probably also plays an essential role in jet ejection by providing ideal conditions for the centrifugal acceleration of material by magnetic fields; the rotating magnetic field structure is most likely responsible for the high collimation observed in many jets (see the contributions by Pudritz & Ouyed, Shu & Shang, Camenzind, and Heyvaerts & Norman in this volume).

A knowledge of mass accretion processes in young stellar objects is essential to developing a fundamental understanding of outflows. Some jets and HH objects show unmistakable evidence for highly time-dependent out-

B. Reipurth and C. Bertout (eds.), Herbig–Haro Flows and the Birth of Low Mass Stars, 391–405.

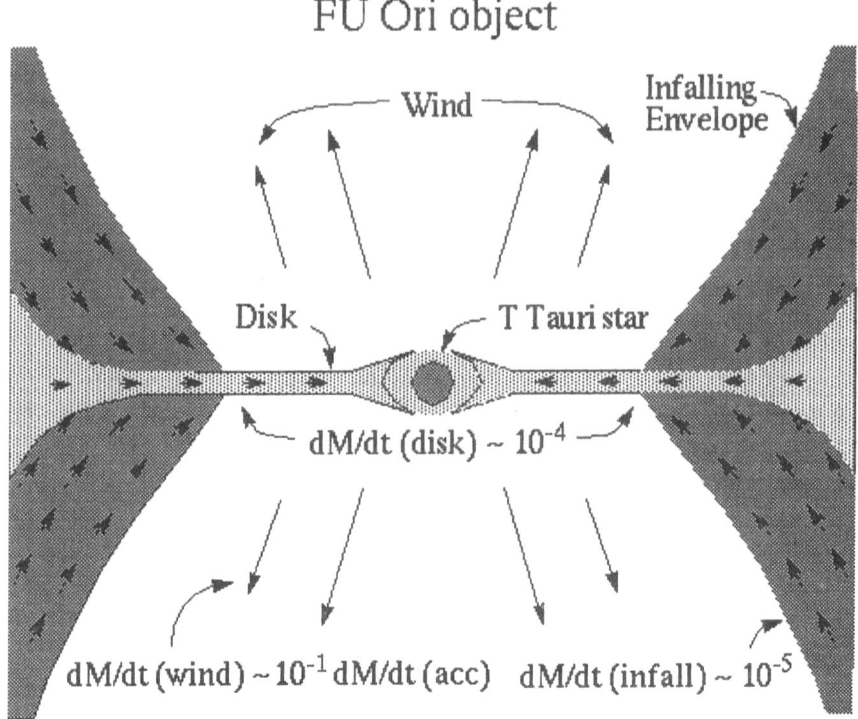

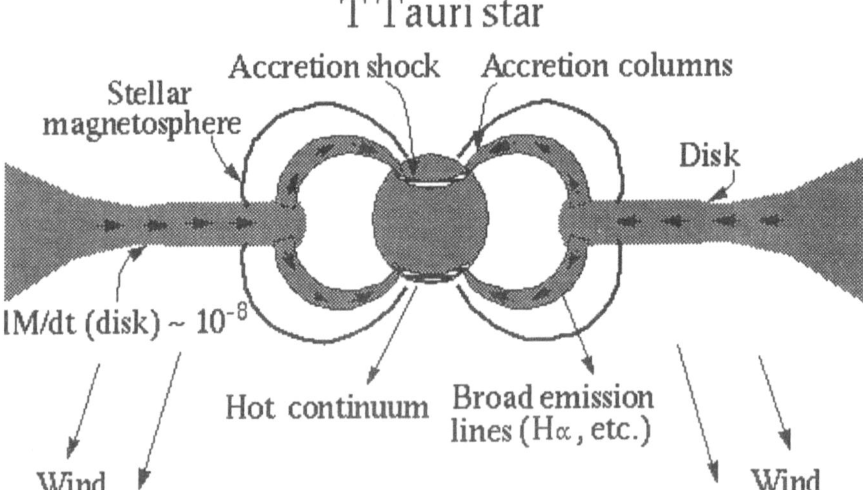

Figure 1. Two main types of accreting pre-main sequence objects (see text). Mass accretion/loss rates are given in units of solar masses per year

flow (see Bally & Devine, this volume), and it has been recognized for some time that the time-dependence of mass accretion might be responsible for this structure (Reipurth 1989, 1991). In addition to the time-variability of accretion, the very large range in accretion rates for objects with outflows provides some important clues to the processes responsible for mass loss.

In this review I shall concentrate on mass accretion in low-mass, pre-main sequence stars. The subject can be conveniently divided into discussions of two types of objects, the FU Ori eruptive variables and the (more numerous) T Tauri stars. FU Ori objects are very young, probably low-mass stars accreting from circumstellar disks at rates as high as $10^{-4} M_{\odot}\,\mathrm{yr}^{-1}$ (Figure 1). This rapid accretion is highly time-variable and probably especially prevalent in the earliest stages of star formation, when material from the infalling natal envelope may still be falling onto the outer disk. The T Tauri stars exhibit much lower mass accretion rates through their disks. Most T Tauri stars are older than FU Ori objects, although it appears quite likely that FU Ori objects oscillate between rapid FU Ori disk accretion and slower, T Tauri-like accretion.

An important distinction between FU Ori and T Tauri accretion is the effect of the stellar magnetosphere (see Edwards, this volume). It now seems likely that the inner disks of T Tauri stars are truncated by the magnetosphere, and that accretion to the central star occurs along magnetic field lines, terminating in a hot accretion shock at the stellar surface (Figure 1). In contrast, the FU Ori objects show no evidence for magnetospheric disk truncation, probably because the magnetosphere of the central T Tauri star is crushed by the high pressures in rapid accretion. Although these objects apparently have very different inner disk boundary conditions, outflows with qualitative similarities are produced in both FU Ori objects and T Tauri stars, and this fact has implications for understanding mass loss mechanisms.

2. FU Ori accretion

The evidence for disk accretion in FU Ori objects has been summarized in a recent review (Hartmann & Kenyon 1996) and so I will not repeat that material here. In outline, the validity of the accretion disk model for FU Ori objects rests on several observational results: that the spectral energy distributions (SEDs) of FU Ori objects can be modeled reasonably well with steady disk spectra, at least shortward of 10μm; that the optical and infrared lines indicate rapid rotation, with doubled line profiles often observed that are expected for a rotating disk; and that the differential rotation observed as a function of wavelength (slower rotation at longer wavelengths) is consistent with quasi-Keplerian disk rotation. The observation of absorption lines in FU Ori objects is consistent with the rapid accretion model; at high accretion rates and optical depths, the viscous energy dissipation is trapped in the disk midplane, helping to create the outwardly-decreasing temperature gradient needed to produce absorption features.

Perhaps the single most important point to be made in the present context is that, during an FU Ori accretion event, the accretion luminosity of the disk far exceeds that of the (original) central star; typical luminosities of FU Ori objects are 200 - 500 $L_{\odot}$, whereas typical T Tauri star luminosities are $\sim 1 L_{\odot}$. This means that the emergent spectrum is dominated by the disk, and can be compared directly with disk models without making direct allowance for radiation of the central star.

The spectral energy distributions (SEDs) of FU Ori objects can be modeled by pure disk models, without including any hot radiating components such as are present in T Tauri stars (Kenyon et al. 1989). The implication is that FU Ori objects have neither the canonical boundary layer between rapidly rotating star and disk (e.g., Lynden-Bell & Pringle 1974) nor the hot magnetospheric accretion shock thought to be present in T Tauri stars (Königl 1991; §3). Both constraints are explicable in terms of the rapid disk accretion rates in FU Ori objects.

The absence of a magnetosphere can be explained simply by the pressure of rapid accretion, as discussed in more detail in §4. The absence of boundary layer emission can also be understood by the following argument. For plausible viscosities, the disks of FU Ori objects are likely to be extremely optically thick (Clarke et al. 1989; Clarke et al. 1990; Bell et al. 1995), resulting in very strong trapping of the internal viscous dissipation of accretion energy. This in turn leads to very high central disk temperatures in the innermost regions, such that the internal sound speed is an appreciable fraction of the local Keplerian velocity (Clarke et al. 1989; Popham et al. 1993). Under these conditions, boundary layers are unlikely to form for some combination of the following reasons. First, since the disk is physically thick, with a ratio of scale height H to cylindrical radius R of $H/R \sim 0.3$, any spatially-concentrated dissipation of energy will be diffused laterally over a region $\Delta R \sim H$, which in turn reduces the maximum temperature observed (Clarke et al. 1989). Second, because of the large optical depths involved, the disk may advect substantial amounts of energy (Clarke et al. 1989); this energy may actually be carried into the central star rather than radiated, possibly explaining why the inner radii of FU Ori disks seem to be a factor of two larger than typical T Tauri star radii (Popham et al. 1993, 1996). Third, gas pressure support in the radial direction is no longer negligible; this means that the disk rotation can be substantially sub-Keplerian over a finite radial distance, implying dissipation of rotational energy over a much larger region of the disk than in a typical boundary layer, and this spreading of dissipating energy also helps prevent the high temperature emission characteristic of boundary layers (Popham et al. 1996).

Accretion rates for FU Ori objects are subject to several uncertainties, in part because of the relatively unknown properties of the central stars. The disk accretion luminosity

$$L_{acc} = \frac{GM_* \dot{M}_{acc}}{(2)R_*} \tag{1}$$

can be used to derive the mass accretion rate $\dot{M}_{acc}$ if the stellar mass M_* and radius R_* are known. The radius can be estimated from the luminosity and peak temperature of disk emission, the latter determined from spectral type matching. In principle, once the radius is known, the mass can be estimated from the rotational velocity, given an inclination; however, application of this method to the FU Ori objects V1057 Cyg and V1515 Cyg indicate very small central masses unless these disks are observed nearly pole-on - which they probably are (Goodrich 1987). In general the inclination uncertainties are large, and it is probably better to simply assume a typical T Tauri stellar mass. The factor of 2 in the denominator of equation (1) is the standard accretion disk result, which assumes that any additional energy release occurs in a boundary layer. As discussed above, there is no boundary layer emission in FU Ori objects, but the factor of 2 remains uncertain because it is not known how much of the accretion energy of the innermost disk is directly radiated and how much is advected into the star.

Within these uncertainties, observations indicate that FU Ori objects must have accretion rates of order $\dot{M}_{acc} \sim 10^{-4} M_\odot \, \mathrm{yr}^{-1}$ at maximum light. This extremely rapid accretion is accompanied by very strong mass ejection from the surface of the disk (Calvet, this volume).

The outburst behavior of FU Ori objects is especially interesting in view of the evidence for multiple jet outbursts (Reipurth 1991; Bally & Devine, this volume). This is very difficult to come by, because we have never observed long enough to see multiple outbursts, or a complete cycle, or even a return to minimum light in any FU Ori object. Event statistics (Hartmann & Kenyon 1996) suggest that the typical low-mass star may undergo 5 - 10 FU Ori outbursts in its lifetime, assuming that we have detected all FU Ori events within 1 Kpc of the Sun over the last 50 years. If we then assume that FU Ori outbursts are confined to the protostellar infall phase, which probably lasts $\sim 10^5$ yr, we find a timescale between outbursts of $\sim 10^4$ yr. If instead the detection statistics are seriously incomplete, so that most of the mass of a typical low-mass star is accreted in FU Ori events, then the frequency increases by a factor of 5, the repetition timescale of ~ 2000 yr appears to be more in line with observed multiple bow shock structures in some jets. My personal guess is that there is a large range in outburst behavior among pre-main sequence stars and FU Ori objects represent one extreme; and that milder, more frequent events, perhaps analogous to the "EXor" outbursts of Herbig (1977), are responsible for the multiple jet bow shocks in most cases.

3. T Tauri accretion

The T Tauri stars have much lower disk accretion rates than FU Ori objects, typically by factors $\sim 10^3 - 10^4$ (see below). The low accretion rates mean that it is difficult to distinguish the local accretion energy heating in the disk from energy input of absorbed light from the central star (Figure 1). It is not generally feasible to determine accretion rates from disk emission. Instead, one must rely on measurements of the hot continuum emission from the accretion shock region (Figure 1).

Several authors have presented estimates of accretion rates using the hot continuum emission of T Tauri stars (Bertout, Basri, & Bouvier 1988; Basri & Bertout 1989; Hartigan et al. 1990, 1991; Valenti, Basri, & Johns 1993; Hartigan, Edwards, & Ghandour 1995). These authors analyzed the hot continuum emission in terms of a boundary layer model, in which the accretion luminosity is related to the mass accretion rate essentially as in equation (1). In the current interpretation of T Tauri emission, material from the disk moves to the star in essentially free-fall from its starting point at the magnetospheric radius R_m. Since the rotational velocities of T Tauri stars are small, one expects the hot continuum emission represents the dissipation of the kinetic energy generated by material falling in from R_m; hence,

$$L_{hot} \sim \frac{GM_* \dot{M}_{acc}}{R_*} \left(1 - \frac{R_*}{R_m}\right) . \qquad (2)$$

While this method appears straightforward in principle, in practice there are problems in evaluating L_{hot} which basically result from the need to separate out the stellar photospheric emission from the accretion shock emission, and to make an appropriate extinction correction. Because the hot continuum is bluer than a typical T Tauri star photosphere, extinction corrections which do not take the added continuum into account may underestimate the extinction, and since much of the excess emission appears in the Balmer continuum, reddening corrections are important. In addition, the Paschen continuum emission is also an important contributor to the total accretion luminosity in many systems, and evaluating this requires careful separation between stellar and accretion shock radiation. As an indication of how difficult the problem is, two recent careful studies, concentrating on differing spectral regions, and making differing assumptions for analysis, differ systematically by an order of magnitude in the inferred hot continuum luminosities and mass accretion rates (Valenti, Basri, & Johns 1993; Hartigan, Edwards, & Ghandour 1995).

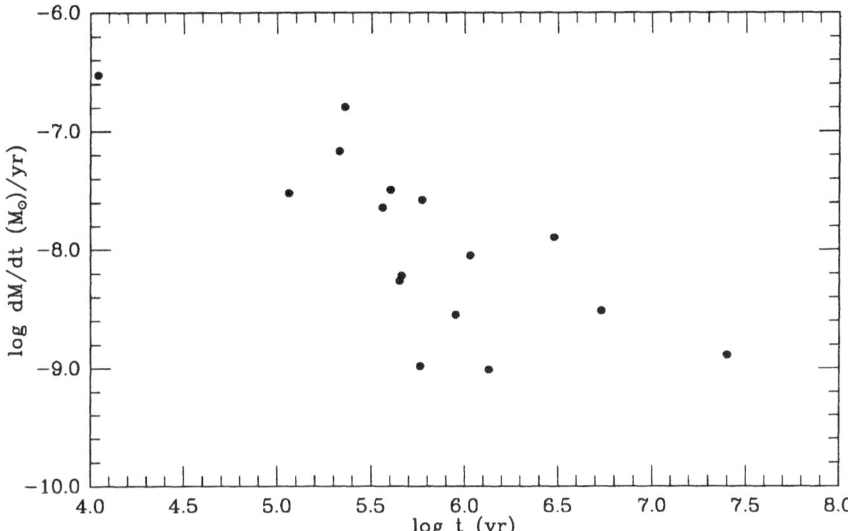

Figure 2. Mass accretion rates for representative T Tauri stars in the Taurus molecular cloud (see text). From Gullbring et al. (1997)

Because of this large systematic discrepancy, and because of the importance of determining mass accretion rates for T Tauri stars, we decided to reexamine this question. Our method (Gullbring et al. 1997) employs elements of both previous works. We obtained intermediate resolution absolute spectrophotometry of a small sample of Taurus classical T Tauri stars spanning the wavelength range 3200 Å to 5200 Å. Thus, we concentrate on the Balmer and upper Paschen continuum emission, like Valenti et al., but we tried to measure the absorption features at intermediate resolution, in the spirit of the high-resolution method of Hartigan et al., to separate between star and hot "veiling" continuum.

Our results, shown in Figure 2, are generally closer to the results of Valenti et al. with substantial scatter. We think that we understand the sources of the discrepancy; basically, Hartigan et al. assumed that the Paschen continuum is stronger than it generally is, and this causes an overestimate of the amount of emission at wavelengths outside of the region of observation, as well as producing a bias toward overestimating the extinction. In addition, there is nearly a factor of two systematic difference in determining mass accretion rates from luminosities using equation (2) (assuming $R_m \sim 5R_*$) rather than using the usual boundary layer equation, as Hartigan et al. assumed.

The results in Figure 2 indicate a wide range in accretion rates among T Tauri stars. The data are biased *against* high accretion rate stars, because the extinction corrections are difficult to determine for these objects, and so both the accretion luminosities and stellar ages are difficult or impossible to determine; thus, the true spread in accretion rates is larger than indicated. There is some evidence for a general decline in mass accretion rates as a function of age, as might be expected if accretion rates depend in part upon the mass remaining in the disk.

4. Implications for magnetospheres and wind ejection

The foregoing discussion indicates that instantaneous mass accretion rates in low-mass stars span a wide range, from up to $\sim 10^{-4} M_\odot \, \mathrm{yr}^{-1}$ in FU Ori events to as low as $\sim 10^{-9} M_\odot \, \mathrm{yr}^{-1}$ in slowly-accreting T Tauri stars. There is evidence for both short-term time variability as well as a general decline in accretion rates as a function of age. In addition to the dramatic variations during FU Ori outbursts, significant changes in accretion onto the central T Tauri star have been observed in classical T Tauri stars. For example, the (currently) strong-emission star DR Tau increased in optical brightness (B magnitude) by about a factor of ten over a period of approximately 10 years (Herbig 1989). Less is known about the variability of accretion in most T Tauri stars. The observed range in mass accretion rate as a function of age in Figure 2 suggests that this variability might be as much as an order of magnitude, but this is likely to be an upper limit given uncertainties in stellar ages and in measuring accretion luminosities, resulting from errors in extinction estimates, differing geometric projections of the accretion shock along the line of sight, rotational modulation of nonaxisymmetric magnetospheric structure, etc.

Figure 3 summarizes the current observational constraints. This figure should be taken as illustrative rather than conclusive. The ages and timescales of the accretion phases are not well known; furthermore, they may overlap. For example, the early phases of accretion probably oscillate between FU Ori outbursts and more typical T Tauri accretion. However, viewed broadly, Figure 3 usefully indicates the wide range of accretion rates present in low-mass stars, and that the high accretion rate phases are probably concentrated to earlier times.

What are the implications of this behavior for wind/jet ejection? In general, we expect that accretion provides the energy for mass ejection

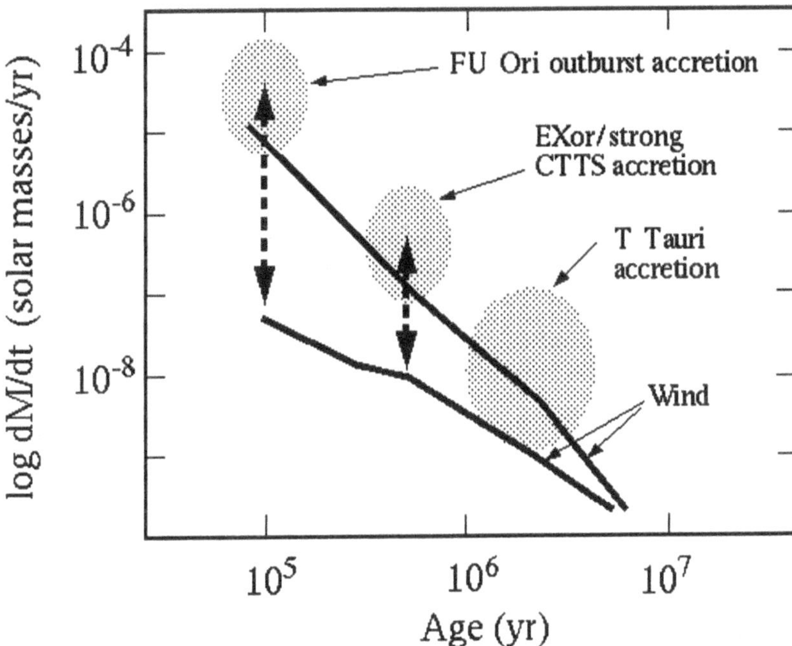

Figure 3. Highly schematic picture of accretion in low-mass stars. FU Ori outbursts are thought to occur mostly at early phases of evolution; the data in Figure 2 suggest that disk accretion rates generally decrease with increasing age. Wind mass loss rates generally should fall below accretion rates for energetic reasons (see text), although the time-variability of accretion, as typified at its extreme end by FU Ori objects, complicates the issue.

(e.g., Shu & Shang, this volume). Therefore, unless there is 100% efficiency of conversion of accretion energy to mass loss energy (which seems unlikely; Hartmann 1995), the kinetic energy flux in the wind must be a fraction of the total accretion energy. It is more useful to consider the mass flux in the wind. In a (quasi) steady state, the energy supplied to the wind or jet comes from the potential energy released as mass moves from infinity down to some radius in the potential well. The mass ejected must escape from this gravitational potential well, and have some extra energy besides to have a finite escape velocity at infinity. Thus, even in the limit where all of the accretion energy goes into accelerating the wind, there must be less mass ejected (at higher energy) than is accreted (at lower energy). Since wind velocities are typically a few times the escape velocity at the point from which mass is ejected, and since not all of the accretion energy goes into the wind (some is radiated away), one might expect that the maximum mass loss rates in winds/jets driven by disk accretion are at most of order 10^{-1} of the mass accretion rates. This is rougly what is observed (Calvet, this volume), which indicates pretty high efficiencies of converting accretion energy into wind energy.

If we take mass loss rates to be 10^{-1} of accretion rates, then we have the curves marked "wind" in Figure 3. The curves must span a range of mass loss rates to accomodate the accretion variability at a given time. Again, one must take these wind limits as schematic rather than highly accurate. Evolutionary timescales are likely to vary from star to star. Given these caveats, Figure 3 makes an important point; namely, that wind ejection is likely to decline steeply with age, and that in terms of impact on the interstellar medium, outbursting/rapidly-accreting young stellar objects are likely to dominate. We should be wary of trying to interpret outflow dynamics in terms of the mass ejection behavior of the most common T Tauri stars, because these objects probably do not dominate the time-integrated outflow properties.

Another way of looking at this is to consider the amount of mass accreted in various phases. One may estimate the mass accreted during the "typical"" T Tauri phase (which I define here as representing ages $\gtrsim 10^6$ yr) from the results in Figure 2;

$$M_{acc} \sim \dot{M}(t) \times t \sim 10^{-8} M_\odot \, \mathrm{yr}^{-1} \times 10^6 \mathrm{yr} \sim 10^{-2} M_\odot \,. \qquad (3)$$

This means that *most* of the mass of a typical T Tauri star of mass $\sim 0.5 M_\odot$ must be accreted at earlier ages. If we assume a more or less universal ratio of $\dot{M}_{wind} \sim 10^{-1} \dot{M}_{acc}$, this suggests that over 90% of the outflow mass ejected occurs at ages $< 10^6$ yr. Moreover, the ram pressure of the flow is also an important quantity in understanding the driving outward of swept-up interstellar medium; this places even more emphasis on possible short-lived events of very rapid mass loss, such as are likely to be associated with outbursts of accretion.

These considerations emphasize the point that "typical" accreting T Tauri stars should not necessarily be taken as the model for outflow sources. It is fairly certain now that stellar magnetospheres play a crucial role in inner disk structure in most T Tauri stars (Königl 1991; Camenzind 1990; see Edwards, this volume). However, FU Ori objects do not show the observational signatures with magnetospheres, and indeed at the high accretion rates during FU Ori objects, one might well expect the ram pressure of the disk to overwhelm the stellar magnetosphere (Shu et al. 1994). Thus it seems to me that the emphasis on the importance of the stellar magnetosphere producing mass ejection in the picture of Shu and collaborators (Shu et al. 1994) may be misplaced.

An issue which is indirectly related to mass accretion and ejection is the location of the magnetospheric radius in T Tauri stars with respect to the co-rotation radius. Magnetic fields which connect to disk regions outside of the co-rotation radius, i.e. to parts of the disk which have a slower angular velocity than that of the star, will help spin down the star by transferring angular momentum outward. However, this angular momentum transfer implies that accretion is not occurring along these field lines. In contrast, stellar magnetic field lines connecting to the disk interior to the corotation radius attach to parts of the disk with higher angular velocities than the star; in this case, the magnetic field lines transfer angular momentum from the Keplerian disk to the star, allowing infall to proceed.

The problem is that we wish to have *both* accretion and stellar spindown. In view of the observation that all T Tauri stars with near-infrared disk emission exhibit accretion diagnostics (Edwards et al. 1994), it appears that nearly all disks of any significance are accreting; on the other hand, it appears that the accreting T Tauri stars are more slowly rotating than the non-accreting T Tauri stars without disks (Bouvier et al. 1993; Edwards et al. 1993). This suggests that disks serve as a rotational brake (e.g., Königl 1991), which happens if magnetic field lines connect to the disk exterior to corotation, at the same time that accretion proceeds, which requires the stellar magnetosphere to connect with disk regions interior to corotation.

The original magnetosphere model of Ghosh & Lamb (1979) solved this problem by having the disk penetrated by a range of magnetic field lines, both interior and exterior to the corotation radius. However, there are difficulties with this model. It is not apparent that the dynamics of the disk can be significantly affected over a wide range of disk radii (Wang 1995, 1996); and there is no guarantee that the magnetic field lines can easily slip through the disk at just the right rate to produce a steady-state configuration (Shu et al. 1994).

For these reasons Shu et al. (1994) developed a model in which the magnetosphere is truncated at the disk at *exactly* the corotation radius. This model is attractive in that it avoids both problems of the Ghosh and Lamb model, and provides a definite inner boundary condition which can be used to calculate the details of the accretion flow and the mass ejection (Ostriker & Shu 1995; Najita & Shu 1994; see Shu & Shang, this volume). However, this solution is obtained at what may be an unacceptable price, namely that the magnetospheric radius is exactly at corotation. For the case of a dipole field, we expect the magnetospheric radius to scale as

$$R_m \propto \dot{M}_{acc}^{-2/7} B_*^{4/7} \qquad (4)$$

(cf. Königl 1991), where B_* is the stellar (dipole) magnetic field. Given the wide range of mass accretion rates in low-mass young stellar objects, and the evidence for some (rapid) time variability, it is not at all clear that the stellar magnetic field will always truncate the disk at the corotation radius. For example, in DR Tau, the accretion rate seems to have increased by an order of magnitude over a period of a decade, which is far too short for the stellar rotation rate to change. Thus, unless the stellar magnetic field changed in precise proportion to the change in disk accretion rate - which seems implausible - an order of magnitude change in $\dot{M}_{acc}$ would yield a change in R_m of a factor of nearly 2, corresponding to a change in the Keplerian angular velocity of the intercepted disk regions of a factor of roughly 2.7. The significance of this result can be emphasized by noting that even if the angular velocities of the star and the disk region where a magnetic field line is attached match to 10%. the field line would still wind up over about 10 rotations - only about 10 weeks for the typical T Tauri star.

It seems necessary to relax the criterion that the magnetospheric truncation occurs precisely at corotation. Indeed, as pointed out by Shu & Shang (this volume), the evidence for non-axisymmetric magnetic field structure implies that not all portions of the truncated disk can be at the corotation radius. It may well be that, under differing conditions of accretion, the central T Tauri stars have a net spin up or spin down torque at any given instant of time, and that an equilibrium is achieved only over relatively long timescales.

Even with this assumption, we still need to consider the likely winding up of field lines in the stellar magnetosphere, as emphasized by Shu in this volume. What this might look like was indicated schematically a few years ago by Van Ballegooijen (1994). He considered a very simple case in which the field lines were force-free outside the star and disk, and were simply rooted in disk regions which rotate at a different angular velocity than the star. This led to magnetic field configurations which varied as indicated very schematically in Figure 4. Van Ballegooijen started with completely poloidal magnetic field lines (radial in projection on the disk plane), as indicated in the upper panel of Figure 4. As the system wound up, the configuration proceeded to something like the lower panel of Figure 4, where the magnetic field lines got twisted and bulged out over the disk. The energy contained in the magnetic fields increased to an apparent catastrophic point, and Van Ballegooijen assumed that reconnection must take place to reduce the system energy - and start the cycle over again.

The interesting feature of this calculation is that the initial magnetic configuration would be qualitatively favorable to accretion, in that disk material could potentially move along field lines always decreasing its distance from the central star, and thus could be falling "downhill". However, after winding up, the field lines bow outward just above the disk, producing a

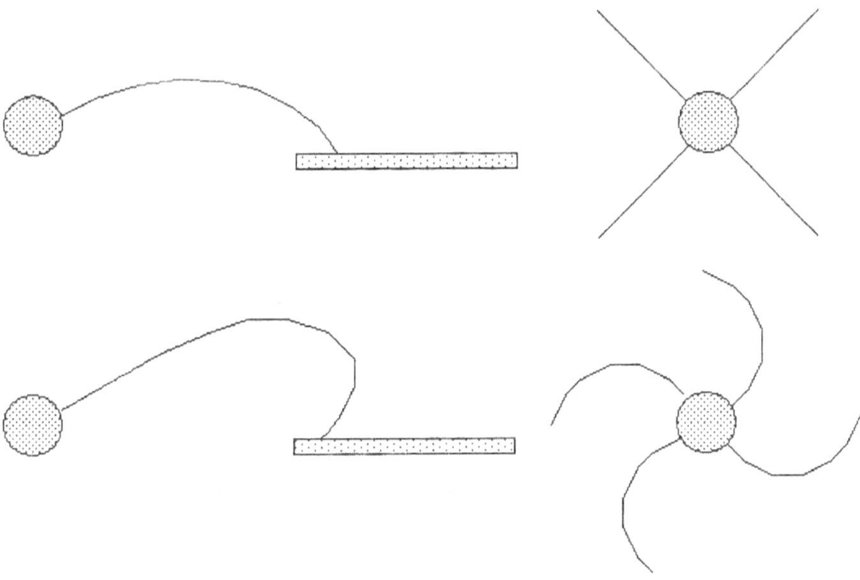

Figure 4. Schematic picture of magnetospheric structure in the situation where the region in the disk in which the stellar magnetic field is rooted has a different angular velocity than the star. The left hand figures show side views, while the right hand figures correspond to top views. Initially poloidal magnetic field lines (top figures) eventually wind up because of the differing angular velocities of the star and the disk region where the magnetic field is fixed (bottom figures). As the field line winds up, the magnetic field strength increases with the increased winding, and the field line bulges out over the disk (lower left). This figure is a schematic representation of the detailed force-free calculations presented by Van Ballegooijen (1994).

potential barrier to accretion (see, e.g., Shu et al. 1994). Thus one might assume that initially untwisted magnetic field lines might allow accretion, but after some twisting accretion would no longer be possible. Moreover, in the configuration shown in the bottom panel of Figure 4, there is an *outward* force of the magnetic field lines near the disk which would tend to push the disk outward. Thus, one might speculate that a time-dependent model of magnetospheric accretion might show complex behavior, with accretion at times along field lines, which then get twisted up, shut off accretion, and even move outward in the disk to couple to regions outside of corotation. In other words, the same magnetic field line might permit accretion and spindown, as in the Shu et al. (1994) model; however, unlike the Shu et al. model, *simultaneous* accretion and spindown does not occur, but rather alternate. The entire magnetosphere might be a complicated, time-dependent structure, with different field lines originating at differing stellar longitudes and latitudes sometimes allowing accretion, sometimes halting accretion and applying a spindown torque, all in a quasi-random pattern. Reconnection should lead to substantial heating and flare activity; this magnetic heating might be responsible in part for producing high enough temperatures to explain the observed permitted line emission from the magnetospheric gas (Muzerolle et al. 1997).

The implications of such complicated and time-dependent structure on mass ejection are not clear. We need to understand the time-dependence of accretion, and separate it from the rotational modulation of non-axisymmetric magnetic field structure. This will require expanded photometric monitoring programs such as undertaken by Bouvier et al. (1993) and Eaton & Herbst (1995) before we are able to determine the true timescales of accretion variability in young stellar objects.

Acknowledgements: This work was supported in part by NASA grant NAGW 2306.

References

Basri, G., & Bertout, C. 1989, ApJ, 341, 340
Bell, K.R., Lin, D.N.C., Hartmann, L.W., & Kenyon S.J. 1995, ApJ, 444, 376
Bertout, C., Basri, G., & Bouvier, J. 1988, ApJ, 330, 350
Bouvier, J., Cabrit, S., Fernandez, M., Martin, E.L., Matthews, J.M. 1993, A&A, 272, 176
Camenzind, M. 1990, Reviews in Modern Astronomy, (Springer-Verlag: Berlin), 3, 234
Clarke, C.J., Lin, D.N.C., Papaloizou, J.C.B. 1989, MNRAS, 236, 495
Clarke, C.J., Lin, D.N.C., Pringle, J.E 1990, MNRAS, 242, 439
Edwards, S., Strom, S.E., Hartigan, P., Strom, K.M., Hillenbrand, L.A., Herbst, W., Attridge, J., Merrill, K.M., Probst, R., & Gatley, I. 1993, AJ, 106, 372
Eaton, N.L., & Herbst, W. 1995, AJ, 110, 2369
Edwards, S., Hartigan, P., Ghandour, L., & Andrulis, C. 1994, AJ, 108, 1056
Ghosh, P., & Lamb, F.K. 1979, ApJ, 234, 296
Gullbring, E., Hartmann, L., Briceño, C., & Calvet, N. 1997, in preparation
Hartigan, P., Hartmann, L., Kenyon, S.J., Strom, S.E., Skrutskie, M.F. 1990, ApJ, 354, 25L
Hartigan, P., Strom, S., Edwards, S., Kenyon, S., Hartmann, L., Stauffer, J., & Welty, A. 1991, ApJ, 382, 617
Hartigan, P., Edwards, S., & Ghandour, L. 1995, ApJ, 452, 736
Hartmann, L., & Kenyon, S.J. 1996, ARAA, 34, 207
Hartmann L. 1995, in Circumstellar Disks, Outflows, and Star Formation, Rev. Mex. Astr. Ap. (Serie de Conferencias), 1, 285.
Herbig, G.H. 1977b, ApJ, 217, 693
Herbig, G.H. 1989, in ESO Workshop on Low-Mass Star Formation and Pre-Main Sequence Objects, ed. B. Reipurth (Garching: ESO), 233
Kenyon, S.J., Hartmann, L., Imhoff, C.L., Cassatella, A. 1989, ApJ, 344, 925
Königl, A. 1989, ApJ, 342, 208
Königl, A. 1991, ApJL, 370, L39
Lynden-Bell, D., & Pringle, J.E. 1974, MNRAS, 168, 603

Muzerolle, J., Calvet, N., & Hartmann, L. 1997, in preparation
Najita, J., & Shu, F.H. 1994, ApJ, 429, 808
Ostriker, E.C., & Shu, F.H. 1995, ApJ, 447, 813
Popham, R., Narayan, R., Hartmann, L., & Kenyon, S. 1993, ApJ, 415, L127
Popham, R., Kenyon, S., Hartmann, L., & Narayan, R. 1996, ApJ, 473, 422
Reipurth, B. 1989, in ESO Workshop on Low Mass Star Formation and Pre-Main Sequence Objects, ed. B. Reipurth (Garching: ESO Conference and Workshop Proceedings No. 33), 247
Reipurth, B. 1991 in The Physics of Star Formation and Early Stellar Evolution, eds. C.J. Lada & N.D. Kylafis (Dordrecht:Kluwer), 497
Shu, F., Najita, J., Ostriker, E., Wilkin, F., Ruden, S., & Lizano, S. 1994, ApJ, 429, 781
Shu, F., Najita, J., Ostriker, E.C., & Shang, H. 1995, ApJ, 455, L155
Valenti, J.A., Basri, G., & Johns, C.M. 1993, AJ, 106, 2024
Van Ballegooijen, A.A. 1994, Space Sci. Rev., 68, 299
Wang, Y.-M. 1995, ApJ, 449, L153
Wang, Y.-M. 1996, ApJ, 465, L111

THE RADIATIVE IMPACT OF FU ORIONIS OUTBURSTS ON PROTOSTELLAR ENVELOPES

K. R. BELL AND K. M. CHICK
Space Sciences Division, NASA Ames Research Center
MS 245-3 Moffett Field, CA 94035, USA
bell@cosmic.arc.nasa.gov, chick@anarchy.arc.nasa.gov

Abstract. We present preliminary results of an investigation into the radiative impact of FU Orionis outbursts on protostellar envelopes. In the thermal accretion disk instability model, the inner portion of the disk is inflated in such a way as to funnel most of the outburst luminosity along the poles of the system. A multidimensional radiative transfer code is employed to derive the thermal equilibrium structure of the surrounding protostellar envelope and the backheated accretion disk during outburst. The radiation emitted during outburst is modeled as a combination of a self-luminous accretion disk and a central point source. We compare the relative heating of circumstellar material due to (1) an isotropic point source and (2) an anisotropic point source in which the emitted radiation is confined to a 30° cone about the polar axis. The isotropic point source creates a spherical dust cavity, while the anisotropic source naturally results in a cavity which is hourglass-shaped. Repeated outbursts may ultimately be responsible for the large-scale polar cavities commonly inferred to exist around young stellar objects. We also show that due to the anisotropy of the radiation expected during outburst, disk annuli within a few *au* will be shielded from the radiation of the outburst.

1. Introduction

Based on evidence suggesting that the driving sources of Herbig Haro (HH) objects are transient but repetitive and that FU Ori outbursts release energy comparable to what is required to power observed HH outflows, it has been suggested that FU Orionis outbursts may be integral to the mechanism which gives rise to HH objects (Reipurth 1989). Several heavily embedded

B. Reipurth and C. Bertout (eds.), Herbig–Haro Flows and the Birth of Low Mass Stars, 407–416.

FU Ori systems which are associated with HH objects include L1551 IRS5 (in which HH objects are spaced at several hundred year intervals), HH57 (for which an HH object with a dynamical age of about 1000 *yr* was discovered before the FU Ori system brightened and was identified), and Z CMa. None of the three "classical" FU Ori systems (FU Ori itself, V1057 Cyg, and V1515 Cyg) are known to have HH objects, but this may be due to a combination of their relatively advanced ages (inferred from small far IR excesses) and the fact that they are all viewed close to pole-on (Kenyon, Hartmann, & Hewett 1988). Recent observations show that in several systems HH ejecta are dynamically clustered at periodic intervals with timescales similar to the expected FU Orionis outburst recurrence rate (HH34: 900 *yr*, Bally & Devine 1994; HH200: 500 *yr*, Bally et al. 1995).

During FU Orionis outbursts, system luminosities increase by a factor of a hundred in a time as short as a few years; outbursts typically last several decades (Herbig 1977). Outbursts are thought to occur repeatedly during the first few hundred thousand years to all low mass, young stellar objects. Statistical arguments suggest that spacing between outbursts is between one and ten thousand years. Numerical calculations (Kawazoe & Mineshige 1994; Bell & Lin 1994) and the observations referred to above favor the smaller number. We refer to Hartmann & Kenyon (1996) for a comprehensive review of the FU Ori phenomenon. In recent years, a detailed theoretical picture of FU Ori outbursts has emerged, in which a flare-up occurs in the central regions of an accretion disk (Lin & Papaloizou 1985; Clarke, Lin & Pringle 1990; Bell & Lin 1994). This modeling suggests that during the outburst, the central disk becomes distorted in such a way that energy is strongly focused upward. In this paper, we apply this theoretical picture to study the effects of an FU Ori outburst on the protostar's placental environment.

One of the on-going puzzles in the field of low mass star formation is the origin of the polar cavities which seem to be ubiquitous among low mass, young stellar objects (Kenyon, Calvet, & Hartmann 1993). Strong winds are common features of these systems, but it is not clear that the gas flow is capable of excavating a cavity. In particular, evidence indicates a high degree of collimation extending down close to the source of these outflows, yet the inferred cavities appear to have wide opening angles (Hartmann, Calvet, & Boss 1996). Although collapse from spherical cores (eg. Cassen & Moosman 1981) may not naturally result in optically thin lines of sight to the central object, systems which collapse from flattened sheets may result in the necessary geometry (Hartmann, Calvet, & Boss 1996). As an alternative, we investigate the possibility that dust-free polar cavities may be the result of dust vaporization by irradiation during FU Orionis outbursts.

Using a radiative transfer code with an assumed density distribution, we present preliminary results into an investigation of the thermal structure of the envelope during an FU Orionis outburst. In §2, we review the FU Orionis mechanism and summarize evidence which suggests that outburst luminosity will be emitted principally along the poles of the system. We also argue that FU Ori systems are deeply embedded objects likely to be surrounded by significant envelope material. We discuss our methods in §3. In §4, we show that the anisotropy of radiation emitted during outburst affects the thermal structure of the envelope and leads to the formation of bipolar, conical cavities. Source directionality will also reduce the irradiation of the surrounding accretion disk and may affect deduced bolometric luminosities. The current results are discussed in §5, and our plans for future work are given in §6.

2. FU Orionis Outbursts

Mass spirals inward through the accretion disk and lands on the central star at a rate which, although slowly declining with time, can be approximated as a radial constant at a given time. Mass transport is accompanied by the local generation of energy via viscous dissipation (Shakura and Sunyaev 1973). When mass flux through the disk is high enough to ionize hydrogen in the inner disk, the system will be subject to a thermal instability caused by the strong dependence of the H^- opacity on the temperature ($\kappa \sim T^{10}$). Numerical calculations (Bell & Lin 1994) suggest that as long as mass is transported into the inner disk at a rate greater than $5 \times 10^{-7}\ M_\odot yr^{-1}$, the inner disk will alternate between long periods of low mass flux quiescence ($\sim 1000\ yr$, $\sim 10^{-8}\ M_\odot yr^{-1}$) and short periods of high mass flux outburst ($\sim 100\ yr$, $\sim 10^{-4}\ M_\odot yr^{-1}$). Time between outbursts varies from 800 yr at an input mass flux of $10^{-6}\ M_\odot yr^{-1}$ to 1200 yr at $10^{-5}\ M_\odot yr^{-1}$ for a value of viscous efficiency of $\alpha = 10^{-4}$. For larger values of α this interval decreases accordingly. Outburst and quiescent mass fluxes depend only on the stellar mass and inner disk cut-off radius ($M_* = 1\ M_\odot$, $R_* = 3\ R_\odot$; Bell & Lin 1994).

Models suggest that for 1 $M_\odot$ systems, only the inner 1/3 au of the disk will be subject to thermal instability (for higher M_* this radius is larger). Light curves and color evolution of modeled outbursts agree well with observations of FU Orionis systems (Bell et al. 1995). A time sequence of B-band images of a model fit to the light curve of V1515 Cyg is shown in Figure 1. The image is viewed at an angle of 30° from pole-on and has a radial extent of 0.3 au. During outburst, the inner disk is depleted in surface density. The slow migration of matter which fills in this diminished region during quiescence can be seen in the first three frames. In 1968,

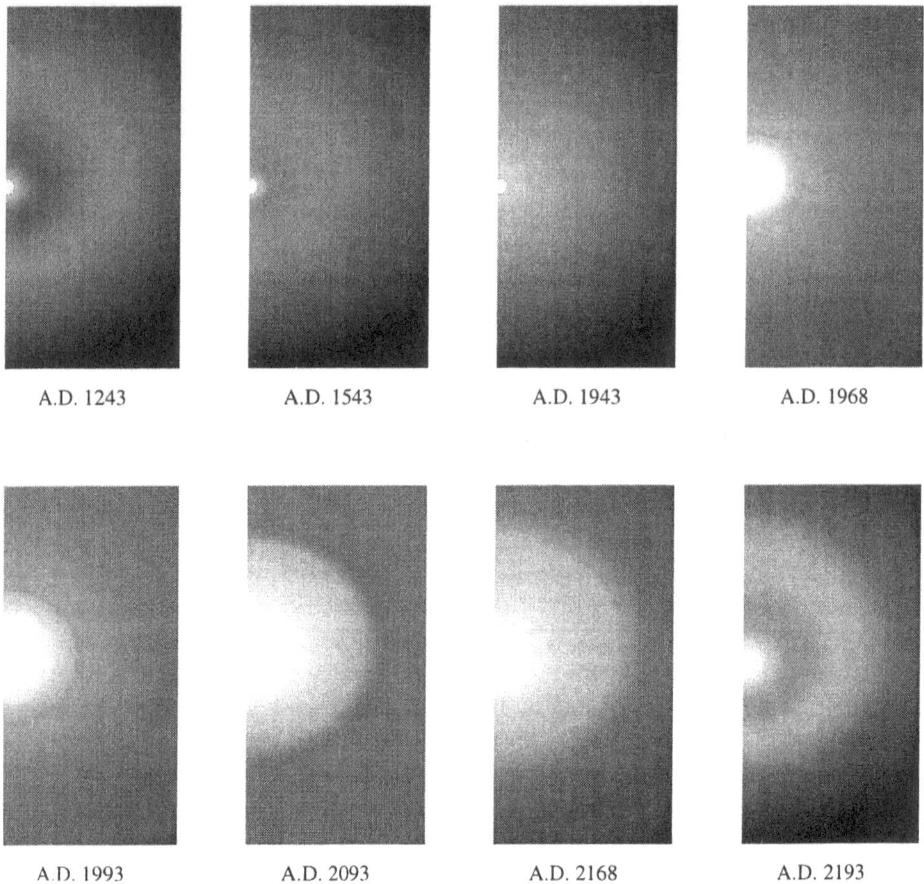

Figure 1. Progression of V1515 Cyg model B images throughout outburst cycle (from Turner, Bodenheimer, & Bell 1997).

the brightening near the star signifies the recent onset of the outburst. The ionization front sweeps radially outward through the inner disk to a distance of about 1/4 *au*. The luminosity then fades uniformly through the inner disk and the system has returned to quiescence by 2193.

Detailed simulations (Clarke , Lin, & Pringle 1990; Bell & Lin 1994; Bell et al. 1995) suggest that the inner disk surface during outburst is inflated inside the radius of hydrogen ionization; the brightest, hottest regions of the disk are consequently confined to a region within a few stellar radii which slopes strongly outward like the inside of a steep bowl. Radial cuts of three outbursting model disk shapes are shown in Figure 2. The labels A1, B1, and C1 refer respectively to light curve fits of the three "classical" FU Ori systems: FU Ori, V1515 Cyg, and V1057 Cyg. For the B1 model

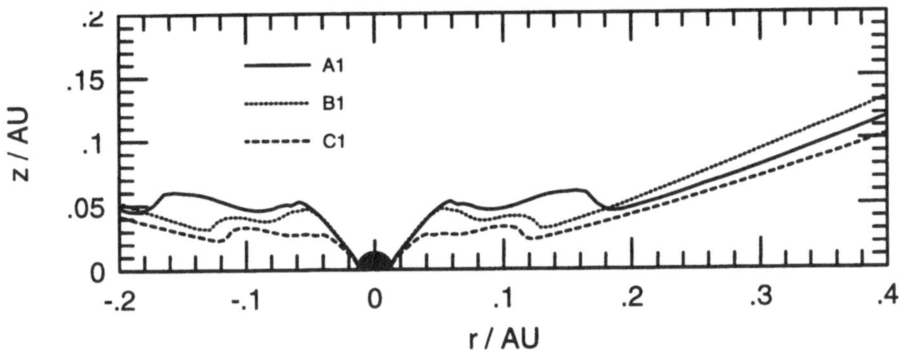

Figure 2. Shape of the inner disk during outburst with central star indicated (from Turner et al. 1997). The strongly sloping inner few stellar radii contain most of the luminosity of the system.

(the same that is shown here in Figure 1), the temperature at the peak in thickness at .05 *au* is only 4750 *K*, down from its maximum of 7900 *K*. The temperature continues to decrease rapidly with radius thereafter. Due to this geometry, we therefore expect that most of the radiation emitted during outburst will be focused upwards along the axis of the system.

Because of the high mass fluxes required to fuel the FU Ori cycle, it is expected that outbursts will be most common in the early evolution of a young stellar object when the remnant molecular cloud core is still dense. The existence of large scale circumstellar material is supported by images of nebulosity around most FU Ori systems (Herbig 1977, Goodrich 1987). Evidence for circumstellar material at much closer radius is given by observations of infrared emission in excess of disk continua (Kenyon & Hartmann 1991). During decay from peak light of the V1057 Cyg outburst, emission from this excess component was seen to track the luminosity of the central object while radiation from intermediate wavelengths, due to emission by the constant mass flux outer disk, remained more nearly constant (Kenyon, Hartmann, & Hewett 1988). Models derived from spectral energy distributions suggest that this material extends down to an *au* or so of the central object (Kenyon & Hartmann 1991; Turner, Bodenheimer, & Bell 1997). Their very association with HH objects further supports the idea that FU Ori systems are enshrouded with circumstellar envelopes, because most HH objects are associated with the heavily embedded Class I or Class 0 protostars.

3. Method

In this contribution, we present preliminary results of an investigation into the thermal impact of FU Ori outbursts on circumstellar envelopes. The radiative equilibrium structure of a model dust envelope is computed using an axisymmetric, multidimensional transfer code. The numerical scheme is summarized in the contributed poster by Chick & Cassen (1997a), and described in detail by Chick (1997) and Chick & Cassen (1997b). Scattering is not included in this procedure; for these relatively opaque models it is not expected to be a factor in the thermal balance. For the results presented here, the dust configuration is roughly toroidal and is based on theoretical models of collapse from a spherical cloud (Cassen & Moosman 1981). The cloud is described by an initial rotation rate of 0.32×10^{-14} sec^{-1}, a mass infall rate of 3×10^{-6} $M_{\odot} yr^{-1}$, and an age of 0.33×10^{6} yr. The cloud infall geometry is characterized by a centrifugal radius of 1.1 au.

In the simulation, energy emanates from two sources: a central point source, and an infinitesimally thin, luminous accretion disk in the equatorial midplane. Energy emitted by the disk is specified by a standard model for accretion disks (Lynden-Bell & Pringle 1974) with the same mass flux rate as the envelope: 3×10^{-6} $M_{\odot} yr^{-1}$. The temperature that the accretion disk surface reaches in radiative equilibrium with the warmed envelope is computed self-consistently. To test the effect of focusing emitted radiation along the axis likely to occur during outburst, heating by an isotropically emitting point source of luminosity L is compared with heating by an anisotropic point source, in which the same luminosity is emitted into a cone of opening angle 30°. Given our choice of cloud and accretion parameters, it is consistent to assign the central source a luminosity of $L = 28$ $L_{\odot}$ which is emitted at optical wavelengths. The disk has an intrinsic luminosity of 9 $L_{\odot}$ which is emitted at thermal wavelengths.

4. Results

The results of the radiative transfer calculation are shown in Figure 3. The right panel shows the temperature structure for an isotropic central source and the left panel for one in which the same luminosity has been confined to a cone of 30°. Only one quadrant is calculated. The gray scale is linear temperature. The radial extent of each image is 3.8 au. Both envelope models have the same density distribution. The regions which are completely white are sufficiently hot such that dust grains have been destroyed and a cavity formed.

Regardless of the density distribution of the envelope, the size and shape of the dust cavity depends principally on the intensity and directionality of the central source. As shown in the right panel of Figure 3, a spheri-

Figure 3. Temperature structure in envelope as a result of radiative transfer calculation. The non-isotropic source is on the left, the isotropic source is on the right. Both figures are plotted on the same linear scale; the lightest point is 1050 K and the darkest point is 350 K. The white regions indicate the dust cavities where grains are vaporized. The scale of the plot is 3.8 au on each side.

cal cavity is produced even in a highly flattened density distribution under illumination by an isotropic source (see also Chick, Pollack, & Cassen 1996). Under illumination by the anisotropic source, however, a conical or hourglass-shaped dust cavity is created.

A further consequence of this anisotropic radiation field is the amount of heating experienced by disk material during outburst. The amount of radiation intercepted by the underlying accretion disk during FU Orionis outbursts may have implications for thermal processing of primitive materials in the solar nebula. The disk in this model has four potential heating sources: intrinsic viscous energy generation, central object radiation which has been captured by the envelope and reprocessed back down onto the disk, direct radiation from the central object, and radiation from other parts of the self-luminous disk. Because in this preliminary calculation, the disk is assumed to be infinitesimally thin, the last two sources are strictly zero.

The disk surface temperature given by the Lynden-Bell & Pringle (1974) model is shown by the dashed line in Figure 4. When radiation is emitted isotropically from the central point source, the envelope is heated uniformly down to the surface of the disk and the reradiation of energy onto the disk's surface is strong. The self-consistent disk temperature profile in this case is the top line in Figure 4. When central object radiation is beamed out the poles, temperatures are reduced globally throughout the cloud; the shape

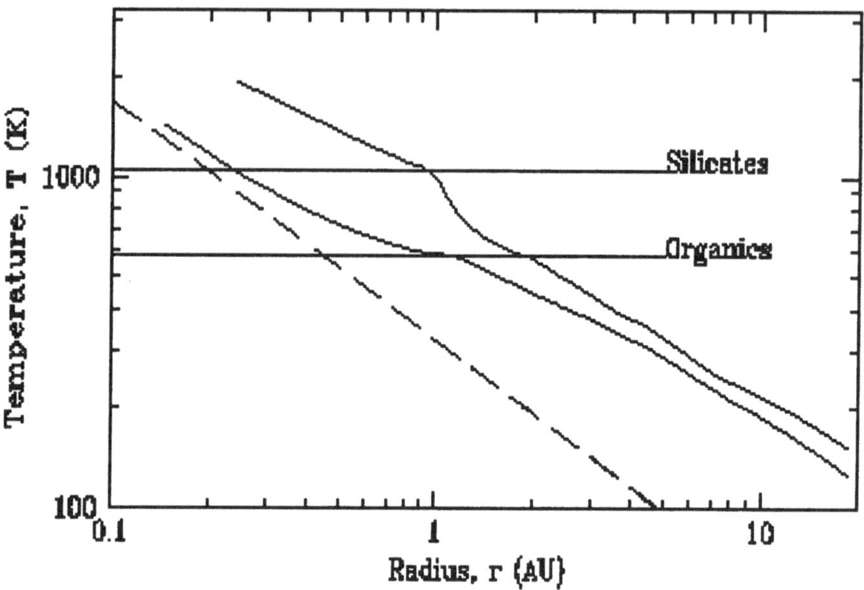

Figure 4. Radial temperature structure of the surface of the disk. The dashed line indicates the intrinsic disk profile due to locally generated accretion energy. The upper solid line includes the effects of radiation from the isotropic central source, the lower solid line from the non-isotropic source. The silicate and organic destruction temperatures are marked.

of the cavity and the disk temperature profile within a few *au* are considerably altered. There is a region where the disk is effectively shielded from the radiation of the central object and approaches the accretional surface temperature. Beyond a few *au*, the thermal radiative transfer washes out the imprint of the anisotropic source. The envelope's temperature structure becomes more uniform, approaching spherical symmetry, and the slope of the disk's surface temperature profile parallels that of the isotropic source case at a somewhat lower value.

5. Discussion

Previous modeling of the impact of FU Ori outbursts on envelope and accretion disk has generally assumed isotropic radiation from the central source. Our modelling shows that far from the anisotropic source the thermal structure does become uniform and is qualitatively similar to the isotropic source case. (Although the thermal structure approaches uniformity, there will still be considerable differences in luminosity, effective temperature etc. between equatorial and axial lines of sight.) Close to the center of the system, isotropic and anisotropic sources give rise to distinctly different thermal structures. Since radiation is beamed out the poles of the system, in the anisotropic case, temperatures are reduced globally throughout the envelope. Equatorial regions of the envelope and the accretion disk surface stay much cooler during outburst than with the isotropic central source. Our results suggest that within a few AU, source anisotropy will significantly reduce the heating by an FU Ori outburst of material in the solar nebula. The overall cooler envelope conditions may enhance the preservation of cometary ices. Due to dust vaporization, an hourglass-shaped cavity is opened along the cloud's axes. This is an efficient mechanism which may account for the ubiquitous polar openings inferred from observations of protostars and is independent of the details of infall physics.

In combination with the strong winds seen during FU Orionis outbursts, radiation emitted preferentially along the poles may help to power and enhance collimation of the jets and HH objects so commonly associated with these systems. Due to the increased heating of envelope material, the onset of outburst would create a sudden vaporization of dust particles. The front edge of new winds may therefore contain unusual and ionized species which may be detectable in some systems (as seen in Chandler & Richer 1997).

The directionality of the FU Ori radiation may cause inaccuracies in estimates of bolometric luminosities. It is generally assumed that a young stellar object is an isotropically radiating system and that one need only integrate out to long enough wavelengths to measure all of its luminosity. Once a strong polar cavity is carved out of the envelope, however, some of the luminosity of the FU Ori outburst escapes entirely from the system along the poles (see also Chick, Pollack, & Cassen 1996). This may account, for example, for the apparently anomalous low luminosity of edge-on system L1551.

6. Future Work

In this preliminary model, the disk has been assumed to be perfectly flat and to have no atmosphere. It is clear that the inner outbursting part of an FU Ori disk is quite complicated in structure. It has also recently been suggested that the outer constant mass flux part of the disk shows subtle behavior; in general the disk will be concave upward (and hence exposed to central object radiation) only out to the radius which marks the formation of dust grains ($\sim 1000\ K$) and concave downward thereafter (Bell et al. 1997). Because photons travel at grazing angles over the surface of the disk, the details of the disk's atmosphere will also be taken into account to properly assess the true impact of FU Ori outbursts on the solar nebula's thermal structure. Rather than using an ad hoc directionally dependent point source as we do now, we will include the detailed structure of both inner and outer disk which results from the full numerical simulations to provide disk temperatures and thicknesses at each radius. We will also consider the effect of different envelope density structures.

Acknowledgements: We would like to thank Neal Turner for his efforts in producing a video of the outbursting models which was shown at the conference and for post script versions of the figures from his paper used in this contribution and in the oral presentation. K.R.B. would like to thank the conference organizing committee and Origins grant NAGW 4456 for funding her travel. K.M.C. acknowledges funding by the NASA Origins of Solar Systems program.

References

Bally, J. & Devine, D. 1994, ApJ, 428, L65
Bally, J., Devine, D., Fesen, R. A., & Laine, A. P. 1995, ApJ, 454, 345
Bell, K. R. & Lin, D. N. C. 1994, ApJ, 427, 987
Bell, K. R., Lin, D. N. C., Hartmann, L. W., & Kenyon, S. J. 1995, ApJ, 444, 376
Bell, K. R., Cassen, P. M., Klahr, H. H. & Henning, Th. 1997 to appear September, ApJ, 486
Cassen, P., & Moosman, A. 1981, Icarus, 48, 353
Chandler, C. J. & Richer, J. S. 1997, in Low Mass Star Formation from Infall to Outflow: Poster Proceedings, Eds. F. Malbet & A. Castets, p. 76
Chick, K. M. 1997, submitted to Journal of Quantitative Spectroscopy and Radiative Transfer
Chick, K. M., & Cassen, P. 1997a, in Low Mass Star Formation from Infall to Outflow: Poster Proceedings, Eds. F. Malbet & A. Castets, p. 207
Chick, K. M. & Cassen, P. M. 1997b, ApJ, 477, 398
Chick, K. M., Pollack, J. B., & Cassen, P. M. 1996, ApJ, 461, 956
Clarke, C. J., Lin, D. N. C., & Pringle, J. E. 1990, MNRAS, 242, 439
Goodrich, R. W. 1987, PASP, 99, 116
Hartmann, L., Calvet, N., & Boss, A. 1996, ApJ, 464, 387
Hartmann, L. & Kenyon, S. J. 1996, ARAA, 34, 207
Herbig, G. H. 1977, ApJ, 217, 693
Kawazoe, E. & Mineshige, S. 1994, PASJ, 45, 715
Kenyon, S. J., Calvet, N., & Hartmann, L. W. 1993, ApJ, 414, 676
Kenyon, S. J. & Hartmann, L. W. 1991, ApJ, 383, 664
Kenyon, S. J., Hartmann, L., & Hewett, R. 1988, ApJ, 325, 231
Lin, D. N. C. & Papaloizou, J. 1993, in Protostars and Planets III, eds. E. H. Levy & J. Lunine (Tucson: Univ. of Arizona Press), 749
Lynden-Bell, D. & Pringle, J. E. 1974, MNRAS, 168, 603
Reipurth, B. 1989, Nature, 340, 42
Shakura, N. I. & Sunyaev, R. A. 1973, A&A, 24, 337
Turner, N. J. J., Bodenheimer, P., & Bell, K. R. 1997, ApJ, to appear May

PROPERTIES OF THE WINDS OF T TAURI STARS

NURIA CALVET
Centro de Investigaciones de Astronomía, Mérida, Venezuela, and Center for Astrophysics, 160 Concord, MA 02138, USA

Abstract. The properties of Classical T Tauri (TTS) winds - their mass loss rate, temperature, origin, and energetics - are reviewed and interpreted in terms of the most recent observational evidence. TTS winds seem to have their origin in the disk and are powered by accretion. Most TTS have $\dot{M}_w \leq 10^{-9} \mathrm{M}_\odot \, \mathrm{yr}^{-1}$ and $\dot{M}_w/\dot{M} \sim 0.1$, with T $\sim 10^4 K$.

1. Introduction

From very early on it was recognized that the characteristic blueshifted absorption component in the emission line profiles of T Tauri stars (TTS) indicated the presence of winds (Kuhi 1964). However, the interpretation of the wind and its origin has changed with the understanding of the origin of the activity in TTS. When TTS were thought to posses an amplified solar type kind of activity, winds were expected to be spherically symmetric, emerging from the stellar surface. As it became apparent that the energy source for the activity was accretion energy, and the presence of an accreting circumstellar disk became the natural scenario, the evidence pointed to the disk as the source of material and power in the wind. Two important points have arisen in this latest picture. First, the wind is far from spherically symmetric, at least at small spatial scales. Second, emission lines as a whole should no longer be seen as wind indicators, i.e., the bulk of the line probably arises in the magnetospheric infalling flow joining the disk and the star (see Edwards, this volume), and do not trace the outflowing material.

In this review, I will refer mostly to the properties of mass ejected near the star, at scales smaller than 100 AU, as inferred from present day observations, and discuss the observational constraints that theoretical models should explain to be applicable to TTS winds. In section 2, I discuss the

B. Reipurth and C. Bertout (eds.), Herbig–Haro Flows and the Birth of Low Mass Stars, 417–432.

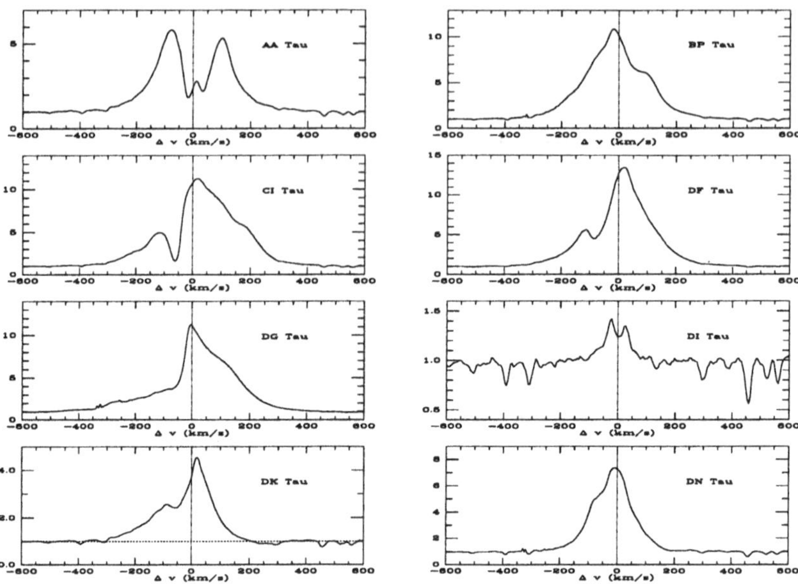

Figure 1. Typical Hα line profiles of TTS

appropriate wind indicators within the present picture of TTS. In the following sections, I review the region of formation (section 3), and the physical conditions of the wind (section 4) based on the information provided by those indicators. I will refer only to the Classical T Tauri stars (CTTS), and call them TTS. The weak line T Tauri stars show no indication of the presence of a wind, at least of the same nature as that of CTTS (Hartigan, Edwards, & Ghandour 1995, HEG hereafter), and will not be discussed.

2. Wind Indicators. Separating Magnetosphere and Wind

Figure 1 shows Hα line profiles for a sample of TTS. The typical features of the line profiles are: (1) An emission peak at nearly zero velocities; (2) A blueshifted absorption component (BAC), at velocities $\sim -50 - 150 \mathrm{km\,s^{-1}}$. A few stars, like AA Tau, show a strong central absorption, which is yet to be explained; (3) Extended blue and red wings, to velocities up to $\pm 250 \mathrm{km\,s^{-1}}$. Stars without accretion, as the WTTS DI Tau in Figure 1, show a narrow line profile.

Original interpretations sought to explain the whole emission line profile as produced in the wind. Spherical wind models (Hartmann, Edwards, & Avrett 1982; Lago 1984, 1986; Natta, Giovanardi, & Palla 1988; Hartmann etal 1990; Giovanardi etal 1991; Johns & Basri 1995a) can successfully explain the fluxes of emission lines, but generally fail in explaining the line profiles. As shown in Calvet, Hartmann, & Hewett (1992, their figure 3) profiles from spherical winds are typically P Cygni type, with deep, below the continuum, BACs and redshifted emission peaks, unlike the observed profiles (c.f. Figure 1). The stochastic spherical wind of Mitskevich, Natta, & Grinin (1993) can explain the observed line profiles, basically because these authors assumed that the wind decelerated outwards. However, it is very difficult to find theoretically a hydrodynamic wind solution with this property (Holzer etal 1983). Spherical winds also fail in explaining the decreasing depth of the BAC going up the Balmer series. Generally, in wind models, the source function falls steeply in the outer regions relative to the Planck function, and this effect gets stronger for lower opacity lines. As a result, the depth of the BAC increases for higher series lines, contrary to the observations.

Given the failure to explain the observed emission line profiles with classical winds, other alternatives were explored. Since the middle 80s, it is generally accepted that TTS are surrounded by disks responsible for the infrared excess, a model originally proposed by Lynden-Bell and Pringle (1974). In this picture, the disk material joined the star through a narrow boundary layer, where it slowed down from Keplerian to the slow rotation of the stellar surface. Nearly half of the accretion energy was supposedly dissipated in this layer; turbulent motions [and shocks] were expected, and winds could potentially be produced (Pringle 1989; Bertout & Regev 1992). Calvet, Hartmann, & Hewett (1992) explored the feasibility of this model by calculating line profiles from a configuration near the star with characteristics appropriate to boundary layers. The resultant line profiles still could not reproduce the observed profiles (c.f. Figure 3 of Calvet, Hartmann, & Hewett [1992]). They were too broad, due to rotation, the peak of the emission was still too redshifted, and at typical inclinations to the line of sight, the BAC was too deep.

In recent years, it became apparent that, given the expected magnetic field strengths in the stellar surface ($\sim$ few KG; Basri, Marcy, & Valenti 1992; Guenther, this volume), and typical mass accretion rates ($\sim 10^{-8} M_{\odot}\,yr^{-1}$, Gullbring etal 1997), the inner disks of TTS would be disrupted by the magnetic field and matter will move into the star along the field lines in nearly free fall (Bertout, Basri, & Bouvier 1988; Krautter, Appenzeller, Jankovics, 1990; Königl 1991; Calvet & Hartmann 1992; Shu etal 1994).

One of the most important evidences in favor of the magnetospheric model is that this model *can successfully reproduce the observed profiles of emission lines.* In the magnetospheric flow, the bulk of the emission comes from material moving at low velocities far from the star, while the wings

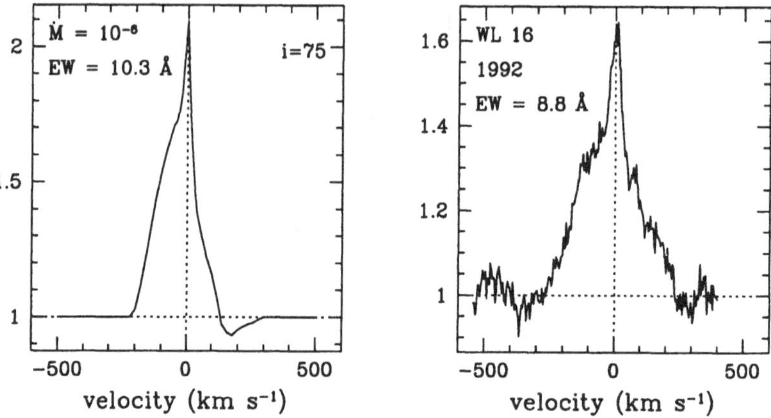

Figure 2. Brγ line profiles. Left: magnetospheric profile from Muzerolle etal (1997). Right: profile of WL 16 from Najita etal (1996)

are produced by the material close to the stellar surface, moving at the free-fall velocity. Therefore, the peak at nearly zero velocity and the line width are naturally explained. Magnetospheric profiles can explain very well the main features of Balmer lines (Hartmann, Hewett, & Calvet 1994). The multilevel, extended Sobolev radiative transfer in the magnetospheric flow by Muzerolle etal (1997) has extended this early treatment to higher Hydrogen series. Figure 2 compares the observed Br γ line profile for WL16 (Najita etal 1996) with a magnetospheric profile of Muzerolle etal (1997). The magnetospheric profile reproduces very well the main features of the observed profile.

The success of the magnetospheric model in explaining the observed line profiles has important consequences. First, the bulk of the line flux comes from the infalling, magnetospheric material, and it is not formed in the wind. Therefore, the hydrogen emission line fluxes are *not* indicators of the wind. The only true wind signature in permitted lines is the blueshifted absorption, which cannot be explained in terms of the magnetospheric model. Second, all the mass loss rate estimated based on the flux of the hydrogen emission lines, both optical and infrared lines (Natta etal 1988; Hartmann etal 1990) are meaningless, since those lines do not arise in the outflowing region.

Only analysis of the BAC can render an estimate of the rate of mass loss in TTS from permitted lines. On the other hand, the low excitation forbidden lines seen in TTS (Hamann 1994; HEG) do probe the outflowing material; in the cases where a spatially extended jet has been resolved,

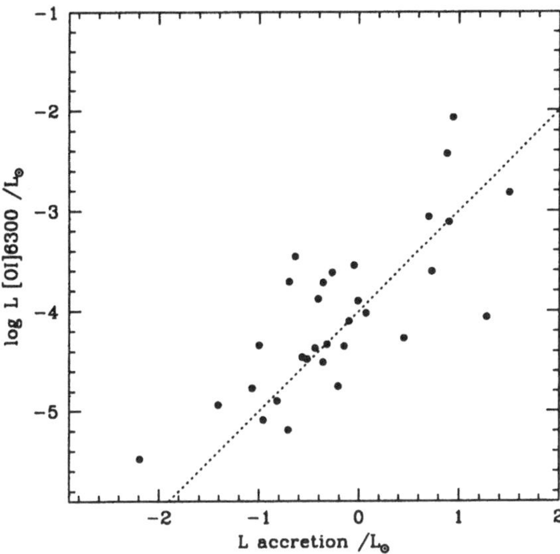

Figure 3. Total luminosity in [OI] 6300 vs. accretion luminosity, from data in HEG

the forbidden emission at the stellar position merges smoothly with the jet emission (Solf & Böhm 1993; Hirth, Mundt, & Solf 1993). Analysis of the forbidden lines provides the best available estimates so far of the mass loss rate in TTS.

3. Origin of the Wind

Since most of the flux of the permitted emission lines comes from the magnetospheric flow, they are no longer a tool to infer the physical characteristics of the wind. The forbidden lines provide the required information at large scales, but probably will be quenched in the densest, innermost zone, and we only have the profile of the BAC to analyze this region. Nonetheless, the analysis of these observational indicators provides clues to discern between the different models that have been discussed for the region where the wind is ejected and powered.

A wind is present whenever the disk is present. Stars that have near infrared excess have forbidden [OI] emission, while stars without excess do not show forbidden lines (Cabrit etal 1990; HEG). A near infrared excess is evidence that a disk with a range of temperatures (in contrast to the single temperature stellar flux) is present (Hartmann, this volume, and references therein), but a correlation with the infrared excess only proves that the disk presence is an essential ingredient for the existence of the wind, because the disk can be heated externally by stellar irradiation (Kenyon & Hartmann 1987; Calvet etal 1992). However, the correlation between the forbidden line luminosity and the accretion luminosity, as measured from the veiling of absorption lines (HEG), shown in Figure 3 strongly suggests that in addition to having a disk origin, TTS winds are powered by accretion energy. The form into which accretion energy is transmitted to the wind in TTS is still a question of debate.

3.1. THE FU ORI WIND

The case of the FU Ori objects can give some insight into the origin of the wind of the less energetic TTS. FU Ori objects are thought to be high mass accretion disks that surround a low mass star (Hartmann & Kenyon 1996; Hartmann, this volume). In one case, a strong line CTTS was observed previous to the disk outburst, so by looking at FU Ori we are in principle observing the same physical system, subject to a higher release of potential energy.

FU Ori objects prove to be the best clear case of a disk wind, in that we can see the wind accelerating region. The energy released by viscous processes in FU Ori disks heats the inner disk regions to temperatures $\sim 7000K$, so the spectrum of the disk atmosphere shows metallic absorption lines with a wide range of strengths. The weak lines are formed deeply in the disk atmosphere, in a region where the Keplerian rotational velocity is larger than the slow expansion velocity, so their line profile shows the characteristic double peak structure expected from a rotating disk. On the other hand, the strong lines are formed high up in the atmosphere, where the expansion velocity is larger than the rotational velocity; their line profiles are single and blueshifted (Calvet etal 1993). The upper panel Figure 4 shows the result of detailed modeling of metallic lines of differing strengths for the case of FU Ori, from Hartmann & Calvet (1995). As the strength of the line increases, the peaks of the double profile merge into one. The lower panel shows observed positions of the peaks of absorption lines in the spectrum of FU Ori (also reported by Petrov & Herbig [1992]). The observations behave as predicted by a rotating accelerating disk wind.

A mass loss rate $\dot{M}$ of $\sim 10^{-5} \mathrm{M}_{\odot}\, \mathrm{yr}^{-1}$ is found for FU Ori (Croswell, Hartmann, & Avrett 1987; Calvet etal 1993), yielding a ratio $\dot{M}_w/\dot{M} \sim 0.1$.

3.2. THE REGION OF ORIGIN OF TTS WINDS

Figure 5 shows observations of the Hα and Na I D lines for a sample of young stellar objects arranged from top to bottom in decreasing order

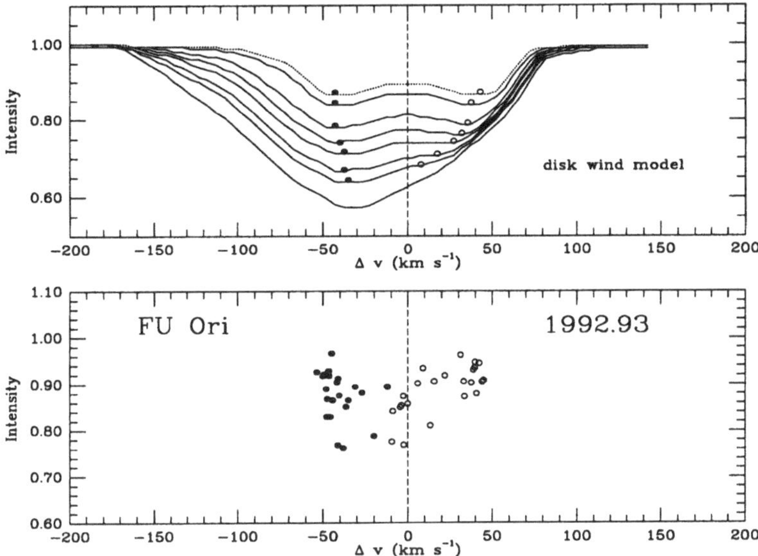

Figure 4. Line profiles for FU Ori disk wind (from Hartmann & Calvet 1995)

of mass accretion rate. At the top, the lines in FU Ori show very strong blueshifted absorption. As the mass accretion rate in the disk decreases, the lines develop emission, probably associated with the emergence of the magnetosphere, which is thought to be crushed out by the rapid accretion rates in FU Ori (Hartmann 1997). At the same time, the strength of the BAC decreases appreciably, to the extent that it is almost absent in the Na I D lines of the lowest mass accretion rate TTS.

Although the profiles shown in Figure 5 have been chosen to show the maximum effect, they agree with the previous conclusion that the winds are powered by accretion, in the sense that they are stronger in the highest mass accretion rate objects. Similarly, since the BAC has to be formed very near the star to absorb the stellar and magnetospheric background light, the profiles shown indicate that the density in the inner accelerating wind region decreases with mass accretion rate.

From the observations of individual BAC, however, it is difficult to deduce the exact region of formation of the wind. Two possibilities have been advanced so far. The wind may arise in the inner disk region, similarly to the case of the FU Ori objects (Blandford & Payne 1982; Pudritz & Norman 1983; Königl 1989; Camenzind 1990; Pelletier & Pudritz 1992; Gómez de Castro & Pudritz 1993; Safier 1993; Paatz & Camenzind 1996).

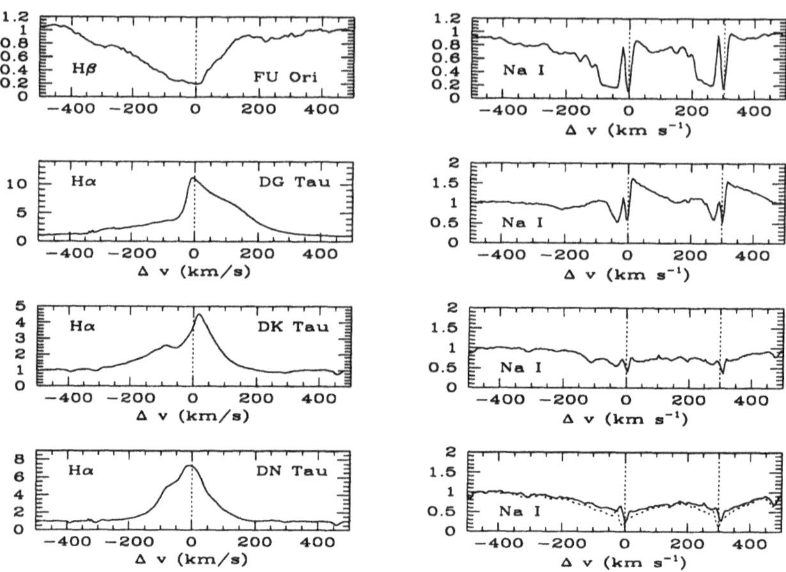

Figure 5. Line profiles of Hα and Na I D for objects arranged by mass accretion rate. From top to bottom: FU Ori, $\dot{M} \sim 10^{-4} M_{\odot}\,yr^{-1}$, DG Tau, $\dot{M} \sim 2 \times 10^{-6} M_{\odot}\,yr^{-1}$, DK Tau, $\dot{M} \sim 4 \times 10^{-7} M_{\odot}\,yr^{-1}$, DN Tau $\dot{M} \sim 3 \times 10^{-8} M_{\odot}\,yr^{-1}$. HEG data. The sharp lines at 0 velocity in the Na I D profiles are of interstellar origin.

Alternatively, the wind may be produced at the same point where the magnetospheric flow is launched, at the radius where the disk corotates with the star, as envisaged by the X-wind model (Shu etal 1994; Najita & Shu 1994; Shu etal 1995). It has proven difficult to device observational tests to distinguish between both kinds of models, especially because the origin of the heating and therefore, the temperature structure, are not well known.

However, variability analyses of line profiles with time baselines covering significantly several stellar periods have provided strong clues as to the location of the wind formation region. The time series analyses comprise several aspects: (a) Search for periodicities in different parts of the line profiles, and compare them with the stellar period; (b) Determine characteristic time scales for variability; (c) Search for correlations between the variations in different parts of the line profiles, to indicate relationships between the physical regions where each arise.

A long time baseline of 37 nights with hundreds of spectra has provided important indications for the characteristics of the BAC and thus of the wind (Johns etal 1992; Giampapa etal 1993; Johns & Basri 1995a,b;

Johns-Krull & Basri 1997). In a sample of 7 stars where the profiles of the strong emission lines have been monitored, the BAC has been found to vary periodically with the same period as the star in one case. This observation supports the view that the wind comes from the corotation radius, in agreement with the predictions of the X-wind model. However, in the rest of the sample, six stars, the BAC is uncorrelated with other parts of the line profile; it is either not periodical, or if periodical, does not vary with the stellar period. Moreover, it shows a longer variation time scale than the rest of the line profile. The authors conclude from these observations that (1) the wind formation region is detached from the region of formation of the main emission, and (2) the wind formation region is exterior to the formation of the main emission. This evidence supports the view that the wind is produced in the disk, in a region exterior to the magnetosphere.

Since the one star where the BAC varies with the stellar period is not a representative TTS, in that it has a higher mass and few evidences of the TTS activity (Edwards, this volume), while the rest of the sample consists of more typical TTS, the variability studies carried out so far indicate a disk origin for the T Tauri wind. It is important, however, to increase the sample of TTS with significant time series coverage to confirm this indication.

3.3. FORBIDDEN LINES. EXTENDED REGION

The forbidden lines probe the extended, lower density regions of the wind. However, the information they provide on the geometry of the wind region is not yet conclusive.

The line profiles of forbidden lines in TTS are generally blueshifted. The overall blueshift of the line has been interpreted as indicative of the presence of a disk, which hides the receding flow of the line (Jankovics, Appenzeller, & Krautter 1983; Appenzeller, Jankovics, & Östreicher 1984; Edwards etal 1987). In addition, the line profiles of forbidden lines in TTS are often double peaked. Following HEG nomenclature, the peak at velocities $\geq -60\mathrm{km\,s^{-1}}$ is called the low velocity component, LVC, while the peak at larger negative velocities is referred to as the high velocity component, HVC.

Figure 6 shows the number of stars where the LVC or the HVC peak is present in the HEG sample, arranged in order of mass accretion rate. The HVC is clearly present only in the high mass accretion stars, while the low velocity component is more ubiquitous. Line ratios indicate that the excitation conditions for both components in a given star are different, with a higher density in the LVC (Hamann 1994; HEG; Solf, this volume). In addition, the HVC is generally spatially resolved, and extends into jets when they can be seen (Solf & Böhm 1993; Hirth, Mundt, & Solf 1993;

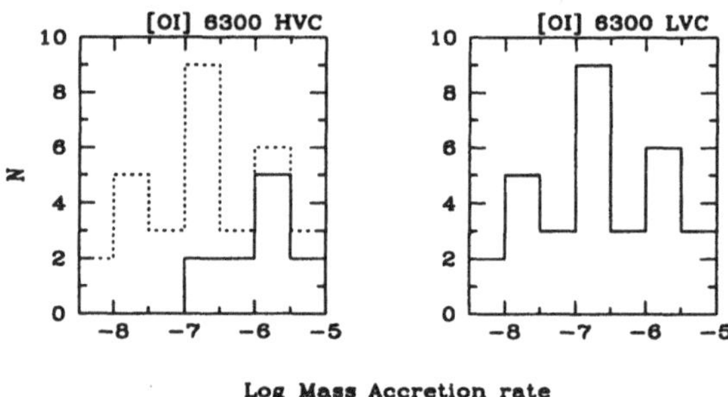

Figure 6. Number of stars with given mass accretion rate where a LVC peak (right panel) or a HVC peak (left panel) is present in [OI] 6300. The dotted line is the total number of stars in the sample, for comparison. Data from HEG

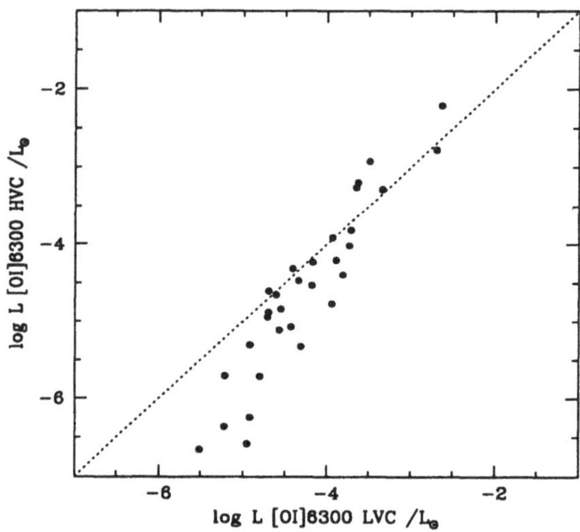

Figure 7. Luminosity of the HVC vs luminosity of the LVC of [OI] 6300. Data from HEG.

Hirth etal 1994). The LVC in general is not resolved, except in [SII] in a few cases (Solf & Böhm 1993).

These observations have led to the suggestion that the two components arise in two distinct physical regions, in contrast to the models that propose geometric or/and inclination effects to explain the profiles (Edwards etal 1987; Raga 1989; Safier 1993; Hartmann & Raymond 1989; Ouyed & Pudritz 1994). In this view, the HVC would correspond to the jet itself while the LVC would come from a wind arising in outer disk regions where the escape velocity is low (Kwan & Tademaru 1988; 1995). However, as shown in Figure 7, the luminosities of the two components are tightly correlated. Moreover, the slope of the correlation is greater than one, suggesting a density effect: at low luminosities, only the higher density low velocity component is present; at high luminosities the wind overall has higher densities, and the low velocity component, produced in the densest regions, may be quenched. This behavior is in agreement with the suggestion that there is only one main emitting region, seen at different scales (Camenzind, this volume). The issue of the origin of the forbidden lines is still an open question.

4. Mass Loss Rates

The determination of mass loss rates in the winds of TTS requires the knowledge of the excitation conditions in the emitting region, as well as of its geometry. As these quantities are very uncertain, mass loss rate determinations are only order of magnitude estimates. In this section, I discuss the mass loss rates from the study of the forbidden lines. In addition, I use the BAC of lines with differing ionization and excitation characteristics to constrain the density and temperature of the inner accelerating region of the wind. With these quantities, I derive crude estimates of the mass loss rate from the permitted lines and find them to be consistent with the forbidden line results. Again, these estimates are only appropriate to order of magnitude, and detailed analysis of simultaneous spectra of individual stars should be made to obtain more accuracy.

4.1. FORBIDDEN LINES

Mass loss rate estimates of TTS have been obtain by HEG. They use only the HVC of [OI] 6300 for their calculations, since this is the component clearly associated with the jet. They include as HVC any emission at large negative velocities, even if there is no line maximum associated with it. They also assume that the forbidden emission is produced in a shock.

The line luminosity is given by

$$L_l \simeq n_u A_{ul} h\nu V \tag{1}$$

where n_u is the upper level of the transition producing the line of frequency ν, A_{ul} is the Einstein coefficient, and V is the emitting volume.

At the same time, the total emitting mass is

$$M \simeq \mu m_H n_H V \tag{2}$$

where μ os the mean atomic weight, and n_H is the total hydrogen density. The usual symbols are used for other atomic constants.

If the observations are limited to a region of size l, in which the material is moving at velocity v, then the mass loss rate can be estimated as

$$\dot{M}_w \simeq M \frac{v}{l} \tag{3}$$

Combining these equations, we get a relationship between the mass loss rate and the line luminosity,

$$\dot{M}_w \simeq \mu m_H \left(\frac{v}{l}\right) \left(\frac{n_H}{n_E}\right) \left(\frac{n_E}{{n_E}^i}\right) \left(\frac{{n_E}^i}{n_u}\right) L_l \tag{4}$$

where n_E is the total density of the element producing the line and ${n_E}^i$ is the density of this element in the corresponding ionization stage. To evaluate this expression, HEG take v from the observed velocity shift and l from the size of the slit. The abundance of the element gives n_H/n_E. To evaluate the ionization stage, $n_E/{n_E}^i$ and the relative population of level u, ${n_E}^i/n_u$, knowledge of the temperature and the electron density is required. HEG assume a temperature of 8000K, arguing that this is the temperature where OI emits the most in shocks. They also assume that most of the oxygen is neutral and most of the sulphur once ionized. To estimate the electron density, they use the ratio of the HVC fluxes of [S II] 6731 and [OI] 6300, and a two level atom formulation.

With these assumptions HEG analyze a sample of 31 stars with accretion, of which 25 have measurable high velocity emission in [OI], for which they obtain mass loss rates in the range between $10^{-10}\mathrm{M}_\odot\,\mathrm{yr}^{-1}$ and $3 \times 10^{-7}\mathrm{M}_\odot\,\mathrm{yr}^{-1}$, with a median value of $\sim 10^{-9}\mathrm{M}_\odot\,\mathrm{yr}^{-1}$. In the case of RW Aur, HEG derive $\dot{M}_w \sim 3 \times 10^{-8}\mathrm{M}_\odot\,\mathrm{yr}^{-1}$, in good agreement with the independent determination of mass loss rate from direct analysis of the resolved jet (Bacciotti, Hirth, & Natta 1996).

The mass loss rate increases with mass accretion rate. Using the mass accretion rate determined by HEG, $\dot{M}_w/\dot{M}$ is $\leq 10^{-2}$. However, more recent determinations of mass accretion rate seem to indicate that the HEG mass accretion rates are overestimated by a factor $\sim$ 10 (Gullbring etal 1997, Hartmann, this volume). If this is the case, then $\dot{M}_w/\dot{M} \sim 0.1$ is a more typical value.

4.2. PERMITTED LINES

Analysis of the BAC of emission lines is difficult because there are many parameters involved: temperature and density, velocity field, geometry, and inclination. Theoretical models give indication on some of these parameters, like the geometry and the velocity field. However, both the basic uncertainty in the excitation characteristics in the theoretical models and the lack of reliable inclination estimates for individual stars make comparison with observations difficult.

However, even with these uncertainties, order of magnitude estimates can be done. The BAC has to form within a few stellar radii from the star, since the atoms producing it have to absorb the stellar plus magnetospheric background. We can safely assume that this may still be the accelerating region of the flow, so that we can use the Sobolev approximation to estimate the optical depth at velocity v:

$$\tau_v = \frac{\pi e^2}{mc} \frac{fc}{\nu_o} \frac{n_l(v)}{dv/dz} \tag{5}$$

where f is the oscillator strength of the line with center frequency ν_o, $n_l(v)$ is the density of the lower level of the transition of atoms moving at velocity v, and dv/dt is the acceleration of the region moving at that velocity.

The mass loss rate can be estimated as

$$\dot{M}_w \simeq \Delta A\, v\, \mu m_H n_H(v) \tag{6}$$

where ΔA is the area crossed by the atoms moving at v. Since $n_l(v)/n_H(v) \sim n_l/n_H$ is a function of n_H and T, if we have an estimate of ΔA and v, we can calculate the optical depth of a given line at that velocity as a function of $\dot{M}_w$ and T, using these two expressions. As the absorption occurs very close to the star, we take $\Delta A \sim \pi(2R_*)^2$, and with typical velocities and radii, $v \sim 100\,\mathrm{km\,s^{-1}}$ and $dv/dz \sim 100\,\mathrm{km\,s^{-1}/R_*}$.

With this procedure, we can calculate optical depths at the center of the BAC, and compare with observations to determine the physical conditions of the region producing the BAC. Several observations can be used: (a) $\tau(H\alpha) \sim 1$ - 10. Natta (1996) measured the optical depth at the center of the BAC in $H\alpha$ for a range of TTS with differing mass accretion rates. She found it to be of order 1 or less, except in the most energetic objects; (b) $\tau(Br\gamma) << 1$. Najita, Carr, & Tokunaga (1996) observed line profiles of Br(γ) for a sample of TTS, generally of high mass accretion rate. They found no hint of the presence of a BAC, even in objects that show it very strongly in Hα; (c) $\tau(Pa\beta) << 1$. Folha etal (1997) find that Paβ does not show the BAC either, except possibly in one case of a star, CW Tau, which has a high mass loss rate as determined from the forbidden lines.

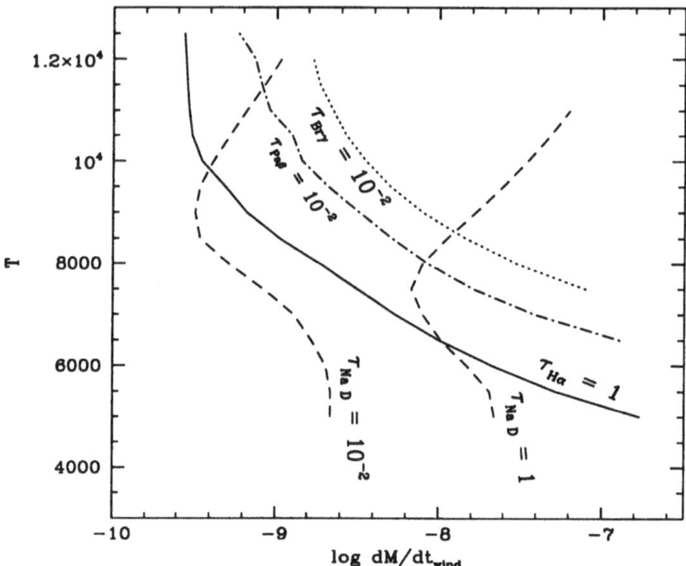

Figure 8. $\dot{M}_w$ vs. T diagram with lines indicating typical values of the optical depth of Hα, Brγ, Paβ, and Na I D.

(d) $\tau(NaID) << 1$ for most T Tauri stars, of order 1 for the high mass accretion rate objects; (c.f. spectra in Edwards etal 1994). These constraints can be incorporated in a ($\dot{M}_w$,T) diagram as shown in Figure 8. $\tau \sim 10^{-2}$, which corresponds to $\sim$ 1% of the neighboring flux, has been adopted as an upper limit when the absorption is not seen.

As noted by Natta & Giovanardi (1990), the Na I D lines are good indicators of density. Objects which show Na BAC have $\dot{M}_w \sim$ few $\times 10^{-8} M_\odot\,yr^{-1}$, while for objects that do not show Na I absorption, $\dot{M}_w \leq 10^{-9} M_\odot\,yr^{-1}$. The hydrogen lines, on the other hand, are good indicators of temperature. In particular, the absence of absorption in the infrared lines, especially Paβ, puts strong constraints on the temperature of the absorbing region. The temperature has to decrease as $\dot{M}_w$ increases, in agreement with the low values of $\tau(H\alpha)$, since a lower temperature is required to have $\tau H(\alpha) \sim 1$. This behavior is expected, since radiative cooling increases with density (c.f. Hartmann, Edwards, & Avrett 1982). For the high $\dot{M}_w$ objects, which show strong BACs in Hα and Na I D, the temperature of the absorbing region has to be lower than $\sim$ 6000K. In contrast, for the lower $\dot{M}_w$ objects to show BAC in Hα, the temperature has to be $\sim$ 8000K or higher. Theoretical models of heating have to be able to explain these constraints.

5. Summary and Conclusions

The present observational evidence suggests that:
– The winds of TTS have their origin in the disk, in a region exterior to the magnetosphere. They are powered by accretion.
– The indicators of the wind - forbidden lines and blueshifted absorption components in permitted emission lines - give consistent results for the mass loss rate: for most TTS, the winds are weak, $\dot{M}_w \leq 10^{-9} M_\odot\, yr^{-1}$. The mass loss rate increases with the rate of accretion in the disk, approximately as $\dot{M}_w/\dot{M} \sim 0.1$. The most massive winds have $\dot{M}_w \sim 10^{-7} M_\odot\, yr^{-1}$.
– Analyses of the blueshifted absorption component of permitted emission lines indicate that the temperature in the inner regions of the wind decreases as $\dot{M}_w$ increases. The temperature must be $\sim 10^4$K for the low $\dot{M}_w$ objects, but less than 6000K for $\dot{M}_w \sim 10^{-7} M_\odot\, yr^{-1}$.
– Although all CTTS have winds insofar as they have a disk, the winds of most of them are too tenuous to produce spatially extended observable jets.

Acknowledgments. I would like to thank the French Ministry of Foreign Affairs for a grant to pay my travel expenses to this conference. I also would like to thank the organizers for their support, L. Hartmann for helpful discussions and for figures 1 and 5, J. Najita for the profile of WL 16, and E. Gullbring for useful comments. This work was supported by NASA Grant 2306.

References

Appenzeller, I., Jankovics, I., Ostreicher, R.: 1984, AA, 141, 108
Basri, G., Marcy, G. W., Valenti, J. A.: 1992, ApJ, 390, 622
Blandford, R.D., Payne, D.G.: 1982, MNRAS 199, 883
Bertout, C., Basri, G., Bouvier, J.: 1988, ApJ, 330, 350
Bertout, C., Regev, O.: 1992, ApJ, 399, L163
Bacciotti, F., Hirth, G. A., Natta, A.: 1996, AA, 310, 309
Cabrit, S., Edwards, S., Strom, S. E., Strom, K. M.: 1990, ApJ, 354, 687
Calvet, N., Hartmann, L.: 1992, ApJ, 386, 239
Calvet, N., Hartmann, L., Hewett, R.: 1992, ApJ, 386, 229
Calvet, N., Magris C.,G., Patiño, A., D'Alessio, P.: 1992, Rev. Mexicana AA, 24, 27
Calvet, N., Hartmann, L., Kenyon, S. J.: 1993, ApJ, 402, 623
Camenzind. M.: 1990, in Rev. Mod. Astron. 3, ed. G. Klare, Sringer-Verlag, 234
Croswell, K., Hartmann, L., Avrett, E. H.: 1987, ApJ, 312, 227
Edwards, S., Cabrit, S., Strom, S., Heyer, I., Strom, K. M., Anderson, E.: 1987, ApJ, 321, 473
Edwards, S., Hartigan, P., Ghandour, L., Andrulis, C.: 1994, AJ, 108, 1056
Folha, D., Emerson, J., Calvet, N.: 1997, Poster Proceedings of the IAU Symposium 182, eds. F. Malbet and A. Castets, Observatoire de Grenoble, 272
Giampapa, M. S., Basri, G., Johns, C. M., Imhoff, C. L. : 1993, ApJS, 89, 321
Giovanardi, C., Gennari, S., Natta, A., Ruggero, S.: 1991, ApJ, 367, 173
Ghosh, P., Lamb, F. K.: 1978, ApJ, 223, 83

Gómez de Castro, A. I., Pudritz, R. E.: 1993, ApJ, 409, 748
Gullbring, E., Hartmann, L., Briceño, C., Calvet, N.: 1997, in preparation
Hamann, F.: 1994, ApJS, 93, 485
Hartmann, L., Edwards, S., Avrett, E.: 1982, ApJ, 261, 279
Hartmann, L., Raymond, J. C.: 1989, ApJ, 337, 903
Hartmann, L., Calvet, N., Avrett, E. H., Loeser, R.: 1990, ApJ, 349, 168
Hartmann, L., Hewett, R., Calvet, N.: 1994, ApJ, 426, 669
Hartmann, L., Kenyon, S. J.: 1996, ARAA, 34, 207
Hartmann, L., Calvet, N.: 1995, AJ., 109, 1846
Hartmann, L.: 1997, Accretion Processes in Star Formation, Cambridge University Press
Hartigan, P., Edwards, S., Ghandour, L.: 1995, ApJ, 452, 736 (HEG)
Hirth, G. A., Mundt, R., Solf, J.: 1993, AA, 285, 929
Hirth, G. A., Mundt, R., Solf, J., Ray, T: 1994, ApJ, L99
Holzer, E., Fla, T., Leer, E.: 1983, ApJ, 275, 808
Jankovics, I., Appenzeller, I., Krautter, J. : 1983, PASP, 95, 883
Johns, C. M., Basri, G.: 1995a, ApJ, 449, 341
Johns, C. M., Basri, G.: 1995b, AJ, 109, 2800
Johns-Krull, C. M., Basri, G.: 1997, ApJ, 474, 433
Kenyon, S. J., Hartmann, L.: 1987, ApJ, 323, 714
Kepner, J., Hartigan, P., Yang, C., Strom, S.: 1993, ApJ, 415, L119
Krautter, J., Appenzeller, I., Jankovics, I.: 1990, AA, 236, 416
Königl, A.: 1989, ApJ, 342, 208
Königl, A.: 1991, ApJ, 370, 39
Kuhi, L. V.: 1964, ApJ, 140, 1409
Kwan, J., Tademaru, E.: 1988, ApJ, 332, L41
Kwan, J., Tademaru, E.: 1995, ApJ, 454, 382
Lago, M. T., V. T.: 1984, MNRAS, 210, 323
Lago, M. T., V. T.: 1986, MNRAS, 222, 213
Lynden-Bell, D., and Pringle, J. E.: 1974, MNRAS, 168, 603
Mitskevich, A. S., Natta, A., Grinin, V. P.: 1993, ApJ, 404, 751
Muzerolle, J., Calvet, N., Hartmann, L.: 1997, in preparation
Najita, J. R., Shu, F. H.: 1994, ApJ, 429, 808
Najita, J., Carr, J. S., Tokunaga, A. T.: 1996, ApJ, 456, 292
Natta, A., Giovanardi, C.: 1990, ApJ, 356, 646
Natta, A., Giovanardi, C., Palla, F.: 1988, ApJ, 332, 921
Natta, A.: 1996, in ASP Conf. Proc. 109, Cool Stars, Stellar Systems and the Sun, eds. R. Pallavicini and A. Dupree, (San Francisco: ASP), 471
Ouyed, R., Pudritz, R. E.: 1994, ApJ, 423, 753
Paatz, G., Camenzind, M.: 1996, AA, 308, 77
Pelletier, G., Pudritz, R.E.: 1992, ApJ, 394, 117
Petrov, P. P., Herbig, G. H.: 1992, ApJ, 392, 209
Pringle, J. E.: 1989, MNRAS, 236, 107
Pudritz, R.E., Norman, C.A.: 1983, ApJ, 274, 677
Raga, A.: 1989, AJ, 98, 976
Safier, P. N.: 1993, ApJ, 408, 148
Shu, F. H., Najita, J. R., Ruden, S. P., Lizano, S.: 1994, ApJ, 429, 797
Shu, F., Najita, J., Ostriker, E.C., Shang, H.: 1995, ApJ, 455, L155
Solf, J., Böhm, K. H.: 1993, ApJ, L31

MAGNETOSPHERICALLY MEDIATED ACCRETION IN CLASSICAL T TAURI STARS

Paradigm for Low Mass Stars Undergoing Disk Accretion?

S. EDWARDS
Smith College
Northampton, MA 01063, USA

Abstract. Observational evidence suggests that disk accretion in low mass stars is mediated by the stellar magnetosphere in the inner few stellar radii of a star/accretion disk system. A strong stellar field appears to disrupt the inner disk and channel accreting material in a funnel flow toward the stellar surface. It is possible that coupling between the stellar magnetosphere and the inner accretion disk plays a profound role in the evolution of a forming star, regulating its spin, establishing its initial angular momentum, and launching the ubiquitous accretion-driven outflows which are the subject of this symposium. Although the evidence that forming low mass stars undergo magnetospherically mediated accretion is considerable, there are a growing number of challenges to this paradigm which require close observational scrutiny.

1. Introduction

You may be wondering why a contribution on magnetospheric accretion has been included in a symposium on Herbig-Haro Flows. There are two compelling reasons. One is that two classes of models use open stellar field lines to launch energetic winds from disk-accreting stars, either from the magnetosphere-disk interface (Shu & Shang, this volume) or from the magnetospheric foot-points on the stellar surface (Camenzind, this volume). The second reason is that it is important to understand which spectroscopic features diagnose the presence of magnetospheric accretion in classical T Tauri stars (cTTS) in order to assess, as near-infrared high resolution spectra become available, whether magnetospheres are important in the more deeply embedded Class I sources (Lada and Wilking, 1984) driving outflows.

B. Reipurth and C. Bertout (eds.), Herbig–Haro Flows and the Birth of Low Mass Stars, 433–442.

Low mass stars first become optically visible as T Tauri stars at an age $t \sim 1$ Myr, at which time approximately half are still surrounded by accretion disks (Strom, Edwards and Skrutskie 1993). Disk accretion rates for cTTS, inferred primarily from optical continuum emission excesses are typically $\dot{M}_{acc} \sim 10^{-7}$ $M_{\odot}$ yr^{-1} (Hartigan, Edwards and Ghandour [HEG], 1995; but see Hartmann this volume). These active young stars also likely generate strong dynamo magnetic fields on the order of 500 to 1500 kG (Guenther, this volume; Montmerle et al., 1993; Basri, Marcy and Valenti, 1992.) which are capable of coupling to the accretion disk and disrupting it within a few stellar radii of the central star (Königl 1991).

There is mounting evidence that magnetosphere-disk coupling plays a vital role in the angular momentum evolution of forming stars, making it essential that the relevance of this mechanism be firmly established. Key areas where magnetosphere-disk coupling may be important are:

• magnetosphere-disk coupling may account for the apparent regulation of the angular velocity of the central star in the presence of disk accretion, countering its tendency to spin up, both from accretion of disk material of high specific angular momentum and from contraction toward the main-sequence (Edwards et al. 1993; Bouvier et al. 1993; Cameron and Campbell, 1993; Ghosh, 1995).

• magnetosphere-disk coupling has the potential to produce a broad dispersion in angular velocities among low mass stars on the ZAMS, if the timescale for disk dissipation extends over timescales from less than 1 Myr up to 10 Myr (Bouvier 1994; Cameron, Campbell and Quaintrell, 1995; Bouvier et al. 1997).

• magnetosphere-disk coupling may not only extract angular momentum from the star, braking its spin; but also may expel this angular momentum from the system via an accretion driven wind using open stellar field lines for a magnetocentrifugal fling (Najita and Shu, 1994).

While details of how the T Tauri magnetosphere couples to the inner accretion disk are not yet established, most investigators agree that the disk will be disrupted somewhere in the vicinity of the co-rotation radius, and the accretion flow will be redirected to free-fall toward the star in a magnetic *funnel flow* (Königl 1991; Hartmann, Hewett and Calvet 1994; Ostriker and Shu, 1995; Paatz and Camenzind 1996; Li, Wickramasinghe and Ruediger, 1996). This paper will review the evidence for funnel flows, and describe some complexities and challenges to this paradigm.

2. Observational Evidence for Magnetospheric Accretion

2.1. SUPPORT FOR FUNNEL FLOWS

Several aspects of cTTS activity are often cited as providing empirical support for the prevalence of magnetically channeled funnel flows:

1. *Kinematic evidence for mass infall at free-fall velocities:* Sensitive spectroscopic surveys of cTTS reveal a high frequency of inverse P Cygni features in hydrogen and permitted metallic lines (Edwards et al. 1994; Najita, Carr, and Tokunaga 1996) with redshifted absorption components indicating infall along the line of sight at velocities comparable to those expected from magnetospheric accretion funnels (Calvet and Hartmann, 1992).

2. *Broad, asymmetric, and centrally peaked hydrogen emission line profiles:* Radiative transfer models of hydrogen lines in a schematic funnel flow can reproduce many peculiarities of the observed H line morphology, suggesting that cTTS emission lines are formed predominantly in accretion flows, not in a wind (Hartmann, Hewett and Calvet 1994; Calvet, this volume). Recent multi-level calculations (Muzerolle, Calvet and Hartmann, in prep.) have successfully reproduced Balmer, Paschen, and Brackett line luminosities as well profile shapes.

3. *Near infrared colors:* Spectral energy distributions of cTTS between 1-5μ are not well fit by a simple combination of a stellar photosphere and a standard accretion disk model; better fits are achieved if the inner disks are truncated with hole sizes about 70% of the co-rotation radius (Bertout, Basri and Bouvier 1988; Kenyon, Yi and Hartmann, 1996; Meyer, Calvet and Hillenbrand 1997). Truncation radii at or inside co-rotation are necessary to allow the accretion flow to fall toward the star.

4. *Periodic photometric modulation from hot spots:* Multiwavelength photometric monitoring reveals that while rotational modulation from many T Tauri stars can be attributed to large cool starspots, some cTTS also reveal signatures of spots hundreds to thousands of degrees hotter than the photosphere (Bouvier et al. 1993; Bertout et al. 1996). These can be accounted for by non-axially symmetric funnel flows if the rotation and magnetic axes are not aligned, allowing accretion shocks on the stellar surface to rotate out of view (Bertout 1989).

5. *Periodic spectroscopic signature of infall and/or outflow?* Time series analysis has revealed one T Tauri star, SU Aur, with periodicity in the absorption strength of both a wind and an infall feature that vary 180° out of phase relative to each other, with a period equal to the rotation period of the star (Johns and Basri 1995a; Petrov et al. 1996). This behavior is consistent with a non-aligned magnetosphere launching both a wind and an accretion flow from the "x-point", which changes its aspect relative to

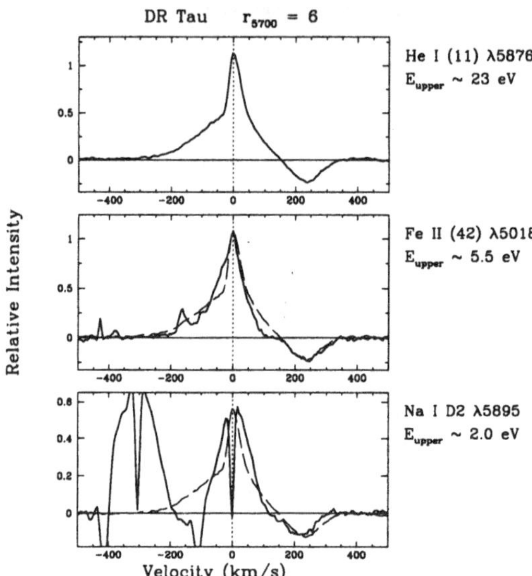

Figure 1. Simultaneous residual line profiles in the high accretion rate cTTs DR Tau. Lines expected to form in very different zones of a magnetic accretion funnel are seen to have nearly identical redshifted absorption. For reference, the normalized He I profile is superposed on the Fe II and Na D2 profile. Intensity is relative to the continuum.

the observer (Shu et al. 1994). Unfortunately this star lacks the definitive signatures for disk accretion that identify classical T Tauri stars among lower mass objects – optical continuum veiling and [O I] λ6300 forbidden emission (HEG) – making the interpretation of these intriguing periodicities unclear.

2.2. CHALLENGES TO FUNNEL FLOWS

While the above phenomena provide strong support for magnetospherically mediated accretion in cTTS, a closer look reveals a complexity that challenges any simple picture of magnetosphere-disk coupling:

1. Some cTTS show very similar inverse P Cygni profiles among lines covering a surprising range of excitation energies, from a low of 2 eV at NaD to a high of 23 eV at He I, as shown in Figure 1. This is not easily explained by current models of the thermal structure of magnetic accretion funnels where the temperature is predicted to increase toward the star due to compressional heating (Martin 1996), with lower excitation lines formed at the base of the flow and higher excitation lines formed near the accretion shock (Martin 1997).

2. Inner disk truncation radii determined from modelling near infrared spectral energy distributions must be viewed with caution. Models to date are relatively simple and do not take into account possible heating of the inner disk from magnetic effects, which can lead to underestimates of the hole sizes (Armitage and Clark 1996).

3. Spectral time series analysis has not revealed periodic behavior in lines or veiling among cTTS that possess definitive accretion signatures, despite considerable time coverage on about 10 stars (Johns and Basri 1995b; Gullbring et al. 1996; Johns-Krull and Basri, 1997). Instead, the H lines are seen to undergo stochastic variations in velocity intervals of 50 to 100 $km\ s^{-1}$ that vary independently of each other, and veiling variations are not well correlated with line variations. This complexity, combined with the fact that luminosities from photometrically derived hot spots are typically only a few percent of the spectroscopically derived veiling luminosity (Bertout et al. 1996), suggests that complex magnetic geometries rather than coherent dipole structures control the accretion flow.

4. Periods of hot spots diagnosed from broad-band photometric studies are observed to vary by up to 25% (Bouvier et al. 1995). Although the reality of this claim has been challenged by Herbst and Wittenmyer (1997), if real, it is not easily accounted for in magnetospheric accretion scenarios (Bertout et al. 1996).

Taken together, the above considerations reveal that our knowledge of magnetic funnel flows in cTTS is superficial, and considerable study is required to achieve the next level of understanding.

3. Permitted Metallic Lines as Funnel Flow Diagnostics?

Emission lines of hydrogen are the strongest features in cTTS spectra. Success in reproducing the general morphology and luminosity of H lines with magnetospheric accretion models lends considerable support to a funnel flow origin for these lines (see Calvet, this volume). However, the optical and near infrared spectra of cTTS also include a rich array of permitted metallic emission lines, which offer additional probes of the near-stellar environment in accretion disk systems. Evidence that lines of He I, Ca II, Mg I, Fe I,II and Si I derive from accretion activity comes from correlations between line strength and accretion diagnostics such as continuum veiling and near infrared excesses, as illustrated in Figure 2.

A growing body of evidence supports the conclusion that the permitted metallic lines originate in *two distinct regimes* characterized by differing kinematic and physical conditions. These lines are typically one to two hundred $km\ s^{-1}$ narrower than H lines in the same star (Edwards et al. 1994) and are characterized by a two component structure, consisting of a

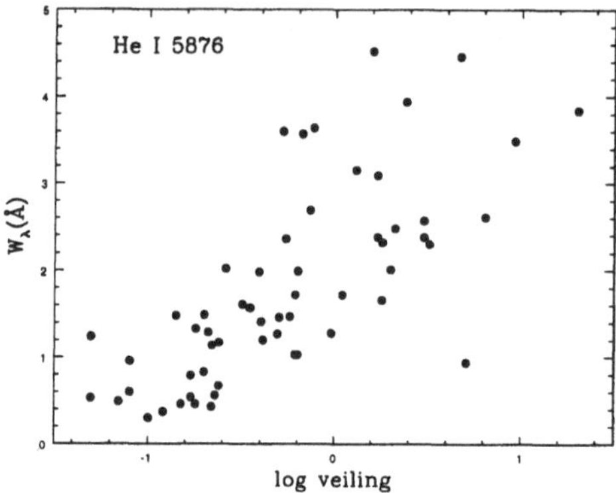

Figure 2. Emission equivalent widths for He I λ5876 for 61 spectra of 30 cTTS from the sample in HEG plotted against the veiling, defined as the ratio of the excess continuum flux to the stellar photosphere.

broad (BC; ≥ 100 $km\ s^{-1}$) and a narrow (NC; < 50 $km\ s^{-1}$) component (Hamann and Persson 1992; Batalha et al. 1996).

An illustration of this morphology is given in Figure 3, which also shows that the relative proportion of BC to NC emission varies considerably among cTTS, ranging between the two extreme cases where the profile is dominated by only one of the two components. Although there is no clear pattern of profile morphology with accretion diagnostics, two-component fits of Gaussian shapes to He I λ5876 profiles in 61 spectra of 30 cTTS reveal that the parameters of the NC are remarkably consistent among the entire sample of stars, with full width at half maxima ~ 48 $km\ s^{-1}$ and centroid velocities at rest relative to the stellar velocity (Beristain, Edwards, Kwan, and Hartigan, 1997; BEKH). The BC is less consistent, ranging in FWHM from 150 to 300 $km\ s^{-1}$, and centroid velocities blueshifted relative to the star by up to 50 $km\ s^{-1}$ in about 1/3 of the sample. Examples of fits to the two kinematic components are illustrated in Figure 4, which also shows that an individual cTTS can transition from metallic lines dominated by BC emission to the opposite on a timescale of weeks, with no correlation with veiling.

Additional evidence that the NC and BC arise in two physically distinct regimes comes from a detailed analysis of unblended Fe I lines (2.5-6 eV) and Fe II lines (5-6 eV) in the active cTTS DR Tau (BEKH). This study demonstrates that a systematic increase in the FWHM of Fe lines from 20 to 80 $km\ s^{-1}$ with increasing line strength is accounted for by differing proportions of NC to BC emission. All Fe lines can be decomposed into two Gaussian contributions with *identical* kinematic components, but stronger

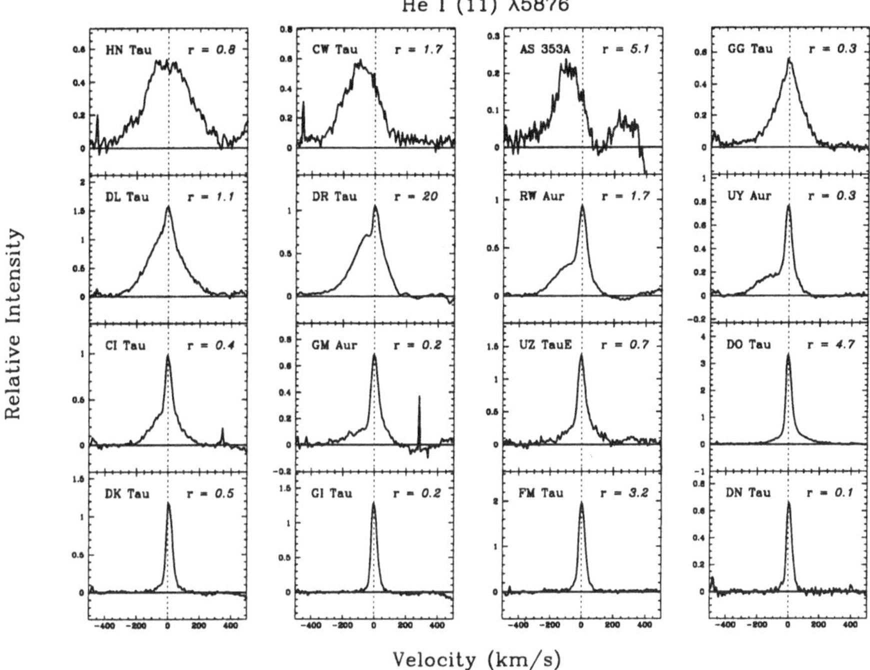

Figure 3. Residual He I λ5876 profiles for 12 cTTS illustrating the variety of relative contributions from the broad component and the narrow component. Profiles are ordered by line width, from broadest to narrowest. Intensity is in units of the continuum.

Fe lines are characterized by a larger contribution from BC relative to NC emission. The simplest explanation of this behavior is that the densities and temperatures of the BC and NC regions are sufficently dissimilar to give rise to Fe lines with differing relative intensities.

It will be important to clarify the origin of each kinematic component in order to fully understand the star-disk interface region. The large line widths of the BC, the tendency for it to be blueshifted, and occasionally to show inverse P Cygni structure, are consistent with this component arising in the magnetospheric accretion flow. In contrast, the NC appears to arise in a region which is hotter and optically thicker than the BC (BEKH), is at rest with respect to the star, and can vary independently of the BC (Gullbring et al., 1996). The near-photospheric line widths of the NC suggest formation on the stellar surface rather than in a Keplerian disk, possibly in the postshock region at the base of accretion flow (Batalha et al., 1996). Additional support for this possibility comes from CIV line profiles in cTTS, which also show strong NC emission and must arise in a region $T \sim 10^5$ K

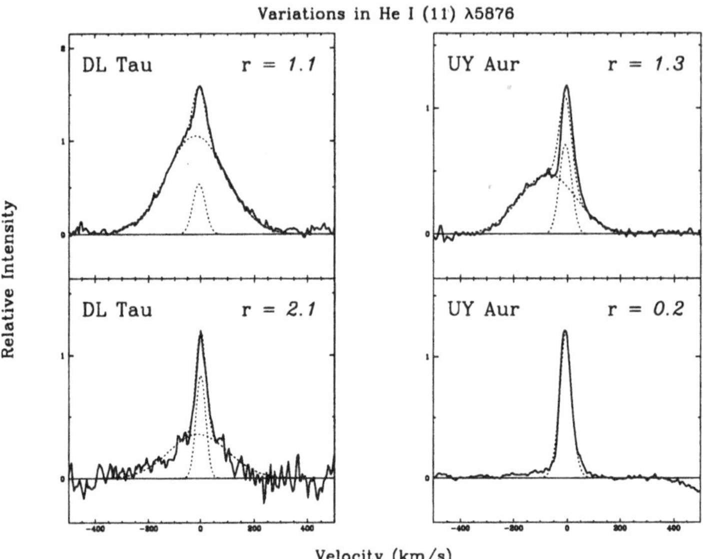

Figure 4. Residual He I λ5876 profiles for 2 cTTS illustrating two component Gaussian fits, which separate the NC and BC contributions. Variations in the relative proportion of NC to BC emission in the same star do not correlate with veiling (r).

(Calvet et al., 1996). If this interpretation is correct, the NC will be a crucial diagnostic of the energetics of the final transfer of mass and angular momentum of material accreting from the disk to the stellar surface.

4. Discussion and Future Prospects

The observational evidence for magnetic funnel flows in disk accreting cTTS is strong, but our understanding of this complex region is still primitive. Most important is the need to establish the relevance of this region to the angular momentum evolution of forming stars. *Is magnetosphere-disk coupling the mechanism which establishes the initial angular momentum of a star?* While the evidence supporting this possiblity is tantalizing, much larger statistical samples of rotation periods of accreting and post-accreting stars must be accumulated. Alternative spin-down mechanisms for accreting stars are being proposed (Popham 1996), and need to be subjected to empirical tests. *Are the accretion-driven winds which characterize all forming stars launched with open field lines from the stellar magnetosphere?* Until definitive observational signatures of the wind origin are identified, the answer to this fundamental question remains uncertain.

The quest to determine the role of the stellar magnetosphere in govern-

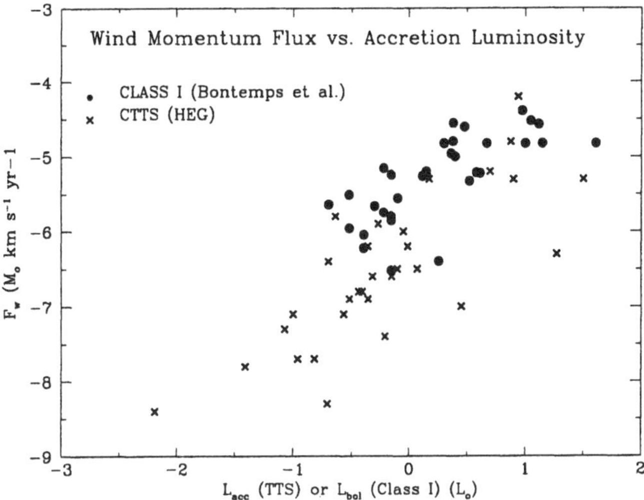

Figure 5. The wind momentum flux plotted against accretion luminosity for two groups of low mass stars. For the Class I sources the wind momentum is determined from $^{12}CO(2-1)$ IRAM data and the bolometric luminosities are assumed to be dominated by accretion (Bontemps et al. 1996). For the cTTS the wind momentum is determined from the high velocity component of the [O I]λ6300 line and the luminosities are the accretion luminosities determined from the veiling flux (HEG).

ing the angular momentum evolution of forming stars also requires scrutiny of the precursors to the optically visible cTTS. Comparison of the overall energetics between cTTS and the more deeply embedded Class I sources suggests a unity of accretion/outflow phenomena. This is illustrated in Figure 5, which shows a similar relation between wind momentum and accretion luminosity for both groups of objects. The advent of high resolution near infrared spectroscopy will allow us to probe this comparison further, making it possible to study the near-stellar regions in Class I sources and to determine the importance of magnetosphere-disk interactions during this earlier stage of star formation.

Acknowledgements: I thank Georgina Beristain for assistance in preparing the figures for this manuscript. Pat Hartigan was instrumental in data reduction and determination of veiling and residual profiles shown here. I also thank Nuria Calvet, Lee Hartmann, John Kwan, James Muzerolle, and Steve Strom for valuable conversations, suggestions, and inspirations.

References

Armitage, P.J. and Clarke, C.J.: 1996, MNRAS, 280, 458
Basri, G., Marcy, G. W., and Valenti, J.A.: 1992, ApJ 390, 622
Batalha, C.C., Stout-Batalha, N.M., Basri, G. and Terra, M.A.O.: 1996, ApJS 103, 211
Beristain, G., Edwards, S., Kwan, J., and Hartigan, P.: 1997, in preparation [BEKH]
Bertout, C.: 1989, ARAA, 27, 351
Bertout, C., Basri, G., and Bouvier, J.: 1988, ApJ 330, 350
Bertout, C., Harder, S., Malbet, F., Mennessier, C., and Regev, O.: 1996, AJ, 112, 2159
Bouvier, J., Cabrit, S., Fenandez M., Martin, E., Matthews, J. 1993, A&A, 101, 495
Bouvier, J.: 1994, in *The Eighth Cambridge Workshop on Cool Stars, Stellar Systems, and the Sun*, ed. J.P. Caillaut, APS Conf Series, 64, 51
Bouvier, J. et al.: 1997, A&A, 318, 495
Bontemps, S., André, P., Terebey, S. and Cabrit, S.: 1996, A&A, 311, 858
Calvet, N, and Hartmann, L. 1992, ApJ, 386, 239
Calvet, N., Hartmann, L., Hewett, R., Valenti, J., Basri, G., and Walter, F., 1996 in *ASP Conf. Proc. 109; Cool Stars, Stellar Systems and the Sun*, eds. R. Pallavicini and A. Dupree, (San Francisco: ASP), 419
Cameron, A., Campbell, C.: 1993; A&A, 274, 309
Cameron, A., Campbell, C., Quaintrell, H.: 1995, A&A, 298, 133
Edwards, S., Strom, S., Hartigan, P., Strom, K., Hillenbrand, L., Herbst, W.Attridge, J., Merrill, M., Probst, R.,and Gatley, I.: 1993, AJ, 106, 372
Edwards, S., Hartigan, P., Ghandour, L., and Andrulis, C.: 1994, AJ, 108, 1056
Ghosh, P.: 1995, MNRAS, 272,763
Ghosh, P. and Lamb, F.: 1979, ApJ, 234, 296
Gullbring, E., Petrov, P.P., Ilyn, I., Tuominen, I., Gahm, G.F. and Loden, K.: 1996, A&A, 314, 835
Hamann, F. and Persson, S.E.: 1992, ApJS 82,287
Hartigan, P., Edwards, S., and Ghandour, L.: 1995, ApJ, 452, 736 [HEG]
Hartmann, L., Hewett, R., and Calvet, N. 1994, ApJ, 426, 669
Herbst, W. and Wittenmyer, R.: 1997 B.A.A.S. 28, 1338
Johns, C. M., and Basri, G. 1995a, ApJ, 449, 341
Johns, C. M., and Basri, G. 1995b, AJ 109, 2800
Johns-Krull, C.M. and Basri, G. 1997, ApJ, 474,433
Kenyon,S., Yi, I. and Hartmann, L. 1996, 462, 439
Königl, A.: 1991, ApJ, 370, L39
Lada, C. J., and Wilking, B. A.: 1984, ApJ, 287, 610
Li, J. Wickramasinghe, D. and Ruediger,G.: 1996 ApJ, 469, 765
Martin S.C.: 1996, ApJ, 473, 1051
Martin S.C.: 1997, ApJ, 478, 33
Montmerle, T., Feigelson, E., Bouvier, J., and André, P. 1993 in *Protostars and Planets, III* ed. E.H. Levy and M.S. Matthews (University of Arizona Press), p. 689
Meyer, M., Calvet, N., and Hillenbrand, L.: 1997 AJ, in press
Najita, J., and Shu, F. H. 1994, ApJ, 429, 808
Najita, J., Carr, J.S., and Tokunaga, A.T. 1996, ApJ, 456, 292
Ostriker, E.C. and Shu, F.H.: 1995, ApJ, 447,813
Paatz, G., Camenzind, M.: 1996, A&A, 308. 77
Petrov, P.P., Gullbring, E., Ilyn,I., Gahm, G., Tuominen, I., Hackman, T. and Loden, K.: 1996, A&A, 314, 821
Popham, R.: 1996, ApJ., 467, 749
Shu, F.H., Najita, J., Ostriker, E., Shang, H.: 1995 ApJ 455, 155
Shu, F.H., Najita, J., Ostriker, E., Wilken, F., Ruden, S., and Lizano, S.: 1994 ApJ, 429,781
Strom, S.E., Edwards, S. and Skrutskie, M.F.: 1993, in *Protostars and Planets III*, ed. E.H. Levy and M.S. Matthews (University of Arizona Press)

JETS, DISK WINDS, AND WARM DISK CORONAE IN CLASSICAL T TAURI STARS

JOHN KWAN
Department of Physics and Astronomy,
University of Massachusetts,
Amherst, MA 01003, USA

Abstract. Arguments, based on analysis of the forbidden line emission, are summarized that point to two components of mass ejection in classical T Tauri stars. A low-speed wind originates from the accretion disk at between ~ 0.05 AU and $\gtrsim 3$ AU from the star, and a high-speed flow, which becomes collimated into a jet within $\lesssim 100$ AU, originates as a wind from either the star or the very inner part of the accretion disk. The [OI] $\lambda5577$ emission in the low-speed component also requires the presence of a warm disk corona, with an electron density of $\sim 10^7$ cm^{-3} and a temperature of ~ 8000 K. Additional indication of a warm disk corona comes from analysis of the central absorptions seen in the profiles of the hydrogen Balmer lines. The need to heat the disk corona implies that a substantial fraction of the energy released in the accretion of matter through the disk may be dissipated at the disk surface.

1. Introduction

In this symposium there are many theoretical discussions on how winds and jets are launched from a classical T Tauri star (cTTS) or its accretion disk (see the papers in this volume by Camenzind; Heyvaerts & Norman; Pudritz & Ouyed; Shu & Shang; and the poster papers by Ferreira, Pelletier, & Appl (1997); and Paatz & Appl 1997). Here the question is posed from the observational point of view: what do the observations tell about the location and properties of the mass ejection? In particular, arguments, based on analysis of the forbidden lines, are presented that indicate there are two components of mass ejection: a high-speed component, originating from the star or the very inner part of the accretion disk, that becomes collimated

B. Reipurth and C. Bertout (eds.), Herbig–Haro Flows and the Birth of Low Mass Stars, 443–454.

into a jet, and a comparatively much slower wind originating from the outer part of the disk up to a distance of several AUs from the star.

The [OI] $\lambda5577$ emission in the low-speed component also points to the presence of a disk corona with a temperature of $\sim$ 8000 K and an electron density of $\sim 10^7$ cm^{-3} . It is this warm corona that extends vertically from the disk into the slow disk wind. Additional indication of a warm disk corona may come from the observed profiles of the hydrogen Balmer lines. It is suggested that the central absorptions seen in those lines are produced by a disk corona within a distance of $\lesssim 10^{13}$ cm from the star.

Throughout this paper I shall constantly refer to the fine sets of [OI] and hydrogen line profiles that Hartigan, Edwards, & Ghandour (1995, hereafter HEG) and Edwards et al. (1994, hereafter EHGA) carefully extract from their echelle spectra. In the next section I shall describe how the observed forbidden line profiles are most readily explained by two separate outflow components. In § 3 the arguments for a warm disk corona, based on the [OI] $\lambda5577$/[OI] $\lambda6300$ intensity ratio and the Balmer line central absorption, are put forward. The implications of a warm disk corona and a disk wind on the process of mass accretion through the disk is discussed in § 4. Many of the thoughts, arguments, and calculations were originally presented in Kwan & Tademaru (1988; 1995, hereafter KT) and Kwan (1997).

2. Two Components of Mass Ejection: a Jet and a Disk Wind

When the OI forbidden lines were first examined, it was realized right away that the generally stronger emission blueward of the stellar velocity indicates the presence of an obscuring disk and that the double-peaked shape of the line profile seen in a fraction of the stars cannot be produced by a geometric structure with spherical symmetry (Appenzeller, Jankovics, & Östreicher 1984; Edwards et al. 1987). The recent high-resolution line profiles of 32 cTTS obtained by HEG provide a large sample for analysis. Referring to their Figure 5, which shows the [OI] $\lambda6300$ profiles, it is seen that a peak located, in the rest frame of the star, at zero velocity or slightly to the blue (by $\sim$ 5 km s^{-1}) is always present. This is often referred to as the low-velocity peak. In about a third of the stars another peak, displaced blueward by $\gtrsim$ 100 km s^{-1} , is prominent. This blue or high-velocity peak is stronger than the low-velocity peak in two stars and comparable to or weaker than the low-velocity peak in the others. In three stars (AS 353A, RW Aur, DF Tau) a third peak, displaced to the red of zero velocity by an amount comparable to the shift of the blue peak, also stands out.

The more frequent appearance of the blue peak than the red peak, and, when the blue peak is not conspicuous, the generally stronger emission blueward of zero velocity strongly suggest the presence of a disk that in most cases obscures the receding part of an outflowing wind. It follows that the red peak is most likely the counterpart of the blue peak. Considering only the gas flow in front of the disk then, the prevalence of the low-velocity peak and the occasional double-peaked line shape need to be explained. The models proposed can be divided according to the two premises: a one-component model in which a high-speed non-isotropic wind produces both the low-velocity and blue peaks as a result of different projections of the velocity vectors along the line of sight (Edwards et al. 1987; Hartmann & Raymond 1989; Safier 1993; Ouyed & Pudritz 1994), and a two-component model in which a high-speed flow produces the blue peak and a much slower wind produces the low-velocity emission.

There are several lines of argument, supported by observational results, that indicate the one-component model is unlikely. In such a model there is a strong relation between the emission properties of the blue and low-velocity peaks, since the blue peak can appear, from a different viewing angle, as a low-velocity peak and vice versa. Given that the viewing angle is arbitrary, the model will be hard pressed to explain the much more frequent appearance of the low-velocity peak than the blue peak. It will also stipulate similar physical conditions for the blue and low-velocity emissions. Observed profiles of other forbidden lines (Hamann 1994; Figs 6, 7, & 8 of HEG), on the other hand, show that, in comparison with the [OI] $\lambda6300$ line, the [OI] $\lambda5577$ line has no or only a weak blue peak, while the [SII] $\lambda6731$ and [NII] $\lambda6583$ lines in quite a few stars have a prominent blue peak but no low-velocity peak. Comparisons of the observed intensity ratios with excitation calculations reveal that the electron densities responsible for the [OI] $\lambda5577$ low-velocity emission and the [SII] $\lambda6731$ blue emission are $\gtrsim 5\times10^6$ cm^{-3} , and $\lesssim 3\times10^4$ cm^{-3} respectively (Hamann 1994; HEG; KT). This large systematic difference in the electron density is not easily effected in a one-component model. The strong contrast of the [SII] $\lambda6731$ and [NII] $\lambda6583$ profiles versus the [OI] $\lambda5577$ profile can actually be used, in those stars in which the two peaks appear blended together in the [OI] $\lambda6300$ profile, to separate out the two contributions (see Fig. 8 of HEG). A third argument against the one-component model comes from high resolution long-slit spectral imaging of several cTTS (Solf & Böhm 1993; Hirth, Mundt, & Solf 1994; see also the paper by Solf in this volume). The emission at the blue peak is found to be more spatially extended along an axis than the low-velocity emission. This systematic dependence of spatial distribution on velocity is not easily reconciled in a one-component model that produces a low observed velocity via geometric projection.

The constraints inherent in a one-component model are lifted in a model with two emission components. The blue and low-velocity peaks can be examined and modelled separately. In this light the gas kinematics needed to produce the two peaks are found to be very different. Assuming a geometry with axial symmetry, and noting that the forbidden lines are optically thin, an observed line profile can be thought of as the sum of the emission profiles from many axisymmetric rings of particles. Let the velocity components of a particle in the rest frame of the star be denoted by v_r, v_ϕ, and v_z in a cylindrical co-ordinate system with the symmetry axis along z. The basic line profile from an axisymmetric ring of particles is symmetric about the observed velocity $v_l = -v_z \; cos \; i$. It extends from this midpoint to a width on each side of $(v^2 - v_z^2)^{0.5} sin \; i$, where i is the inclination of the line of sight from the z-axis, and v is the velocity magnitude. The amplitude of the line is lowest at the midpoint and highest at the two edges, so the basic profile has a double-peaked appearance.

In several cTTS the blue peak is prominently displaced (by ≥ 80 km s^{-1}) from zero velocity and is fairly sharp (see the [OI] λ6300 line in AS 353A, DO Tau, and CW Tau, and the [SII] λ6731 line in DO Tau, DD Tau, DL Tau, and DK Tau from Figs. 5 & 7 of HEG), with the half-width at half-maximum being less than $\sim$ $^1/_4$ of the displacement of the peak. To produce it from a superposition of many basic profiles, a substantial contribution is needed from axisymmetric rings of particles with v_z large and within several percent of v. This condition indicates a high-speed flow that becomes collimated into a jet. The result from long-slit spectral imaging that the emission at the blue peak is spatially elongated along the direction marked by an extended optical jet and/or Herbig-Haro objects (Solf & Böhm 1993; Hirth, Mundt, & Solf 1994) provides further corroboration.

The fraction of cTTS in HEG's sample showing a blue peak is $\sim$ $^1/_2$ if one includes the stars DD Tau, HN Tau, AA Tau, and Lk Ca 8 in which the blue and low-velocity peaks (while blended together in the [OI] λ6300 profile) can be separated out by utilizing the [SII] λ6731, [NII] λ6583, and [OI] λ5577 profiles. Then, given the possibility that physical conditions amenable to producing strong forbidden line emission may not always be present, particularly if excitation of the gas in the jet is via internal shocks produced by a time-dependent flow, the above fraction suggests that a high-speed flow that becomes collimated within $\lesssim$ 100 AU of the star may be a common characteristic of cTTS.

Examining the low-velocity peak in the [OI] λ6300 profile (see Fig. 5 of HEG), it is seen that the peak is blue-shifted from the stellar velocity by $\lesssim$ 5 km s^{-1} and the line intensity falls off roughly symmetrically and fairly sharply from the peak to values of $\sim\pm$100 km s^{-1} . The fact that the low-velocity peak is always present and blue-shifted, but only slightly, indicates that, whatever the viewing angle, the particles producing the low-velocity peak has a positive but small v_z ($\lesssim$10 km s^{-1}). To produce the line shape from a superposition of many basic profiles, it is also necessary that the particles have a broad range in v from a value $\lesssim$10 km s^{-1} (the instrumental resolution of the spectra) to $\sim$100 km s^{-1} . A casual comparison of observed profiles with theoretical ones calculated assuming that the number of particles per unit velocity interval in the upper state of a forbidden line is proportional to v^α finds that the index α lies between 0 and -1, indicating that a substantial fraction of the emitting particles have low speeds. Furthermore, it is difficult to avoid a dip in the intensity right at line center unless there are particles with speeds about 10 km s^{-1} or lower (see Fig. 2 of KT).

The broad distribution in v (from $\lesssim$10 to $\sim$100 km s^{-1}) needed to produce the line shape of the low-velocity peak suggests that Keplerian rotation from $\sim$0.05 AU out to $\gtrsim$3 AU is a likely cause. The small but positive v_z of the gas then signals a wind emanating from the disk. This hypothesis is consistent with the observational result that, when the low-velocity peak is clearly distinct from the blue peak in the [OI] λ6300 profile and [SII] λ6731 is also observed to have low-velocity emission (in DG Tau, CW Tau, DQ Tau, UY Aur), the progression of the blue-shifts among the [OI] λ5577, [OI] λ6300, and [SII] λ6731 lines, which have successively lower critical electron densities for collisional de-excitation, indicates a higher v_z at a lower electron density (Solf & Böhm 1993; Hirth, Mundt, & Solf 1994; HEG).

In CW Tau and DG Tau the [OI] λ6300 luminosity in the low-velocity component is high, $\sim 4\times10^{-3}\ L_\odot$ (including a factor of two to account for obscuration by the disk), and it places a constraint on the size of the emission region. Assuming an [OI] λ6300 excitation temperature given by $T_{\lambda6300}$ in the range between 6×10^3 and 10^4 K, oxygen to be all neutral, and a solar composition, the mass needed to produce the above luminosity is $M \simeq 5\times10^{-8}(8\times10^3\ \mathrm{K}/T_{\lambda6300})^3\ M_\odot$. For a mass loss rate given by $\dot{M}$, an outflow speed given by v_t, the linear size along the outflow direction is then $l \sim Mv_t/\dot{M} \sim 1.6\times10^{13}(8\times10^3\ \mathrm{K}/T_{\lambda6300})^3(v_t/10\ \mathrm{km\ s^{-1}})(10^{-7}M_\odot\ yr^{-1}/\dot{M})$ cm. This result indicates that, at least for these two stars, the low-velocity emission cannot originate from the neighborhood of the star. It adds credence to the hypothesis of an extended disk wind.

The above calculation, in the disk wind hypothesis, determines the vertical height of the [OI] λ6300 emission from the disk if v_t is replaced by v_z. A similar manipulation using the [SII] λ6731 luminosity in CW Tau and DG Tau, which is $\sim 7\times10^{-4}$ and $\sim 2\times10^{-3}\ L_\odot$ (HEG), obtains, assuming sulphur is all SII, the corresponding vertical height of $\sim 25\ (8\times10^3\ \mathrm{K}/T_{\lambda6731})^{2.6}(v_z/10\ \mathrm{km\ s^{-1}})(10^{-7}M_\odot\ yr^{-1}/\dot{M})$ AU and $\sim$70 AU respectively, where $T_{\lambda6731}$ is the [SII] λ6731 excitation temperature. Given the large uncertainties in $T_{\lambda6731}$, v_z, and $\dot{M}$, these heights are roughly compatible with the observed values of $\sim$150 and $\sim$50 AU for the [SII] λ6731 emission in CW Tau and DG Tau respectively (Hirth, Mundt, & Solf 1994; Solf & Böhm 1993).

If the origin of the low-velocity emission is a disk wind, that of the high-velocity emission is less clear. Observed Balmer lines of hydrogen sometimes show blue absorptions at $\lesssim -100$ km s^{-1} (EHGA). The need to maintain population in the n=2 excited level of hydrogen in order to produce the absorption indicates that the high-speed outflowing gas responsible probably lies close to the star. Theoretically it also makes sense that the gas attaining a terminal speed of $\sim$300 km s^{-1} starts originally at a gravitational potential from where the escape velocity is not much smaller, otherwise the stresses built up to produce the acceleration will be unnecessarily excessive. In the case of the disk wind producing the low-velocity emission, for example, $v_z \cos i$, as revealed by the spectral imaging studies (Hirth, Mundt, & Solf 1994; Solf & Böhm 1993), reaches a terminal value of only $\sim$12 and $\sim$43 km s^{-1} in CW Tau and DG Tau respectively. It is likely, therefore, that the high-speed gas that ultimately becomes collimated originates as a wind from either the star or the very inner part of the accretion disk.

3. A Warm Disk Corona

Besides conveying information on the kinematics of the gas through their profiles, the forbidden lines also reveal through their intensities the physical conditions giving rise to their excitation. The intensity ratio [OI] λ5577/[OI] λ6300, in particular, strongly constrains the electron density n_e and temperature T. It will be shown that its value in the low-velocity component suggests the presence of a warm disk corona ($n_e \sim 10^7$ cm^{-3} , $T \sim 8000$ K) out to a distance of $\gtrsim$3 AU from the star. In addition, analysis of the central absorptions seen in the Balmer lines of hydrogen points to the absorption gas being within 10^{13} cm from the star and at a height above the disk that is large in relation to its distance (i.e., $z/r \gtrsim 0.3$). The presence of high density gas ($\sim 10^{10}$ cm^{-3}) at that location is also indicative of a warm disk corona. These two lines of evidence are presented sequentially below.

3.1. [OI] $\lambda\lambda$6300, 5577 EMISSION

The [OI] λ5577 line originates from an excited level that is $\sim$2.2 eV above the upper level of [OI] λ6300. It has a spontaneous emission rate of $\sim$1.3 s^{-1} versus that of $\sim 5 \times 10^{-3}$ s^{-1} for [OI] λ6300. The conditions favoring λ5577 over λ6300 emission are then a high electron density and a high temperature. Observationally, in the low-velocity component [OI] λ5577/[OI] λ6300 lies between 0.1 and 1 (HEG). To produce a ratio of 0.2, for example, the electron density needed is $\sim 5 \times 10^6$ cm^{-3} at $T = 10^4$ K, and increases to $\sim 4 \times 10^7$ cm^{-3} at $T = 6000$ K (see Fig. 3 of KT). For ease of reference, $T = 8000$ K and $n_e = 10^7$ cm^{-3} shall be taken as representative values.

In the disk wind hypothesis the low-velocity emission originates at a distance out to $\gtrsim 3$ AU from the star. To produce an n_e of 10^7 cm^{-3} that far from the star places a taxing demand on the ionization source. It is readily shown that photoionization by the stellar or veiling continuum is inadequate. If $r \sim 2.5 \times 10^{13}$ cm denotes roughly half the extent of the low-velocity emission along the disk, the number of ionizing photons incident per second per unit area at r needs to be $\sim n_e^2 \alpha_r r$, where $\alpha_r \simeq 2.6 \times 10^{-13}$ cm^3 s^{-1} is the radiative recombination coefficient. The corresponding luminosity, assuming the photons to have energy $h\nu$, is then $\sim n_e^2 \alpha_r 4\pi r^3 h\nu$ $\sim 2.8 \times 10^{-2}(n_e/10^7 \text{ cm}^{-3})^2(r/2.5 \times 10^{13} \text{ cm})^3(h\nu/13.6\ eV)\ L_\odot$. It is stronger than the Lyman continuum of the veiling flux, if the latter is represented by a blackbody of luminosity $\leq 1\ L_\odot$ and temperature $\leq 2 \times 10^4$ K. The Lyman continuum will also suffer severe intervening attenuation by the gas infalling onto the star along stellar magnetic field lines. The required ionizing luminosity also greatly exceeds the observed X-ray luminosity of $\sim 3 \times 10^{-5}\ L_\odot$ from cTTS (Neuhäuser et al. 1995).

Assuming that the energy source is heat, hydrogen excitation calculations have been carried out that include collisional processes involving the ground and n=2 levels, Balmer photoionization, two-photon radiative decay, and electron-proton recombinations. The proton density is sufficiently high to ensure that the 2s and 2p level populations are in the thermal-equilibrium ratio of 1:3, and the decay route through escape of Lyα photons has been neglected in order to obtain a conservative estimate of the physical conditions. To produce an n_e of 10^7 cm^{-3} at a distance of $r = 2.6 \times 10^{13}$ cm, for example, it is found that a temperature of greater than 10^4, equal to 9000, and slightly less than 8000 K is needed if the hydrogen density n_H equals 10^8, 10^9, and 10^{10} cm^{-3} respectively (see Fig. 4 of Kwan 1997).

The local heating rate needed to excite hydrogen and other elements, such as OI and FeII, that produce strong cooling, is high. In the above example of $n_H = 10^9$ cm^{-3} , $r = 2.6 \times 10^{13}$ cm, and $T = 9000$ K, it is $\sim 1.5 \times 10^{-8}$ erg s^{-1} cm^{-3} . This heating rate has to be maintained over a z height that is at least as large as the product of the sound speed and the electron-proton recombination time, which is $\sim 3.6 \times 10^{11}$ cm. The required energy input rate per unit area on the disk at $r = 2.6 \times 10^{13}$ cm is then $\sim 1.1 \times 10^4$ erg s^{-1} cm^{-2}. Presumably the energy is ultimately derived from accretion. Assuming a stellar mass of 0.7 $M_\odot$, the corresponding mass accretion rate needed is $\sim 1.3 \times 10^{-7}\ M_\odot\ yr^{-1}$.

3.2. BALMER LINE CENTRAL ABSORPTION

In 5 of the 15 cTTS that EHGA obtained profiles of the hydrogen Balmer lines a strong absorption at line center is prominent, even in Hδ. This absorption must occur within 3×10^{14} cm from the star, otherwise, if the n=2 level is populated by electron-proton recombinations, the Hα luminosity from the absorption region will itself exceed the observed value, or, if the n=2 level is populated by collisional excitation, the heating luminosity needed will be prohibitively high (Kwan 1997). This constraint, together with the fairly common occurrence of the central absorption, argues strongly for an origin that is intrinsic to the star and its accretion disk.

The deep central absorption in Hβ, Hγ, or Hδ extends far below the continuum in AA Tau, DR Tau, and RW Aur (see Fig. 3 of EHGA). The veiling continua in these three stars are, respectively, about 0.2, 10, and 3.4 times the stellar photospheric continua (EHGA). In AA Tau then the absorption gas must be able to cover, along the line of sight, a large fraction of the stellar surface, while in DR Tau and RW Aur, assuming that the veiling continuum is produced near the polar region of the star, some absorption gas must lie at least one stellar radius above the disk plane.

There are two possible explanations for the incidence of 5 out of 15 stars showing a central absorption. One is that the 5 stars are the only ones in the sample with an absorption region, which covers the entire 4π solid angle about the star. The other is that the absorption region is present in each of the 15 stars but it covers, on average, only a fraction $f \sim 5/15$ of the 4π solid angle. The first explanation is less likely because of the following argument. It is reasonable to assume that all the Lyman transitions are ineffective decay routes owing to their very large opacities. Absorption of an Hα photon then leads to emission of another Hα photon. If the absorption gas covers the whole 4π solid angle, there will be no net loss of Hα photons in traversing the absorption region. The Hα photons can be re-distributed in frequency (as in the standard case of a P-Cygni profile being produced by a spherical-symmetric outflow), but here the absorption is centered at the stellar velocity in three stars and only slightly shifted to the blue in the other two, so the scattering will not produce a broad frequency re-distribution. There can be a large bulk motion transverse to the line of sight, but in this case the motion, most probably rotation, will itself point to a geometric structure that only partially covers the 4π solid angle. Furthermore, following absorptions of Hβ, Hγ, Hδ, etc. within the region, there is net Hα emission. Thus it is difficult to comprehend the strong Hα absorptions seen in AA Tau, GK Tau, and RW Aur if the absorption gas covers the star completely. It seems more likely that the absorption gas is present in each star, but the covering fraction probably varies substantially from star to star.

In DK Tau and GK Tau the central absorption in Hβ extends below the continuum, but that in Hγ and Hδ are successively weaker (EHGA). These results, together with the fact that the Hβ, Hγ, and Hδ opacities decrease in order, suggest that the line profile, in the absence of the absorption, is centrally peaked. The absorption gas will then need to lie at a distance from the star that is beyond the emission region of the Balmer lines.

Of the five stars with a Balmer line central absorption four, among a group of eight in the sample, also show a red absorption in one or more Balmer lines. The remaining, DR Tau, has a red absorption in its Na D line (EHGA). The red absorption is interpreted as being produced by infalling gas along stellar magnetic field lines that intercept the disk (Hartmann, Hewett, & Calvet 1994; EHGA). It appears that when a central absorption is seen, a red absorption is also present.

Considering all the arguments and interpretations presented above, it is believed that the gas producing the central absorption is located beyond the magnetospheric accretion column in order to be able to cover the stellar surface, the veiling continuum, and the region producing the hydrogen emission lines. The moderate covering fraction subtended by this absorption region, the small radial motion of the gas, and the possible correlation of the central absorption with the red absorption indicate that the gas is confined more to the equatorial than to the polar region. The need for some of the absorption gas to lie one stellar radius or more above the disk in order to cover the veiling continuum source suggests that a disk corona is present.

To determine the physical conditions needed to produce the Balmer line central absorption, calculations of the n=2 level population have been carried out. The Lyman continuum is assumed to be completely absorbed by the gas in the magnetospheric accretion column. X-ray photoionization is also judged to be ineffective (Kwan 1997). In addition to thermal collisions and Balmer photoionization, two processes that utilize the continuum between 12 and 13.6 eV are included (Kwan & Alonso-Costa 1988). The first is ionization from the $2p^3$ ^2D state of NI. That state, lying at ~ 2.4 eV above the ground state, can be ionized by photons more energetic than 12.16 eV. Charge-exchange of NII with hydrogen will then produce hydrogen ionization. The second process is absorption of Lyβ in the damping wings, followed by a spontaneous decay to the 2s level. These two processes reduce the heat input needed to produce the hydrogen excitation if the veiling continuum at those photon energies is not too weak.

The details of the calculations are presented in Kwan(1997). Here the main results are summarized. Assuming a continuum luminosity of 0.3

$L_\odot$ between 3.4 and 13.6 eV, and an energy distribution (erg s^{-1} Hz^{-1}) that is proportional to ν^{-3}, it is found that, over the distance range 3.2×10^{11} cm $\leq r \leq 10^{13}$ cm, the temperature needed to produce an Hδ opacity near unity is $\sim$ 3300, $\sim$ 4500, and $\gtrsim$ 7000 K if the hydrogen density n_H equals 10^{11}, 10^{10}, and 10^9 cm^{-3} respectively. For the same luminosity of 0.3 $L_\odot$ but a ν^{-5} frequency dependence, the required temperature is about 4000 and 5500 K for $n_H = 10^{11}$ and 10^{10} cm^{-3} , respectively, while the corresponding temperature required in the absence of NI excited-state ionization is about 5600 and 6500 K.

The above conditions of a fairly high density and temperature, together with the need for the gas to be present at a substantial height above the disk (i.e., $z/r \gtrsim 0.3$) in order to account for the incidence of absorption, corroborates the argument for a warm disk corona.

4. Discussion

The disk-wind origin for the low-velocity forbidden line emission and the presence of a warm disk corona have important implications on the accretion process through the disk, so it is important to establish their validity first. The determination of $n_e \sim 10^7$ cm^{-3} and $T \sim 8000$ K needed to produce the low-velocity [OI] λ5577 emission is firm. It is the location of this emission that has to be ascertained. Spatial imaging with the Hubble Space Telescope may reveal the extent of the emission along the disk. Observations with a very high spectral resolution to compare the [OI] λ5577 and [OI] λ6300 profiles near the low-velocity peak will provide strong constraints on the detailed modelling.

On the interpretation of the Balmer line central absorption as being caused by a warm disk corona, several self-consistency checks can be made. Even though a strong continuum between 12 and 13.6 eV is not essential, it facilitates the hydrogen excitation with a minimum need for heat input. It is useful to measure the continuum shape beyond the Balmer edge, not only to make the excitation calculations more definite, but also to determine the veiling continuum luminosity more accurately. Another check lies in the shape of the Paschen lines. These lines are optically thin in the absorption region, so they should not possess central absorptions. Recent observations of Paβ in T Tauri stars show that indeed they do not (Folha, private communication; see also the paper by Folha, Emerson, & Calvet in this symposium). A third check is the correlation between the central and red absorptions. A large sample of cTTS to examine the central, red, and blue absorptions will not only elucidate the geometrical distributions of the different gas flows close to the star, but may even help determine whether the high-speed outflow, which becomes collimated into a jet, originates from the star or the very inner part of the accretion disk.

If the postulate of a warm disk corona is correct, then the energy input rate needed to maintain it is considerable. As estimated in §3.1, it entails a mass accretion rate of $\sim 1.3 \times 10^{-7}\ M_\odot\ yr^{-1}$. While this estimate is quite uncertain, it nonetheless points to a substantial fraction of the energy released from the accretion of matter through the disk being dissipated at the disk surface. This is contrary to the usual assumption of dissipation of all the energy in the form of blackbody radiation from the bulk of the disk. One way for this to occur, in light of the recent emphasis by Balbus & Hawley (1991) of a magnetohydrodynamic instability, may be the damping at the disk surface of Alfven waves that are generated as a result of the instability.

Even if most of the energy dissipation occurs at the disk surface, half of the radiation emitted in the disk corona will propagate toward the disk. It will be absorbed by dust grains and re-radiated in the infrared. The local infrared spectrum will be broader than that of a single blackbody producing an equivalent loss, however, as the dust temperature will be a function of vertical distance from the corona.

The need to heat the corona points to an accretion process that is driven by magnetohydrodynamic turbulence. The extension of the disk corona into a disk wind indicates that angular momentum transport via a magnetohydrodynamic wind is also taking place. It will be interesting to ascertain if both processes are integral to the accretion of matter through a disk.

References

Appenzeller, I., Jankovics, I., & Östreicher, R. 1984, A&A, 141, 108
Balbus, S. A., & Hawley, J. F. 1991, ApJ, 376, 214
Edwards, S., Cabrit, S., Strom, S. E., Heyer, I., Strom, K. M., & Anderson, E. 1987, ApJ, 321, 473
Edwards, S., Hartigan, P., Ghandour, L., & Andrulis, C. 1994, AJ, 108, 1056 (EHGA)
Ferreira, J., Pelletier, G., Appl, S.: 1997, in *Low Mass Star Formation - from Infall to Outflow*, poster proceedings of IAU Symp. No. 182, eds. F. Malbet & A. Castets, Observ. de Grenoble, p. 112
Hamann, F. 1994, ApJS, 93, 485
Hartigan, P., Edwards, S., & Ghandour, L. 1995, ApJ, 452, 736 (HEG)
Hartmann, L., Hewett, R., & Calvet, N. 1994, ApJ, 426, 669
Hartmann, L., & Raymond, J. 1989, ApJ, 337, 903
Hirth, G., Mundt, R., & Solf, J. 1994, A&A, 285, 929
Kwan, J. 1997, ApJ, submitted
Kwan, J., & Alonso-Costa, J. L. 1988, ApJ, 330, 870
Kwan, J., & Tademaru, E. 1988, ApJ, 332, L41
Kwan, J., & Tademaru, E. 1995, ApJ, 454, 382 (KT)
Neuhäuser, R., Sterzik, M. F., Schmitt, J. H. M., Wichmann, R., & Krautter, J. 1995, A&A, 299, 391
Ouyed, R., & Pudritz, R. E. 1994, ApJ, 423, 753
Paatz, G., Appl. S.: 1997, in *Low Mass Star Formation - from Infall to Outflow*, poster

proceedings of IAU Symp. No. 182, eds. F. Malbet & A. Castets, Observ. de Grenoble, p. 303
Safier, P. N. 1993, ApJ, 408, 148
Solf, J., & Böhm, K. H. 1993, ApJ, 410, L31

THERMAL STRUCTURE OF MAGNETIC FUNNEL FLOWS

STEVEN C. MARTIN
Department of Astronomy & Astrophysics
The University of Chicago
5640 South Ellis Ave.
Chicago, Illinois 60637
USA

Abstract. The thermodynamic structure of gas that is channeled by stellar magnetic fields onto a young (pre–main-sequence) star is presented. In this model, the star possesses a dipole magnetic field which disrupts the inner regions of a geometrically thin accretion disk and channels the inflowing gas onto the stellar surface, thereby forming an accretion funnel. The temperature and ionization degree of the inflowing gas is calculated by solving the heat equation coupled to statistical rate equations for hydrogen. It is found that for typical accretion rates of $\sim 10^{-7}\ M_{\odot}\ yr^{-1}$, temperatures of $\sim$ 7000 K and hydrogen ionization fractions (n_{H^+}/n_H) of $\sim 10^{-2}$ can be attained in the funnel flow. The principal heat source is found to be adiabatic compression, and coolants include bremsstrahlung radiation as well as line emission from the Ca II and Mg II ions. The relatively hot and ionized funnel flow lead to observational signatures such as inverse P Cygni line profiles seen in upper Balmer and near-infrared lines. In addition, carbon monoxide bandhead emission may be an important tracer of the outer portions of the funnel flow.

1. Introduction

Many pre–main-sequence stars are known to have inflowing gas as well as outflowing gas. Both of these flows can manifest themselves simultaneously and appear to be physically related (e.g., Cabrit et al. 1990). Tracers of the outflows include P Cygni profiles of permitted lines (e.g., Mundt 1984), blueshifted forbidden emission lines (e.g., Hartigan, Edwards, & Ghandour 1995), and well collimated optical jets which possess (many times multiple)

B. Reipurth and C. Bertout (eds.), Herbig–Haro Flows and the Birth of Low Mass Stars, 455–463.

Herbig-Haro objects (e.g., Mundt, Brugel, & Bührke 1987). The inflowing gas is inferred to be in the form of a geometrically thin accretion disk (e.g., Adams, Lada, & Shu 1988). Disk properties have been deduced by detailed modeling of both ultraviolet and infrared continuum emission (e.g., Bertout, Basri, & Bouvier 1988; Hartigan, Edwards, & Ghandour 1995).

In the event that the star is unmagnetized (or has a sufficiently weak magnetic field), the accretion disk joins onto the star through an equatorial boundary layer (e.g., Lynden-Bell & Pringle 1974). However, in the presence of a strong stellar magnetic field, the inner accretion disk can become disrupted and the gas from the disk can be channeled onto the star along magnetic field lines forming accretion funnels (e.g., Königl 1991; Hartmann, Hewett, & Calvet 1994). At the stellar surface, the infalling gas (essentially in free-fall) must be halted leading to strong shocks where the kinetic energy of the infalling gas is thermalized and re-radiated (mainly in the ultraviolet). This scenario may be ubiquitous among, for example, classical T Tauri stars as implied by rotating "hot spots" (e.g., Kenyon et al. 1994), emission line profiles (e.g., Edwards et al. 1994; and Edwards this volume), and inferred stellar magnetic field strengths (Basri, Marcy, & Valenti 1992, see also Guenther this volume).

The radiative properties of magnetic funnel flows are determined by both the kinematic and thermodynamic characteristics of the infalling gas. By solving the thermal balance equation coupled to rate equations for hydrogen, the temperature and ionization structure for funnel flows can be computed. These calculations can form the basis for interpreting the observed tracers of funnel flows from young stars, in particular P Cygni line profiles and possibly carbon monoxide bandhead emission.

2. Gas Dynamics

To compute the dynamics of the gas, the star is assumed to possess a dipole magnetic field which disrupts a geometrically thin accretion disk at several stellar radii (typically at ~ 5 stellar radii). The entire flow is therefore axisymmetric and specified a priori. Although the geometry adopted here is dipolar, the calculations and results would be valid for generic converging inflows (the model would also be applicable even if the field were not continuous, e.g., magnetic flux tubes). Since this model is most appropriate for T Tauri stars, rotation is not important on the dynamics of the infalling gas (T Tauri stars are slow rotators, see Hartmann & Stauffer 1989).

In general, the inflowing material consists of several components: neutral gas, ionized gas, and dust. If the neutrals were not well coupled to the ions through collisions, the neutrals would fall toward the star along radial trajectories while the ions would follow the magnetic field lines. At the other extreme, perfect ion-neutral coupling would ensure that both components flow inward along the magnetic field lines. Using the equation of motion for the neutrals, it is possible to estimate the relative drift between these two components and therefore assess how good the coupling is (see Martin 1996). It turns out the coupling is sufficient throughout the flow to treat the ions and neutrals as effectively coupled. Dust could have strong effects both on the emergent radiation from the funnel flow and on its thermodynamic structure. However, it is most likely that funnel flows are dust free based on the small values of A_V ($\sim$ few at most) deduced for T Tauri stars (see Appendix A of Martin 1996). Thus, the inflowing material can be treated as a single fluid and both the velocity and density of the gas can be derived at any point along the flow (see Hartmann, Hewett, & Calvet 1994).

3. Gas Thermodynamics

The temperature and ionization degree of the gas can be determined by solving the first law of thermodynamics coupled to rate equations for hydrogen (Martin 1996). The first law reduces to the standard heat equation in steady state (in a fixed reference frame) and can be written as

$$\frac{3}{2} n_{\rm H} k \boldsymbol{v} \cdot \nabla T = kT \boldsymbol{v} \cdot \nabla n_{\rm H} - n_{\rm H} \sum_j I_j \boldsymbol{v} \cdot \nabla f_j + \Gamma - \Lambda, \tag{1}$$

where $n_{\rm H}$ is the total number density of hydrogen nuclei, T is the kinetic temperature, I_j is the binding energy of hydrogen in state j, f_j is the fractional particle density (n_j/n_H) and k is Boltzmann's constant. Note the meaning of the terms on the right hand side of equation (1): the first term represents adiabatic cooling for an expanding flow and compressional heating for a converging flow, the second term allows for changes in the chemistry along the flow, the third term (Γ) is the heating rate per unit volume and the last term (Λ) is the cooling rate per unit volume. In practice, the most difficult part of solving equation (1) is specifying the last two terms for the densities and temperatures of interest in the funnel flow (typically, $T \sim 5000$ K and $n_H \sim 10^{12}$ cm^{-3} for an accretion rate $\sim 10^{-7}$ $M_\odot$ yr^{-1}). The required rate equations for hydrogen may be written simply as

$$\boldsymbol{v} \cdot \nabla f_j = \sum_{j' \neq j} R_{j',j} - f_j \sum_{j' \neq j} R_{j,j'}, \tag{2}$$

where R is the creation or destruction rate (s^{-1}) due to collisional and radiative processes. This system of equations is closed by requiring particle conservation. Once the model parameters are specified (e.g., stellar mass, radius, and temperature; the mass accretion rate; and the initial temperature) the above system of equations is integrated from a point just above

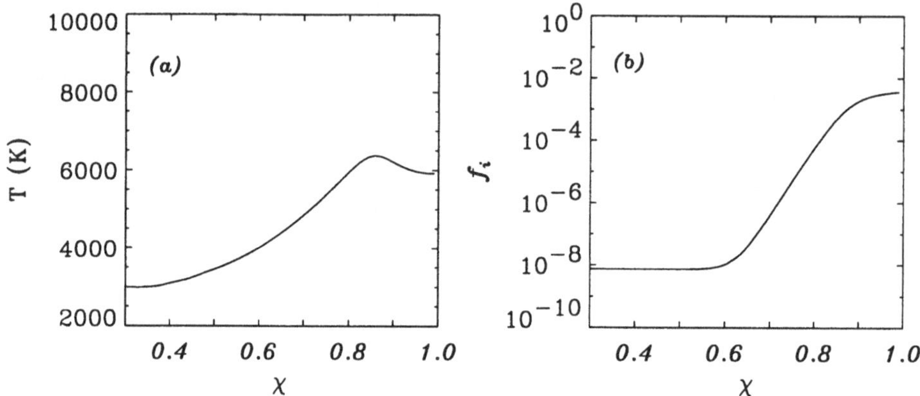

Figure 1. Thermal structure of gas along a flow line for parameters of a T Tauri star and a mass accretion rate of 10^{-7} $M_\odot$ yr^{-1}. The abscissa is the fraction along the flow line measured in units of total flow line length. The equations are integrated from $\chi = 0.3$ to $\chi \approx 1$ where the shock forms. (*a*) Temperature and (*b*) ionization fraction $f_i = n_{H^+}/n_H$ (from Martin 1996).

the disk (where the gas is loaded onto the field lines) to the stellar surface. The results from a sample calculation are displayed in Figure 1. Near the end point of the flow (i.e., the shock at $\chi \approx 1$), the largest temperature and ionization degree is attained. The high densities encountered in the funnel flow prevent Lyman photons from effectively penetrating and ionizing the gas. However, Balmer continuum photons can easily penetrate the gas flow and are the main ionizing source. The heat sources and coolants for this calculation can be identified from Figure 2. The principal heat source is compressional, due to the converging nature of the flow. This heating leads to a continual rise in temperature and ionization degree until $T \sim 6500$ K is reached, at which point the Ca II and Mg II ions efficiently cool the gas. In fact, these ions behave as a thermostat that regulates the gas temperature (see Appendix B of Martin 1996).

4. Observational Implications

Recent observations of young stars have found several pieces of evidence for magnetically channeled gas accretion (e.g., Kenyon et al. 1994, Edwards et al. 1994, and see especially Edwards this volume). Emission line profiles provide the most useful diagnostics for these types of flows as they give information on both the gas kinematics and thermodynamics. Inverse P Cygni line profiles were first recognized in some of the upper Balmer lines (Edwards et al. 1994), but now lines in the infrared (e.g., Paβ) are routinely

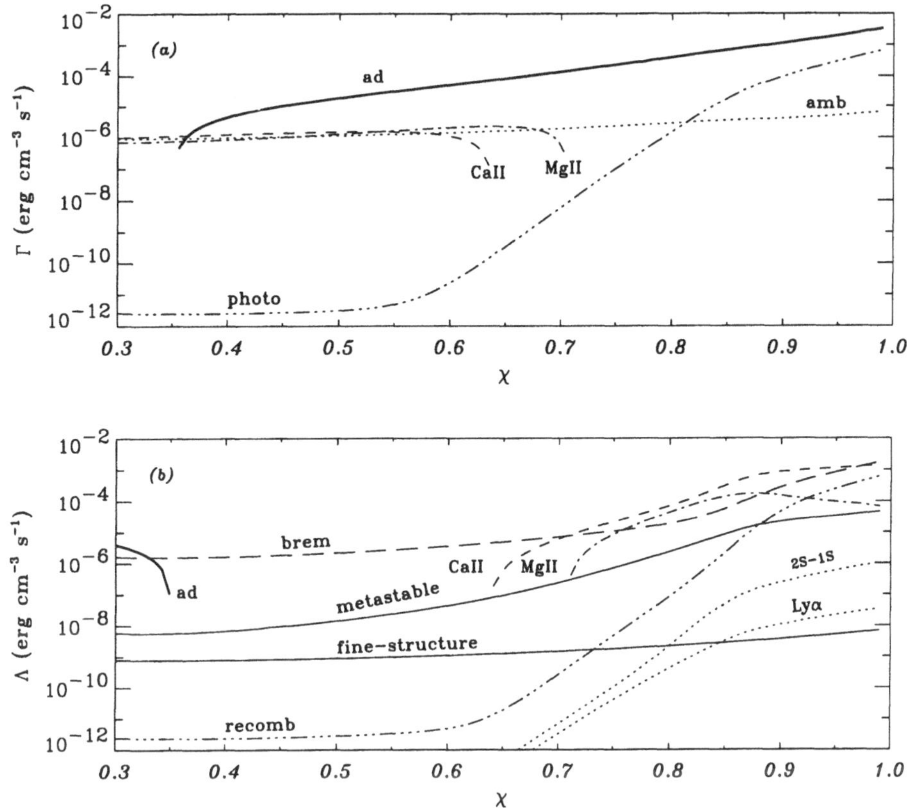

Figure 2. Heating and cooling rates for the calculation shown in Figure 1. (*a*) Heating rates. *Heavy solid line*: adiabatic; *dotted line*: ambipolar diffusion. (*b*) Cooling rates. *Heavy solid line*: adiabatic; *long dashed line*: bremsstrahlung; *short dashed line*: Ca II; *dash-dotted line*: Mg II; *triple-dot dashed line*: recombination. See Martin (1996) for details.

being observed (Folha, Emerson, & Calvet 1997). In addition, since some regions of the funnel flow are cool enough to allow molecules to exist (e.g., carbon monoxide), they may provide another important tracer of funnel flows.

The highest temperatures and ionization degrees are found in the inner regions of the funnel flow (see Fig. 1). For a typical accreting T Tauri star, temperatures $\sim$ 6500 K and ionization fractions of a few percent can be established. Thus, this model predicts that inverse P Cygni line profiles seen in the higher Balmer lines should form in the inner flow regions, where the temperatures are sufficient to populate the upper levels. The kinematics of the flow can be understood from Figure 3. This figure displays the constant-

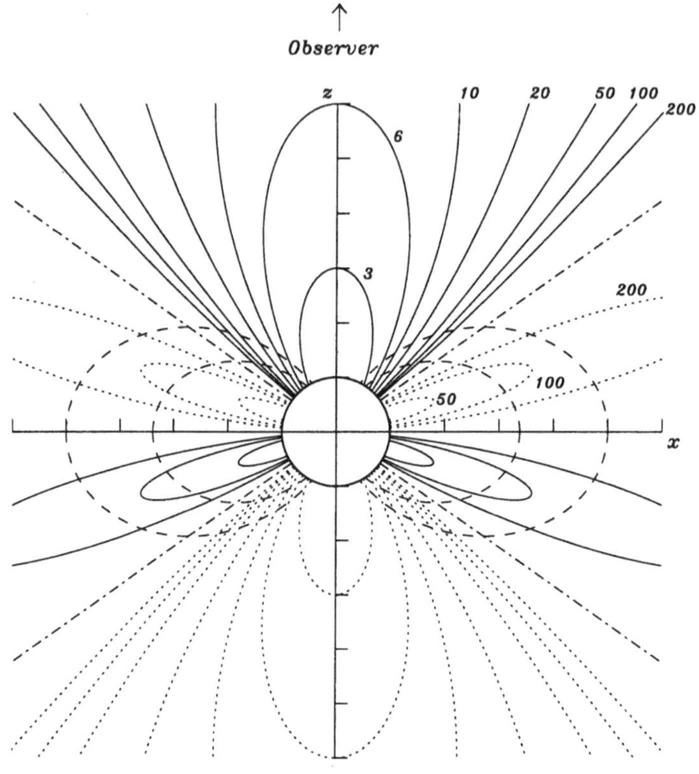

Figure 3. Constant-projected-velocity curves for an inclination of 0° in a dipole flow geometry. The Curves are labeled by $\Upsilon = (v_{ff}/v_{los})^2$ where v_{ff} is the free-fall velocity and v_{los} is the line-of-sight velocity. Positive velocities (redshifted) are represented by solid curves and negative velocities (blueshifted) by dotted curves (Martin 1996).

projected-velocity curves in a dipole flow geometry. The largest velocities occur in the inner regions (near the shock) and thus redshifted absorption within a line would occur at relatively large velocities. This prediction is borne out by observations where redshifted absorption features are seen at velocities $\gtrsim$ 200 km s^{-1}. Furthermore, blueshifted absorption features (at times seen simultaneously) typically occur at velocities $\lesssim$ 100 km s^{-1} and are most likely due to absorption in an outflowing wind.

Molecular lines, such as the first overtone transitions of carbon monoxide (CO), probe moderate-temperature ($T \sim 3000$ K), high-density ($n_H \gtrsim$

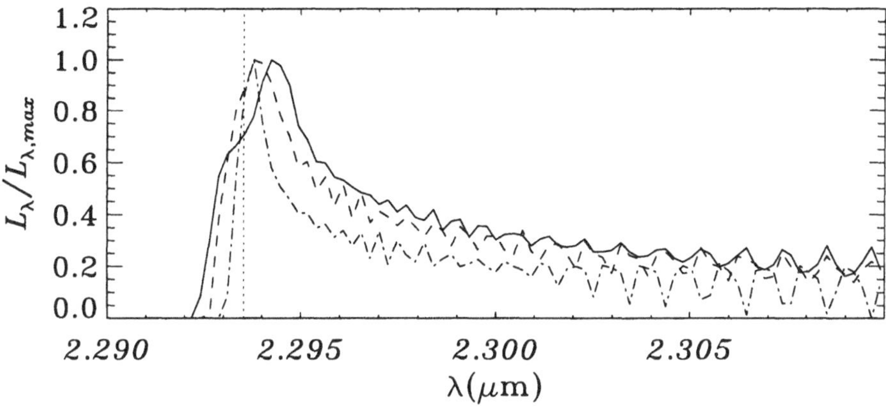

Figure 4. Bandhead profiles of the $v = 2 - 0$ transition of CO for the thermodynamic model shown in Figure 1. The *solid line*, *dashed line*, and *dash-dotted line* refer to viewing angles (with respect to the rotation axis) of 80°, 45°, and 10°, respectively. *Dotted line*: rest wavelength of the bandhead. From Martin 1997.

10^{10} cm^{-3}) gas. These conditions can be found in funnel flows and thus CO could be an important probe of the outer regions. High-resolution line profiles of CO bandheads from young stars indicate emission up to relatively large velocities ($\pm \sim 200$ km s^{-1}; e.g., Chandler et al. 1993). These profile characteristics have led most authors to favor an accretion disk as the origin of the CO lines (e.g., Chandler, Carlstrom, & Scoville 1995). However, it turns out that in a dipole flow geometry the kinematics and thermodynamics in the outer portions of the funnel flow can give the observed line profiles (Martin 1997). Shown in Figure 4 are a sample of the $v = 2 - 0$ bandhead profiles for CO using the computed thermal structure shown in Figure 1. Hence, the observed features (such as the blue shoulder and the redshifted peak) in this model depend mainly on the observers inclination angle. In addition, the computed luminosities of the bandheads fall near the low end of the observed range. Since the luminosity is directly proportional to the density and therefore the mass accretion rate (see Martin 1997), only those sources with the largest accretion rates should have detectable CO bandhead emission.

5. Conclusions

In the funnel flow model, stellar magnetic field lines disrupt the innermost regions of an accretion disk and channel the gas onto the surface at relatively high latitudes (see Figure 5). Since the flow is converging, compressional heating operates and increases the gas temperature to values ~ 7000 K. At

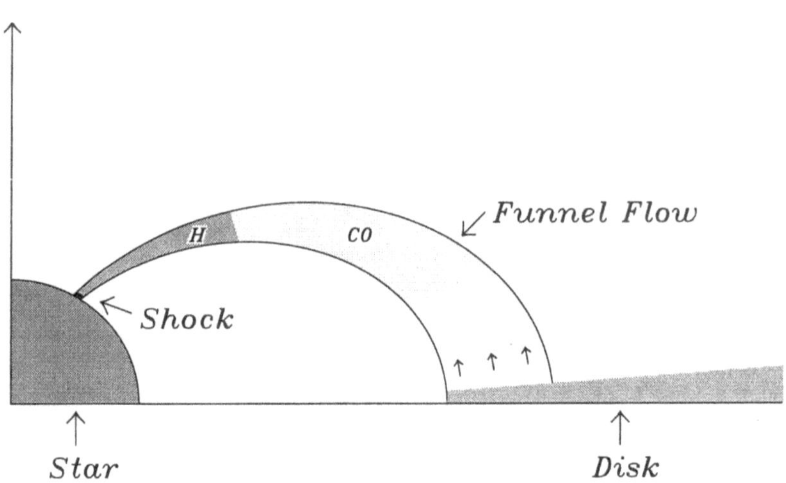

Figure 5. Dipolar funnel flow model. Only one quadrant is displayed, since the model is axisymmetric and reflection symmetric. Temperature increases towards the shock. *Light-shaded* region represents the portion of the flow that can produce carbon monoxide bandhead emission. *Dark-shaded* region is where inverse P Cygni profiles form (e.g., higher Balmer lines). From Martin 1997.

the stellar surface, strong shocks form where the gas in free-fall is halted and its kinetic energy thermalized, leading to excess ultraviolet radiation. Many pre–main-sequence stars display signatures of gas inflow along stellar magnetic field lines. For example, some T Tauri stars display inverse P Cygni line profiles and ultraviolet excess. The kinematics and thermodynamics of the infalling gas imply that inverse P Cygni line profiles (of, e.g., hydrogen lines) form close to the accretion shock, where the gas velocities are largest. In fact, the observed profiles display absorption at relatively large redshifted velocities, as expected from the model. In addition, the outer regions of the flow are cool enough to allow molecular emission, and thus may explain the carbon monoxide bandhead profiles seen for some young stars.

Acknowledgements: The author would like to thank the symposium organizers for the invitation to present this work. This research was supported in part by NASA grants NAG5-2766 and NAG5-3687. Also, funding to attend this symposium was provided in part by a AAS international travel grant and an IAU travel grant.

References

Adams, F. C., Lada, C. J., & Shu, F. H. 1988, ApJ, 326, 865
Basri, G., Marcy, G. W., & Valenti, J. A. 1992, ApJ, 390, 622
Bertout, C., Basri, G., & Bouvier, J. 1988, ApJ, 330, 350
Cabrit, S., Edwards, S., Strom, S. E., & Strom, K. M. 1990, ApJ, 354, 687
Chandler, C. J., Carlstrom, J. E., Scoville, N. Z., Dent, W. R. F., & Geballe, T. R. 1993, ApJ, 412, L71
Chandler, C. J., Carlstrom, J. E., & Scoville, N. Z. 1995, ApJ, 446, 793
Edwards, S., Hartigan, P., Ghandour, L., & Andrulis, C. 1994, AJ, 108, 1056
Folha, D., Emerson, J., & Calvet, N. 1997, in *Low Mass Star Formation - from Infall to Outflow*, poster proceedings of IAU Symp. No. 182, eds. F. Malbet & A. Castets, p. 272
Hartigan, P., Edwards, S., & Ghandour, L. 1995, ApJ, 452, 736
Hartmann, L., Hewett, R., & Calvet, N. 1994, ApJ, 426, 669
Hartmann, L., & Stauffer, J. R. 1989, AJ, 97, 873
Kenyon, S. J., Hartmann, L., Hewett, R., Carrasco, L., Cruz-Gonzalez, I., Recillas, E., Salas, L., Serrano, A., Strom, K. M., Strom, S. E., & Newton, G. 1994, AJ, 107, 2153
Königl, A. 1991, ApJ, 370, L39
Lynden-Bell, D., & Pringle, J. E. 1974, MNRAS, 168, 603
Martin, S. C. 1996, ApJ, 470, 537
————. 1997, ApJ, 478, L33
Mundt, R. 1984, ApJ, 280, 749
Mundt, R., Brugel, E. W., & Bührke, T. 1987, ApJ, 319, 275

MAGNETIC FIELDS OF T TAURI STARS

EIKE W. GUENTHER
Thüringer Landessternwarte Tautenburg
D-07778 Tautenburg, Germany

Abstract. The magnetic field strengths of several T Tauri stars are derived by measuring the width of unblended Fe I lines of high and low values of $g_{eff} \cdot \lambda^2$ using the autocorrelation function. The T Tauri stars were selected for their low values of $v \cdot sin\, i$, and large strengths of the Ca II emission component. The derived magnetic field strength are 2.0 ± 0.6 kG and 2.6 ± 0.8 kG for the classical T Tauri stars Lk Ca 15 and T Tau, respectively. An upper limit of 0.6 ± 0.8 kG is found for the weak-line T Tauri star Lk Ca 16. The method is tested by analysing two non-magnetic main sequence stars, and a late-type star that is known to have a strong magnetic field.

1. Introduction

Large observational and theoretical efforts have been invested in trying to understand how classical T Tauri stars (CTTS) can keep their low rotation rates while accreting matter. Currently, the most popular model is that strong magnetic fields ($\approx 1kG$) couple star and disk, so that angular momentum is transported outward, while matter is flowing inward (see reviews of F. Shu, and M. Camenzind in this volume).

Although strong magnetic fields might be highly important for CTTS, no field strength has been measured yet. However, there are some indications for the presence of magnetic fields in CTTS: The strongest evidence is the detection of circular polarized radio emission in HD 283447 (Phillips *et al.*, 1996), and T Tauri (Phillips, 1993). However, both stars are binaries, and HD 283447 is intermittent between a weak-line T Tauri star (WTTS) and a CTTS. Although many more CTTS have been detected in the radio regime, the radiation is in general thermal (Bieging *et al.*, 1984). As pointed out by André (1987), the non-detection of polarized radio emission

B. Reipurth and C. Bertout (eds.), Herbig–Haro Flows and the Birth of Low Mass Stars, 465–474.

in many CTTS does not necessarily exclude magnetic fields, because the ionized winds of CTTS are expected to free-free absorb any nonthermal radio emission produced near the surface of the star. Strong circular polarization of opposite helicity has also been detected in the two lobes of T Tau S (Ray *et al.*, 1997). This is expected if the outflows are magnetically collimated. Additionally, there is some evidence that magnetic fields are important for the acceleration of the winds (Paatz & Camenzind, 1996). Another argument for the presence of magnetic fields is that CTTS are relatively bright in X-rays, and that they occasionally show X-ray flares (Preibisch *et al.* 1993; Montmerle *et al.* 1983). Although some of the short time variability in the optical could be flares too, Gahm(1994) argues that most events are probably due to variations of the accretion rate and not due to flares. Another hint are the narrow emission line components (Ca II, He I) that originate most likely in plage regions on the surface of the star (Batalha *et al.*, 1996). There are thus some indications that many CTTS have strong magnetic fields but this does not necessarily mean that the star and disk are magnetically coupled. Evidence for that comes from line profiles: (1) the observed large infall velocities can best be explained in the context of magnetic accretion (Edwards *et al.*, 1994), and (2) the shape of the profiles can best be reproduced with models that assume that these lines are formed in magnetospheric infall zones (Hartmann *et al.*, 1994).

The evidence for magnetic fields in WTTS is much stronger than for CTTS: Basri *et al.* (1992) detected a magnetic field of 1000 ± 500 gauss in the weak-line T Tauri star TAP35, and found an upper limit of 700 G in Tap 10. Broad-band photometry implies that there are dark spots on WTTS. These spots cover areas of typically 5-40% of the stellar surface (Bertout 1989; Bouvier *et al.* 1993). In good agreement with this data are the results from Doppler imaging. Reconstructions of the surface markings clearly show cool spots on these stars (Joncour *et al.* 1994a, 1994b; Strassmeier *et al.* 1994; Rice & Strassmeier 1996). More evidence comes from observations in the radio regime: At least some WTTS are non-thermal radio sources. The non-thermal nature of the radiation is evident from the circular polarization and from the large brightness temperature of $\geq 10^7 K$. In general, the radio-emission of WTTS is consistent with gyrosynchroton radiation from a dipolar magnetic structure, and with a strength of the magnetic field of the order of one kG at the surface of the star (White *et al.*, 1992). Additionally, VLBI-observations show that the sizes of the magnetospheres are of the order of $R_{mag} \approx 25 R_*$ in diameter (André *el al.*, 1992). The correlation between the X-ray luminosity and the stellar rotation rate implies presence of a stellar dynamo (Bouvier, 1990), and flares have been observed in the radio X-ray and optical regime (White *et al.* 1992; Feigelson *et al.* 1985; Guenther & Emerson 1997). Since CTTS are rather similar to WTTS, we may argue that CTTS should have strong magnetic fields, if WTTS have strong magnetic fields. However, Neuhäuser & Preibisch (1994) argue that the larger X-ray fluxes (and faster rotation) of WTTS compared to CTTS implies a larger magnetic activity of WTTS. It can thus be concluded that there is some evidence for the presence of magnetic fields in CTTS and WTTS but it remains to be shown that CTTS and WTTS have magnetic fields with typical strength of kG. The aim of this work is to present measurements of the magnetic field strength in T Tauri stars.

2. Methods to measure magnetic fields

Since the magnetic field of a TTS will have its maximum magnetic flux density near the surface of the star and will decrease rapidly outwards, the strongest fields can be expected close to the star. Thus, magnetic field measurements should be carried out using photospheric lines, rather than emission lines. Since the photospheric spectrum of CTTSs is veiled, it will be easier to detect the fields in WTTSs. If the magnetic field structure is very complicated, observation of Stokes V will show no signal, because the signal from each magnetic north pole is canceled out by a magnetic south pole. However, the presence of many regions of opposite polarity will still lead to a broadening of the Stokes I-profile (Robinson *et al.*, 1980), and to a change in the equivalent width of the lines.

2.1. CHANGES OF THE EQUIVALENT WIDTH

The presence of a magnetic field changes slightly the equivalent width of photospheric lines. Magnetic fields can thus be measured by measuring the equivalent width and modeling of the stellar atmosphere. The clear advantage of this method is that the equivalent width can be measured even in stars with relatively high values of $v \cdot sin\, i$, and even when the lines are slightly blended, if only the whole spectrum is modeled. The disadvantage of this method is that a full radiative transfer code is needed and that the gf-values have to be precisely known. The gf-values are usually determined by measuring the equivalent widths in non-magnetic main sequence stars. The magnetic field is then derived by measuring the equivalent widths in the magnetic star by using the gf-values from the non-magnetic stars. Basri *et al.* (1992) used this method to measure a magnetic field of 1000 ± 500 gauss in the WTTS TAP35, and derived an upper limit for Tap 10 of 700 G.

2.2. THE LINE-BROADENING METHOD

Another method is to measure the width of photospheric lines. Magnetic fields can simply be measured from the enhanced broadening of Zeeman sensitive lines over Zeeman insensitive lines. This is usually done by plotting $g_{eff} \cdot \lambda^2$ against the line width for lines from the flat part of the curve of growth (Preston 1971; Hensberge & De Loore 1974). The slope of this curve then gives the magnetic field strength. The problem of the line-broadening method is that width of the lines has to be derived extremely accurately, and that the non-magnetic broadening of the lines could in principle be different for lines with different heights of formation. This would be the case, if the turbulence in the stellar atmosphere has a strong gradient. Although it can be expected that this effect is small, because the amplitude of the vertical component of the micro-turbulence decreases from just 2.5km/s to 1km/s when going from $\tau = 1$ to $\tau = 10^{-3}$ (Hollweger, 1967), and the macro-turbulence is of the order of 2km/s only (Stix, 1989), it is probably safer to use lines originating in the same layers of the atmosphere (Robinson *et al.* 1980; Guenther & Mattig 1991). This can be done by using lines of the same ion, of the same equivalent width, and in about the same spectral region. A disadvantage of this method is that it works better for slowly rotating stars. The most accurate way to measure the width of the lines is to calculate the autocorrelation functions for lines with large values of $g_{eff} \cdot \lambda^2$ and of lines with small values of $g_{eff} \cdot \lambda^2$ (Queloz *et al.*, 1996). The magnetic field strength can then be derived from the difference of the width of the two autocorrelation functions. In the following, this method will be applied to unblended Fe I lines from the flat part of the curve of growth to derive magnetic field strengths of T Tauri stars.

3. Selection criteria of the objects

Since the aim of this work is detecting magnetic fields in WTTS and CTTS, it is better to analyse stars with a high probability for the presence of a strong magnetic field. For solar like stars, Schrijver *et al.* (1989) found a correlation between the flux of the emission core of the Ca II K line and the magnetic field strength. The flux of this component is related to the magnetic field strength, because the component is formed in plage regions. Since the emission core of the Ca II 8542 Å-line is formed in plage regions too, there should be a similar correlation for this line. Thus, stars with strong emission cores in the Ca II 8542 Å-line can be expected to have a strong magnetic field. The stars for this work where thus selected using the following criteria:

- small $v \cdot sin\, i$

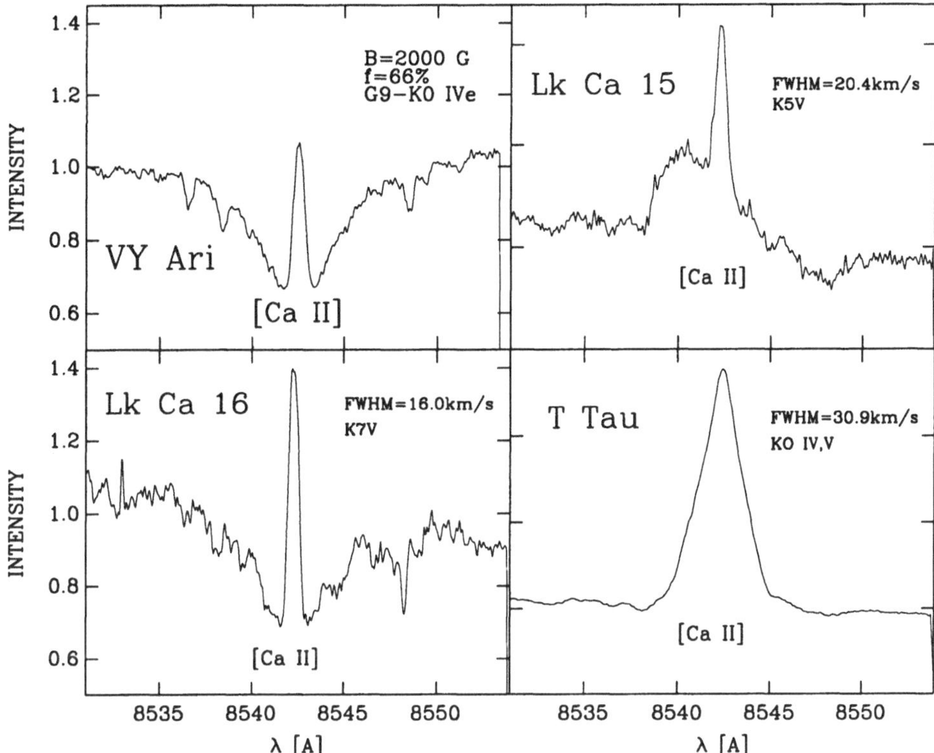

Figure 1. Ca II lines of the magnetic template star VY Ari, the WTTS Lk Ca 16, and the classical T Tauri stars Lk Ca 15 and T Tau. The narrow emission component seen in VY Ari originates in plage regions on the surface of the star, and is thus related to the strength of the magnetic field. This component is clearly visible in Lk Ca 15 and Lk Ca 16 too, making these stars suitable targets for the search of magnetic fields. Only a broad Ca II line is visible in T Tau.

- a large Ca II emission line component
- spectral type earlier than K7
- relatively large apparent brightness
- low veiling

Figure 1 shows the profiles of the Ca II 8542 Å-line of the sample. VY Ari is a star with a magnetic field strength of 2.8 kG and a filling factor of 66% (Bopp *et al.*, 1989). The emission line core is thus quite strong. It is interesting to note that the emission core of the Ca II lines is even stronger in the WTTS Lk Ca 16 than in VY Ari. Lk Ca 15 is intermittent between a CTTS and a WTTS. At the time of the observation we measured an equivalent width in $H\alpha$ of 24 Å, making it a CTTS at the time of the observations. The narrow component of the Ca II 8542 line is quite large.

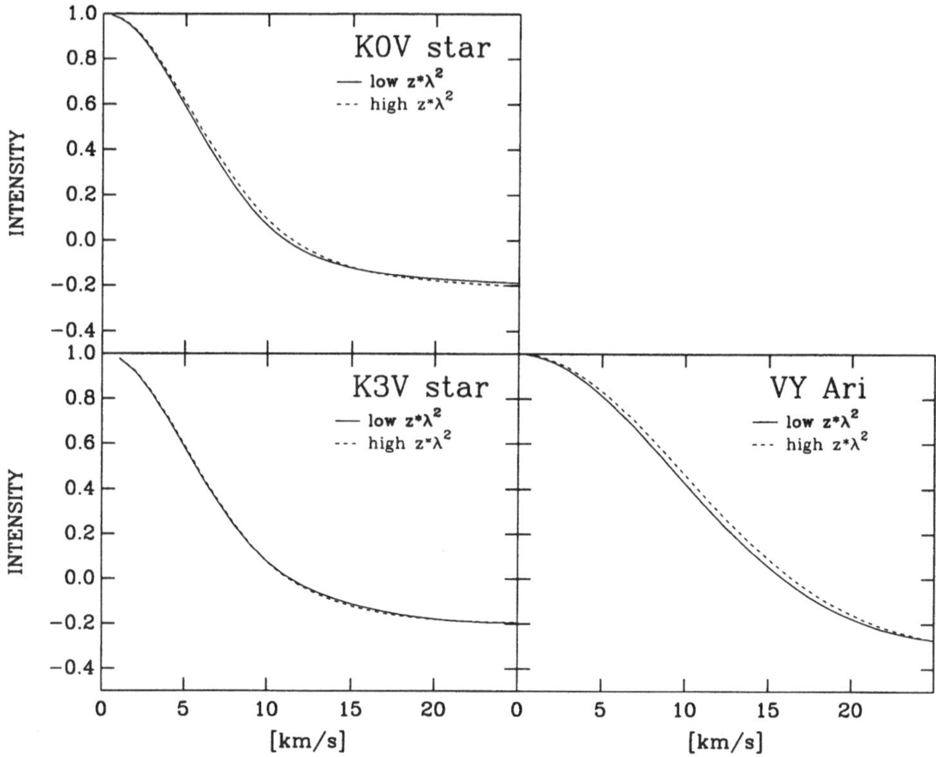

Figure 2. Autocorrelation functions of unblended Fe I lines with low (full line) and high (dashed line) values of $g_{eff} \cdot \lambda^2$ of the main sequence stars HD 10780 (spectral type K0V), HD 16160 (spectral type K3V) and the magnetic template star VY Ari. The autocorrelation function of low and high $g_{eff} \cdot \lambda^2$-values have exactly the same width in the case of HD 16160 indicating that no magnetic field could be detected in this star. In the case of HD 10780, the autocorrelation function for lines with high values of $g_{eff} \cdot \lambda^2$ is slightly broader than that of the lines with low values of $g_{eff} \cdot \lambda^2$, indicating that there is a small magnetic field. In the case of VY Ari, the autocorrelation function for lines with high values of $g_{eff} \cdot \lambda^2$ is significantly broader than that of the low $g_{eff} \cdot \lambda^2$-values.

T Tauri was selected because of the relatively low veiling and the small $v \cdot sin\, i$. The Ca II emission in T Tauri is just a broad emission line, and the narrow Ca II emission core is thus not visible.

4. Measurements of the magnetic field strength

Spectra of these stars were taken with the William Herschel Telescope using the University of Utrecht spectrograph. The resolution of the spectra is $\lambda/\Delta\lambda = 55000$, and the spectral-range 5260-9240Å. The signal to noise ratio is one-hundred or higher. The magnetic field strength is derived by measuring the width of the autocorrelation function of unblended Fe I lines

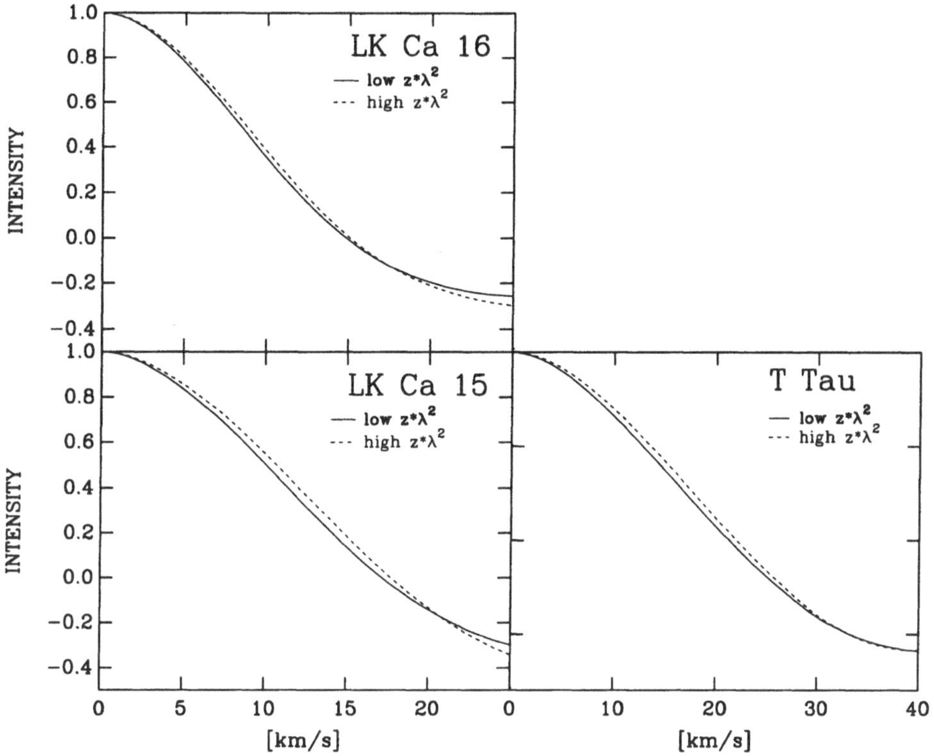

Figure 3. Like Figure 2 but for the WTTS Lk Ca 16, and the CTTS Lk Ca 15 and T Tauri. The autocorrelation function for lines with high value of $g_{eff} \cdot \lambda^2$ is significantly broader than that of the the lines with low values of $g_{eff} \cdot \lambda^2$ in Lk 15 and T Tau, indicating that a magnetic field has been detected in theses stars.

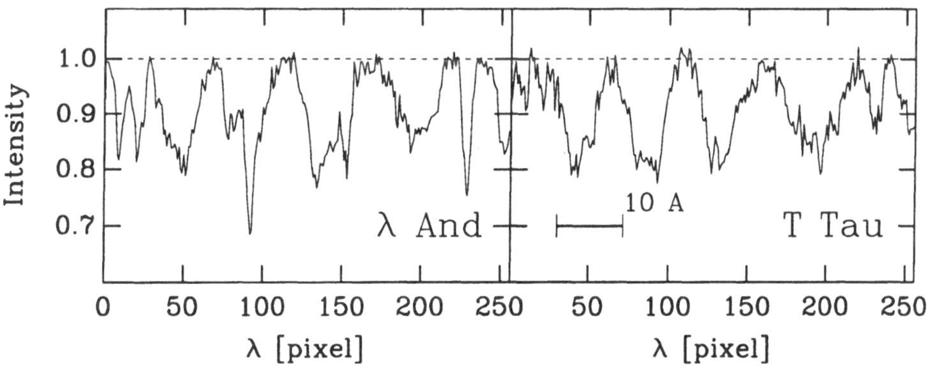

Figure 4. Spectrum of T Tau (right) and of the comparison star λ And (left). Photospheric lines are clearly detected in T Tau. These lines could be used to measure the magnetic field strength in CTTS and WTTS.

with a relatively high, and relatively low values of $g_{eff} \cdot \lambda^2$. A difference of 0.60 km/s between the two autocorrelation functions corresponds to a magnetic field of one kG. In order to test the method the width of the autocorrelation functions for the two main sequence stars HD 10780 and HD 16160 were measured. These stars are not expected to have strong magnetic fields. Additionally, the magnetic field of VY Ari was derived. Figure 2 shows the results. The autocorrelation function for low and high $g_{eff} \cdot \lambda^2$-values have exactly the same width in the case of HD 16160, indicating that no magnetic field could be detected in this star. In the case of HD 10780, the autocorrelation function for lines with high value of $g_{eff} \cdot \lambda^2$ is slightly broader than that of lines with low values of $g_{eff} \cdot \lambda^2$. Thus, HD 10780 seems to have a small magnetic field of the order of 0.7 ± 0.3 kG. In the case of VY Ari, the autocorrelation function for lines with a high value of $g_{eff} \cdot \lambda^2$ is significantly broader than for lines with low $g_{eff} \cdot \lambda^2$-values. The field strength of VY Ari is 1.3 ± 0.6 kG. This corresponds well to the value of B=2000 and f=66% from Bopp *et al.*(1989). Figure 3 shows the autocorrelation functions for the WTTS Lk Ca 16, and the CTTS Lk Ca 15 and T Tauri. The autocorrelation function for lines with high value of $g_{eff} \cdot \lambda^2$ is significantly broader than that for lines with low values of $g_{eff} \cdot \lambda^2$ in Lk Ca 15 and T Tau, indicating that a magnetic field has been detected in these stars. The corresponding field strength are 2.0 ± 0.6 kG and 2.6 ± 0.8 kG, for Lk Ca 15 and T Tau, respectively. The errors of the values of the magnetic fields are derived from using different sets of lines and different algorithms in finding the continuum. No magnetic field could be detected in Lk Ca 16. Table 1 summarizes the results.

TABLE 1. Magnetic field strength

star	spec type	fB [G]
HD 16160	K3V	$\leq 0.2 \pm 0.2$ kG
HD 10780	K0V	0.7 ± 0.3 kG
VY Ari	G9-K0IVe	1.3 ± 0.6 kG
Lk Ca 16	K7V	$\leq 0.6 \pm 0.8$ kG
Lk Ca 15	K5	2.0 ± 0.6 kG
T Tau	K0IV,V	2.6 ± 0.8 kG

5. Conclusions

Magnetic fields with a strength of 2.0 ± 0.6 kG and 2.6 ± 0.8 kG have been detected for Lk Ca 15 and T Tau by measuring the width of unblended Fe I lines with different values of $g_{eff} \cdot \lambda^2$. These are the first direct measurements of magnetic fields in classical T Tauri stars. It can thus be concluded that at least some classical T Tauri stars have strong magnetic fields. Although Lk Ca 16 has a strong Ca II emission line, no magnetic field has been detected.

6. Outlook

Since the magnetic broadening of the lines is $\propto g_{eff} \cdot \lambda^2$, it might be better to use lines in the infrared spectral region instead of the optical (although in practise, because of the increase of the intrinsic absorption line width with λ, the real gain is only $\propto \lambda$ (Giampapa *et al.*, 1993)). Besides the gain in magnetic sensitivity, the larger brightness of these late type stars in the infrared will lead to relatively short exposure times, and the low extinction in the infrared might even allow to measure the magnetic field strength of some embedded sources.

The line broadening method would ideally be used on pairs of unblended lines originating from the same levels of the atmosphere, where one is a Landé factor $g_{eff} = 0$ line, and the other is highly sensitive to the magnetic field. The only $g_{eff} = 0$ lines in the near infrared region are the Fe I 1.00733μm (multiplet 1993) and the Fe I 1.01519μm (multiplet 1861) line (Nave *et al.*, 1994). There are indeed two transitions from the same multiplets close by: the Fe I 1.00925μm (multiplet 1993), and the Fe I 1.01399μm (multiplet 1861) line. However, the Fe I 1.00925μm line has $g_{eff} = 1.375$ (7π, 12σ components), and the Fe I 1.01399μm line has $g_{eff} = 0.75$ (3π, 6σ components) only (Ramsauer *et al.* 1995; Nave *et al.* 1994).

The potential of the line pair Fe I 1.56572μm ($g_{eff} = 1.802$) and Fe I 1.56528μm ($g_{eff} = 3.00$) was pointed out by Solanki *et al.* (1992) in the context of the Sun. These two lines have very similar properties but differ in their sensitivity to the magnetic field. Guenther & Emerson (1996) demonstrated that these lines are present in the spectra of WTTSs, and that the magnetic field strength can be determined with an accuracy better than 500G in relatively short integration times if $v\sin i < 20$ km/s. Figure 4 shows a spectrum of the CTTS T Tauri and a comparison star in the same spectral region taken with the infrared camera MAGIC and the Coudé-spectrograph of the 2.2m telescope on Calar Alto. The spectrum clearly shows that these photospheric lines are present in CTTS too. Thus, the potential of infrared lines for measuring magnetic fields in CTTS and WTTS is high, and infrared lines could be used for this kind of work in the future, especially if infrared echelle spectrographs become available.

References

André, P.: 1987, in *Protostars and Molecular Clouds*, eds. Montmerle & C. Bertout

(Saclay: CEA-Doc), 143
André, P., Deeney, B.D., Phillips, R.B., Lestrade, J.-F.: 1992, ApJ 401, 667
Basri, G., Marcy, G. W., Valenti, J.A.: 1992, ApJ 390, 622
Batalha, C.C, Stout-Batalha, N.M., Basri, G., Terra, M.A.O.: 1996, ApJ Suppl. 103, 211
Bertout, C.: 1989, ARA&A 27, 351
Bieging, J.H., Cohen, M., Schwartz, P.R.: 1984, ApJ 282, 699
Bopp, B.W., Saar, S.H., Ambruster, C., Feldman, P., Dempsey, R., Allen, M., Barden, S.P.: 1989, ApJ 339, 1059
Bouvier, J.: 1990, AJ 99, 946
Bouvier, J., Cabrit, Fernandez, M., Martin, E.L., Matthews, J.M.: 1993, A&A 272, 176
Edwards, S., Hartigan, P., Ghandour, L., Andrulis, C.: 1994, AJ 108, 1056
Feigelson, E.D., Montmerle, T.: 1985, ApJ 289, L19
Gahm G.F: 1994, in *Flares and Flashes, Proceedings of IAU Colloqium No.151*, J. Greiner, H. Duerbeck, R.E. Gershberg (eds.), Springer Heidelberg, p. 203
Giampapa, M.S., Colub, L., Worden, S.P.: 1983, ApJ 268, L121
Guenther, E. Mattig, W.: 1991, A&A 243, 244
Guenther, E.W., Emerson, J.P.: 1996, A&A 309, 777
Guenther, E.W., Emerson, J.P.: 1997, A&A in press
Hartmann, L., Hewett, R., Calvet, N.: 1994, ApJ 426, 669
Hatzes, A.P.: 1995, ApJ 451, 784
Hensberge, H., De Loore, C.: 1974, A&A 37, 367
Hollweger, H.: 1967, Z. Astrophys. 22, 265
Joncour, I., Bertout, C., Bouvier, C.: 1994a, A&A 291, L19
Joncour, I., Bertout, C., Bouvier, C.: 1994b, A&A 285, L25
Montmerle, T., Koch-Miramond, L., Flaarone, E., Grindlay, J.E.: 1983, ApJ 269, 182
Nave G., Johansson, S., Learner, R.C.M., Thorne, A.P.: 1994, ApJS 94, 221
Neuhäuser, R., Preibisch, T.: 1994, in *Flares and Flashes, Proceedings of IAU Colloqium No.151*, J. Greiner, H. Duerbeck, R.E. Gershberg (eds.), Spinger Heidelberg, p. 216
Paatz G., Camenzind M.: 1996 A&A 308, 77
Phillips, R., Lonsdale, C.J., Feigelson, E.D.: 1993, ApJ 403, L43
Phillips, Lonsdale, C.J., Feigelson, E.D., Deeney, B.D.: 1996 AJ 111, 918
Preibisch, Th., Zinnecker, H., Schmitt, J.H.M.M.: 1993, A&A 279, L33
Preston, G.W.: 1971, ApJ 164, 309
Queloz, D., Babel, J., Mayor, M.: 1996, in *Cool Stars, Stellar Systems, and the Sun 9th Cambridge Workshop*, ASP Conference Series Vol 109, R. Pallavicini, A.K. Dupree (eds.), p. 627
Ramsauer, J., Solanki, S.K., Biémont, E.: 1995, A&AS, 113, 71
Ray T.P., Muxlow, T.W.B., Axon, D.J., Brown, A., Corcoran, D., Dyson, J., Mundt, R.: 1997, Nature 385, 415
Rice, J.B, Strassmeier, K.G.: 1996, A&A 316, 164
Robinson, R.D., Worden, S.P., Harvey, J.W.: 1980, ApJ 236, L155
Schrijver, C.J., Coté, J., Zwaan, C., Saar, S.H.: 1989, ApJ 337, 964
Solanki, S.K., Rüdi, I., Livingston, W.: 1992, A&A 263, 312
Stix, M.: 1989 *The Sun*, Springer Heidelberg, p.142
Strassmeier, K.G., Welty, A.D., Rice, J.B.: 1994, A&A 285, L17
White, S.M., Pallavicini, R., Kundu, M.R.: 1992, A&A 259, 149

EVIDENCE FOR MAGNETIC FIELDS IN THE OUTFLOW FROM T TAU S

T.P. RAY
Dublin Institute for Advanced Studies

T.W.B. MUXLOW
National Radio Astronomy Laboratory, Jodrell Bank

D.J. AXON
Space Sciences Division of ESA,
Space Telescope Science Institute, Baltimore

A. BROWN
Center for Astrophysics and Space Astronomy,
University of Colorado

D. CORCORAN
Dublin Institute for Advanced Studies

J. E. DYSON
Department of Physics and Astronomy, University of Leeds

AND

R. MUNDT
Max-Planck-Institut für Astronomie, Heidelberg

Abstract. We have observed at 5GHz the T Tau system with high resolution ($\lesssim 0''.1$) using the Multi-Element Radio Linked Interferometer (MERLIN) based at Jodrell Bank. Both the optical star (T Tau N) and its well-known infrared companion (T Tau S) were detected. The radio emission from T Tau S was found to be roughly extended in the direction of what is thought to be its outflow axis. More importantly we discovered that this radio emission split up into two spatially separated lobes of opposite helicity in the left and right-circular polarization channels. The circularly polarized lobes appear to straddle the star so that the "flow" and the "counterflow" were of opposite helicity. Such observations are the first direct evidence for the presence of magnetic fields in extended outflows that we are aware of. The radio flux appears to be due to gyrosynchrotron emission from mildly

B. Reipurth and C. Bertout (eds.), Herbig–Haro Flows and the Birth of Low Mass Stars, 475–480.

relativistic electrons ($\gamma \approx 2-3$). These electrons may have been accelerated in shocks close to the source. Using reasonable assumptions, the inferred magnetic fields strengths are surprisingly large ($\gtrsim$ several gauss) at distances of approximately 10–20 AU from their source. This is consistent with the magnetic fields being part of a collimated flow.

1. Introduction and Observational Details

T Tau is in many ways not typical of the class of stars to which it gives it name and it is perhaps because of its enigmatic nature that it remains an object of intense study. It is associated not only with a well-known reflection nebula (Hind's Nebula or NGC 1555) but it is at the centre of a diffuse HH region (Burnham's Nebula) powered by shocks (Solf *et al.*, 1988; Böhm & Solf, 1994) and a complex system of interlocking loops is seen in the near-infrared (see the poster by Herbst *et al.*, 1997). T Tau has been shown using infrared speckle interferometry to have a companion which lies approximately $0''.7$ south (Ghez *et al.*, 1991) of the visible star. Proper motion studies (Ghez *et al.*, 1995) have implied that the optical star and T Tau S almost certainly form a gravitationally bound system, the projected separation of which is decreasing at a rate of approximately 8 mas/yr. Both the optical star and its infrared companion have been detected at radio wavelengths (Schwartz *et al.*, 1984; Skinner & Brown, 1994) and these radio sources are referred to as T Tau N and T Tau S respectively. The centimeter emission from T Tau S is known to be variable on timescales of a few days and to be circularly polarized (Philips *et al.*, 1993; Skinner & Brown, 1994).

In November 1992, we used the Multi-Element Radio Interferometer Network (MERLIN) based at Jodrell Bank to map T Tau at 6 cm in phase referencing mode. Details are contained in Ray *et al.* (1997). Our final smoothed beam size was $0''.2$ and the rms noise level was 170 μJy/beam. The total field of view was $7''.68\text{x}7''.68$.

2. Results

Our 6cm radio map of the T Tau region is shown in Figure l. The two previously known radio sources T Tau N and T Tau S can be seen, along with extended emission and a third source which we have labeled T Tau R. It is not clear whether T Tau R is another star in the T Tau system or perhaps part of an outflow. If it is a highly embedded star (note that it was not seen using near-infrared speckle interferometry) then it may have been confused in the past with T Tau N (Ray *et al.*, 1997). The positions

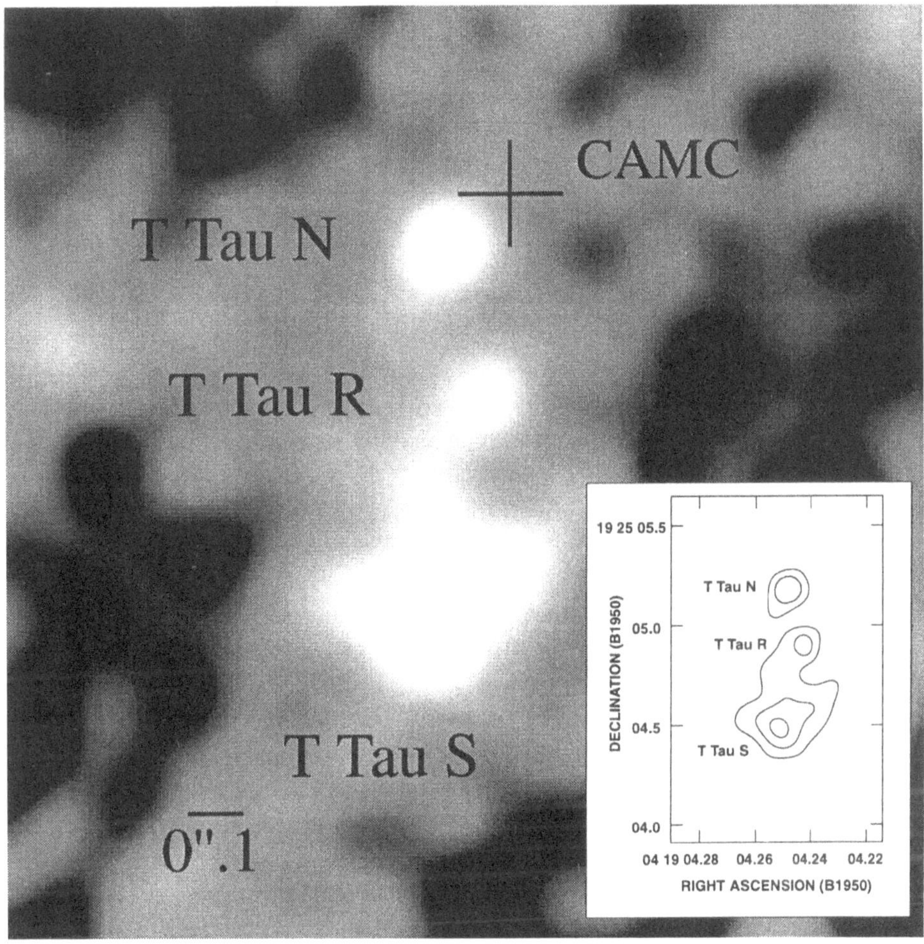

Figure 1. MERLIN 6 cm intensity image of T Tau from Ray *et al.* (1997). Apart from the known radio sources T Tau N and T Tau S, extended radio emission is observed between these two stars. There is also a third source which may have been wrongly classified as the optical star in the original VLA discovery image (see text and Ray *et al.*, 1997). The radio emission from T Tau S appears extended in the northwest–southeast direction due to its outflow (see also Bally & Devine, this volume). The position of the optical star, as derived from the Carlsberg Automatic Meridian Circle, and allowing for its known proper motion, is also shown. The 1σ errors in the optical position are indicated by the cross. Inset is a contour plot of the same data of the T Tau region; contours are at 0.4, 0.7 and 1.2 milliJanskys.

of the various radio sources, along with that of the optical star, are listed in Table 1. The optical position of T Tau, allowing for its proper motion, is derived from Carlsberg Automatic Meridian Circle data (Ray *et al.*, 1997) and it is also marked in Figure 1. From the overlap in their positions, it is

Source	RA(B1950)	Dec(B1950)	I(mJy)	R(mJy)	L(mJy)
T Tau N	04 19 04.246	+19 25 05.17	0.9	0.5	1.3
T Tau R	04 19 04.243	+19 25 04.88	0.8	0.6	1.1
T Tau S	04 19 04.246	+19 25 04.54	2.1	2.4	1.8
T Tau (Optical)	04 19 04.240	+19 25 05.26	–	–	–

TABLE 1. MERLIN determined positions and 6 cm flux densities for sources in the T Tau region. All positions quoted are Equinox B1950.0 and Epoch J1992.9 i.e. the epoch of the radio observations

clear that the radio source T Tau N can be unambiguously identified with the optical star. We should mention that the standard errors in the radio and optical positions are $\pm 0.''035$ and $\pm 0''.1$ respectively. Flux densities for the 3 sources are also given in Table 1 not only for total intensity (I) but right (R) and left (L) circular polarization components as well (see below). The typical errors in the flux densities are $\pm$0.15mJy.

We have already referred to the fact that the radio emission from T Tau S has been known to be circularly polarized on occasions. VLA observations by Philips *et al.* (1993) of T Tau at 18cm, although not resolving the emission, showed that its degree of circular polarization to be very high. Interestingly nearly simultaneous VLBI observations by the same authors overresolved the emission. From this fact Philips *et al.* (1993) argued that the size of the T Tau S radio emitting region must lie somewhere between the resolutions afforded by the VLBI and VLA systems i.e. from tens of stellar radii to a few tens of AU. Somewhat to our surprise, examination of the left and right circularly polarized channels of our radio image, showed that the emission from T Tau S divided up into two spatially separated lobes of opposite helicity (see Figure 2). The emission centroids of the two lobes were approximately 0''.15 apart corresponding to a distance of just over 20AU if we assume the Taurus-Auriga Cloud is 140pc away.

As far as we are aware, this is the first time that extended i.e. resolved polarization has been observed at radio wavelengths in the vicinity of a young stellar object (YSO). Moreover it clearly implies the presence of magnetic fields on scales of tens of AU rather than a few stellar radii. As pointed out by Skinner & Brown (1994) the only feasible mechanism for generating the observed emission is gyrosynchrotron radiation from electrons with mildly relativistic Lorentz factors i.e. $\gamma \approx 2$–3. As shown by Ray *et al.* (1997), the inferred magnetic field strengths are then of order several gauss at 10–20AU. Such fields are likely to be dynamically important and are certainly consistent with magnetic collimation models (see, for example, Ouyed *et al.*, 1997 and Pudritz & Ouyed, this volume) in which

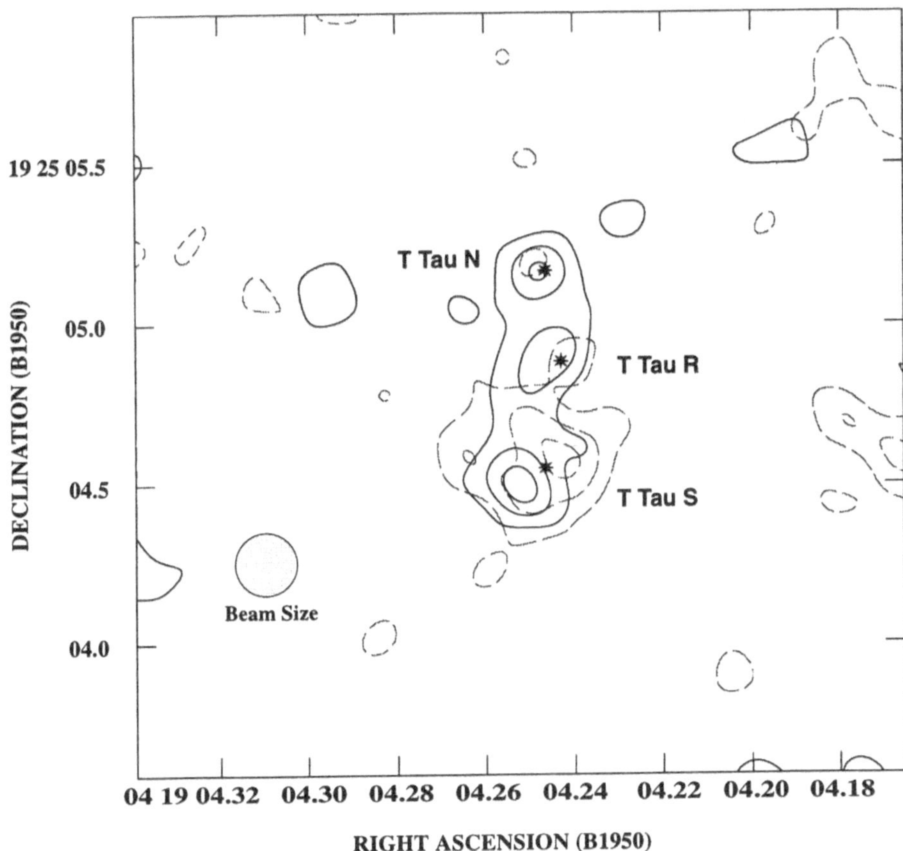

Figure 2. Our MERLIN contour map of the 6cm radio emission in the T Tau region. Right and left circularly polarized components are shown as dashed and continuous lines respectively. Contour levels are in each case 0.4, 0.8 and 1.2 mJy. Note the spatially resolved separation of the emission from T Tau S into two lobes of opposite helicity and the net left-hand circular polarization of T Tau N. Thus it appears that the 'flow' and 'counterflow' from T Tau S are both circularly polarized but in opposite senses. The source positions are shown as asterisks. Adapted from Ray *et al.* (1997).

outflows are produced, through the generation of a wind, by the winding up of magnetic fields anchored in an accretion disk. It is not clear at this stage whether the mildly relativistic electrons we observed in the outflow from T Tau S, were accelerated in reconnection events close to the star (i.e. at distances of at most a few $R_{\star}$) or whether they received their energy in the outflow through diffusive shock acceleration.

The strength of the inferred fields at distances of 10–20 AU clearly points to them being part of a collimated flow (Ray *et al.*, 1997) although we also emphasize that the fields may have been to some degree amplified through radiative shocks. It is also worth pointing out that T Tau S underwent what was probably an EX Lupi type outburst in 1990/91 due, presumably, to an increase in its accretion rate. In this regard it is interesting that while T Tau S appeared point-like in radio maps taken just before and around the time of the 1990 outburst (Skinner & Brown, 1994), it is extended along the outflow direction in our radio image two years later. Thus the radio "lobes" may have had their origin in the EX Lupi type outburst.

Finally, we remark that a small number of other YSO outflows are known to have non-thermal radio spectra (see, for example, Curiel *et al.*, 1993). Whereas linear polarization has been looked for but not found in these flows (e.g., the poster paper by Wilner *et al.*, 1997) it could be the case that some outflows are emitting gyrosynchrotron rather than synchrotron radiation. If this is so then we should be looking for circular as opposed to linear polarization.

References

Böhm, K. H. and Solf, J., 1994, ApJ 430, 277
Curiel, S., Rodríguez, L. F., Moran, J. M. and Canto, J., 1993, ApJ 415, 191
Dyck, H. M., Simon, T. and Zuckerman, B., 1982 ApJ 255, L103
Ghez, A. M., Neugebauer, G., Gorham, P. W., Haniff, C. A., Kulkarni, S. R., Matthews, K., Koresko, C. and Beckwith, S., 1991, AJ 102, 2066
Herbst, T. M., Robberto, M. and Beckwith, S. V. W., 1997, Low Mass Star Formation - from Infall to Outflow, Poster Proceedings of IAU Symp. No. 182, eds. F. Malbet & A. Castets, p215
Ghez, A. M., Weinberger, A. J., Neugebauer, G., Matthews, K. and McCarthy, D. W., JR., 1995, AJ 110, 753
Königl, A. and Ruden, S. P., 1993, Protostars and Planets III, eds. Levy, E. & Lunine, J., University of Arizona Press, 641
Ouyed, R., Pudritz, R. E. and Stone, J. M., 1997, Nature, 385, 409
Philips, R. B., Lonsdale, C. J. and Feigelson, E. D., 1993, ApJ 403, L43
Ray, T. P., Muxlow, T. W. B., Axon, D. J., Brown, A., Corcoran, D., Dyson, J. and Mundt, R., 1997, Nature, 385, 415
Schwartz, P. R., Simon, T., Zuckerman, B. and Howell, R.R., 1984, ApJ 280, L23
Skinner, S.L. and Brown, A., 1994, AJ, 107, 1461
Solf, J., Böhm, K. H. and Raga, A. C., 1988, ApJ 334, 229
Wilner, D.J., Reid, M.J., Menten, K.M. and Moran, J.M., 1997, Low Mass Star Formation - from Infall to Outflow, Poster Proceedings of IAU Symp. No. 182, eds. F. Malbet & A. Castets, p193

V. Low- and High-Mass Protostars and their Environment

THE EVOLUTION OF FLOWS AND PROTOSTARS

PHILIPPE ANDRÉ
CEA Service d'Astrophysique, Centre d'Etudes de Saclay
F-91191 Gif-sur-Yvette Cedex, France

Abstract. In this paper, I summarize the key properties of Class 0 protostars, discuss the observational evidence for a decline of outflow/inflow power with evolutionary stage, point out a possible connection with the initial conditions of fast protostellar collapse, and suggest a new collapse scenario in regions of induced, multiple star formation. The main thesis put forward here is that Class 0 objects have unusually powerful outflows because they accrete at a substantially higher rate than their Class I descendants.

1. Introduction

Despite recent observational and theoretical progress, the initial conditions of star formation and the first phases of protostellar collapse remain poorly understood. The purpose of this paper is to show that studying the evolution of molecular outflows through the embedded phase can shed light on this problem by providing important clues to both the mass-loss and the mass-accretion history of protostars.

1.1. TRACKING YSO EVOLUTION

In order to track protostellar evolution one needs a practical age indicator. The evolutionary indicator advocated here is the circumstellar mass derived from (sub)millimeter continuum measurements (André, Ward-Thompson, & Barsony 1993 – hereafter AWB93; André & Montmerle 1994 – hereafter AM94; Saraceno et al. 1996). Independently of the details of any protostellar theory, and in a statistical sense at least, one expects larger amounts of circumstellar material to surround younger stellar objects. In the case of embedded protostars, this decrease of circumstellar mass with time results

B. Reipurth and C. Bertout (eds.), Herbig–Haro Flows and the Birth of Low Mass Stars, 483–494.

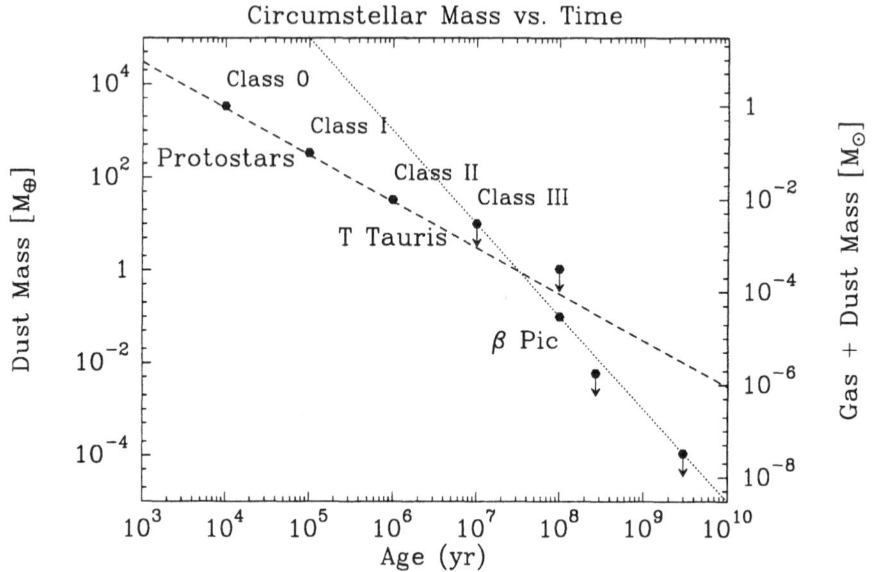

Figure 1. Circumstellar dust mass (in earth masses – left axis) as a function of age for low-mass stars, based on (sub)millimeter continuum measurements by AM94 and Zuckerman & Becklin (1993). The right axis gives estimates of the corresponding total masses (in solar masses), assuming a constant gas-to-dust mass ratio of 100. For comparison, the dashed and dotted lines show t^{-1} and t^{-2} variations of mass with time, respectively.

from the progressive dissipation of the protostellar envelope through accretion and ejection. For instance, in the standard theory of Shu et al. (1987), the mass enclosed within a given radius R of the infalling envelope scales approximately as $M_{env}(r < R) \propto t^{-1/2}$ with time.

The advent of sensitive bolometer arrays on large radiotelescopes such as the IRAM 30 m and the JCMT provides a very effective method of measuring the circumstellar mass around YSOs of any type by mapping their dust continuum emission in the (sub)millimeter range. Thanks to the small dust optical depth at these wavelengths, the measured emission is directly proportional to the circumstellar dust mass. The total (gas + dust) circumstellar mass may be derived if the gas-to-dust ratio is assumed or can be calibrated. The dust absorption coefficient is not exactly known, but the uncertainties are much reduced when the appropriate dust model is used for each type of sources (see Henning, Michel, & Stognienko 1995).
The effectiveness of this time clock is illustrated in Figure 1 which shows a global decline in the median circumstellar mass from protostars to main-

sequence stars, going approximately as $(\text{age})^{-1}$ or $(\text{age})^{-2}$. In this figure, the ages of Class 0 and Class I YSOs have been assumed to equal their respective lifetimes as derived from statistical arguments (e.g. AM94), while those of optically visible stars are estimated from positions in the HR diagram.

1.2. DEFINING PROPERTIES OF CLASS 0 PROTOSTARS

Following the above ideas, AWB93 introduced the concept of Class 0 protostars and proposed an age ordering of embedded YSOs based on the submillimeter to bolometric luminosity ratio $\mathrm{L}_{\mathrm{submm}}^{\lambda>350\mu}/\mathrm{L}_{\mathrm{bol}}$. While the (integrated) submillimeter luminosity $L_{submm}^{\lambda>350\mu}$ provides a relative measure of the (total) circumstellar mass $\mathrm{M}_{\mathrm{env}}$, the bolometric luminosity $\mathrm{L}_{\mathrm{bol}}$ may be used to constrain the central stellar mass $M_\star$ (see AM94). In the youngest sources, $\mathrm{L}_{\mathrm{submm}}^{\lambda>350\mu}/\mathrm{L}_{\mathrm{bol}}$ should tend to reproduce the variations of the mass ratio $M_{env}/M_\star$, which is expected to decrease with protostellar age.
AWB93 defined Class 0 sources as those young stellar objects which have $\mathrm{L}_{submm}^{\lambda>350\mu}/\mathrm{L}_{bol} > 5 \times 10^{-3}$. [A roughly equivalent, but more practical, criterion is $\mathrm{S}_{1.3mm}^{int}(d/160pc)^2/L_{\mathrm{bol}} \sim 0.2\ \mathrm{Jy/L_\odot}$.]
Since such sources are often undetected shortward of 10 μm, the existence of the central YSO is only indirectly inferred from, e.g., the detection of compact VLA radio continuum emission (e.g. Bontemps et al. 1995), or the presence of a collimated CO outflow (see Bachiller 1996). The formal $\mathrm{L}_{submm}^{\lambda>350\mu}/\mathrm{L}_{bol}$ boundary given above between Class 0 YSOs and the more evolved Class I sources of Lada (1987) was set so as to correspond to a mass ratio $M_{env}/M_\star = 1$, assuming the most plausible relations between L_{bol} and $M_\star$ on the one hand and between $\mathrm{L}_{\mathrm{submm}}^{\lambda>350\mu}$ and M_{env} on the other hand (see AWB93 and AM94 for details). *Class 0 sources are thus excellent candidates for being very young protostars in which a hydrostatic core has already formed but not yet accreted the bulk of its final mass.* As most of their mass is still in the form of a dense circumstellar envelope, Class 0 objects can potentially tell us a lot about the physics of protostellar collapse.

2. Evidence for a Decline of Outflow/Inflow Power with Time

Class 0 protostars tend to drive powerful, "jet–like" CO molecular outflows (see review by Bachiller 1996). In contrast, the CO outflows from Class I sources tend to be poorly collimated and much less powerful.

In an effort to quantify this evolution of molecular outflows during the protostellar phase, Bontemps et al. (1996 – hereafter BATC) have obtained and analyzed a homogeneous set of CO(2–1) data around a large sample of low-luminosity ($L_{bol} < 50\ L_\odot$), nearby ($d < 450$ pc) embedded YSOs,

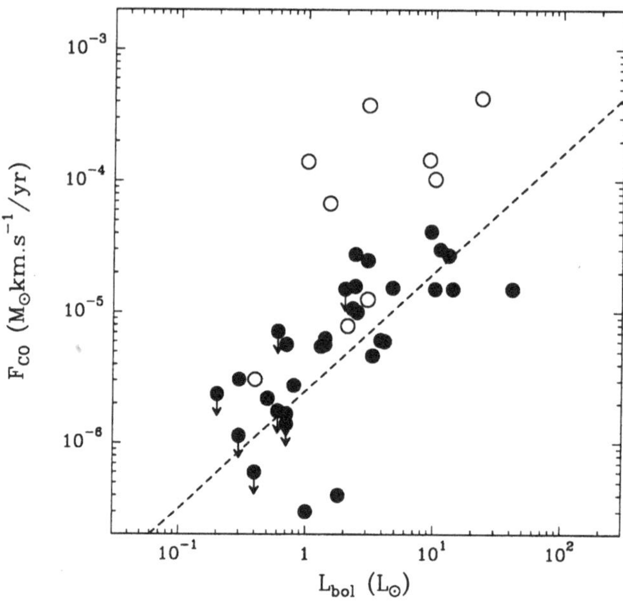

Figure 2. CO momentum flux versus bolometric luminosity for a sample of Class 0 (open circles) and Class I (filled circles) sources. The 'best fit' F_{CO}–L_{bol} correlation found for Class I sources is plotted as a dashed line. Taken from Bontemps et al. (1996).

including 36 Class I sources and 9 Class 0 sources. The results of these observations show that essentially *all* embedded YSOs have some degree of outflow activity, suggesting the outflow phase and the infall/accretion phase coincide. This is consistent with the idea that accretion cannot proceed without ejection and that outflows are directly powered by accretion (e.g. Pudritz et al. 1991, Shu et al. 1994, Ferreira & Pelletier 1995).

More importantly, in the F_{CO}–L_{bol} diagram shown in Figure 2, Class 0 objects lie an order of magnitude above the well-known correlation between outflow momentum flux (F_{CO}) and bolometric luminosity (L_{bol}) that holds for Class I sources. Since lower projected velocities, and thus lower apparent momentum fluxes, are expected from outflows more inclined to the line of sight, this indicates that Class 0 objects are not merely highly obscured Class I sources viewed edge-on: Class 0 objects differ qualitatively from Class I sources *independently of inclination effects.*

Furthermore, BATC find that outflow momentum flux is well correlated with circumstellar envelope mass in their *entire* sample (which includes both Class I and Class 0 sources). Bontemps et al. argue that this new correlation is independent of the F_{CO}–L_{bol} correlation and most likely results from a progressive decrease of outflow power with time during the accretion phase.

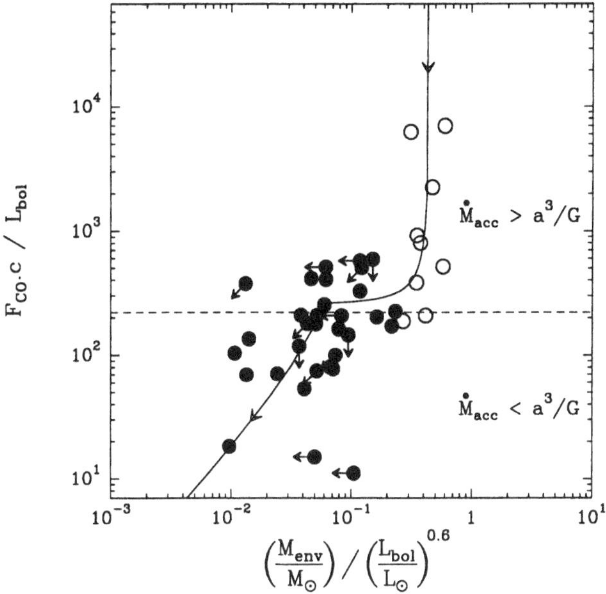

Figure 3. $F_{CO}\,c/L_{bol}$ versus $M_{env}/L_{bol}{}^{0.6}$ for the sample of Class 0 (open circles) and Class I (filled circles) sources studied by Bontemps et al. (1996). The normalized momentum flux $F_{CO}\,c/L_{bol}$ can be taken as an empirical tracer of the accretion rate $\dot{M}_{acc}$ providing there is no significant luminosity evolution (see § 2.1); $M_{env}/L_{bol}{}^{0.6}$ is an evolutionary indicator which decreases with time (see § 1). This diagram should therefore mainly reflect the evolution of $\dot{M}_{acc}$ during the embedded phase. The solid curve shows the accretion rate history predicted by the model of Henriksen et al. (1997) (see § 4).

This is illustrated in the $F_{CO}\,c/L_{bol}$ versus $M_{env}/L_{bol}{}^{0.6}$ diagram of Figure 3, which should be essentially free of any luminosity effect.

2.1. INTERPRETATION OF THE OBSERVED OUTFLOW EVOLUTION

Since magneto-centrifugal accretion/ejection models of bipolar outflows (e.g. Shu et al. 1994, Ferreira & Pelletier 1995) predict that the mass-loss rate is a definite fraction of the mass-accretion rate (typically, $\dot{M}_{jet} \sim 0.1\,\dot{M}_{acc}$), BATC suggest that the decline of outflow power with evolutionary stage seen in Figure 3 reflects a corresponding decrease in the mass-accretion/infall rate.

In the picture of accretion-driven jets/outflows, the CO outflow momentum flux F_{CO} is related to the accretion rate $\dot{M}_{acc}$ by

$$F_{CO} = f_{ent}\,\dot{P}_{jet} = [f_{ent}\,(\dot{M}_{jet}/\dot{M}_{acc})\,V_{jet}] \times \dot{M}_{acc},$$

where $f_{\rm ent}$ is the efficiency of entrainment of the flow by the jet, $\dot{M}_{\rm jet}$ is the mass-loss rate associated with the jet, and $V_{\rm jet}$ is the jet velocity. Using typical values ($f_{\rm ent} \sim 1$, $\dot{M}_{\rm jet}/\dot{M}_{\rm acc} \sim 0.1$, $V_{\rm jet} \sim 100\,{\rm km\,s^{-1}}$), the results of BATC suggest that, *on average*, $\dot{M}_{\rm acc}$ declines from $\sim 10^{-5}\,{\rm M_\odot yr^{-1}}$ for the youngest Class 0 protostars to $\sim 2 \times 10^{-7}\,{\rm M_\odot yr^{-1}}$ for the most evolved Class I sources, corresponding to a decrease of $\dot{M}_{\rm jet}$ from $\sim 10^{-6}\,{\rm M_\odot yr^{-1}}$ to $\sim 2 \times 10^{-8}\,{\rm M_\odot yr^{-1}}$. Note, however, that Ophiuchus YSOs and Taurus YSOs seem to follow quite different accretion/ejection histories (see § 4.2, Fig. 6, and Fig. 7 below).

3. Link to the Initial Conditions of Star Formation

As shown by Henriksen (1994), the beginning of the accretion phase depends quite critically on the initial conditions of fast protostellar collapse. In the standard theory of low-mass star formation, the initial conditions correspond to singular isothermal spheres (SISs), which have $\rho_{SIS}(r) = (a^2/2\pi G)r^{-2}$ where a is the isothermal sound speed. This leads to a constant accretion rate $\dot{M}_{\rm acc} \sim a^3/G$ (Shu 1977), which is apparently in conflict with the conclusions of § 2 above. Now, the SIS is obviously an idealized model of the actual pre-collapse conditions. Recent results suggest that, in the real world, the initial conditions differ sometimes significantly from a singular isothermal sphere.

In particular, submillimetre dust continuum mapping shows that the radial density profiles of isolated pre-stellar cores are relatively steep towards their edges (i.e., sometimes steeper than $\rho(r) \propto r^{-2}$), but *flatten out* near their centers, becoming less steep than $\rho(r) \propto r^{-2}$ (Ward-Thompson et al. 1994; André, Ward-Thompson, & Motte 1996 – AWM96). A representative, well-documented example of an isolated pre-stellar core is provided by L1689B, which is located in the ρ Ophiuchi complex but outside the main cloud. In this case, the radial density profile approaches $\rho(r) \propto r^{-2}$ between $\sim$ 4000 AU and $\sim$ 15000 AU, and is as flat as $\rho(r) \propto r^{-0.4}$ or $\rho(r) \propto r^{-1.2}$ (depending on the deprojection hypothesis) at radii less than $\sim$ 4000 AU (see AWM96 and Fig. 4).
By contrast, *protostellar* envelopes are always found to be strongly centrally peaked and do *not* exhibit the inner flattening seen in *pre-stellar* cores (Ladd et al. 1991, Motte et al. 1996).

Fragmentation introduces an additional complication in regions of multiple star formation. For instance, the ρ Ophiuchi main cloud is a star-forming cluster where the fragmentation size scale is clearly much smaller than 0.1 pc (e.g. Motte et al. 1997). In this case, the radius of the 'sphere of influence' of any given protostar must be less than $\sim$ 4000 AU, and the scale-free approximation associated with the SIS must quickly break down.

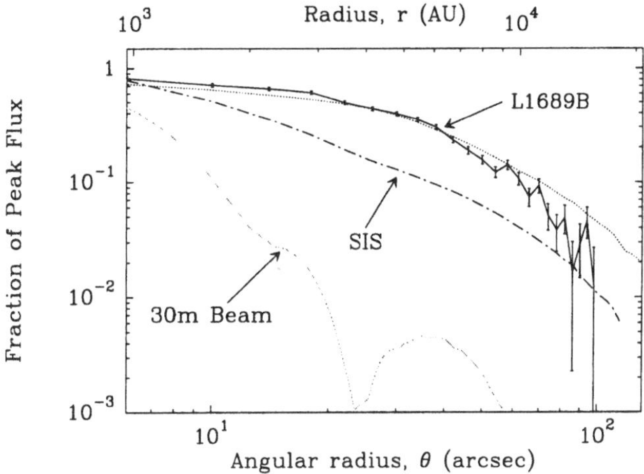

Figure 4. Azimuthally averaged intensity profile of the pre-stellar core L1689B (solid line), as observed with the IRAM 30 m telescope at 1.3 mm (adapted from AWM96). For comparison, the dash-dotted curve shows a spherically symmetric, isothermal model with $\rho(r) \propto r^{-2}$ such as a SIS. The dotted curve represents a spherical model with $\rho(r) \propto r^{-1.3}$ for $r < 4000$ AU (i.e., $\theta < 25''$) and $\rho(r) \propto r^{-2}$ for $r \geq 4000$ AU. The telescope beam profile is shown as a light dotted curve. The model profiles result from a full simulation of the continuum dual-beam mapping technique (see Motte et al. 1996).

Interestingly, recent large-scale, high-angular resolution imaging with the mid-infrared camera on board the ISO satellite (Abergel et al. 1996) suggests that the ρ Oph dense cores are characterized by very sharp edges (i.e., steeper than $\rho \propto r^{-3}$ or $\rho \propto r^{-4}$), possibly produced by external pressure. (The ρ Oph cores are seen as deep absorption structures by ISOCAM.)

4. Suggested Collapse Scenario

In order to improve our understanding of the accretion phase history in the case of non-singular initial conditions, Henriksen, André, & Bontemps (1997 – hereafter HAB97) have recently carried out simplified, analytical calculations of protostellar collapse based on the following assumptions. The initial conditions are represented by an idealized pre-stellar core which consists of a strictly flat inner plateau up to a radius r_N, an r^{-2} 'envelope' up to r_B, and a steeper power-law 'environment' farther out (see Fig. 5). The core collapse is initiated by an external disturbance at $t = t_o < 0$. It is supposed that, after a transitional period (leading to, e.g., magnetic decoupling), supersonic velocities develop *prior to* stellar core formation. The flat inner core region is then free to collapse, essentially homologously due to

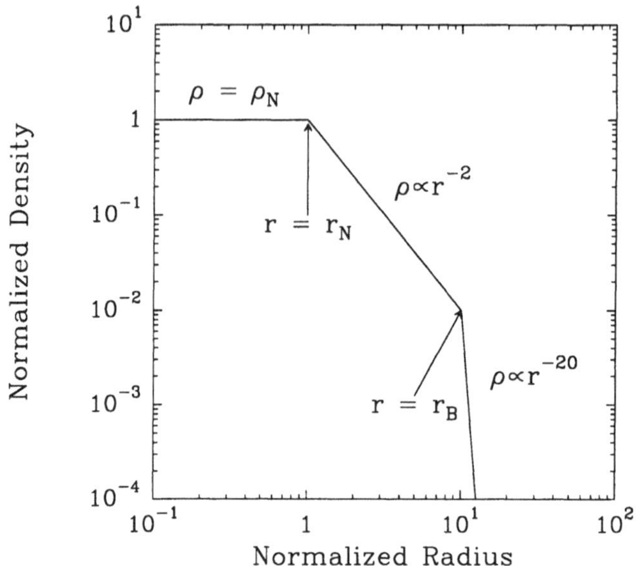

Figure 5. Density *vs.* radius in log–log format for an idealized pre-stellar core corresponding to the initial conditions of the collapse model proposed by HAB97.

the absence of significant pressure gradients. In this case, pressure-free calculations are justified and give qualitatively correct answers, as confirmed by comparison with recent numerical computations which fully account for pressure effects (Foster & Chevalier 1993, Tomisaka 1996). In the calculations of HAB97, the beginning of free core collapse is taken to be at $t = t_1 \equiv -t_{ff}(N)$. The core collapse is thus complete at $t = 0$, where $t_{ff}(N)$ is the free-fall time of a uniform sphere from rest. For all times $t > t_1$, the evolution is assumed to be of the 'gravity-dominated' self-similar form (see HAB97 for details), given uniquely by the density profile at the $t = t_1$ epoch.

In this scenario, the collapse occurs in various stages. The first stage ($t_1 < t < 0$) corresponds to the nearly isothermal, dynamical collapse of the pre-stellar flat inner region, which ends with the formation of a *finite-mass* stellar nucleus. At this stage, the flow is characterized by an inward-going compression wave (see Whitworth & Summers 1985). Observationally, this initial phase, which does not exist in the standard Shu picture, should correspond to 'isothermal protostars', i.e., supersonically collapsing cloud fragments with no central YSOs (see Mezger et al. 1992). It is followed at $t > 0$ by the main accretion/ejection phase, during which the nonzero central point mass accretes the surrounding envelope. The evolution can then be described by an inside-out expansion wave reminiscent of the standard Shu (1977) solution. However, because of the significant infall velocity field achieved during the first collapse stage, the accretion rate is initially higher than in the Shu theory. Enhanced accretion persists as long as the gravitational pull of the initial point mass remains significant. The accretion rate then quickly converges towards the characteristic value $\sim a^3/G$, which is the constant rate found by Shu (1977). At late times, accretion of the outer environment leads to a terminal phase of residual accretion/ejection, during which the accretion rate declines below the Shu value.

When the boundary radius r_B is not much larger than the radius r_N of the flat inner plateau (i.e. $r_B \sim 1.6\, r_N$), the simplified model of HAB97 provides a good overall fit of the diagram of outflow efficiency $F_{\rm CO}\, c/L_{\rm bol}$ versus normalized envelope mass $M_{\rm env}/L_{\rm bol}{}^{0.6}$ observed by BATC (see Figure 3 above). This requires that the fraction of cloud mass in the central plateau region be relatively large, $M_N/M_{cloud} \sim 30$ %.

Based on the fit shown in Figure 3, HAB97 tentatively associate the short initial period of energetic accretion/ejection predicted by their model at the beginning of the accretion phase with the observationally-defined Class 0 stage (AWB93, see § 1.2). In this view, Class I objects are more evolved and correspond to the longer period of moderate accretion/ejection when the accretion rate approaches the Shu value ($\dot{M}_{\rm acc} \lesssim a^3/G$).

4.1. ACCRETION LUMINOSITIES OF CLASS 0 AND CLASS I SOURCES

If, as proposed by BATC and HAB96, $\dot{M}_{\rm acc}$ is a factor of ~ 10 larger for Class 0 sources than for Class I sources, one may naively expect the former to have much higher accretion luminosities than the latter. This would be in apparent contradiction with observations which indicate that Class 0s are not significantly over-luminous compared to Class Is. For instance, in the embedded YSO sample of BATC, the ratio of the average bolometric luminosities for the two classes is $< L_{\rm bol} >_0 / < L_{\rm bol} >_I \sim 1.6$.

In reality, several factors contribute to reduce the theoretical accretion luminosity ($L_{acc} = GM_\star \dot{M}_{acc}/R_\star$) at the Class 0 stage:
(1) The central stellar mass $M_\star$ is smaller for Class 0 sources;
(2) The stellar radius $R_\star$ is likely to be larger if $\dot{M}_{\rm acc}$ is higher, since one expects the rough scaling $R_\star \propto \dot{M}_{\rm acc}{}^{1/3}$ (see Fig. 7 of Stahler 1988);
(3) The amount of accretion energy dissipated in the wind can be expected to be larger for Class 0s than for Class Is (and could be a significant fraction of the total accretion energy).

As shown in HAB97, the combined effects of (1), (2), (3) above are likely to render the luminosities of Class 0 sources similar to those of Class I.

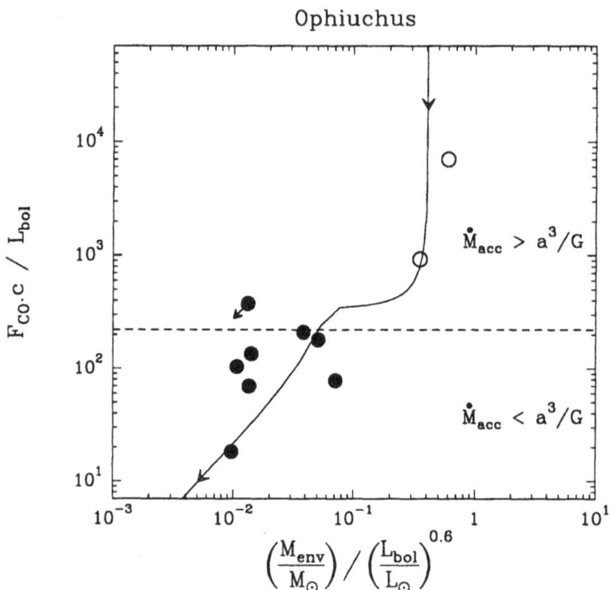

Figure 6. Same $F_{CO}\,c/L_{bol}$ versus $M_{env}/L_{bol}{}^{0.6}$ diagram as Fig. 3, but for the sub-sample of Class 0 (open circles) and Class I (filled circles) sources observed by BATC in Ophiuchus. The solid curve shows the model accretion history advocated by HAB97 (see text).

4.2. DIFFERENCES BETWEEN OPHIUCHUS AND TAURUS

In ρ Ophiuchi, the model fit of Figure 6 suggests that the Class 0 sources are in the initial phase of vigorous accretion with $\dot{M}_{acc} \sim 10^{-4}\,M_\odot yr^{-1}$, while the Class I sources may be in their terminal accretion phase with $\dot{M}_{acc} \lesssim 10^{-5}\,M_\odot yr^{-1}$. The latter is not surprising since, due to the small fragmentation size scale in this cloud (e.g. Motte et al. 1997), the expansion wave characterizing the collapse at $t > 0$ will reach the boundary of any given protostellar envelope/core in a time $t_B = r_B/a \lesssim 5\times10^4$ yr (assuming $a = 0.35\,\mathrm{km\ s^{-1}}$), shorter than the typical Class I lifetime of $\sim 2 \times 10^5$ yr (estimated by, e.g., Greene et al. 1994).
The clear contrast seen in Figure 6 between Class 0 and Class I objects strongly supports the contention that, in star-forming clusters at least, these two classes of YSOs differ qualitatively (AWB93).

On the other hand, all the Taurus sources of Figure 7 appear to be in the 'asymptotic' phase during which the accretion rate is approximately that predicted by the Shu theory, i.e., $\dot{M}_{acc} \sim a^3/G \sim 2 \times 10^{-6}\,M_\odot yr^{-1}$ (see also Kenyon, Calvet, & Hartmann 1993). In this case there is a much

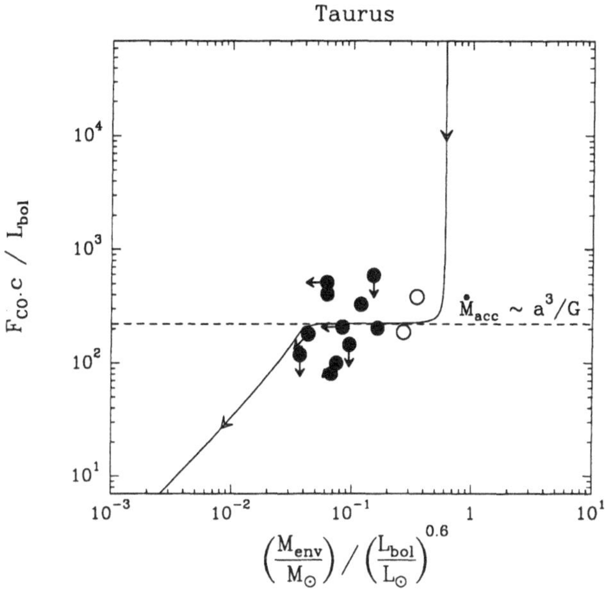

Figure 7. Same as Fig. 6 but for the sub-sample of candidate Class 0 (open circles) and Class I (filled circles) protostars observed by BATC in Taurus.

better continuity between the two classes, suggesting the candidate Class 0 protostars of Taurus may be viewed as "extreme Class I" sources.

5. Conclusions

Measuring outflow strength as a function of envelope mass in large, representative samples of embedded YSOs is a powerful, albeit indirect, tool to gain insight into the mass-accretion history of protostars.

Our recent work (BATC and HAB97) suggests that the protostellar accretion rate is roughly constant in Taurus, in agreement with the $\sim a^3/G$ value predicted by the 'standard' theory of Shu and collaborators. This is consistent with the idea that most Taurus stars form in relative isolation, following the *self-initiated* contraction/collapse of dense cores due to ambipolar diffusion (e.g. Mouschovias 1991). The initial conditions of fast protostellar collapse are then effectively close to a singular isothermal sphere (e.g. Li & Shu 1996).

However, our results also suggest that Class 0 protostars in star-forming clusters such as ρ Ophiuchi accrete at a significantly higher rate than the standard a^3/G value, and much more vigorously than their Class I descendants. In our view, this is because, in these regions, star formation is *induced* by the impact of (slow) shock waves (e.g. Loren & Wootten 1986, Boss 1995), and the collapse initial conditions depart significantly from a singular isothermal sphere. In this case, the collapse scenario proposed by HAB97 (see § 4), which predicts a strong decline of the accretion rate at the beginning of the main accretion phase, is likely to provide a better description of early protostellar evolution than the standard theory.

Acknowledgements: It is a pleasure to thank my colleagues S. Bontemps, R. Henriksen, F. Motte, and D. Ward-Thompson for their contributions to the work presented in this paper.

References

Abergel, A., Bernard, J.P., Boulanger, F. et al. 1996, A&A, 315, L329
André, P., Montmerle, T. 1994, ApJ 420,837 - AM94
André P., Ward-Thompson D., Barsony M., 1993, ApJ, 406, 122 - AWB93
André P., Ward-Thompson D., Motte, F. 1996, A&A, 314, 625 - AWM96
Bachiller, R. 1996, A.R.A.A., 34, 111
Bontemps, S., André, P., & Ward-Thompson, D. 1995, A&A, 297, 98
Bontemps, S., André, P., Terebey, S. & Cabrit, S. 1996, A&A, 311, 858 - BATC
Boss, A.P. 1995, ApJ 439, 224
Ferreira, J., Pelletier, G. 1995, A&A 295,807
Foster, P.N., Chevalier, R.A. 1993, ApJ 416,303 - FC93
Greene T.P., Wilking B.A., André P., Young E.T., Lada C.J. 1994, ApJ 434, 614
Henning, Th., Michel, B., & Stognienko, R. 1995, Planet. Space Sci., 43, 1333
Henriksen, R.N. 1994, in: Montmerle T., Lada C.J., Mirabel I.F., Trân Thanh Vân J. (eds.) The Cold Universe. Editions Frontières, p.241 - H94
Henriksen, R.N., André, P., & Bontemps, S. 1997, A&A, in press
Kenyon, S.J., Calvet, N., & Hartmann, L. 1993, ApJ 414, 676
Lada, C.J. 1987, in: Peimbert M., Jugaku J. (eds.) Star Forming Regions. IAU 115, p.1
Ladd, E.F., Adams, F.C., Casey, S. et al. 1991, ApJ 382, 555
Li, S., & Shu, F.H. 1996, ApJ, 472, 211
Loren, R.B., Wootten, A., 1986, ApJ, 306, 142
Mezger, P.G., Sievers, A.W., Haslam, C.G.T., et al. 1992, A&A, 256, 631
Motte, F., André, P., Neri, R. 1996, in: Siebenmorgen, R., Kaufl, H.U. (eds.) The Role of Dust in the Formation of Stars. ESO Astrophysics Symposia. Springer, Berlin, p. 47
Motte, F., André, P., Neri, R. 1997, A&A, in preparation (see Poster Proceedings)
Mouschovias, T.M. 1991, in The Physics of Star Formation and Early Stellar Evolution, Eds. Lada & Kylafis (Kluwer), p. 449
Pudritz, R.E., Pelletier, G, & Gomez de Castro, A.I. 1991, in The Physics of Star Formation and Early Stellar Evolution, Eds. Lada & Kylafis (Kluwer), p. 539
Saraceno P., André P., Ceccarelli C., Griffin M., Molinari S. 1996, A&A, 309, 827
Shu F. 1977, ApJ, 214, 488
Shu, F.H., Adams, F.C., Lizano, S. 1987, ARA&A 25,23
Shu F., Najita J., Ostriker E. et al. 1994, ApJ 429, 781
Stahler S.W. 1988, ApJ 332, 804
Tomisaka, K. 1996, PASJ, 48, L97
Ward-Thompson, D., Scott, P.F., Hills, R.E., & André, P. 1994, MNRAS, 268, 276
Whitworth, A., & Summers, D. 1985, MNRAS, 214, 1
Zuckerman, B., & Becklin, E.E. 1993, ApJ, 414, 793

CIRCUMSTELLAR MOLECULAR ENVELOPES

G. A. FULLER
Physics Department, UMIST
P.O. Box 88, Manchester, M60 1QD, UK

AND

E. F. LADD
Five College Radio Astronomy Observatory
University of Massachusetts, Amherst, MA 01003, USA
and
Department of Physics, Bucknell University
Lewisburg, PA 17837, USA

Abstract. The results of $C^{17}O$ and $C^{18}O$ observations of low luminosity young stars are discussed and the derived properties of the circumstellar molecular envelopes described. Much of the circumstellar material is quiescent and cold. The youngest sources have envelopes which contain up to $\sim 0.3\ M_{\odot}$ of material within 3000 – 4000 AU of the central star. From a survey of sources in Taurus, both the timescale and mechanism of clearing of the circumstellar region are investigated. Neither the observed outflows nor standard infall models alone are capable of clearing the circumstellar regions quickly enough. Maps of the circumstellar material around two very young sources suggest that the stars form from flattened configurations of material.

1. Introduction

The structure and dynamics of circumstellar material play a large role in determining how a protostar forms and evolves. The study of the circumstellar regions of protostars therefore provides key details for understanding how stars and circumstellar disks form and evolve.

The millimetre and submillimetre continuum emission from young stars clearly shows the presence of significant amounts of circumstellar mate-

B. Reipurth and C. Bertout (eds.), Herbig–Haro Flows and the Birth of Low Mass Stars, 495–506.

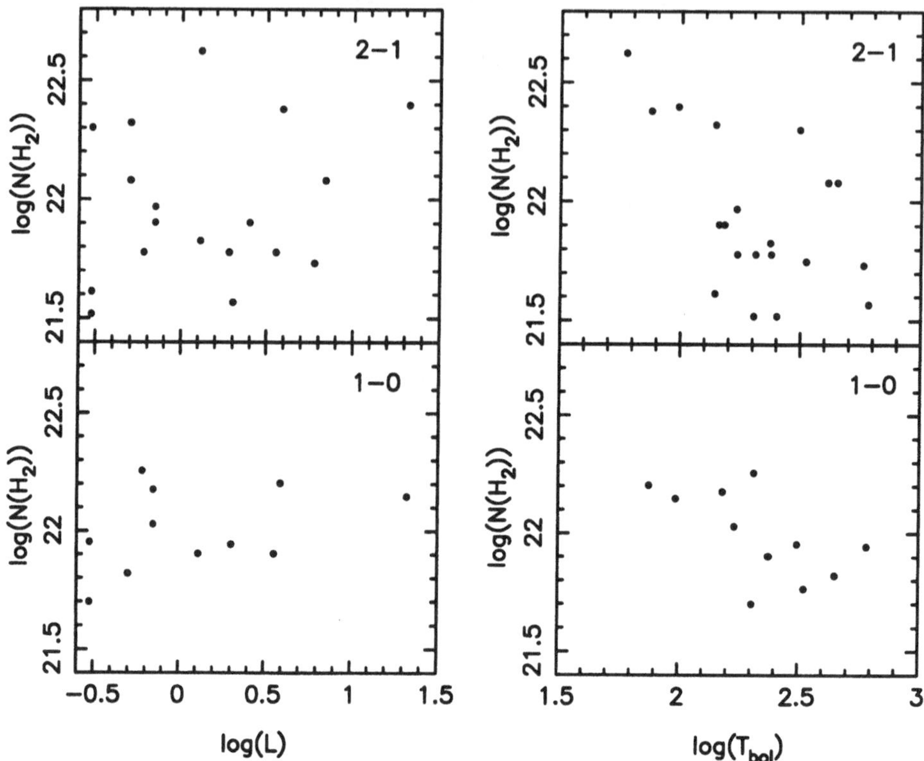

Figure 1. Column density derived from $C^{18}O$ and $C^{17}O$ observations versus source luminosity (left) and bolometric temperature (right). The upper figures show the results derived from the J= 2 → 1 transitions and the lower figures show the results from the J= 1 → 0 transitions.

rial associated with these stars (Ladd et al. 1991a,b) and modeling of this emission has provided constraints on the density and temperature of circumstellar material (e.g. Terebey et al., 1993; André & Montmerle, 1994). However, there has been little systematic work on molecular line tracers of this material. Molecular line probes of these regions provide a different view of the circumstellar material, one which is less biased by the temperature structure of the regions. The most significant advantage of molecular probes over dust is of course that the velocity and line width of spectral lines are direct probes of the dynamics of the circumstellar material.

We have examined the circumstellar regions of young stars in molecular lines using two complementary approaches. First, we have completed a single-beam survey of sources in Taurus using the J= 1 → 0 and J= 2 → 1 transitions of $C^{18}O$ and $C^{17}O$ using FCRAO and CSO. In parallel with this survey, we are mapping selected sources in these lines and the J= 3 → 2

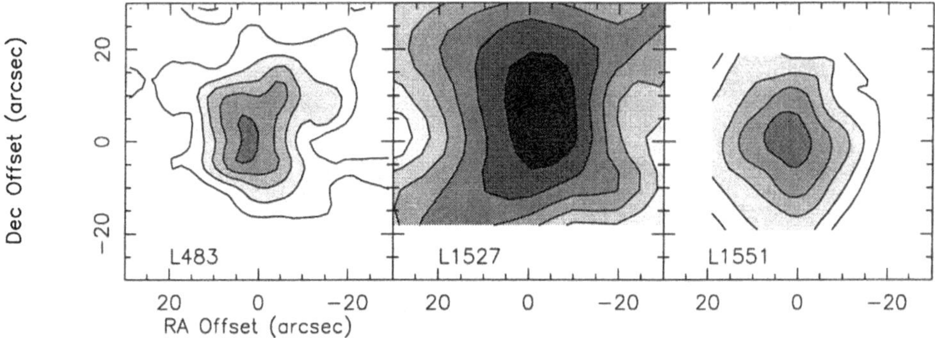

Figure 2. Maps of the dense circumstellar material around three young sources. Left: Map of the integrated intensity of the $C^{18}O$ J= 3 → 2 around the embedded source in L483. Centre: Integrated $C^{17}O$ J= 2 → 1 emission towards the the embedded source in L1527 (Fuller et al., 1996). Right: Integrated $C^{17}O$ J= 2 → 1 emission towards the embedded source in L1551 (Fuller et al., 1995).

transitions of the same species using FCRAO and JCMT.

2. Rare Species of CO As Probes

Our choice of $C^{17}O$ and $C^{18}O$ as probes of the circumstellar material is based on the fact that these isotopomers of carbon monoxide typically have low optical depth toward young stellar objects, that because of their low dipole moment the low-J energy levels are thermalized in these regions, and that the abundances of these isotopomers appear to be little affected by chemical or dynamical processes within the cloud. $C^{17}O$ is a factor of $\sim 10^3$ less abundant than the most common CO isotope, ^{12}CO, and its lines are optically thin for beam-averaged column densities in excess of 10^{23} cm^{-2}. By observing the low optical depth lines of two isotopomers, we can calculate the optical depth, column density, and excitation temperature of the emitting material.

Figure 1 shows that these CO isotopomers trace the circumstellar material around young sources. The hydrogen column densities derived from observations of the J= 1 → 0 and the J= 2 → 1 lines are not correlated with the source luminosity but are correlated with the source bolometric temperature, $T_{\rm bol}$ (Myers & Ladd, 1993). As a characteristic of the spectral energy distribution of an embedded source, $T_{\rm bol}$ measures the amount of circumstellar dust around a young star; sources with higher values of $T_{\rm bol}$ have less circumstellar dust. Since the column density derived from the CO observations decreases with increasing $T_{\rm bol}$, the CO is tracing the same circumstellar material as the dust measured by $T_{\rm bol}$. The correlation also

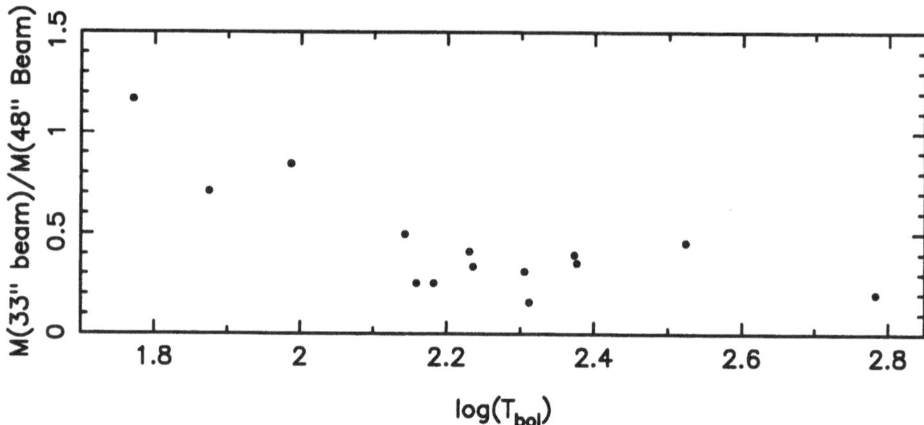

Figure 3. Ratio of beam mass derived from the J= 2 → 1 observations to that derived from the J= 1 → 0 observations versus source bolometric temperature. Two sources, L1536 and IRAS04181B, are not shown on this figure as the column densities derived towards these sources are thought to be unreliable (Ladd et al., 1997).

shows that the decrease in gas column density is not due to depletion of the gas phase molecules onto dust grains.

It has been argued that the 1.3mm dust continuum emission from young stellar objects is directly proportional to the mass of the circumstellar envelope (André et al., 1993; Saraceno et al., 1996). However, the 1.3mm flux of both outflow sources and HH exciting sources is strongly correlated with the source luminosity (Saraceno et al., 1996; Reipurth et al., 1993). This correlation suggests that the 1.3mm emission is not a function of mass alone but is determined by a combination of both the mass and temperature of the circumstellar material. The absence of such a correlation in the CO data shows the column density and mass results (and the observed line integrated intensities) to be relatively insensitive to temperature.

The most direct evidence that $C^{18}O$ and $C^{17}O$ trace the circumstellar material comes from the sources which have been mapped. Three of these sources are shown in Figure 2. The maps show that $C^{17}O/C^{18}O$ emission is centred on the young stars and have a typical radial extent of $\sim$ 3000AU.

3. Mass of Circumstellar Material

Both the survey and maps show that the circumstellar regions contain a substantial mass of material, with the youngest sources having the most circumstellar material. The Taurus survey shows that within about 3400 AU of the central stars the mass of the circumstellar material ranges from about 0.3 $M_\odot$ for the youngest sources to 0.02 $M_\odot$ for the older sources. Similarly the objects which have been mapped, which are the most embedded sources, show 0.2 to 0.5 $M_\odot$ within a region 3000 to 4000 AU in radius around the sources.

These masses are significantly higher than derived from other molecular line observations. An analysis of the CS emission from a sample containing some of the same young sources in Taurus as observed in CO, found circumstellar masses ranging from $0.003M_\odot$ to $0.015M_\odot$ (Moriarty-Schieven et al., 1995). The CS emission also has considerably larger velocity dispersion and higher temperature than the $C^{17}O$. This suggests that the CS emission is a relatively poor tracer of the quiescent circumstellar material and may instead be associated with the outflows from the young stars. Maps of some sources show that there is considerable CS emission associated with the outflows, for example in NGC 1333 IRAS2 (Sandell et al., 1994). Also, towards L1527 a high resolution map of CS $J=2 \rightarrow 1$ by Ohashi et al. (1996) shows the emission peaking 20″ from the location of the embedded source whereas the $C^{17}O$ $J=2 \rightarrow 1$ emission is centred on the source (Figure 2).

The $J=1 \rightarrow 0$ and $J=2 \rightarrow 1$ data have been analysed independently and therefore produce independent estimates of the column density towards the sources. Figure 3 shows the ratio of the mass within the two beams. Since the beams have different FWHM, 48″ for the $J=1 \rightarrow 0$ observations and 33″ for the $J=2 \rightarrow 1$ observations, this ratio depends on the density profile of the material around each source. If the material is uniformly distributed around a source the ratio of mass in the smaller $J=2 \rightarrow 1$ beam to the larger $J=1 \rightarrow 0$ would just scale as the beam area and so be 0.47. On the other hand, if the material is concentrated near the star, the ratio should increase, and in the limit approach unity, as the larger beam would contain only a slightly larger mass of material than the smaller beam. Figure 3 shows that there is a inverse correlation between this ratio and $T_{\rm bol}$.

The younger sources have higher values of this ratio and are consistent with their circumstellar material being strongly centrally condensed. The older, high $T_{\rm bol}$ sources have low values of the ratio and are consistent with uniform distributions of material. This change from centrally condensed to consistent with a uniform distribution of circumstellar material suggests that as material is removed from the circumstellar regions, it is removed first, and most significantly, from the region closest to the forming star.

4. The Temperature of the Circumstellar Material

Both the $J=1 \rightarrow 0$ and $J=2 \rightarrow 1$ survey data indicate excitation temperatures of 5 to 14 K for the circumstellar material. These values are consistent with the 7 to 15 K found from the $J=2 \rightarrow 1$ maps. The ratio of $J=2 \rightarrow 1$ to $J=3 \rightarrow 2$ emission suggests somewhat higher excitation temperature for

TABLE 1. Fits to lines towards L1527

Line	dv ($km\,s^{-1}$)	
	Comp. 1	Comp. 2
$C^{18}O$ 1-0	0.47	1.7
$C^{17}O$ 1-0	0.41	1.2
$C^{18}O$ 2-1	0.57	1.87
$C^{17}O$ 2-1	0.44	...
$C^{18}O$ 3-2	0.54	2.01
$C^{17}O$ 3-2	0.50	...

some sources; this result is not unexpected as the J= $3 \rightarrow 2$ emission is likely to contain a contribution from warmer material. The comparison of these two lines is, however, more uncertain due to the different beam sizes of the observations. These temperatures are lower than has been estimated from either CS J= $5 \rightarrow 4$ (Moriarty-Schieven et al., 1995) or the dust continuum emission, even for Class 0 sources which have spectral energy distributions which are well matched by single temperature models (Moriarty-Schieven et al., 1994; André et al., 1993).

The excitation temperatures are in reasonable agreement with the models of Ceccarelli, Hollenbach and Tielens (1996). Scaling the results from their 20 $L_{\odot}$ model to 1 $L_{\odot}$ (which is typical of the Taurus sources), yields a temperature of 9.5 K at a radius of 4.2×10^{16}cm (2800 AU) for the forming star.

5. The Velocity Dispersion of the Circumstellar Material

Most of the sources show a narrow line width component ($\sim$ 0.5–0.7 $km\,s^{-1}$), indicating that much of the circumstellar material is relatively quiescent. This narrow velocity component is not seen in CS J= $5 \rightarrow 4$ (Moriarty-Schieven et al., 1995).

Some sources also show a second broader component which is more apparent in $C^{18}O$ than in $C^{17}O$ and in the higher J transitions than in the lower J transitions, suggesting that the broad component has lower optical depth and is warmer than the narrow component. An example of Gaussian fits to the line profiles towards the embedded source in L1527 is given in Table 1. Notice the consistency of the linewidth of the narrow component and that for a given transition the $C^{17}O$ linewidth is narrower, consistent with the $C^{18}O$ line being more optically thick than the $C^{17}O$ line.

6. Variation of Column Density with T_{bol}

It has been suggested that sources may have low values of T_{bol}, not because they have large amounts of circumstellar material, but because they are viewed edge-on. The column density-T_{bol} correlation shown in Figure 1 shows that this is not the case. The derived column densities are independent of inclination angle and the low T_{bol} sources have the highest column densities of material, and the most massive circumstellar envelopes.

T_{bol} is an age ordering quantity and T_{bol} can be correlated with age estimates for the sources (Chen et al., 1995; Ladd et al., 1997). Using this correlation together with the column density-T_{bol} correlation it is possible to investigate how the amount of circumstellar material decreases with source age.

The column density is found to vary with source age, t_{yrs}, as

$$\begin{array}{ll} 48'' \text{ beam} & \log(N_{H_2}) = 2.9 - 0.42\log(t_{yrs}) \pm 0.29 \\ 33'' \text{ beam} & \log(N_{H_2}) = 4.3 - 0.74\log(t_{yrs}) \pm 0.45 \end{array}$$

for the J= 1 → 0 and J= 2 → 1 data respectively. These relations can be converted to the rate at which mass is lost from the respective beams by multiplying the column density by the respective beam size and differentiating with respect to the source age. This leads to beam mass loss rates of

$$\begin{array}{ll} 48'' \text{ beam} & \log(\dot{M}_{beam}) = 0.7 - 1.42\log(t_{yrs}) \pm 0.41 \\ 33'' \text{ beam} & \log(\dot{M}_{beam}) = 2.0 - 1.74\log(t_{yrs}) \pm 0.60 \end{array}$$

These show that both beam mass loss rates are decreasing functions of source age. For the youngest source, L1527, $\dot{M}_{beam} > 10^{-5}$ $M_{\odot}yr^{-1}$ while for older sources such as HL Tau, $\dot{M}_{beam} < 10^{-7}$ $M_{\odot}yr^{-1}$.

There are two ways in which the mass within the beam could decrease. Material could be moved out of the beam by the outflows from the star, alternatively material could collapse to a compact optically thick region to which the observations are insensitive.

6.1. OUTFLOW

The role of outflows in removing material from our beam can be assessed using the mass loss rates derived from the observed bipolar outflows associated with the sources. Of the sources observed into the J= 2 → 1 transitions, 11 have outflows detected in CO.

In order to determine whether the outflows could remove the circumstellar material at the rate required by the $C^{17}O/C^{18}O$ observations, it is important to understand the origin of the CO in the observed outflows. If all the CO in the bipolar outflows is material that has been entrained

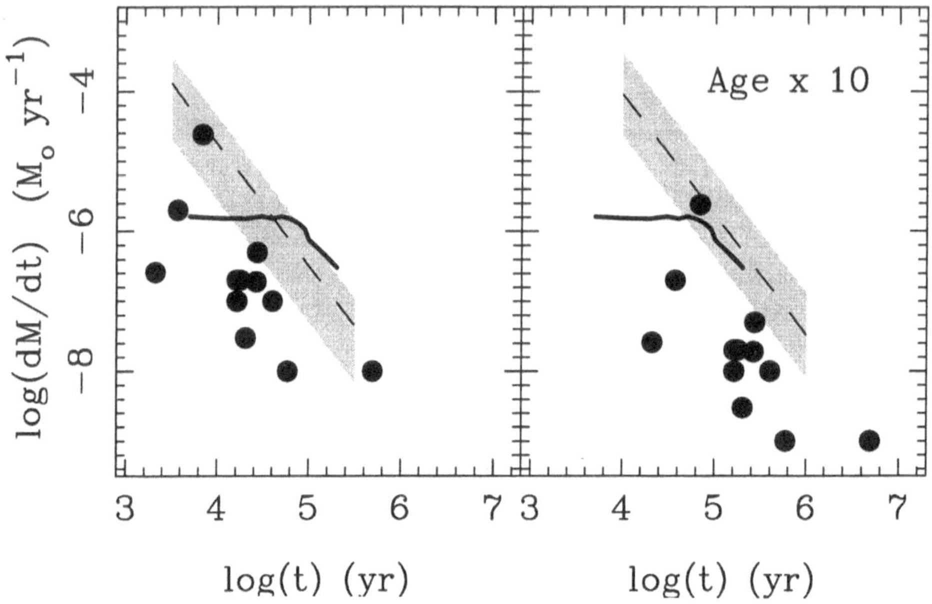

Figure 4. Mass loss rates derived from the J= 2 → 1 observations versus source age. In each panel the dashed line and gray region indicate the best fit beam mass loss rate and the estimated uncertainty from the J= 2 → 1 $C^{17}O/C^{18}O$ data. The points indicate the mass loss rates due to the outflows from the sources. The solid line is the beam mass loss rate for a collapsing singular isothermal sphere at 10 K. The panel to the left shows the results for using the T_{bol} - source age correlation discussed in the text. The other panel shows the results if the sources are assumed to be 10 times older than implied by the T_{bol} correlation.

from the circumstellar regions around the young stars, then the mass loss rate from the circumstellar region is just equal to the CO outflow mass loss rate. If, however, the CO is only entrained at large distances from the central star, outside the beams used to make the $C^{17}O/C^{18}O$ observations, the circumstellar mass loss rate is equal to the stellar wind mass loss rate. This is smaller than the observed CO outflow mass loss rate by the ratio of the outflow speed to the wind speed, a factor typically assumed to be 0.1.

The presence of shocks and HH objects at large distances from the central sources together with high angular resolution observations of outflows within a few arcseconds of young stars (e.g. Chandler et al. 1996) suggest that entrainment takes place over much of the length of an outflow. Therefore in comparing the outflow mass loss rates with the $C^{17}O/C^{18}O$ data we have assumed that the circumstellar mass loss rate is one third of the outflow mass loss rate.

The outflow mass loss rates are compared with the observed beam mass loss rates derived from the J= 2 → 1 $C^{17}O/C^{18}O$ observations in Figure 4. Except for one source, L1551 IRS5 (Fuller et al. 1995), the outflow mass loss rates fall below the $C^{17}O/C^{18}O$ rates, in many cases by an order of magnitude. The outflows therefore do not appear capable of removing the circumstellar material sufficiently quickly to match the observed decline in the mass of circumstellar material. Although only the results from the J= 2 → 1 observations are shown in Figure 4, analysis of the J= 1 → 0 results lead to the same conclusions about role of outflows (and infall) in clearing the circumstellar regions (Ladd et al., 1997).

6.2. INFALL

In the standard singular isothermal sphere (SIS) model for the collapse of a dense core, the central regions of the core collapse first and an expansion wave marking the onset of collapse propagates outwards at the local sound speed. The mass accretion rate during the collapse is constant and given by

$$\dot{M} = 0.975\frac{a^3}{G} \tag{1}$$

where a is the isothermal sound speed (Shu 1977). For a gas temperature of 10K, the mass accretion rate is therefore $1.6 \times 10^{-6} M_{\odot} yr^{-1}$. The rate of loss of mass within the beam used for the J= 2 → 1 observations due to a 10K SIS infall is shown as the solid curve in Figure 4. The left panel of the figure shows that at early times the mass loss rate from the beam is constant at $1.6 \times 10^{-6} M_{\odot} yr^{-1}$ as material collapses into the central region where it becomes optically thick and invisible to our observations. At a time 5.8×10^4 years after the start of the collapse, the expansion wave radius is equal to the radius of the J= 2 → 1 beam. After this time the beam mass loss rate decreases since once the expansion wave radius is larger than the beam, material initially outside the beam starts to infall and compensate for material which is collapsing at centre of the core and becoming undetectable.

The beam mass loss rate for the 10K SIS model is significantly smaller than the rate derived from the observations for the youngest sources. But for sources older than about 4×10^5 ($10^{4.6}$) years the SIS beam mass loss rate exceeds that observed. For the older sources this infall model could therefore account for the observed decrease of circumstellar material. A SIS model with a temperature of 20K, for which the sound speed is consistent with the observed line widths of the $C^{18}O$, has a higher beam mass loss rate than the 10 K model at all times. Such a model has a mass loss rate equal to the observed beam mass loss rate for slightly younger sources than the 10K model, but still can not match the very high mass loss rate for the youngest sources.

Perhaps the largest uncertainty in comparing the infall model and the observations is the ages of the sources. The effect of artificially increasing the age of all the sources by a factor of 10 is shown in the right hand panel of Figure 4. In this figure all the sources are assumed to be 10 times older than in the other panel. The best fit to the beam mass loss rate is shown as the dashed line and its uncertainty is indicated by the shaded area. The solid points indicate the outflow mass loss rates. The assumed greater age of the sources has moved these points down by an order of magnitude. This occurs because these values have been derived using the outflow dynamical age as an estimate of the source age and therefore would decrease if the sources were in fact older than their dynamical ages suggest (Parker et al., 1991; Padman & Richer, 1994; Masson & Chernin, 1993). Increasing the source ages therefore does not affect the relative locations of the outflow mass loss rates and observed beam mass loss rates.

The solid line shows again the 10K SIS model. Since the location of this curve is fixed by the properties of the infall, its location is unaffected by the larger age of the sources. However, the greater age of the sources has moved the best fit beam mass loss rate further from the SIS model. For the assumed factor of ten increase in source ages, this 10 K SIS model never exceeds the best fit beam mass loss rate for the sources.

6.3. OUTFLOW AND INFALL ?

The discussion above shows that the outflows from the sources are unlikely to alone account for the observed clearing of the circumstellar molecular cores. On the other hand, SIS infall models can not match the observations for the youngest sources. This suggests two possible conclusions. The first is that at early times the sources have mass accretion rates much higher than in the standard SIS model. Infall models based on models of cores other than as singular isothermal spheres such as Bonner-Ebert spheres predict higher initial (and time variable) mass accretion rates (Foster & Chevalier, 1993; Henriksen et al., 1997) and Bontemps et al. (1996) have invoked high mass accretion rates for the youngest embedded sources to explain the more energetic outflows from these sources.

The alternative possibility for the observed beam mass loss rates is that early in the evolution of young sources the stellar winds are very efficiently coupled to the dense circumstellar material. The outflows are therefore responsible for clearing the circumstellar material for the youngest sources but later in their evolution, at times greater than about 4×10^4 years, infall becomes the dominant process by which the remaining circumstellar material is cleared. It might indeed be expected that the stellar winds are much better coupled to the dense circumstellar material if they initially have broken out of the dense circumstellar environment, but later escape the circumstellar regions predominantly through these cleared paths.

7. The Structure of the Circumstellar Environment

Maps showing the distribution of the material traced by $C^{17}O$ and $C^{18}O$ around three young stars are shown in Figure 2. Of these three sources, the oldest as measured by its bolometric temperature of 96K, is L1551 IRS-5. Towards this source the $C^{17}O$ emission traces two components of circumstellar material: a cross of material with the arms of the cross aligned in the cardinal directions and a more extended cooler, more quiescent component. Both these components are also visible in the submillimetre continuum emission. The continuum and line emission give somewhat different relative masses of these two components with the line emission suggesting that not more than half the total mass is in the cross component whereas the continuum emission is consistent with a much smaller fraction of the total mass in a hot cross component. The origin of this cross structure has been discussed in detail by Ladd et al. (1995) and Fuller et al. (1995) who conclude that it is tracing the warm region of interaction between the outflow from the source and the walls of the outflow cavity.

The two younger sources shown in Figure 2, L483 and L1527 which have bolometric temperatures of 46K and 59K respectively, both show similar circumstellar structures. The outflows from both sources are aligned close to east-west and in each case the rare isotopes of CO show a 2000 – 3000 AU radius structure elongated perpendicular to the outflow axis. These structures have masses of $\sim 0.5\ M_\odot$ and $\sim 0.3\ M_\odot$ for L483 and L1527 respectively.

Flattened structures of this physical size around other sources, most notably HL Tau (Hayashi et al., 1993), have been interpreted as examples of pseudo-disks resulting from the collapse of a dense core (Galli & Shu, 1993a,b). However, there is a problem with describing the structures around the sources in L483 and L1527 as pseudo-disks. The pseudo-disks of Galli & Shu result from the collapse of an initially spherical dense core and only produce flattened structures as large and massive as seen in L1527 and L483 at times greater than 10^5 years. This is at least an order of magnitude older than the estimated ages of these sources, which for L1527 is $\sim$ 5000 years.

It seems unlikely that these structures are the remnants of initially spherical distributions as this would require the outflows to have cleared material from a very large solid angle around the sources in a very short time. Perhaps the simplest interpretation of these flattened structures is that they predate the formation of the central star, so that the young stars have formed from flattened distributions of dense gas. This would be consistent with the modeling of reflection nebulae associated with embedded sources which are best explained if the stars form from such flattened structures (Hartmann et al., 1996). The formation, structure and evolution of such flattened initial configurations has recently been investigated by Li & Shu (1996; 1997). If flattened structures are the initial configurations out of which stars form, all young sources should be surrounded by such structures and there should be little correlation between the properties of the structures and the properties of the outflows from the stars. Observations of other extremely young sources should allow us to better understand the properties and origins of these structures.

References

André, P., and Montmerle, T. 1994, ApJ, 420, 837
André, P., Ward-Thompson, D., and Barsony, M. 1993, ApJ, 406, 122
Bontemps, S., Andre, P., Terebey, S., and Cabrit, S. 1996, AA, 311, 858
Ceccarelli, C., Haas, M. R., Hollenbach, D. J., and Rudolph, A., L. 1997, ApJ, 476, 771
Chandler, C. J., Terebey, S., Barsony, M., Moore, T. J. T., and Gautier, T. N. 1996, ApJ, 471, 308
Chen, H., Myers, P. C., Ladd, E. F., and Wood, D. O. S. 1995, ApJ, 445, 377
Foster, P. N., and Chevalier, R. A. 1993, ApJ, 416, 303
Fuller, G. A., Ladd, E. F., and Hodapp, K. -W. 1996, ApJL, 463, L97
Fuller, G. A., Ladd, E. F., Padman, R., Myers, P. C., and Adams, F. C. 1995, ApJ, 454, 862
Galli, D., and Shu, F. H. 1993a, ApJ, 417, 220
Galli, D., and Shu, F. H. 1993b, ApJ, 417, 243
Hartmann, L., Calvet, N., and Boss, A, 1996, ApJ, 464, 387
Hayashi, M., Ohashi, N., and Miyama, S. M. 1993, ApJ, 418,71
Henriksen, R., André, P., and Bontemps, S. 1997, AA, in press
Ladd, E. F., Fuller, G. A., and Deane, J. R. 1997, ApJ in press
Ladd, E. F., Fuller, G. A., Padman, R., Myers, P. C., and Adams, F. C. 1995, ApJ, 439, 771
Li, Z-Y, and Shu, F. H. 1996, ApJ 472, 211
Li, Z-Y, and Shu, F. H. 1997, ApJ 475, 237
Masson, C., R., and Chernin, L. M. 1993, ApJ, 414, 230
Moriarty-Schieven, G. H., Wannier, P. G., Mangum, J. G., Tamura, M., and Olmsted, V. K. 1995, ApJ, 455, 190
Moriarty-Schieven, G. H., Wannier, P. G., Tamura, M., and Keene, J. 1994, ApJ, 400, 260
Myers, P. C., and Ladd, E. F. 1993, ApJL, 413, 47
Ohashi, N., And Hayashi, M., Kawabe, R., And Ishiguro, M. 1996, ApJ, 466, 317
Padman, R., and Richer, J. S. 1994, APSS, 216, 129
Parker, N. D., Padman, R., and Scott, P. F. 1991, MNRAS, 252, 442
Reipurth, B., Chini, R., Krugel, E., Kreysa, E., and Sievers, A. 1993, AA, 273, 221
Sandell, G., Knee, L. B. G., Aspin, C., Robson, I. E., and Russell, A. P. G. 1994, AA, 285, 1
Saraceno, P., André, P., Ceccarelli, C., Griffin, M., and Molinari, S. 1996, AA, 309, 827
Shu, F. H. 1977, ApJ, 214, 488
Terebey, S., Chandler, C. J., and André, P. 1993, ApJ, 414, 759

MILLIMETER INTERFEROMETRY OF CLASS 0 SOURCES: ROTATION AND INFALL TOWARDS L1448N

SUSAN TEREBEY
Extrasolar Research Corporation
720 Magnolia Ave., Pasadena CA 91106, USA

AND

DEBORAH L. PADGETT
Jet Propulsion Laboratory
IPAC 100-22, Caltech, Pasadena CA 91125, USA

Abstract. In order to study the kinematics of protostellar collapse we present high-spatial resolution millimeter interferometer data of the class 0 source L1448 N. The L1448 N cloud core has fragmented into two rapidly rotating subcores containing three protostars. One subcore is a protobinary system with 2000 AU separation. Strong evidence for infall is seen towards the N(B) component.

1. Introduction

Since their introduction Class 0 sources have been an interesting but controversial category of protostellar sources (André *et al.*, 1993). By definition class 0 sources have narrow "cold" spectral energy distributions, exhibit large circumstellar masses through their strong millimeter continuum fluxes, and show no visible or near-infrared emission at the position of the embedded star. Advocates propose them to be an earlier evolutionary phase than class I sources, having typical ages on order 2×10^4 yr. In a general way their properties match the predictions of theoretical models for young protostars (Shu *et al.*, 1987). It is encouraging that recent observations of collapse motions (Zhao *et al.*, 1993; Zhao, 1995) involve targets which are primarily class 0 sources. However some authors argue the definition is flawed, and that many or all of the class 0 sources may instead be class I

B. Reipurth and C. Bertout (eds.), Herbig–Haro Flows and the Birth of Low Mass Stars, 507–514.

sources which have large inclination angles, i.e. are seen principally along the bulk of their flattened circumstellar structures.

By happenstance a high fraction of class 0 sources are also molecular jet sources, although the reason for this association is unclear. Will class 0 sources become typical class I sources as they age, or is there a more direct connection to Herbig-Haro jets? Are there other physical parameters of the collapsing cloud, such as rapid rotation (Goodman *et al.*, 1993) or high mass, which are important factors in the molecular jet phenomenon?

To gain a better understanding of the kinematics and density of the collapse phase we are studying class 0 sources using the high-spatial resolution provided by millimeter interferometry. Class 0 sources are excellent candidates for study: they are likely to be young, and by definition their large circumstellar masses ensure that they are strong millimeter targets. In this paper we present a case study of the class 0 source L1448 N, sometimes called L1448 IRS3.

2. Observations

The data presented are millimeter spectral line and continuum observations from the Owens Valley Millimeter Interferometer. The spatial resolution is 8″ for the long wavelength $C^{18}O$(1-0) and 2.7 mm continuum data. A higher spatial resolution of 2.5″ was achieved for the shorter wavelength $C^{18}O$(2-1), ^{13}CO(2-1), and 1.3 mm continuum observations. Previous continuum results and system performance are reported elsewhere (Terebey *et al.*, 1993).

3. Results

3.1. OVERVIEW OF L1448 CLOUD CORE

The L1448 cloud shows fragmentation and star formation activity in several regions. The cloud, as defined by NH_3 data is $10' = 1$ pc across and contains 50 $M_\odot$ of gas (Bachiller and Cernicharo, 1986). L1448 is located in the Perseus molecular cloud at a distance of 300 pc.

Figure 1 shows the distribution of the dense gas in integrated $C^{18}O$(1-0) emission towards the central 1′ of L1448 N. The well known outflow source L1448 C lies outside the field about 1′ to the southeast (Bachiller *et al.*, 1990). It is apparent in Figure 1 that there are two cores, one centered on L1448 N, and a second located 20″ northwest, which we call L1448 NW.

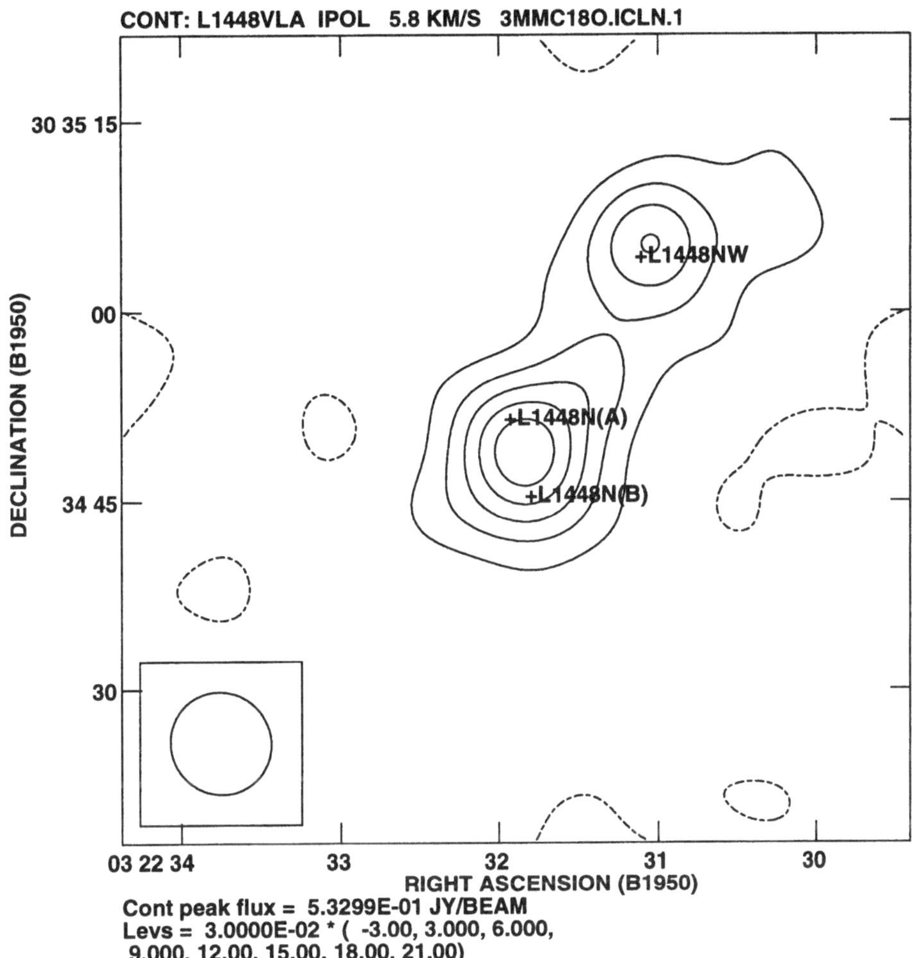

Figure 1. The integrated $C^{18}O$(1-0) map shows the dense gas in the L1448 N cloud core has fragmented into two subcores. Plus symbols show the positions of embedded protostellar sources.

3.2. EMBEDDED PROTOSTELLAR SOURCES

Millimeter continuum emission is sensitive to heated circumstellar dust and has proven to be a good indicator of protostars. The OVRO data show three sources of continuum emission whose positions are shown by plus symbols in Figure 1. Curiel *et al.* (1990) previously detected the VLA sources A and B in L1448 N. At millimeter wavelengths the continuum emission is dominated by the southernmost source (Terebey *et al.*, 1993), whose position coincides with the VLA source L1448 N(B). The separation of the components is about 6″ = 1800 AU. The third embedded protostellar source is located in the center of the L1448 NW core, about 20″ northwest.

The strong millimeter continuum emission from N(B) combined with the IRAS fluxes show L1448 N(B) has a cold spectral energy distribution which fits the definition of a class 0 source. What then is the status of the two other millimeter continuum sources? Unfortunately the sources are spatially unresolved by IRAS. However near-infrared images have the requisite spatial resolution and sensitivity to detect class I sources at the distance of Perseus. Although near-infrared images show H_2 emission and reflection nebulosity in the region there is no near-infrared emission detected from the protostellar positions above the detection threshold of 18th magnitude at K. The lack of near-infrared emission suggests all three protostars may be class 0 sources.

3.3. FRAGMENTATION AND ROTATION

Both the L1448 N and NW subcores show large velocity gradients which are consistent with rotation. The position angle of the rotational axis is approximately -45° for both subcores. The suggested geometry is that of a narrow rotating cylinder which has fragmented along its length. The magnitude of the gradient is about 100 km s^{-1} pc^{-1}, representing a rotational period of about 60,000 yr. The implied dynamical mass is about 1.5 $M_\odot$ which is consistent with the spectral line mass estimate. The projected distance between the two subcores is about 20″, which suggests the fragmentation length scale is comparable at 20″ = 10^{17} cm.

The L1448 N subcore shows a flattened structure resembling a disk which extends over 30″ in diameter. Figure 2 shows a position-velocity slice taken along the major axis in the optically thin $C^{18}O$(1-0) line. The P-V diagram shows a solid-body rotation curve in the inner 10″. Outside this radius the velocity is flat or slightly declining. We constructed a simple thin disk model to explore the relative importance of density, inclination, and rotation law on the P-V diagram. Figure 3 shows the best fit model generated for a 60° inclined disk (approaching edge-on) with 16″ radius and r^{-1} surface density profile. The rotation curve is solid body inside a 5″ radius, switching to Keplerian outside that radius.

3.4. THE 2000 AU PROTOBINARY

The 30″ flattened rotating structure contains two protostellar sources, N(A) and N(B), which are separated by 6″ = 1800 AU. The dense gas forms a bridge which varies slowly in velocity between the two sources. We interpret the system to be a rotating and collapsing cloud which is in the process of

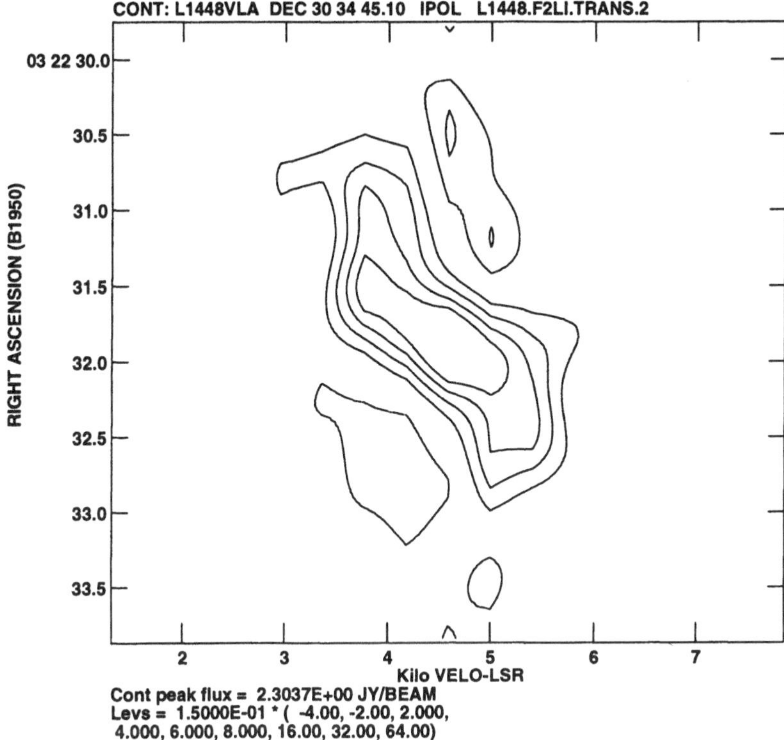

Figure 2. A position-velocity diagram shows evidence for rotation in L1448 N.

forming a protobinary. Corroborating dynamical evidence is the fact that the rotation curve looks like a solid-body in the inner $5''$, suggesting that the dynamical mass is distributed, i.e. non-pointlike, and this occurs over the same scale as the binary separation.

The two sources are presumably coeval but differ markedly in their circumstellar mass. The continuum flux differs by a factor of ten between the two components. In addition evidence for an outflow is only seen toward N(B), the source with the high circumstellar mass. At centimeter wavelengths the N(A) component dominates (Curiel *et al.*, 1990). This suggests that N(A) has already accumulated most of its mass, while N(B) is in the midst of a rapid accretion phase.

3.5. INFALL

Infall motions exhibit a systematic pattern which can be searched for and in a few cases detected (Zhao *et al.*, 1993; Zhao, 1995). Spectral line profiles

Figure 3. A simple model with solid body rotation in the interior, and Keplerian rotation exterior provides a good fit to the position-velocity data.

should broaden in progressively smaller beams centered on the source. The magnitude of the velocity should be consistent with gravitationally bound motion. In the case of optically thick lines, the line will be self-absorbed, but with a stronger blue than red shoulder.

Figure 4 shows the $^{13}CO(2\text{-}1)$ profile toward L1448 N(B) with 2.5″ spatial resolution. The profile shows the characteristic infall signature for an optically thick spectral line. The magnitude of the velocity is consistent with gravitationally bound motion around a central mass of 0.75 $M_{\odot}$.

4. Conclusions

Millimeter interferometric observations of the class 0 source L1448 N were made in $C^{18}O(1\text{-}0)$, $C^{18}O(2\text{-}1)$ and $^{13}CO(2\text{-}1)$ in order to study the kinematics (rotation and infall) of a young protostellar source. L1448 N shows a variety of behaviors which are shared in part by other class 0 sources:

- The L1448 N cloud core has fragmented into two subcores containing three protostars. The lack of near-infrared emission suggests all three protostars may be class 0 sources.

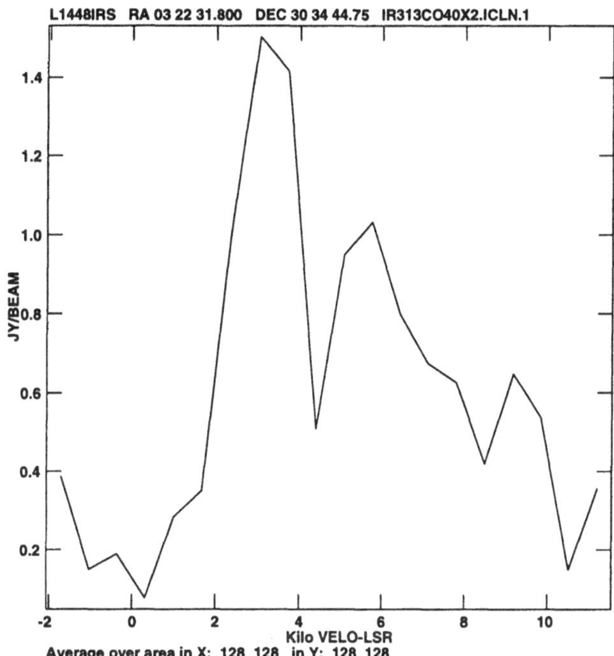

Figure 4. The ^{13}CO(2-1) spectrum directly towards the N(B) protostar shows the asymmetric profile characteristic of infall motions in an optically thick line tracer.

• The two subcores are rapidly rotating. In addition, the rotational axes are aligned. The suggested geometry is that of a narrow rotating cylinder which has fragmented along its length. The precollapse fragmentation scale in the Perseus cloud is about 6000 AU = 10^{17} cm. The rotation rate is about 100 km s^{-1} pc^{-1}, for a rotational period of about 60,000 yr.

• The L1448 N subcore is a protobinary system, a dynamically rotating and collapsing core which is currently forming the two protostars N(A) and N(B). The two components are connected by a continuous gas bridge. The binary separation is 2000 AU, compatible with observed separations of stars in wide binary systems.

• The two sources are presumably coeval but differ markedly in their circumstellar mass and outflow properties. The continuum flux, which is proportional to the circumstellar mass, differs by a factor of ten between the two components. An outflow is seen only toward L1448 N(B), the source with the higher circumstellar mass. At centimeter wavelengths the N(A) component dominates.

• Toward the most massive/strongest continuum source L1448 N(B), the optical depth is sufficiently high to show the characteristic spectral signature of infall motions. The broadening of the spectral line due to infall is seen in the central 3″, whereas the velocity gradient due to rotation extends over 30″.

High sensitivities can be achieved by millimeter interferometers studying Class 0 objects, which are by definition strong millimeter continuum sources. A promising direction for future research is the detailed comparison of observations of protostellar density and kinematics with theoretical collapse models.

Acknowledgements: The authors gratefully acknowledge support from the NASA Origins of Solar Systems Program. They thank Alberto Noriega-Crespo and Dave Van Buren for their help.

References

André, P., Ward-Thompson, D., and Barsony, M. 1993, ApJ, 406, 122
Bachiller, R., and Cernicharo, J. 1986, A&A, 168, 262
Bachiller, R., Cernicharo, J., Martin-Pintado, J., Tafalla, M. and Lazareff, B. 1990, A&A, 231, 174
Curiel, S., Raymond, J. C., Rodriguez, L. F., Canto, J. and Moran, J. M. 1990, ApJ, 365, L85
Goodman, A. A., Benson, P. J., Fuller, G. A., and Myers, P.C. 1993, ApJ, 406, 528
Shu, F. H., Adams, F. C., and Lizano, S. 1987, ARAA, 25, 23
Terebey, S., Chandler, C. J., and André, P. 1993, ApJ, 414, 759
Zhao, S., 1995, ApJ, 442, 685
Zhao, S., Evans, N. J. II, Kompe, C. and Walmsley, C. M. 1993, ApJ, 404, 232

WATER MASERS TOWARD LOW-LUMINOSITY YOUNG STELLAR OBJECTS

MARK J CLAUSSEN
National Radio Astronomy Observatory, Socorro, NM, USA
KEVIN B. MARVEL
Owens Valley Radio Observatory, Big Pine, CA, USA
H. ALWYN WOOTTEN
National Radio Astronomy Observatory
Charlottesville, VA, USA
AND
BRUCE A. WILKING
University of Missouri, St. Louis, MO, USA

Abstract. A review of the importance of water maser observations toward young stellar objects (YSOs) is presented. Also, we present recent, differing types of observations of water masers near YSOs. Single antenna observations, taken regularly, characterize the variability of the masers and allow estimates of time and spatial scales. High resolution ($\sim$ 1 mas) multi-epoch observations allow proper motions to be studied. Detailed analysis of such proper motions will allow the placement of the masers in the circumstellar (a disk) or near-stellar environment at the base of the outflow. Radio interferometric techniques are the best method of making estimates of the kinematics of the gas in these regions.

1. Introduction

Water maser emission at 22 GHz is found to be a tracer of mass-loss activity in young stars of all masses (e.g., Rodríguez *et al.* 1980, Genzel & Downes 1977). This mass loss occurs during an early, embedded phase of pre-main-sequence stars characterized by large quantities of circumstellar dust which completely absorbs visible light from the central object. The emergent energy from such YSOs is photospheric and disk radiation reprocessed by

B. Reipurth and C. Bertout (eds.), Herbig–Haro Flows and the Birth of Low Mass Stars, 515–524.

circumstellar dust and radiated at mid- to far-infrared wavelengths. But all too often the interest of astronomers is focussed on the spectacularly strong masers, with high-velocity features, that arise in flows near very massive YSOs, those which are forming O or B stars, e.g. the Orion water masers, or the masers near W49. The low-mass YSOs and their associated gas outflows have been well-studied, both observationally and theoretically (note the beautiful results at this meeting). So it is important to study the water masers around embedded sources of low luminosity. Although they may be generally weak, the sensitivity of large radio telescopes are such that telescope time is no longer prohibitive in the study of weak maser sources.

Water masers, whether near high-mass or low-mass stars, have long been known to be variable. Recent surveys of low-mass YSOs for H_2O maser emission have demonstrated that while such emission is common, it is also highly variable; e.g., Comoretto *et al.* (1990); Felli, Palagi, & Tofani (1992); Wilking *et al.* (1994a); Xiang & Turner (1992); Xiang & Turner (1995). To characterize the frequency and timescales for maser emission in these YSOs requires multi-epoch observations.

The organization of this contribution will be as follows: first, what can we learn about the stellar or outflow sources by studying water masers ? In particular, what can water masers arond low-luminosity sources reveal about the outflow or the central star ? Second, a review of a program of a monthly monitoring survey of water masers toward low-luminosity YSOs and the results of that survey will be presented. This is important because it characterizes the time scale of water maser activity in the embedded sources. Third, some assorted VLA observations of water masers around these objects will be presented, with a discussion of the position of the water masers with respect to the central energy sources. Finally, a preliminary presentation of VLBA data of the water masers around the YSO IRAS05413-0104, which is the driving source of the molecular hydrogen jet and H-H object HH 212 will be made. We will use data from IRAS05413-0104 for examples of all the water maser observations reported here. Multiple epoch high-resolution observations of these water masers can tell us about the motions of the gas near these young stellar objects.

2. Why Study Water Masers toward YSOs ?

Generally speaking, the radiative transfer problem for masers is difficult enough so that one cannot expect to use the observations to determine densities or temperatures of the masing gas, except within some broad bounds. Even if these parameters could be determined, the masers are likely to be found in clumps of gas so that those parameters couldn't be used to make statements about the general physical parameters of the environment of the embedded object (at least for the densities). On the other hand, we know that for water masers at 22 GHz to even exist there must be some lower limit to the temperature and densities for the energy levels to be significantly populated. The necessary temperature is 200 — 300 K, and the H_2 densities in the clumps must be in the range 10^8 - 10^9 cm^{-3}.

But water masers at 22 GHz do give us the ability to do at least three things:

1) Observe the kinematics of the gas where the masers reside, with high velocity resolution (≤ 0.1 km s^{-1}) via radio spectroscopy.

2) Observe the kinematics of the gas with high angular resolution using radio interferometry (scales of 1 arcsecond down to 1 mas).

Masers can be used to do these two things for the same reason that we have difficulties in interpreting their physical properties. They are non-linear processes, so they are very bright and small.

3) Masers also tell us that there *are* places where water molecules reside at fairly high densities and warm temperatures. These are necessary but not sufficient conditions for water masers, that is, there may be warm, dense clumps of water molecules where masers are not found.

So if we would like to use water masers to study the kinematics of the environment of YSOs at high angular and spectral resolution, we must characterize the emission from a number of objects which we intend to study.

Since their discovery it has been well known that water masers in the interstellar medium are highly variable — why they are is another question which is tied in with the way they are pumped, the radiative transfer, and beaming effects — but they are, and we need to characterize the variability if we are to study kinematics using them.

A number of groups have used single antennas over the years to conduct surveys of star-forming regions in order to characterize the frequency and luminosity with which water masers occur in these regions. Some of these are: Reinhard Genzel and Dennis Downes in the late 1970s generally looked at high-mass star formation regions; Ed Churchwell and his collaborators toward compact and ultra-compact HII regions; and most recently, Marcello Felli and his group have observed low- to mid-luminosity sources. Only in the case of Felli's group has there been a concerted effort to characterize the variability of sources by observing a sample more than once.

3. Single Antenna Water Maser Monitoring Survey

After the discovery of water masers toward low-luminosity objects in the ρ Ophiuchus cloud in 1987, and follow-up observations at the VLA by Wilking & Claussen (1987) and by Terebey, Vogel, & Myers (1989), we realized by the difference in spectra, that to better understand the water masers around low-luminosity objects, we would have to monitor them on monthly time scales over many months to characterize their variability. So we embarked on what we called a monitoring survey; to observe as many sources as possible once per month for as long as we could get time. We have reported on this survey in three recent papers: Wilking *et al.* (1994a), Wilking *et al.* (1994b), and Claussen *et al.* (1996). Here we will summarize and show some data from this survey.

The sample of sources was selected from the IRAS PSC which were known to be associated with molecular outflows and/or dense molecular cores. We also required sources to have rising νS_ν throughout the IRAS bands, to lie in nearby clouds, and to have IRAS in-band luminosities $\leq$ 120 $L_\odot$. Other embedded sources were included such as those revealed by IRAS Pointed Observations, the CPC, and some sources that were confused in the IRAS PSC. We were clearly selecting toward Class I and Class 0 objects.

A list of 45 sources were thus compiled for observations. From these, 27 were monitored on a regular basis which we defined as a minimum of seven observations with no gaps larger than three months. Observations were made using the 37-m radio telescope at Haystack Observatory and the NRAO 43-m radio telescope at Green Bank. Figure 1 shows the spectra of water masers toward IRAS05413-0104 taken over the 13-month period.

The results of the monitoring survey are 1) nine new water maser sources were found, resulting in 24 YSOs with L$\leq$ 120 $L_\odot$ known to display water maser activity; 2) the maser phase in low-mass embedded sources is estimated to occupy about one-third the duration of the embedded phase of evolution; 3) some water maser features appear and disappear on timescales less than about 2 months — this characteristic timescale is consistent with shock crossing times for reasonable values of the shock velocity and size; and 4) the isotropic luminosity of the water masers in any given source can vary by at least 2 orders of magnitude over the course of a year — this implies that a single-epoch measurement of the luminosity of the masers may provide a misleading result.

4. Assorted VLA Observations

We have used the VLA irregularly, in different configurations making use of bits of telescope time available at the end of software time, etc. in order to refine the positions of the water masers and relate them to other known objects in the field of view such as mm-wave continuum sources, cm-wave continuum sources, IRAS positions, CO sources, etc. Details of these observations and more specific results from them have been and will be reported

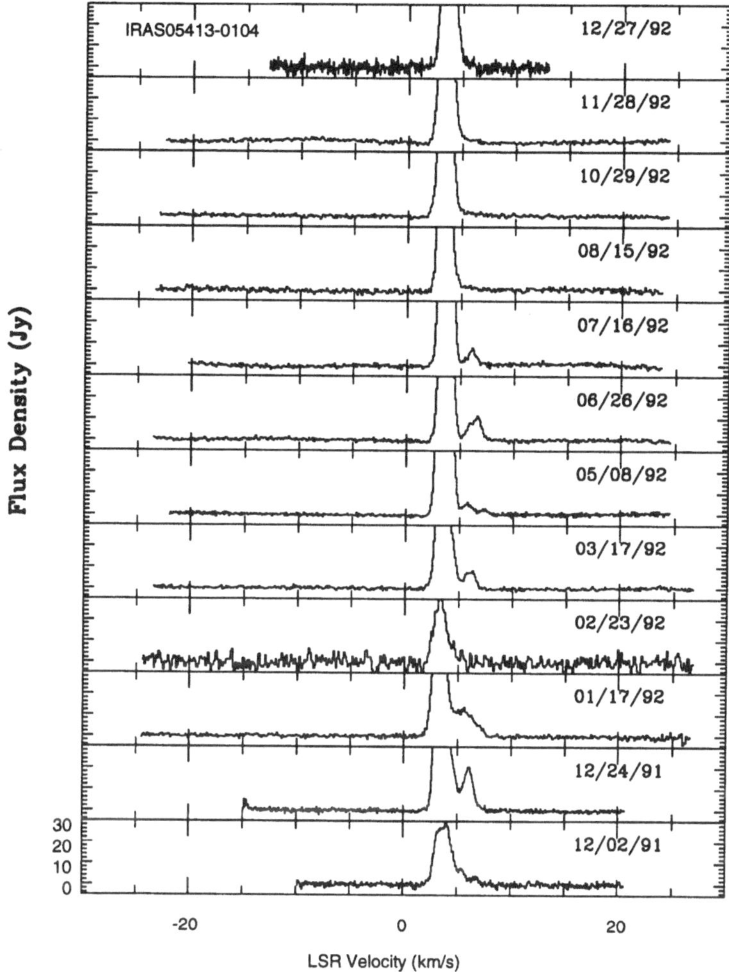

Figure 1. Multi-Epoch Spectra of Water Masers in IRAS05413-0104. The peak flux density for the bright feature is upwards of 300 Jy.

elsewhere (Rogers & Gottschalk 1993; Wootten *et al.* 1997).

As an example of our VLA data, we show in Figure 2 a picture of IRAS05413-0104 , showing the positions of the water masers, the IRAS source, the 3 mm continuum source, and the centroid of the molecular hydrogen jet. Figure 2 shows that the water masers are clearly associated with the centroid of the molecular hydrogen jet and the 2.7 mm OVRO continuum source. The water masers, of course give the best positional accuracy of any of these.

The results of these VLA observations point up the fact that, in most

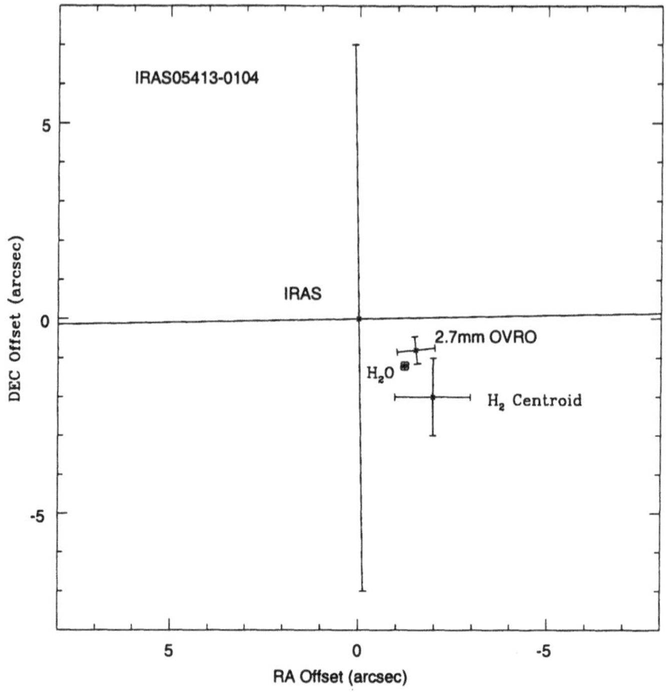

Figure 2. Objects Near the Embedded Source IRAS 05413-0104

sources, the masers all lie within a few hundred A.U. of the stellar object. This suggests that the masers lie in high density gas near the driving source of the outflow, and likely produced in shocks near the star.

5. VLBA Observations of Water Masers in IRAS05413-0104

The water masers toward IRAS05413-0104 are one of three maser sources that we detected all the time in our monitoring survey. The masers were usually fairly strong, but IRAS05413-0104 is **not** one of the most luminous sources in the survey. At 15 $L_\odot$, the bolometric luminosity of the embedded source is more nearly the median in the survey. IRAS05413-0104 is located at the center of a beautiful H_2 jet (Zinnecker, McCaughrean, & Rayner, 1997). There is a 3 mm radio source at the center as well, and the water masers coincide with the 3mm source, midway between the two central H_2 knots, which are separated by about 8 arcseconds.

Our VLBA data was taken at four epochs: July 4, July 26, August 16, and September 15, 1996. The entire 10-station array was used, as well as one antenna from the VLA. At 22 GHz the beam size is about 1 mas,

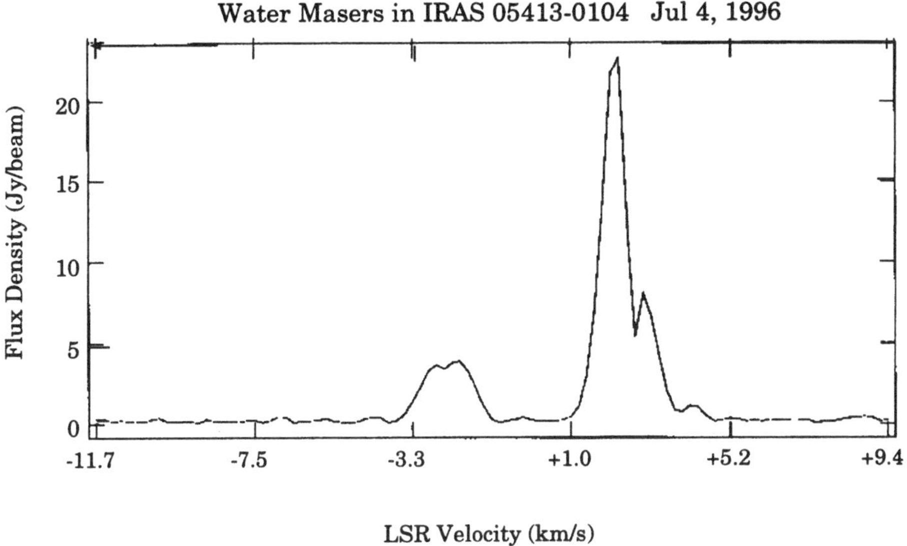

Figure 3. The Water Maser Spectrum in IRAS 05413-0104 on July 4, 1996

which at the distance of IRAS05413-0104 is 0.5 A.U. At the initial epoch, the spectrum of the water masers looked like Figure 3. Note that the feature that is most blue-shifted is new since 1992, but since we don't have monitoring data since then, we don't know when it appeared or how long it has been around.

When we made the observations, we expected to see several maser "spots", or features. What we saw is shown in Figure 4 — there are more than 30 distinct features here. The first thing to notice from Figure 4 is that the masers lie mostly along a line that has a position angle (N through E) almost exactly the same as the position angle of the jet (Zinnecker, McCaughrean, & Rayner, 1997). This suggests that the masers are tracing the outflow very near to the star, rather than lying in a disk around the star. The blue-shifted features (at ~ -3 km s^{-1}; see Figure 3) are all found in the group of masers in the northeast of the map of Figure 4. The most extreme red-shifted emission (at ~ 5 km s^{-1}) are found as the compact clump of features just south and east of the (0,0) position. The strong, intermediate velocity feature relates to the linear string of masers at the (0,0) position in Figure 4.

In Figure 5, maps of the blue-shifted masers are shown for all four epochs. The very high resolution available by observing water masers with the VLBA is clear from this figure, as well as the definite changes in the details of the structure, although the general structure over 130 A.U. is

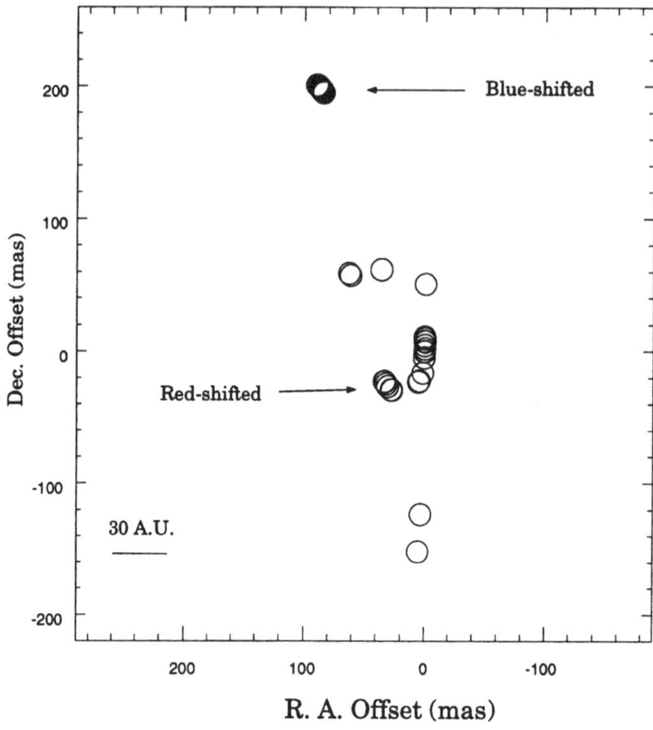

Figure 4. Water Maser Positions near IRAS 05413-0104 on July 4, 1996.

very similar over all four epochs. The proper motion detectable by relative measurements over one month interval is 25 km s^{-1} for a distance of 500 pc. We are currently analysing the details of the proper motions of the maser spots and the results for IRAS05413-0104 will be published soon (Claussen *et al.*, 1997).

6. Summary

We have discussed the importance of maser observations, both as single antenna monitoring observations, and high resolution ~1 mas monitoring observations, to understand the structure of the gas within a few A.U. of the embedded sources. Water masers are a very valuable probe of the circumstellar or near stellar environment. Further studies will allow us to place the masers in a circumstellar disk or at the very base of the outflow.

Acknowledgements: The VLA and the VLBA are instruments operated by the National Radio Astronomy Observatory, which is a facility of the National Science Foundation under cooperative agreement by As-

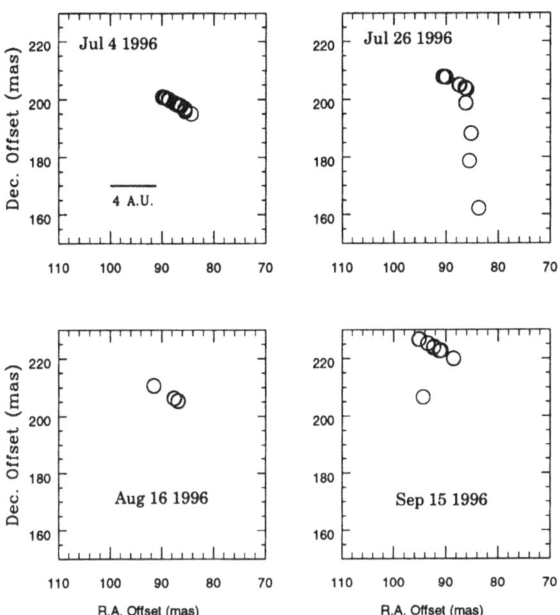

Figure 5. Blueshifted Water Maser Positions near IRAS 05413-0104 four Four Epochs

sociated Universities, Inc. The authors would like to thank the Scientific Organizing Committee for providing the opportunity to present these results at the conference.

References

Claussen, M. J, Wilking, B. A., Benson, P. J., Wootten, H. A., Myers, P. C., & Terebey, S. 1996, ApJ Supp. Series, 106, 111
Claussen, M. J, Marvel, K. B., Wootten, H. A., & Wilking, B. A. 1997, *in preparation*
Comoretto, G., it et al. 1990, A&A Supp. Series, 84, 179
Genzel, R., & Downes, D. 1977, A&A Supp. Series, 30, 145
Felli, M., Palagi, F., & Tofani, G. 1992, A&A, 255, 293
Rodríguez, L. F., Moran, J. M., Ho, P. T. P., & Gottlieb, E. W., 1980, ApJ 235, 845
Rogers, C. B. & Gottschalk, J. 1993, B.A.A.S., 25, 1367
Terebey, S., Vogel, S. N., & Myers, P. C. 1989, ApJ, 340, 472
Wilking, B. A. & Claussen, M. J. 1987, ApJL, 320, L133
Wilking, B. A., Claussen, M. J., Benson, P. J., Myers, P. C., Terebey, S., & Wootten, H. A. 1994, ApJ, 431, L119
Wilking, B. A. & Claussen, M. J., Benson, P. J., Myers, P. C., Terebey, S., & Wootten, H. A. 1994b, in *Clouds, Cores, and Low-Mass Stars*, eds. D. P. Clemens & R. Barvainis (ASP: San Francisco), p. 299

Wootten, H. A., Wilking, B. A., Meehan, L. & Claussen, M. J 1997, *in preparation*
Xiang, D. & Turner, B. E. 1992, Acta Astronomica, 33, 87
Xiang, D. & Turner, B. E. 1995, ApJ Supp. Series, 99, 121
Zinnecker, H., McCaughrean, M., & Rayner, J., in *Low Mass Star Formation from Infall to Outflow*, ed. F. Malbet & A. Castets 1997, IAU Symposium No. 182.

MASSIVE STAR FORMATION: OBSERVATIONAL CONSTRAINTS

ED CHURCHWELL
Dept. of Astronomy, University of Wisconsin
475 N. Charter St.,Madison, WI, 53706, USA

1. Introduction

Observations during the past several years strongly imply that virtually every star, independent of final mass, goes through a phase of rapid outflow simultaneously with rapid accretion during formation. The structure and properties of outflows and accretion disks associated with low-mass star formation has received intensive observational attention during the past several years (see the reviews and references in Lada 1985; Edwards, Ray, and Mundt 1993; Fukui *et al.* 1993; and this symposium). Young stellar objects (YSOs) with $L_{bol} < 10^3$ $L_\odot$ will be referred to as "low-mass" stars in this review. The range of physical properties of CO outflows associated with YSOs of all masses are enormous, see Fukui *et al.* (1993). I will focus attention in this review on what we know about massive YSOs and their environments.

2. Stellar Luminosity and Outflow Properties

Edwards, Ray, and Mundt (1993), Cabrit and Bertout (1992, hereafter CB92), and others have shown that the bolometric luminosity (L_{bol}) of the central star is correlated with various outflow properties such as the outflow mechanical luminosity (L_f), the momentum flux ($\dot{P}_f$) or equivalently the driving force (F_f), and the mass outflow rate ($\dot{M}_f$). Clearly, the outflow properties are not independent of the luminosity (or the mass) of the central star. However, the mechanism responsible for driving and collimating the outflows remains unresolved. CB92 suggested that the flows may be driven by an ionized stellar wind based on the fairly tight correlation of the intrinsic 6cm continuum flux density ($(S_6/d^2) \propto N_c$) with L_f and $\dot{P}_f$. Other suggestions have also been proposed including several versions of accretion

B. Reipurth and C. Bertout (eds.), Herbig–Haro Flows and the Birth of Low Mass Stars, 525–536.

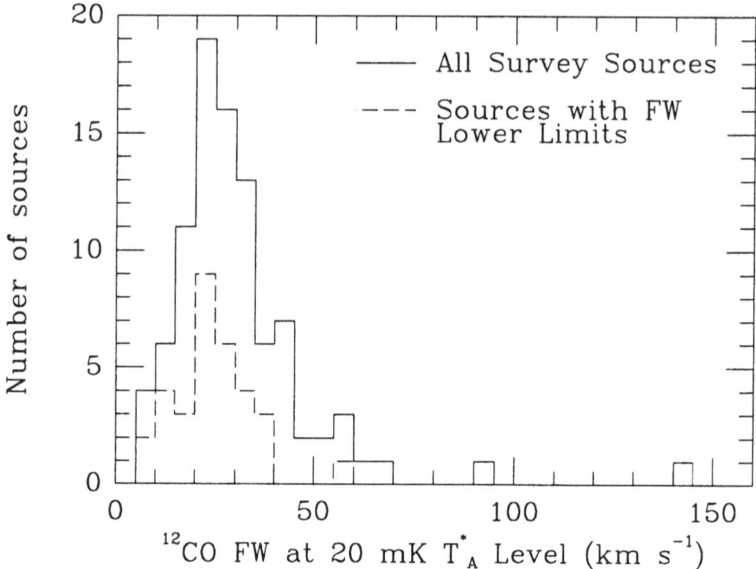

Figure 1. Distribution of measured ^{12}CO (1-0) line full-widths at 20mK from SC95.

driven outflows involving magnetic fields and rotation, photoevaporation of accretion disks coupled with radiative acceleration, etc. I will return to this issue later in regard to massive star outflows.

3. The Frequency of Massive Star Outflows

Motivated by the fact that no systematic study of the frequency of outflows associated with massive stars existed, Shepherd and Churchwell (1995; hereafter SC95) undertook a CO (1-0) line profile survey of 94 massive star formation regions using the NRAO 12m telescope (HPBW~60") to identify potential outflow sources. They found fully 90% of their sample was associated with high velocity (HV) gas. At the 1σ T_A^* level, only 10% had a full width FW<15 km s^{-1}; 49% had 15<FW<30 km s^{-1}; 30% had 30<FW<45 km s^{-1}; and 11% had FW>45 km s^{-1}. The distribution of FWs at 20 mK is shown in Fig. 1 which shows that most ^{12}CO (1-0) FWs lie in the range 20-30 km s^{-1} toward massive star formation regions when observed with a spatial resolution of ~60".

From line profiles alone, one cannot distinguish outflows from other bulk motions such as accretion, rotation, infall, shocks, etc. Shepherd and

Churchwell (1996; hereafter SC96) selected 10 potential outflow sources from the survey of SC95 and made OTF maps in the ^{12}CO (1-0) line and high S/N profiles of ^{13}CO (1-0) using the NRAO 12m telescope. The ^{13}CO data were used to correct for optical depth effects in the line wings. Of the 10 sources mapped, 5 have well defined bipolar outflows, 3 had broad line wings but had S/N ratios too low to confidently map the line wing emission distributions (due either to bad weather or instrumental problems), and 2 had multiple velocity components in the beam but no evidence for bipolar outflows. For this small sample, we conclude that at least 50% of the potential outflow sources selected on the basis of broad line wings are indeed produced by outflows. This could be as high as 80% depending on the status of the 3 sources for which our maps were inadequate to map the line wings.

4. Outflow Properties

SC95 and SC96 determined the outflow properties for 7 massive star formation regions using ^{13}CO to correct for optical depth effects in the ^{12}CO line wings. The outflow properties were derived assuming an inclination angle of 45° to the line of sight. In Table 1 below, the average values of a few of the outflow parameters are compared. Average values for 11 outflows driven by stars with $L_{bol} < 750\ L_\odot$ were taken from Lada (1985) and those for $L_{bol} > 10^3\ L_\odot$ were taken from SC95 and SC96, which includes only B-stars. Values with a superscript $*$ do not include upper limits in the average. As we will see below, one expects the average parameters for outflows driven by more luminous O-stars to be substantially larger than those listed in Table 1.

TABLE 1. Mean Outflow Properties: Low- and High-Mass Stars

Parameter	Mean $L_{bol} > 10^3\ L_\odot$	Mean $L_{bol} < 750\ L_\odot$
R_f(pc)	0.5	0.5
$M_f(M_\odot)$	44*	1.8*
$F_f(10^{-3} M_\odot km\ s^{-1} yr^{-1})$	8.3*	0.8*
$L_f(L_\odot)$	8*	1.1*

Clearly, the outflow mass, force, and mechanical luminosity of luminous stars are an order of magnitude or more greater than those of low-luminosity stars. This is also true for the mass flow rate $\dot{M}_f$ as illustrated in the $\dot{M}_f$

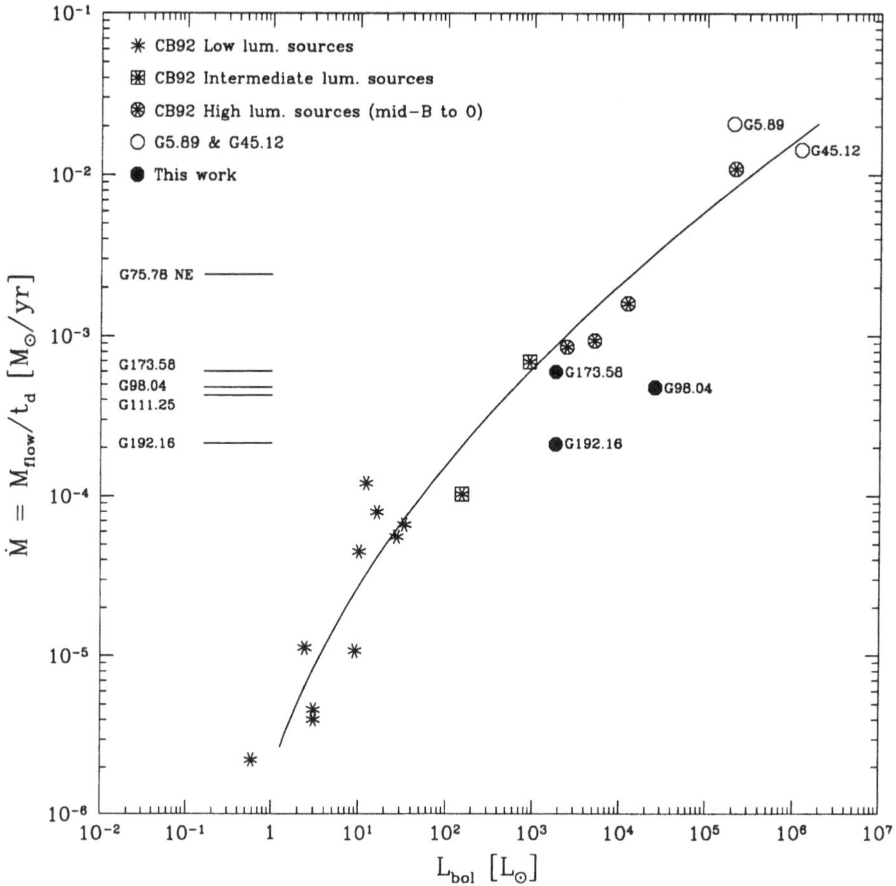

Figure 2. Mass outflow rates versus bolometric luminosity. The solid curve is a second order least squares fit to the data which are indicated by symbols identified in the figure. Figure is from SC96.

versus L_{bol} plot shown in Fig. 2 below. Here, in addition to the CB92 sources, those of SC96 and G5.89 and G45.12 from Acord *et al.* (1997) and Hunter *et al.* (1997), respectively, are included.

Fig. 2 shows that $\dot{M}_f$ is a tight function of L_{bol} in the range $1 \leq L_{bol} \leq 10^6$ $L_\odot$. The solid curve in Fig. 2 is a second order least squares fit to the data which has the form:

$$log(L_{bol}) = 11.2 + 3.35 log(\dot{M}_f) + 0.24(log(\dot{M}_f))^2 \quad (1)$$

with a residual standard error of $\sigma = 0.36$. The small dispersion of $\dot{M}_f$

with L_{bol} can be used to infer the bolometric luminosity of the central star as pointed out by SC96 and used by them to argue that the G192.16 and G98.04 outflows are driven by lower luminosity stars than those responsible for the nearby UC HII regions. SC96 also showed that within a given luminosity range, the outflow masses generally increase with age (as expected) and outflows driven by luminous stars are substantially more massive than those of lower luminosity stars of the same age. For example, the Orion A-IRc2 outflow is three orders of magnitude more massive than that of L1448-C which is an order of magnitude older than the Orion outflow. This is a consequence of the much greater mass flux of O-stars relative to lower luminosity stars.

5. Interferometric Observations

The study of massive star formation regions suffers from three important disadvantages: none are closer than 450 pc, they generally occur in clusters with many lower mass stars close by, and they are deeply embedded in their natal molecular clouds and therefore suffer large visual extinctions. As a consequence, radio (or FIR) interferometric observations are required to determine morphologies on scale sizes <0.1 pc and to identify the source of the outflow. The observations of G75.78NE by Shepherd, Churchwell, and Willner (1997; hereafter SCW97) and G45.12 by Hunter, Phillips, and Menten (1997; hereafter HPM97) are instructive. Single dish observations indicate a single massive outflow toward G75.78NE which is offset from the UC HII region and apparently is not driven by the ionizing star of the HII region (SC96). BIMA observations of the region with a resolution of $\sim$5" by SCW97 reveals at least 3 separate massive outflows in the single dish beam. One of them is centered on a millimeter continuum source, but the other two have no detected radio continuum or IR emission that could be construed as the origin of the outflows. Similarly, single dish observations of G45.12 by HPM97 show a massive outflow centered on the UC HII region. OVRO observations with a resolution of $\sim$2.5" reveals at least 2 very massive outflows (see Table 2) only one of which is centered on a compact radio continuum source. Some of the properties of the individual outflows resolved toward G75.78 NE and G45.12 are given in Table 2 below.

Why do we not see a radio continuum source toward all massive outflows if they are driven by massive stars, as claimed? A possible explanation is that during the phase of rapid accretion, matter falls so rapidly onto the YSO that its Strömgren radius is basically coincident with the stellar photosphere and therefore not detectable as a UC HII region during this phase. The FIR emission is probably dominated by the nearby UC HII region and poor spatial resolution prevents one from recognizing multiple sources in

TABLE 2. Molecular Outflows Associated with OB Stars

Source	Sp. Type	M_* $M_\odot$	$\log(\dot{M}_f)$ $M_\odot yr^{-1}$	M_f $M_\odot$	τ_d 10^4 yr	Notes	Refs.
G5.89	O6.5	28	-1.59	77	0.3	1,2	1
G45.12+0.13							
	O5.5	35	-1.62	~4800	~20	1,2,3	2
G45.12+0.13W							
G75.78NE-C	B0	16	-2.80	58	3.7	2	3
G75.78NE-N			-3.45	19	5.3	2	3
G75.78NE-E			-3.36	20.7	4.7	2	3
G98.04+1.45	O6.5	28	-3.36	40	9.1	1	4
G111.25-0.77	B0.5	13	-3.36	16	3.7	1	4
G173.58+2.45	B2.5	7	-3.22	32	5.3	1	4
G192.16-3.82	B2.5	7	-3.67	58	27	1	4
W75N	B1-B2	11,8	-2.92	48	4.4	3	5
MonR2(IRS1)	B0	16	-2.95	170	15	1,3	7
DR21	O4-O5	>40	-1.2	~3000	5	3	6

Notes
1. Single dish results
2. Interferometer results
3. Multiple sources present

References
1. Acord *et al.* 1997
2. HPM97
3. SCW97
4. SC96
5. Hunter *et al.* 1994
6. Garden *et al.* 1991
7. Tafalla *et al.* 1997

the beam. High resolution NIR and intermediate IR observations might be able to resolve this issue. At any rate, high resolution CO images clearly illustrate that massive star formation regions have multiple, physically close outflows most of which are not driven by the UC HII region in the field.

6. Origin of the Mass in Massive Outflows

Outflows driven by luminous stars are often, if not always, more massive than that of the star responsible for driving the outflows. In Table 2, taken from Churchwell (1997), the properties of several massive outflows are tabulated. Churchwell (1997) lists 4 outflows toward G45.12, however, after further analysis HPM97 finds only two outflows. Table 2 has been modified to account for this change; the entries for G45.12 are single dish results which are integrated parameters for all outflows in this region.

What is the origin of the mass in massive star outflows? Four possibilities are: 1) accumulated stellar winds; 2) entrained ISM in stellar bipolar jets; 3) swept-up ISM; and 4) infalling matter diverted into bipolar outflows. I will consider each possibility in the following.

6.1. ACCUMULATED STELLAR WINDS

The winds from O and B stars are known to be radiation driven (see reviews and references in Abbott 1985; Lamers and Groenewegen 1990; Pauldrach *et al.* 1990; and Castor 1993). For stellar winds to supply an outflow of 50 $M_\odot$ over a typical period of 10^4 yr, the average mass loss rate would have to be $\sim 5\times10^{-3}$ $M_\odot$ yr^{-1}. This is about two orders of magnitude greater than the highest mass loss rate observed toward O-stars and WR stars. A mass loss rate of this magnitude, if it is radiation driven, cannot be greater than $\frac{L_*/c}{v_\infty}$ where L_* is the luminosity of the star, c is the speed of light, and v_∞ is the wind terminal velocity. This requires that the wind terminal velocities must be in the range 0.4 to 4 km s^{-1} for $L_*=10^{5-6}$ $L_\odot$. Such velocities are several times lower than observed molecular outflow velocities. The mass accumulation could be greater if the dynamical lifetimes under-estimate the actual ages of the outflows due to confinement by ambient pressure and the mass loss rates could be greater if multiple scattering is important. In a bipolar outflow, multiple scattering is unlikely to be important because of scattering losses perpendicular to the flow axis. Also, a consideration of the momentum and kinetic energies involved in massive star outflows leads one to conclude that confinement by the ambient interstellar medium is unlikely to increase actual lifetimes by more than a factor of 2 or so beyond that implied by dynamical ages. Thus, neither of these effects appear capable of increasing the accumulated outflows by large enough factors to account for the observed values. It is, therefore, highly unlikely that the outflows are driven by stellar winds which are flowing slower than the outflows themselves. Thus, it also follows that the outflow masses are unlikely to be the result of accumulated stellar winds.

6.2. ENTRAINED ISM IN STELLAR BIPOLAR JETS

Let us now consider entrained interstellar matter (ISM) in stellar bipolar jets. The issue here is how much ISM can be entrained in a bipolar jet. The mass entrainment rate per unit area is:

$$R_e = \epsilon\rho_0 c_0 \ , \tag{2}$$

(Cantó and Raga 1991) where ϵ is the entrainment efficiency (<<1), ρ_0 is the ambient ISM density, and c_0 is the speed of sound in the ambient ISM. For a temperature T=20-50 K and density $n_{H_2} \approx 10^5$ cm^{-3},

$$R_e(gms^{-1}cm^{-2}) = \epsilon(0.9 - 1.4)\times10^{-14} \ , \tag{3}$$

If we take the central flow in G75.78NE as an example (see Table 2), we have a flow radius of 0.69 pc and an age of 3.7×10^4 yr. To estimate its entrainment area, an approximate cone shape is assumed. For a flow Mach number in the range 10-20, the jet opening angle is expected to be $\theta = 2/\sigma \approx 0.04 rad = 2.3°$ from experimental results for a "mixing layer limited" entrainment regime (see Fig. 2 in Cantó and Raga 1991). This gives a total surface area (both lobes) of about $5.8 \times 10^{35} cm^2$ and an entrained mass:

$$M_f = R_e A_f \tau_d \approx \epsilon(3-5) M_\odot \, , \tag{4}$$

where τ_d is the dynamical lifetime of the outflow and A_f is the entrainment area of the jets. In the case of stellar jets with temperature $T_j = 10^4$ K and an ambient temperature $T_0 < 100$ K, Cantó and Raga (1991) find that the "mixing layer limited" regime is appropriate and that the entrainment efficiency in this regime is $\epsilon < c_0/2c_j$ where c_0 and c_j are the sound speed in the ambient medium and the jet, respectively. For the above conditions, and $\epsilon < 0.1$, the total mass of entrained ISM would be $M_f < 1$ $M_\odot$. One could increase the entrained mass by assuming that the stellar mass loss occurs not as a jet but as a flow of matter with much larger opening angles. For example, if $\theta = 30°$ (effectively increasing the mixing layer area by more than an order of magnitude), $M_f = \epsilon(52-81)M_\odot < 8M_\odot$. Thus, even in the large opening angle case, one can only entrain a few solar masses in outflows; certainly nothing like the several tens of solar masses observed. Since entrained mass is proportional to the third power of the outflow lobe separation, an under-estimate of this parameter would under-estimate the entrained mass. Failure to correct for inclination of the flow axis to the line of sight under-estimates the lobe separation on average by a factor of $\sqrt{2}$ which results in an average under-estimate of the entrained mass by a factor of ~ 0.35, not enough to account for the observed masses. It, therefore, appears that neither accumulated stellar winds nor entrained ISM in stellar jets are capable of providing the mass observed in the outflows from young OB stars.

6.3. SWEPT-UP ISM

It is also possible to set a large mass of ISM into motion along an outflow axis if the density of ISM in the immediate neighborhood of a forming star is large, and if the opening angle of the outflow is large (i.e. not a jet). In this case, the leading boundary of the outflows could act like pistons sweeping up ISM in front of the expanding outflow lobes rather than punching a hole through the ambient medium and sweeping aside the matter as would be the case for a narrow-opening angle jet. There is ample evidence that the densities of ISM is large in the neighborhood of forming massive stars (see Cesaroni *et al.* 1991; Churchwell, Walmsley, and Wood 1992; Cesaroni, Walmsley, and Churchwell 1992; Hofner *et al.* 1995; and others). Thus, the main issue here is whether the opening angles of massive star outflows are narrow and jet-like or wide. In fact, there is evidence for both.

The Orion A-IRc2 outflow appears to have a fairly wide opening angle as traced by the shock-excited H_2 emission obtained by Allen and Burton (1993). The interpretation of the origin and nature of the outflow is controversial, but whatever the correct interpretation, it is clear that the opening angle of the flow is broad on the scale of the H_2 emission.

In contrast, the outflows in G45.12, which are among the most massive outflows known, are well collimated. The greater distance of this source, however, prevents us from viewing it with the same resolution possible for Orion. Nonetheless, the high degree of collimation of these very massive outflows produce small cross-sections for interaction with the ISM and are unlikely to sweep-up significant amounts of mass relative to that in the flows.

At this juncture, perhaps all one can say is that swept-up ISM is a possible origin for outflow masses. High resolution images will be required to resolve the issue of whether wide opening angles are common among massive outflows and if they continue to spread with distance from the origin.

6.4. ACCRETION DRIVEN OUTFLOWS

A fundamental problem associated with star formation is how they shed angular momentum in order for infalling matter to actually reach the star. The discovery of optical jets and radio bipolar molecular outflows associated with YSOs suggests that this is how they shed angular momentum and permit stars to form. However, the process by which accreting matter is diverted into outflows is not understood, especially for very massive stars. Several models which rely on magnetic fields and rotation have been proposed in the literature (Shu 1991; Pudritz 1988; Shu *et al.* 1994; Shu 1995; and others at this symposium) for low-mass stars. It is not clear whether any of these are applicable to massive stars. Although the mechanism itself is not understood, it appears likely that deflection of infalling matter into bipolar outflows may be a primary mechanism that feeds outflows of massive YSOs. Some implications of this are examined in the following.

First, it implies that massive YSOs must go through a very rapid accretion phase with a mass accretion rate:

$$\dot{M}_{acc} > \dot{M}_f \approx 10^{-2} - 10^{-3} M_{\odot} yr^{-1} . \tag{5}$$

$\dot{M}_{acc}$ must be larger than $\dot{M}_f$ since some of the accreting matter ultimately ends up on the central star. $\dot{M}_{acc}$ also places restrictions on the density structure of the accreting matter. Since the infall rate cannot exceed the free-fall rate, if gravity is the only inward force, we have:

$$\rho(r) \geq \frac{\dot{M}_{acc}}{4\pi r^{3/2}(2GM_*)^{1/2}}, \tag{6}$$

where r is the distance from the central star and M_* is the mass of the central YSO. For example, for $\dot{M}_{acc} = 10^{-2} M_\odot yr^{-1}$ toward an O5 ZAMS star (M_* ≈40$M_\odot$; Straizys and Kuriliene 1981), the density of infalling matter at 1000 AU would be ≥10^8 cm^{-3} and ≥10^5 cm^{-3} at ~ 0.5 pc. This requirement on the density of infalling matter would provide a natural explanation for the very high densities detected in the immediate vicinity of very compact millimeter continuum sources such as G9.62E and F (Hofner *et al.* 1995) and CS cores with FIR colors consistent with UC HII regions (Bronfman, Nyman, and May 1996).

Second, if the outflow has achieved its maximum mass (i.e. old and near the end of the bipolar outflow phase), then the outflow mass plus stellar mass gives the minimum mass that a molecular cloud core must have to form a star of a given mass. For example, the central outflow in the G75.78NE region is apparently driven by an early B-star (about 16 $M_\odot$) and has an outflow mass of about 58 $M_\odot$. Thus, the fraction of the infalling matter actually incorporated into the star is only ~ 20% and ~80% is diverted back outward into the outflow. In this sense, the efficiency of getting matter onto the central star is only about 20% and one concludes that a cloud core with a mass ≥5 times that of the final star mass is required to form the star. If massive star formation is accompanied by an initial mass function distribution in which most of the mass is in lower mass stars, as appears to be the situation from deep NIR imaging of massive star formation regions (see McCaughrean 1993; McCaughrean and Stauffer 1994; Aspin and Walther 1990, and others), then the mass of the natal molecular cloud core would have to be much greater than the most massive star formed in the association. This may be partly the reason why massive stars are so rare; typical cloud cores generally do not achieve such large masses.

Third, some massive outflows such as NGC6334I (Bachiller and Cernicharo 1990) have flow velocities as high as 100 km s^{-1} or greater. From escape velocity considerations, one can show that these flows must originate from deep in the gravitational well of the central stellar object (≤7 to ≤180 AU for a 40 $M_\odot$ central star and a flow velocity in the range 100 to 20 km s^{-1}, respectively). The inequality arises because $v_{flow} \geq v_{escape}$. Thus, observation of the critical region where outflows originate will require spatial resolutions better than 10 to 100 AU.

Fourth, accretion rates of 10^{-2} to 10^{-3} $M_{\odot}$ yr^{-1} will result in most of the stellar UV radiation being absorbed close to the central star, thus delaying the formation of a detectable UC HII region until after this phase of star formation. For example, a density greater than 10^8 cm^{-3} of infalling matter toward an O5 star would result in a Strömgren radius $\leq$100 AU. This implies that the massive outflows observed to date are likely to be associated with YSOs that have not yet formed a detectable UC HII region. Those that are associated with a UC HII region, such as G5.89, are probably no longer being driven but are relics of an earlier rapid accretion phase. This has the further implication that detection of massive YSOs in their rapid accretion phase or earlier can only be achieved either via their molecular outflows or their thermal dust emission at FIR to mm wavelengths or high excitation molecular line probes, but not from radio free-free emisssion. This does not rule out hot ($>10^5$ K), shocked, ionized gas produced by infall onto an accretion disk or in the outflow which might be detectable in very high resolution radio continuum observations.

7. Summary

About 90% of all massive star formation regions have associated high velocity gas. In at least 50% of these, the high velocity gas is due to bipolar outflows. Molecular outflow properties are tightly correlated with stellar luminosity for L_{bol} in the range ~1-10^6 $L_{\odot}$. The correlation of $\dot{M}_f$ with L_{bol} is tight enough that one can predict with reasonable accuracy the stellar L_{bol} from a measurement of the outflow rate. The energetic properties of outflows associated with massive YSOs are systematically larger than those of low-mass stars. One also finds that massive star formation regions when observed with high spatial resolution tend to have multiple outflows. In most cases the origin of the outflows are not the UC HII region in the field. The origin of the mass in massive outflows is not likely to be due to accumulated stellar winds or to entrainment of ambient ISM. It could possibly be due to swept-up ISM if the outflow opening angles are large and the flows continue to spread with distance from the origin. It is argued that highly collimated massive outflows are likely to be fed by accreting matter that has been diverted into bipolar outflows. Interesting consequences of this idea are: 1) that massive YSOs must go through a very rapid mass accretion phase with the density structure of the infalling matter dependent on distance from the YSO as $\rho(r) \propto r^{-3/2}$; 2) the efficiency of getting infalling matter onto a central star is only $\sim$ 20% with the rest of the matter being diverted back out into bipolar outflows; 3) the high speed outflows must originate within 10-100 AU of the central YSO; and, 4) during the rapid accretion phase, radio free-free emission is unlikely to be detected.

Acknowledgements: I would like to thank Deborah Shepherd for her invaluable help in preparing the figures for this review and for many helpful discussions of this subject. I thank Joe Cassinelli for several insightful discussions on stellar winds and stellar mass accretion.

References

Abbott, D. C. 1985, in *Radio Stars*, eds. R. M. Hjellming and D. M. Gibson, Reidel, Dordrecht, p. 61
Acord, J. M., Walmsley, C. M., Churchwell, E. 1997, Ap. J., in press
Allen, D. A., Burton, M. G. 1993, Nature, 363, 54
Aspin, C., Walther, D. M. 1990, A&A, 235, 387
Bachiller, R., Cernicharo, J. 1990, A&A, 239, 276
Bronfman, I., Nyman, L.-Ä., May, J. 1996, A&A Suppl., 115, 81
Cabrit, S., Bertout, C. 1992, A&A, 261, 274 (CB92)
Cantó, J., Raga, A. C. 1991, Ap. J., 372, 646
Castor, J. I. 1993, in *Massive Stars: Their Lives in the Interstellar Medium*, eds. J. P. Cassinelli and E. B. Churchwell, A. S. P. Conf. Series, 35, 297
Cesaroni, R., Walmsley, C. M., Churchwell, E. 1992, A&A, 256, 618
Cesaroni, R., Walmsley, C. M., Kömpe, C., Churchwell, E. 1991, A&A, 252, 278
Churchwell, E. 1997, Ap. J. Lett., in press
Churchwell, E., Walmsley, C. M., Wood, D. O. S. 1992, A&A, 253, 541
Edwards, S. Ray, T,, Mundt, R. 1993, in *Protostars and Planets III*, eds. E. H. Levy and J. L. Lunine, Univ. Arizona Press, Tucson&London, p. 567
Fukui, Y., Iwata, T., Mizuno, A. Bally, J., Lane, A. P. 1993, in *Protostars and Planets III*, eds. E. H. Levy and J. L. Lunine, Univ. Arizona Press, Tucson&London, p. 603
Garden, R. P., Hayashi, M., Gatley, I., Hasegawa, T., Kaifu, N. 1991, Ap. J., 374, 540
Hofner, P., Kurtz, S., Churchwell, E., Walmsley, C. M., Cesaroni, R. 1995, Ap. J., 460, 359
Hunter, T. R., Phillips, T. G., Menten, K. M. 1997, Ap. J., in press (HPM97)
Hunter, T. R., Taylor, G. B., Felli, M., Tofani, G. 1994, A&A, 284, 215
Lada, C. J. 1985, Ann. Rev. A&A, 23, 267
Lamers, H. G. J. L. M. & Groenewegen, M. A. T. 1990, in *Properties of Hot Luminous Stars*, ed. C. D. Garmany, A. S. P. Conf. Series, 7, 189
McCaughrean, M. 1993, in *Massive Stars: Their Lives in the Interstellar Medium*, eds. J. P. Cassinelli and E. B. Churchwell, A. S. P. Conf. Series, 35, 80
McCaughrean, M., Stauffer, J. R. 1994, A. J., 108, 1382
Pauldrach, A. W. A., Puls, J., Gabler, R., Gabler, A. 1990, in *Properties of Hot Luminous Stars*, ed. C. D. Garmany, A. S. P. Conf. Series, 7, 171
Pudritz, R. E. 1988, in *Galactic and Extragalactic Star Formation*, eds. R. E. Pudritz and M. Fich, Kluwer Acad. Pub., Netherlands, p. 135
Shepherd, D. S., Churchwell, E., Willner, D. J. 1997, Ap. J., in press (SCW97)
Shepherd, D. S., Churchwell, E. 1996, Ap. J., 472, 225 (SC96)
Shepherd, D. S., Churchwell, E. 1995, Ap. J., 457, 267 (SC95)
Shu, F. H. 1991, in *The Physics of Star Formation and Early Evolution*, eds. C. J. Lada and N. D. Kylafis, Kluwer Acad. Pub., Dordrecht, Netherlands, p. 365
Shu, F. 1995, Rev. Mex. A. A. (Serie de Conf), 1, 375
Shu, F., Najita, J., Ostricker, E., Wilkin, F., Ruden, S., Lizano, S. 1994, Ap. J., 429, 781
Straizys, V., Kuriliene, G. 1981, Ast.&Sp. Sci., 80, 353
Tafalla, M., Bachiller, R., Wright, M. C. H., Welch, W. J. 1997, Ap. J., 474, 329

LOW-MASS VERSUS HIGH-MASS STAR FORMATION

T. W. HARTQUIST
*Max-Planck-Institute für extraterrestrische Physik,
D-85740 Garching, Germany*

AND

J. E. DYSON
*Department of Physics and Astronomy,
The University of Leeds,
Leeds. LS2 9JT, UK*

Abstract. Structures like the clumps identified in the CO maps of the Rosette Molecular Cloud and the dense cores such as those in B5, a cluster of cores and young low-mass stars, are key to considerations of star formation. Whether star formation is a self-inducing process or one that causes itself to turn off depends greatly on whether the responses of the interclump and intercore media to young stars cause the collapse of clumps or cores to be faster than their ablation. We present a naive introduction to the lengthscales over which such responses are significant, mention ways in which the responses might induce collapse, review some of the little that is known of how flows of media around clumps and cores ablate them, and then return to the issue of the lengthscales over which such responses are significant by considering the global properties of mass-loaded flows in clumpy star forming regions.

1. Introduction

Regions of recent and/or ongoing high-mass star formation are amongst the most brilliant diffuse optical sources in the sky, in contrast to regions in which only low-mass stars are born. The dramatic nature of the response of the interstellar medium around high-mass stars and work on the ages of massive stars in Orion led eventually to the realisation that that type of

B. Reipurth and C. Bertout (eds.), Herbig–Haro Flows and the Birth of Low Mass Stars, 537–549.

response induces further massive stars to form and that such stars can be produced sequentially (Elmegreen & Lada 1977).

In contrast, the formation of a low-mass star is generally treated as a process occurring in isolation (e.g. Mouschovias 1987; Shu, Adams, & Lizano 1987), an approach which may be valid in many cases. However, young low-mass stars are sources of radiation and winds sufficiently powerful to induce significant responses in the interstellar matter near them, and we must address the possibility that low-mass star formation is often self-regulating.

Much of the issue of differences between and similarities of high-mass and low-mass star formation is, therefore, concerned with the ways in which gas in an interstellar cloud responds to the formation of a high-mass star or the birth of a low-mass star. In section 2 we consider briefly the lengthscales on which the responses to low-mass and high-mass stars are significant. The ways in which clumps react to a newly formed nearby star are important in determining whether star formation induces further stellar birth; in sections 3 and 4 respectively, we mention mechanisms by which clump collapse might be triggered or retarded by a star. Section 5 returns to the topic of section 2 and contains comments about what is known of how the clumpiness of a medium might alter its global response to stellar sources. Section 6 concludes the review.

2. The Scale of Response

Cioffi, McKee, & Bertschinger (1988) have shown that a supernova occurring in a uniform non-magnetic medium will increase the pressure of the surrounding medium within a radius $R_{\rm SN}$, given by

$$R_{\rm SN} \simeq 70\ E_{51}^{31/98} {n_{\rm H}}^{-18/49} \left\{ \frac{\beta C_1}{10 \text{ km s}^{-1}} \right\}^{-3/7} \text{ pc.} \tag{1}$$

E_{51}, $n_{\rm H}$, C_1, and β respectively are the supernova energy in units of 10^{51} erg, the hydrogen nuclei number density in cm^{-3}, the sound speed of the ambient medium, and a constant near unity. The use of equation (1) to estimate the lengthscale $R_{\rm SNGMC}$ over which a supernova will have a significant effect in a giant molecular cloud (GMC) is naive because a cloud is magnetized and clumpy. Even so, to make a rough estimate of $R_{\rm SNGMC}$, we will use $R_{\rm SN}$, but to do so we must first specify $n_{\rm H}$ and C_1. We assume that the supernova occurs in the interclump medium of a GMC of which the Rosette Molecular Cloud (RMC) is a particularly well-studied example. Williams, Blitz, & Stark (1995) argued that the mean interclump neutral atomic number density in the RMC is 4 cm^{-3}. Williams et al. gave the mean H_2 number density in clumps as about 220 cm^{-3} and found the velocity dispersions in the clumps to be $2 - 3$ km s^{-1} typically. For an interclump medium of atomic hydrogen with a number density of about 4 cm^{-3} to confine such a clump (which seems required because according to Williams et al., a good fraction of the clumps in the RMC are not massive enough to be bound by the components of the gravitation field that they generate) the effective interclump sound speed must be roughly 30 km s^{-1} (which may be the case if the interclump magnetic field is of the corresponding strength). For $n_{\rm H} \simeq 4$ cm^{-3} and $C_1 \simeq 30$ km s^{-1}, equation (1) gives $R_{\rm SN} \simeq 26\ E_{51}^{31/98} \beta^{-3/7}$ pc, a particularly interesting result because it suggests that a single supernova can affect the pressure over a reasonable fraction, but not all, of a GMC. If it were to cause a large increase in the pressure over a distance small compared to the typical distance between clumps in which high-mass stars form one might expect high-mass star formation not to be self-triggered at all. If a single supernova could cause a large increase in the pressure through the interclump medium of an entire GMC, one might expect a burst of high-mass star formation over the entire GMC rather than high-mass star formation to be sequential.

The ^{13}CO and ^{12}CO observations of the RMC that Williams et al. analysed do not readily lead to the discovery of dense cores, which are more easily identified from NH_3 data but can be mapped in CO emissions (e.g. Myers 1990). Young low-mass stars are associated with about half of the dense cores indicating that they are often progenitors of stars. Those in which low-mass stars form usually have molecular hydrogen number densities of about $10^4 - 10^5$ cm^{-3} and masses within a factor of a few of 10 M$_\odot$. While some cores are more isolated, a sizeable fraction are found to exist in clusters somewhat like Barnard 5 (B5). The CO map of B5 of Goldsmith, Langer, & Wilson (1986) shows the presence of 5 dark cores as well as the locations of young stellar objects which are IRAS sources. The CO data provide clear evidence that the winds of the young stars affect the dynamics of the intercore medium of the B5 core cluster on scales comparable to the intercore separations. Obviously, at least in some cases, the formation of a low-mass star cannot be considered to be occurring in isolation. Rather, sometimes on smaller scales and during later stages of collapse the births of low-mass stars affect subsequent low-mass star formation in manners that may be similar to or may contrast interestingly with the ways in which the births of high-mass stars affect subsequent high-mass star formation. The degree of similarity or contrast most likely has very much to do with the nature of the responses of clumps and of dense cores, to flows around them driven by the stars.

3. Triggering Clump or Core Collapse

Mouschovias & Spitzer (1976) showed that gravitational collapse of a magnetized clump will occur only if its mass exceeds a critical mass M_c and the external pressure is greater than a minimum value P_m. For a clump with the same distribution of mass on magnetic field lines as a spherical clump of uniform density and with a constant magnetic field

$$M_c = \frac{0.53}{\pi} \left\{ \frac{5}{9G} \right\}^{1/2} \Phi, \tag{2}$$

and

$$P_m = \frac{1.89 C_{cl}^8}{G^3 M_{cl}^2 (1 - (M_c/M_{cl})^{2/3})^3}, \tag{3}$$

where M_{cl} is the clump mass, C_{cl} is the isothermal sound speed in the clump, Φ is the magnetic flux through the clump, and G is the gravitational constant. Low-mass star formation is generally supposed to occur in clumps in which initially $M_{cl} < M_c$ but ambipolar diffusion gradually reduces Φ until, eventually, $M_{cl} \geq M_c$ (e.g. Mouschovias 1987; Shu et al. 1987). High-mass star formation is more likely triggered in clumps in which $M_{cl} > M_c$ by an increase in external pressure (e.g. Elmegreen & Lada 1977; Mouschovias 1987; Shu et al. 1987) to a value above P_m from one that is initially below P_m.

Since the RMC clumps identified by Williams et al. (1995) have masses varying by over two orders of magnitude and average number densities that are restricted (perhaps, due to the sensitivity of CO observations to the range of measured densities) to within about a factor of four of the mean, some RMC clumps probably are magnetically supercritical (i.e. have $M_{cl} > M_c$) while others are magnetically subcritical (i.e. have $M_{cl} < M_c$). In the standard view of low-mass star formation, a core in B5 is either magnetically subcritical but becoming marginally supercritical due to ambipolar diffusion or is marginally supercritical. An increase of the pressure in the RMC interclump medium or the B5 intercore medium caused by a star compresses the clumps or the cores but only in the RMC could it possibly directly induce the gravitational instability of some (but certainly not all) objects.

In the last sentence we used the phrase 'possibly . . . induce' rather than simply 'induce' because even a huge increase of the external pressure of a magnetically supercritical cloud will not induce it to collapse if C_{cl} is a too sensitively increasing function of the external pressure (c.f. equation (3)). The turbulent pressure within a typical RMC clump exceeds the thermal pressure by more than an order of magnitude, and a characteristic speed associated with the turbulence should replace the isothermal sound speed in equation (3) when it is employed in considerations of RMC clump stability. The turbulence within an RMC clump probably consists of a superposition of Alfvén waves (e.g. Arons & Max 1975; Mouschovias & Psaltis 1995), but the excitation mechanism remains unknown implying that the dependence upon the external pressure of the characteristic speed associated with the turbulence to be used in equation (3) is also unknown. In our view, until this excitation mechanism is identified, the theory of sequential high-mass star formation and the theory of starbursts in external galaxies remain distant objectives (Hartquist, Dyson, & Williams 1997).

However, knowledge of some of the mechanisms damping the waves comprising the turbulence does exist. One of importance is the friction between ions and neutrals which is easily analysed for small amplitude waves superposed on a uniform medium (Kulsrud & Pearce 1969). The damping rate due to ion-neutral friction is inversely proportional to the fractional ionisation for waves of fixed angular frequency as long as that angular frequency is much less than the inverse of the time required for collisions with ions to transfer to a neutral particle a momentum of magnitude comparable to the neutral particle's initial momentum. Hence, the ionization structure of an RMC clump is likely of importance in determining the properties of the turbulence in it.

Hartquist et al. (1993) have suggested that S^+ is the most abundant ion throughout RMC clumps that do not contain infrared sources (the presence of which indicates that the clumps have collapsed). Because the simple hydrogen abstraction reactions $SH^+ + H_2 \rightarrow SH_2^+ + H$ and $SH_2^+ + H_2 \rightarrow SH_3^+ + H$ are slow at low temperatures and the ionization potential of S is lower than that of C, gas phase sulphur remains primarily in S^+ at several times greater visual extinction than that to which carbon remains in C^+. Hartquist et al. proposed that the RMC clumps in which stars have formed have at one time or another been compressed sufficiently for their visual extinctions to exceed a critical visual extinction above which most of the sulphur in them has become neutral, whereas the RMC clumps have not been compressed sufficiently exceed this critical value. In clumps in which the sulphur is primarily neutral, the fractional ionization is probably one to two orders of magnitude lower than that in clumps in which most gas phase sulphur is in S^+. Thus, in the former class of clumps, the ion-neutral frictional damping of turbulence is much more rapid than in the latter class of clumps, probably resulting in the reduction of the turbulent pressure in a clump belonging to the former class and its collapse primarily along the magnetic field lines in it; of course, some collapse perpendicular to the field lines also occurs. Thus, high-mass star formation may (in a case in which the clump is initially supported in part by turbulence and is magnetically supercritical) be induced directly by an increase in the interclump pressure (in response to a recently formed star) if the pressure increase gives rise to a sufficient rise in visual extinction that the fractional ionization at the clump centre drops significantly. Of course, partial collapse of a magnetically subcritical and initially turbulent clump may be induced in the same way, but ambipolar diffusion must then act for star formation to occur leading probably to low-mass star formation.

The principal ions at the centres of the cores in an object like B5 are most likely Mg^+, Na^+, and molecular ions, and compression by the increase of the pressure in the intercore medium probably does not alter the fractional ionization radically (i.e. by more than a factor of a couple or a few). In addition, in each of many dense cores, the line widths of emissions indicate that the turbulent pressure is negligible compared to the thermal pressure (Myers & Benson 1983). Therefore, the specific sort of induced collapse picture presented above for RMC clumps does not apply to dense cores.

Though we have expressed reservations above about the idea that the compression of a dense core causes major changes in the fractional ionization, the possibility that desorption of elements heavier than helium from grains ceases abruptly once a critical visual extinction (differing from the critical visual extinction mentioned above in the consideration of sulphur chemistry and varying from one star forming region to another) has been exceeded, has been advanced (Williams, Hartquist, & Whittet 1992). Freeze-out of heavy elements (and in particular, sodium, magnesium and other heavy metals) may have a major effect on the fractional ionization. When the fractional ionization is much above about 10^{-8}, grain-neutral drag is negligible and the ambipolar diffusion time scale of a magnetically subcritical core is proportional to the fractional ionization (e.g. Mouschovias 1987). Unfortunately, the desorption and freeze-out of elements in cores is not understood, and until it is, a number of important issues concerning ionization structure and ambipolar diffusion in dense cores and the possibility that low-mass star formation may be sped up significantly by an increase in the pressure of the intercore medium, will also remain unclear.

4. Ways in Which Star Formation May Hinder Further Star Formation

As mentioned at the end of the previous section, the time scale for ambipolar diffusion, important for the formation of stars in clumps and cores that are initially magnetically subcritical, is proportional to the fractional ionization. Oppenheimer & Dalgarno (1974) and subsequently Elmegreen (1979) and Umebayashi & Nakano (1980) have shown that at high visual extinctions where ionization is induced primarily by cosmic rays, the most abundant ions are those of heavy metallic atoms (e.g. Na^+ and Mg^+) but that most of the heavy metallic atoms may be neutral. McKee (1989) has argued that the radiation of a young low-mass star in a core cluster like B5 can penetrate a dense core sufficiently to maintain a sizeable fraction of the gas phase metallic atoms in ionized states. Therefore, the birth of a low-mass star in a core cluster may retard the formation of other low-mass stars by causing an increase in the ambipolar diffusion time scale.

As seen in Goldsmith et al. (1986), the winds of the young stars in B5 move material around on lengthscales comparable to the intercore separations, and as the cores undergo collapse they are also ablated by the stellar winds. Further stellar birth would not occur if the collapse time scale were long compared to the ablation time scale.

Nittmann, Falle, & Gaskell (1982) argued that the evaporation of a non-magnetic clump initially bound by the pressure of a more diffuse static non-magnetized medium through which a hypersonic shock passes, takes place within several clump sound crossing times. The results of the numerical simulations of Klein, McKee, & Colella (1994) are consistent with those arguments and the earlier simulations. Evaporation is reduced if the compression causes the clump to become Jeans unstable (c.f. Woodward 1976). Evaporation also occurs but at a lowered rate if the flow in the medium surrounding the clump is subsonic in the frame of the clump (in which the Mach number of the tenuous medium's flow is measured to be M_T). Hartquist et al. (1986) have estimated the timescale for the mass of the clump to be incorporated into the wind to be of the order of the sound crossing timescale multiplied by the cube root of the ratio of the sound speeds in the surrounding medium and the clump and divided by $M_T^{5/3}$; the timescale on which the coherence of the clump is destroyed may be smaller by a factor of the ratio of the sound speeds in the clump and surrounding medium divided by M_T. Clearly, if M_T is not considerably less than unity, the ambipolar diffusion timescale is much longer than the sound crossing timescale, and these estimates of evaporation timescales are relevant, then initially magnetically subcritical dense cores will be ablated before ambipolar diffusion allows them to collapse due to gravity. (The numerical simulations of MacLow et al. (1994) of shocks in magnetized media overrunning magnetized clumps suggest that the evaporation timescales appropriate for non-magnetized media and clumps may underestimate those relevant when magnetic fields are important.)

The timescale for ionization equilibrium to be reached in clumps is short compared to any dynamical timescale, and the damping timescale is likely to be less than the sound crossing timescale if S^+ is not a significantly abundant ion. Thus, ablation is probably less important in the collapse of magnetically supercritical RMC clumps induced by an increase in external pressure and in high-mass star formation than it is in low-mass star formation from magnetically subcritical dense cores. If this occurs throughout a region of star formation, an IMF biased towards more massive stars may obtain.

We would like to study the ablation of clumps and cores observationally. TMC-1 is a ridge of dense cores with a young low-mass star near the core thought to be the densest. That core has a measured H_2 number density of about 5×10^5 cm^{-3} (Hirahara et al. 1992). The interaction of the stellar wind with it creates a turbulent boundary layer with unknown properties. Charnley et al. (1990), Nejad & Hartquist (1994), and Rawlings & Hartquist (1997) attempted to identify atomic and molecular diagnostics at such a boundary layer. The models of Charnley et al. (1990) and of Nejad & Hartquist (1994) are based on the assumptions that a boundary layer is isobaric and that H^+ and He^+ from shocked stellar wind and heat are injected into core gas at specified rates. The models of Rawlings & Hartquist (1997) are based on the assumptions that the boundary layer is isobaric and that diffusive transport between the core and the wind takes place in the boundary layer; for an optically thin boundary layer the free parameter is the diffusion timescale and the molecular, atomic, and ionic densities are specified to be appropriate for a shocked stellar wind and a dense core on either side of the boundary layer. The chemistry in all their models was restricted to simple species. An observation of CO ($J = 6 \rightarrow 5$) emission towards core B (in the Hirahara et al. (1992) nomenclature) in TMC-1 would serve as a useful diagnostic of the thermal structure of the boundary layer. Observations of CH and OH emissions with Arecibo might yield useful data on whether diffusive mixing in the boundary layers occurs. (Because the boundary layers are magnetized, mixing of stellar wind ions with dense core neutrals to the extent that they react together chemically might be negligible, but the issue is difficult to address theoretically.) A major problem is that a boundary layer may contain only a few percent of the total material of a core so that emissions arising in it might be hard to detect against the core emissions; this difficulty makes the existing data for CI emission from TMC-1 (Schilke et al. 1995) irrelevant for boundary layer diagnosis even though C^0 and C^+ could acquire high fractional abundances in a boundary layer if mixing is thorough.

However, the fact that CI emission is detectable from a dense core may imply that mixing does take place in boundary layers in low-mass star forming regions (Williams, Hartquist, & Caselli 1996; Rawlings & Hartquist 1997; references therein). For the moment assume that dense cores form when magnetically subcritical RMC clumps dissipate their turbulence and undergo partial collapse. Williams et al. (1995) have concluded that only a few percent of the hydrogen in a clump in the RMC is not in molecular form; at RMC clump densities, the H_2 formation time exceeds the chemical timescale on which most carbon is converted to CO and in addition H_2 shields CO from photodissociation (e.g. van Dishoeck 1987) so that nearly all the carbon in an RMC clump may be contained in CO. Rawlings & Hartquist (1997) argued against clump-scale diffusive mixing being a means of maintaining high C abundances in clumps or cores, and the low atomic hydrogen to H_2 abundances in the RMC clump provides some evidence that clump-scale diffusion mixing is chemically unimportant.) Thus, collapse to form a dense core from a clump rich in atomic carbon on a timescale short compared to the chemical timescale can be ruled out as the source of atomic carbon in dense cores. However, CI emission can arise from a dense core in a core cluster if mixing of ions and neutrals occurs in the boundary layer between shocked stellar wind and a dark core of a previous generation and the ablated material - stellar wind mixture passes through a low enough speed termination shock and then undergoes collapse to form a new generation of cores on a timescale short compared to the chemical timescale ($\simeq 10^6$ years) required to convert almost all carbon to CO. We suggest that CI emission and cyanopolyyne emission (cyanopolyyne formation is rapid when atomic carbon and simple carbon bearing molecules are simultaneously abundant) from dense cores implies that ablation of dense cores is significant in some regions and may prevent some cores from forming stars and that multiple generations of core formation occur in some regions.

The issue of the relative importance of photoabsorption versus purely mechanical effects makes the interpretation of some observed stellar wind-clump interactions in regions of high-mass star formation somewhat uncertain. Observations of cometary globules containing young low-mass stars in Orion show many of them to have long thin pointy tails (Sutherland et al. 1997). One possibility is that the tails form as the shocked wind of a nearby high-mass star flows subsonically past the clumps (Dyson, Hartquist & Biro 1993); if this mechanism is responsible for their formation, spectroscopic observations made with HST or the best ground-based adaptive optics systems should lead eventually to a great deal of insight into wind-clump interactions in star forming regions. Alternatively, a tail may form if a D-type ionization front drives a small amount of material from the parts of the clump initially most distant from the line connecting the clump's centre of mass and the star towards the back of the clump (that side away from the star) and towards that line. This material is then shadowed by the main part of the clump as it flows away from the star to make a tail. The numerical simulations performed to demonstrate tail formation as a consequence of photoionization (Lefloch & Lazareff 1994; 1995) include no diffuse component of the radiation field, and until simulations including a reasonable diffuse component are performed, we will remain somewhat sceptical about whether this mechanism actually forms tails.

The subject of the photoevaporation of clouds into a surrounding medium has a long history. However, so far, the photoevaporation of magnetized clumps has received little attention. Redman et al. (1997) have shown that the density jump across a D-critical ionization front is significantly reduced compared to the non-magnetic case when the ratio of the magnetic pressure to the gas pressure in the gas ahead of the IF exceeds a factor of about unity. The consequences for the evaporation rates of clumps requires a self-consistent treatment analogous to those carried out for non-magnetized clumps (Dyson 1968; Kahn 1969; Bertoldi & McKee 1990).

Other examples of clumps being affected by the winds of high-mass stars are the clumps in the Orion Kleinmann - Low region producing much of the H_2 emission there (Tedds et al. 1996). The absence of double peaks in the line profiles implies that the H_2 emission is probably not arising in the interclump material as it passes through bow shocks around the clumps unless the clumps are themselves highly fragmented (which appears unlikely from the observational data). However, the great symmetry of the line profiles about a central velocity seems inconsistent with the emission arising in gas being ablated from the clumps. The NH_3 observations of Wiseman & Ho (1996) implying that the clumps they observed in Orion are surrounded by sheaths that are warmer than the more central parts of the clumps in principle contain information about the energy dissipation in boundary layers between the clumps and a wind.

5. Global Properties of Mass-Loaded Flows in Clumpy Star Forming Regions

Most studies of the effects of mass, momentum, and energy injection by stars on their surroundings are based on the assumption that the circumstellar material is smoothly distributed. Almost by definition, star forming regions are clumpy. In clumpy media, there is mass, momentum and energy interchange in boundary layers set up at global flow-clump interfaces. The global flow properties, including the scales over which they operate, are affected by this interchange; conversely, the interchange and the clumps themselves are affected by the global flows. Clump material can be fed into a global flow by hydrodynamic ablation, photoevaporation and conduction.

Hydrodynamic ablation is considerably reduced in very subsonic as opposed to supersonic flows (Section 4). Williams, Hartquist & Dyson (1995) have analysed the dynamics of spherically symmetric, isothermal mass-loaded winds. Both sub and supersonic flow regions are possible, depending on the distribution of mass-loading centres. However, self-consistent solutions with Mach number dependent mass injection rates have not yet been obtained. Clearly the response of clumps to stellar winds can be critically affected by their location within the global flow. Unless the wind undergoes a supersonic-subsonic transition and the subsonic region is thick, the range of influence of the wind is not appreciably affected by mass loading since momentum is conserved in the flow

Williams, Dyson & Redman (1996 and references therein) have considered models for ultracompact HII regions (UCHIIR) in which mass is fed into the HII region by either hydrodynamic ablation or by photoionization. In these models, the HII region is bounded by a stationary recombination front rather than by the usual expanding ionization front. The typical radius of such a region is that characteristic of an UCHIIR i.e. ~ 0.1 pc, which is about two orders of magnitude less than the radius of the initial Strömgren sphere which would be set up in the interclump medium of the RMC. Mass-loading greatly reduces the range of influence of photoionization. If the effects of a stellar wind are included, the scale sizes are unaltered but the Mach number variation in the flow - and therefore the ablation rate - is. If the mass-loaded wind exits supersonically from the UCHIIR (Dyson, Williams & Redman 1995), any subsequent supersonic pressurisation is effected by a neutral wind moving at most at a few times the ionized gas sound speed rather than at the much higher wind terminal velocity. Mixing at boundary layers between clumps and the global flow involves neutral gas and not highly ionized gas.

The internal structures of supernova remnants are strongly affected by interior mass addition (e.g. McKee & Ostriker 1977; Chièze and Lazareff 1981; Cowie, McKee & Ostriker 1981; Dyson & Hartquist 1987; White & Long 1991; Arthur & Henney 1996). Mass addition can enhance radiative cooling. For example, a remnant in the standard uniform density Sedov phase has an integrated energy loss $\propto t^{1.5}$, whereas a remnant hydrodynamically mass-loaded from a uniform distribution of mass-loading sources has an integrated energy loss $\propto t^{4.4}$ (Dyson & Hartquist 1987). In deriving these dependences, Kahn's (1976) cooling law approximation has been used, which almost certainly has led to an underestimate of the effects of mass-loading on radiative losses as, in reality, the ionization structure in picked-up material requires a significant amount of time to reach equilibrium resulting in an increase in the cooling rate coefficient.

6. Conclusions

The presence of stars can induce additional stellar birth only if the range of the ambient medium's response is greater than the distances between the progenitor clumps *and* the gravitational collapse of these clumps occurs on timescales shorter than any evaporation timescales. Unless it is embedded in a very subsonic flow, the ratio of the gravitational collapse and evaporative timescales for a magnetically subcritical dense core, similar to those in which low-mass stars form, is likely to be much greater than the corresponding ratio for a magnetically supercritical clump, like those in which high-mass stars are born. Evaporation of cores or clumps into global flows can lead to supersonic to subsonic transitions in the global flows thus creating regions in which the ablation rate is suppressed resulting in a decreased ratio of the gravitational collapse and evaporation timescale and hence an increased chance of star formation. However, in many cases, mass-loading greatly diminishes the range of response of the ambient medium to a star. The consequences for phenomena such as sequential star formation and superwind generation in starburst galaxies are yet to be explored.

References

Arons J., Max C. E., 1975, *Astrophys.J.*, **196**, L77

Arthur S. J., Henney W. J., 1996, *Astrophys.J.*, **457**, 752

Bertoldi F., McKee C. F., 1990, *Astrophys.J.*, **354**, 529

Charnley S. B., Dyson J. E., Hartquist T. W., Williams D. A., 1990, *Mon.Not.R.ast.Soc.*, **243**, 405

Chièze J. P., Lazareff B., 1981, *Astron.&Astrophys.*, **95**, 194

Cioffi D. F., McKee C. F., Bertschinger E., 1988, *Astrophys.J.*, **334**, 252

Cowie L. L., McKee C. F., Ostriker J. P., 1981, *Astrophys.J.*, **247**, 908

Dyson J. E., Hartquist T. W., 1987, *Mon.Not.R.ast.Soc.*, **228**, 453

Dyson J. E., 1968, *Astrophys.Space Sci.*, **1**, 388

Dyson J. E., 1994, in Ray T. P., Beckwith S. V. W., eds, Star Formation and Techniques in Infrared and mm Astronomy. Springer-Verlag: Berlin, 93

Dyson J. E., Hartquist T. W., Biro S., 1993, *Mon.Not.R.ast.Soc.*, **261**, 430

Dyson J. E., Williams R. J. R., Redman M. P., 1995, *Mon.Not.R.ast.Soc.*, **277**, 700

Elmegreen B. G., Lada C. J., 1977, *Astrophys.J.*, **214**, 725

Elmegreen B. G., 1979, *Astrophys.J.*, **232**, 729

Goldsmith P. F., Langer W. D., Wilson R., 1986, *Astrophys.J.*, **303**, L11

Hartquist T. W., Dyson J. E., Williams R. J. R., 1997, *Astrophys.J.*, **In press**

Hartquist T. W., Dyson J. E., Pettini M., Smith L. J., 1986, *Mon.Not.R.ast.Soc.*, **211**, 715

Hartquist T. W., Rawlings J. M. C., Williams D. A., Dalgarno A., 1993, *Quart.J.R.ast.Soc.*, **34**, 213

Hirahara Y., Suzuki H., Yamamoto S., Kawaguchi K., Ohishi M., Takano S., Ishikawa S. I., Masuda A., 1992, *Astrophys.J.*, **394**, 539

Kahn F. D., 1969, *Physica*, **41**, 172

Klein R. I., McKee C. F., Colella P., 1994, *Astrophys.J.*, **420**, 213

Kulsrud R., Pearce W. A., 1969, *Astrophys.J.*, **156**, 445

Lefloch B., Lazareff B., 1995, *Astron.&Astrophys.*, **289**, 559

Lefloch B., Lazareff B., 1996, *Astron.&Astrophys.*, **301**, 552

MacLow M. M., McKee C. F., Klein R. I., Stone J. M., Norman M. L., 1994, *Astrophys.J.*, **433**, 757

McKee C. F., Ostriker J. P., 1977, *Astrophys.J.*, **215**, 213

McKee C. F., 1989, *Astrophys.J.*, **345**, 782

Mouschovias T. C., Psaltis D., 1995, *Astrophys.J.*, **444**, L105

Mouschovias T. C., Spitzer Jr. L., 1976, *Astrophys.J.*, **210**, 326
Mouschovias T. C., 1987, in Morfill G. E., Scholer M., eds, Physical Processes in Interstellar Clouds. Reidel: Dordrect, 413
Myers P. C., Benson D. J., 1983, *Astrophys.J.*, **266**, 309
Myers P. C., 1990, in Hartquist T. W., ed, Molecular Astrophysics - A Volume Honouring Alexander Dalgarno. Cambridge University Press: Cambridge, 328
Nejad L. A. M., Hartquist T. W., 1994, *Astrophys.Space Sci.*, **220**, 253
Nittmann J., Falle S. A. E. G., Gaskell P. H., 1982, *Mon.Not.R.ast.Soc.*, **201**, 833
Oppenheimer M., Dalgarno A., 1974, *Astrophys.J.*, **192**, 29
Rawlings J. M., Hartquist T. W., 1997, *Astrophys.J.*, **In press**
Redman M. P., Williams R. J. R., Dyson J. E., Hartquist T. W., Fernandez B. R., 1997, *Astron.&Astrophys.*, **Submitted**
Schilke P., Keene J., Le Bouret J., Pineau des Forêts G., Roueff E. G., 1995, *Astron.&Astrophys.*, **294**, L17
Shu F. H., Adams F. C., Lizano S., 1987, *Ann.Rev.Astron.Astrophys.*, **25**, 84
Sutherland R. S., Hartquist T. W., Bally J., Dyson J. E., 1997, *Astrophys.J.*, **Submitted**
Tedds J. A., Brand P. W. J. L., Brand M. G., Chrysostomou A., Fernandes A. J. L., 1996, in Millar T. J., Raga A. C., eds, Shocks in Astrophysics. Kluwer: Dordrect, 39
Umebayashi T., Nakano T., 1980, *PASJ*, **32**, 405
van Dishoek E. F., 1987, in Vardya M. S., Tarafdar S. P., eds, IAU Symposium 120 - Astrochemistry. Reidel: Dordrect, 51
White R. L., Long K. S., 1991, *Astrophys.J.*, **373**, 543
Williams J. P., Blitz L., Stark A. A., 1995, *Astrophys.J.*, **451**, 252
Williams R. J. R., Dyson J. E., Redman M. P., 1996, *Mon.Not.R.ast.Soc.*, **280**, 667
Williams D. A., Hartquist T. W., Caselli P., 1996, *Mon.Not.R.ast.Soc.*, **282**, 900
Williams R. J. R., Hartquist T. W., Dyson J. E., 1995, *Astrophys.J.*, **446**, 759
Williams D. A., Hartquist T. W., Whittet D. C. B., 1992, *Mon.Not.R.ast.Soc.*, **258**, 599
Wiseman J. J., Ho P. T., 1996, *Nature*, **382**, 139
Woodward P. R., 1976, *Astrophys.J.*, **196**, 555

ORION PROPLYDS AND THE EAGLE'S EGGS

Observing the effects of high-mass star formation on low-mass neighbours

MARK MCCAUGHREAN
Max-Planck-Institut für Radioastronomie,
Auf dem Hügel 69, 53121 Bonn, Germany

Abstract. We describe the effects of environment on low-mass star formation, from the perspective that most stars in the galaxy form in dense clusters and aggregates in GMCs, often in the presence of massive OB stars. First, we discuss the observational evidence for circumstellar disks around stars of the Trapezium Cluster, and how they are affected by their environment. Second, we examine the more distant M 16, where photo-evaporation by OB stars may disrupt the star formation process, and may inhibit the development of circumstellar disks.

1. Introduction

Jets and outflows from young stars play an important role in regulating angular momentum removal and dissipation of the outer envelope, and thus perhaps defining parameters such as the final stellar mass. In addition, the outflow may stir up the surrounding molecular cloud on scales up to parsecs, thus affecting further collapse and star formation in the cloud. However, our impression of the global importance of such effects can be biased by the fact that the most beautiful (and thus most studied) examples of collimated jets and outflows are generally associated with more or less isolated stars, in relatively quiescent regions such as the Taurus-Auriga dark clouds, or in the farther reaches of the Orion giant molecular clouds. In those regions, star formation is occuring at a low density, with only $\sim$0.3–3 stars/pc^2 (*cf.* Gomez *et al.* 1993 for Taurus-Auriga). In contrast, most stars in the galaxy form in GMCs in dense aggregates or clusters of 10–1000 members, with only a minority spread throughout the cloud in a "pedestal" population (Lada *et al.* 1993; Zinnecker *et al.* 1993). Within these clusters, the stellar density is much greater, $\sim 10^3$–10^4 stars/pc^3, implying interactions between stars in the protostellar phase, with adjacent cloud cores compet-

B. Reipurth and C. Bertout (eds.), Herbig–Haro Flows and the Birth of Low Mass Stars, 551–560.

ing for material, and later, when star-disk systems may collide. Also, many clusters contain OB stars which, with their ionizing radiation and strong winds, add another level of disruption to their low-mass neighbours, typically only 0.01–1 pc away. The cumulative effects of cluster life are, as yet, relatively poorly understood, but departure from the relative quiescence of the isolated star formation paradigm may, in fact, play a crucial role in determining important global parameters such as the stellar inital mass function and the fraction of stars which form planetary systems.

2. Circumstellar disks in the Orion Nebula

The most prominent nearby site of on-going high- and low-mass star formation is in the Orion GMCs at ~ 500 pc. There are concentrated sites of star formation in the clouds, as well as a pedestal, distributed population of young stars. Orion B (L 1630) contains dense clusters of up to 200 stars associated with NGC 2024, NGC 2023, NGC 2068, and NGC 2071, and a near-IR survey of the cloud showed that at least 60% (and up to 90% after field star corrections) of the total stellar population is in those clusters (Lada *et al.* 1991). Similarly, a survey of much of the southern half of the Orion A (L 1641) cloud concluded that roughly half of the stars are in dense aggregates or small clusters of up to 20 members (Strom *et al.* 1993; Allen 1996). The Orion A cloud also includes the Orion Nebula H II region, inside which is the Trapezium Cluster of up to 1000 young stars (Herbig & Terndrup 1986; Zinnecker *et al.* 1993; Prosser *et al.* 1994; McCaughrean & Stauffer 1994; Hillenbrand 1997), and thus, averaged over the whole cloud, cluster-mode star formation again dominates. The Trapezium Cluster is a particularly important laboratory for studying the effects of dense clusters on circumstellar disks.

2.1. DETECTION VIA IR EXCESS

Far-IR and millimetre excesses are a clear signature of a circumstellar disk, but the limited spatial resolution of space-based IR satellites and single dish millimetre telescopes makes it impossible to measure conventional SEDs for stars in the Trapezium Cluster, typically only a few arcsec apart. However, the hot inner edges of disks radiate at 2–10 μm, where ground-based IR imaging can separate the stars. Infrared imaging of the cluster from 1–4 μm shows that $\sim$60–80% (depending on the reddening vector for the region) of the Trapezium Cluster stars show excess 3.4 μm emission typical of the inner edge of a disk (McCaughrean *et al.* 1996; Kenyon & Hartmann 1995). Mid-IR imaging of a few select sources confirms that these excesses extend to 12 μm, and that the optical-IR SEDs can be fit with a standard disk model (Hayward & McCaughrean 1997). The implication then is that disks are common in the cluster, but we must keep in mind that 2–10 μm emission only probes material within 1–2 AU of the central star: a canonical circumstellar disk capable of producing a planetary system similar to our own (30 AU radius) would need to be much larger.

2.2. DETECTION AS IONIZED PROPLYDS

The core of the Trapezium Cluster is dominated by the eponymous Trapezium OB stars, the most massive of which, θ^1Ori C, is an O6 star responsible for ionizing the Orion Nebula, as well as a number of compact ($\sim$1 arcsec, 500 AU diameter) nebulae discovered through ground-based optical and radio imaging (Laques & Vidal 1979; Garay *et al.* 1987; Churchwell *et al.* 1987; Felli *et al.* 1993), and since studied in great detail with the HST (O'Dell *et al.* 1993; Stauffer *et al.* 1994; O'Dell & Wen 1994; Bally *et al.* 1995; O'Dell & Wong 1996).

The mass of these nebulae can be crudely estimated by converting the radio continuum flux into a mass-loss rate ($\sim 10^{-7}\ M_\odot$/yr), and multiplying by the age of the cluster ($\sim 10^6$ yrs) to give a reservoir of $\sim 0.1\ M_\odot$ (Churchwell *et al.* 1987; Stauffer *et al.* 1994). Since optical-IR stars are visible in almost every one of these nebulae (McCaughrean & Stauffer 1994; O'Dell & Wong 1996), the underlying reservoir of material must be in the form of a disk (Churchwell *et al.* 1987). Accordingly, these sources are thought to be externally ionized proto-planetary disks or "proplyds" (O'Dell *et al.* 1993). Initial models of such sources are able to reproduce some of the key observational features (Henney *et al.* 1996; Johnstone *et al.* 1996).

The disks are not generally directly observed in the proplyds; their presence is inferred by contrasting the mass and size of the reservoir with the visibility of the central star. However, the reservoir mass may be much lower than presumed. First, the mass-loss rate is calculated by assuming that the ionized gas is flowing freely away: if it is actually trapped, the mass-loss rate will be lower. Second, the timescale over which a proplyd is exposed to the ionizing radiation from θ^1Ori C may be much shorter than 10^6 yrs: the OB stars may have formed later than the surrounding low-mass stars and may be only $\sim 10^5$ yrs old, while the 1–2 km/s velocity dispersion of the cluster will cycle the low-mass stars in and out of the core on timescales of 10^5 yrs. If the mass-loss rate and duration of exposure to ionizing radiation were both an order of magnitude lower, a reservoir mass as small as 0.001 $M_\odot$ would be inferred, more consistent with direct observations at millimetre wavelengths which indicate masses $\sim 0.005\ M_\odot$ or lower (Mundy *et al.* 1996; Lada, personal communication). Such low reservoir masses would not necessarily invalidate the conclusion that they are in the form of disks, since the optical visibility of the central stars places a stringent constraint on the

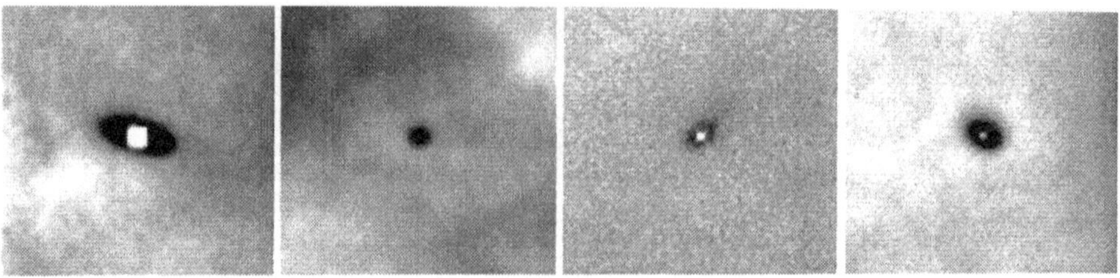

Figure 1. HST images of four silhouette disks in the Orion Nebula. From left to right: Orion 218-354, 167-231, 121-1925, and 183-405. All are seen against the bright Hα background of the nebula. Each panel is 4.1 arcsec or 1800 AU square, *cf.* the 60 AU diameter of the Solar System.

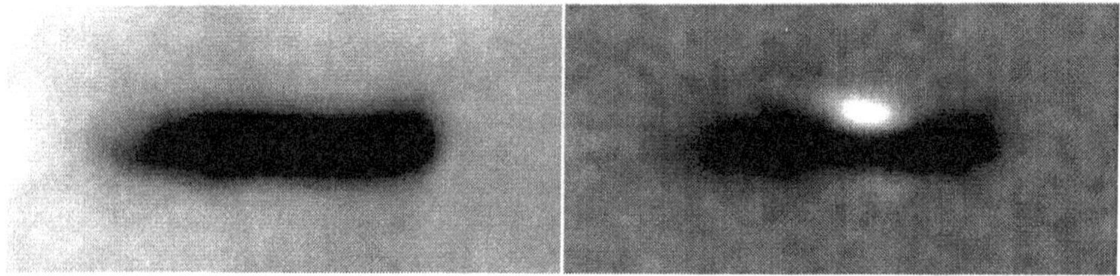

Figure 2. HST images of the Orion 114-426 near edge-on disk. The right panel shows the disk clearly silhouetted against the nebular Hα background. The left image shows an F547M continuum image, in which faint reflection nebulae are seen above and below the plane of the disk. Each panel is 1800×900 AU in size.

geometry. However, these disks would be an order of magnitude less massive than the minimum mass solar nebula, and thus unlikely to form planetary systems similar to our own. Therefore, although the ionized proplyds probably contain disks, it is not clear if they are true proto-planetary disks: more accurate mass estimates are needed via millimetre interferometry.

2.3. DETECTION AS SILHOUETTES

In special cases, a disk may be imaged *directly* as a silhouette against the bright background emission of the Orion Nebula, and a small number of such sources have been found (Figures 1 and 2; McCaughrean & O'Dell 1996). These sources seem to be too far from the OB stars to be strongly ionized, although they are still likely bona fide members of the Trapezium Cluster. The silhouettes show a range in shapes, from circular to elliptical to cigar-shaped, consistent with a family of disks seen at various orientation angles from face-on to near edge-on. These disks range from 50–1000 AU in diameter, and each case, a young (1–3 Myr), low-mass (0.3–1.5 $M_{\odot}$) central star is seen either directly or, in Orion 114-426, indirectly via faint polar reflection nebulae above and below the plane of the near edge-on disk.

Since the silhouette disks are resolved, their structure can be probed directly, at least at the edges. Figure 3 shows model disks placed in silhouette against a bright background, and their major axis profiles compared to that of one source, Orion 183-405. The first point to notice is that the edge of the disk *is* resolved, *i.e.*, a model disk with an abrupt edge results in a profile that is sharper than observed. Conversely however, the observed profile is sharper than predicted for standard disk models, with surface densities $\Sigma \propto r^{-p}$ and p=0.75–1.5, depending on the mass of the disk relative to the central star. The observed profile is in fact most consistent with a truncation at some radius followed by an exponential decay beyond.

It is possible that this exponential edge is simply an effect of disk evolution, independent of environment: time-dependent models predict the development of transition zones between an inner quasi-equilibrium disk and an outer infalling envelope (Saigo & Hanawa 1997), or between disk material that is predominantly spiralling inward, and that which is moving outward, carrying away excess angular momentum (Hartmann, personal communication). It is also possible that the disks are truncated by their local environment, *i.e.*, a hot H II region full of stars. First, the ionizing radiation and winds from the OB stars could evaporate and ablate the disks at some radius as presumably occurs for the ionized proplyds. Second, pressure equilibrium will be established at some radius (and thus density) between the cold, high-density disk material and the hot, low-density H II region. Third, a rotating star-disk system moving through the H II region would induce compression and stripping on the leading and trailing edges of the disk respectively, thus leading to truncation and distortion. Finally, interactions between pairs of star-disk systems moving within the cluster could truncate the disks (Clarke & Pringle 1991; Heller 1995; Larwood *et al.* 1996). Indeed, recent models predict just the sort of exponential truncation seen in the silhouette disks (Hall *et al.* 1996). To clarify which mechanisms are responsible for disk truncation, further high-resolution multi-wavelength surface density profiles obtained with the HST will have to be compared with models incorporating parameters including the distance from θ^1Ori C, H II region density and temperature, local stellar density, *etc.*

3. Star formation in the M 16 elephant trunks

The now familiar HST image of M 16, the Eagle Nebula, shows molecular cloud cores being photo-evaporated by OB stars in the NGC 6611 cluster to form tall pillars or "elephant trunks" (Hester *et al.* 1996). Small-scale crenellations on the pillars were dubbed "evaporating gaseous globules", or EGGs, an acronym which also implies that these globules are the hatching sites of young stars. If low-mass stars are indeed forming in these EGGs,

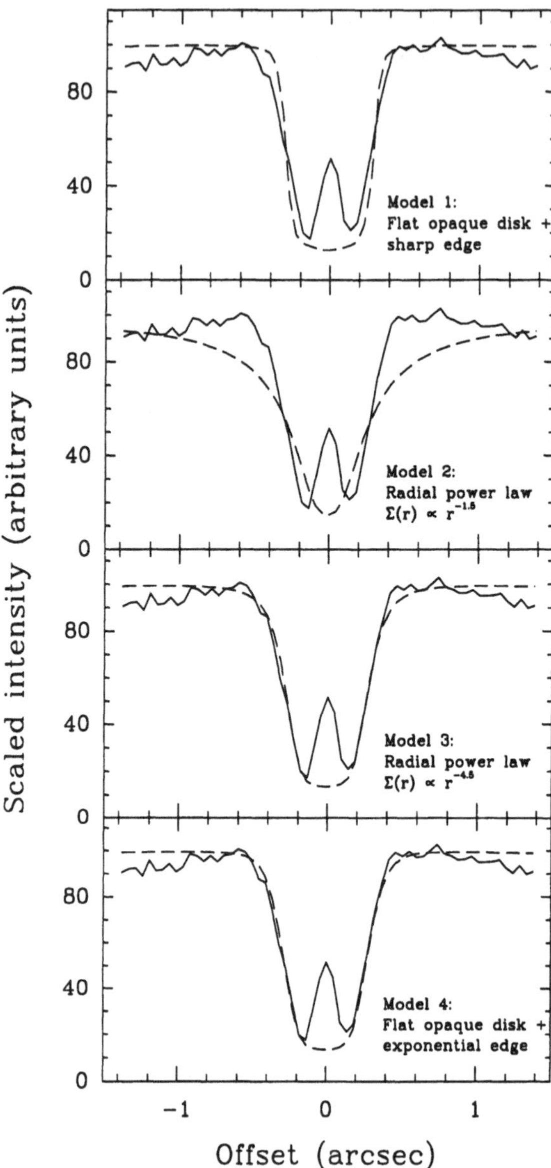

Figure 3. A major axis intensity profile across an [O III] image of the Orion 183-405 silhouette, including the central star (solid curve). Four disk models are superimposed (dashed curves): the model disk is thin, tilted at 45°, and convolved with a model HST+WFPC2 PSF to match the data. Model 1 is a completely opaque disk with sharp edges, and is too sharp to fit the observed disk edge. Model 2 has a canonical disk model power-law surface density profile with $\Sigma(r) \propto r^{-1.5}$: now the model profile is much too shallow. Model 3 has a steeper power-law dependence, $\Sigma(r) \propto r^{-4.5}$, and while the fit is good, such a steep power-law is non-physical. Finally, Model 4 is an opaque central disk modified to have an exponentially decaying edge: this model fits well, and is physically the most plausible.

there might be considerable ramifications, since the massive stars would apparently be responsible for terminating infall onto the low-mass stars, and thus determining the stellar masses and evolution of circumstellar disks. However, only limited evidence for stars embedded in the EGGs was presented by Hester *et al.* (1996): their HST continuum images showed one or two EGGs to have stars near their tips, while a few more coincidences were found in the near-IR images of Hillenbrand *et al.* (1994). However, the relatively poor sensitivity of the latter data limited the search to higher-mass stars ($>3\,M_{\odot}$), and the coarse image sampling (1.3 arcsec/pixel) made it difficult to be sure of the one-to-one correspondence between an IR source and an EGG: at near-IR wavelengths, the M 16 field is heavily contaminated by background field stars in the inner galactic bulge.

Clearly, more sensitive higher-resolution near-IR imaging is required. This has been attempted using a ground-based adaptive optics system (Currie *et al.* 1996), and images will also be taken with the HST under the NICMOS GTO program. However, both systems have a limited FOV (typically $\sim 20 \times 20$ arcsec), allowing only targeted observations of specific interesting EGGs to be carried out. This makes it difficult to carry out a statistically unbiased survey of all the EGGs, and equally importantly, prevents a comparably deep wide-field survey of the background population, a vital part of assessing field star contamination in this very crowded region.

We have carried out a multi-colour near-IR imaging survey of a 13×13 arcmin region of M 16, covering the elephant trunks and a large surrounding area, including the nearby NGC 6611 cluster and the dense field star population in the galactic bulge beyond. Only the central section, corresponding to the WFPC2 FOV, is shown in Figure 4. The integration time of 2 minutes each at J, H, and K' gave 5σ point source detection limits of $19^{\mathrm{m}}.2$, $18^{\mathrm{m}}.1$, and $17^{\mathrm{m}}.5$ respectively. The corresponding physical limits are complicated to define, since they are a product of the distance to M 16, the mass and age of a young star, and the extinction being penetrated. Roughly speaking however, $K'=17^{\mathrm{m}}.5$ is equivalent to a 0.5 Myr old, $0.5\,M_{\odot}$ star at 2.0 kpc, seen through $A_V \sim 30^{\mathrm{m}}$. Assuming a typical diameter for an EGG of 1 arcsec or 2000 AU, this extinction corresponds to a constant uniform volume density of $\sim 2 \times 10^{6}\,\mathrm{cm}^{-3}$, appropriate for a small dense cloud core. We have made a careful search for near-IR stars associated with the optical EGGs, aided by an accurate astrometric link between the two images (0.06 arcsec RMS). In the 73 EGGs of Hester *et al.* (1996), we have identified only 8 near-IR sources, *i.e.*, roughly 10% of the EGGs have stars along their line-of-sight. Some of these are chance alignments with field stars: our measurements of the background field density indicate that two or three of the alignments will be coincidental. Therefore, roughly 7% of the EGGs are physically associated with low-mass stars down to $0.5\,M_{\odot}$.

Figure 4. The elephant trunks in M 16 at near-IR wavelengths. The image is a black-and-white rendering of a true-colour *JHK'* composite that only partly captures the almost transparent appearance of the elephant trunks and the dominant population of background field stars. The image covers roughly 1.5×1.5 pc at the 2.0 kpc to M 16. The data were obtained using Omega-Prime on the Calar Alto 3.5-m telescope in June 1996. With a 1024×1024 pixel HgCdTe array and an image scale of 0.4 arcsec/pixel, the field is 6.7×6.7 arcmin: only a 2.5×2.5 arcmin extract is shown here.

Assuming a standard Miller-Scalo IMF, roughly one-third of all stars have masses $0.5\,M_\odot$ or greater, and thus our survey could be missing two-thirds of an embedded low-mass population. Correcting for this, it is plausible that some 20% of all the EGGs might contain young stars.

Thus, the EGGs are indeed plausible sites of star formation, but only a minority of the EGGs appear to have been able to form stars before being disrupted. An interesting question is whether or not the stars have disks, following the suggestion by Hester *et al.* (1996) that the same ionization and winds from the OB stars in NGC 6611 responsible for illuminating and eroding the elephant trunks will also destroy any disks as soon as the young stars are exposed. In this regard, we are reminded of the ionized proplyds in Orion, and indeed, Hester *et al.* (1996) have suggested that the EGGs and proplyds are similar objects, simply viewed at different angles. This is a tricky issue to address here, but we can at least make two points. First, the EGGs and proplyds are *different* in one important regard: well over 90% of all proplyds in Orion *do* have associated near-IR stars, which cannot be said of the EGGs. Second, disks *do* appear to have survived in the ionized proplyds at least on small scales, as inferred from their near-IR excesses.

Nevertheless, the effects of environment are again found to be important in M 16: the OB stars ablate the columns, revealing the small population of embedded stars 'prematurely', and thus may play a role in determining the stars' masses (Hester *et al.* 1996). However, the limited WFPC2 FOV covered only a small and selective part of the region. The NGC 6611 cluster, centred ~ 1 pc to the north-west, contains thousands of low-mass stars that may have formed relatively quiescently *before* the OB stars turned on and disrupted their surroundings: in the elephant trunks, we may only be seeing 'secondary' star formation around the edges of the main cluster.

4. Summary and prospects

In the limited space available, we have only been able to briefly mention the effects of the external environment on star formation in the Orion and Eagle nebulae. The ionizing radiation and strong winds from the OB stars, as well as dynamical interactions in the dense cluster environment itself, appear to shape the evolution of low-mass stars and their circumstellar disks, truncating the reservoir for further accretion onto a star, thus perhaps setting its mass, and possibly destroying disks before planets have had a chance to form. Until recently, observations of such phenomena have been difficult, and it is only now possible to study the more distant star-forming regions in detail with the advent of high-spatial resolution optical, infrared, and millimetre instrumentation. The next generation of instrumentation including NICMOS on the HST, adaptive optics on 8-m class telescopes, SIRTF, FIRST, the NGST, and large sub-millimetre arrays promises further important insights. At the same time, theoretical work is beginning to take on the difficult task of understanding the complex energetic effects of high-mass stars on their low-mass neighbours, the effects of competition and interactions between the low-mass stars, and the cumulative effects on disks and planets around those stars.

Acknowledgements: I would like to thank my collaborators, including John Bally, Bob O'Dell, John Stauffer, John Rayner, Hans Zinnecker, and Tom Hayward. I would also like to thank Jeff Hester for enlightening discussions and Bo Reipurth, Claude Bertout, *et al.* for organising the meeting.

References

Allen, L. E. 1996, PhD thesis, University of Massachussetts-Amherst
Bally, J., Devine, D., & Sutherland, R. 1995, *Rev. Mex. A. A. (Serie de Conferencias)*, 1, 19
Clarke, C. J., & Pringle, J. E. 1991, *Mon. Not. R. astr. Soc.*, 249, 584
Currie, D., Kissell, K., Shaya, E., Avizonis, P., Dowling, D., & Bonaccini, D. 1996, ESO Messenger, 86, 31
Felli, M., Taylor, G. B., Catarzi, M., Churchwell, E. B., & Kurtz, S. 1993, *Astr. Ap. Suppl.*, 101, 127
Garay, G., Moran, J. M., & Reid, M. J. 1987, *Ap. J.*, 314, 535
Gomez, M., Hartmann, L. W., Kenyon, S. J., & Hewett, R. 1993, *A. J.*, 105, 1927
Hall, S. M., Clarke, C. J., & Pringle, J. E. 1995, *Mon. Not. R. astr. Soc.*, 278, 303
Hayward, T. L., & McCaughrean, M. J. 1997, *A. J.*, 113, 346
Heller, C. H. 1995, *Ap. J.*, 455, 252
Henney, W. J., Raga, A. C., Lizano, S., & Curiel, S. 1996, *Ap. J.*, 465, 216
Herbig, G. H. & Terndrup, D. M. 1986, *Ap. J.*, 307, 609
Hester, J. J., *et al.* 1996, *A. J.*, 111, 2349
Hillenbrand, L. A. 1997, *A. J.*, in press
Johnstone, D., Hollenbach, D., Storzer, H., Bally, J., & Sutherland, R. 1996, BAAS, 189, #49.12
Kenyon, S. J., & Hartmann, L. 1995, *Ap. J. Suppl.*, 101, 117
Lada, E. A., Evans, N. J., II, DePoy, D. L., & Gatley, I. 1991, *Ap. J. (Letters)*, 371, 171
Lada, E. A., Strom, K. M., & Myers, P. C. 1993, in *Protostars and Planets III*, eds. E. H. Levy & J. I. Lunine, (Tucson: Univ. of Arizona Press), p245
Laques, P., & Vidal, J.-L. 1979, *Astr. Ap.*, 73, 97
Larwood, J., Nelson, R. P., Papaloizou, J. C. B., & Terquem, C. 1996, *Mon. Not. R. astr. Soc.*, 282, 597
Levy, E. H. & Lunine, J. I. 1993, editors of *Protostars & Planets III*, (Tucson: Univ. of Arizona Press)
McCaughrean, M. J., & Stauffer, J. R. 1994, *A. J.*, 108, 1382
McCaughrean, M. J., Rayner, J. T., Zinnecker, H., & Stauffer, J. R. 1996, in *Disks and Outflows around Young Stars*, eds. S. V. W. Beckwith, J. Staude, A. Quetz, & A. Natta, (Heidelberg: Springer), p33
McCaughrean, M. J., & O'Dell, C. R. 1996, *A. J.*, 111, 1977
Mundy, L. G., Lada, E. A., & Looney, L. 1995, *Ap. J.*, 452, 137
O'Dell, C. R., Wen, Z., & Hu, X. 1993, *Ap. J.*, 410, 696
O'Dell, C. R., & Wen, Z. 1994, *Ap. J.*, 436, 194
O'Dell, C. R., & Wong, S.-K. 1996, *A. J.*, 111, 846
Prosser, C. F., Stauffer, J. R., Hartmann, L., Soderblom, D. R., Jones, B. F., Werner, M. W., & McCaughrean, M. J. 1994, *Ap. J.*, 421, 517
Saigo, K., & Hanawa, T. 1997, in *Low Mass Star Formation from Infall to Outflow*, eds. F. Malbet & A. Castets, IAU Symposium 182, p247
Stauffer, J. R., Prosser, C. F., Hartmann, L., & McCaughrean, M. J. 1993, *A. J.*, 108, 1375
Strom, K. M., Strom, S. E., & Merrill, K. M. 1993, *Ap. J.*, 412, 233
Zinnecker, H., McCaughrean, M. J., & Wilking, B. A. 1993, in *Protostars and Planets III*, eds. E. H. Levy & J. I. Lunine, (Tucson: Univ. of Arizona Press), p429

TWO-WIND INTERACTION MODELS OF THE PROPLYDS IN THE ORION NEBULA

W. J. HENNEY AND S. J. ARTHUR
Instituto de Astronomía, UNAM, Unidad Morelia,
J. J. Tablada 1006, 58090 Morelia, Michoacán, México

Abstract. Many low-mass stars in the Orion nebula are associated with very compact ($\simeq 1$ arcsec) emission knots, known variously as proplyds, PIGs or LV knots. Some of these knots are teardrop-shaped, with "tails" pointing away from the massive star θ^1 Ori C, which is the principal exciting star of the nebula. We discuss models of such knots, which invoke the interaction of the fast stellar wind from θ^1 Ori C with a transonic photoevaporated flow from the surface of an accretion disk around a young low-mass star. We review previous analytic work and compare the results of the model with the observed brightnesses, morphologies and emission line profiles of the knots, as well as presenting new results from numerical hydrodynamical simulations.

1. Introduction

The proplyds are bright compact emission line knots, with sizes of order 0.5–2.0 arcseconds, that are found in the inner region of the Orion nebula (Laques and Vidal, 1979; Garay et al., 1987; Churchwell et al., 1987; Felli et al., 1993; O'Dell et al., 1993; O'Dell and Wen, 1994; McCaughrean, 1997) and nearly all of which contain an embedded low-mass star (Meaburn, 1988; McCaughrean and Stauffer, 1994). Many of the proplyds show a head/tail morphology, in which the tail points away from the star θ^1 Ori C, the most massive star of the Trapezium cluster. Emission line spectroscopy of the proplyds in the [O III] 5007Å line (Massey and Meaburn, 1993; Massey and Meaburn, 1995; Henney et al., 1997) show a bright central core with full width half maximum (FWHM) of $\simeq 50\,\mathrm{km\ s^{-1}}$, together with faint wings extending out to $> 100\,\mathrm{km\ s^{-1}}$ from the line center.

B. Reipurth and C. Bertout (eds.), Herbig–Haro Flows and the Birth of Low Mass Stars, 561–570.

The obvious explanation for these objects is that the radiation and stellar wind from θ^1 Ori C is interacting with the circumstellar material around the low-mass stars. In our models we assume that this circumstellar material is in the form of an optically thick, geometrically thin accretion disk. The effect of the radiation from θ^1 Ori C will be to ionize the material in the disk, which will produce a photoevaporated flow away from the disk and *towards* the ionizing source. Hence, to produce the tails pointing away from θ^1 Ori C, this flow must somehow be confined and redirected. Two possible candidates for this confinement mechanism are the exciting star's radiation pressure and the ram pressure of its stellar wind. However, only the second of these is feasible, as is shown in § 2. In § 3 a simple analytic model of the resultant two-wind interaction is briefly described and the successes and failures of the model in explaining observed properties of the proplyds are outlined in § 4. In § 5, preliminary results of numerical hydrodynamic simulations are presented, which remove some of the arbitrary assumptions of the model. Complications such as the possible existence of a neutral photodissociated flow are critically discussed in § 6.

2. Confinement Mechanisms — Radiation vs. Ram Pressure

The gas pressure at the base of the ionized flow can be calculated simply by equating the numerical flux, F_0, of Lyman continuum (Ly-c) photons arriving at the ionization front (IF) with the numerical flux of newly-ionized ions entering the photoevaporated wind:

$$F_0 = n_0 \, u_0 \ , \tag{1}$$

where n_0 is the ion number density and u_0 the ion velocity at the base of the wind. This leads to the following expression for the gas pressure

$$P_{\rm gas} \equiv \mu m_{\rm H} n_0 \, c_0^2 = \mu m_{\rm H} c_0 F_0 \, / \mathcal{M}_0 \ , \tag{2}$$

where μ is the mean atomic mass ($\simeq 1.3$), $m_{\rm H}$ is the mass of hydrogen, c_0 is the sound speed in the ionized gas ($\simeq 12\,{\rm km\ s^{-1}}$) and $\mathcal{M}_0$ is the Mach number at the base of the flow, which will be of order 1–2 (Dyson, 1968; Kahn, 1969; Bertoldi, 1989).

The *unattenuated* ionizing flux is given by $F_\star = \dot{S}_* / 4\pi d^2$, where $\dot{S}_*$ is the stellar ionizing photon rate, for which estimates vary between $7 \times 10^{48}\,{\rm s^{-1}}$ (Panagia, 1973) and $3 \times 10^{49}\,{\rm s^{-1}}$ (Bertoldi and Draine, 1996), and d is the distance of the proplyd from the exciting star. However, at the distances of the proplyds from θ^1 Ori C, most of this flux is used up in maintaining the ionization state of the photoevaporated flow against recombination. With the assumption that $F_0 \ll F_\star$, one can write

$$F_\star e^{-\tau_0} = f(\tau_0) \, n_0^2 \alpha_{\rm B} r_{\rm d} \ , \tag{3}$$

where τ_0 is the dust absorption optical depth of the flow, α_B is the Case B recombination coefficient (2.6×10^{-13} cm^3 s^{-1}), r_d is the disk radius and $f(\tau_0) \simeq (3+\tau_0)^{-1}$ depends slightly on the assumed geometry (Henney et al., 1996). Assuming $\tau_0 = 0$, one finds (Eqs. 1 and 3) that the *percentage* of ionizing photons reaching the ionization front is

$$\beta_{\%} = \frac{100\,F_0}{F_\star} = 1.5\,\mathcal{M}_0\,d_{17}\,r_{15}^{-1/2}\,\dot{S}_{49}^{-1/2}\,, \tag{4}$$

where d_{17} and r_{15} are the proplyd distance and disk radius measured in units of 10^{17} cm ($\simeq 0.3$ pc) and 10^{15} cm ($\simeq 66$ au) respectively and $\dot{S}_{49}$ is the stellar ionizing photon rate in units of 10^{49} s^{-1}. Allowing for the effects of dust makes little difference to this estimate.

The ionizing radiation pressure from θ^1 Ori C that acts on the photoevaporated flow can be written as

$$P_{\rm rad} = \frac{F_\star \langle h\nu \rangle}{c}\,, \tag{5}$$

where c is the speed of light, $F_\star$ is the *unattenuated* ionizing flux from θ^1 Ori C, and $\langle h\nu \rangle$ is the mean energy of ionizing photons absorbed in the flow ($\simeq 13.6$eV). Hence,

$$\frac{P_{\rm rad}}{P_{\rm gas}} \simeq \frac{h\nu_0 F_\star \mathcal{M}_0}{\mu m_{\rm H} c_0 c F_0} = 0.033\,\mathcal{M}_0\,\beta_{\%}^{-1} = 0.022\,d_{17}^{-1}\,r_{15}^{1/2}\,\dot{S}_{49}^{1/2}\,. \tag{6}$$

This ratio is always significantly less than unity, therefore the ionizing radiation pressure is **incapable** of confining the photoevaporated flow. If there were enough dust opacity at the base of the flow, then it is conceivable that the non-ionizing radiation from θ^1 Ori C (FUV, optical) may make a significant contribution to the radiation pressure. However, the bolometric luminosity of θ^1 Ori C is only ~ 3 times its Ly-c luminosity, so the above conclusion is unchanged and radiation pressure falls an order of magnitude short of the thermal pressure even for the closest proplyds ($d_{17} \simeq 0.5$).

Turning now to the ram pressure, $P_{\rm hyd}$, of the stellar wind from θ^1 Ori C, this will be given by

$$P_{\rm hyd} = \rho_{\rm w} v_{\rm w}^2\,, \tag{7}$$

where $\rho_{\rm w}$ and $v_{\rm w}$ are respectively the stellar wind density and velocity. Since the wind is radiation-driven, one would expect the ratio $P_{\rm hyd}/P_{\rm rad}$ to be of order unity and, using the observed parameters of θ^1 Ori C (Howarth and Prinja, 1989; Panagia, 1973), one finds (Henney et al., 1996) that this is indeed the case. However, although the radiation pressure must act on the base of the wind where the majority of the recombinations occur, the ram pressure need not do so, but will act on the surface of contact between

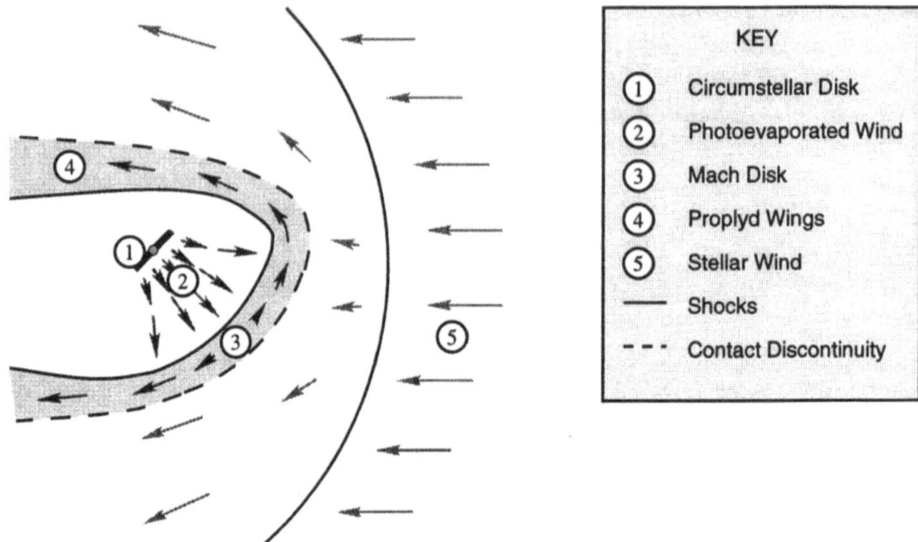

Figure 1. Schematic diagram of the two-wind interaction model. Ionizing photons from θ^1 Ori C drive a photoevaporated wind from one face of the circumstellar disk, which interacts supersonically with θ^1 Ori C's stellar wind.

the evaporated flow and the stellar wind, wherever that may be. Hence, the photoevaporated gas will flow divergently away from the disk until its pressure (reduced by geometric dilution) falls to that of $P_{\rm hyd}$, at which point it can be confined by the stellar wind.

3. Analytic Two-Wind Models

The analytic model (Henney et al., 1996) depends chiefly on the dimensionless parameter $\lambda \equiv P_{\rm gas}/P_{\rm hyd}$. From the discussion of the previous section, it is evident that $\lambda > 1$, in which case the photoevaporated flow, which is initially mildly supersonic (Dyson, 1968; Kahn, 1969; Bertoldi, 1989), will begin to flow freely away from the disk. It is assumed that the streamlines are straight and that the initial flow diverges with a half-opening angle of 45°. If the velocity remained constant, the density would fall as $(1+z)^{-2}$, where z is the height above the disk in units of the disk radius, but a pressure gradient causes the flow to accelerate.

The flow will shock at the point where its pressure has fallen to that of $P_{\rm hyd}$, which occurs at a distance

$$D \simeq \frac{1.19\,(\ln\lambda)^{1/4}\lambda^{1/2}}{\cos^2\theta_0}\,r_{\rm d} = 4\text{–}20\,r_{\rm d}\ , \qquad (8)$$

where θ_0 is the inclination angle of the disk normal with respect to the direction to θ^1 Ori C and, for the second equality, $\lambda = 10$–100 is assumed. The shock will be radiative, so can be treated as isothermal, although the radiation from the shock itself makes a negligible contribution to the proplyd luminosity. A shock will also form in the wind from θ^1 Ori C, but this will be non-radiative, hence the assumption of ram pressure balance used to derive equation 8 is not strictly valid (see § 5).

A thin, almost flat, layer (Mach disk) of shocked photoevaporated flow material forms parallel to the circumstellar disk and gas flows outwards along this layer, reaching a velocity at the edge of

$$v_\ell \simeq 3.8(\ln \lambda)^{1/2} c_0 = 70\text{–}100 \,\mathrm{km\ s^{-1}} \,. \tag{9}$$

The gas is then swept back by the wind of θ^1 Ori C to form the proplyd wings and tail. Figure 1 illustrates the components of the model in a schematic form.

For reasonable values of λ, the photoevaporated disk wind is the brightest component of the model (with a luminosity $\simeq 0.5\lambda^{1/2}$ times that of the Mach disk plus tail) and also the smallest, leading to a core-halo morphology (Henney et al., 1996, Fig. 11).

4. Comparison with Observations

The ensemble properties of the proplyds are quite well reproduced by the analytic model. The models show good agreement with the observed trends of proplyd size and luminosity vs. distance from θ^1 Ori C, the former increasing and the latter decreasing (McCullough et al., 1995; Henney et al., 1996, Figs. 9 and 10). The implied circumstellar disk radii are between 20 and 60 au ($r_{15} \simeq 0.3$–1.0). These correlations, however, are rather insensitive to the details of the model.

The morphologies of individual proplyds are compared with model predictions in Figure 2 (a larger sample is given in Fig. 14 of Henney et al., 1996), where it can be seen that the models successfully reproduce single and double tails, both of which are observed (O'Dell and Wen, 1994; O'Dell and Wong, 1996; Johnstone et al., 1996). However, the double tails may be merely the result of absorption in the core of the tail, which is not consistent with the models as they stand. The crescent head observed in many proplyds would correspond to the Mach disk in the models, but this is rather problematic since the models predict that this should be less bright than the photoevaporated wind component (§ 3), which is not the case for most proplyds, although dust absorption at the base of the wind ($\tau \simeq 0.5$–1 for the closest proplyds) would alleviate this problem (this is included in the fit to OW 158–327).

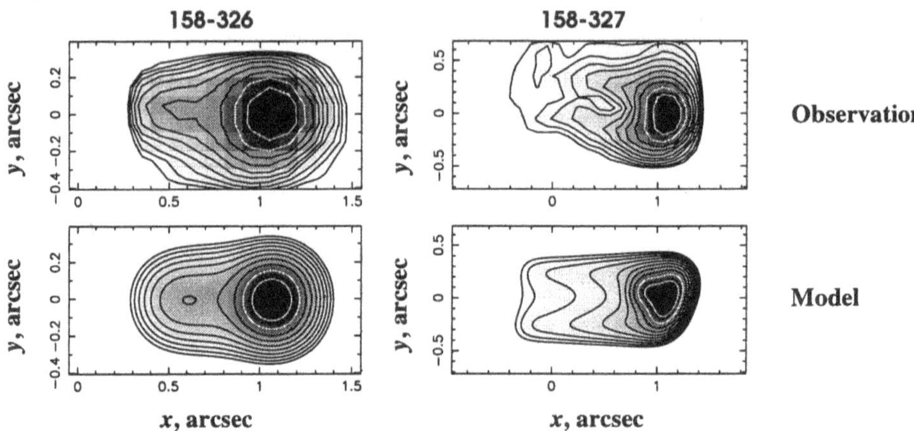

Figure 2. Comparison of morphological predictions of the two-wind interaction model with observations. Contours and greyscales show Hα surface brightness (logarithmic scale) for *HST* observations and model images. The interval between successive contours is $2^{1/2}$.

Detailed comparisons between model predictions and high resolution [O III] 5007Å spectra of individual proplyds are presented in Henney et al. (1997). The evaporated wind produces the bright core of the line, with width of a few times the sound speed in the ionized gas, while the Mach disk and tail produce the high-velocity line wings that are observed. However, in order to reproduce the $\simeq 100\,\mathrm{km\ s^{-1}}$ widths of the line wings seen in LV 5 (OW 158–323) and LV 2 (OW 167–317), values of $\lambda = 50$–200 are required, which are 3–4 times larger than those found in fitting the morphologies of the same objects (Henney et al., 1996).

5. Hydrodynamical Simulations

Figure 3 shows the results of an example numerical simulation of the two-wind interaction (Henney and Arthur, 1997). In this simulation, the circumstellar disk (oriented vertically in the figure) is inclined by 45° with respect to the direction of the stellar wind from θ^1 Ori C (other parameters are described in the figure caption). The transfer of ionizing radiation in the photoevaporated flow is not calculated self-consistently in the models, but the boundary conditions are assumed constant over the disk surface and are taken from the analytic model. Also, the simulation parameters correspond to a rather small value of λ ($\simeq 8$), since a larger value would require an unfeasibly large computational grid.

The main differences with respect to the analytic calculation are due to the relaxation of two arbitrary assumptions of the model. Firstly, the pho-

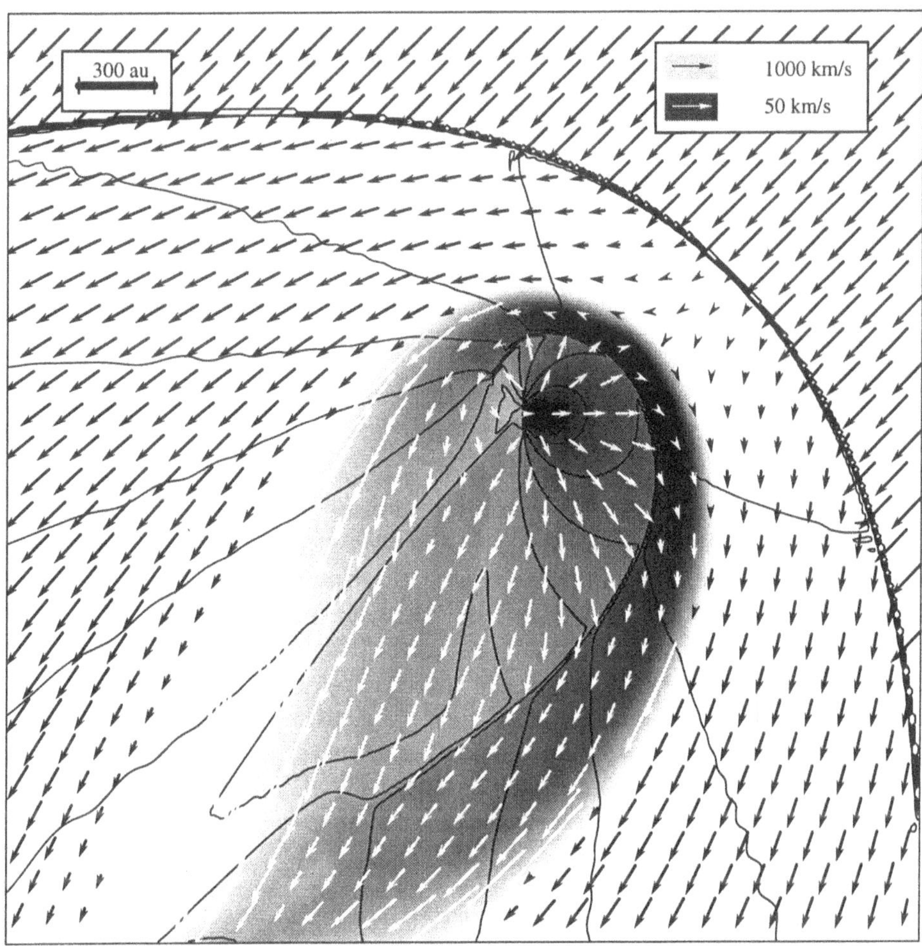

Figure 3. Hydrodynamical simulation of the two-wind interaction, calculated in 2D slab symmetry with a grid size of 300 × 300 cells. Greyscale shows the gas density while contours show the gas pressure (both logarithmic scale). Arrows show gas velocity. Photoevaporated disk material (white arrows) has an isothermal equation of state. Stellar wind material (black arrows) has an adiabatic equation of state. Note the different velocity normalizations of the black and white arrow lengths. The parameters of the model are $r_d = 30$ au, $\theta_0 = 45°$, $n_0 = 2 \times 10^5 \mathrm{cm}^{-3}$, $n_w = 1 \mathrm{cm}^{-3}$, $u_w = 1000\,\mathrm{km\ s}^{-1}$.

toevaporated flow is calculated self-consistently, instead of being assumed to follow straight streamlines with opening angle 45°. Pressure gradients in the mildly supersonic wind in fact cause the flow to diverge increasingly with distance from the disk. Secondly, the (non-radiative) shock in the stellar wind is treated properly, instead of merely assuming ram pressure balance between the two winds as was done in the analytic model. Both these factors affect the shape of the dense layer of shocked photoevaporated wind material, which is more curved than in the analytic calculation. The flat "Mach disk" of the analytic model is no longer apparent and there is no sharp distinction between the Mach disk and tail. The gas velocities reached in the shocked photoevaporated wind are also slightly smaller than in the analytic model. Unfortunately, the use of slab symmetry, which allows the asymmetric interaction to be modelled in two dimensions, means that it is not possible to produce emission maps or spectra from the simulations. Nonetheless, the morphology and kinematics of the simulations are rather similar to those of the analytic calculation modulo the differences noted above. The arguments for and against the two-wind interaction model are hence little affected.

6. Discussion and Speculation

Despite the success of the two-wind interaction models in reproducing the observed morphologies and kinematics of the proplyds, some problems remain. In particular, the apparent absorption in the tails of some objects and the [O III]/IR arcs (Bally et al., 1995; Hayward et al., 1994) that are seen between the closer proplyds and θ^1 Ori C are both hard to explain with the two-wind model. An alternative view (Bally et al., 1995; Johnstone et al., 1996) is that the disk evaporation is controlled by non-ionizing FUV photons, with the ionization front occurring away from the disk. The proplyd morphology would then be determined by the shape of the ionization front. The hydrodynamic interaction with the stellar wind from θ^1 Ori C would, on this view, still occur, but farther out in the flow, perhaps producing the [O III]/IR arcs. This model has had most success in explaining the characteristics of HST 10 (OW 182–413), but this object does not seem to be typical of proplyds as a class (in particular, its tail does not point exactly away from θ^1 Ori C and it may not contain a central star). Johnstone et al. (1996) compare the mass-loss rates from photodissociated and photoionized disk winds and conclude that the former will dominate for all proplyds. However, they *assume* that the warm ($\simeq 1000\,\mathrm{K}$) photodissociated gas will be able to flow freely away from the disk at its sound speed ($\simeq 2.5\,\mathrm{km\ s^{-1}}$), but this is not necessarily the case.

If one allows, for the sake of argument, that a free-flowing photodissociated wind, with a density at its base of $10^6 n_{\mathrm{n},6}\,\mathrm{cm^{-3}}$, is initially established, then, once the ionizing radiation from θ^1 Ori C is switched on, an R-type ionization front will be driven rapidly into the flow. For proplyds closer than

$$d'_{17} \simeq 0.55\, n_{\mathrm{n},6}^{-1}\, r_{15}^{-1/2}\, \dot{S}_{49}^{1/2} \,, \tag{10}$$

the flow will be immediately ionized all the way down to the disk. For proplyds further away, the ionization front undergoes a transition to D-type some way out from the disk, at which point its progress will slow and it will begin to drive a shock into the atomic flow. The density of newly ionized gas n_0 will adjust itself to that given by equations 1 and 4 of § 2, but with r_d replaced by the radius of the ionization front. The pressure of the ionized gas will be roughly 20 times that of neutral gas of the same density, so that for proplyds closer than

$$d''_{17} \simeq 20\, n_{n,6}^{-1}\, r_{15}^{-1/2}\, \dot{S}_{49}^{1/2} \,, \tag{11}$$

the shock will reach the surface of the disk before stalling, hence quenching the neutral flow in a time of 30–500 years. In proplyds farther away than d''_{17} from θ^1 Ori C, the shock will stall at a distance $z_0 r_d$ from the disk, where

$$1 + z_0 \simeq 0.136\, n_{n,6}^{2/3}\, d_{17}^{2/3}\, r_{15}^{1/3}\, \dot{S}_{49}^{-1/3} \,. \tag{12}$$

The chief uncertainty in the estimates of d'_{17} and d''_{17} is the density at the base of the neutral flow. However, taking the parameters of HST 10 (Johnstone et al., 1996), $r_{15} = 1.3$, $z_0 = 2.3$, $d_{17} = 5$, and assuming $n_{n,6}$ is the same for all proplyds, one finds that $d'_{17} \simeq 0.03\, r_{15}^{-1/2}$ and $d''_{17} \simeq r_{15}^{-1/2}$. No proplyds are observed with $d_{17} < d'_{17}$ but a substantial fraction have $d_{17} < d''_{17}$, although the exact number depends on the distribution of disk radii. This can only be determined directly for the dark silhouette disks (McCaughrean and O'Dell, 1996), which show $r_{15} = 0.4$–7.6, but the bright proplyds are likely to have smaller disks ($r_{15} \simeq 0.1$–1.0, Henney et al., 1996; Johnstone et al., 1996). Hence, roughly half of all bright proplyds will **not** have an extended neutral evaporated flow.

The real situation is undoubtedly much more complicated than portrayed above (c.f. Bertoldi and Draine, 1996), but the basic argument, that the neutral flow must have a higher pressure than the overlying ionized flow in order to exist, should remain valid. A further problem for the neutral flow is gravity: the escape speed from the circumstellar disk will equal the sound speed in the neutral gas at a disk radius of $r_{esc,15} \simeq 2.1 M_\star$, where $M_\star$ is the mass of the central star in solar masses ($\simeq 0.1$–2, McCaughrean and Stauffer, 1994). Hence, except for the proplyds with the lowest mass central stars, gravity will dominate the dynamics of the photodissociated region.

Note that the argument against radiation pressure in § 2 applies a fortiori to a neutral photodissociated flow since its pressure would have to be larger than the ionized flow. However, the confinement problem could be circumvented if it were maintained that the material in the tail, instead of having been redirected from an initial flow towards θ^1 Ori C, was instead part of a flow from the back side of the disk, possibly driven by the diffuse radiation field. This could also explain the absorption seen in the core of some tails, but whether the flow would be dense enough for this is not clear. Alternatively, the tails could be formed from the remnants of a dense slow wind from a massive star (Sutherland et al., 1997).

In conclusion, the two-wind interaction model has had qualified success in explaining the observed properties of the proplyds closer to θ^1 Ori C. Various discrepancies remain, however, and further work is necessary both in extending this model and in developing alternatives.

Acknowledgements: We are very grateful to Alex Raga, Susana Lizano and John Meaburn for their contributions to the work discussed in this paper. WJH also acknowledges useful discussions with John Dyson, Dave Hollenbach, Mark McCaughrean and Bob O'Dell. Financial support for this research has been provided by DGAPA-UNAM under project number IN105295 and by Cray Research, Inc.

References

Bally, J., Devine, D. and Sutherland, R. 1995, In: S. Lizano and J. M. Torrelles (eds.): *Circumstellar Disks, Outflows and Star Formation*, RMAA (Serie de Conferencias) 1, 19
Bertoldi, F. 1989, ApJ 346, 735
Bertoldi, F. and Draine, B. T. 1996, ApJ 458, 222
Churchwell, E., Felli, M., Wood, D. O. S. and Massi, M. 1987, ApJ **321**, 516
Dyson, J. E. 1968, Ap&SS 1, 388
Felli, M., Taylor, G. B., Catarzi, M., Churchwell, E. and Kurtz, S. 1993, A&AS 101, 207
Garay, G., Moran, J. M. and Reid, M. J. 1987, ApJ, 314, 535
Hayward, T. L., Houck, J. R. and Miles, J. W. 1994, ApJ, 433, 157
Henney, W. J. and Arthur, S. J. 1997, in preparation
Henney, W. J., Meaburn, J., Raga, A. C. and Massey, R. 1997, A&A in press
Henney, W. J., Raga, A. C., Lizano, S. and Curiel, S. 1996, ApJ 465, 216
Howarth, I. D. and Prinja, R. K. 1989, ApJS 69, 527
Johnstone, D., Hollenbach, D., Storzer, H., Bally, J. and Sutherland, R. 1996, BAAS 189, 4912 (Available at http://www.cita.utoronto.ca/~johnston/orion/abstract.html)
Kahn, F. D. 1969, Physica 41, 172
Laques, P. and Vidal, J. L. 1979, A&A 73, 97
Massey, R. M. and Meaburn, J. 1993, MNRAS 262, L48
Massey, R. M. and Meaburn, J. 1995, MNRAS 273, 615
McCaughrean, M. J. 1997, this volume
McCaughrean, M. J. and O'Dell, C. R. 1996, AJ, 111, 1977
McCaughrean, M. J. and Stauffer, J. R. 1994, AJ, 108, 1382
McCullough, P. R., Fugate, R. Q., Christou, J. C., Ellerbroek, B. L., Higgins, C. H., Spinhirne, J. M., Cleis, R. A. and Moroney, J. F. 1995, ApJ 438, 394
Meaburn, J. 1988, MNRAS 233, 791.
O'Dell, C. R. and Wen, Z. 1994, ApJ 436, 194
O'Dell, C. R., Wen, Z. and Hu, X. 1993, ApJ 410, 696
O'Dell, C. R. and Wong, S. K. 1996, AJ 111, 8460
Panagia, N. 1973, AJ 78, 929
Sutherland, R. S., Hartquist, T. W., Bally, J. and Dyson, J. E. 1997, in preparation

CONFERENCE SUMMARY

A. C. RAGA
Instituto de Astronomía, UNAM, 04510 México, D. F., México

Abstract. The present paper reviews the talks and poster papers presented at the IAU 182 symposium. The papers are divided into different subjects, highlighting their contributions to the field of outflows from young stars. The observational contributions were notable because of the results recently obtained with new instruments (e. g., with radio interferometers and space based observatories). The theoretical contributions were noted by the increased understanding of the production of collimated winds from young stars, and by numerical and analytic work on the propagation of the outflows.

1. Introduction

The IAU 182 symposium is the first conference dedicated to Herbig-Haro objects since the symposium on HH objects in Mexico in 1983. This unusually long time-span between conferences (at least, for astronomical subjects) has resulted in an extremely exciting meeting. In particular, very interesting new observational results have been presented, driven by new instrumental developments : mm interferometers (Plateau de Bure, Nobeyama, Owens Valley, BIMA, and now even the VLA !), ISO, infrared cameras and spectrographs on many optical telescopes and HST. Notably, the UV wavelength range has been mostly absent from the meeting.

From the theoretical point of view, this meeting has been the ground for discussions on detailed models of HH jets and molecular outflows, as well as models for the production and collimation of MHD winds from young, low mass stars. The interaction between the observers and the theoreticians that occurred during this meeting was very positive, and hopefully will result in important progress in the near future.

B. Reipurth and C. Bertout (eds.), Herbig–Haro Flows and the Birth of Low Mass Stars, 571–586.

The present paper is organized as follows. Section 2 describes a more or less complete tabulation of the poster papers and talks presented at the meeting. Section 3 discusses some of the important observational work presented at the meeting, and section 4 discusses part of the theoretical work. Finally, section 5 is dedicated to conjectures about the (near) future evolution of the field of outflows from young stars.

2. Papers presented at the meeting

One of the most interesting novelties in the organization of the meeting was the production of a "Poster Proceedings" book, edited by Malbet and Castets (1997), which was distributed to the participants upon arrival. This book is most useful and, together with these Proceedings should provide a more or less complete coverage of the work presented at the meeting.

Tables 1-5 list all of the poster papers and talks given at the meeting, divided into a number of different categories. These categories have been chosen so as to highlight the contributions of the papers to the field of outflows from young stars, which was the main topic of the conference.

I would like to apologize to all second authors for the fact that in order to save space both the talks and the poster papers are listed exclusively under the name of the first author (in the "first author" column of the tables). The talks are labeled with a "t" in the "reference" column, and the poster papers are labeled with the page number of the corresponding paper in the poster proceedings book (Malbet and Castets 1997). Some authors presented both talks and posters on similar subjects, and the two contributions are then listed together in the same line (with both a "t" and a page number in the "reference" column). Most of the papers corresponding to the talks can of course be found in these Proceedings.

Finally, there were a number of "late poster papers" presented at the meeting. Some of these are included in tables 1-5, with an "lp" entry in the "reference" column. However, a probably high percentege of the late poster papers has not been included in the tables.

I have risked to introduce a column of "comments", in which I have tried to describe the different papers in 2-5 words. It is of course impossible to do a fair description of any of the papers in such a way. These comments only represent the things that came to my mind as particularly interesting from looking at the papers in the poster proceedings book and at my notes. I feel that they might be useful as a guide to the subjects covered in the meeting.

The observational papers have been divided into :

Outflows : subdivided into categories depending mostly on wavelength range, instrumental technique, or particular instruments,

T Tauri, FU Ori stars : subdivided into variability, spectra and disks/ envelopes around T Tauri stars,

Cores and globules (no subdivision)

Star formation : mostly includes papers which do not fit in the other categories,

Orion M 42 objects : papers on OMC 1 HH objects, proplyds, etc. .

The theoretical papers are listed together under the **Theory** heading, which is subdivided into : wind production/collimation; jet/outflow models; disks, envelopes; collapse; chemistry of outflows; and other topics.

The division into these categories is of course quite arbitrary, since many papers include many different observational results, and many are also both observational and theoretical. This is particularly true of the longer talks, which in many cases covered a broad range of subjects.

If we take the classification of the papers in tables 1-5 seriously, we would conclude that out of a total of 166 papers, 127 were observational, and 39 were theoretical. Of the 166 theoretical and observational papers, 87 were directly concerned with HH objects and/or molecular outflows (keeping the 3:1 observational to theoretical paper ratio of the total sample). The majority of the remaining papers are concerned with observations and theory of the near environment of the outflow sources (35 papers) and cloud cores (18 papers).

Tables 1-5 do not include a very interesting late poster paper by Reipurth, which predicted the appearance of a new version of his HH object catalogue (Reipurth 1994). This new version will include approximately twice as many objects as the previous 1994 edition, and I personally look forward to having a copy soon !

The following two sections describe a few of the more notable aspects of the observational and theoretical results (respectively) presented at the meeting. The discussion is centred on the work directly related to the formation and propagation of outflows from young stars.

3. Observational results

The probably most exctiting reported observations of outflows were the ones resulting from the new, space based instruments ISO and HST. The ISO observations (see table 1) have opened up a previously little observed region of the spectrum of outflows, showing lines of [O I] and [C II], and in some objects also lines of different molecules.

The HST images presented in this meeting (see table 2) give a surprising new view of HH jets, with the well known chains of knots resolved into small bow-shaped structures, shocks in sharp bends, and less well defined structures in other cases. Proper motion determinations have been carried out, and new knots have been seen to appear close to the outflow sources.

Some of the HST images of HH objects (e. g., of large heads of jets) appear to finally have spatially resolved the cooling regions behind HH shock waves in an unambiguous way. This has been shown in a quite dramatic way, e. g., in the Hα–[S II] subtracted images of Reipurth and Heathcote (1997). Very interesting structures are also observed in some bow shocks, where a complex fragmentation appears to be occurring.

The mm interferometrical observations of molecular outflows presented at the meeting (see table 1, and Guilloteau *et al* 1997) are possibly even more interesting than the ISO and HST observations. We now see that at least some molecular outflows are highly collimated, and have structures that look like shells at the walls of elongated cavities, or like well collimated jets. In some objects, both structures coexist, with higher velocities in the jet-like, central component. Also, different molecules are seen in each component. These observations represent a dramatic improvement in spatial resolution over previous observations of molecular outflows, and should provide a very strong impulse for further modelling.

Also, single dish mm observations of outflows appear to be giving most interesting results (see table 1, and Bachiller and Pérez Gutiérrez 1997). Information is now available of the spatial distribution of the emission of several different molecules along some outflows. These observations appear to finally provide similar constraints on molecular outflow models to the ones that have been available for many years for HH objects.

New radio continuum interferometric observations of the jet-like structures associated with the sources of HH objects were reported in 6 papers. Rodríguez (1997) pointed out that now there are $\sim$ 80 HH sources detected in radio continuum, $\sim$ 30 of which have been resolved spatially (mostly into jet-like structures). Maps with the new 7 mm detectors at the VLA provide an unprecedented (for HH flows) resolution of $\sim 0.04''$. Ray *et al* (1997) reported the detection of polarization offset from the position of an outflow source (from which the magnetic field can be estimated). Such observations clearly provide important constraints for wind collimation mechanisms.

It is interesting to note that the same number of papers has been presented on ground based IR imaging/spectroscopy as on optical imaging and spectroscopy. This is a measure of the extremely lively progress of IR observations of HH objects (see table 2, and Eislöffel 1997). Most of the IR observations have been restricted to imaging, but we also see that some spectroscopic work is being done (giving both line ratios and radial velocities, see, e. g., Noriega-Crespo 1997). These observations of course focus on studying the very rich IR spectrum of the H_2 molecules.

These IR observations show H_2 structures associated with optically detected HH objects or with optically invisible "H_2 jets". A clearly important issue is to see to what extent these two kinds of objects correspond to similar phenomena. Also, it appears to be most interesting to compare the spatial distribution of the H_2 emission with the emission of other molecules. This has been done, e. g., by Chandler & Richer (1997) for the case of HH 211 (see also Zinnecker *et al.*, 1997; Guilloteau *et al* 1997).

The optical imaging papers focused on searches for new HH objects, and on the study of the so-called "superjets" (Bally and Devine 1997). On the first subject, the work presented in the conference shows that searches of HH objects based on optical imaging still yield a number of interesting new outflows. On the latter subject, the 3 papers presented at this meeting (see table 2) show exciting identifications of jets extending over many parsecs. Even out to these unprecedently large distances from the source, a very striking jet/counterjet symmetry appears to be preserved.

The paper of Böhm and Goodson (1997) presented a new analysis of spectrophotometric data of HH objects, concluding that the presently available shock wave models appear to have quite surprising problems. The remaining papers on optical spectroscopy focussed on 20 to 300 km/s resolution spectroscopy with long-slit or spectro-imaging spectrographs (the lower resolutions corresponding to the latter technique). Four papers (Solf 1997; Mundt 1997; Corcoran and Ray 1997; Lavalley *et al.* 1997) were dedicated to the analysis of the so-called "microjets", extending for distances of only 2-4$''$ from the source. These results appear to be highly interesting in that they provide some of the most important constraints available for wind production/collimation models.

Such constraints are also provided by the observations of the properties of T Tauri stars, which show evidence for the co-existence of inflow and outflow in many objects (Hartmann 1997; Calvet 1997; Edwards 1997). The observation of variabilities in timescales ranging from hours to decades (see table 3), as well as the determination of magnetic fields (Guenther 1997), also provide important constraints on wind collimation mechanisms as well as on models of jet structures resulting from variabilities of the source.

Finally, a number of observational papers on dense cores were presented. These studies ranged from radio wavelenghts (e. g., Terebey and Padgett 1997) to the IR (Gredel 1997), and included in many cases detailed comparisons with kinematic+radiative transfer models. From such comparisons, a quite clear picture of the collapse and rotation of the clouds can be obtained. In very intriguing papers, André (1997) and Fuller and Ladd (1997) set out the basis of how to obtain an evolutionary sequence for cores and very young stars.

4. Theoretical results

The theoretical papers were to some extent dominated by the wednesday session on MHD wind models. Camenzind, Heyvaerts, Kwan, Pudritz and Shu participated in a round-table discussion chaired by Hartmann, and there were also a number of talks during the day. Let me give a short description of the main issues that arose in the talks of the participants :

- Camenzind (1997) presented a model of a wind from a strongly magnetized star, and discussed the possible (though non-intuitive !) importance of the light cylinder in such flows. He also showed predictions of line-profiles from the wind model (based on the introduction of a somewhat arbitrary heating term),
- Heyvaerts (1997) gave a very technical talk which described the general properties of asymptotic wind solutions (for large distances from the source). He also talked about the possibility of having a "polar boundary layer" to prevent on-axis crushing. This of course is an intriguing idea in the context of observations of HH flows,
- Kwan (1997) talked about coronae around disks, and presented kinematic models of these regions,
- Pudritz and Ouyed (1997) presented numerical simulations of a jet ejected from a magnetized disk. These models assumed an arbitrary mass loading from the disk into the wind. Interesting predictions were made of the conditions under which the jets are stable or unstable, the latter jets having chains of knots,
- Shu and Shang (1997) presented the X-wind model, which appears to be the more complete model to date. A complete solution of the dynamical structure of the wind has been computed by Shu and collaborators, as well as an analytic solution for the asymptotic regime of large distances from the source. Shu also presented X-wind driven thin shell solutions into a stratified environment.

There were also a number of other talks on MHD wind production and collimation (see table 4), describing important aspects of these flows. One of the important remaining problems of wind models appears to be the determination of the temperature structure of the base of the wind, which is most interesting as it would enable a comparison with the T Tauri microjets. This issue was addressed in two papers (Martin 1997; Paatz and Appl 1997). A different issue was raised by Frank and Mellema (1997), who discussed the collimation of a gasdynamic wind through the interaction with a stratified environment. Finally, the paper of Ultchin *et al.* (1997) was very interesting, studying the possibility of producing a "blobby" outflow through the interaction between a star with "starspots" and an accretion disk.

There were a number of interesting papers on the dynamics of HH jets and molecular outflows. Many issues were described in the review of Cabrit *et al* (1997). The talks of Stone (1997) and Massaglia *et al* (1997) as well as several poster papers (see table 4) addressed the effects of Kelvin-Helmholtz instabilities in radiative jets. Smith *et al* (1997) presented numerical simulations of jets from variable sources, which included a simplified treatment of the chemistry.

More detailed discussions of the chemistry (but of course for dynamically more simple situations !) were presented by Hollenbach (1997) and Pineau des Forêts *et al* (1997). It would be most interesting to see if it is possible to implement a comparable chemical sophistication in 2 or 3D calculations. Also intriguing was the talk of Taylor (1997) who stressed the importance of the radiative field of HH objects in evaporating grain mantles and modifying the chemistry of the surrounding environment.

There were a number of papers on modelling disks (e. g., Bell and Chick 1997), envelopes (e. g., Berger and Ménard 1997) and collapse of cores (e. g., Hanawa *et al.* 1997, see also table 5). Particularly interesting was the paper of Toropin *et al.* (1997) which presented a full MHD calculation of an infalling envelope leading to the production of an outflow.

Finally, I would like to commend the papers of Hartquist and Dyson, Henney and Arthur, Wilkin *et al* and Gvaramadze (1997) for having described very clear analytic work. Interestingly, these papers are all on the "other topics" section of table 5. I hope that this does not show that we will not see any future purely analytic papers directly concerned with outflows from young stars !

5. Final speculations

As a final section of this paper, let me give a series of speculations about what will possibly be interesting ways forward in the field of outflows from young stars. These possibilities are mostly not my original ideas, as they attempt to reflect some of the most fruitful discussions between observers and theoreticians that occurred during the meeting.

1. Do the outflows directly reflect the time-history of the ejected wind ?

From a purely naive point of view, one would think that the amazingly symmetric structures observed in the two lobes of many outflows directly imply an origin for the structures that is associated with time-dependent phenomena of the ejection. Furthermore, the strong symmetry appears to imply that the interaction with the environment (which is observed to be highly inhomogeneous) cannot have a strong effect on the propagation of the outflow. These arguments would lead us to believe that, at least in some of the "superjets", the outflows behave in an approximately ballistic way until many parsecs away from the source !

Of course a possibility would be that the first "outflow event" clears out the inhomogeneous environment, leaving behind a less dense, more homogeneous medium. Successive ejection events could then travel into and interact with this homogeneous medium and still preserve symmetries between the two lobes even if the flow is not ballistic. Appropriate numerical simulations should be able to tell us whether or not this alternative scenario is a working possibility.

A clearly also important point is to attempt to directly relate the observed structures in HH jets with the variability of the source. To this effect, both the HST images of regions close to HH sources and the studies of variabilities of T Tauri stars appear to be most important.

2. What do the outflows with high lobe-to-lobe asymmetries tell us ?

There are certainly objects which show very strong lobe-to-lobe asymmetries. It is then of course quite straightforward to interpret these asymmetries as evidence for interaction with the (highly inhomogeneous) environment. However, as we have seen in this meeting, even radio continuum jets with lengths of a few arcseconds sometimes show important asymmetries. If these asymmetries are indeed due to strong interactions with the environment, it appears to be difficult to reconcile this with the striking symmetries observed on much larger scales. Are we then actually seeing asymmetric ejections from the sources ? Could the same source produce both asymmetric (for small knots) and symmetric (for large working surfaces) ejection events ?

3. Does the cross section of an outflow reflect the cross section of the ejected wind ?

From direct radial velocity, proper motion and line flux measurements it is possible to obtain an idea of the cross section of a collimated flow. It would be most interesting to see more work in this direction being carried out in the future. The important question then is whether or not this cross section directly reflects the angular dependence of the wind ejected from the source. The work presented in this meeting of the interaction of initially isotropic winds with a stratified circumstellar environment appears to indicate that the answer might be negative.

4. The microjets

Of course, the situation is more hopeful for the microjets (described in this meeting) or the thermal radio jets observed to extend for a few arcseconds away from the source. Do these "short" jets preserve more faithfully the structure of the wind ejected from the source ? At what distances from the source does the lateral interaction with the environment produce shocks that modify the jet cross section ?

5. HST

The new observations of HH objects with the HST provide a set of data which is ideal for comparisons with numerical simulations. It is now both important and straightforward to produce numerical models with more sophisticated treatment of the ionization structure in order to obtain predictions that can be directly compared to the high resolution HST data. In the near future, we will undoubtedly be seeing "second epoch" HST images, which will give a most intriguing picture of the time-evolution of HH flows.

6. Molecular outflows, chemistry

In this conference, we have seen an explosion in the available information about molecular outflows. A large increase in the spatial resolution, together with an increase in the number of different lines and chemical species that are studied provide suitable data for very detailed comparisons with models. Clearly, more sophisticated chemical/dynamical models will have to be produced to interpret this data. Possibly, in this way the real nature of the relation between HH jets and molecular outflows (two winds, lateral, or head entrainment) will finally be established.

7. Instabilities, turbulence

We have seen several papers about K-H instabilities in radiative jets (a field in which little effort had been done in the past). Do these instabilities lead to fully turbulent jets ? How does the turbulence affect the propagation of the jet ? Can we empirically determine whether or not HH jets and/or molecular outflows are turbulent ?

To conclude, I would like to point out that these seven points are limited to the topics that interest me more closely (as evidenced by the fact that I was able to follow the discussions) and clearly do not reflect the wide variety of discussions which occurred during this most fruitful meeting.

Acknowledgements: I would like to thank Claude Bertout and Bo Reipurth for inviting me to give this talk.

References

André, P. 1997, these Proceedings.
Bachiller, R., Pérez Gutiérrez, M. 1997, these Proceedings.
Bally, J., Devine, D. 1997, these Proceedings.
Bell, K.R., Chick, K.M. 1997, these Proceedings.
Berger, J.-P., and Ménard, F. 1997, in *Low mass star formation - from infall to outflow*, Poster Proceedings of the IAU Symposium n° 182, eds. F. Malbet and A. Castets (Observatoire de Grenoble), p. 201.
Böhm, K. H., Goodson, A.P. 1997, these Proceedings.
Cabrit, S., Raga, A.C., Gueth, F. 1997, these Proceedings.
Calvet, N. 1997, these Proceedings.
Camenzind, M. 1997, these Proceedings.

Chandler, C. J., and Richer, J. S. 1997, in *Low mass star formation - from infall to outflow*, Poster Proceedings of the IAU Symposium n° 182, eds. F. Malbet and A. Castets (Observatoire de Grenoble), p. 76.
Corcoran, M., and Ray, T. P. 1997, in *Low mass star formation - from infall to outflow*, Poster Proceedings of the IAU Symposium n° 182, eds. F. Malbet and A. Castets (Observatoire de Grenoble), p. 82.
Edwards, S. 1997, these Proceedings.
Eislöffel, J. 1997, these Proceedings.
Frank, A., and Mellema, G. 1997, these Proceedings.
Fuller, G., and Ladd, E. F. 1997, these Proceedings.
Gredel, R. 1997, in *Low mass star formation - from infall to outflow*, Poster Proceedings of the IAU Symposium n° 182, eds. F. Malbet and A. Castets (Observatoire de Grenoble), p. 284.
Guenther, E. 1997, these Proceedings.
Guilloteau, S., Dutrey, A., Gueth, F. 1997, these Proceedings.
Gvaramadze, V. 1997, in *Low mass star formation - from infall to outflow*, Poster Proceedings of the IAU Symposium n° 182, eds. F. Malbet and A. Castets (Observatoire de Grenoble), p. 129.
Hanawa, T., Matsumoto, T., and Nakamura, F. 1997, in *Low mass star formation - from infall to outflow*, Poster Proceedings of the IAU Symposium n° 182, eds. F. Malbet and A. Castets (Observatoire de Grenoble), p. 147.
Hartmann, L. 1997, these Proceedings.
Hartquist, T.W., Dyson, J.E. 1997, these Proceedings.
Henney, W.J., Arthur, S.J. 1997, these Proceedings.
Heyvaerts, J. 1997, these Proceedings.
Hollenbach, D. 1997, these Proceedings.
Kwan, J. 1997, these Proceedings.
Lavalley, C., Dougados, C., and Cabrit, S. 1997, in *Low mass star formation - from infall to outflow*, Poster Proceedings of the IAU Symposium n° 182, eds. F. Malbet and A. Castets (Observatoire de Grenoble), p. 147.
Malbet, F., and Castets, A. (eds.) 1997, *Low mass star formation - from infall to outflow*, Poster Proceedings of the IAU Symposium n° 182 (Observatoire de Grenoble).
Martin, S. 1997, these Proceedings.
Massaglia, S., Micono, M., Ferrari, A., Bodo, G., Rossi, P. 1997, these Proceedings.
Mundt, R. 1997, these Proceedings [manuscript not submitted]
Noriega-Crespo, A. 1997, these Proceedings.
Paatz, G. and Appl, S. 1997, in *Low mass star formation - from infall to outflow*, Poster Proceedings of the IAU Symposium n° 182, eds. F. Malbet and A. Castets (Observatoire de Grenoble), p. 303.
Pineau des Forêts, G., Flower, D.R., Chièze, J.-P. 1997, these Proceedings.
Pudritz, R., Ouyed, R. 1997, these Proceedings.
Ray, T. P. *et al* 1997, these Proceedings.
Reipurth, B. 1994, *A general catalogue of Herbig-Haro objects*, electronically published via anon. ftp to ftp.hq.eso.org, directory /pub/Catalogs/Herbig-Haro.
Reipurth, B., Heathcote, S. 1997, these Proceedings.
Rodríguez, L. F. 1997, these Proceedings.
Shu, F., Shang, H. 1997, these Proceedings.
Smith, M., Völker, R., Suttner, G., Yorke, H.W. 1997, these Proceedings.
Solf, J. 1997, these Proceedings.
Stone, J. 1997, these Proceedings.
Taylor, S. 1997, these Proceedings.
Terebey, S., Padgett, D.L. 1997, these Proceedings.
Toropin, Y., Saveljev, V., and Chechetkin, V. 1997, in *Low mass star formation - from infall to outflow*, Poster Proceedings of the IAU Symposium n° 182, eds. F. Malbet and A. Castets (Observatoire de Grenoble), p. 254.

Ultchin, Y., Regev, O., and Bertout, C. 1997, in *Low mass star formation - from infall to outflow*, Poster Proceedings of the IAU Symposium n° 182, eds. F. Malbet and A. Castets (Observatoire de Grenoble), p. 318.
Wilkin, F., Cantó, J., Raga, A.C. 1997, these Proceedings.
Zinnecker, H., McCaughrean, M., and Rayner, J., in *Low mass star formation - from infall to outflow*, Poster Proceedings of the IAU Symposium n° 182, eds. F. Malbet and A. Castets (Observatoire de Grenoble), p. 199.

TABLE 1. Poster papers and talks

Topic	First author	Reference	comments
Outflows			
General			
	Padman	t	cavities, jets, changes in coll., precession
	André	t	evolution of inflow/outflow
Radio (single dish)			
	Cernicharo	t	IRAM, HH 111, CO tube, bullets, momentum estimates
	Bachiller	t	IRAM, spatial progression of chemistry
	Bence	57	JCMT, wind-driven shells
	Dent	88	JCMT, molecules in HH 2
	Gibb	120	JCMT, CO in HH 25, 26
	Hogerheijde	138	JCMT, CO, evolutionary sequence ?
	Ogura	166	SEST, CO in HH 135, 136
Radio (interferometry)			
	Guilloteau	t	PdBI obs. of outflows, cavities, jets, precession
	Rodríguez	t	VLA thermal jets (~ 80), geometries, sources
	Ray	t	Merlin, polariz., determination of B off the source
	Bloemhof	60	VLA continuum of known HH objects, also IR
	Bontemps	63	VLA continuum, new jet (asymm.)
	Cesaroni	73	PdBI, new flow, also single dish+IR
	Chandler	76	VLA, inner SiO jet in HH 211, also JCMT
	Dutrey	101	PdBI, SiO bow shocks
	Girart	123	BIMA, high vel. HCO^+
	Gueth	126	PdBI, superimposed cavities
	Hughes	141	VLA continuum, Cepheus A
	Ladd	144	BIMA, strong changes in position angle
	Martí	160	VLA continuum, proper motions of HH 80-81 jet
	Nagar	163	BIMA, CO sheath of HH 111
	Shepherd	175	OVRO, CO in flow from B star, also IR images
	Wilner	193	VLA continuum, non-thermal emission in jet
	Zhang	195	BIMA, jet-like flow, SiO, CO, also IR (H_2)
	Estalella	266	VLA, NH_3 in quadrupolar outflow
	Wiseman	lp	NH_3 in flows in the Orion Ridge region
ISO			
	Liseau	t	typical HH spectrum : [O I] 63μ, [C II] 158μ
	Ceccarelli	66	LWS spectra of outflow, [O I], CO, H_2O
	Mazzini	lp	spectra of 17 HH objects

TABLE 2. Poster papers and talks, continued

Topic	First author	Reference	comments
IR observations			
	Eislöffel	t	H_2 in embedded flows, line profiles, entrainment
	Noriega-Crespo	t	Excitation of H_2, mixing layers, fluorescence
	Ayala	5	H_2, new flow, also spectra
	Massi	21	H_2, new flow, known HH objects
	Palacios	30	H_2, known HH objects
	Schultz	39	", shocked cloudlet
	Eiroa	103	H_2 close to Serpens radio jet
	Herbst	135	H_2, new flow, also spectra
	Fernandes	109	spectra, contribution of H_2 fluorescence
	Richer	172	images and spectra of RNO 43 precessing jet
	Smith	178	Fabry-Perot, ortho/para H_2 ratios
	Tabone	184	H_2 in HH 24, 135, 136
	Zinnecker	198	point-symmetric H_2 bow shocks in HH 212
	Herbst	215	H_2 Fabry-Perot image of T Tau environment
	Miralles	227	H_2, new flow
	Nisini	lp	FIR spectroscopy of outflow
	Zinnecker	lp	unbiased H_2 jet survey of Orion A
	Quirrenbach	lp	H_2 in T Tauri, adaptive optics
Optical imaging			
	Bally	t	superjets, timescales, symmetries
	Cernicharo	8	new flow, also mm continuum map
	Alten	51	new flows
	Corporon	85	", flow in a cavity
	Devine	91	new flows
	Devine	95	HH 34 superjet, also radial vel.+proper motions
	Hojaev	218	environment of pre-FU Ori star
	Eislöffel	lp	superjets, evidence for precession
Optical spectroscopy			
	Böhm	t	high/low excitation, problems with shock modelling
	Solf	t	micro-jet long-slit spectra, two components, accel.
	Mundt	t	micro-jets in many stars
	Corcoran	82	long-slit, micro-jet, Herbig Ae star
	Fridlund	117	long-slit, 2D coverage, HH 29
	Lavalley	147	2D spectro-imaging of DG Tau micro-jet
	Magakian	158	long slit+ 2D spectro-imaging of R Mon jet
	Movsessian	lp	2D spectro-imaging of Haro 6-5B
HST			
	Reipurth	t	HH 46/47, 111, 34, 80/81, 1/2, Hα−[S II]
	Fridlund	t	L 1551 IRS 5 jet, reflection neb., new knots
	Stapelfeldt	t	HL, XZ, DG, DG B, and T Tauri, HH 30, reflection neb.

TABLE 3. Poster papers and talks, continued

Topic	First author	Reference	comments
T Tauri, FU Ori stars			
Variability			
	Semkov	42	optical spectrophot., possible FU Ori star
	Beskrovnaya	204	polarization, Hα profile
	Pogodin	244	line profiles, outflow/inflow
	Chelli	263	Br γ variability in DF Tau
	Fernández	269	line profiles, Hα, He I, Na I D
	Folha	272	Pa β line profile, infalling material
	Gameiro	275	line profiles, Hα, He I, Na I D
	Castro	278	UV (IUE) continuum variability
	Grankin	281	photometry of WTTS
	Bertout	288	photometry, spot model, acc. luminosity
	Heines	294	linear polarization, wavelenght dependence
	Ibrahimov	297	1981-1995 photometry of FUors
	Reipurth	309	FUor BBW 76, photometry, line profiles
	Smith	315	DR Tau, photometric var. with $\tau \sim 1$ hr
Spectra			
	Hartmann	t	magnetospheres, smaller (1/10) values for $\dot{M}$
	Calvet	t	T_e from blue absorpotion comp. of permitted lines
	Edwards	t	accretion columns, cTTS with short rotation periods
	Guenther	t	Ca II K lines, determination of B, 2-3 kG
	Mennessier	300	Doppler imaging, spots
	Pedrosa	306	classification of Hα profiles
Disks, envelopes			
	Kitamura	t	NMA maps of disks, L 1551 IRS 5, DG and TM Tauri
	Meyer	224	6-12 μ spectra of disk
	Monin	230	1-5 μ adaptive optics, disks in binaries
	Harvey	291	SSV 13, 47-95 μ scans, photometric variab.
	Dutrey	lp	2.7 mm cont. survey of T Tauri disks
	Dutrey	lp	DM Tau and GC Tau disks, mm interf., many molecules
Cores and globules			
	Fuller	t	evol. sequence for cores, time indicators
	Terebey	t	mm interf., infall+rotation, modelling of line prof.
	Choi	11	radio interf., HH 1 source
	Mangum	18	radio, single dish
	Morato	24	"
	Motte	27	"
	Panis	33	"
	Sugitani	45	radio interf.
	Wang	48	IR imaging
	Anglada	54	radio, single dish, sources of outflows
	Olberg	169	CS, DCO^+ in environment of HH 110, 111
	Ward-Thompson	257	ISO maps of pre-stellar core
	Wiesemeyer	260	mm interf. of infalling envelopes, also models
	Gredel	284	HH 100 IRS, observations and models of IR cont.

TABLE 4. Poster papers and talks, continued

Topic	First author	Reference	comments
Star formation			
	Churchwell	t	high mass star formation
	Claussen	t	water masers, kinematics with $10^{-3''}$ resolution
	Giovanetti	14	optical and IR photometry of young stars
	Plazy	36	IR photometry of young stars
Orion M 42 objects			
	McCaughrean	t	proplyds, young low mass stars in trouble
	O'Dell	t	IR fingers, proper motions, large bow shocks
	Everett	106	H_2 fingers in OMC-1
	Persi	238	H_2, [S II] images of IRAS sources
	Tedds	lp	H_2 fingers in OMC-1
	Burton	lp	HH knots in Orion
Theory			
Wind production/coll.			
	Shu	t	X-wind, asymptotic flow, wind-driven shell, X-rays
	Camenzind	t	dynamos in disk, wind from star, emission lines
	Heyvaerts	t	asymptotic wind solutions, on-axis boundary condition
	Appl	t	K-H and current driven instabilities
	Ferreira	113, t	braking of star by outflow, saturation mechanism
	Lery	152, t	slow vs. fast rotators
	Frank	t	gasdynamic wind/environment interaction, radiative
	Martin	t	thermal structure of funnel flows, low temperatures
	Paatz	303	heating/cooling mechanisms in wind
	Shang	312	cosmic rays produced in MHD wind
Jet/outflow models			
	Cabrit	t	acceleration, bubbles, cavities, mix. layers, bow sh.
	Pudritz	t	knots in magnetized jet, currents loop back along bow
	Smith	t	simulations with chemistry, velocity cross section
	Massaglia	t	K-H in radiative, axisym. jets, comparisons with obs.
	Stone	t	K-H in slab and 3D radiative jets, twisting jets
	Rubini	t	crossing shocks in initially overpressured jet
	Bacciotti	t	plasma diagnostics, ion. state of jets, observations
	Chiuderi	79	linear+nonlinear MHD, shock-less
	Leeuwin	150	travelling knots, radiative, axisym.
	Suttner	181	internal working surfaces, H_2 and CO emission
	Thiele	187	adiabatic, 3D with B, nose cones
	Cerqueira	70	radiative, 3D, stability
	Downes	98	radiative K-H, slab jet
	Frank	115	adiabatic K-H with B, axisym.
	Hardee	132	K-H in radiative, 3D jet with B
	Toropin	254	production of MHD jet from collapse onto disk

TABLE 5. Poster papers and talks, continued

Topic	First author	Reference	comments
Chemistry of outflows			
	Hollenbach	t	chemistry in J and C shocks, instabilities, masers
	Taylor	t	"searchlight chemistry" in clumps near HH objects
	Pineau des Forêts	7, t	C shocks with chem., time-dep.: no critical points
Disks, envelopes			
	Bell	t	FU Ori disks, thermal instab., values of α
	Kwan	t	disk coronae, interpretation of line profiles
	Berger	201	polarisation modulations, circumbinary envelopes
	Chick	207	rad. transfer in envelopes, nested grid
	D'Alessio	210	irradiated accretion disks
	Men'shchikov	221	dusty, reprocessing disk around HL Tau
	Nakamura	235	thin disk simulation, formation of binaries
	Pickett	241	3D, ad./isoth. disks, formation of structure
	Ultchin	318	ejection of diamagnetic blobs from disks
Collapse			
	Hanawa	212	similarity sol. for collapse to form disk+star
	Nakamura	232	axisym. simulation of collapse with B
	Saigo	247	similarity, collapsing core+disk solution
	Tomisaka	250	accretion rate in collapsing core with B
Other topics			
	Hartquist	t	high mass stars, compact H II regions
	Henney	t	models for proplyds, two-wind interactions
	Wilkin	t	wind bow shocks, radiated energy, momentum transfer
	Gvaramadze	129	fingers in OMC-1
	Mac Low	155	stability of 3D C-shocks
	Wilkin	190	stellar wind b. s., non-uniform media, analytic

LIST OF PARTICIPANTS

Allain, Stéphanie; Obs. de Grenoble, *allain@obs.ujf-grenoble.fr*
André, Philippe; CEA-SAp, Saclay, *andre@sapvxg.saclay.cea.fr*
Anglada, Guillem; Inst. de Astron., UNAM , Mexico, *guillem@astroscu.unam.mx*
Appl, Stefan; Obs. de Strasbourg, *appl@astro.u-strasbg.fr*
Ayala, Sandra; Inst. de Astron., UNAM, Mexico, *sayala@astroscu.unam.mx*
Bacciotti, Francesca; Univ. di Firenze, Arcetri, *fran@arcetri.astro.it*
Bachiller, Rafael; Obs. Astron., Alcala de Henares, *bachiller@oan.es*
Bacmann, Aurore; CEA-SAp, Saclay, *bacmann@discovery.saclay.cea.fr*
Bally, John; Center for Astroph. and Space Astron., *bally@nebula.colorado.edu*
Bell, K. Robbins; NASA/Ames & UC Santa Cruz, *bell@cosmic.arc.nasa.gov*
Bence, Stephen; MRAO, Cambridge, *sjb39@mrao.cam.ac.uk*
Berger, Jean-Philippe; Obs. de Grenoble, *berger@obs.ujf-grenoble.fr*
Bertout, Claude; Institut d'Astrophysique de Paris, *bertout@iap.fr*
Beskrovnaya, Nina; Pulkovo Obs., St-Petersbourg, *beskr@pulkovo.spb.su*
Bloemhof, Eric; Calif. Inst. of Technology, Pasadena, *eeb@astro.caltech.edu*
Böhm, Karl-Heinz; Univ. of Washington, Seattle, *bohm@astro.washington.edu*
Bontemps, Sylvain; Stockholm Obs., *bontemps@astro.su.se*
Burton, Michael; Univ. New South Wales, Sydney, *mgb@newt.phys.unsw.edu.au*
Cabrit, Sylvie; DEMIRM, Obs. de Paris, *sylvie.cabrit@obspm.fr*
Calvet, Nuria; CIDA, Merida, *ncalvet@cfa.harvard.edu*
Camenzind, Max; Landesternwarte, Heidelberg, *mcamenzi@lsw.uni-heidelberg.de*
Castets, Alain; Obs. de Grenoble, *castets@obs.ujf-grenoble.fr*
Ceccarelli, Cecilia; Obs. de Grenoble, *ceccarel@obs.ujf-grenoble.fr*
Cernicharo, José; IEM, Madrid, *cerni@astro.iem.csic.es*
Chalabaev, Almas; Obs. de Grenoble, *chalabae@obs.ujf-grenoble.fr*
Chandler, Claire; MRAO Cambridge, *cjc@mrao.cam.ac.uk*
Chelli, Alain; Obs. de Grenoble, *chelli@obs.ujf-grenoble.fr*
Chick, Kenneth; NASA/Ames, *chick@anarchy.arc.nasa.gov*
Chiuderi, Claudio; Univ. of Firenze, Arcetri, *chiuderi@arcetri.astro.it*
Choi, Minho; BIAA, Taipei, *minho@biaa3.biaa.sinica.edu.tw*
Churchwell, Ed; Univ. of Wisconsin, Madison, *churchwell@astro.wisc.edu*
Clark, Stuart; Univ. of Hertfordshire, *sclark@stour.herts.ac.uk*
Claussen, Mark; NRAO, Socorro, *mclausse@nrao.edu*
Corcoran, Myles; CEA-Sap, Saclay, *corcoran@discovery.saclay.cea.fr*
Curiel, Salvador; Inst. de Astron., UNAM, Mexico, *scuriel@astroscu.unam.mx*
D'Alessio, Paola; Inst. de Astron., UNAM, Mexico, *dalessio@astroscu.unam.mx*

Dent, Bill; Joint Astronomy Centre, Hawaii, *dent@jach.hawaii.edu*
Devine, David; Univ. of Colorado, *devine@starburst.colorado.edu*
Dougados, Catherine; Obs. de Grenoble, *dougados@obs.ujf-grenoble.fr*
Downes, Turlough; Dublin Inst. for Advanced Studies, *tpd@cp.dias.ie*
Dutrey, Anne; IRAM, Grenoble, *dutrey@iram.fr*
Duvert, Gilles; Obs. de Grenoble, *duvert@obs.ujf-grenoble.fr*
Dyson, John; Univ. of Leeds, *jed@ast.leeds.ac.uk*
Edwards, Suzan; FCRAO, Massachusetts, *sedwards@smith.edu*
Eiroa, Carlos; Univ. Autonoma de Madrid, *carlos@xiada.ft.uam.es*
Eislöffel, Jochen; IfA, Hawaii, *jochen@ifa.hawaii.edu*
Estalella, Robert; Univ. de Barcelona, *robert@mizar.am.ub.es*
Everett, Mark; The Ohio State Univ., *everett@payne.mps.ohio-state.edu*
Felli, Marcello; Osservatorio di Arcetri, *felli@arcetri.astro.it*
Fernandes, Amadeu; Center for Astroph., Oporto Univ., *amadeu@astro.up.pt*
Fernandez, Matilde; MPIA, Heidelberg, *matilde@mpia-hd.mpg.de*
Ferreira, Jonathan; Obs. de Grenoble, *ferreira@obs.ujf-grenoble.fr*
Flower, David; Univ. of Durham, *david.flower@durham.ac.u*
Folha, Daniel; Queen Mary & Westfield College, London, *d.f.m.folha@qmw.ac.uk*
Frank, Adam; Univ. of Rochester, *afrank@holly.pas.rochester.edu*
Franqueira, Mercedes; Univ. Complutense de Madrid, *mf@orion.mat.ucm.es*
Fridlund, C.V.M; ESTEC/ESA, Noordwijk, *mfridlun@astro.estec.esa.nl*
Fuller, Gary; UMIST, Manchester, *gfuller@umist.ac.uk*
Gameiro, Jorge; Center for Astroph., Oporto Univ., *jgameiro@astro.up.pt*
Garcia, Paulo; Obs. de Lyon, *pgarcia@obs.univ-lyon1.fr*
Geoffray, Hervé; Obs. de Grenoble, *geoffray@obs.ujf-grenoble.fr*
Giannini, Teresa; IFSI, Frascati, *teresa@vega.ifsi.fra.cnr.it*
Gibb, Andy; Univ. of Kent, Canterbury, *agg@starlink.ukc.ac.uk*
Giovanetti, Philippe; CESR, Toulouse, *giovanetti@cesr.cnes.fr*
Girart, José; Harvard-Smithsonian CfA, Cambridge, *jgirart@cfa.harvard.edu*
Gomez de Castro, Ana; Univ. Complutense de Madrid, *aig@eucmos.sim.ucm.es*
Gougeon, Samuel; ESRF, Grenoble, *gougeon@esrf.fr*
Gredel, Roland; ESO, Santiago, *rgredel@eso.org*
Grewing, Michael; IRAM, Grenoble, *grewing@iram.fr*
Guenther, Eike; TLS, Tautenburg, *guenther@tls-tautenburg.de*
Gueth, Frédéric; IRAM, Grenoble, *gueth@iram.fr*
Guilloteau, Stéphane; IRAM, Grenoble, *guillote@iram.fr*
Gvaramadze, Vasili; Abastumani Astr. Obs., Moscow, *vgvaram@mx.iki.rssi.ru*
Hanawa, Tomoyuki; Nagoya Univ., *hanawa@a.phys.nagoya-u.ac.jp*
Hardee, Philip; Univ. of Alabama, *hardee@venus.astr.ua.edu*
Harder, Stéphane; Obs. de Grenoble, *Stephan.Harder@obs.ujf-grenoble.fr*
Hartmann, Lee; CfA, Cambridge, *hartmann@cfa.harvard.edu*
Harvey, Paul; Univ. of Texas, Austin, *pmh@astro.as.utexas.edu*
Hatchell, Jennifer; UMIST, Manchester, *jjh@ast.phy.umist.ac.uk*
Heines, Anke; Max Planck Research Unit, Jena, *anke@astro.uni-jena.de*
Henney, William; Inst. de Astron., UNAM, Mexico, *raga@astroscu.unam.mx*
Herbst, Tom; MPIA, Heidelberg, *herbst@mpia-hd.mpg.de*
Heyvaerts, Jean; Obs. de Strasbourg, *heyvaert@cdsxb6.u-strasbg.fr*
Hogerheijde, Michiel; Leiden Obs., *michiel@strw.leidenuniv.nl*
Hojaev, Alisher; Ulugbek Astron. Inst., Tashkent, *hojaev@saturn.silk.glas.apc.org*
Hollenbach, David; NASA/Ames, *hollenbach@warped.arc.nasa.gov*

Hughes, Victor; Queen's Univ., Kingston, *hughes@qucdn.queensu.ca*
Kitamura, Yoshimi; ISAS, Sagamihara, Kanagawa, *kitamura@pub.isas.ac.jp*
Kumar, Nanda; Physical Research Laboratory, *nanda@prl.ernet.in*
Kwan, John; UMass, Amherst, *kwan@donald.phast.umass.edu*
Ladd, Edwin; FCRAO, Massachusetts, *ladd@wayback.phast.umass.edu*
Lavalley, Claudia; Obs. de Grenoble, *lavalley@obs.ujf-grenoble.fr*
Leeuwin, Francine; Osservatorio di Arcetri, *leeuwin@arcetri.astro.it*
Lefloch, Bertrand; IRAM, Granada, *lefloch@gra-ux1.iram.es*
Lery, Thibaut; Obs. de Strasbourg, *lery@astro.u-strasbg.fr*
Liseau, René; Stockholm Obs., *rene@astro.su.se*
Lorenzetti, Dario; Osservatorio di Roma, *dloren@coma.mporzio.astro.it*
Mac Low, Mordecai-Mark; MPIA, Heidelberg, *mordecai@mpia-hd.mpg.de*
Magakian, Tigran Yu; Byurakan Astroph. Obs., *tigmag@sci.am*
Malbet, Fabien; Obs. de Grenoble, *malbet@obs.ujf-grenoble.fr*
Mangum, Jeff; NRAO, Tucson, *jmangum@nrao.edu*
Marti, Josep; CEA-SAp, Saclay, *jmarti@discovery.saclay.cea.fr*
Martin, Steven; The Univ. of Chicago, *smartin@jets.uchicago.edu*
Massaglia, Silvano; Univ. di Torino, *massaglia@ph.unito.it*
Massi, Fabrizio; Osservatorio di Roma, *massi@coma.mporzio.astro.it*
McCaughrean, Mark; MPIFR, Bonn, *mjm@mpifr-bonn.mpg.de*
Mellema, Garrelt; Stockholm Obs., *garrelt@astro.su.se*
Men'shchikov, Alexander; Max Planck Res. Unit, Jena, *sascha@astro.uni-jena.de*
Ménard, François; Obs. de Grenoble, *menard@obs.ujf-grenoble.fr*
Mennessier, Catherine; Obs. de Grenoble, *mennessi@obs.ujf-grenoble.fr*
Meyer, Michael; MPIA, Heidelberg, *meyer@mpia-hd.mpg.de*
Miralles, Mari Paz; CfA, Cambridge, *mmiralles@cfa.harvard.edu*
Miranda, Luis; Univ. Complutense de Madrid, *lfm@ucmast.fis.ucm.es*
Moneti, Andrea; VILSPA, Madrid, *amoneti@iso.vilspa.esa.es*
Monin, Jean-Louis; Obs. de Grenoble, *monin@obs.ujf-grenoble.fr*
Montmerle, Thierry; CEA-SAp, Saclay, *montmerle@sapvxg.saclay.cea.fr*
Morata, Oscar; Univ. de Barcelona, *oscar@fareb1.am.ub.es*
Motte, Frédérique; Obs. de Grenoble, *motte@obs.ujf-grenoble.fr*
Movsessian, Tigran ; Byurakan Astroph. Obs., *tmov@helios.sci.am*
Mundt, Reinhard; MPIA, Heidelberg, *mundt@mpia-hd.mpg.de*
Nakamura, Fumitaka; Niigata Univ., *fnakamur@ed.niigata-u.ac.jp*
Neri, Roberto; IRAM, Grenoble, *neri@iram.fr*
Nisini, Brunella; IFSI, Frascati, *bruni@taurus.ifsi.fra.cnr.it*
Noriega-Crespo, Alberto; IPAC, Pasadena, *alberto@ipac.caltech.edu*
O'Dell, Robert; MPIA, Heidelberg, *cro@mpia-hd.mpg.de*
O'Sullivan, Stephen; Dublin Inst. for Advanced Studies, *so@cp.dias.ie*
Ogura, Katsuo; Kokugakuin Univ., Tokyo, *k30943@m-unix.cc.u-tokyo.ac.jp*
Olberg, Michael; Onsala Space Obs., *olberg@oso.chalmers.se*
Padman, Rachael; MRAO Cambridge, *rachael@mrao.cam.ac.uk*
Palacios, Javier; Univ. Autonoma de Madrid, *carlos@xiada.ft.uam.es*
Panis, Jean-François; BIAA, Taipei, *panis@biaa.sinica.edu.tw*
Pedrosa, António; Center for Astroph., Oporto Univ., *apedrosa@astro.up.pt*
Perrier, Christian; Obs. de Grenoble, *perrier@obs.ujf-grenoble.fr*
Persi, Paolo; IAS, Frascati, *persi@saturn.ias.fra.cnr.it*
Pickett, Brian; NASA/Ames, *pickett@cosmic.arc.nasa.gov*
Pineau des Forêts, Guillaume; DAEC, Obs. de Paris, *forets@obspm.fr*

Plazy, Frédéric; Obs. de Grenoble, *plazy@obs.ujf-grenoble.fr*
Pogodin, Mikhail; Pulkovo Obs., St-Petersbourg, *pogodin@pulkovo.spb.su*
Pudritz, Ralph; McMaster Univ., *pudritz@physics.mcmaster.ca*
Raga, Alejandro; Inst. de Astron., UNAM, Mexico, *raga@astroscu.unam.mx*
Ray, Tom; Dublin Inst. for Advanced Studies, *tr@cp.dias.ie*
Regev, Oded; Technion, Haifa, *regev@phastro1.technion.ac.il*
Reipurth, Bo; ESO, Santiago, *reipurth@eso.org*
Richer, John; MRAO, Cambridge, *jsr@mrao.cam.ac.uk*
Rodriguez, Luis Felipe; Inst. de Astron., UNAM, Mexico *luisfr@astroscu.unam.mx*
Rubini, Francesco; Univ. di Firenze, Arcetri, *rubini@arcetri.astro.it*
Saigo, Kazuya; Nagoya Univ., *saigo@a.phys.nagoya-u.ac.jp*
Saraceno, Paolo; IFSI, Frascati, *saraceno@auriga.ifsi.fra.cnr.it*
Schultz, Angie; NASA/Ames, *schultz@ssa1.arc.nasa.gov*
Schuster, Karl-Friedrich; IRAM, Grenoble, *schuster@iram.fr*
Semkov, Evgeni; Inst. of Astron., Sofia, *evgeni@wfpa.acad.bg*
Sepulveda, Inma; Univ. de Barcelona, *inma@fareb1.am.ub.es*
Shang, Hsien; Univ. of California, Berkeley, *shang@astron.berkeley.edu*
Shepherd, Debra; Calif. Inst. of Technology, *dss@astro.caltech.edu*
Shu, Frank; Univ. of California, Berkeley, *shu@vshu.berkeley.edu*
Smith, Kester; Queen Mary College, London, *K.W.Smith@qmw.ac.uk*
Smith, Michael; Univ. of Wuerzburg, *smith@astro.uni-wuerzburg.de*
Solf, Josef; TLS, Tautenburg, *solf@tls-tautenburg.de*
Stanke, Thomas; AIP, Postdam, *tstanke@aip.de*
Stapelfeldt, Karl; JPL, Pasadena, *krs@wfpc2-mail.jpl.nasa.gov*
Stone, James; Univ. of Maryland, *jstone@astro.umd.edu*
Strom, Stephen; UMass, Amherst, *sstrom@donald.phast.umass.edu*
Sugitani, Koji; Nagoya City Univ., *sugitani@nsc.nagoya-cu.ac.jp*
Tabone, Ann-Maree; Univ. of Wollongong, *tabone@davinci.sci.uow.edu.au*
Taylor, Stephen; Univ. College, London, *sdt@star.ucl.ac.uk*
Tedds, Jonathan; Univ. of Edinburgh, *jat@roe.ac.uk*
Terebey, Susan; Extrasolar Research Corp., *terebey@extrasolar.com*
Thiele, Markus; Landesternwarte, Heidelberg, *M.Thiele@lsw.uni-heidelberg.de*
Tomisaka, Kohji; Niigata Univ., *tomisaka@ed.niigata-u.ac.jp*
Toropin, Yuriy; Sternberg Astronomical Inst., Moscow, *toropin@sai.msu.su*
Wang, Jun-Jie; Beijing Astronomical Obs., *wangjj@bao01.bao.ac.cn*
Ward-Thompson, Derek; Royal Obs., Edinburgh, *dwt@roe.ac.uk*
Wiesemeyer, Helmut; MPIFR, Bonn, *hwiesemeyer@mpifr-bonn.mpg.de*
Wilkin, Francis; Univ. of Calif., Berkeley, *fwilkin@astron.berkeley.edu*
Wilner, David; Harvard-Smithsonian CfA, *dwilner@cfa.harvard.edu*
Wiseman, Jennifer; NRAO, Charlottesville, *jwiseman@nrao.edu*
Zhang, Qizhou; Harvard-Smithsonian CfA, *qzhang@cfa.harvard.edu*
Zinnecker, Hans; AIP, Potsdam, *hzinnecker@aip.de*

LIST OF POSTER PAPERS

A. Star Forming Regions

B. Herbig-Haro objects, jets and outflows

C. Circumstellar environments, infall and accretion

D. Central stellar objects and their winds

E. Additional posters

Available only in the Web version

How to get the poster papers

The poster book has been distributed to all the participants at the conference and to most of the main astronomical libraries. No further copies are available. However, the poster book is available on the Web at the following address:

`http://www-laog.obs.ujf-grenoble.fr/meetings/iau182`

I am an unquenchable fire, the centre of all energy, the stout heroic heart;
I am truth and light, I hold power and glory in my sway;
My presence disperses dark clouds, I have been chosen to tame the fates;
I am the Dragon

— *T. Low, The Handbook of Chinese Horoscopes*

This molecular hydrogen image (v=1–0 S(1) line at 2.122 μm) shows HH 288 in Cassiopeia, dubbed the Dragon Jet after its resemblance to the mythical fire-breather. The jet emanates from a deeply embedded young source and drives an associated molecular outflow; at the 2 kpc kinematic distance of the ambient cloud, the jet is over 3 parsec long.

Data taken by Mark McCaughrean using Omega-Prime on the Calar Alto 3.5-m; 0.8 arcsec seeing; 33 minutes integration time; 4.3×4.6 arcmin field; continuum not subtracted.

Zeitfracht Medien GmbH
Ferdinand-Jühlke-Straße 7
99095 Erfurt, Deutschland
produktsicherheit@kolibri360.de